Membranes and Transport

Volume 1

Membranes and Transport

Volume 1

Edited by

Anthony N. Martonosi

State University of New York
Syracuse, New York

Plenum Press • *New York and London*

Library of Congress Cataloging in Publication Data

Main entry under title:

Membranes and transport.

Bibliography: p.
  Includes index.
    1. Membranes (Biology). 2. Biological transport. I. Martonosi, Anthony, 1928–
    . [DNLM: 1. Membranes — Metabolism. 2. Biological transport. QH 509
M5335]
QH601.M4817   574.87′5   82-3690
ISBN 0-306-40853-9 (v. 1)   AACR2

© 1982 Plenum Press, New York
A Division of Plenum Publishing Corporation
233 Spring Street, New York, N.Y. 10013

Printed in the United States of America

Contributors

Adolph Abrams
Departments of Biochemistry/Biophysics/
   Genetics
University of Colorado School of Medicine
Denver, Colorado 80262

R. Akeroyd
Laboratory of Biochemistry
State University of Utrecht
Transitorium III
NL-3584CH Utrecht, The Netherlands

Angelo Azzi
Medizinisch-chemisches Institut der Universität
CH-3012 Bern, Switzerland

Donner F. Babcock
Institute for Enzyme Research and the Department of
   Biochemistry
University of Wisconsin
Madison, Wisconsin 53706

J. Baddiley
Microbiological Chemistry Research Laboratory
The University
Newcastle upon Tyne NE1 7RU, England

Hagan Bayley
Department of Biochemistry
College of Physicians & Surgeons
Columbia University
New York, New York 10032

Jon Beckwith
Department of Microbiology and Molecular Genetics
Harvard Medical School
Boston, Massachusetts 02115

Helmut Beinert
Institute for Enzyme Research
University of Wisconsin
Madison, Wisconsin 53706

Gary Bellus
Department of Molecular, Cellular, and Developmental
   Biology
University of Colorado
Boulder, Colorado 80309

T. Berglindh
Laboratory of Membrane Biology
University of Alabama
Birmingham, Alabama 35294

Konrad Bloch
The James Bryant Conant Laboratories
Harvard University
Cambridge, Massachusetts 02138

M. R. Block
Laboratoire de Biochimie (INSERM U. 191 et
   CNRS/ERA 903)
Département de Recherche Fondamentale
Centre d'Etudes Nucléaires,
38041 Grenoble Cedex, France, et Faculté de
   Médecine de Grenoble, France

F. Boulay
Laboratoire de Biochimie (INSERM U. 191 et
   CNRS/ERA 903)
Département de Recherche Fondamentale
Centre d'Etudes Nucléaires,
38041 Grenoble Cedex, France, et Faculté de
   Médecine de Grenoble, France

G. Brandolin

Laboratoire de Biochimie (INSERM U. 191 et
    CNRS/ERA 903)
Département de Recherche Fondamentale
Centre d'Etudes Nucléaires,
38041 Grenoble Cedex, France, et Faculté de
    Médecine de Grenoble, France

Arnold F. Brodie

Department of Biochemistry
University of Southern California School of Medicine
Los Angeles, California 90033

Fyfe L. Bygrave

Department of Biochemistry
Faculty of Science
The Australian National University
Canberra, Australian Capital Territory 2600, Australia

Roderick A. Capaldi

Institute of Molecular Biology
University of Oregon
Eugene, Oregon 97403

Ernesto Carafoli

Laboratory of Biochemistry
Swiss Federal Institute of Technology (ETH)
8092 Zürich, Switzerland

D. Chapman

Department of Biochemistry and Chemistry
Royal Free Hospital School of Medicine
University of London
London WC1N 1BP, England

Richard J. Cherry

Eidgenössische Technische Hochschule
Laboratorium für Biochemie
ETH-Zentrum
CH-8092 Zürich, Switzerland

John H. Collins

Department of Pharmacology and Cell Biophysics
University of Cincinnati College of Medicine
Cincinnati, Ohio 45267

Graeme B. Cox

Biochemistry Department
John Curtin School of Medical Research
Australian National University
Canberra, Australian Capital Territory 2601, Australia

John E. Cronan, Jr.

Department of Microbiology
University of Illinois
Urbana, Illinois 61807

P. R. Cullis

Department of Biochemistry
University of British Columbia
Vancouver, British Columbia, Canada

Harry A. Dailey

Department of Biochemistry
University of Connecticut Health Center
Farmington, Connecticut 06032

K. van Dam

Laboratory of Biochemistry
University of Amsterdam
1018TV Amsterdam, The Netherlands

J. F. Danielli

Danielli Associates and Life Sciences Department
Worcester Polytechnic Institute
Worcester, Massachusetts 01609

Bernard D. Davis

Bacterial Physiology Unit
Harvard Medical School
Boston, Massachusetts 02115

Jean Davoust

Institut de Biologie Physico-Chimique
75005 Paris, France

R. A. Demel

Laboratory of Biochemistry
State University of Utrecht
Transitorium III
NL-3584CH Utrecht, The Netherlands

Philippe F. Devaux

Institut de Biologie Physico-Chimique
75005 Paris, France

Bernhard Dobberstein

European Molecular Biology Laboratory
6900 Heidelberg, West Germany

J. Allan Downie

Biochemistry Department
John Curtin School of Medical Research
Australian National University
Canberra, Australian Capital Territory 2601, Australia

Michael Edidin

Department of Biology
The Johns Hopkins University
Baltimore, Maryland 21218

Maria Erecińska

Department of Pharmacology and Department of
    Biochemistry and Biophysics
University of Pennsylvania
Philadelphia, Pennsylvania 19104

Ernesto Freire

Department of Biochemistry
University of Tennessee
Knoxville, Tennessee 37916

M. Fry

Institute for Enzyme Research
University of Wisconsin
Madison, Wisconsin 53706

Stephen D. Fuller

Institute of Molecular Biology
University of Oregon
Eugene, Oregon 97403

Masamitsu Futai

Department of Microbiology
Faculty of Pharmaceutical Sciences
Okayama University
Okayama 700, Japan

Peter B. Garland

Biochemistry Department
Dundee University
Dundee DD1 4HN, Scotland.

Jeffery L. Garwin

Department of Microbiology and Molecular Genetics
Harvard Medical School
Boston, Massachusetts 02115

James L. Gaylor

Central Research and Development
Glenolden Laboratory
E.I. Du Pont and Company
Glenolden, Pennsylvania 19707.

Robert B. Gennis

Department of Chemistry
School of Chemical Sciences
University of Illinois
Urbana, Illinois 61801

Frank Gibson

Biochemistry Department
John Curtin School of Medical Research
Australian National University
Canberra, Australian Capital Territory 2601, Australia

I. M. Glynn

Physiological Laboratory
University of Cambridge
Cambridge CB2 3EG, England

D. E. Green

Institute for Enzyme Research
University of Wisconsin
Madison, Wisconsin 53706

Charles M. Grisham

Department of Chemistry
University of Virginia
Charlottesville, Virginia 22901

Richard Grosse

Biomembrane Section
Central Institute of Molecular Biology
Academy of Sciences of the German Democratic
  Republic
1115 Berlin–Buch, East Germany

Charles R. Hackenbrock

Department of Anatomy
Laboratories for Cell Biology
University of North Carolina School of Medicine
Chapel Hill, North Carolina 27514

Bruce A. Haddock

Biogen S.A.
1227 Carouge/Geneva, Switzerland

Lowell P. Hager

Department of Biochemistry
School of Chemical Sciences
University of Illinois
Urbana, Illinois 61801

I. C. Hancock

Microbiological Chemistry Research Laboratory
The University
Newcastle upon Tyne NE1 7RU, England

Noam Harpaz

Research Institute, Hospital for Sick Children,
  and Department of Biochemistry
University of Toronto
Toronto, Ontario M5G 1X8, Canada

Tom R. Herrmann

Department of Biological Chemistry
University of Maryland School of Medicine
Baltimore, Maryland 21201

Lowell E. Hokin

Department of Pharmacology
University of Wisconsin Medical School
Madison, Wisconsin 53706

C. Huang

Department of Biochemistry
University of Virginia School of Medicine
Charlottesville, Virginia 22908

Masayori Inouye

Department of Biochemistry
State University of New York
Stony Brook, New York 11794

Oleg Jardetzky

Stanford Magnetic Resonance Laboratory
Stanford University
Stanford, California 94305

William P. Jencks

Graduate Department of Biochemistry
Brandeis University
Waltham, Massachusetts 02254

Peter Leth Jørgensen

Institute of Physiology
University of Aarhus
8000 Aarhus C, Denmark

Yasuo Kagawa

Department of Biochemistry
Jichi Medical School
Minamikawachi
Tochigi-ken 329-04, Japan

Vijay K. Kalra

Department of Biochemistry
University of Southern California School of Medicine
Los Angeles, California 90033

Hiroshi Kanazawa

Department of Microbiology
Faculty of Pharmaceutical Sciences
Okayama University
Okayama 700, Japan

S. J. D. Karlish

Department of Biochemistry
Weizmann Institute
Rehovoth, Israel

Ann-Louise Kerner

Department of Biochemistry
University of Connecticut Health Center
Farmington, Connecticut 06032

M. Klingenberg

Institute for Physical Biochemistry
University of Munich
8000 Munich 2, West Germany

H. R. Koelz

Laboratory of Membrane Biology
University of Alabama
Birmingham, Alabama 35294

John G. Koland

Department of Chemistry
School of Chemical Sciences
University of Illinois
Urbana, Illinois 61801

B. de Kruijff

Department of Molecular Biology, and Department of
    Biochemistry
University of Utrecht
3584CH Utrecht, The Netherlands

Henry A. Lardy

Institute for Enzyme Research and the Department of
    Biochemistry
University of Wisconsin
Madison, Wisconsin 53706

Peter K. Lauf

Department of Physiology
Duke University Medical Center
Durham, North Carolina 27710

G. J. M. Lauquin

Laboratoire de Biochimie (INSERM U. 191 et
    CNRS/ERA 903)
Département de Recherche Fondamentale
Centre d'Etudes Nucléaires,
38041 Grenoble Cedex, France et Faculté de Médecine
    de Grenoble, France

Richard M. Leimgruber

Departments of Biochemistry, Biophysics, and
    Genetics
University of Colorado School of Medicine
Denver, Colorado 80262

K. R. Leonard

European Molecular Biology Laboratory
6900 Heidelberg, West Germany

Anthony W. Linnane

Department of Biochemistry
Monash University
Clayton, Victoria 3168, Australia

Harvey F. Lodish

Department of Biology
Massachusetts Institute of Technology
Cambridge, Massachusetts 02139

Gregory D. Longmore

Research Institute, Hospital for Sick Children,
    and Department of Biochemistry
University of Toronto
Toronto, Ontario M5G 1X8, Canada

David H. MacLennan

Banting and Best Department of Medical Research
Charles H. Best Institute
University of Toronto
Toronto, Ontario M5G 1L6, Canada

Henry R. Mahler
Department of Chemistry and the Molecular, Cellular, and Developmental Biology Program
Indiana University
Bloomington, Indiana 47405

Anthony N. Martonosi
Department of Biochemistry
SUNY, Upstate Medical Center
Syracuse, New York 13210

Sangkot Marzuki
Department of Biochemistry
Monash University
Clayton, Victoria 3168, Australia

J. T. Mason
Department of Biochemistry
University of Virginia School of Medicine
Charlottesville, Virginia 22908

Michael W. Mather
Department of Chemistry
School of Chemical Sciences
University of Illinois
Urbana, Illinois 61801

Alan C. McLaughlin
Biology Department
Brookhaven National Laboratory
Upton, New York 11973

Stuart McLaughlin
Department of Physiology and Biophysics
Health Sciences Center
State University of New York
Stony Brook, New York 11794

David I. Meyer
European Molecular Biology Laboratory
6900 Heidelberg, West Germany

Kurt Mühlethaler
Swiss Federal Institute of Technology
Institute for Cell Biology
CH-8093 Zürich, Switzerland

Saroja Narasimhan
Research Institute, Hospital for Sick Children, and Department of Biochemistry
University of Toronto
Toronto, Ontario M5G 1X8, Canada

Robert C. Nordlie
Department of Biochemistry
University of North Dakota School of Medicine
Grand Forks, North Dakota 58202

Eric Oldfield
School of Chemical Sciences
University of Illinois at Urbana
Urbana, Illinois 61801

Sergio Papa
Institute of Biological Chemistry
Faculty of Medicine
University of Bari
Bari, Italy

Peter L. Pedersen
Laboratory for Molecular and Cellular Bioenergetics
Department of Physiological Chemistry
The Johns Hopkins University School of Medicine
Baltimore, Maryland 21205

Robert O. Poyton
Department of Molecular, Cellular, and Developmental Biology
University of Colorado
Boulder, Colorado 80309

E. Rabon
Laboratory of Membrane Biology
University of Alabama
Birmingham, Alabama 35294

Shmuel Razin
Department of Membrane and Ultrastructure Research
The Hebrew University–Hadassah Medical School
Jerusalem, Israel

Peter H. Reinhart
Department of Biochemistry
Faculty of Science
The Australian National University
Canberra, Australian Capital Territory 2600, Australia

Reinhart A. F. Reithmeier
Department of Biochemistry
University of Alberta
Edmonton, Alberta T6G 2H7, Canada

Kurt R. H. Repke
Biomembrane Section
Central Institute of Molecular Biology
Academy of Sciences of the German Democratic Republic
1115, Berlin-Buch, East Germany

Harald Reuter
Department of Pharmacology
University of Bern
3010 Bern, Switzerland

Henry Roberts

Department of Biochemistry
Monash University
Clayton, Victoria 3168, Australia

Charles O. Rock

Department of Biochemistry
St. Jude Children's Research Hospital
Memphis, Tennessee 38101

Ben Roelofsen

Laboratory of Biochemistry
State University of Utrecht
Transitorium 3
NL-3584CH Utrecht, The Netherlands

G. Saccomani

Laboratory of Membrane Biology
University of Alabama
Birmingham, Alabama 35294

G. Sachs

Laboratory of Membrane Biology
University of Alabama
Birmingham, Alabama 35294

Milton R. J. Salton

Department of Microbiology
New York University School of Medicine
New York, New York 10016

Harry Schachter

Research Institute, Hospital for Sick Children,
    and Department of Biochemistry
University of Toronto
Toronto, Ontario M5G 1X8, Canada

H. J. Schatzmann

Department of Veterinary Pharmacology
University of Bern
3012 Bern, Switzerland

Heinz Schneider

Department of Anatomy
Laboratories for Cell Biology
University of North Carolina School of Medicine
Chapel Hill, North Carolina 27514

Arnold Schwartz

Department of Pharmacology and Cell Biophysics
University of Cincinnati College of Medicine
Cincinnati, Ohio 45267

Adil E. Shamoo

Department of Biological Chemistry
University of Maryland School of Medicine
Baltimore, Maryland 21201

Hans Sigrist

Institute of Biochemistry
University of Bern
CH-3012 Bern, Switzerland

Thomas P. Singer

Molecular Biology Division
Veterans Administrations Medical Center
San Francisco, California 94121,
    and Department of Biochemistry and Biophysics
University of California
San Francisco, California 94143

Carolyn W. Slayman

Departments of Human Genetics and Physiology
Yale University School of Medicine
New Haven, Connecticut 06510

Clifford L. Slayman

Department of Physiology
Yale University School of Medicine
New Haven, Connecticut 06510

Brian Snyder

Department of Biochemistry
University of Virginia School of Medicine
Charlottesville, Virginia 22908

Philipp Strittmatter

Department of Biochemistry
University of Connecticut Health Center
Farmington, Connecticut 06032

Phang-C. Tai

Bacterial Physiology Unit
Harvard Medical School
Boston, Massachusetts 02115

Wayne M. Taylor

Department of Biochemistry
Faculty of Science
The Australian National University
Canberra, Australian Capital Territory 2600, Australia

Henry Tedeschi

Department of Biological Sciences
State University of New York
Albany, New York 12222

David D. Thomas

Department of Biochemistry
University of Minnesota Medical School
Minneapolis, Minnesota 55455

H. Ti Tien

Department of Biophysics
Michigan State University
East Lansing, Michigan 48824

Yuji Tonomura
Department of Biology
Faculty of Science
Osaka University
Toyonaka, Osaka 560, Japan

Victor Darley-Usmar
Institute of Molecular Biology
University of Oregon
Eugene, Oregon 97403

A. J. Verkleij
Department of Molecular Biology, and Department of
   Biochemistry
University of Utrecht
3584CH Utrecht, The Netherlands

Donald A. Vessey
Liver Studies Unit
Veterans Administration Medical Center
San Francisco, California 94121,
   and Departments of Medicine and Pharmacology
University of California
San Francisco, California 94143

P. V. Vignais
Laboratoire de Biochimie (INSERM U. 191 et
   CNRS/ERA 903)
Département de Recherche Fondamentale
Centre d'Etudes Nucléaires,
38041 Grenoble Cedex, France, et Faculté de Médecine
   Grenoble, France

Alan S. Waggoner
Department of Chemistry
Amherst College
Amherst, Massachusetts 01002

Robert H. Wasserman
Department of Physiology
New York State College of Veterinary Medicine
Cornell University
Ithaca, New York 14853

Janna P. Wehrle
Laboratory for Molecular and Cellular Bioenergetics
Department of Physiological Chemistry
The Johns Hopkins University School of Medicine
Baltimore, Maryland 21205

H. Weiss
European Molecular Biology Laboratory
6900 Heidelberg, West Germany

H. V. Westerhoff
Laboratory of Biochemistry
University of Amsterdam
1018TV Amsterdam, The Netherlands

John S. White
Department of Biochemistry
School of Chemical Sciences
University of Illinois
Urbana, Illinois 61801

Mårten Wikström
Department of Medical Chemistry
University of Helsinki
SF-00170 Helsinki 17, Finland

David F. Wilson
Department of Biochemistry and Biophysics
University of Pennsylvania
Philadelphia, Pennsylvania 19104

K. W. A. Wirtz
Laboratory of Biochemistry
State University of Utrecht
Transitorium III
NL-3584CH Utrecht, The Netherlands

Henry C. Wu
Department of Microbiology
Uniformed Services University of the Health Sciences
Bethesda, Maryland 20014

Taibo Yamamoto
Department of Biology
Faculty of Science
Osaka University
Toyonaka, Osaka 560, Japan

Peter Zahler
Institute of Biochemistry
University of Bern
CH-3012 Bern, Switzerland

David Zakim
Liver Studies Unit
Veterans Administration Medical Center
San Francisco, California 94121,
   and Departments of Medicine and Pharmacology
University of California
San Francisco, California 94143

H. Vande Zande
Institute for Enzyme Research
University of Wisconsin
Madison, Wisconsin 53706

Asher Zilberstein
Department of Biology
Massachusetts Institute of Technology
Cambridge, Massachusetts 02139

Preface

This work is a collection of short reviews on membranes and transport. It portrays the field as a mosaic of bright little pieces, which are interesting in themselves but gain full significance when viewed as a whole. Traditional boundaries are set aside and biochemists, biophysicists, physiologists, and cell biologists enter into a natural discourse.

The principal motivation of this work was to ease the problems of communication that arose from the explosive growth and interdisciplinary character of membrane research. In these volumes we hope to provide a readily available comprehensive source of critical information covering many of the exciting, recent developments on the structure, biosynthesis, and function of biological membranes in microorganisms, animal cells, and plants. The 182 reviews contributed by leading authorities should enable experts to check up on recent developments in neighboring areas of research, allow teachers to organize material for membrane and transport courses, and give advanced students the opportunity to gain a broad view of the topic. Special attention was given to developments that are expected to open new areas of investigation. The result is a kaleidoscope of facts, viewpoints, theories, and techniques, which radiates the excitement of this important field. Publication of these status reports every few years should enable us to follow progress in an interesting and easygoing format.

I am grateful to the authors, to Plenum Publishing Corporation, and to several of my colleagues for their thoughtful suggestions and enthusiastic cooperation, which made this work possible.

Anthony N. Martonosi

Lovell, Maine

Contents

I · The Molecular Architecture of Biological Membranes

II · The Physical Properties of Biological and Artificial Membranes

III · *Biosynthesis of Cell Membranes: Selected Membrane-Bound Metabolic Systems*

IV · The Structure, Composition, and Biosynthesis of Membranes in Microorganisms

V · Bioenergetics of Electron and Proton Transport in Mitochondria

VI · *Energy-Transducing ATPases and Electron Transport in Microorganisms*

VII · Ion Transport Systems in Animal Cells

Contents of Volume 2

X · Channels, Pores, Intercellular Communication

XI · Excitable Membranes

XV · *Membrane-Linked Metabolic and Transport Processes in Plants*

Membranes
and Transport

Volume 1

I

The Molecular Architecture of Biological Membranes

1

Experiment, Hypothesis, and Theory in the Development of Concepts of Cell Membrane Structure 1930–1970

J. F. Danielli

1. INTRODUCTION

Many of the basic hypotheses concerning membrane structure were developed in the period 1900–1945. From 1945 to the present, intensive experimental study of these hypotheses has left our concepts of the role of lipids in the membrane largely unchanged. It has also greatly extended the evidence for a variety of roles for proteins in the membrane. Information about the way in which proteins enter into membrane structure, and how they carry out their roles, has been remarkably extended since 1960. In this review, I shall analyze the development of the basic hypotheses in terms of the interplay between hypothesis, experiment, and theory, as it occurred during the period 1930–1970.

The functions of hypothesis, experiment, and theory are quite different. *Hypothesis* is an expression of intuition, which serves as a guide for development of experiment and of theory. Experiment and theory are alternative and complementary methods of testing the validity of a hypothesis. Without such testing, although a hypothesis may be esthetically satisfying, it is largely useless and by providing a degree of satisfaction may actually retard the development of true understanding.

Experiment provides a test for hypothesis by establishing quantities that can be seen to be compatible, or incompatible, with the hypothesis. For example, from experiment we may be able to say, e.g., 60% of the lipid in a particular membrane is in the bilayer configuration, or receptors are present for norepinephrine, and so on.

This article is an extended version of my article in *The Role of Membranes in Metabolic Regulation* (Academic Press, New York, 1972).

J. F. Danielli • Danielli Associates and Life Sciences Department, Worcester Polytechnic Institute, Worcester, Massachusetts 01609.

Theory provides a definition of process and relationship, preferably in quantifiable terms that permit prediction, such as: if the lipid bilayer hypothesis is correct, permeation of the membrane must be by activated diffusion and will be given by, e.g.,

$$p = ae/(nb + 2e)$$

or, for example, a particular model is impossible because its free energy is so much higher than that of other configurations of the same molecules.

In many fields a period of observation precedes hypothesis formation, and the observations that are made lead to the first hypotheses. In cell membrane biology this period of observation lies prior to 1930.

2. EARLY WORK

Over the period 1900–1930, substantial amounts of qualitative evidence indicated that the surface layers of the cell must be predominantly lipid. This arose partly from the permeability studies of such investigators as Overton and Osterhaut, and partly from studies of the conductance of the membrane carried out, for example, by Hober, Fricke, and Cole. The impedance studies also showed clearly that the membrane could not be more than a few molecules thick. For reviews of this work, see for example, Hober (1945) and Harvey and Danielli (1938). Although there was a considerable measure of agreement that the membrane must contain lipid components, there was much doubt as to whether the membrane was a homogeneous lipid layer, or a mosaic of different structures, or could perhaps contain substantial pores. It was not possible to resolve this problem at the time, because although quantitative data were available, methods for analyzing the data were in the main qualitative. In 1934, drawing upon work by Harvey, I was able to show that protein adsorbed strongly to lipid surfaces, even when these lipid surfaces are of low initial surface free energy (Danielli and Harvey, 1935). These observations suggested that protein was an essential component of the membrane. The concept that proteins might be an essential part of cell membranes was novel, and the question thus arose as to how both the proteins and the lipid molecules were arranged to constitute membranes. To solve this problem, I turned to the theory of amphipathic molecules. Thanks to the work of investigators such as Langmuir, Adam, and Rideal, the general behavior of amphipathic molecules was quite well understood at that time, and led readily to appropriate hypotheses.

3. THE DEVELOPMENT OF THE PAUCIMOLECULAR LAYER THEORY

The first hypothesis I developed concerning the arrangements of the molecules was put forward in 1934, and is shown in Fig. 1. In this hypothesis the interaction between lipid and protein was postulated to be polar. It must be remembered that at this time extraordinarily little was known of the structure of proteins. The α-helix structure, for example, had not been suggested at that time. In Fig. 2 are shown a number of the other possible arrangements of molecules that were considered in a second paper (Danielli, 1936). Calculation of surface free energies indicated that structures such as (b) and (h) were the most probable as they would have the lowest surface free energies. Structures such as (h) were based on the supposition that part of the membrane could be made of protein molecules that were held together by nonpolar forces, forming a bilayer structure in which the

EXTERIOR

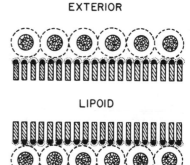

LIPOID

Figure 1. The first paucimolecular layer model suggested for plasma
membranes (1934).

INTERIOR

bilayer component was protein. Little attention was paid to this suggestion at the time, but
it has now become of interest for reasons that will appear later.

In 1934, when I put forward my first hypothesis of membrane structure, I was un-
aware of the work of Gorter and Grendal (1926), and so had no evidence on whether the
lipid layer was bimolecular or thicker, though the impedance studies showed that the total
thickness was not much greater than 100 Å. In 1935, N. K. Adam let me search through

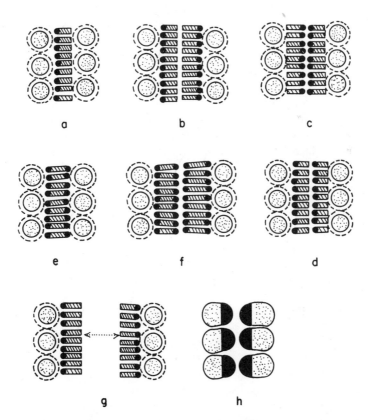

Figure 2. Some alternative arrangements of lipids and proteins (1936). Protein molecules are shown either as
spheres with polar surfaces or as amphipathic molecules (h).

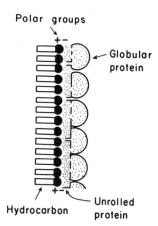

Polar groups

Globular protein

Unrolled protein

Hydrocarbon

Figure 3. The second paucimolecular model: general organization of a half-membrane (1938).

his reprint collection, and I found Gorter and Grendal's paper, which made it highly probable that the red cell membrane was bimolecular in lipid. It was reasonable to make the assumption that other cell membranes were also bimolecular, and from 1936 onward I made this assumption, which was eventually justified by a variety of experimental work.

Once this first hypothesis (Fig. 1) had been advanced, the question immediately arose: Was the configuration of proteins in the membrane, as shown for example in Fig. 1, correct, or was there some other configuration of proteins that would be more appropriate? Studies of the surface properties of proteins I made between 1934 and 1937 showed that the suggested interaction between protein and lipid was not likely to be correct (Harvey and Danielli, 1938; Danielli, 1938). It was found that when proteins such as ovalbumin were absorbed at surfaces, a surface free energy was available from nonpolar forces that was of the order of 100,000 cal/molecule. The nonpolar groups responsible for this nonpolar energy were normally concentrated in the interior of the protein, with the polar groups on the surface of the protein molecule. The nonpolar free energy was thus seen to provide a stabilizing force for the globular protein (Danielli, 1939a). However, when adsorption took place on the surface, which promoted unrolling of the protein structure, these forces became available for stabilizing the adsorbed protein at the interface (Danielli, 1939a,b). Thus, the hypothesis proposed in 1934 was replaced by 1937 by a second hypothesis, in which the organization of the lipid components remained as in the 1934 hypothesis but protein was postulated to be arranged in a primary layer that involved nonpolar interactions between lipid and protein and a secondary layer in which the protein molecules of the second layer were mainly attached through nonpolar forces. Figure 3 shows the general arrangement postulated for molecules at the membrane surface, and Fig. 4 shows the arrangement of individual molecules in the lipid and the first protein layer (taken from Davson and Danielli, 1943). Thus, the first hypothesis put forward for membrane structure was almost immediately disproved by a set of experimental observations combined with a set of calculations of free energies.

The question now arose: Was it possible to show that such structures are indeed stable by making artificial membranes having analogous structure? Artificial membranes having these structures were made as spherical shell membranes, in which the lipid phase was a mixture of triolein, oleic acid, and lecithin and the protein component was egg albumin (Danielli, 1936). These membranes were stable for several days and were demonstrated at Princeton at a meeting of the Society for General Physiology in 1935 (Fig. 5).

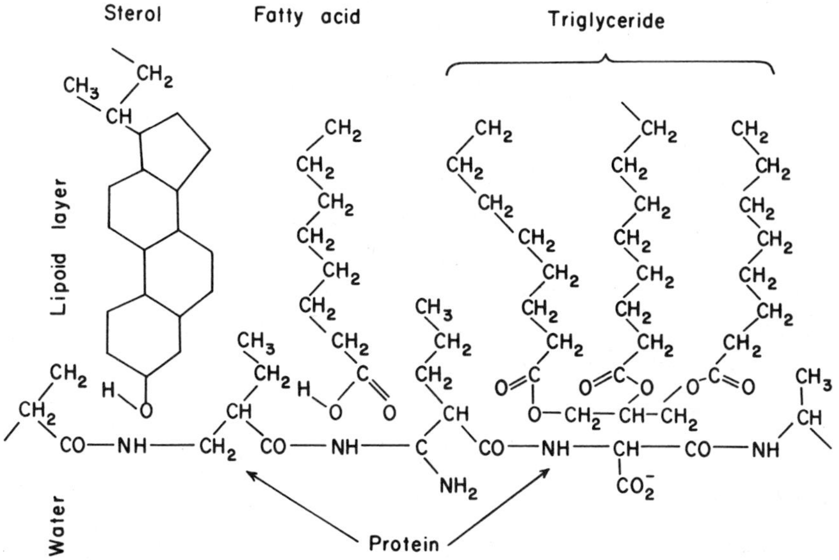

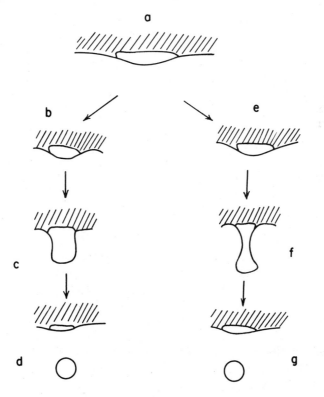

Figure 4. The second paucimolecular model, revealing nonpolar interactions between lipid and protein molecules (1938).

Figure 5. Experimental formation of lipoprotein spherical shell membranes (1936).

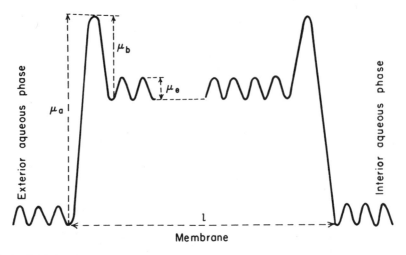

Figure 6. Model of the energy barriers to diffusion across the second paucimolecular model (1941).

4. COULD THESE STRUCTURES ACCOUNT FOR CELL MEMBRANE PERMEABILITIES?

The fact that model membranes having the postulated structure were in fact stable was gratifying, but the extent to which this model really corresponded to the structure of cell membranes needed much further examination. For this I turned to a quantitative study of the permeability properties of natural membranes. To do this I utilized the theory of activated diffusion, which was itself developed in the 1930s in the fields of physics and physical chemistry. To obtain a quantitative treatment of permeability data, it was necessary to obtain a model of the energy barriers to diffusion that would correspond to the postulated structure of the membrane. In developing this model I was particularly helped by the work being done at the time by J. D. Bernal on hydrogen bonding, for this enabled me to envisage the diffusion process as a discontinuous phenomenon, and to propose a physical explanation for the major discontinuity in diffusion rate at the water–lipid interface. The model that was used is shown in Fig. 6 (Danielli, 1941). It led to the following general equation for the permeability of such membranes:

$$p = ae/(nb + 2e)$$

In this equation p is the permeability, a, b, and e are functions of the free energies of activation for diffusion through the corresponding energy barrier, and n is a measure of the thickness of the membrane. It was shown using data especially of Jacobs and of Collander that this equation correctly described the rate of permeation of most molecules. This was true for a range of 10^2 in molecular weight, 10^4 in permeability, and 10^5 in oil–water partition coefficients. In the model, it is assumed that the only significant barrier to diffusion was caused by the hydrocarbon layer of the bilayer, and that the free energies restricting the rate of diffusion arise from the existence of this hydrocarbon layer. The fact that the great majority of molecules behave in a way that fits the equation indicates that these molecules see only hydrocarbon groups when they pass through the membrane. The evidence for this was published in 1943 (Davson and Danielli, 1943). However, although most molecules behaved as though they saw only a hydrocarbon layer, certain other types of molecules, e.g., sugars, phosphate, and amino acids, behaved differently. They diffused faster than would be predicted by the above equation; they behaved as though they were involved in an enzymelike process, interacting with a polar component of the membrane. And it was found that, whereas with molecules that obey the above equation structural detail was of secondary importance, in the case of molecules that permeate faster than predicted by this equation, structural detail was of outstanding importance so that even methylation of a single hydroxyl group at times made a profound difference to the permeability constant.

Thus, by 1943, the evidence was quite clear that the membrane must contain at least two types of structures that are fundamentally different. One of these structures is essentially hydrocarbon in nature. The other structure is essentially polar in nature. It was suggested that the closest analogy known at the time to the polar structures was enzymes, and that these structures should in fact be regarded as enzymelike in nature. As extremely little was known of protein structure at that time, particularly lipophilic proteins, all that could be said of the protein structure was that it must extend through the thickness of the membrane (Davson and Danielli, 1943), as was indicated in diagrammatic form by Fig. 7. Because the nature of the membrane proteins was so obscure, development of the basic physical chemistry of those proteins was not possible. On the other hand, much could be done about the structure of the lipid bilayer. A particularly interesting fact was that, using the equation given above, it was possible to calculate the relative viscosity of cell membranes. It was shown, for a variety of cells, that the viscosity was only 10^2 to 10^5 times that of water, thus showing that (as was to be expected from monolayer studies) the lipid component was liquid, except where solidified locally by, e.g., interaction with membrane proteins (Davson and Danielli, 1943). These figures for membrane lipid viscosity have since been confirmed by a variety of experimental studies.

The theoretical study of permeability constants showed quite clearly that the second hypothesis that had been developed for the structure of the membrane was wrong and that the membrane must in fact contain at least two strikingly different structures, i.e., by that

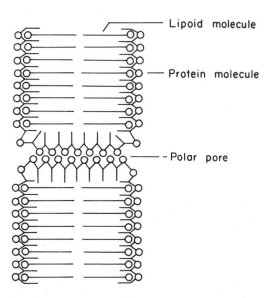

Figure 7. Third paucimolecular model, revealing protein extending through the membrane (1943).

time the interaction between hypothesis, theory, and experiment had led to the discarding of two hypotheses and the generation of a third hypothesis. It is this third hypothesis that is largely substantiated today.

We may summarize the situation by saying that up to about 1930 experiment had been well ahead of theory, that over the period 1930–1940 theoretical examination of data and hypotheses moved ahead of experiment. Over the period approximately 1940–1970 the theoretical studies remained more effective than experimental studies, and that only over the last 15 years have the experimentalists substantially broken new ground.

5. DEVELOPMENTS 1940–1970

Over this period there was remarkable development of new techniques and a corresponding development of experimental studies of cell membranes. Particularly outstanding was the demonstration by electron microscopy that the type of membrane that had hitherto been supposed to exist only at the surfaces of cells was also an integral part of many organelle structures. Taking the period as a whole, it was characterized by certain qualities:

1. A wide-ranging development of new techniques, including electron microscopy, spin resonance, nuclear magnetic resonance, calorimetry, freeze-fracture, and X-ray diffraction studies.
2. A great elaboration of wildcat hypotheses that were not very satisfactorily related either to theory or to experiment, and which I shall not catalog here.
3. A general failure to realize that the membrane consists of more than one structure, i.e., is a mosaic.
4. An undue concentration on the bilayer aspect of membranes, which led to a considerable struggle to demonstrate that the model that I had put forward in 1934 was or was not right, when in fact it had been shown to be incorrect already by 1940.
5. A rather desperate struggle over the nature and role of membrane proteins, which had no very effective outcome by 1970 (but has been attended by much success in the period 1970–1980).
6. An intensive development of the theory of bilayers and of experimental study of lipid bilayers.
7. A clarification of the differences between simple diffusion, facilitated diffusion, active transport, and exchange diffusion.
8. A rapid development of understanding of the wide range of receptor molecules present in membrane structure, culminating in the understanding of the role of cyclic AMP.

So far as membrane structure is concerned, the outcome of these studies has been to produce a satisfactory volume of evidence that the uniform component of all or most cell membranes is a lipid bilayer. That is to say, the continuous phase of the membranes is a lipid bilayer, in which the proteins are the discontinuous elements of a two-dimensional solution that are embedded in, or attached to, the lipid phase. Thus, the theory of amphipathetic molecules, as it was used in the 1930s, did in fact give the right answer for the main lipid component of membranes. The question of the structures into which proteins enter in membranes still remained open, though from the X-ray data it seemed reasonably clear that in many membranes there was a substantial protein component on either side of the bilayer, and from the diffusion studies and from freeze-fracture studies it seemed evident that protein molecules in some cases extended through the thickness of the membrane.

6. THE USE OF FREE-ENERGY CALCULATIONS

When a new structure is suggested as a component of cell membranes, it is often difficult to find an experimental approach to discover whether the suggestion has any validity. Under these circumstances it is very often of value to consider the free energy of the proposed arrangement of molecules in comparison with the free energies of other possible structures. Indeed, this is a valuable procedure, even when experimental techniques are also available for testing a hypothesis. For example, consider the hypothesis that the lipid layer of cell membrane is micellar, as was proposed by some electron microscopists. If one considers neutral phospholipids, three alternative structures are available: the bilayer structure, cylindrical micelles, and spherical micelles. Calculations of the differences in free energy for these structures show that the cylindrical micelle has a free energy in aqueous phases about 6 kcal greater than that of the bilayer structure and the spherical micelle has a free energy about 12 kcal greater than the bilayer structure. Thus, the conclusion emerges very clearly that, in aqueous media, bilayers have the least free energy and consequently bilayer structures will predominate. On the other hand, it also follows that because the free-energy differences between these structures are not huge, all three structures will coexist. If no other molecular components are present, the bilayer will predominate. It is also clear from free-energy calculations that dehydration of a bilayer system will cause tremendous changes in the free energies of the different structures and that consequently it is extremely difficult from studies of anhydrous systems to make determinations about the details of the same systems in the aqueous state.

In this particular case the conclusion that was reached was that the "either/or" approach to the structure of a lipid component of cell membranes is wrong. The question is not whether the membrane consists of bilayer or of micelle, but how much of each type of component is present at a particular time. Furthermore, because the free-energy differences are not huge, we can ask what sorts of molecules can provide a free-energy term that will stabilize an alternative structure at particular locations in the membrane.

It is probable that the cyclic antibiotics and other "small" molecules that greatly increase the permeability of membranes to ions do so by stabilizing alternative arrangements of the lipid molecules, the necessary free energy being derived from interaction between the "small" molecules and the lipid.

7. CONSIDERATIONS BASED UPON EVOLUTION BY NATURAL SELECTION

Insight about cell membranes, and what we may expect to find in them, can also arise from consideration of the process of evolution. Let us first ask the question: Why is it that over a period of about 3 billion years of evolution the bilayer element has come to be selected as a universal membrane component? The first important consideration is probably that it provides a remarkably efficient barrier to free diffusion, i.e., it provides a very economical restraint from free mixing of the interior of a cell with the external medium. I calculated in 1940 that a membrane 100 times thicker would not be significantly more efficient than a bilayer structure. A second and perhaps more important consideration is that a lipid bilayer, being liquid crystalline in nature at physiological temperatures, provides an undemanding two-dimensional matrix in which a great variety of functional molecules may be embedded and associate with one another without significant need for isomorphous properties, except insofar as they are required by necessary physiological interactions between macromolecules. If the membranes were made of proteins only, or of some other type of macromolecule, in order to derive a satisfactory diffusion barrier it would be necessary for the molecules to be substantially isomorphous or complementary in structure. Most membranes probably contain over 100 macromolecular species, and it is rare for more than 10% of the macromolecules composing a membrane to be of any one species. Therefore, if the membrane proteins were to be isomorphous or complementary in structure, it would be a great constraint on evolution, for whenever a mutation occurred that modified one of the macromolecular species, it would be necessary for others to be modified simultaneously. Consequently, in order to get a significant change in membrane function, it would be necessary to have simultaneous mutations involving a number of macromolecular species. Now the rate of favorable mutation for any one gene is probably not greater than 10^{-10}, and probably lower. Favorable mutations are in themselves uncommon events. If it were necessary for, say, five species of molecules to change simultaneously, we could expect a favorable change in macromolecular structure to occur at the rate of, say, 10^{-50}. Thus, we very soon reach the point at which no change is possible in geological time. It is evident that the fact that macromolecules may embed in the lipid bilayer, and thereby avoid the necessity for being isomorphous or complementary, confers an extraordinary evolutionary advantage upon the cell, as independent molecular evolution is then possible for each membrane function.

We may illustrate this point by saying that because the bilayer structure is an undemanding matrix, it can readily incorporate structures that selectively modify the barrier to diffusion that is initially imposed by the presence of the bilayer. This permits selective transport, including active transport, and the development of excitability, without the necessity for profound modification of membrane structure, except locally in terms of specific macromolecules that are embedded in the membrane.

We can now turn to the question that has sometimes been asked as to which family of molecules is primarily responsible for establishing the structure of the membranes, i.e., is the prime mover a protein or is the prime mover a lipid? When we think of this question from the point of evolution, we see it is not really a sensible question. Over the 2–3 billion years of evolution, natural selection must have acted continuously to produce complementarity between protein and lipid. And so by this time the protein and lipid found in membranes must be closely adapted to one another and one can no more assign the function of prime mover to protein or to lipid than one can assign the function of prime mover to chicken or egg.

These same considerations, which limit the likelihood of membranes in general being composed of isomorphous or complementary macromolecules, also limit the likelihood of cooperative phenomena occurring in membranes. A cooperative process, involving dozens of types of macromolecules, is quite improbable. Thus, if we are to see cooperative changes in membranes, they are likely to occur only in regions of membranes that are composed of a very small variety of macromolecules.

8. CELL ASSEMBLY TECHNIQUES

I wrote in 1971 that artificial cell assembly techniques were moving to the point at which it would be possible to use these techniques to study some of the dynamic properties of cell membranes. One of the most interesting of such properties is the rate of replacement of molecules in cell membranes and the control of membrane composition. The techniques essentially involve taking the components of cells and reorganizing them so as to obtain a new cell. The simplest technique of this type is cell fusion, which, for example, permits the study of the rate of mingling of the antigens of the separate cells after fusion has taken place. The second set of techniques involves the addition of nuclei, viruses, or single chromosomes to cells or cytoplasms. The addition is made in such a way as to change the genetic control of cell behavior and composition. A third type of techniques involves taking a membrane, natural or artificial, and filling it with those components whose function it is desired to test in relation to, e.g., a membrane composition. Work on these lines has been slower than I anticipated in 1971, but during the next 10 years we can expect to see great advances based upon the combination of techniques such as these and immunochemical analytic techniques.

9. CONCLUSION

To sum up, I would say that at the present time the general existence of the bilayer component of cell membranes is well established. The *presence* of protein (and glycoprotein) components is equally well established. But the details of the way in which proteins form part of the membrane can be described only in the most general terms, e.g., by saying that some proteins extend through the thickness of the membrane, some are embedded in one surface, some are absorbed on a surface, and that their behavior indicates that the continuous phase of the membrane is the liquid-crystalline bilayer, and the proteins are the disperse phase. The interactions between proteins and lipids involve polar bonding and nonpolar bonding to an extent that varies from protein to protein.

Little theoretical work can be done on the membrane proteins until the three-dimensional structure of some of these proteins are known. From studies on protein folding in aqueous media (see, e.g., Goel and Ycas, 1979), it now seems probable that the conformation of a protein in aqueous solution can be calculated, provided its amino acid sequence is known. I anticipate that similar studies will make it possible to calculate the conformation of a polypeptide chain in hydrocarbon phases, and that the final step will be to combine the two procedures to calculate the conformation of a protein the individual molecules of which are partly in an aqueous phase and partly in a lipid phase, as is the case with transmembrane proteins.

Optimal progress, and optimal use of resources, will require careful study of the relation between hypothesis, theory, and experiment, and if this is done, much of the confusion that characterized the field over the period 1950–1970 can be avoided. Looking back on the period 1945–1970, the most surprising thing to me was that, although my 1934 model was widely quoted and used, few people actually read the subsequent papers, or understood that, in the book that I published with Davson (Davson and Danielli, 1943), the original simplistic model of Fig. 1 had been proved to be wrong and that my third model (Fig. 7) had emerged. This model has been modified in the light of modern knowledge of protein structure, but its main features are retained, and seem certain to remain as our knowledge of the protein components becomes more sophisticated.

REFERENCES

Danielli, J. F. (1936). *J. Cell. Comp. Physiol.* **7,** 393–408.
Danielli, J. F. (1938). *Cold Spring Harbor Symp. Quant. Biol.* **6,** 190.
Danielli, J. F. (1939a). *Proc. R. Soc. London Ser. B* **127,** 34–35.
Danielli, J. F. (1939b). *Proc. R. Soc. London Ser. B* **127,** 73.
Danielli, J. F. (1941). *Trans. Faraday Soc.* **37,** 121–124.

Danielli, J. F., and Davson, H. (1935). *J. Cell. Comp. Physiol.* **5,** 495–508.

Danielli, J. F., and Harvey, E. N. (1935). *J. Cell. Comp. Physiol.* **5,** 483–494.

Davson, H., and Danielli, J. F. (eds.) (1943). In *The Permeability of Natural Membranes,* Cambridge University Press, London.

Goel, N., and Ycas, M. (1979). *J. Theor. Biol.* **77,** 253.

Gorter, E., and Grendal, F. (1926). *Proc. K. Acad. Wetensoh. Amsterdam* **29,** 314.

Harvey, E. N., and Danielli, J. F. (1938). *Biol. Rev. Cambridge Philos. Soc.* **13,** 319–341.

Hober, R. (1945). *Physical Chemistry of Cells and Tissues,* Churchill, London.

2

Complementary Packing of Phosphoglyceride and Cholesterol Molecules in the Bilayer

C. Huang and J. T. Mason

1. INTRODUCTION

It is well established for phospholipids that the molecular structure and packing properties of the hydrophobic region of the molecule are intimately coupled to the associated structural properties of the polar head group of the phospholipid. For example, phosphatidylethanolamine molecules extracted from biological sources such as hen egg yolk form both lamellar (or bilayer) and hexagonal type II phases when exposed to excess water at physiological temperatures (Reiss-Husson, 1967; Junger and Reinauser, 1969). In contrast, phosphatidylcholine molecules form only the bilayer structure under the same conditions (Luzzati and Husson, 1962; Luzzati, 1968). The large shift in the main endothermic transition temperature, T_m, of the lipid hydrocarbon chains from the crystalline gel to the liquid-crystalline state serves as another example. The value of T_m for synthetic [$C_{16 : 0}$–$C_{16 : 0}$] dipalmitoylphosphatidylcholine is 41.4°C (Mabrey and Sturtevant, 1976), whereas phosphatidylethanolamine with the same saturated hydrocarbon chains has a T_m of about 63.8°C (Mabrey and Sturtevant, 1977). Even within the same phospholipid such as dipalmitoylphosphatidylcholine, an alteration in polar-head-group conformations or properties can be shown to affect the structural and packing properties of the hydrophobic portion of the lipid bilayers. For example, upon the addition of about 4 molecules of water per molecule of dipalmitoylphosphatidylcholine, the conformation and the mobility of the lipid polar head group in bilayers are found to be distinctly different from those in anhydrous solids (Griffin, 1976; Bush et al., 1980a). Any marked change in the structural motional properties of the polar head group of phospholipids upon hydration is not unexpected, for

C. Huang and J. T. Mason • Department of Biochemistry, University of Virginia School of Medicine, Charlottesville, Virginia 22908.

it is well known that the water molecules can associate with the polar head group of phospholipid molecules. Interestingly, the addition of about four water molecules to the head group of phosphatidylcholine also alters significantly the packing properties of the hydrophobic region of the lipid molecule in bilayers in terms of the degree of intrachain (*trans/gauche*) and interchain (lattice) disorders (Bush *et al.*, 1980a).

The propagation of a conformational change in one part of a phospholipid molecule to the rest of the same molecule and also to the neighboring molecules in bilayers is in certain ways analogous to many phenomena observed with other biological assemblies such as the allosteric effect in enzymes and other macromolecules. The structural properties of hexokinase and phosphoglycerate kinase, as revealed by X-ray crystallographic studies, serve as an example. These kinases are known to consist of two large domains connected by a "hinge region" but separated spatially by an open and deep cleft (Anderson *et al.*, 1979; Banks *et al.*, 1979). Upon substrate binding, the kinase undergoes a large-scale "hinge-bending" conformational change, resulting in a closing of the deep cleft. The point to be made here is that phospholipids in bilayers can also undergo relatively large conformational change upon addition of other amphipathic molecules. This is particularly true when the bilayer is in the liquid-crystalline state. Here, we shall discuss the interaction of phospholipids with cholesterol in membranes of the lamellar phase, and emphasis will be given to their complementary recognition of each other in membrane assembly.

2. THE CHEMISTRY AND BIOSYNTHESIS OF CHOLESTEROL

Cholesterol is an amphiphilic molecule with a quasi-planar and frayed structure. As a membrane lipid, cholesterol is unique in that it has a rather small polar head group, the 3β-hydroxy group (see Fig. 1); in fact, it is the only major membrane lipid in which the size of the polar head group is smaller than that of a water molecule.

X-Ray crystal structures of anhydrous cholesterol and cholesterol monohydrate have been reported (Craven, 1976; Shieh *et al.*, 1977). In both cases, crystallographic data indicate that there are eight independent cholesterol molecules packed in a triclinic unit cell. The arrangement of these eight molecules is unusually complex with the lowest possible crystal symmetry (space group $P1$). The total thickness of the cholesterol bilayer is 33.9 Å (Shieh *et al.*, 1977). The cross-sectional area of cholesterol, estimated from the maximally condensed monolayer at the air–water interface, is 38 Å^2 (Jones, 1975).

The hydrophobic region of the cholesterol molecule consists of a fused tetracyclic ring system (the steroid nucleus) and a branched isooctyl side chain (Fig. 1). The tetracyclic ring system is stereochemically rigid with a length of 9 Å. In contrast, the isooctyl side chain is quite flexible with considerable thermal motion at room temperature; the length

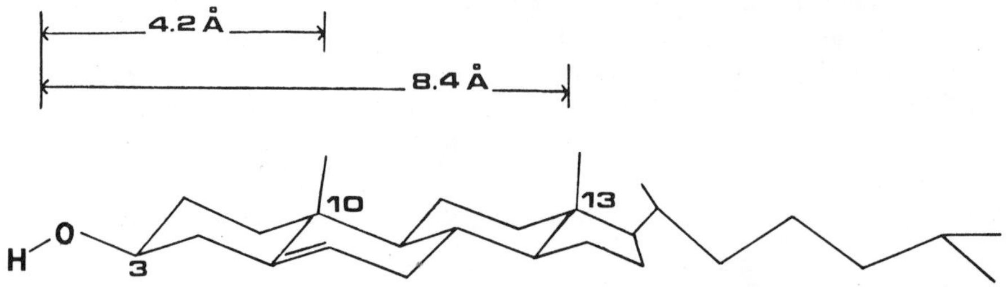

Figure 1. Hydrophobic region of the cholesterol molecule.

can vary from 5.51 to 6.60 Å, depending on the crystal environment (Shieh *et al.*, 1977). The α face or underface of the fused tetracyclic ring system is relatively flat, because the seven axial hydrogen atoms at C-1, C-3, C-7, C-9, C-12, C-14, and C-17 are roughly coplanar. The β face, however, has greater relief due to the presence of two angular methyl groups at C-10 and C-13. In contrast to the alignment of the two angular methyl groups, C-18 and C-19, which are perpendicular to the plane of steroid nucleus, the projecting methyl group C-21 lies nearly parallel to the C-16–C-17 bond in the crystal structure of cholesterol monohydrate (Craven, 1976).

The pathway for the biosynthesis of cholesterol is reasonably well understood (Bloch, 1965); here, a brief outline of the biosynthetic pathway of cholesterol will be described and then discussion will be centered on those points that may be related to the cholesterol–phospholipid interaction in bilayers. The biosynthesis of cholesterol can be conveniently divided into three stages. The first stage is the formation of the active isoprene unit, the C_5 compound 3-isopentenylpyrophosphate, from acetyl CoA via a number of enzymatic reactions. The second stage of the pathway is a series of successive condensations of six isoprene units to yield an acyclic C_{30} compound, squalene. The final stage is the aerobic cyclization of squalene to a 3β-OH-bearing sterol, lanosterol, and the aerobic conversion of lanosterol to cholesterol, a C_{27} compound. In addition to the saturation of a double bond in the side chain and the removal of three methyl groups from the C-4 and C-14 atoms, the last stage of the biosynthetic pathway involves the shift of a double bond from the Δ^8 to the Δ^5 position by way of the sequence $\Delta^8 \rightarrow \Delta^7 \rightarrow \Delta^{5,7} \rightarrow \Delta^5$.

The dehydrogenation step of cholest-7-en-3β-ol to cholesta-5,7-dien-3β-ol, or $\Delta^7 \rightarrow \Delta^{5,7}$, has been reported to involve an electron transfer chain that includes cytochrome b_5 (Reddy *et al.*, 1977). It should be emphasized that the same membrane-bound hemeprotein cytochrome b_5 is well known to serve as an electron carrier in the multicomponent electron transport chain of the stearyl-CoA desaturase system (Shimakata *et al.*, 1972), which is responsible for the production of unsaturated fatty acyl CoA. Perhaps this common dehydrogenation step in the final stage of both cholesterol and natural phospholipid biosynthetic pathways reflects some structural similarities of the relatively flat α face of the sterol molecule and a linear configuration of the saturated chain, as they are the substrates of similar desaturase systems.

3. CONFORMATION OF MEMBRANE PHOSPHOGLYCERIDES

Phospholipids isolated from biological membranes are primarily mixed-chain lipids. Usually, the fatty acid esterified at the *sn*-carbon 2 of the glycerol backbone is unsaturated with varying chain lengths (usually from C_{16} to C_{24}) and degrees of unsaturation. The unsaturated double bond possesses a *cis* configuration. If there is more than one *cis* double bond, the double bonds are aways separated from each other by one methylene group (Kunau, 1976). The position along the fatty acyl chain where a double bond most frequently occurs is between the C-9 and the C-10 atoms. The fatty acyl chain at the *sn*-1 of the glycerol backbone is usually saturated, again with varying chain lengths. For example, the most abundant phospholipid isolated from egg yolk is 1-palmitoyl-2-oleoyl-phosphatidylcholine.

Because of the heterogeneity of the fatty acyl chains, single species of phospholipids such as phosphatidylcholine isolated from biological sources generally do not form crystals. Nevertheless, X-ray fiber diagrams of egg phosphatidylcholine in the crystalline state have been reported (Sakurai *et al.*, 1977). The three-dimensional crystal structure of egg phosphatidylcholine is not unambiguous, because it is based on a one-dimensional analysis. However, X-ray crystal structures of synthetic saturated phospholipids such as dilauroyl-DL-phosphatidylethanolamine–acetic acid complex and dimyristoyl-Lα-phosphatidylcholine dihydrate at atomic resolution have been reported (Hitchock *et al.*, 1974; Pearson and Pascher, 1979). The phosphatidylcholine crystals are monoclinic with two lipid molecules in the asymmetric unit (space group $P2_1$). Before we discuss some of the common structural features of the saturated phosphatidylcholine and phosphatidylethanolamine observed in the crystals, two stereochemical characteristics of the lipid molecule should be borne in mind. (1) The five atoms in the

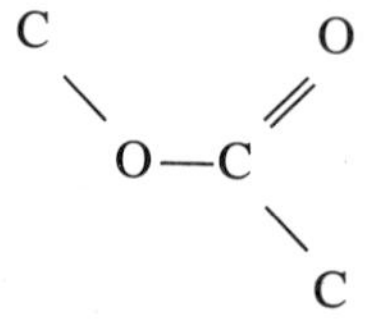

$$
\begin{array}{ccc}
\text{C} & & \text{O} \\
\diagdown & & \diagup\!\diagup \\
& \text{O}\!-\!\text{C} & \\
& & \diagdown \\
& & \text{C}
\end{array}
$$

ester groups are coplanar, because the ester O–C bond has partial double-bond character (Pauling, 1968). (2) The ${}^{\delta}\bar{\text{O}}$–C–C–N${}^+$ group has a predominant *gauche* configuration to minimize the distance between charged groups with opposite signs and the phosphodiester linkage is *gauche–gauche* (Sundaralingam, 1972).

From X-ray crystal structures of synthetic phosphatidylcholine and phosphatidylethanolamine (Hitchock *et al.*, 1974; Pearson and Pasher, 1979), one can conclude that the plane occupied by the primary ester containing the *sn-1* glycerol carbon atom is roughly perpendicular to the plane of the secondary ester as shown in Fig. 2. Although the initial segment of the *sn-2* fatty acyl chain extends parallel to the bilayer surface, the chain bends at the C-2 atom so that the rest of the chain runs parallel to the *sn-1* fatty acyl chain. The sharp bend is caused by a *gauche* rotation occurring at the carbon–carbon bond next to the carbonyl carbon:

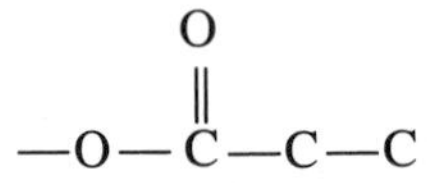

$$
\begin{array}{c}
\text{O} \\
\|\| \\
-\text{O}-\text{C}-\text{C}-\text{C}
\end{array}
$$

One important structural feature resulting from the bend is that the two terminal methyl groups of the two alkyl chains are not in register, being separated by a distance of approximately 3.7 Å.

Two conformations of the polar head group are revealed in the crystal of dipalmitoyl-phosphatidylcholine dihydrate. The phosphate–choline dipoles are inclined to the plane of

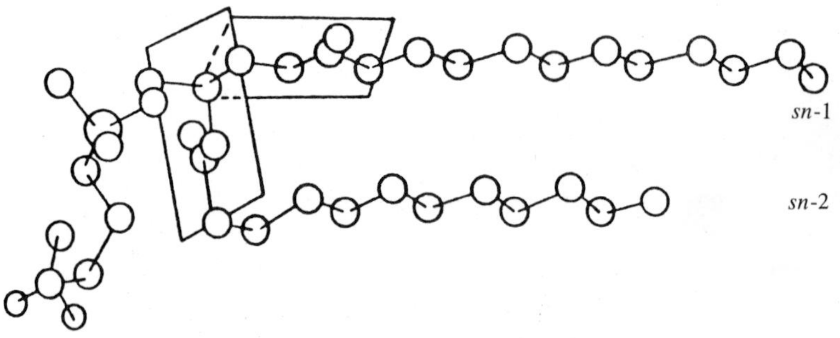

Figure 2. Model revealing stereochemistry of lipid molecules.

the bilayer by 17° and 27°, respectively, for the two conformations. In the case of dilauroylphosphatidylethanolamine, the orientation of the P–N vector is approximately parallel to the bilayer surface.

As discussed in the Introduction, phospholipid molecules in membrane bilayers, especially in the liquid-crystalline state, are very flexible. In fact, they are dynamic structures with considerable degrees of motion (Thompson and Huang, 1980). It is, therefore, incorrect to assume that the X-ray crystallographic solution is either a unique or a correct one, even for the averaged conformation of phospholipids in the oriented two-dimensional lamellar arrays of the bilayer. In fact, no reliable X-ray crystallographic data are yet available for the naturally occurring mixed-chain phospholipids. Nevertheless, X-ray crystallographic data do provide some guidelines for interpretation of the results obtained with phospholipids in bilayers. Some of the fundamental structural features of phospholipids revealed by X-ray studies such as the planarity of the ester group

$$C-O-\overset{\overset{\displaystyle O}{\parallel}}{C}-C$$

and the orientation of the head group appear to be invariant for phospholipids in bilayers in the liquid-crystalline state (Franks, 1976; Worcester and Franks, 1976). From a conformational standpoint, the major difference between the phospholipid molecule in the crystal and that in the lamellar liquid-crystalline bilayer is the difference in the arrangement of carbon atoms of the fatty acyl chain that results from rotation of groups about the single carbon–carbon bonds. For instance, with the exception of C-2 at the *sn*-2 position of the glycerol backbone, all carbon atoms along the carbon–carbon bond axis in the fatty acyl chain of crystalline saturated phospholipids are in the *trans* (*t*) conformation. In contrast, appreciable amounts of *gauche* conformers can be detected directly by monitoring the carbon–carbon stretching region of Raman bands from bilayers of saturated phospholipids in the liquid-crystalline state (Lippert and Peticolas, 1971; Spiker and Levin, 1975). Moreover, ^{2}H NMR studies of deuterated dipalmitoylphosphatidylcholine bilayers in the liquid-crystalline state and similar studies of deuterated stearic acids intercalated in lamellar dispersion of egg phosphatidylcholine both show that the molecular order parameter along the long molecular axis of the fatty acyl chain remains constant with high molecular order to the level of the C-10 atom and then decreases monotonically to a small value at the terminal methyl carbon atom (Seelig, 1977; Smith *et al.*, 1977). Because the molecular order parameter represents the average orientation of the chain segment to which the deuterium is attached, ^{2}H NMR results together with the theoretical calculations of Marcelja (1974) can be taken as evidence to imply that the *gauche* conformers ($g^\pm$) in the first 10-carbon segment are β-coupled to form g^+tg^- and g^-tg^+ kinks and these kinks are not stationary but migrate rapidly up and down the chain. Also, ^{2}H NMR data suggest that the probability of the formation of isolated *gauche* conformers ($tg^\pm t$) increases distinctly for positions along the acyl chain beyond C-10 toward the terminal methyl group. As previously discussed, for naturally occurring phospholipids the fatty acyl chain at the *sn*-2 carbon position is usually unsaturated, and the *cis* double bond is most frequently found between C-9 and C-10. The *cis* bond makes the long chain axes on each side of the double bond form an angle of 130° (Lagaly *et al.*, 1977). However, if a single *gauche* configuration occurs at the β position on either side of the *cis* double bond together with a 30° rotation of the carbon–carbon single bond adjacent to the *cis*-double bond on the same side of the double bond, which may be designated as the Δ*tg* kink, then the overall configuration of this unsaturated chain will become linear, but laterally displaced (see Fig. 3). In contrast to

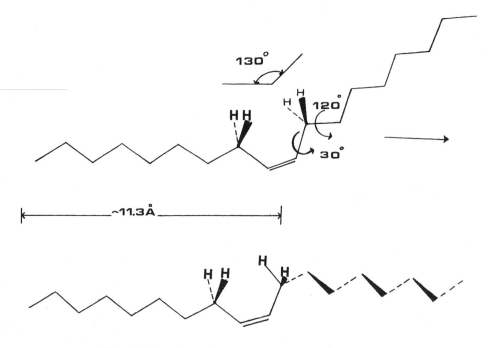

Figure 3. Effect of Δtg kink formation in cholesterol bilayer.

g^+tg^- kinks, the migration of the Δtg kink is limited due to the fixed position of double bonds. There is strong evidence to suggest that Δtg kinks may indeed occur in bilayers of natural phospholipids (Huang, 1977a,b). An important consequence of the formation of $\Delta^9 tg$ kinks in bilayers is that the resulting hydrophobic pocket with a length of approximately 11.3 Å (Fig. 3) provides a perfect complementary locality for the two protruding angular methyl groups on the β face of cholesterol, which are about 4.2 and 8.4 Å, respectively, away from the oxygen atom at the polar head group (Fig. 1).

4. CHOLESTEROL IN MIXED-CHAIN PHOSPHATIDYLCHOLINE BILAYERS

Various NMR techniques have shown that the presence of cholesterol in egg phosphatidylcholine bilayers has selective effects on the motions of the various regions of the phospholipid molecules (Thompson and Huang, 1980; Wennerström and Lindblom, 1977). For instance, the motions of the head group and the methyl group of the acyl chain terminus are not affected by cholesterol, whereas the dynamics of methylene groups near the polar-head region and the olefinic carbon near the center of the acyl chain are markedly reduced. These data strongly suggest that the first 11-carbon segment of egg phosphatidylcholine including the unsaturated double bond between C-9 and C-10 carbon atoms is in direct van der Waals contact with the rigid steroid nucleus of cholesterol. If this is indeed the case, then the rapid kink migration of the first 11-carbon segment must be severely damped by cholesterol and an induced specific configuration to complement the stereospecific and rigid conformation of the steroid nucleus will be adapted by the acyl chains. Therefore, cholesterol can be expected to have an ordering (or condensing) effect on the bilayer of natural phospholipids.

Based on electron density profiles and neutron-scattering amplitude density profiles, it is now well established that the long molecular axis of cholesterol is perpendicular to the bilayer surface and that the position of the cholesterol 3β-OH groups is in proximity to the carbonyl group of phospholipids in the bilayer (Franks and Lieb, 1979). Experiments with cholesterol and its analogs on the physical properties of bilayers indicate that cholesterol has the exact length required to maximize interactions between neighboring chains without disturbing the bilayer structure of egg phosphatidylcholine (Craig *et al.*, 1978; Suckling *et al.*, 1979). Thus, the structure of cholesterol as a whole, not just the steroid nucleus, appears to be critical for the maximum sterol–phospholipid interaction in bilayers.

Two structural configurations of egg phosphatidylcholine have been suggested as possible models to fit the X-ray electron density profile of egg phosphatidylcholine–cholesterol bilayers (Franks, 1976). In one model, the phospholipid molecule has exactly the same configuration as the one derived from the X-ray crystal structure in which the two carbonyl groups are arranged differently at the *sn*-1 and *sn*-2 chains (Hitchock *et al.*, 1974). In the second model, no sharp bend is proposed to occur at the C-2 atom in the *sn*-2 chain; hence, the two carbonyl groups are more or less equivalent. More recent work of Zacci *et al.* (1979) indicates that the two terminal methyl groups of the two alkyl chains of the dipalmitoylphosphatidylcholine in the $L_{\beta'}$ phase are more nearly in register in the center of the bilayer. Instead of 3.7 Å as revealed by the X-ray crystal work, a separation of only 1.8 Å is observed for the two methyl groups in bilayers in the $L_{\beta'}$ phase. Recent Raman work of Bush *et al.* (1980b) suggests that cholesterol and water have a combined effect in inducing the *sn*-2 chain to assume a conformation more equivalent to the *sn*-1 chain. Taken together with the NMR data discussed earlier, these results suggest that the conformational change of the phospholipid induced by cholesterol in the bilayer can be concluded to be quite substantial; it extends from the carbonyl region at the glycerol backbone all the way down along the chain to the methyl terminus, a domain equivalent in length to the overall dimension of the cholesterol molecule.

5. A MODEL FOR CHOLESTEROL–MIXED-CHAIN PHOSPHOLIPID PACKING ASSOCIATIONS IN BILAYERS

In 1977, a structural model for cholesterol–natural phospholipid complexes in the bilayer was proposed by Huang based on structural information available at the time. Basically, the model states that the plane of the cholesterol molecule may occur at a random angle relative to the bilayer normal in the two-dimensional array of the phospholipids provided that the two faces of the plane are preferentially oriented with respect to the fatty acyl chain in the bilayer such that:

1. The 3β-OH group of cholesterol can hydrogen bond to the carbonyl oxygen of the *sn*-1 chain of the natural phospholipids in the bilayer.
2. The β face of the steroid nucleus will be packed in close contact with the unsaturated *sn*-2 fatty acyl chain of the phosphatidylcholine molecule. This close contact is configurationally possible because the hydrophobic pocket generated by the $\Delta^9 tg$ kink (Fig. 3) has the proper dimensions to fit the two angular methyl groups of the sterol β face (Fig. 1).
3. The α face of cholesterol will preferentially pack in close contact with the saturated (*sn*-1) chain in order to maximize the van der Waals contact. This close contact may be a necessary condition for the possible formation of a hydrogen bond between the cholesterol OH group and the *sn*-1 carbonyl oxygen, if such a bond does exist.

In the light of many recent findings on cholesterol–phospholipid interactions (Bush *et al.*, 1980a,b; Clejan *et al.*, 1979), it appears that the proposed hydrogen bond between the cholesterol OH and the *sn*-1 carbonyl oxygen of natural phospholipids in bilayers is, perhaps, geometrically too restricted to occur. In fact, one of the experimental results that was used as evidence to support the hydrogen-bond hypothesis (Yeagle *et al.*, 1975), namely, a substantial downfield shift of the ^{13}C carbonyl resonance of egg phosphatidylcholine upon incorporation of cholesterol (Keough *et al.*, 1973), cannot be reproduced (Wennerström and Lindblom, 1977). Nevertheless, the cholesterol molecule is known to intercalate into the bilayer with its 3β-OH group anchoring at the bilayer interface. In the interface, the water molecules and the fatty acid ester groups are both present (Franks and Lieb, 1979). It is thus not entirely unreasonable to assume that the 3β-OH group of cholesterol can hydrogen bond with either carbonyl oxygens of the ester groups or perhaps with water molecules. However, spectroscopic detection of the hypothesized hydrogen bond between the 3β-OH group of cholesterol and the *sn*-1 carbonyl oxygen of the phospholipid will be very difficult, if not impossible, due to the presence of water molecules in the bilayer interface.

The proposed preferential interactions of the two faces of cholesterol with the saturated and the unsaturated chain of natural phosphoglycerides have not been subjected rigorously to experimental tests. The stereospecific association, however, does explain many experimental results (Huang, 1977a,b). Moreover, the potential usefulness of the model lies in the predictions that result from it. Based on the stereospecific model, one would predict that lanosterol, a precursor of cholesterol in its biosynthetic pathway, cannot pack as nicely as cholesterol in the egg phosphatidylcholine bilayer because the α-methyl group at the steroid nucleus C-14 would prevent the close contact between the linear *sn*-1 chain of the phospholipid and the α face of the lanosterol ring system. This perturbative α face would have the effect of making the neighboring linear segment of saturated acyl chain vulnerable to β-coupled rotations of single C–C bonds. Consequently, lanosterol would be predicted to be more mobile in the bilayer. Experimental data indeed bear this out (Yeagle *et al.*, 1977). Another virtue of the stereospecific model is its explanation of the cholesterol ordering (or condensing) effect on the bilayer of natural phospholipids and of the absence of cholesterol condensing effects on saturated phospholipids in the gel state (Thompson and Huang, 1980). Finally, this model also emphasizes that while the cholesterol molecule is stereochemically rigid, the natural phosphoglycerides can undergo considerable degrees of conformational changes along the long axis of the molecule in the bilayer; whether or not the cholesterol is involved in hydrogen bonding with the ester carbonyl oxygen is not critical to the basic premise of the stereospecific model.

ACKNOWLEDGMENTS. This work was supported in part by Research Grant GM-17452 from the National Institute of General Medical Sciences, USPHS. We thank Professor Robert W. McGilvery for critical reading and enlightening discussions of the manuscript.

REFERENCES

Anderson, C. M., Zucker, F. H., and Steitz, T. A. (1979). *Science* **204**, 375–380.

Banks, R. D., Blake, C. C. F., Evans, P. K., Haser, R., Rice, D. W., Hardy, G. W., Marrett, M., and Phillips, A. W. (1979). *Nature (London)* **279**, 773–777.

Bloch, K. (1965). *Science* **150**, 19–28.

Bush, S. F., Adams, R. G., and Levin, I. W. (1980a). *Biochemistry* **19**, 4429–4436.

Bush, S. F., Levin, H., and Levin, I. W. (1980b). *Chem. Phys. Lipids* **27**, 101–111.

Clejan, S., Bittman, R. R., Deros, P. W., Isaacon, Y. A., and Rosenthal, A. F. (1979). *Biochemistry* **18**, 2118–2125.

Craig, I. F., Boyd, G. S., and Suckling, K. E. (1978). *Biochem. Biophys. Acta* **508**, 418–421.

Craven, B. M. (1976). *Nature (London)* **260**, 727–729.

Franks, N. P. (1976). *J. Mol. Biol.* **100**, 345–358.

Franks, N. P., and Lieb, W. R. (1979). *J. Mol. Biol.* **133**, 469–500.

Griffin, R. G. (1976). *J. Am. Chem. Soc.* **98**, 851–853.

Hitchock, P. B., Mason, R., Thomas, K. M., and Shipley, G. G. (1974). *Proc. Natl. Acad. Sci. USA* **71**, 3036–3040.

Huang, C. (1977a). *Lipids* **4**, 348–356.

Huang, C. (1977b). *Chem. Phys. Lipids* **19**, 150–158.

Jones, M. N. (1975). *Biological Interface,* Elsevier, Amsterdam.

Junger, E., and Reinauser, H. (1969). *Biochem. Biophys. Acta* **183**, 304–308.

Keough, K. M., Oldfield, E., Chapman, D., and Beynon, P. (1973). *Chem. Phys. Lipids* **10**, 37–50.

Kunau, W. H. (1976). *Angew. Chem. Int. Ed. Engl.* **15**, 61–74.

Lagaly, G., Weiss, A., and Stuke, E. (1977). *Biochim. Biophys. Acta* **470**, 331–341.

Lippert, J. L., and Peticolas, W. L. (1971). *Proc. Natl. Acad. Sci. USA* **68**, 1572–1576.

Luzzati, V. (1968). In *Biological Membranes* (D. Chapman, ed.), pp. 71–123, Academic Press, New York.

Luzzati, V., and Husson, F. (1962). *J. Cell Biol.* **12**, 207–219.

Mabrey, S., and Sturtevant, J. M. (1976). *Proc. Natl. Acad. Sci. USA* **73**, 3862–3866.

Mabrey, S., and Sturtevant, J. M. (1977). *Biochem. Biophys. Acta* **486**, 444–450.

Marcelja, S. (1974). *Biochim. Biophys. Acta* **367**, 165–176.

Pauling, P. (1968). In *Structural Chemistry and Molecular Biology* (A. Rich and N. Davidson, eds.), pp. 553–565, Freeman, San Francisco.

Pearson, R. H., and Pascher, I. (1979). *Nature (London)* **281**, 499–501.

Reddy, V. V. R., Kupfer, D., and Caspi, E. (1977). *J. Biol. Chem.* **252**, 2797–2801.

Reiss-Husson, F. (1967). *J. Mol. Biol.* **25**, 363–382.

Sakurai, I., Iwuyanaji, S., Sakwrai, T., and Sato, T. (1977). *J. Mol. Biol.* **117**, 285–291.

Seelig, J. (1977). *Q. Rev. Biophys.* **10**, 353–418.

Shieh, H. S., Hoard, L. G., and Nordman, C. E. (1977). *Nature (London)* **267**, 287–289.

Shimakata, T., Mihava, K., and Sato, R. (1972). *J. Biochem.* **72**, 1163–1174.

Smith, I. C. P., Stockton, G. W., Tulloch, A. P., Polnaszek, C. F., and Johnson, K. G. (1977). *J. Colloid Interface Sci.* **58**, 439–451.

Spiker, R. C., and Levin, I. W. (1975). *Biochim. Biophys. Acta* **388**, 361–373.

Suckling, K. E., Blair, H. A. F., Boyd, G. S., Craig, I. F., and Malcolm, B. R. (1979). *Biochem. Biophys. Acta* **351**, 10–21.

Sundaralingam, M. (1972). *Ann. N.Y. Acad. Sci.* **195**, 324–355.

Thompson, T. E., and Huang, C. (1980). In *Membrane Physiology* (T. E. Andreoli, J. F. Hoffman, and D. D. Fanestil, eds.), pp. 27–48, Plenum Press, New York.

Wennerström, H., and Lindblom, G. (1977). *Q. Rev. Biophys.* **10**, 67–96.

Worcester, D. L., and Franks, N. P. (1976). *J. Mol. Biol.* **100**, 359–375.

Yeagle, P. L., Hutton, W. C., Huang, C., and Martin, R. B. (1975). *Proc. Natl. Acad. Sci. USA* **72**, 3477–3481.

Yeagle, P. L., Martin, R. B., Laka, A. K., Lin, H. K., and Bloch, K. (1977). *Proc. Natl. Acad. Sci. USA* **74**, 4924–4926.

Zaccai, G., Büldt, G., Seelig, A., and Seelig, J. (1979). *J. Mol. Biol.* **134**, 693–706.

3

Sterols and Membranes

Konrad Bloch

> *Nature is the end, and what each thing is when fully*
> *developed, we call nature.*
> *Politics,* Aristotle

1. INTRODUCTION

Research described in this article was prompted by some speculations dealing with the ultimate origin and evolution of the sterol molecule and the corollary inquiry whether various structural features characteristic of cholesterol and its cyclic biosynthetic precursors can be rationalized in terms of function (Bloch, 1976). The riddle why certain organic molecules occur in cells and others do not has received surprisingly little attention. This Aristotelian question is perhaps unanswerable given the fact that knowledge of the chemical components comprising the primordial soup is beyond reach. Yet as I will attempt to show, in the instance of cholesterol the inquiry into the motives of nature has met with some success and the hypothesis verified experimentally to some extent. We can invoke evolutionary pressures as the driving force that shaped precursor molecules to the structure that ultimately came to reside and function competently in cell membranes.

Cholesterol and the related cholestane derivatives of yeast, fungi, and plants occur ubiquitously, though not universally, in the membrane envelope of eukaryotic cells. Opinions as to the role of cholesterol in membranes as distinct from its role as a precursor of hormones, vitamins, and bile acids range from statements that "this role is unknown" (Tanford, 1978) to explicit views that "sterols are important for the regulation or modulation of membrane fluidity" (Demel and DeKruyff, 1976). In assessing the evidence, a distinction must be made between sterol effects that can be demonstrated with artificial membranes or isolated cells and the significance of fluctuating membrane sterol levels *in vivo*. Whatever the ultimate answer, it is a reasonable supposition that the universal occurrence of sterols is not fortuitous but has functional significance.

Konrad Bloch • The James Bryant Conant Laboratories, Harvard University, Cambridge, Massachusetts 02138.

We have elsewhere reasoned that squalene, the acyclic sterol precursor, must be an ancient molecule (Bloch, 1979). The occurrence of the hydrocarbon in the primitive and anaerobic *Archaebacteria* (Tornabene *et al.*, 1978) and several other prokaryotes (Goldberg and Shechter, 1978) supports this view. For squalene per se, no physiological role has so far been described. However, cells seem to have explored various ways for exploiting the chemically facile transformation of squalene to cyclized products. In the more primitive mode, proton-initiated and OH⁻-terminated cyclization invariably produces pentacyclic alcohols, e.g., the fully cyclized hopane derivatives diplopterol in *Acetobacter* species (Ourisson *et al.*, 1979) and tetrahymanol in *T. pyriformis* (Mallory *et al.*, 1963). For this protozoan, tetrahymanol and cholesterol appear to be functionally equivalent (Mallory and Conner, 1971), while bacterial diplopterol duplicates the condensing effects of cholesterol in phospholipid monolayers (Poralla *et al.*, 1980). In contrast to the products of oxidative cyclization to be discussed below, pentacyclic triterpenes seem to function per se, without subsequent nuclear modifications. They did not evolve further, we believe, because their rigid, relatively planar structures can be readily accommodated in phospholipid bilayers (Fig. 1).

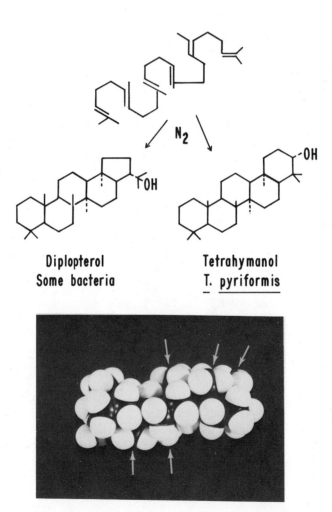

Figure 1. Structural formulas and space-filling models of tetrahymanol and diplopterol. The pentacyclic triterpenes are formed anaerobically from squalene and are not altered metabolically.

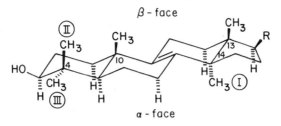

Figure 2. The two faces, α and β, of lanosterol. Encircled numbers indicate the sequence of enzymatic methyl group removal.

In the oxidative and therefore more modern pathway (Corey *et al.*, 1966; van Tamelen *et al.*, 1966), formation of squalene epoxide precedes cyclization, leading ordinarily to tetracyclic products (lanosterol or cycloartenol) with retention of an aliphatic eight-carbon side chain. Lanosterol and probably also cycloartenol are transient intermediates, not major membrane components.

In this discussion the three nuclear demethylations occurring at the sterol α face of lanosterol will be stressed as the significant metabolic events that afford the sterol structures normally found in membranes associated with phospholipid bilayers. These demethylations occur sequentially and invariably in the order shown in Fig. 2 (Gautschi and Bloch, 1959; Sharpless *et al.*, 1969). The consequences are twofold. Methyl groups are replaced by hydrogen, reducing the overall bulk of the molecule, and second, the finished, fully demethylated, tetracyclic sterol ring structure is flat. It displays two planar, roughly parallel surfaces, comprised at the α face (bottom) by six axial hydrogen atoms at C-3, C-5, C-1, C-7, C-12, and C-14 and at the other (β face) by coplanar methyl groups at C-19, C-18, and C-21.

As space-filling models show, the flat, biplanar cholesterol ring system when properly positioned in phospholipid bilayers can engage in multiple nonpolar interactions with contiguous fatty acyl chains. No such fit is possible with lanosterol because the 14α-CH_3 group projects from the α plane. We therefore attribute the oxidative removal of this substituent, the first of three modification steps, to evolutionary pressures. The sterol molecule becomes more competent for expression of membrane function. In support of this rationalization, the physical responses of artificial membranes to insertion of cholesterol and lanosterol respectively can be shown by three independent methods to differ strikingly. Unlike cholesterol, lanosterol fails to impede solute permeation (Lala *et al.*, 1978) or to raise microviscosity of artificial membranes (C. Dahl *et al.*, 1980). Moreover, ^{13}C NMR data show that while membrane-embedded cholesterol is fully immobilized, various sterol carbon resonances appear when lanosterol replaces cholesterol (Yeagle *et al.*, 1977). The membrane-associated 4,4′,14-trimethyl sterol is fully mobile. Clearly, mobility measures the degree of sterol–phospholipid chain interactions, i.e., the strength of packing or condensing effects.

In the biosynthetic pathway, 14α-demethylation is followed by removal of the axial 4β-methyl group to yield 4α-CH_3-Δ^7-cholestenol (lophenol). In the final demethylation step, lophenol affords Δ^7-cholestenol (Nes and McKean, 1977). The progressive increases in the microviscosity of lecithin vesicles caused by these sequential demethylations are shown in Fig. 3. With the departure of each methyl group, the $\bar{\eta}$ values become larger, reaching a maximum with cholesterol itself (C. Dahl *et al.*, 1980). In fact, by this criterion, sterols more effective than cholesterol have not so far been found.

Because the bulk if not all of the sterols residing in natural membranes contain a double bond in ring B of the sterol nucleus, microviscosities of some stanols and Δ^5-stenols have

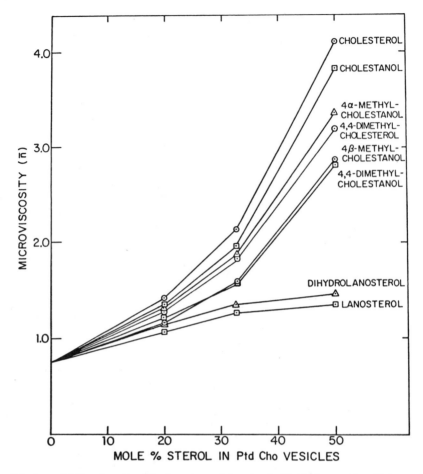

Figure 3. Left panel: Microviscosities ($\bar{\eta}$) of sterol-containing phosphatidylcholine vesicles. Right panel: Effect of sterols on growth rates (mass doubling times) and cell yields of *M. capricolum*.

been compared (C. Dahl *et al.*, 1980). Invariably, B-ring unsaturated stenols produce larger $\bar{\eta}$ values than the corresponding stanols, suggesting a positive contribution of this structural feature to membrane function.

That modulation of membrane fluidity is one of the physiologically significant sterol functions is indicated by growth studies with sterol-requiring organisms. In eukaryotic cells, e.g., anaerobic yeast (Nes *et al.*, 1978; M. Sobus and K. Bloch, unpublished results), pupating insects (Clark and Bloch, 1959a), and an animal cell mutant deficient in demethylating enzymes (Chang *et al.*, 1977), lanosterol fails to satisfy the nutritional requirement for a sterol source, the presumption being that membranes containing this molecule as the only sterol are too fluid. Detailed information on the physiological consequences of sequential lanosterol demethylation has become available from studies with *Mycoplasma capricolum,* a prokaryotic sterol and fatty acid auxotroph (Razin, 1973). By two criteria, growth rates (or cell yields) and effects on the physical state of isolated membranes, sterol effectiveness increases in the order: lanosterol < 4,4-dimethylcholestanol ≤ 4β-methylcholestanol < 4α-methylcholestanol < cholestanol < cholesterol (C. Dahl *et al.*, 1980). This is precisely the order in which this series of sterols induces fluidity changes in model membranes (Fig. 3). Moreover, because the intermediary steps in the enzymatic (eukar-

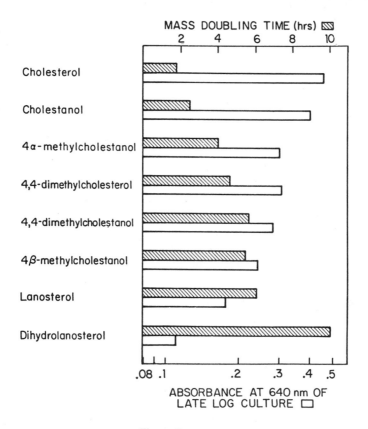

MASS DOUBLING TIME (hrs)

ABSORBANCE AT 640 nm OF
LATE LOG CULTURE

Figure 3. continued

yotic) lanosterol demethylation occur in the same temporal sequence, it is reasonable to argue that the respective nuclear modifications did not arise as chance events; they occurred in response to evolutionary pressures improving the fitness of the sterol molecule for membrane function.

That lanosterol supports moderate *Mycoplasma* growth and does so without being metabolically altered was unexpected, in view of its incompetence for eukaryotic cells. Because *Mycoplasma* species grow only on complex media, the question arose whether some adventitious exogenous sterol not removed by exhaustive delipidation of the medium contributed to or was essential for growth on lanosterol. Small amounts of cholesterol, clearly medium derived, were indeed encountered occasionally in lanosterol-grown cells even though no growth occurs without sterol supplementation. In experiments prompted by these concerns, the combination of low levels of external cholesterol (0.5 μg/ml) and a 20-fold greater amount of lanosterol (10 μg/ml) stimulated the bacterial growth rate more than twofold (J. Dahl *et al.*, 1980). The synergistic character of the growth response suggested a dual or multiple rather than single function for sterols in membranes.* Thus, apart from regulating the bulk physical state of the phospholipid bilayer—the role traditionally ascribed to cholesterol—sterols might act more selectively in localized membrane domains. In exploring the basis, possibly metabolic, of the synergistic effect, we have recently varied the concentration of the fatty acid supplement, an approach that proved profitable. A two- to fourfold increase in the concentrations of palmitate and elaidate in

*Over 20 years ago, during a study of the sterol requirement for pupating insects, we observed a similar dual membrane role for the sterol molecule (Clark and Bloch, 1959b).

the *Mycoplasma* medium supplemented with lanosterol as the sole sterol source raised the bacterial growth rate to levels obtained either with nonlimiting cholesterol or with the synergistic combination of cholesterol + lanosterol (Dahl *et al.*, 1981). Thus, high fatty acid levels spare cholesterol, but only when a bulk sterol (e.g., lanosterol) is provided. Expressed differently, in cells growing suboptimally on lanosterol, the rate of fatty acid uptake appears to be limiting. Cholesterol, in small amounts, specifically facilitates this process. The specific, cholesterol-promoted step in the synthesis of membrane phospholipids remains to be elucidated.

As noted earlier, control of bulk fluidity is believed to be the principal role that sterols play in membranes. Yet under certain chosen conditions, *Mycoplasma* growth rates and membrane fluidity are not necessarily correlated. It is true that when cholesterol serves as the sole sterol source, the bacterial cell yield parallels the cholesterol content of the cell membranes and in turn correlates positively with the bacterial membrane microviscosity (or inversely with fluidity) (J. Dahl *et al.*, 1980). On the other hand, cells grown under synergistic conditions, containing lanosterol and cholesterol in a ratio of approximately 10 : 1, display very low, lanosterol-determined microviscosity (J. Dahl *et al.*, 1980). Yet their rate of growth and cell yield approach those of cholesterol-rich cells. It is therefore possible to achieve nearly identical, optimal *Mycoplasma* growth under conditions that produce cells exhibiting either high or low bulk fluidity. One might argue that in membranes whose bulk sterol is provided by lanosterol there exist less fluid, localized cholesterol-rich regions which are essential. However, this seems unlikely because 3α-methylcholesterol, a sterol that duplicates the synergistic growth effect of cholesterol in combination with lanosterol (Dahl *et al.*, 1981), does not as such alter the fluidity of either artificial lecithin vesicles (Lala *et al.*, 1978) or *Mycoplasma* membranes (Odriozola *et al.*, 1978). For *M. capricolum* at least, it seems doubtful that the exclusive role of sterol is to ensure membrane stability or cell viability by fluidity control.

In discussing alternative functional modes, we have pointed out that lanosterol shares with cholesterol the property of increasing the distance between phospholipid head groups, as judged from $^{31}P\{'H\}$ nuclear Overhauser effects observed with lecithin vesicles (Yeagle *et al.*, 1977). This effect is seen even though lanosterol interacts much less strongly than cholesterol with fatty acyl chains in the bilayer. Bulk and rigidity rather than planarity of the inserted polycyclic molecule may therefore be the more critical structural feature for sterol function in membranes.

2. SELECTIVE DEMETHYLATION

It deserves emphasis that nuclear demethylation is a highly selective process, confined to the lanosterol α face (CH$_3$ groups at C-14 and C-4). Sterols lacking the bridge-head methyl groups at C-18 or C-19 or the C-21 methyl carbon of the side chain are only rarely found in nature. The apparent absence of selective pressures to dealkylate the sterol β face can again be rationalized in terms of productive or nonproductive interactions of the steroidal ring system with its phospholipid partners in the membrane bilayer. In model membranes, 19-norcholesterol raises the microviscosity somewhat but definitely less than cholesterol. It is inferior to cholesterol in condensing the bilayer phase (Lala *et al.*, 1978). Similarly, 19-norcholesterol is a slightly poorer sterol source for the sterol-requiring *M. capricolum* (J. Dahl, unpublished results) and significantly less active than cholesterol in supporting the growth of a yeast mutant deficient in sterol biosynthesis (T. Buttke, unpublished results). In the space-filling model, the methyl groups at C-18, C-19, and C-21 are seen to be roughly coplanar, an alignment that is favorable as such for interactions with fatty acyl chains. Therefore, their departure would not be expected to improve van der Waals contacts and, it may be surmised, did not occur for that reason. Compelling arguments why the retention of β-face alkyl groups may in fact be advantageous, particularly for interactions with *cis*-olefinic fatty acyl chains, have been presented by Huang (1977). In sum, the late stages of sterol biosynthesis, i.e., the conversion of lanosterol to cholesterol, appear to be designed to remove only substituents that weaken or interfere with sterol–fatty acid interactions but to retain those that are either neutral or beneficial.

3. RATIONALIZATION OF THE 4α-CH₃–4β-CH₃ DEMETHYLATION SEQUENCE

While the adverse membrane effects of the 14α-CH_3 substituent can be ascribed to steric interference, the reasons for the ordinarily compulsory removal of the two methyl groups at C-4 are less apparent, at least on steric grounds. Neither of the two CH_3 groups is sufficiently exposed to interfere with fatty acyl chain packing. Judging from fluidity measurements of model membranes (C. Dahl *et al.*, 1980) and the ability to support the growth of sterol auxotrophs (C. Dahl *et al.*, 1980), 4,4'-dimethyl and 4β-methyl (axial) sterols have the same relatively low membrane competence, while the behavior of 4α-methylstanols or stenols approaches that of cholesterol. This fact alone suggests that the axial 4β-methyl group renders the sterol molecule less competent than the equatorial 4α-CH_3 epimer. In line with this reasoning, the oxidative enzymes catalyzing dealkylation of C-4 produce the 4α-monomethyl epimer, not the 4β epimer, as an intermediate (Nes and McKean, 1977). Elimination of the 4β-methyl group has the highest priority. As seen in space-filling models, the 4β-methyl group is sufficiently close to the angular CH_3 group at C-10 to provoke *syn*-diaxial interactions. Conceivably, such interactions disturb the contacts with phospholipid acyl chains at the sterol β face. That the 4β epimer does not occur in nature fits gratifyingly with the hypothesis that each metabolic step selected by evolution affords a product with membrane properties superior to those of its antecedent.

4. STEROL STRUCTURE AND FUNCTION IN YEAST

The important discovery by Andreasen and Stier that growth of *Saccharomyces cerevisiae* under strictly anaerobic conditions requires supplementation with a sterol and an unsaturated fatty acid (Andreasen and Stier, 1954) paved the way for numerous studies on sterol structure–function relationships in a eukaryotic cell. Some of the relevant literature reports must, however, be viewed with caution because total anaerobiosis is technically difficult to achieve, for the apparent reason that the oxygen tension required for certain of the oxygenase-catalyzed reactions in the sterol pathway is exceedingly low. For example, there are conflicting reports whether or not lanosterol can support anaerobic growth of yeast. According to the most recent findings (Nes *et al.*, 1978; M. Sobus and K. Bloch, unpublished results), this point has been settled in the negative. It may be recalled that another eukaryotic cell, a demethylase mutant of Chinese hamster ovary cells blocked in lanosterol metabolism, is likewise unable to grow unless supplied with cholesterol (Chang *et al.*, 1977).

For extending results obtained with anaerobic wild-type yeast, a mutant recently described by Gollub *et al.* (1977) has proven exceptionally useful. The mutant, GL7, is deficient in squalene-epoxide cyclase and also in heme biosynthesis and consequently devoid of all cytochromes. 14α-Demethylation, a P_{450}-dependent process (Ohba *et al.*, 1978), can therefore not occur. In contrast to wild-type yeast, GL7 grows well on lanosterol anaerobically, without modifying it (Buttke and Bloch, 1980), as expected from the P_{450} deficiency. Gollub and colleagues mention related observations showing that for yeast strain 587, lanosterol is an adequate sterol source (Gollub *et al.*, 1977). Conceivably, the membranes or organellar structures of these yeast mutants are also altered, allowing growth to occur on a structurally more primitive sterol.

Aerobically, also, lanosterol will support growth of GL7 but curiously only after suitable adaptation. In the presence of air, adapted cells efficiently demethylate lanosterol at C-4 to produce 14α-methyl-4-desmethyl cholestenols (Gollub *et al.*, 1977; Buttke and Bloch, 1980).* The growth-promoting molecule for aerobic GL7 is therefore not lanosterol itself but presumably a 4-desmethyl sterol. An incidental conclusion to be drawn from these results is that the oxygenase-catalyzed demethylations at C-4, in contrast to C-14 demethylation, are not heme dependent. The positive growth response of GL7 to sterols containing a nonmetabolizable 14α-CH_3 group suggests in any event a greater tolerance of certain yeast cells to sterol structures that are either "incompetent" in model membranes (Lala *et al.*, 1978) and in a demethylase-deficient mammalian cell (Chang *et al.*, 1977), or poor sterol sources for *Mycoplasma* (C. Dahl *et al.*, 1980). This raises the question whether the spatial sterol–phospholipid relationships as generally postulated apply to the membranes of yeast. According to the space-filling model, sterols containing an exposed 14α-CH_3 group are poorly accommodated. It should be added that whatever little we know about the role of membrane sterols in animal cells, there is even less certainty about sterol function in yeast, or for that matter about intracellular sterol localization in this organism.

5. STRUCTURAL REQUIREMENT FOR C-4 DEMETHYLATION

As mentioned above, given the proper conditions, yeast mutant GL7 readily demethylates lanosterol to a 14α-methyl cholestenol in aerobic cultures. It was therefore predictable that 4,4-dimethyl or 4-monomethyl sterols would be similarly metabolized. This prediction has been experimentally verified (T. Buttke and K. Bloch, unpublished results). The role of the B-ring double bond in the substrates to be dealkylated deserves some comment. 4,4-Dimethylcholest-7-enol supported growth and was readily converted to cholesta-5,7-dienol, as was 4α-methylcholest-7-enol (lophenol). The Δ^7 isomers are normal intermediates at this stage of sterol biosynthesis. By contrast, the corresponding 4-alkyl-Δ^5 isomers, which do not occur naturally, were metabolically inert. Interestingly, the 4-alkyl stanols were also active sterol sources, yielding cholestanol in the growing cultures. Clearly, the presence of the Δ^7 double bond is incidental and dispensable for C-4 dealkylation. On the basis of these observations, we can rationalize the fact that during sterol biosynthesis, introduction of the Δ^5 double bond is invariably delayed until both C-4 demethylation steps have occurred. We cannot, however, offer any mechanistic explanation why a Δ^5 double bond should interfere with the oxygenase-catalyzed C-4 demethylations. Nonetheless, the specific sequence of events as it normally occurs during the final stage of lanosterol demethylation can be viewed as a mechanistically determined rather than fortuitous choice.

---

*The evidence suggests that in animal tissues, removal of the 14α-CH_3 group prior to C-4 demethylation is compulsory (Sharpless *et al.*, 1968).

6. *SIDE-CHAIN ALKYLATION OF YEAST STEROLS*

The sterols of yeast and fungi, in contrast to those of animal tissues, typically contain an extra methyl group attached to C-24 of the sterol side chain. In plants, an ethyl substituent occupies the same position. We have recently raised the question whether and why such side-chain modifications confer a demonstrable benefit on the cells that produce them. Yeast, under conditions of sterol auxotrophy (either wild-type anaerobically, or various mutants), invariably fare better when supplied with ergosterol, stigmasterol, or β-sitosterol than cells grown in the presence of cholesterol or 7-dehydrocholesterol (Nes *et al.*, 1978; Buttke *et al.*, 1980). In studies addressing this question with the mutant GL7 (an unsaturated fatty acid as well as sterol auxotroph), we find that relative growth rates depend not only on sterol structure but also on the unsaturated fatty acid supply (Buttke *et al.*, 1980). In the presence of linoleic acid, linolenic acid, or a mixture of palmitoleic and oleic acids, growth was as rapid with cholesterol or 7-dehydrocholesterol as with ergosterol. On media supplemented with oleic acid alone, the cells grew more slowly with 7-dehydrocholesterol than with ergosterol; on petroselenic acid (Δ^6-C_{18}), the cell yield on 7-dehydrocholesterol dropped to about 5% of that seen with ergosterol. Regardless of the unsaturated fatty acid supplied, i.e., over a wide range of fatty acid fluidities, GL7 grows well on ergosterol, the sterol yeast normally produces. When sterols lacking the C-24 alkyl substituent replace ergosterol, good growth occurs only with the more unsaturated, lower melting fatty acids. It appears that yeast normally compensates for its limited capacity to synthesize unsaturated, especially dienoic or trienoic fatty acids by alkylating the sterol side chain. Both devices appear to be designed to disorder the membrane bilayer interior.

The fatty acid patterns of cells grown on 7-dehydrocholesterol or ergosterol respectively likewise display an intimate interdependence between fatty acid and sterol structure. GL7 supplied with a $C_{16\,:\,1}$–$C_{18\,:\,1}$ mixture (1 : 4) incorporates only half as much palmitate into membrane phosphatidylethanolamine when 7-dehydrocholesterol rather than ergosterol is the sterol source (Buttke *et al.*, 1980). The correspondingly increased levels of cellular unsaturated fatty acids may represent an attempt to raise membrane fluidity in the absence of the disordering effect that the bulky C-24 methyl group of ergosterol brings about normally (Semer and Gelerinter, 1979). We find that GL7 responds in essentially the same manner (conditions for optimal growth, more saturated phospholipid fatty acid patterns) to β-sitosterol and stigmasterol as to ergosterol (Buttke *et al.*, 1980). For yeast physiology, the presence of an alkyl group at C-24 rather than its size is therefore the important structural feature. Why plants invariably introduce the larger ethyl substituent into the sterol side chain by a second alkylation remains an intriguing question.

7. *FUNCTION OF STEROL HYDROXYL*

Regardless of structural diversity elsewhere in the molecule, membrane sterols invariably possess an unsubstituted equatorial hydroxyl group, introduced at the stage of squalene epoxide–lanosterol cyclization. Significance has been attached to the fact that this OH group is free and therefore potentially available if not essential for H-bonding to some electron-rich region of the glycerophosphoryl-X moiety of membrane phospholipid (Brockerhoff, 1974; Huang, 1976; Yeagle *et al.*, 1976). Whether such H-bonded interactions contribute to membrane structure and stability has not been settled.

In the instances of both *M. capricolum* (Odriozola *et al.*, 1978) and in the anaerobically grown yeast mutant GL7 (Lala *et al.*, 1979), cholesteryl methylether as such satisfies the essential sterol requirement for growth. It accounts for essentially all of the recovered cellular sterol. At least judged by growth, blocking the free OH group irreversibly does not impair membrane function. Clearly, esterification with fatty acids or ketone formation is not essential for the two sterol auxotrophs investigated.

8. SUMMARY

The objective of the studies described here was to offer a functional explanation for the various enzymatic modifications of the sterol molecule during the lanosterol–cholesterol (or ergosterol) transformation. Obviously, sterol auxotrophs are best suited for answering such questions and their evolutionary implications. *M. capricolum,* exceptional as a bacterial sterol auxotroph, has properties ideal for our purposes. First, the organism is structurally simple, containing only a single membrane. Second, it incorporates and uses a wide variety of sterols without structural alterations. Whatever physiological changes ensue from modifications in the sterol structure are therefore caused by the test sterol per se. Third, the genetic deficiencies in the synthesis of both saturated and unsaturated fatty acids provide a handle for manipulating phospholipid composition as well. Beneficial or adverse relationships between the two lipid classes can therefore be assessed. Examples of such relationships have been given.

With the aid of the yeast mutant GL7, similar information has been obtained for a unicellular eukaryotic cell. Aerobically, the mutant metabolizes sterols to some extent and therefore has been useful for outlining some of the steps in yeast sterol biosynthesis. In future work with the mutant it should be possible to address the unsolved question whether in yeast sterol association or content is functionally important for organelles other than the plasma membrane.

We have provided evidence that allows us to view the selective sequential demethylation of lanosterol at C-14 and C-4 as beneficial events for stepwise improvement of membrane function. Correlations between sterol-controlled membrane fluidity and microbial growth rates have been established. Nevertheless, some of our findings raise doubts that bulk membrane fluidity is indeed the single factor controlling or determining the physiological response. Finally, using pairs of sterols, we have encountered a synergistic effect of sterols on microbial growth, best explainable by a dual role of sterols in membranes.

ACKNOWLEDGMENTS. This work was supported by grants-in-aid from the National Institutes of Health, the National Science Foundation, and the Eugene P. Higgins Fund of Harvard University.

The author is indebted to Drs. Jean Dahl, Charles Dahl, and Thomas M. Buttke for valuable discussions.

REFERENCES

Andreasen, A. A., and Stier, T. J. B. (1954). *J. Cell. Comp. Physiol.* **41,** 23–27.
Bloch, K. (1976). In *Reflections on Biochemistry* (A. Kornberg, B. L. Horecker, L. Cornudella, and J. Oro, eds.), pp. 143–150, Pergamon Press, Oxford.
Bloch, K. (1979). *Crit. Rev. Biochem.* **7,** 1–5.
Brockerhoff, H. (1974). *Lipids* **9,** 645–650.
Buttke, T., and Bloch, K. (1980). *Biochem. Biophys. Res. Commun.* **92,** 229–236.
Buttke, T., Jones, S., and Bloch, K. (1980). *J. Bacteriol.* **144,** 124–138.

Chang, T. Y., Telakowsky, C., vanden Heuvel, W., Alberts, A. W., and Vagelos, P. R. (1977). *Proc. Natl. Acad. Sci. USA* **74,** 832–836.

Clark, A. J., and Bloch, K. (1959a). *J. Biol. Chem.* **234,** 2578–2582.

Clark, A. J., and Bloch, K. (1959b). *J. Biol. Chem. 234,* 2583–2588.

Corey, E. J., Russey, W. E., and Ortiz de Montellano, P. R. (1966). *J. Am. Chem. Soc.* **88,** 4750–4751.

Dahl, C., Dahl, J., and Bloch, K. (1980). *Biochemistry* **19,** 1462–1467.

Dahl, J., Dahl, C., and Bloch, K. (1980). *Biochemistry* **19,** 1468–1472.

Dahl, J., Dahl, C., and Bloch, K. (1981). *J. Biol. Chem.* **256,** 87–91.

Demel, R. A., and DeKruyff, B. (1976). *Biochim. Biophys. Acta* **457,** 109–132.

Gautschi, F., and Bloch, K. (1959). *J. Biol. Chem.* **243,** 1343–1347.

Goldberg, I., and Shechter, I. (1978). *J. Bacteriol.* **135,** 717–720.

Gollub, E. G., Liu, K., Dayan, J., Adlersberg, M., and Sprinson, D. B. (1977). *J. Biol. Chem.* **252,** 2846–2854.

Huang, C. (1976). *Nature (London)* **259,** 242–244.

Huang, S. H. (1977). *Lipids* **12,** 348–356.

Lala, A. K., Lin, H. K., and Bloch, K. (1978). *Bioorg. Chem.* **7,** 437–445.

Lala, A. K., Buttke, T. M., and Bloch, K. (1979). *J. Biol. Chem.* **254,** 10582–10585.

Mallory, F. B., and Conner, R. L. (1971). *Lipids* **6,** 149–153.

Mallory, F. B., Gordon, J. T., and Conner, R. L. (1963). *J. Am. Chem. Soc.* **85,** 1362–1363.

Minale, L., and Sodano, G. (1974). *J. Chem. Soc. Perkin Trans. 1* **1974,** 1888–1892.

Nes, R. W., and McKean, M. L. (1977). In *Biochemistry of Steroids and Other Isopentenoids,* p. 375, University Park Press, Baltimore.

Nes, R. W., Sekula, B. C., Nes, W. D., and Adler, J. H. (1978). *J. Biol. Chem.* **253,** 6218–6225.

Odriozola, J. M., Waitzkin, E., Smith, T. L., and Bloch, K. (1978). *Proc. Natl. Acad. Sci. USA* **75,** 4107–4109.

Ohba, M., Sato, R., Yoshida, Y., Nishino, T., and Katsuki, H. (1978). *Biochem. Biophys. Res. Commun.* **85,** 21–27.

Ourisson, G., Albrecht, P., and Rohmer, M. (1979). *Pure Appl. Chem.* **51,** 709–729.

Popov, S., Carlson, R. M. K., Weymann, A., and Djerassi, C. (1975). *Tetrahedron Lett.* **31,** 758–760.

Poralla, K., Kannenberg, E., and Blume, A. (1980). *FEBS Lett.* **113,** 107–110.

Razin, S. (1973). In *Advances in Microbial Physiology* (A. H. Rox and D. W. Tempest, eds.), pp. 1–80, Academic Press, New York.

Semer, R., and Gelerinter, E. (1979). *Chem. Phys. Lipids* **23,** 201–211.

Sharpless, K. B., Snyder, T. E., Spencer, T. A., Maheshwari, K. K., Guhn, G., and Clayton, R. B. (1968). *J. Am. Chem. Soc.* **90,** 6874–6875.

Sharpless, K. B., Snyder, T. E., Spencer, T. A., Maheshwari, K. K., Nelson, J. A., and Clayton, R. B. (1969). *J. Am. Chem. Soc.* **91,** 3394–3396.

Tanford, C. (1978). *Science* **200,** 1012–1018.

Tornabene, T. G., Wolfe, R. S., Balch, W. E., Holzer, G. E., and Oro, J. (1978). *J. Mol. Evol.* **11,** 259–266.

van Tamelen, E. E., Willet, J. D., Clayton, R. B., and Lord, K. (1966). *J. Am. Chem. Soc.* **88,** 4752–4754.

Yeagle, P. L., Hutton, W. C., Huang, C., and Martin, R. (1976). *Biochemistry* **15,** 2121–2124.

Yeagle, P. L., Martin, R., Lala, A. K., Lin, H. K., and Bloch, K. (1977). *Proc. Natl. Acad. Sci. USA* **74,** 4924–4926.

4

Compositional Domain Structure of Lipid Membranes

Ernesto Freire and Brian Snyder

1. INTRODUCTION

During the past few years it has become evident that the molecular constituents of biological membranes are not randomly organized within the bilayer matrix and that many of their physical and functional properties are sensitive to the particular way in which lipid and protein molecules are distributed within the bilayer (Taylor *et al.*, 1971; Shimshick and McConnell, 1973; Hui and Parsons, 1975; Fishman and Brady, 1976; Papahadjopoulos *et al.*, 1976; van Dijck *et al.*, 1978; Thompson, 1978; Correa-Freire *et al.*, 1979). There are various types of ordered molecular arrangements, ranging from nonspecific aggregation or lateral phase separation processes to highly specific molecular interactions leading to the formation of complex structural patterns (Satir, 1976; Lee, 1977; Caspar *et al.*, 1977; Makowski *et al.*, 1977; Wallace and Engelman, 1978). In general, the lateral distribution of molecules in a lipid bilayer is dictated by the energetics of the interactions between the various components and as such is susceptible of being altered by changes in temperature, pH, ionic strength, concentration of ligand molecules, or other physicochemical variables (Wallace and Engelman, 1978; Pearson *et al.*, 1979). The compositional domain structure of lipid membranes has been investigated by direct visualization using electron microscopy (Caspar *et al.*, 1977; Pearson *et al.*, 1979; Hui, 1981) or by appropriately transforming spectroscopic and physicochemical data into the parameters describing the lateral organization of the membrane (von Dreele, 1978; Wolber and Hudson, 1979; Klausner *et al.*, 1980; Kleinfeld and Solomon, 1982). Recently we have developed a system of Monte Carlo calculations (see Binder, 1979, for a review on Monte Carlo methods) that allows us to generate with the computer lipid and protein distributions and to relate these distributions with various physical and functional properties of the membrane (Freire and Sny-

Ernesto Freire • Department of Biochemistry, University of Tennessee, Knoxville, Tennessee 37916.
Brian Snyder • Department of Biochemistry, University of Virginia School of Medicine, Charlottesville, Virginia 22908.

der, 1980a,b, 1982; Snyder and Freire, 1980). These computer methods are directed to evaluate the number of contacts between molecular species, the size distribution of compositional domains, their topological localization, geometry, and physical extension within the membrane. In this review we will briefly summarize the distributional parameters that have been obtained for various phospholipid and cholesterol–phospholipid mixtures, and their correlation with transport processes across and along the plane of the membrane.

2. ENERGETICS OF LATERAL ORGANIZATION

Compositional domains arise from the process of distributing molecules of different species within the lateral plane of the bilayer. In general, the size, shape, and number of these domains are a function of the interaction energies and the molar composition of the mixtures (see Lee, 1977, for a review). For two-component systems the lateral distribution of molecules can be described in terms of a single affinity constant, P, defined as the excess intrinsic probability for establishing a contact between two molecules of the same type (Freire and Snyder, 1980a). The magnitude of P is proportional to the degree of nonideality in the mixing between components. A value of $P = 1$ defines the ideal mixture in which the two components are randomly distributed within the bilayer. A $P > 1$ reflects a tendency for the individual components to undergo lateral separation into compositionally distinct domains. Conversely, a $P < 1$ is indicative of repulsive interactions between molecules of the same type. The actual value of P can be estimated from phase diagrams (Freire and Snyder, 1980a) or from the compositional dependence of a physical observable that is sensitive to the lateral distribution of molecules within the bilayer (Snyder and Freire, 1980). Typical P values for mixtures of phosphatidylcholines range from 1.25 for DMPC–DPPC in the liquid-crystalline phase to 3.0 for DMPC–DSPC in the gel phase. These P values correspond to excess mixing energies ranging between 150 and 700 cal/mole and indicate a greater affinity between lipid molecules of the same type. For phosphatidylcholines the P values are larger in the gel phase, indicating that the gel-to-liquid-crystal phase transition of these mixtures is coupled to a lateral reorganization process. For cholesterol–phospholipid mixtures the P values range from 1.0 for cholesterol–DMPC and cholesterol–palmitoylsphingomyelin to 4.0 for cholesterol–lignoceroylsphingomyelin. These P values can be used to generate molecular distributions with the computer as described by Freire and Snyder (1980a). The resulting computer-generated distributions provide a sequence of instantaneous bilayer configurations that can be directly analyzed to obtain the dependence of the lateral organization parameters on the composition, the bilayer size, and the magnitude of the interaction energies.

3. CONCENTRATION DEPENDENCE OF COMPOSITIONAL DOMAIN STRUCTURE

As the molar fraction of a membrane component increases, its characteristic domain size within the bilayer also increases. It must be noted that there are two distinct mechanisms accounting for this increase in domain size: (1) addition of new molecules to preexisting domains, and (2) fusion of two or more domains into larger ones. The analysis of the domain distribution functions for mixtures of phosphatidylcholines (Freire and Snyder, 1980b) and cholesterol–phospholipid mixtures (Snyder and Freire, 1980) has revealed that the first mechanism predominates at both low and high concentrations, and that there is a very narrow critical concentration range in which the increase in domain size occurs almost exclusively through fusion of preexisting domains. For binary mixtures of phospholipids the critical concentration range is centered at $\sim$ 50 mole% (Freire and Snyder, 1980b), whereas for cholesterol–phospholipid mixtures the critical concentration for the cholesterol-rich lipid domains is $\sim$ 20 mole% cholesterol. Even though the critical concentration range is very narrow, the fusion process is quite massive and involves most of the existent domains. This process, technically called percolation (Stoll and Domb, 1978; Coniglio and Russo, 1979), results in the formation of a single-domain network that covers the entire bilayer surface. Percolation processes like the above can be defined for domains of homogeneous compositions and also for domains of heterogenous composition characterized by some common property, e.g., it is possible to define the percolation of fluid domains, the percolation of fast-diffusion domains, etc. The common characteristic is that at the percolation threshold these domains become connected with each other, forming a network that covers the entire bilayer surface. This change is the lateral connectivity of the compositional domains induces very dramatic changes in the physical properties of the bilayer.

4. LATERAL CONNECTIVITY AND TRANSPORT PROCESSES

The lateral connectivity of compositional domains can be affected by the chemical and physical environment of the bilayer provided that the molar composition of the membrane is close to a percolation point. This is illustrated in Fig. 1. Panel A represents a hypothetical lateral configuration for a relatively high P value (high affinity between like molecules). There are a few, very compact (low perimeter-to-surface ratio) compositional domains resembling isolated islands within the bilayer surface. A decrease in the value of P will cause an increase in the number of contacts between unlike molecules, the compositional domains will become ramified and they will touch one another, giving rise to a very intricate network as shown in panel B.

Percolation processes like the one described above might dramatically affect several properties of the membrane depending on the characteristics of the percolating domains. Perhaps one of the most important and biologically relevant cases is that in which the

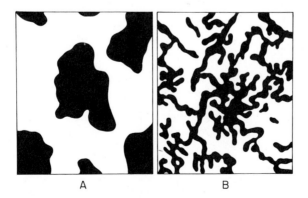

A   B

Figure 1. Schematic illustration of a hypothetical percolation process induced by decreasing the affinity between black molecules. For the high-affinity state (panel A), the compositional domains are large, compact, and disconnected; a decrease in the affinity (panel B) results in the formation of a highly ramified and connected network that extends over the entire membrane surface.

percolating domains have a much larger diffusion coefficient than the rest of the membrane. In this case the existence of a highly connected fast-diffusion network ensures a very rapid and efficient transport of molecules throughout the entire membrane surface. Below the percolation point lateral transport is severely restricted because the motion of molecules must occur through alternating fast- and slow-diffusion regions. Above the percolation point the spreading of molecules over the whole membrane surface can be done entirely through fast-diffusion regions. Recently, fluorescence photobleaching recovery studies by Rubenstein *et al.* (1979) on cholesterol–phosphatidylcholine mixtures below their phase transition have shown an order of magnitude jump in the lateral diffusion coefficient at $\sim$ 20 mole% cholesterol, i.e., at the same cholesterol concentration in which the cholesterol-rich domains become connected.

For cholesterol–phospholipid mixtures the formation of a highly connected network of cholesterol-rich domains is accompanied by the isolation of the pure lipid domains into small areas incapable of long-range cooperative behavior. This phenomenon is reflected in the disappearance of the sharp transition peak at 20 mole% cholesterol (Estep *et al.*, 1978). Consequently, membrane properties that are maximized by the existence of a sharp phase transition show a dramatic decrease at $\sim$ 20 mole% cholesterol. For example, Pownall *et al.* (1979) have shown that the ability of dimyristoylphosphatidylcholine to associate with human high-density apolipoprotein A-I is maximal at the lipid phase transition temperature and that it abruptly disappears between 17 and 20 mole% cholesterol. Swaney (1980) has recently extended these studies to binary mixtures of phosphatidylcholines and has also found that the binding ability is abruptly lost between 19 and 22 mole% cholesterol. These studies have shown that transport processes both across and along the plane of the bilayer are susceptible of being modulated by the compositional domain structure of the bilayer and that this domain structure may in turn be affected by the chemical and physical environment.

ACKNOWLEDGMENT. This work was supported by Grants GM-27244 and GM-26894 from the National Institutes of Health.

REFERENCES

Binder, K. (1979). *Top. Curr. Phys.* **7**, 1–45.

Caspar, D. L. D., Goodenough, D. A., Makowski, L., and Phillips, W. C. (1977). *J. Cell Biol.* **74**, 605–628.

Coniglio, A., and Russo, L. (1979). *J. Phys. A Math. Nucl. Gen.* **12**, 545–550.

Correa-Freire, M. C., Freire, E., Barenholz, Y., Biltonen, R., and Thompson, T. E. (1979). *Biochemistry* **18**, 442–445.

Estep, T. N., Mountcastle, D. B., Biltonen, R. L., and Thompson, T. E. (1978). *Biochemistry* **17**, 1984–1989.

Fishman, P. H., and Brady, R. O. (1976). *Science* **194**, 906–915.

Freire, E., and Snyder, B. (1980a). *Biochemistry* **19**, 88–94.

Freire, E., and Snyder, B. (1980b). *Biochim. Biophys. Acta* **600**, 643–654.

Freire, E., and Snyder, B. (1982). *Biophys. J.,* in press.

Hui, S. W. (1981). *Biophys. J.* **34**, 383–395.

Hui, S. W., and Parsons, D. F. (1975). *Science* **190**, 383–384

Klausner, R. D., Kleinfeld, A. M., Hoover, R. L., and Karnovsky, M. J. (1980). *J. Biol. Chem.* **255**, 1286–1295.

Kleinfeld, A. M., and Solomon, A. K. (1982). In preparation.

Lee, A. G. (1977). *Biochim. Biphys. Acta* **472**, 285–344.

Makowski, L., Caspar, D. L. D., Phillips W. C., and Goodenough, D. A. (1977). *J. Cell Biol.* **74**, 629–645.

Papahadjopoulos, D., Vail, W. J., Pangborn, N. A., and Poste, G. (1976). *Biochim. Biophys. Acta* **448**, 265–283.

Pearson, R. P., Hui, S. W., and Stewart, T. P. (1979). *Biochim. Biophys. Acta* **557**, 265–282.

Pownall, H. J., Massey, R. B., Kussnow, S. K., and Gotto, A. M., Jr. (1979). *Biochemistry* **18**, 574–579.

Rubenstein, J. L. R., Smith, B. A., and McConnell, H. M. (1979). *Proc. Natl. Acad. Aci. USA* **76**, 15–18.

Satir, B. (1976), *J. Supramol. Struct.* **5**, 381–389.

Shimshick, E. J., and McConnell, H. M. (1973). *Biochemistry* **12**, 2351–2360.

Snyder, B., and Freire, E. (1980). *Proc. Natl. Acad. Sci. USA* **77**, 4055–4059.

Stoll, E., and Domb, C. (1978). *J. Phys. A Math. Nucl. Gen.* **11**, L57–L61.

Swaney, J. B. (1980). *J. Biol. Chem.* **255**, 8791–8797.

Taylor, R. B., Duffus, W. P. H., Raff, M. D., and De Petris, S. (1971). *Nature (London)* **233**, 225–230.

Thompson, T. E. (1978). In *Molecular Specialization and Symmetry in Membrane Function* (A. K. Solomon and M. Karnovsky, eds.), pp. 78–98, Harvard University Press, Cambridge, Mass.

van Dijck, P. W. M., de Kruijff, B., Verkleij, A. J., van Deenen, L. L. M., and de Gier, J. (1978). *Biochim. Biophys. Acta* **512**, 84–96.

von Dreele, P. H. (1978). *Biochemistry* **17**, 3939–3943.

Wallace, B., and Engelman, D. (1978). *Biochim. Biophys. Acta* **508**, 431–449.

Wolber, P. K., and Hudson, B. P. (1979). *Biophys. J.* **28**, 197–210.

5

Structural and Functional Aspects of Nonbilayer Lipids

B. de Kruijff, P. R. Cullis, and A. J. Verkleij

1. INTRODUCTION

The bilayer model of the lipid part of biological membranes has been very popular over the past few decades owing to the fact that it provided a rationale for many observations made in the field of membrane biology. In recent years, however, it has become increasingly clear that this bilayer model is incomplete as it does not explain two basic properties of membranes. The first is that next to bilayer-forming lipids there is an abundant occurrence of membrane lipids that after isolation and dispersion at physiological temperatures in aqueous buffers do not adopt a bilayer organization (nonbilayer lipids). Second, during many membrane processes like fusion and lipid flip-flop, (part of) the lipids will (temporarily) leave the bilayer organization (nonbilayer processes). The hypothesis that nonbilayer lipids and the structures they form are actively involved in nonbilayer processes had lead to the proposal of an alternative model of biological membranes (for reviews see Cullis and de Kruijff, 1979; de Kruijff *et al.*, 1980a). In this article we will summarize the present knowledge of the structural and functional properties of these nonbilayer lipids.

2. PROPERTIES OF NONBILAYER LIPIDS

2.1. Lipid Polymorphism

The ability of membrane lipids to adopt a variety of phases on hydration has been known for some time (Luzzatti *et al.*, 1968). X-ray analysis (for review see Shipley, 1973)

B. de Kruijff and A. J. Verkleij • Department of Molecular Biology, and Department of Biochemistry, University of Utrecht, 3584CH Utrecht, The Netherlands.   *P. R. Cullis* • Department of Biochemistry, University of British Columbia, Vancouver, British Columbia, Canada.

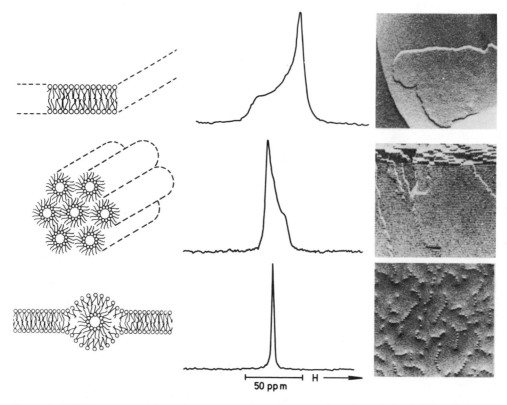

Figure 1. ^{31}P NMR spectra and freeze-fracture morphology of aqueous dispersions of phospholipids. Hydrated erythrocyte PE adopts the bilayer phase below 8°C and the H_{II} phase above 8°C (Cullis and de Kruijff, 1978). As an example of an ''isotropic'' phase, the ^{31}P NMR spectrum and freeze-fracture morphology (lipidic particles) of a PC/CL (Ca^{2+}) dispersion (Verkleij et al., 1979b; de Kruijff et al., 1979) are shown. As a model of this phase, the intrabilayer inverted micelle is shown.

and in more recent years ^{31}P NMR and freeze-fracture studies on fully hydrated preparations of the major membrane lipids have shown that as a rule either the bilayer or the hexagonal (H_{II}) phase is preferred. In the case of ^{31}P NMR the lineshape obtained from model systems as well as biological membranes is a sensitive indicator of the phase(s) adopted by the phospholipids (Cullis and de Kruijff, 1979). As is shown in Fig. 1, an asymmetrical lineshape with a low-field shoulder is characteristic of bilayer phospholipids, whereas H_{II}-phase phospholipids exhibit a narrower lineshape with reversed asymmetry. Alternatively, lipids in structures that allow isotropic motion ($\tau_C < 10^{-4}$) give rise to narrow symmetric signals. Typical bilayer-forming lipids are phosphatidylcholines (PC), sphingomyelin, diglucosyl- and digalactosyldiglycerides. The main H_{II} lipids are unsaturated phosphatidylethanolamines (PE), monoglucosyl- and monogalactosyl diglycerides, and cardiolipin (CL) in the presence of Ca^{2+} (for references see Cullis and de Kruijff, 1979). Phosphatidylserines (PS) and phosphatidylglycerols prefer the bilayer phase at neutral pH and ambient temperatures. At lower pH (Hope and Cullis, 1980) or at elevated temperatures (Harlos and Eibl, 1980), a hexagonal phase can be adopted. Cholesterol can induce formation of hexagonal phases from bilayer systems.

The phase preferences of a lipid can be understood in terms of the dynamic molecular shape of the molecule (Cullis and de Kruijff, 1979; Israelachvili et al., 1977). H_{II} lipids will exhibit a "cone" shape where the polar head group is at the smaller end of the cone. Lipids that have an inverted cone shape (e.g., lysophospholipids) organize themselves in micelles, whereas lipids with a more cylindrical shape pack optimally into a bilayer. The observation that an equimolar mixture of lysophosphatidylholine (inverted cone shape) and cholesterol (cone shape) is organized in a bilayer (Rand *et al.*, 1975) nicely illustrates this molecular shape concept.

2.2. Modulation of Bilayer–Nonbilayer Transitions

An important characteristic of the phase preferences of membrane lipids is that transitions between the H_{II} and the bilayer phase can occur. For example, in the case of unsaturated PEs, bilayer $\rightarrow$ H_{II} transitions occur as the temperature is increased through a characteristic value, which is dependent on the fatty acid composition. This transition temperature is about 8°C for human erythrocyte PE (Fig. 1). These transitions can also be induced isothermally by changes in divalent cation concentration. For example, Ca^{2+} can induce a bilayer $\rightarrow$ H_{II} phase transition for beef heart cardiolipin (Fig. 2) (Cullis *et al.*, 1978) and for PE/PS (Cullis and Verkleij, 1979) or PE/CL (de Kruijff and Cullis, 1980a) mixed systems. Most interestingly, lipid–protein interactions also can modulate the phase behavior of the lipids. Cytochrome *c* specifically induces the H_{II} and an isotropic phase in CL-containing systems (Fig. 2) (de Kruijff and Cullis, 1980a), whereas poly-L-lysine causes this effect on mixed PE/CL bilayers (de Kruijff and Cullis, 1980b). That integral mem-

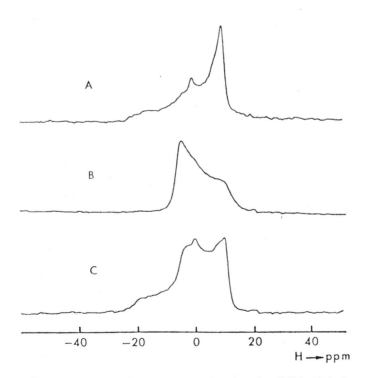

Figure 2. 81-MHz ^{31}P NMR spectra at 30°C of an aqueous dispersion of cardiolipin (A) in the presence of Ca^{2+} (B) and cytochrome *c* (C). Beef heart cardiolipin (50 μmoles) was dispersed in 1.0 ml 100 mM NaCl, 10 mM Tris–HCl, 0.2 mM EDTA, pH 7.0. In (B) 0.1 ml 1 M $CaCl_2$ and in (C) 0.2 ml buffer containing 36 mg oxidized cytochrome *c* was added.

brane proteins stabilize bilayer structures may be inferred from the observation that in total lipid extracts of the rod outer segment (de Grip *et al.*, 1979) and *E. coli* cytoplasmic membrane (Burnell *et al.*, 1979), the H_{II} and an isotropic phase are observed whereas the lipids in the intact membranes are mainly organized in bilayers.

2.3. Lipidic Particles

Because biological membranes contain a mixture of bilayer and H_{II} lipids, the phase behavior of such mixtures is of obvious interest. ^{31}P NMR studies have demonstrated that instead of gradually going from an H_{II} phase to a bilayer structure when a bilayer lipid is titrated into an H_{II} type of lipid structure, a new spectral component is observed indicating fast isotropic motion of the lipid molecules (Fig. 1). This behavior was observed for many different lipid mixtures including the hydrated total lipid extracts from the rod outer segment (de Grip *et al.*, 1979), *E. coli* (Burnell *et al.*, 1979) and inner mitochondrial membranes (Cullis *et al.*, 1980a). The organization of this intermediate phase is therefore of particular interest. However, a variety of lipid structures can give rise to narrow symmetric ^{31}P NMR spectra characteristic of isotropic averaging, including small bilayer vesicles, micelles, inverted micelles, or phases like the cubic ones. Freeze-fracture analysis of these preparations has shown that on the fracture face of these pure lipid systems numerous particles and pits are present that often are organized in a stringlike fashion (Verkleij *et al.*, 1979a; de Kruijff *et al.*, 1979). All present evidence suggests that these lipidic particles represent intrabilayer inverted micelles that are located either inside one bilayer or at the nexus of intersecting bilayers. This tendency of nonbilayer lipids to prefer an inverted micellar structure instead of the H_{II} phase in mixtures with bilayer lipids opens intriguing functional possibilities.

2.4. Biological Membranes

From the previous data it is clear that nonbilayer structures, in particular inverted micellar ones, might occur in biological membranes. ^{31}P NMR investigations of human red cell membranes have shown that the lipids are organized in extended bilayers (Fig. 3). This organization is extremely stable, being unaffected by extensive treatment with phospholipases and proteolytic enzymes (Cullis and de Kruijff, 1979). This stability might be related to the large mechanical stress that the erythrocyte encounters in the circulation or alternatively might reflect the relatively low metabolic activity of this plasma membrane. A very different situation is observed in metabolically more active membranes such as the endoplasmic reticulum of the rat liver (Fig. 3). At physiological temperatures a lineshape is observed that demonstrates isotropic motion for a large fraction of the endogenous phospholipids. At lower temperatures the spectrum indicates bilayer structure. This temperature-dependence behavior has been observed for isolated rat (de Kruijff *et al.*, 1980c), rabbit (Stier *et al.*, 1978), and beef (de Kruijff *et al.*, 1978) liver microsomes, as well as for the rat liver inner mitochondrial membrane (Cullis *et al.*, 1980b) and the *E. coli* inner membrane (Burnell *et al.*, 1979). ^{31}P (de Kruijff *et al.*, 1980b) and ^{13}C (de Kruijff *et al.*, 1980c) NMR studies on intact rat liver also demonstrated isotropic motion of part of the membrane phospholipids at 37°C. Although the isotropic motion could originate from rapid lateral diffusion of the lipids over curved bilayer surfaces, the data would be consistent with the transient occurrence of nonbilayer lipid structures, such as inverted micelles.

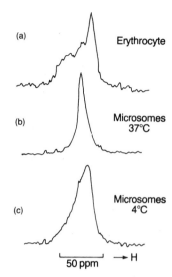

Figure 3. ^{31}P NMR spectra of various biological membranes. Reproduced with permission from *TIBS* (1980) **5**, 79–81.

2.5. Functional Aspects

Fusion and lipid flip-flop are clear examples of nonbilayer processes in which nonbilayer lipids have been shown to play an active role. For example, chemical fusogens that fuse erythrocytes also induce the H_{II} phase in the erythrocyte membrane (Cullis and Hope, 1978), the extent of fusion being proportional to the amount of H_{II} phase formed (Hope and Cullis, 1981). Furthermore, in different lipid systems vesicle fusion is accompanied by the appearance of lipidic particles at the site of fusion (Verkleij *et al.*, 1979b, 1980). Similarly, phospholipid flip-flop in model membrane systems is greatly enhanced by the presence of nonbilayer lipid structures (Gerritsen *et al.*, 1980). The fast flip-flop of lipids in the endoplasmic reticulum membrane at 37°C (Zilversmit and Hughes, 1977; van den Besselaar *et al.*, 1978) and in bacterial membranes (Rothman and Leonard, 1977) also is fully consistent with the transitory formation of inverted micelles.

Nonbilayer lipids and the structures they form appear to be especially important for the inner mitochondrial membrane, as this membrane is particularly rich in nonbilayer lipids. At 37°C in the presence of Ca^{2+}, both CL and PE, which amount to 60% of the total lipids, prefer the H_{II} phase. That inverted micelles of the cardiolipin–Ca^{2+} complex play a role as ionophore in Ca^{2+} transport across the inner mitochondrial membrane is indicated by the following observations: (1) Ca^{2+} induces the formation of lipidic particles in CL-containing membranes (Verkleij *et al.*, 1979a; de Kruijff *et al.*, 1979); (2) these particles facilitate divalent cation transport (de Kruijff *et al.*, 1979; Gerritsen *et al.*, 1980); (3) ruthenium red, a potent inhibitor of Ca^{2+} transport in mitochondria, blocks the formation of nonbilayer phases by Ca^{2+} in CL-containing membranes and also blocks the uptake of Ca^{2+} into an organic phase by CL (Cullis *et al.*, 1980b).

Literature data as well as our own studies on cytochrome *c*–CL systems (for review see de Kruijff *et al.*, 1981) suggest that nonbilayer lipids may play an important role in posttranslational protein insertion, as the formation of nonbilayer lipid structures can provide a low-energy pathway for the translocation of polar head groups into or across a lipid bilayer. Finally, the observations that cytochrome *c* specifically induces nonbilayer structures in CL-containing bilayers (de Kruijff and Cullis, 1980b) and that CL is absolutely required for the enzymatic activity of cytochrome oxidase (Fry and Green, 1980) suggest that an intramembrane cytochrome c–CL complex is important for electron transport between cytochrome c and cytochrome oxidase.

3. CONCLUDING REMARKS

Although our understanding of the structural and functional aspects of nonbilayer membrane lipids is still at a primitive level, we feel that the results obtained so far are exciting and appear to be leading to a new appreciation of the role of lipids in membranes. Further research will have to be directed toward a characterization of the molecular structure of the "isotropic" phases observed in biological membranes. New areas of research will include the chloroplast membranes, which like the mitochondrion are particularly rich in nonbilayer lipids. Furthermore, it can be expected, and to some extent has already been documented, that a detailed understanding of the phase characteristics of functionally very important nonbilayer lipids such as phosphatidic acids, diglycerides, and phosphatidylinositol (phosphates) may lead to a molecular interpretation of biological signal transmission.

REFERENCES

Burnell, E., van Alphen, L., de Kruijff, B., and Verkleij, A. J. (1979). *Biochim. Biophys. Acta* **597**, 492–501.

Cullis, P. R., and de Kruijff, B. (1978). *Biochim. Biophys. Acta* **513**, 31–42.

Cullis, P. R., and de Kruijff, B. (1979). *Biochim. Biophys. Acta* **559**, 399–420.

Cullis, P. R., and Hope, M. J. (1978). *Nature (London)* **271**, 672–674.

Cullis, P. R., and Verkleij, A. J. (1979). *Biochim. Biophys. Acta* **552**, 545–550.

Cullis, P. R., Verkleij, A. J., and Ververgaert, P. H. J. Th. (1978). *Biochim. Biophys. Acta* **513**, 11–20.

Cullis, P. R., de Kruijff, B., Hope, M. J., Nayar, R., Rietveld, A., and Verkleij, A. J. (1980a). *Biochim. Biophys. Acta* **600** 625–635.

Cullis, P. R., de Kruijff, B., Hope, M. J., Nayar, R., and Schmidt, S. (1980b). *Can. J. Biochem.* **58**, 1091–1101.

de Grip, W. J., Drenth, E. H. S., van Echteld, C. J. A., de Kruijff, B., and Verkleij, A. J. (1979). *Biochim. Biophys. Acta* **558**, 330–337.

de Kruijff, B., and Cullis, P. R. (1980a). *Biochim. Biophys. Acta* **601**, 235–240.

de Kruijff, B., and Cullis, P. R. (1980b). *Biochim. Biophys. Acta* **602**, 477–490.

de Kruijff, B., van den Besselaar, A. M. H. P., Cullis, P. R., van den Bosch, H., and van Deenen, L. L. M. (1978). *Biochim. Biophys. Acta* **514**, 1–8.

de Kruijff, B., Verkleij, A. J., van Echteld, C. J. A., Gerritsen, W. J., Mombers, C., Noordam, P. C., and de Gier, J. (1979). *Biochim. Biophys. Acta* **555**, 200–209.

de Kruijff, B., Cullis, P. R., and Verkleij, A. J. (1980a). Trends in Biochem. Sci. **5**, 79–81.

de Kruijff, B., Rietveld, A., and Cullis, P. R. (1980b). *Biochim. Biophys. Acta* **600**, 343–357.

de Kruijff, B., Rietveld, A., and van Echteld, C. J. A. (1980c). *Biochim. Biophys. Acta* **600**, 597–606.

de Kruijff, B., Verkleij, A. J., van Echteld, C. J. A., Gerritsen, W. J., Noordam, P. C., Mombers, C., Rietveld, A., de Gier, J., Cullis, P. R., Hope, M. J., and Nayer, R. (1981). In *Cell Biology 1980–1981* (H. Schweiger, ed.), pp. 559–572, Springer-Verlag, Heidelberg.

Fry, M., and Green, D. E. (1980). *Biochem. Biophys. Res. Commun.* **93**, 1238–1246.

Gerritsen, W. J., de Kruijff, B., Verkleij, A. J., de Gier, J., and van Deenen, L. L. M. (1980). *Biochim. Biophys. Acta* **598**, 554–560.

Harlos, K., and Eibl, H. (1980). *Biochemistry* **19**, 895–899.

Hope, M. J., and Cullis, P. R. (1980). *Biochem. Biophys. Res. Commun.* **92**, 846–852.

Hope, M. J., and Cullis, P. R. (1981). *Biochim. Biophys. Acta* **640**, 82–90.

Israelachvili, J. N., Mitchell, D. J., and Ninham, B. W. (1977). *Biochim. Biophys. Acta* **470**, 185–201.

Luzzatti, V., Gulik-Krzywicki, T., and Tardieu, A. (1968). *Nature (London)* **218**, 1031–1034.

Rand, R. P., Pangborn, W. A., Purdon, A. D., and Tinker, D. O. (1975). *Can. J. Biochem.* **53**, 189–195.

Rothman, J. E., and Leonard, J. (1977). *Science* **195**, 743–753.

Shipley, G. C. (1973). In *Biological Membranes* (D. Chapman and D. F. H. Wallach, eds.), Vol. 2, pp. 1–89, Academic Press, New York.

Stier, A., Finch, S. A. E., and Bösterling, B. (1978). *FEBS Lett.* **91,** 109–112.

van den Besselaar, A. M. H. P., de Kruijff, B., van den Bosch, H., and van Deenen, L. L. M. (1978). *Biochim. Biophys. Acta* **510,** 242–255.

Verkleij, A. J., Mombers, C., Gerritsen, W. J., Leunissen-Bijvelt, J., and Cullis, P. R. (1979a). *Biochim. Biophys. Acta* **555,** 358–361.

Verkleij, A. J., Mombers, C., Leunissen-Bijvelt, J., and Ververgaert, P. H. J. Th. (1979b). *Nature (London)* **279,** 162–163.

Verkleij, A. J., van Echteld, C. J. A., Gerritsen, W. J., Cullis, P. R., and de Kruijff, B. (1980). *Biochim. Biophys. Acta* **600,** 620–624.

Zilversmit, D. B., and Hughes, M. E. (1977). *Biochim. Biophys. Acta* **469,** 99–110.

6

Divalent Cations, Electrostatic Potentials, Bilayer Membranes

Stuart McLaughlin

"How many? How tightly? Where? Why?" these were the questions posed by Scatchard (1949) in his elegant study of the adsorption of ions to proteins. These succinct questions also define what we wish to know about the adsorption of divalent cations to phospholipid bilayer membranes. The first two questions, which relate to the stoichiometry of the binding and the magnitude of the binding constants, will be addressed in this chapter. The third question, which relates to the exact location of the bound divalent cations in the phospholipid head group, requires a molecular approach such as nuclear magnetic resonance (NMR) and will be addressed in the following chapter. The fourth question, which relates to the physiological relevance of the adsorption of divalent cations to the bilayer component of biological membranes, cannot yet be completely answered. We shall, however, consider two phenomena: (1) the effect of divalent cations on the electrostatic potential at the surface of a nerve membrane and (2) the effect of the adsorption of calcium to intracellular membranes on the diffusion coefficient of this ion.

Scatchard (1949) was well aware that the adsorption of ions to a macromolecule changes the electrostatic potential at the remaining binding sites and that "there may be an electrostatic effect but there may also be additional effects which cannot be explained by any simple electrostatic theory." The situation is similar with bilayers. There is certainly an electrostatic effect but there are additional effects as well. The ability of divalent cations to induce phase transitions and phase separations of the lipids as well as aggregation and fusion of the membranes cannot be described in terms of any simple theory and will be ignored in this chapter (e.g., Träuble and Eibl, 1974; Jacobson and Papahadjopoulos, 1975; McDonald *et al.*, 1976; Forsyth *et al.*, 1977; Papahadjopoulos *et al.*, 1977; van Dijck *et al.*, 1978; Portis *et al.*, 1979; Ohki and Duzgunes, 1979; Cohen *et al.*, 1980).

It is difficult to describe the electrostatic potential adjacent to either a protein (Edsall and Wyman, 1958) or a polyelectrolyte molecule (Rice and Nagasawa, 1961). Fortunately,

Stuart McLaughlin • Department of Physiology and Biophysics, Health Sciences Center, State University of New York, Stony Brook, New York 11794.

there is a simple way to describe the potential adjacent to a charged planar surface such as a phospholipid bilayer. The description was due originally to Gouy (1910) and to Chapman (1913), who combined the Poisson and Boltzmann equations, and has come to be known as the theory of the aqueous diffuse double layer. The four major assumptions inherent in the Gouy–Chapman theory of the diffuse double layer are: (1) that the ions in solution are point charges, (2) that the surface charges are uniformly smeared over the interface, (3) that the dielectric constant of the aqueous phase is constant, and (4) that image charge effects may be ignored. References to the literature in which these and other assumptions are discussed can be found in reviews by Grahame (1947), Verwey and Overbeek (1948), Haydon (1964), Mohilner (1966), Barlow (1970), Bockris and Reddy (1970), Sparnaay (1972), Aveyard and Haydon (1973), and McLaughlin (1977). A theory with so many obviously incorrect assumptions might be thought to be of little practical value. Indeed, after reviewing the evidence obtained from mercury dropping electrodes, Bockris and Reddy (1970) concluded that "the Gouy–Chapman theory might best be described as a brilliant failure." For reasons that are not yet well understood, the theory works much better at the surface of a bilayer membrane than it does at the surface of a mercury electrode. The earlier work on bilayer membranes has been reviewed (McLaughlin, 1977) and will not be discussed here. More recently, the potential adjacent to a phospholipid surface has been investigated by direct measurements above monolayers (Ohki and Sauve, 1978); by capacitance and conductance measurements on planar bilayers (Schoch et al., 1979; J. Cohen, personal communication); by zeta potential measurements on multilamellar bilayer vesicles (Eisenberg et al., 1979; McLaughlin et al., 1981; Lau et al., 1981); and by spin-label and fluorescent probe measurements on sonicated vesicles (Castle and Hubbell, 1976; Eisenberg et al., 1979). Furthermore, the adsorption of divalent cations to phospholipid bilayer membranes has been measured using electron spin resonance (ESR) (Puskin, 1977; Puskin and Coene, 1980), NMR (Grasdalen et al., 1977; McLaughlin et al., 1978; Lau et al., 1981), equilibrium dialysis (Newton et al., 1978), and Ca-sensitive electrodes (McLaughlin et al., 1981). The data obtained from each of these experimental approaches agree remarkably well with the Gouy–Chapman theory, if it is assumed that monovalent and divalent cations exert not only a nonspecific, or "screening," effect in the aqueous diffuse double layer adjacent to the surface, but also adsorb specifically, or chemically, to the phospholipid molecules. Stern (1924) was the first to modify the Gouy–Chapman theory to account for such specific adsorption. In essence, he combined the Gouy–Chapman theory with the Langmuir adsorption isotherm. The use of the Langmuir adsorption isotherm is, of course, also an oversimplification. As noted by Gileadi (1967): "The Langmuir adsorption isotherm may now be regarded a classical law in physical chemistry. It has all the ingredients of a classical equation; it is based on a clear and simple model, can be derived easily from first principles, is very useful now, about fifty years after it was first derived and will probably be useful for many years to come, and is rarely ever applicable to real systems, except as a first approximation." Fortunately, the Gouy–Chapman–Stern theory appears to be a good first approximation for describing the adsorption of cations to bilayer membranes.

In the simplest form of the Gouy–Chapman–Stern theory the relationship between the surface charge density, σ, and the surface potential, ψ_0, is given by the Grahame (1947) equation from the Gouy–Chapman theory of the diffuse double layer:

$$\sigma = \pm \{2\epsilon_r \epsilon_0 RT \Sigma_i C_i [\exp(-z_i F \psi_0 / RT) - 1]\}^{1/2} \qquad (1)$$

where C_i is the concentration of ions of valence z_i in the bulk aqueous phase, ϵ_r is the dielectric constant of the aqueous phase, ϵ_0 is the permittivity of free space, F is the

Faraday constant R is the gas constant, and T is the absolute temperature. The concentrations of monovalent and divalent cations in the aqueous phase at the surface of the membrane, $C^+(0)$ and $C^{2+}(0)$, are related to the bulk aqueous concentrations, C^+ and C^{2+}, by the Boltzmann equation:

$$C^+(0) = C^+ \exp(-F\psi_o/RT)$$
$$C^{2+}(0) = C^{2+} \exp(-2F\psi_o/RT) \tag{2}$$

The adsorption of the monovalent cation to the negative lipid, P^-, to form $1:1$ complexes, C^+P^-, can be described by the Langmuir adsorption isotherm (Eisenberg *et al.*, 1979):

$$\{C^+P^-\} = K_1\{P^-\}C^+(0) \tag{3}$$

Where K_1 is an intrinsic association constant and the braces denote a surface concentration. The divalent cation forms $1:1$ complexes, $C^{2+}P^-$, with the negative lipid (see below):

$$\{C^{2+}P^-\} = K_2\{P^-\}C^{2+}(0) \tag{4}$$

where K_2 is an intrinsic association constant. When zwitterionic lipids, P, are present in the membrane, the divalent cation is again assumed to form $1:1$ complexes and the Langmuir isotherm is assumed to be valid:

$$\{C^{2+}P\} = K_3\{P\}C^{2+}(0) \tag{5}$$

where K_3 is an intrinsic association constant. The total surface concentration of negative lipid, $\{P^-\}^{tot}$, is the sum of the free and bound concentrations, and the total surface concentration of the zwitterionic lipid, $\{P\}^{tot}$, is also the sum of the free and bound concentrations. Algebraic manipulation of these equations yields

$$\sigma = \frac{-\{P^-\}^{tot}[1 - K_2 C^{2+}(0)]}{[1 + K_1 C^+(0) + K_2 C^{2+}(0)]} + \frac{2K_3\{P\}^{tot}C^{2+}(0)}{[1 + K_3 C^{2+}(0)]} \tag{6}$$

The combination equations (1), (2), and (6) is referred to as a Stern equation.

The stoichiometry of the binding of divalent cations, such as Ca and Mg, to negative lipids, such as phosphatidylserine (PS) and phosphatidylglycerol (PG), is a topic of some debate. However, there is now strong evidence that these cations form at least some $1:1$ complexes with the lipids (McLaughlin *et al.*, 1981; Lau *et al.*, 1981). The magnitude of the intrinsic $1:1$ association constant for the Ca–PS complex, for example, can be determined by measuring the concentration of Ca at which the charge on a PS vesicle reverses sign, $[Ca]^{rev} = 0.08M$ (McLaughlin *et al.*, 1981). At $[Ca]^{rev}$, the number of Ca ions adsorbed in positively charged $1:1$ complexes is equal to the number of negatively charged PS molecules. The intrinsic association constant is thus $K_2 = 1/[Ca]^{rev} = 12\ M^{-1}$. The calculation is independent of the number of electroneutral $2:1$ complexes formed between PS molecules and Ca, the number of electroneutral $1:1$ complexes formed between PS and Na, the distance of the hydrodynamic plane of shear from the membrane, the area occupied by the lipid molecules, the viscosity and dielectric constant of the aqueous phase adjacent to the membrane, etc. Values of the $1:1$ association constants of a variety of divalent cations with brain PS are shown in Table I.

Although these electrophoretic charge reversal measurements do not rule out the possibility that PS could also form a large number of electroneutral $2:1$ complexes with the

Table I. The Intrinsic 1 : 1 Association Constants (M⁻¹) of Some Divalent Cations with Four Common Phospholipids

	PS[a]	PG[b]	PC[a]	PE[a]
Ni	40	7.5		
Co	28	6.5		
Mn	25	11.5		
Ba	20	5.5		
Sr	14	5.0		
Ca	12	8.5	3	3
Mg	8	6.0	2	2

[a] Data from McLaughlin *et al.* (1981).
[b] Data from Lau *et al.* (1981).

divalent cations, I believe that the available evidence favors the argument that these lipids form mainly 1 : 1 complexes with the divalent cations. This simple postulate is sufficient to explain the loss of divalent cations from the aqueous phase on addition of lipid vesicles, as measured by electrodes, equilibrium dialysis, NMR and ESR techniques; it can also explain the effect of low concentrations of divalent cations on the surface potential, as determined from zeta potential, compensation potential, and conductance probe measurements (see McLaughlin *et al.*, 1981, for references). Thus, there is no experimental evidence for the formation of 2 : 1 complexes in the absence of membrane aggregation. When aggregation of the bilayers occurs, the effective lipid : divalent cation stoichiometry must be 2 : 1 in the region where the membranes are juxtaposed because of electroneutrality. In this case there may be *trans* complexes formed between lipids on adjacent membranes (Portis *et al.*, 1979).

The Stern equation can describe the adsorption of Ca and Mg not only to PS membranes but also to bilayers formed from mixtures of PS and the zwitterionic lipids phosphatidylcholine (PC) and phosphatidylethanolamine (PE) (McLaughlin *et al.*, 1981). It also describes adequately the adsorption of divalent cations to PG membranes (Lau *et al.*, 1981). As discussed in the next chapter, the association of the alkaline earth cations with each of these four lipids appears to involve mainly outer-sphere rather than inner-sphere complxes.

Divalent cations exert a stabilizing effect on the membranes of excitable cells: when the calcium concentration is increased, a larger depolarization is required to reach the threshold potential and elicit an action potential. This stabilizing effect can be understood in terms of the ability of calcium to change the electrostatic potential at the surface of the membrane (see Hille *et al.*, 1975, for references). It is not yet known whether the negative surface charges adjacent to the sodium and potassium channels in nerves arise from the anionic lipids in the bilayer portion of the membrane or from the proteins constituting the channel. The former possibility is not ruled out by selectivity experiments with lipids: the ability of divalent cations to bind to a PC : PS bilayer (McLaughlin *et al.*, 1981) parallels the ability of the cations to affect the electrical properties of the sodium channel (Hille *et al.*, 1975).

Calcium also controls and triggers a variety of intracellular events. To perform these functions, calcium must diffuse from one part of the cell to another. It is not widely recognized that the reversible adsorption of calcium to phospholipids in intracellular membranes will reduce the diffusion coefficient of calcium in the cytoplasm. For example,

McLaughlin and Brown (1981) have calculated that the diffusion coefficient of calcium in rod cells should be reduced by 1–2 orders of magnitude because of the adsorption of calcium to the intracellular disk membranes.

REFERENCES

Aveyard, R., and Haydon, D. A. (1973). *An Introduction to the Principles of Surface Chemistry*, Cambridge University Press, London.

Barlow, C. A., Jr. (1970). In *Physical Chemistry, An Advanced Treatise*, pp. 167–246, Academic Press, New York.

Brockris, J. O'M., and Reddy, A. K. N. (1970). *Modern Electrochemistry*, Plenum Press, New York.

Castle, J. D., and Hubbell, W. L. (1976). *Biochemistry* **15**, 4818–4831.

Chapman, D. L. (1913). *Philos Mag.* **25**, 475–481.

Cohen, F. S., Zimmerberg, J., and Findelstein, A. (1980). *J. Gen. Physiol.* **75**, 251–270.

Edsall, J. T., and Wyman, J. (1958). *Biophysical Chemistry*, Academic Press, New York.

Eisenberg, M., Gresalfi, T., Riccio, T., and McLaughlin, S. (1979). *Biochemistry* **18**, 5213–5223.

Forsyth, P. A., Marcella, S., Mitchell, D. J., and Ninham, B. W. (1977). *Biochim. Biophys. Acta* **469**, 335–344.

Gileadi, E. (1967). *Electrosorption*, Plenum Press, New York.

Goüy, M. (1910). *J. Phys. (Paris)* **19**, 457–468.

Grahame, D. C. (1947). *Chem. Rev.* **41**, 441–501.

Grasdalen, H., Eriksson, L. E. G., Westman, J., and Ehrenberg, A. (1977). *Biochim. Biophys. Acta* **469**, 151–162.

Haydon, D. A. (1964). *Recent Prog. Surf. Sci.* **1**, 94–158.

Hille, B., Woodhull, A. M., and Shapiro, B. I. (1975). *Philos. Trans. R. Soc. London Ser. B* **270**, 301–318.

Jacobson, K., and Papahadjopoulos, D. (1975). *Biochemistry* **14**, 152–161.

Lau, A., McLaughlin, A., and McLaughlin, S. (1981). *Biochim. Biophys. Acta* **645**, 279–292.

McDonald, R. C., Simon, S. A., and Baer, E. (1976). *Biochemistry* **15**, 885–891.

McLaughlin, A. C., Grathwohl, C., and McLaughlin, S. (1978). *Biochim. Biophys. Acta* **513**, 338–357.

McLaughlin, S. (1977). *Curr. Top. Membr. Transp.* **9**, 71–144.

McLaughlin, S., and Brown, J. (1981). *J. Gen. Physiol.* **77**, 475–487.

McLaughlin, S., Mulrine, N., Gresalfi, T., Vaio, G., and McLaughlin, A. (1981). *J. Gen. Physiol.* **77**, 445–473.

Mohilner, D. M. (1966). *Electroanal. Chem.* **1**, 241–409.

Newton, C., Pangborn, W., Nir, S., and Papahadjopoulos, D. (1978). *Biochim. Biophys. Acta* **506**, 281–287.

Ohki, S., and Duzgunes, N. (1979). *Biochim. Biophys. Acta* **552**, 438–449.

Ohki, S., and Sauve, R. (1978). *Biochim. Biophys. Acta* **511**, 377–387.

Papahadjopoulos, D., Vail, W. J., Newton, C., Nir, S., Jacobson, K., Poste, G., and Lazo, R. (1977). *Biochim. Biophys. Acta* **465**, 579–598.

Portis, A., Newton, C., Pangborn, W., and Papahadjopoulos, D. (1979). *Biochemistry* **18**, 780–790.

Puskin, J. S. (1977). *J. Membr. Biol.* **35**, 39–55.

Puskin, J., and Coene, M. T. (1980). *J. Membr. Biol.* **52**, 69–74.

Rice, S. A., and Nagasawa, M. (1961). *Polyelectrolyte Solutions*, Academic Press, New York.

Scatchard, G. (1949). *Ann. N.Y. Acad. Sci.* **51**, 660–672.

Schoch, P., Sargent, D. F., and Schwyzer, R. (1979). *J. Membr. Biol.* **46**, 71–89.

Sparnaay, M. J. (1972). *The Electrical Double Layer*, Pergamon Press, Oxford.

Stern, O. (1924). *Z. Electrochem. Angew. Phys. Chem.* **30**, 508–516.

Träuble, H., and Eibl, H. (1974). *Proc. Natl. Acad. Sci. USA* **71**, 214–219.

van Dijck, P. W. M., de Kruijff, B., Verkleij, A. J., van Deenen, L. L. M., and de Gier, J. (1978). *Biochim. Biophys. Acta* **512**, 84–96.

Verwey, E. J. W., and Overbeek, J. Th. G. (1948). *Theory of the Stability of Lyophobic Colloids*, Elsevier, Amsterdam.

7

Nuclear Magnetic Resonance Studies of the Adsorption of Divalent Cations to Phospholipid Bilayer Membranes

Alan C. McLaughlin

1. INTRODUCTION

Thermodynamic aspects of the adsorption of divalent cations to phospholipid bilayer membranes can be reasonably well described by the Gouy–Chapman–Stern theory (S. McLaughlin, this volume). To proceed beyond this thermodynamic description requires molecular information about the structure of the divalent cation–phospholipid complexes. Because most experimental techniques, i.e., equilibrium dialysis (Portis *et al.*, 1979) or ion-sensitive electrodes (McLaughlin *et al.*, 1981), measure only the loss of divalent cations from the aqueous medium, they give no information on the bound complexes. For example, they cannot be used to determine which groups in the phospholipid molecule provide ligands for the divalent cation, or to distinguish between inner-sphere complexes, where the ligand is inserted into the first coordination sphere of the divalent cation, and outer-sphere complexes, where the ligand and the fully hydrated cation form an "ion pair" (Basolo and Pearson, 1967; Hewkin and Prince, 1970; Ahland, 1972; Beck, 1968). This review briefly discusses how nuclear magnetic resonance (NMR) can provide this type of molecular information and can also be used to quantitatively test one of the major assumptions of the Gouy–Chapman–Stern theory.

Alan C. McLaughlin • Biology Department, Brookhaven National Laboratory, Upton, New York 11973.

2. EFFECTS OF DIVALENT CATIONS ON THE NMR SPECTRA OF SONICATED PHOSPHOLIPID BILAYER MEMBRANES

2.1. Diamagnetic Divalent Cations

In sonicated phospholipid vesicles the rapid isotropic Brownian rotation averages the residual anisotropic interactions present in unsonicated membranes (McLaughlin *et al.*, 1975) and the observed signals are narrow enough to produce essentially "high resolution" ^{13}C NMR (Batchelor *et al.*, 1972; Metcalfe *et al.*, 1971) and ^{31}P NMR (Michaelson *et al.*, 1973; Berden *et al.*, 1974) spectra. Diamagnetic divalent cations, such as calcium, have small effects on the chemical shifts and relaxation times of these signals (Hutton *et al.*, 1977; Grasdalen *et al.*, 1977). However, the theoretical basis for the shifts is unclear, and an analysis of the relaxation times is complicated by the anisotropic motion (McLaughlin *et al.*, 1973). It is therefore difficult to use these small effects to investigate the structure of complexes formed between phospholipids and diamagnetic divalent cations. Fortunately, this problem can be circumvented by the use of paramagnetic divalent transition metal ions.

2.2. Paramagnetic Divalent Cations

2.2.1. Theory

Paramagnetic divalent transition metal cations have large effects on the NMR signals from groups that form inner-sphere complexes. In contrast to the small effects of diamagnetic divalent cations, these effects are well understood theoretically (Shulman *et al.*, 1965). Because the effects of cobalt are the easiest to quantitate, it is the most useful paramagnetic divalent cation for this purpose. The effects of cobalt on the linewidth, $1/T_{2P}$, and the shift, $\Delta\omega_P$, are given by the following equations:

$$1/T_{2P} = f[\Delta\omega_M^2 \tau_M/(1 + \Delta\omega_M^2 \tau_M^2)] \tag{1}$$
$$\Delta\omega_P = f[\Delta\omega_M/(1 + \Delta\omega_M^2 \tau_M^2)] \tag{2}$$

where f is the fraction of groups involved in inner-sphere complexes with cobalt, τ_M is the lifetime of the complex, and $\Delta\omega_M$ is the shift of the NMR signal for the bound ligand. Equations (1) and (2) are simplified forms of more general equations (Shulman *et al.*, 1965), but are valid for analyzing the effects of cobalt on the ^{31}P NMR spectra (McLaughlin *et al.*, 1978a) and ^{13}C NMR spectra (A. C. McLaughlin, unpublished observations) of phospholipid membranes. As $\Delta\omega_M$ can be determined from studies on model compounds (McLaughlin *et al.*, 1978b), a simultaneous measurement of $1/T_{2P}$ and $\Delta\omega_P$ allows a calculation of f and τ_M.

2.2.2. Inner-Sphere Complexes

In sonicated phospholipid vesicles the observed effects of cobalt on the NMR signals can be used to determine which groups form inner-sphere ligands for the divalent cation. For phosphatidylglycerol there are no effects on the ^{13}C NMR signals from the carbonyl or the hydroxyl groups, but there are large effects on the [13]P NMR signal (Lau *et al.*, 1981). It can thus be concluded that only the phosphodiester group forms detectable amounts of inner-sphere complexes with cobalt. The number of inner-sphere complexes can be calculated using equations (1) and (2).

For phosphatidylcholine it was also concluded that only the phosphodiester group forms inner-sphere complexes with cobalt (Lau *et al.*, 1981). However, with phosphatidylserine, preliminary results indicate both the phosphodiester group and the carboxyl group form inner-sphere complexes with cobalt (A. C. McLaughlin, unpublished observations).

2.2.3. Outer-Sphere Complexes

It is difficult to directly measure the number of divalent cations involved in outer-sphere complexes (Beck, 1968). An indirect approach is to calculate the difference between the number of bound divalent cations and the number of divalent cations involved in inner-sphere complexes. When the amount of cobalt sequestered in the "double layer" is negligible (McLaughlin *et al.*, 1981), the number of bound cobalt ions can be calculated from the difference between the total and the free cobalt concentrations. The free cobalt concentration can be calculated from the observed ^{31}P NMR linewidth by using a calibration curve determined from dialysis experiments.

For phosphatidylglycerol membranes the calculated number of phosphodiester groups involved in inner-sphere complexes with cobalt is approximately one-fifth the number of bound cobalt ions (Lau *et al.*, 1981). If the stoichiometry of the inner-sphere complex is 1 : 1 (S. McLaughlin, this volume), this finding implies that only one-fifth of the bound cobalt ions are involved in inner-sphere complexes with phosphodiester groups. Because ^{13}C NMR results show that the remaining bound cobalt ions are not involved in inner-sphere complexes with other membrane groups, they must be involved in outer-sphere complexes. The outer-sphere complexes presumably involve an ion pair between the fully hydrated cobalt ion and the phosphodiester group.

3. DISCRETE CHARGE EFFECTS

In the Gouy–Chapman–Stern theory it is assumed that the charges on the phospholipid molecules and the bound divalent cations are uniformly smeared over the surface of the membrane. This assumption is not correct. For a negatively charged membrane the magnitude of the micropotential, the potential at the divalent cation binding site, is larger than the average value of the surface potential. Furthermore, when divalent cations bind to a small fraction of the phospholipid molecules, they exert less effect on the micropotential than on the average value of the surface potential. These phenomena are usually termed "discrete charge" effects (Grahame, 1958; Levine, 1971; Nelson and McQuarrie, 1975; Tsien, 1978; Sauve and Ohki, 1979).

Cobalt can be used as a probe to study the effect of diamagnetic divalent cations such as calcium on the micropotential, ψ_m. Very low concentrations of cobalt substantially broaden the ^{31}P NMR signal but do not change the micropotential (Lau *et al.*, 1981). The effect on the ^{31}P NMR linewidth is proportional to the number of cobalt ions bound in inner-sphere complexes. The number of inner-sphere complexes is proportional to the free cobalt concentration in the aqueous phase adjacent to the phosphodiester group, and this free concentration is, in turn, proportional to the Boltzmann factor, $\exp[-2F\psi_m/RT]$. The ratio of the ^{31}P NMR linewidths in the presence, $1/T_{2P}^{Ca}$, and absence, $1/T_{2P}^{0}$, of calcium is given by the expression

$$(1/T_{2P}^{Ca})/(1/T_{2P}^{0}) = \exp[-2F\,\Delta\psi_m/RT] \tag{3}$$

where $\Delta\psi_m$ is the change in the micropotential on the addition of calcium. For phosphatidylglycerol membranes the effects of calcium on the micropotential, calculated using equation (3), are very similar to the effects on the surface potential calculated from the Gouy–Chapman–Stern theory (Lau *et al.*, 1981). The close agreement between the two values suggests that discrete charge effects do not play a significant role in the binding of divalent cations to bilayer membranes.

4. DISCUSSION

We have briefly outlined how the paramagnetic effects of cobalt, a divalent transition metal ion, can be used to obtain quantitative molecular information about the bound phospholipid complexes. The conclusion that a substantial fraction of bound cobalt ions are involved in outer-sphere complexes with phosphatidylglycerol is consistent with previous studies of cobalt binding to model oxyanions. For example, using ultrasound absorption techniques, Eigen and Tamm (1962) demonstrated that sulfate ions form predominantly outer-sphere complexes with cobalt. Using spectroscopic techniques, Smithson and Williams (1958) came to the same conclusion for both sulfate and nitrate ions, and using infrared techniques Brintzinger (1963) concluded that in cobalt-ATP approximately one-half of the terminal phosphate groups were involved in outer-sphere complexes with the divalent cation.

Although we have no direct experimental evidence that the alkaline earth cations form mainly outer-sphere complexes with phospholipids, two circumstantial lines of evidence are consistent with this suggestion. First, Phillips (1966) has pointed out that when an anion forms predominantly outer-sphere complexes with a series of divalent cation, the association constants are very similar. The intrinsic association constants for the phosphatidylglycerol complexes of the alkaline earth cations are very similar (S. McLaughlin, this volume). The same is true for phosphatidylserine, phosphatidylcholine, and phosphatidylethanolamine (McLaughlin *et al.*, 1978a, 1981). Second, the order of magnitude of the association constant for this type of complex is predicted theoretically to be 10 M^{-1} (Basolo and Pearson, 1967; Hewkin and Prince, 1970), which is very similar to the intrinsic association constants found for the alkaline earth complexes of the four phospholipids.

The suggestion that the alkaline earth cations form mainly outer-sphere complexes with phospholipids is also consistent with the effects of calcium on the ^{31}P NMR signal from unsonicated phosphatidylglycerol membranes. Above the phase transition of the calcium–phosphatidylglycerol complex, the observed phosphorus chemical shift anisotropy is very similar to the value found in the absence of calcium (Cullis and deKruijff, 1976). As the phosphorus chemical shift anisotropy is sensitive to the details of the rapid internal motion of the phosphodiester group (Seelig, 1978), this similarity implies that, under these conditions, calcium does not significantly perturb the motion of the phosphodiester group.*

In conclusion, the NMR results we have obtained with cobalt appear to be generally relevant to the interaction of a wide range of divalent cations with phospholipid bilayer membranes. In conjunction with EPR (Pushkin, 1977), diffraction (Lis *et al.*, 1981; Pangborn, 1980), and other NMR techniques (Cullis and deKruijff, 1976; Hope and Cullis, 1980), they provide detailed molecular information about the complexes formed between phospholipid molecules and divalent cations.

---

* Because the intensities of the spectra were not calibrated, this conclusion can be applied only to those phospholipid molecules that contribute to the observed ^{31}P NMR spectra.

ACKNOWLEDGMENTS. The author's work described in this review was supported by Grant GM 24971 from the USPHS and the United States Department of Energy.

REFERENCES

Ahland, S. (1972). *Coord. Chem. Rev.* **8,** 21–29.

Basolo, F., and Pearson, R. G. (1967). *Mechanisms of Inorganic Reactions*, pp. 34–38, Wiley, New York.

Batchelor, J. G., Prestegard, J. H., Cushley, R. J., and Lipsky, S. R. (1972). *Biochem. Biophys. Res. Commun.* **48,** 70–75.

Beck, M. T. (1968). *Coord. Chem. Rev.* **3,** 91–115.

Berden, J. A., Cullis, P. R., Hoult, D. I., McLaughlin, A. C., Radda, G. K., and Richards, R. E. (1974). *FEBS Lett.* **46,** 55–58.

Brintzinger, J. (1963). *Biochim. Biophys. Acta* **77,** 343–345.

Cullis, P. R., and de Kruijff, B. (1976). *Biochim. Biophys. Acta* **436,** 523–540.

Eigen, M., and Tamm, K. (1962). *Z. Elektrochem.* **66,** 107–121.

Grahame, D. C. (1958). *Z. Elektrochem.* **62,** 264–274.

Grasdalen, H., Eriksson, L. E. G., Westman, J., and Ehrenberg, A. (1977). *Biochim. Biophys. Acta* **469,** 151–162.

Hewkin, D. J., and Prince, R. H. (1970). *Coord. Chem. Rev,* **5,** 45–73.

Hope, M. J., and Cullis, P. R. (1980). *Biochem. Biophys. Res. Commun.* **92,** 846–852.

Hutton, W. C., Yeagle, P. L., and Martin, R. B. (1977). *Chem. Phys. Lipids* **19,** 255–265.

Lau, A., McLaughlin, A. C., and McLaughlin, S. G. A. (1981). *Biochim. Biophys. Acta* **645,** 279–292.

Levine, S. (1971). *J. Colloid Interface Sci.* **37,** 619–634.

Lis, L. J., Lis, W. T., Parsegian, V. A., and Rand, R. P. (1981). *Biochemistry* **20,** 1771–1777.

McLaughlin, A. C., Podo, F., and Blasie, J. K. (1973). *Biochem. Biophys. Acta* **330,** 109–121.

McLaughlin, A. C., Cullis, P. R., Berden, J. A., and Richards, R. E. (1975). *J. Magn. Reson.* **20,** 146–165.

McLaughlin, A. C., Grathwohl, C., and McLaughlin, S. (1978a). *Biochim. Biophys. Acta* **513,** 338–357.

McLaughlin, A. C., Grathwohl, C., and Richards, R. E. (1978b). *J. Magn. Reson.* **31,** 283–293.

McLaughlin, S., Mulrine, N., Gresalfi, T., Vaio, F., and McLaughlin, A. C. (1981), *J. Gen. Physiol.,* in press.

Metcalfe, J. C., Birdsall, N. J. M., Feeney, J., Lee, A. G., Levine, Y. K., and Partington, P. (1971). *Nature* (*London*) **233,** 199–201.

Michaelson, D. M., Horwitz, A. F., and Klein, M. P. (1973). *Biochemistry* **12,** 2637–2645.

Nelson, A. P., and McQuarrie, D. A. (1975). *J. Theor. Biol.* **55,** 13–27.

Pangborn, W. A. (1980). *Fed. Proc.* **39,** 2191.

Phillips, R. (1966). *Chem. Rev.* **66,** 501–527.

Portis, A., Newton, C., Pangborn, W., and Papahadjopoulos, D. (1979). *Biochemistry* **18,** 780–790.

Pushkin, J. S. (1977). *J. Membr. Biol.* **35,** 39–55.

Sauve, R., and Ohki, S. (1979). *J. Theor. Biol.* **81,** 157–179.

Seelig, J. (1978). *Biochim. Biophys. Acta* **436,** 523–540.

Shulman, R. G., Sternlicht, H., and Wyluda, B. J. (1965). *J. Chem. Phys.* **43,** 3116–3122.

Smithson, J. M., and Williams, R. J. P. (1958). *J. Chem. Soc.* **1958,** 457–462.

Tsien, R. Y. (1978). *Biophys. J.* **24,** 561–567.

8

Intrinsic Protein–Lipid Interactions in Biomembranes

D. Chapman

An area of recent interest in research on biomembranes is that concerned with the mutual effect and perturbation of the dynamics and movement of intrinsic proteins and their surrounding lipids, i.e., protein–lipid interactions. A number of questions have been posed, and this article will consider some of these questions.

1. *How do the dynamics of those lipids adjacent to intrinsic proteins compare with the dynamics of the bulk lipid?*

These adjacent lipids have sometimes been termed *boundary lipid*. A considerable number of reports have been published concerning "boundary-layer" lipids. Several authors have studied spin-labeled lipids with mixtures of cytochrome oxidase and phospholipids, either natural (Jost *et al.*, 1973) or synthetic (Marsh *et al.*, 1978). Analyses of the ESR spectra from a series of mixtures of varying lipid-to-protein ratio are *interpreted in terms of two distinct lipid environments* for the spin probe. In one environment the probe is mobile and the spectral component is similar to that found for the probe in pure lipid bilayers. The second spectral component corresponds to a highly immobilized probe, and this has been interpreted as a fraction of the lipid bilayer tightly bound to the cytochrome oxidase in a monomolecular layer. A source of uncertainty in the interpretation of these results is that cytochrome oxidase contains six subunits whose functions and relationship to the membrane are not yet fully understood (Eytan and Schatz, 1975).

Dahlquist and co-workers (1977) studied the effects of cytochrome *c* oxidase upon the specifically deuterated lipid 1-(16^1,16^1,16^1)trideuteropalmitoyl-2-palmitoleoylphosphatidylcholine and suggested that the protein converted a sizeable fraction of the lipid molecules into a more ordered component.

Various terms have been used to describe this perturbed lipid, including boundary-layer lipid (Jost *et al.*, 1973), halo lipid (Träuble and Overath, 1973; Stier and Sackmann,

D. Chapman • Department of Biochemistry and Chemistry, Royal Free Hospital School of Medicine, University of London, London WC1N 1BP, England.

1973), and annulus lipid (Warren *et al.*, 1975). In all these views it was assumed that a rigid or immobilized lipid shell exists separating the hydrophobic intrinsic protein from the adjacent fluid bilayer regions. It was suggested by some workers that this shell is a single rigid lipid layer (Jost *et al.*, 1973); Hesketh *et al.* (1976) went further and argued that this single rigid lipid shell is long-lived and excludes cholesterol. They argued that the rate of exchange between annular lipid and bulk lipid is slow even when the bulk lipid is fluid. The annulus lipid is said to correspond to a stoichiometric complex of lipid and protein, to control enzyme activity, and to be sensitively responsive to anesthetics.

Recent studies (Oldfield *et al.*, 1978; Rice *et al.*, 1979) do not support these conclusions. *No evidence is found in the spectra for two components, bilayer plus "boundary lipid."* Furthermore, the presence of the cytochrome oxidase causes a *decrease* in the quadrupole splitting (or order parameter) of the methyl group of the lipid chains. These authors argue that the cytochrome oxidase has the effect of preventing lecithin hydrocarbon chains from crystallizing below the lipid T_c transition temperature, while *above T_c* it causes a *disordering* of the lipid chains.

A number of workers have studied rhodopsin and boundary-layer lipid effects associated with this protein. Rhodopsin was tested with spin-labeled fatty acids. Pontus and Delmelle (1975) found evidence of a rigid boundary layer around this hydrophobic protein. However, previously Hong and Hubbell (1972) had reached a different conclusion from spin-label experiments with rhodopsin reincorporated into phospholipids; they suggested that the *average* viscosity of the membranes was dependent on the ratio of lipid to protein. This is in agreement with recent results put forward by Cherry *et al.* (1977) from very different experiments involving bacteriorhodopsin. Using proton NMR and rod outer segment membranes, Brown *et al.* (1977) concluded that, although the rotational motion of hydrocarbon chains was affected by the proximity of the proteins, *all phospholipids can diffuse rapidly* in the plane of the membrane, *suggesting a homogeneous lipid phase*.

Devaux and co-workers (Baroin *et al.*, 1978; Favre *et al.*, 1979) recently carried out an extensive number of experiments using spin-label methods where they studied the immediate environment of the lipid near rhodopsin in the disc membranes and without changing the lipid-to-protein ratio. They concluded that, with a hydrophobic spin label covalently bound to rhodopsin via a hydrophobic SH group, but able to locate in its near environment, it gives rise to an ESR spectrum not very different from the spectrum of a spin-labeled fatty acid diffusing freely in the lipid bilayer, i.e., a high degree of disorder or fluidity is observed. (On the other hand, when the membranes are partially delipidated using phosopholipase A_2, an immobile component is always observed. A difference of four orders of magnitude occurs between the correlation times of the probe situated in the boundary layer of intact disc membranes and membranes with low lipid content.)

In general, the NMR studies using ^1H, ^{19}F, or ^2H (time scale 10^{-3} to 10^{-5} sec) on various biomembranes and reconstituted systems reveal only one type of lipid mobility. This is the case with ^{19}F studies of *E. coli* membranes (Brown *et al.*, 1977; Gent and Ho, 1978), rhodopsin in disc membranes, and ^2H NMR studies of cytochrome oxidase (Kang *et al.*, 1979) and sarcoplasmic reticulum (Rice *et al.*, 1979). The ^2H NMR spectra of reconstituted Ca^{2+}-ATPase systems are shown in Fig. 1. The data indicate a continuity between the bulk lipid phase and the boundary layer, and ready diffusion of all the lipids takes place. The exchange must be fast on a time scale of 10^4 Hz. (See Chapman *et al.*, 1979, for a review.)

In summary, the ^2H NMR spectra of reconstituted membranes reveal four interesting features (see Seelig and Seelig, 1980). (1) A continuity between the bulk and boundary lipid occurs. (2) Membranes with protein exhibit a small, but finite decrease (10–25%) in the quadrupole splitting compared to pure lipid samples. (3) The deuterium T_1 relaxation times are shorter by about 20–30% in reconstituted membranes. (4) The apparent linewidth of both the ^{2}H and the ^{31}P NMR spectra increases.

The reduction in the deuterium quadrupole splitting has been ascribed to a disordering effect of the protein interface. In most membrane models presented in the literature, the membrane proteins are drawn as smooth cylinders or rotational ellipsoids. This is probably not very realistic. Even if the protein backbone is arranged in an α-helical configuration, the protrusion of amino acid side chains will lead to an uneven shape of the protein surface. Though already disordered in the pure lipid bilayer, the fatty acid acyl chains may become even more distorted by contact with the hydrophobic site. Another feature to consider is the effect of the protein on the lipid polar group packing within the lipid bilayer (Pink and Chapman, 1979). The spatial disorder of the hydrocarbon chains may be further augmented by density fluctuations on the protein surface itself. It has been suggested that membrane proteins have a ''fluidlike'' outer region that provides an approximate fluid mechanical match with the liquid-crystalline phospholipid membrane (Bloom, 1979).

From the increase in *spatial disorder* it cannot be concluded that the membrane is *more fluid*. On the contrary, deuterium T_1 relaxation time measurements suggest a decrease in the rate of segment reorientation in the presence of protein. Such deuterium T_1 measurements have been performed with cytochrome c oxidase and reconstituted sarcoplasmic reticulum. The addition of protein decreases the relaxation time in both cases (Seelig and Seelig, 1980).

2. What determines the distribution and movement of intrinsic proteins and are they randomly arranged in the plane of the lipid bilayer matrix?

Freeze-fracture electron microscopy of model and natural biomembranes sometimes reveals particles that appear randomly distributed in the plane of the lipid matrix. [Experiments with bacteriorhodopsin have shown that it is not possible to relate each particle to a specific intrinsic protein (Fisher and Stoeckenius, 1977).] However, there have been few studies that prove statistically that true random arrangements of proteins do occur, and work is still required on this aspect.

Studies that are consistent with a random arrangement of intrinsic components occurring in *model* biomembranes are those recently carried out using the fluorescent probe 1,6-diphenyl-1,3,5-hexatriene (Hoffmann *et al.*, 1980). The polarization of the probe was studied in reconstituted systems of an intrinsic polypeptide (gramicidin A) or of various intrinsic proteins. As the concentration of the intrinsic molecule was increased, the value of the polarization P reached a limiting value. Empirically, each of the curves was observed to fit a simple exponential equation. The value of the exponent is related to the number of probe molecules or lipids that surround the intrinsic molecule.

A careful analysis using probability theory shows that the occurrence of such exponential curves is consistent with the occurrence of a random arrangement of intrinsic molecules. The probe molecule is markedly affected by the presence of the intrinsic molecule according to $P_1 = e^{-Mx}$, where M is the number of DPH molecules that can be accommodated around the intrinsic molecule in half of the lipid bilayer, and x is related to the concentration of intrinsic molecules in the lipid bilayer matrix. The dependence of $P(M, x)$ upon x reflects the fact that as the concentration of polypeptide or protein increases, so does the probability of (say) protein–protein contacts, while the number of lipid molecules and probe molecules that, can contact the protein decreases. A comparison of calculated polarization values, at various concentrations of intrinsic molecules in a fluid lipid bilayer, agrees well with the experimental values (Hoffman *et al.*, 1981).

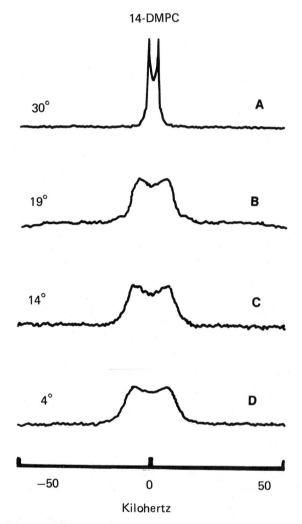

14-DMPC

30°   A

19°   B

14°   C

4°   D

−50   0   50

Kilohertz

Figure 1. Deuterium NMR spectra of DMPC-d_3 and DMPC-d_3/ATPase (sarcoplasmic reticulum, ATP phospho-hydrolase, EC 3.6.1.3) complexes as a function of temperature. (A) Pure DMPC-d_3, 30°C, 100-kHz effective spectral width, 0.54-sec recycle time, 2048 data points, $\tau_1 = \tau_2 = 50$ μsec, 7-μsec 90° pulse widths, 20,000 scans, 150-Hz line broadening. (B) Same as (A) except 19°C, $\tau_1 = \tau_2 = 70$ μsec, 15,000 scans. (C) Same as (B) except

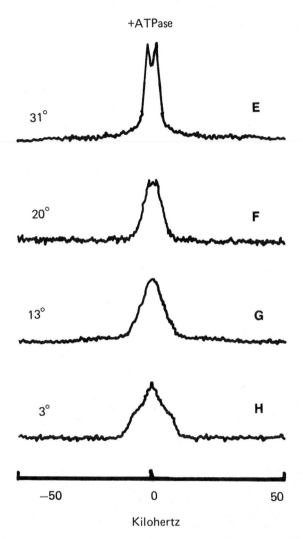

+ATPase

31°   E

20°   F

13°   G

3°   H

−50   0   50

Kilohertz

14°C. (D) Same as (B) except 4°C. (E) DMPC-d_3/ATPase (41 : 1), 31°C, other conditions as in (A) except $\tau_1 = \tau_2 = 90$ μsec, 40,000 scans. (F) Same as (E) except 20°C. (G) Same as (E) except 13°C, 80,000 scans. (H) Same as (E) except 3°C, 80,000 scans. The protein–lipid complex samples contained ∼10 mg phospholipid, which was 25% ^{2}H-labeled (Rice *et al.*, 1979).

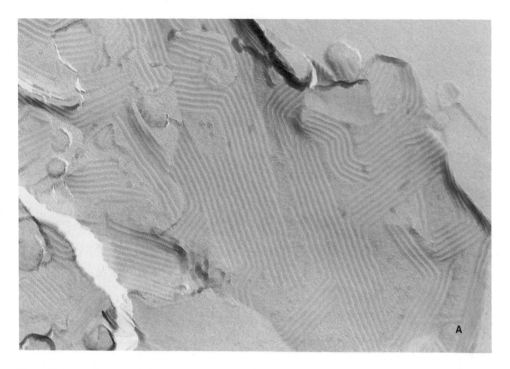

Figure 2. Freeze-fracture micrographs of dimyristoylphosphatidylcholine recombinants with Ca^{2+}-ATPase. The molar ratio of lipid to protein is 237. Sample (A) was quenched from 18°C and (B) from 4°C (Gomez-Fernandez *et al.*, 1980).

The distribution of proteins in the plane of the lipid matrix can be affected by various means so that migration of the proteins can take place. An example of this is observed with the sarcoplasmic reticulum Ca^{2+}-ATPase reconstituted into pure-lipid–water systems (Gomez-Fernandez *et al.*, 1980). In this case, freeze-fracture micrographs of Ca^{2+}-ATPase reconstituted into dimyristoylphosphatidylcholine–water systems show that the particles occur in patches when a preliminary cooling has occurred so that the lipid chains crystallize prior to the rapid quenching to liquid-nitrogen temperature. This is shown in Fig. 2 where in (a) the sample was quenched from 18°C and in (b) from 4°C. The coexistence of the rippled phase and the high protein to lipid patches shows the extent to which migration and phase separation occurs (Gomex-Fernandez *et al.*, 1980).

The mobility of the intrinsic protein in systems where high protein–lipid patches occur can be determined by the melting of the lipids in these patches. This can mean that protein rotation increases dramatically (Hoffman *et al.*, 1980), as does enzymatic activity some-times $\sim$ 10°C below the T_c of the pure lipid, e.g., at 30°C with ATPase incorporated with dipalmitoyl–lecithin–water systems ($T_c = 41$°C).

ACKNOWLEDGMENT. We wish to thank the Wellcome Trust for financial support.

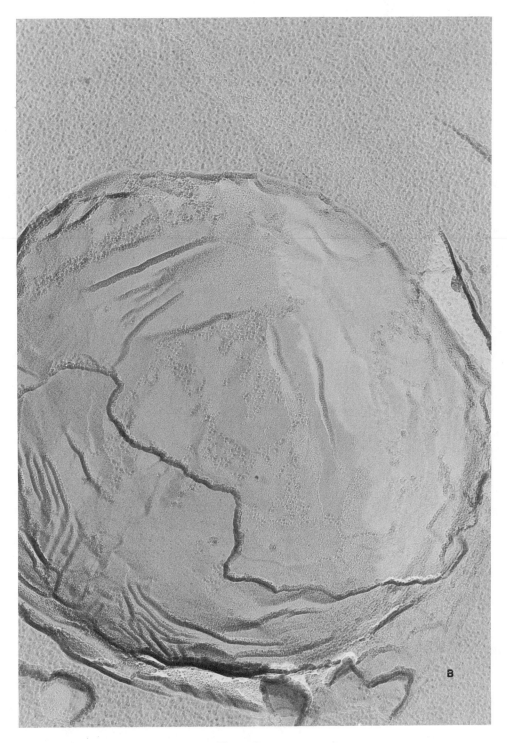

Figure 2. continued

REFERENCES

Baroin, A., Bienvenue, A., and Devaux, P. (1978). *Biochemistry* **18**, 1151–1155.

Bloom, M. (1979). *Can. J. Phys.* **57**, 2227–2230.

Brown, M. F., Miljanich, G. P., and Dratz, E. A. (1977). *Biochemistry* **16**, 2640–2648.

Chapman, D., Gomez-Fernandez, J. C., and Goni, F. M. (1979). *FEBS Lett. Rev.* **98**, 211–223.

Cherry, R. J., Müller, U., and Schneider, G. (1977). *FEBS Lett.* **80**, 465–469.

Dahlquist, F. W., Muchmore, D. C., Davis, J. H., and Bloom, M. (1977). *Proc. Natl. Acad. Sci. USA* **74**, 5435–5439.

Favre, E., Baroin, A., Bienvenue, A., and Devaux, P. (1979). *Biochemistry* **18**, 1156–1162.

Fisher, K. A., and Stoeckenius, W. (1977). *Science* **197**, 72–74.

Gent, M. P. N., and Ho, C. (1978). *Biochemistry* **17**, 3023–3038.

Gomez-Fernandez, J. C., Goni, F. M., Bach, D., Restall, C. J., and Chapman, D. (1980). *Biochim. Biophys. Acta* **598**, 502–516.

Hesketh, T. R., Smith, G. A., Houslay, M. D., McGill, K. A., Birdsall, N. J. M., Metcalfe, J. C., and Warren, G. B. (1976). *Biochemistry* **15**, 4145–4151.

Hoffman, W., Sarzala, M. G., Gomez-Fernandez, J. C., Goni, F. M., Restall, C. J., Chapman, D., Heppeler, G., and Kreutz, W. (1980). *J. Mol. Biol.* **141**, 119–132.

Hoffmann, W., Pink, D. A., Restall, C. J., and Chapman, D. (1981). *Eur. J. Biochem.* **114**, 585–589.

Hong, K., and Hubbell, W. L. (1972). *Proc. Natl. Acad. Sci. USA* **69**, 2617–2621.

Jost, P. C., Griffith, O. H., Capaldi, R. A., and Vanderkooi, G. (1973). *Proc. Natl. Acad. Sci. USA* **70**, 480–484.

Kang, S., Gutowsky, H. S., Hshung, J. C., Jacobs, R., King, T. E., Rice, D., and Oldfield, E. (1979). *Biochemistry* **18**, 3257–3267.

Marsh, D., Watts, A., Maschke, W., and Knowles, P. F. (1978). *Biochem. Biophys. Res. Commun.* **81**, 397–402.

Oldfield, E., Gilmore, R., Glaser, M., Gutowski, H. S., Hshung, J. C., Kang, S., Meadows, M., and Rice, D. (1978). *Proc. Natl. Acad. Sci. USA* **75**, 4657–4660.

Pontus, M., and Delmelle, M. (1975). *Biochim. Biophys. Acta* **401**, 221–230.

Rice, D. M., Meadows, M. D., Scheinman, A. O., Goni, F. M., Gomez-Fernandez, J. C., Moscarella, M. A., Chapman, D., and Oldfield, E. (1979). *Biochemistry* **18**, 5893–5903.

Seelig, J., and Seelig, A. (1980). *Q. Rev. Biophys.* **13**, 19–61.

Stier, A., and Sackmann, E. (1973). *Biochim. Biophys. Acta* **311**, 400–408.

Träuble, H., and Overath, P. (1973). *Biochim. Biophys. Acta* **307**, 491–512.

Warren, G. B., Houslay, M. D., Metcalfe, J. C., and Birdsall, N. J. M. (1975). *Nature (London)* **255**, 684–687.

9

Essential Structural Features and Orientation of Cytochrome b_5 in Membranes

Philipp Strittmatter and Harry A. Dailey

1. INTRODUCTION

It is clear the cytochrome b_5 serves as a mobile electron carrier or shuttle to provide reducing equivalents for a number of oxidation–reduction reactions that utilize cytoplasmically generated reduced pyridine nucleotides as electron sources. Thus, two flavoproteins, NADH cytochrome b_5 reductase (Spatz and Strittmatter, 1973) and NADH cytochrome P-450 reductase (Oshino et al., 1971; Enoch and Strittmatter, 1979a), reduce the membrane-bound heme protein rapidly, the former with a turnover number of 30,000 min^{-1} at 30°C (Spatz and Strittmatter, 1973) and the latter at 7000 min^{-1} at 33°C (Enoch and Strittmatter, 1979a). Reduced cytochrome b_5 is the direct electron donor for stearyl-CoA desaturase (Δ^9 fatty acyl CoA desaturase) (Holloway and Wakil, 1970; Oshino and Omura, 1973; Holloway and Katz, 1972; Strittmatter et al., 1974; Enoch et al., 1976), and may participate in the reduction of cytochrome P-450 as well (Hildebrandt and Estabrook, 1972; Lau et al., 1974; Saesame et al., 1974; Imai and Sato, 1977). In addition, the microsomal heme protein has been implicated as the reductant in the Δ^6 desaturation of fatty acids (Okayasu et al., 1977; Lee et al., 1977), cholesterol biosynthesis (Reddy et al., 1977), plasmalogen biosynthesis (Paultanf et al., 1974), and more recently in the elongation of fatty acids (Keyes et al., 1979). Both the relatively large and nonstoichiometric amount of cytochrome b_5 in endoplasmic reticulum relative to any one of the enzymes with which it interacts, and direct evidence that interaction with NADH cytochrome b_5 reductase in prepared phospholipid and microsomal vesicles is diffusion limited (Strittmatter and Rogers,

Philipp Strittmatter and Harry A. Dailey • Department of Biochemistry, University of Connecticut Health Center, Farmington, Connecticut 06032. Present address of Harry A. Dailey: Department of Microbiology, University of Georgia, Athens, Georgia 30602.

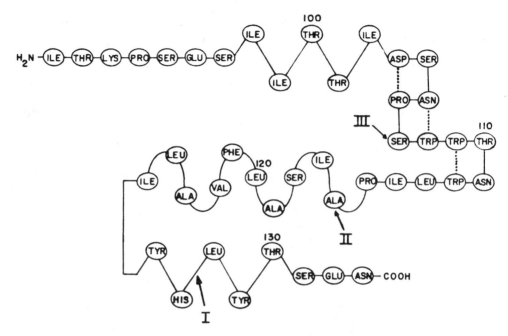

Figure 1. Primary and predicted secondary structure of the nonpolar peptide of cytochrome b_5. The arrows show the limits of cleavage by carboxypeptidases for the three derivatives. Secondary structure features are: β sheet, ⋈ ; β turn, ⊏ ; α helix, ϻ . (From Dailey and Strittmatter, 1978, with permission of the American Society of Biological Chemists, Inc.)

1975; Rogers and Strittmatter, 1974) indicate that the translational movement of the heme protein within membrane bilayers is a crucial functional feature of this electron carrier.

Our focus here is to examine the structural aspects of cytochrome b_5 required for its catalytic functions as well as binding and orientation in membranes. The 16,000-dalton heme protein (Spatz and Strittmatter, 1971) is amphipathic, containing a hydrophilic, catalytic heme peptide of approximately 80 residues joined by a short apparently unstructured sequence to a hydrophobic COOH-terminal segment of approximately 40 residues that is required for binding to membranes (Strittmatter *et al.*, 1972). The detailed crystallographic studies of Mathews (1976) on the compact tertiary structure of the heme peptide segment (11,000 daltons), obtained by mild proteolysis, revealed a tight heme-binding crevice that sequesters the iron atom in the interior and orients the heme edge containing the propionyl side chains at the surface of the molecule. In contrast, the nonpolar segment (Fleming *et al.*, 1978) (Fig. 1) is characterized by an extensive sequence of nonpolar residues (residues 98–129) that follow the short joining region (residues 91–97). Figure 1 also shows that the method of Chou and Fasman (1977a,b) predicts a high degree of secondary structure in this region, consistent with circular dichroism measurements, involving two short segments of β sheet a section of α helix, and a cluster of three β turns. The absence of predicted structure in the joining region is consistent with the susceptibility of this region to proteolytic digestion (Strittmatter and Rogers, 1975).

In the following sections we will examine more recent information that bears first on the definition of the catalytic surface and then turn to data on the structure and orientation of the nonpolar segment in membranes. These considerations form the basis for a proposed relatively detailed structural model for functional cytochrome b_5 *in situ*.

2. THE ANIONIC ACTIVE SITE OF CYTOCHROME b_5

Early studies on the interaction between the catalytically active soluble proteolytic fragments of cytochrome b_5 and NADH cytochrome b_5 reductase (Loverde and Strittmatter, 1968) showed that acylation of only seven lysyl residues of the reductase results in almost complete loss of catalytic activity, the heme protein acting as electron acceptor. This implied involvement of charge interactions between lysyl residues on the reductase and complementary negatively charged carboxyl groups on cytochrome b_5 also became apparent in the rapid oxidation of reduced cytochrome b_5 by cytochrome c, a rapid nonphysiological reaction. Salemme (1976), on the basis of the tertiary structure of the two proteins determined by X-ray crystallography and best-fit computer programs, proposed complementary charge interactions between cytochrome c lysyl residues 13, 27, 72, and 79 and carboxyl residues Glu 52, Glu 48, Asp 64, and one exposed propionate heme group of cytochrome b_5. Ng et al. (1977) experimentally tested this hypothesis by characterizing the effects of single lysyl residue modification of cytochrome c. In this manner they implicated lysyl residues 13, 25, 27, and 72 or 79 in interactions with cytochrome b_5.

We have utilized a water-soluble carbodiimide and various nucleophiles to produce derivatives of both the soluble heme peptide fragment and intact cytochrome b_5 (Dailey and Strittmatter, 1979, 1980). A loss of carboxylate charge was achieved with methylamine as the nucleophile, and a similar charge effect was produced by the more reactive, but bulkier, glycine ethyl ester. Charge retention was effected with taurine. Individual homogeneous methylamide derivatives were isolated by ion-exchange chromatography, and, with the other nucleophile derivatives containing various levels of carboxylate derivatization, were examined as electron acceptors for the two microsomal flavoproteins and as electron donor for stearyl-CoA desaturase.

The initial studies with the soluble catalytic segments of the cytochrome and NADH cytochrome b_5 reductase (Dailey and Strittmatter, 1979) with methylamine derivatives implicated Glu 47, Glu 48, Glu 52, and a heme propionate in interactions with the reductase. Modification of these residues in various combinations resulted in an increase in the K_m (from 8.8 to 91 μM) for the heme peptide of cytochrome b_5. Asp 64 was also tentatively identified as a fifth anionic interaction site, there being a further increase in K_m in a derivative carrying five methylamide groups. With the larger glycine ethyl ester as nucleophile, K_m increases of more than 2 orders of magnitude (250–350 μM) were observed. In both cases the effect of charge elimination at these sites was on the apparent K_m values alone; the V_{max} values remained unaltered, suggesting that the methylamide and the glycine ethyl ester groups do not significantly effect the cytochrome reductase orientation required for rapid electron transfer. In contrast, taurine-modified heme peptide, with retention of all possible charge pair interactions, yields an unaltered apparent K_m. This modification, however, results in a movement of each negative charge, now a sulfonate, approximately 4 Å from the previous carboxylate anion, thus increasing the nearest approach distance between the heme edge of the cytochrome and the reduced flavin of the reductase. This results in an expected marked decrease in V_{max} to as little as 2% of that of the native protein. In electron-transfer reactions this additional distance may be critical, for the efficiency of electron tunneling drops rapidly as the distance between donor–acceptor pairs increases (Hopefield, 1974).

Figure 2 (Dailey and Strittmatter, 1979) shows an outline of the interacting surface of the heme peptide (A) and the backbone structure from this view (B) constructed using the coordinates of Mathews (1976). The carboxyl groups identified as charge pairing sites with NADH cytochrome b_5 reductase, as well as those postulated by Salemme (1976) in electron

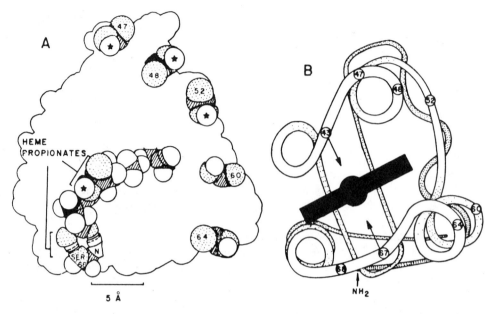

Figure 2. (A) Outline and specific features of a Cory–Pauling–Koltun space-filling model of the heme peptide of cytochrome b_5. For clarity of presentation, only the exposed heme edge and side-chain carboxyl groups are shown. In the drawing, oxygen atoms are stippled, carbon atoms are cross-hatched, and hydrogen atoms are outlined. The heme propionate hydrogen bonded to the Ser 68 side-chain hydroxyl and amide nitrogen (N) is also shown. Residues modified and identified in this study are Glu 47, 48, and 52, and the exposed heme propionate. The model was constructed using the coordinates of Mathews (1976). (B). Backbone tracing of the heme peptide of cytochrome b_5. This figure is to the same scale as (A). The NH_2 terminus is marked, and the edge of the heme is shown as a rectangle with the position of the iron represented by the solid circle. Residues marked are: 43 and 67, histidines that are coordinated with the heme iron; 68, serine hydrogen-bonded to one heme propionate; Glu 47, 48, 52, 60, and Asp 64. The segment from residues 43 to 47 loops slightly outward toward the observer, thus accounting for the apparent two-dimensional nearness of these two marked residues. (From Dailey and Strittmatter, 1979, with permission of the American Society of Biological Chemists, Inc.)

transfer to cytochrome c, are all present on the face of the molecule that contains the exposed edge of the heme. This surface, which lacks any positively charged amino acid residues, forms a nearly planar surface of anionic charge that is perpendicular to the plane of the sequestered protoheme macrocycle. The distribution of charges is such that the essential carboxyl groups form a semicircle about the periphery of this heme edge face with the exposed heme 12 Å distant from any carboxyl group. In the case of electron transfer between the heme groups of cytochrome c and cytochrome b_5, Salemme (1976) proposed complex formation in which the hemes of the two proteins assume a nearly coplanar orientation. The participation of essentially the same charge constellation in electron transfer from the reductase to the heme peptide of cytochrome b_5 may produce a similar specific orientation of the flavin to the heme to facilitate electron transfer.

The view that a single anionic site on cytochrome b_5 is utilized in the spectrum of microsomal oxidation–reduction reactions is strengthened by similar studies in which derivatives of intact cytochrome b_5 were used to characterize the interactions with NADPH cytochrome P-450 and stearyl-CoA desaturase (Dailey and Strittmatter, 1980). Again the same four carboxylate residues 47, 48, 52, and the heme propionate were implicated in electron transfer by observed increases of at least an order of magnitude in the K_m values of various methylamine derivatives. It is not yet clear whether quantitative variation in the effects on the apparent K_m values for the two reductases and desaturase reflects a variation in contributions of other nonionic contacts in these protein–protein interactions or more subtle differences in reaction sequences leading to electron transfer. This situation is emphasized by the fact that taurine derivatives of cytochrome b_5, in which there is a forced 4-Å change in charge pair positions, do not show a decreased V_{max} with the P-450 reductase as was observed with NADH cytochrome b_5 reductase. This ability to accommodate the imposed displacement of crucial anionic charges indicates either a more flexible arrangement of residues forming charge pairs or a less restrictive alignment of the flavin electron donor and heme. In this regard, the presence of both FAD and FMN in the P-450 reductase (Vermilion and Coon, 1978), with different sites for reduction by NADPH and oxidation by either cytochrome c or cytochrome P-450, brings added functional complexity to the active site(s) of this flavoprotein.

3. STRUCTURE AND ORIENTATION OF NONPOLAR SEGMENT IN PHOSPHOLIPID BILAYERS

The ready isolation of the short joining region (residues 91–97) with the nonpolar segment of cytochrome b_5 (Fleming and Strittmatter, 1978), representing residues 91–133 of the intact protein, permitted direct binding studies to phospholipid bilayers, initially to small dimyristyl lecithin vesicles. These experiments determined that the binding of the COOH-terminal nonpolar segment mimics the complete protein and that only one of the three tryptophanyl residues is fluorescent. The fluorescence emission of this residue, Trp 109, increases approximately twofold upon insertion of the peptide in the lipid vesicles. The subsequent development of a method, utilizing excitation energy transfer from a donor tryptophanyl residue in bound proteins to trinitrophenyl or dansyl acceptor groups on the surface of phospholipid bilayers (Koppel *et al.*, 1979), revealed that the intramembrane position of this indole ring is located approximately 20–22 Å below the surface of the bilayer (Fleming *et al.*, 1979).

Because this membrane-binding segment is at the COOH terminus of the intact heme protein, it was also possible to use limited digestion with carboxypeptidases A and Y to determine which portions of the entire sequence are required for membrane binding and catalytic interactions with the desaturase that cannot utilize the soluble heme peptide segment as electron donor (Fleming and Strittmatter, 1978). The Roman numerals in Fig. 1 indicate the three derivatives generated by limited carboxypeptidase Y digestion (I), carboxypeptidase A digestion (II), and sequential digestion of II with carboxypeptidase Y to remove the proline and then carboxypeptidase A (III). Both derivatives I and II lacking either 6 or 18 COOH-terminal residues bind to dimyristyllecithin vesicles and are completely active in the stearyl-CoA reductase system reconstituted in egg lecithin vesicles. In contrast, derivative III, derived from II by removal of the next 9 COOH-terminal residues, does not bind to vesicles or act as electron shuttle in the desaturase reaction sequence. This directly identifies the nine residues (107–115), of which seven are nonpolar, as the crucial structural feature for functional attachment to membranes. It is of particular interest that the following sequence of 11 residues (116–126) is not essential for this type of protein–lipid interaction. Circular dichroism data indicate that both derivatives I and II retain the secondary structure of the residual sequence that was originally discerned in the entire sequence of the nonpolar peptide (Fig. 1). Thus, only a portion of the nonpolar segment yields a minimal structure that results in insertion and provides the orientation and presumably translational mobility required for interaction with both vesicle-bound NADH cytochrome b_5 reductase and stearyl-CoA desaturase.

A more subtle variation in membrane attachment of the heme protein was discerned in experiments that utilized a method for preparing uniformly large vesicles (1000-Å diameter) (Enoch and Strittmatter, 1979b) to examine, in more detail, protein transfer and carboxypeptidase Y digestion of vesicle-bound cytochrome (Enoch *et al.*, 1979). Gel filtration was employed to separate equilibrated large and small vesicles or microsomes and measure the extent of cytochrome b_5 transfer between these vesicle types. As expected from our earlier work, cytochrome b_5 was not transferred from dimyristyllecithin or microsomal vesicles to either of the other two vesicle preparations at an appreciable rate ($< 10\%$ in several hours at 30 °C). However, cytochrome b_5 bound to either small (250–300Å) or large egg lecithin vesicles would readily transfer between these two sizes of vesicles or to either microsomes or dimyristyl lecithin vesicles. This distinction in binding was also reflected in the release of the soluble heme peptide segment by carboxypeptidase Y digestion. The terms "tight" and "loose" were used to distinguish the more stable binding, representing the biological situation, from the mobile protein–vesicle interactions. The latter "loose" binding appears to be a problem of insertion into the stable egg lecithin vesicles, which are less leaky than preparations of dimyristyllecithin and do not contain membrane proteins that provide the localized disruption of regular bilayer packing in biological membranes and therefore foci for normal cytochrome insertion. The observation that normal "tight" binding to egg lecithin vesicle preparations could be achieved by sonication or by inclusion of deoxycholate, peptides, or proteins in the vesicles, is consistent with this view.

With this additional sensitive test for detecting essential structural features required for proper orientation in the membrane as well as functional binding, we reexamined the interactions of both derivatives of cytochrome b_5, shortened at the COOH terminus, and nonpolar peptide derivatives produced by modification of the carboxylate residues of the isolated nonpolar segment, using phospholipid vesicles (Dailey and Strittmatter, 1981a). The removal of only six COOH-terminal residues, which include the carboxyl groups of Glu 132 and COOH-terminal Asn 133, by carboxypeptidase digestion resulted in the loss of the characteristic "tight" binding to either synthetic phospholipid vesicles or isolated microsomes. Moreover, chemical modification of the four carboxyl groups of the isolated nonpolar peptide segment with carbodiimide and methylamine to produce a derivative with no anionic charged residues also resulted in a loss of this type of stable membrane interaction exhibited by intact cytochrome b_5 and unmodified nonpolar peptide. Thus, although the nine-residue sequence that includes the fluorescent tryptophanyl residue (residues 107–115) is sufficient for membrane interaction in a functional manner, the short polar COOH-terminal segment, containing two of the four carboxyl groups of the membrane-binding domain of cytochrome b_5, is essential for the lipid–protein interactions that lead to the normal "tight" orientation both *in situ* and in reconstituted phospholipid bilayer systems.

All of the binding data led us to a consideration of experimental approaches that might distinguish between two contrasting models considered previously (Enoch *et al.*, 1979) that would yield such a stable bound cytochrome. One involves a transmembrane orientation of the nonpolar peptide segment placing the COOH terminus on the side opposite to the NH₂-terminal segment, including the catalytically functional heme peptide. Presumably, initial binding in this fashion at a perturbed region of the bilayer would then result in a structure in which the anionic and polar COOH-terminal sequence would face the opposing lipid–water interfaces to provide the thermodynamically favorable situation for stable binding. The second model further develops the original suggestion of Tanford and Reynolds (1976) that the COOH terminus is on the same side of the bilayer as the NH_2 terminus. In this case, ionic interactions of Glu 132 and Asn 133, perhaps by charge pairing with cations in the polar head group region, would provide a degree of stabilization of the bound structure. We were successful in using chemical probes for the two tyrosyl residues near the COOH terminus (Tyr 126 and Tyr 129) to determine that both the NH_2-terminal residues of the nonpolar peptide and the tyrosyl residues present at positions 5 and 8 from the COOH terminus are oriented on the outer leaflet in closed phosphatidylcholine vesicles (Dailey and Strittmatter, 1981b). Both of these latter residues ionize rapidly upon addition of sodium hydroxide to the outer aqueous phase at a rate at least an order of magnitude greater than ionization of an indicator dye trapped within the vesicles. This is consistent with positions of both tyrosyl residues at or near the outer hydrated region of the lipid vesicles. Only one of these (Tyr 129) will react with the polar reagent diazotized sulfanilic acid, again added to the outer aqueous phase and impermeant to the vesicle preparations. This places the tyrosyl residue nearest the COOH terminus within the reach of this reagent, within the head-group region, and the second phenolic group, three residues beyond, more submerged in the outer leaflet of the bilayer. Thus, the model featuring a loop of the nonpolar segment into and out of the bilayer appears to describe normal cytochrome b_5 binding.

4. A STRUCTURAL MODEL FOR CYTOCHROME b₅ IN MEMBRANES

The basic two-domain structure of cytochrome b_5 utilizes the nonpolar, terminal segment to anchor the protein in phospholipid bilayers with the catalytically functional heme peptide in the aqueous phase within several angstroms of the bilayer surface as determined by a short, apparently unstructured, joining segment of approximately nine residues (residues 89–97). This binding provides both the required proximity and orientation of this electron carrier for its highly efficient interactions with a spectrum of electron donors and acceptors arranged on the cytosolic side of the endoplasmic reticulum of mammalian cells. The identification of a single anionic surface forming a crescent-shaped reactive site that appears to involve charge pairing with at least two flavoproteins, the iron-containing desaturase, and cytochrome c is not only consistent with our earlier view that the heme protein is distributed randomly in the endoplasmic reticulum and undergoes random translational motion, but also that the interactions resulting from collisions with the variety of other microsomal oxidative enzymes will be effective in producing rapid electron transfer only if the protein–protein contacts result in these precise ionic interactions. The orientation of the electron carrier pair in each case must be optimal for a catalytic event.

The constraints of this requirement are clear. Because all of these proteins are attached to a membrane, their degrees of freedom are largely limited to rotational and translational movement, and any segmental motion that each protein structure or its interactions with the bilayer permit. The amphipathic structure of both cytochrome b_5 and NADH cytochrome b_5 reductase, in which each domain is joined by a short sequence that is susceptible to cleavage by a number of endopeptidases at several sites in these sequences, led us to the original suggestion that this region is in fact flexible (Strittmatter *et al.*, 1972). This would permit a degree of wobble that might be particularly significant in permitting optimal cytochrome b_5 interactions with a variety of other proteins in which the electron acceptor or donor groups are aligned differently with respect to the bilayer surface. In an initial measurement of the rotational diffusion of cytochrome b_5 in vesicles, however, Vaz *et al.* (1979) found that the rotational motion of the heme was sensitive to the lipid phase tran-

Figure 3. (A) Cory–Pauling–Koltun space-filling model of the nonpolar peptide segment of cytochromes, based on the primary and secondary structure shown in Fig. 1, the placement of both NH_2 and COOH termini on the same side of the bilayer, the known position of Trp 109, and the probable positions of Tyr 126 and Tyr 129. (B) Basic backbone structure of model and selected residue positions.

sition. In their model, which assumes a heme orientation parallel to the plane of the bilayer as Mathews suggest (1976), the entire molecule would therefore be more rigid. It is possible, however, that the measurements of the triplet-state absorbance anisotropy decay of the rhodium protoporphyrin IX derivative of cytochrome b_5 did not detect the contribution of the wobble arising from this flexible region either, e.g., because this wobble is relatively limited by proximity of the membrane surface or involves a time scale outside of the resolution of the method.

Some of the structural features of the nonpolar binding domain are now more clearly

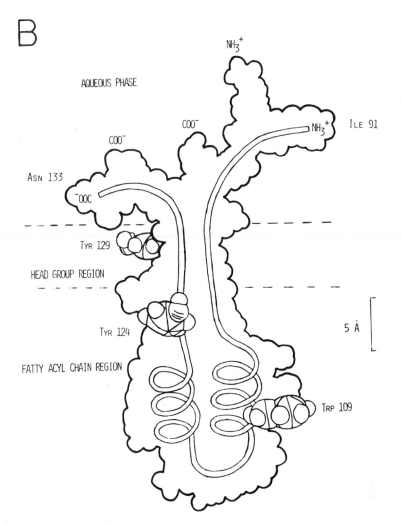

B

AQUEOUS PHASE

NH$_3^+$

COO⁻

COO⁻

ASN 133

⁻OOC

NH$_3^+$ ILE 91

TYR 129

HEAD GROUP REGION

TYR 124

5 Å

FATTY ACYL CHAIN REGION

TRP 109

Figure 3. continued

defined. A relatively short sequence of nine residues is sufficient to anchor the cytochrome in place so that it can participate effectively in a number of electron transfer reactions. Nevertheless, only the complete segment, including the five polar residues at the COOH terminus, is required for the stable binding that mimics the *in vivo* situation. The placement of this final polar sequence of the nonpolar domain on the same side of the bilayer as the heme peptide domain in a loop arrangement can be combined with the accumulated data on the binding and structure of the nonpolar peptide segment to generate a relatively detailed model for the tertiary structure of "tightly" bound nonpolar peptide segment. Figure 3A shows a space-filling model constructed on the basis of the present data and several assumptions. Trp 109 is 20–22 Å from the bilayer surface whereas both the NH$_2$-terminal polar and unstructured residues and the COOH-terminal polar sequence are on the same outer surface of the bilayer. Tyr 129 is present in the outer bilayer phospholipid head-group region and Tyr 126 just slightly below in the fatty acyl side-chain region. In addition, the model utilizes the known primary structure and the predicted secondary structure to construct the loop that gives rise to such residue positions. With this latter assumption regarding secondary structure, which is consistent with the circular dichroism data, the loop permits the required hydrogen bonding of residues 98–102 with 126–130 to form an antiparallel β-sheet structure. Trp 109 is then brought to its known position in the hydrocarbon region in a three consecutive β-turn structure that closely resembles a 3_{10} type of helix, which lies next to the predicted α helix of residues 115–125, forming a largely nonpolar foot to anchor the protein in the hydrophobic milieu of the bilayer. The final COOH-terminal anionic residues are placed in the head-group region to contribute to stabilization of both the overall tertiary structure and membrane position by ion pairing with phospholipid head groups. Figure 3B diagrammatically emphasizes selective components of this structure for clarity. This model results both in the presentation of a highly nonpolar surface to maximize van der Waals contacts in the hydrocarbon region and the opportunity for significant stabilization of the tertiary structure by hydrogen bonding. Table I shows that this latter type of bonding arises mainly from interactions expected for the segments of secondary structure, but polar side-chain residues placed in the hydrocarbon region of the bilayer also readily form hydrogen bonds in this structural model.

The value of this detailed and speculative model extends beyond the attempted synthesis of existing data to present the type of stable tertiary structure that can serve as an anchoring foot for the heme peptide. Hopefully, probes that reach only selected regions of

Table I. Hydrogen Bonds Stabilizing Nonpolar Peptide Structural Model of Fig. 3

Structure	Residues forming hydrogen bonds[a]			
β Sheet	Ile	98 (NH)	Thr	130 (CO)
	Thr	130 (NH)	Ile	98 (CO)
	Thr	100 (NH)	Leu	128 (CO)
	Leu	128 (NH)	Thr	100 (CO)
	Ile	102 (NH)	Tyr	126 (CO)
	Tyr	126 (NH)	Ile	102 (CO)
α Helix	Ser	118 (NH)	Ile	114 (CO)
	Ala	119 (NH)	Pro	115 (CO)
	Leu	120 (NH)	Ala	116 (CO)
	Phe	121 (NH)	Ile	117 (CO)
	Val	122 (NH)	Ser	118 (CO)
	Ala	123 (NH)	Ala	119 (CO)
	Leu	124 (NH)	Leu	120 (CO)
	Ile	125 (NH)	Phe	121 (CO)
β Turn	Asp	103 (NH)	Pro	106 (CO)
	Asn	105 (NH)	Trp	108 (CO)
	Trp	109 (NH)	Trp	112 (CO)
	Ser	107 (NH)	Thr	110 (CO)
Side chains	Thr	100 (OH)	Tyr	126 (OH)
	Thr	101 (OH)	His	127 (N)
	Ser	107 (OH)	Ser	118 (OH)
	Tyr	129 (OH)	Ser	131 (OH)
	Trp	112 (N)	Trp	109 (CO)

[a]Groups involved are shown in parentheses.

the bilayer, whether designed for cross-linking or as optical reporters, will test the predicted locations of a number of other residues and gradually define the true outline of this peptide segment. Additional distance measurements to residues such as the single imidazole, the other two tryptophanyl residues, and polar residues would provide particularly restrictive data. At the same time it will be necessary to determine more precisely the degree to which wobble arising from flexibility in the joining region contributes to the catalytic efficiency of this heme protein, and the extent to which protein–protein interactions involving segments immersed in the bilayer facilitate the most effective alignment of the active sites of cytochrome b_5 and any other microsomal enzyme. In this sense the relatively simple structure of cytochrome b_5 represents the experimental model for effectively pursuing similar molecular details of the spectrum of enzymes of the endoplasmic reticulum and other biological membranes.

ACKNOWLEDGMENT. This research was supported by Grant GM-15925 from the United States Public Health Service.

REFERENCES

Chou, P. Y., and Fasman, G. D. (1977a). *Trends Biochem. Sci.* **2,** 128–131.
Chou, P. Y., and Fasman, G. D. (1977b). *J. Mol. Biol.* **115,** 135–175.
Dailey, H. A., and Strittmatter, P. (1978). *J. Biol. Chem.* **253,** 8203–8209.
Dailey, H. A., and Strittmatter, P. (1979). *J. Biol. Chem.* **254,** 5388–5396.
Dailey, H. A., and Strittmatter, P. (1980). *J. Biol. Chem.* **255,** 5184–5189.
Dailey, H. A., and Strittmatter, P. (1981a). *J. Biol. Chem.* **256,** 1677–1680.
Dailey, H. A., and Strittmatter, P. (1981b). *J. Biol. Chem.* **256,** 3951–3955.
Enoch, H. G., and Strittmatter, P. (1979a). *J. Biol. Chem.* **254,** 8976–8981.
Enoch, H. G., and Strittmatter, P. (1979b). *Proc. Natl. Acad. Sci. USA* **76,** 145–149.
Enoch, H. G., Catala, A., and Strittmatter, P. (1976). *J. Biol. Chem.* **251,** 5095–5103.
Enoch, H. G., Fleming, P. J., and Strittmatter, P. (1979). *J. Biol. Chem.* **254,** 6483–6488.
Fleming, P. J., and Strittmatter, P. (1978). *J. Biol. Chem.* **253,** 8198–8202.
Fleming, P. J., Dailey, H. A., Corcoran, D., and Strittmatter, P. (1978). *J. Biol. Chem.* **253,** 5359–5372.
Fleming, P. J., Koppel, D. E., Lau, A. L. Y., and Strittmatter, P. (1979). *Biochemistry* **18,** 5458–5464.
Hildebrandt, A., and Estabrook, R. W. (1971). *Arch. Biochem. Bophys.* **143,** 66–79.
Holloway, P. W., and Katz, J. T. (1972). *Biochemistry* **11,** 3689–3696.
Holloway, P. W., and Wakil, S. J. (1970). *J. Biol, Chem.* **245,** 1862–1865.
Hopfield, J. J. (1974). *Proc. Natl. Acad. Sci. USA* **71,** 3640–3644.
Imai, Y., and Sato, R. (1977). *Biochem. Biophys. Res. Commun.* **75,** 420–426.
Keyes, S. R., Alfano, J. A., Jansson, I., and Cinti, D. L. (1979). *J. Biol. Chem.* **254,** 7778–7784.
Koppel, D. E., Fleming, P., and Strittmatter, P. (1979). *Biochemistry* **18,** 5450–5457.
Lau, A. Y. H., West, S. B., Vore, M., Ryan, D., and Levin, W. (1974). *J. Biol. Chem.* **249,** 6701–6709.
Lee, T. C., Baker, R. C., Stephens, N., and Snyder, F. (1977). *Biochim. Biophys. Acta* **489,** 25–31.
Loverde, A., and Strittmatter, P. (1968). *J. Biol. Chem.* **243,** 5779–5787.
Mathews, F. S. (1976). In *The Enzymes of Biological Membranes* (A. Martonosi, ed.), Vol. 4, pp. 143–198, Plenum Press, New York.
Ng, S., Smith, M.B., Smith, H.T., and Millett, F. (1977). *Biochemistry* **16,** 4975–4978.
Okayasu, T., Ono, T., Shinojima, K., and Imai, Y. (1977). *Lipids* **12,** 267–271.
Oshino, N., and Omura, T. (1973). *Arch. Biochem. Biophys.* **157,** 395–404.
Oshino, N., Imai, Y., and Sato, R. (1971). *J. Biochem. (Tokyo)* **69,** 155–167.
Paultanf, F., Prough, R. A., Masters, B. S. S., and Johnston, J. M. (1974). *J. Biol. Chem.* **249,** 2661–2662.
Reddy, V. R., Kupfer, P., and Caspi, E. (1977). *J. Biol. Chem.* **252,** 2797–2801.
Rogers, M. J., and Strittmatter, P. (1974). *J. Biol. Chem.* **249,** 5565–5569.
Salemme, F. R. (1976). *J. Mol. Biol.* **102,** 563–568.
Sasame, H. A., Thorgeirsson, S.S., Mitchell, J.R., and Gillette, J. R. (1974). *Life Sci.* **14,** 35–46.
Spatz, L., and Strittmatter, P. (1971). *Proc. Natl. Acad. Sci. USA* **68,** 1042–1046.

Spatz, L., and Strittmatter, P. (1973). *J. Biol. Chem.* **248,** 793–799.

Strittmatter, P., and Rogers, M. J. (1975). *Proc. Natl. Acad. Sci. USA* **72,** 2658–2661.

Strittmatter, P., Rogers, M. J., and Spatz, L. (1972). *J. Biol. Chem.* **247,** 7188–7194.

Strittmatter, P., Spatz, L., Corcoran, D., Rogers, M. J., Setlow, B., and Redline, R. (1974). *Proc. Natl. Acad. Sci. USA* **71,** 4565–4569.

Tanford, C., and Reynolds, J. A. (1976). *Biochim. Biophys. Acta* **457,** 133–170.

Vaz, W. L. C., Austin, R. H., and Vogel, H. (1979). *Biophys. J.* **26,** 415–426.

Vermilion, J. L., and Coon, M. J. (1978). *J. Biol. Chem.* **253,** 8812–8819.

10

Activation of Pyruvate Oxidase and Interaction with Membrane Components

John G. Koland, Michael W. Mather, Robert B. Gennis, John S. White, and Lowell P. Hager

1. INTRODUCTION

Pyruvate oxidase is one of several flavoprotein dehydrogenases that feed electrons into the *E. coli* respiratory chain. The enzyme catalyzes the oxidative decarboxylation of pyruvate to acetic acid and CO_2 (Hager, 1957). Although usually a minor component in *E. coli,* it can be induced to higher levels in mutant strains and provides a pathway for pyruvate catabolism alternative to that of the pyruvate dehydrogenase complex. Pyruvate oxidase is a tetramer with identical subunits of 60,000 daltons (Williams and Hager, 1960; O'Brien *et al.*, 1976). Each subunit has a tightly bound flavin adenine dinucleotide (FAD), and the addition of a second cofactor, thiamin pyrophosphate, is necessary to elicit catalytic activity. Enzymatic activity is easily monitored spectrophotometrically using ferricyanide as an artificial oxidant.

   Purified pyruvate oxidase has been particularly useful in the study of protein–lipid interactions. Unlike the other flavoprotein dehydrogenases from *E. coli,* it is a peripheral membrane protein and can be released from the membrane by sonication (O'Brien *et al.*, 1976). The purified enzyme is free of lipids and is stable in the absence of detergents or chaotropic agents.

John G. Koland, Michael W. Mather, and Robert B. Gennis. • Department of Chemistry, School of Chemical Sciences, University of Illinois, Urbana, Illinois 61801.   *John S. White and Lowell P. Hager.* • Department of Biochemistry, School of Chemical Sciences, University of Illinois, Urbana, Illinois 61801.

2. AMPHIPHILIC ACTIVATION OF PYRUVATE OXIDASE

2.1. Amphiphilic Activation

The central focus of research on pyruvate oxidase over the past several years has been the strong effect of amphiphiles, including phospholipids, on enzymatic activity (Cunningham and Hager, 1971a,b; Blake *et al.*, 1978). Figures 1 and 2 illustrate that a wide

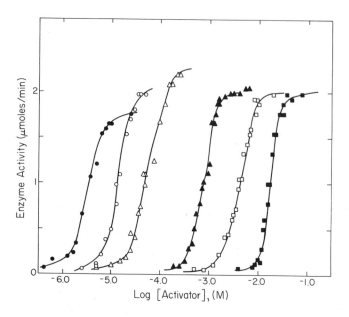

Figure 1. Activation of pyruvate oxidase by charged detergents and fatty acids. Pyruvate oxidase was assayed at room temperature in the presence of the indicated concentrations of various monomeric amphiphiles. Enzyme activity is expressed in μmoles of pyruvate consumed per 8.8 μg of enzyme. The representative amphiphiles are octadecyltrimethylammonium bromide (●), hexadecyltrimethylammonium bromide (o), tetradecyltrimethylammonium bromide (△), decanoic acid (▲), nonanoic acid (□), and octanoic acid (■). (From Blake *et al.*, 1978.)

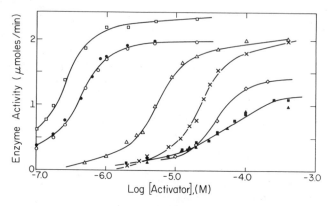

Figure 2. Activation of pyruvate oxidase by representative aggregated amphiphiles. As in Fig. 1, pyruvate oxidase was assayed in the presence of various activators: cardiolipin (□), phosphatidylglycerol (o), phosphatidylserine (●), dihexadecylphosphatidic acid (△), gangliosides (X), dihexadecylphosphatidylcholine (◇), ditetradecylphosphatidylethanolamine (■), and dihexadecylphosphatidylethanolamine (▲). (From Blake *et al.*, 1978.)

variety of phospholipids, charged detergents, and fatty acids are capable of activating this enzyme approximately 25-fold. Although the oxidase interacts most favorably with negatively charged amphiphiles, it is also activated by cationic, uncharged, and zwitterionic amphiphiles (Blake *et al.*, 1978). The ability of an amphiphile to activate pyruvate oxidase increases with the length of its hydrocarbon chain up to 14–16 carbon atoms, suggesting that the primary mode of interaction between the protein and the amphiphiles is hydrophobic rather than electrostatic (Blake *et al.*, 1978).

2.2 Kinetic Consequences of Activation

Table I shows some of the steady-state kinetic parameters obtained when the behavior of the enzyme is compared in the presence and absence of a phospholipid activator. In addition to the dramatic effect on the turnover number, the presence of the lipid alters the K_m values and the Hill coefficients for substrates and cofactors (Cunningham and Hager, 1971b). Equilibrium binding studies with thiamin pyrophosphate (O'Brien *et al.*, 1977) have clearly shown that the influence of lipids and detergents on the K_m and Hill coefficient of this cofactor directly reflects changes in its equilibrium binding isotherm. Hence, it is evident that the activity of this membrane-bound enzyme can be modulated by membrane lipids not only through changes in the turnover number, but also by affecting the strength and extent of cooperativity of substrate and cofactor binding.

Although the effect of lipid activators on the overall turnover number of the enzyme is conveniently monitored by steady-state kinetic techniques, the microscopic consequences of lipid activation must be determined by other methods, such as rapid kinetics. The catalytic cycle of pyruvate oxidase may be broken down into the sequence of steps shown in Fig. 3, by analogy with other thiamin pyrophosphate-requiring enzymes (Krampitz, 1969; Hager and Krampitz, 1963). Rapid kinetic techniques have been used to study the effect of lipid activators on isolated portions of this cycle. The rate constant describing the oxidation of the reduced flavoprotein (V) by ferricyanide is not increased in the presence of the lipid activator (Blake, 1977). Hence, the entire effect of lipids on the steady-state turnover number is due to an increase in the rate by which the oxidized flavoprotein is reduced by pyruvate (steps 1–4). By using rapid chemical quench techniques with [1-^{14}C]pyruvate, the rate of CO_2 release catalyzed by the enzyme can be measured directly (steps 1–3). The rate of CO_2 release is identical to the rate of enzymatic turnover in the absence of lipids (M. W. Mather and R. B. Gennis, unpublished observations) and is

Table I. Steady-State Kinetic Parameters for the Pyruvate Oxidase Reaction [a]

Activator	k_{cat}[b] (sec^{-1})	$K_{0.5}$[c] for pyruvate[d] (mM)	$K_{0.5}$[c] for TPP (mM)	Hill coefficient of TPP	$K_{0.5}$[c] for activator (μM)	Hill coefficient of activator
None	7	100	6	1.0	—	—
Phosphatidylglycerol[e]	160	11	3	1.5	0.4	1.0
Sodium dodecyl sulfate[f]	180	14	7	1.0	8.0	2.8

[a] From Blake *et al.* (1978).
[b] k_{cat} is the catalytic center activity or turnover number per FAD.
[c] $K_{0.5}$ is the concentration of the indicated species at which half-maximal activity is observed.
[d] The Hill coefficient for pyruvate is in all cases equal to 1.0.
[e] Activator exists in an aggregated form.
[f] Activates below its critical micelle concentration as a monomeric species.

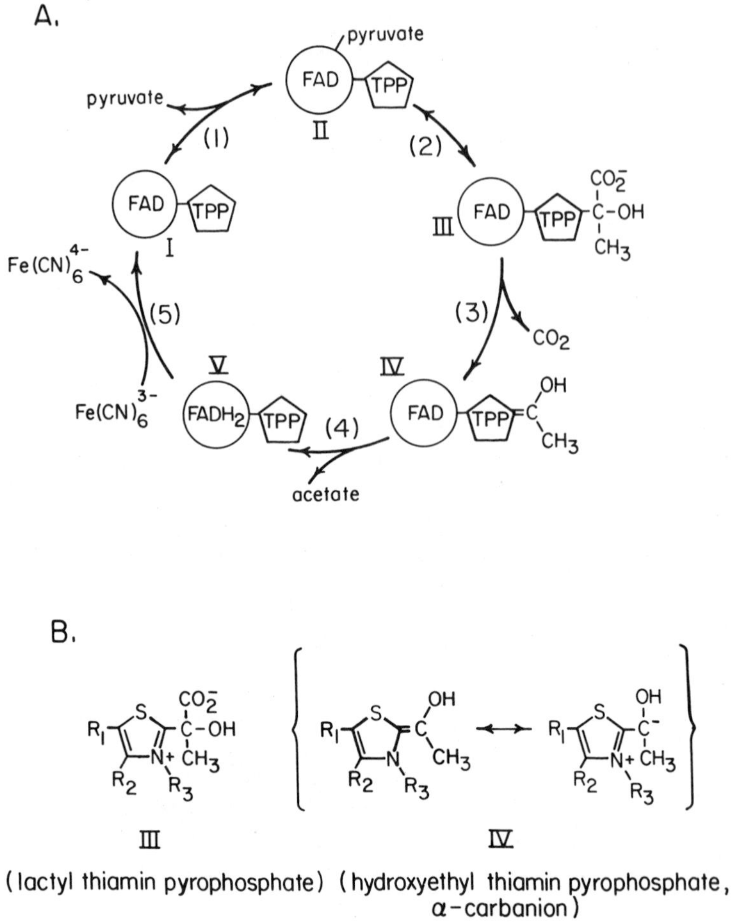

A.

B.

III   IV

(lactyl thiamin pyrophosphate)   (hydroxyethyl thiamin pyrophosphate, α-carbanion)

Figure 3. Steady-state catalytic cycle of pyruvate oxidase. (A) Catalytic cycle. (B) Detailed structures of proposed intermediate species. Based on the scheme of Breslow (1958).

dramatically accelerated in the presence of lipids (Blake, 1977). The data indicate that the rate of reduction of the flavoprotein (step 4) by the intermediate species (IV) is fast, and that activation must be due to the acceleration of a step prior to the flavin reduction (step 4).

Heavy atom isotope experiments (M. O'Leary and R. B. Gennis, unpublished observations), comparing the rates of decarboxylation of pyruvate containing ^{12}C and ^{13}C in the carboxyl position, clearly show that the step involving the actual decarboxylation (step 3) is not rate limiting either in the presence or the absence of the lipid activator. These data are most easily interpreted by postulating that a single step in the mechanism is accelerated by lipids and that this step is rate limiting in the overall cycle. This step is most likely to be the formation of the covalent adduct between thiamin pyrophosphate and pyruvate bound at the active site of the enzyme (step 2). Alternatively, a slow conformational change associated with the binding of the substrate could also be responsible for the observed effects. A summary of the various rate constants that have been measured is presented in Table II.

Table II. Experimentally Measured Second-Order Rate Constants of Pyruvate Oxidase [a,b]

Activator	Rapid mixing experiments			Steady-state parameters [c]	
	CO_2 evolution	Flavin reduction	Flavin oxidation	$k_{cat}/K_{0.5}$ (pyruvate)	$k_{cat}/K_{0.5}$ (ferricyanide)
None	69	67	150,000	70	3,500
Phosphatidylglycerol	—	>7,200	70,000	15,000	50,000
Sodium dodecyl sulfate	>17,000	>6,400	83,000	13,000	60,000

[a] From Blake (1977) and M. Mather and R. B. Gennis (unpublished observations).
[b] All rate constants are in units of $M^{-1} sec^{-1}$.
[c] Steady-state data are shown for comparison. (See Table I for definition of parameters.) These quantities represent the minimum theoretical values for any second-order rate constant characterizing a portion of the reductive half and the oxidative half, respectively, of the catalytic cycle.

2.3. Activation as an Allosteric Process

Whereas the influence of lipids on events at the catalytic site has been explored in considerable detail, perhaps the most significant experiments with pyruvate oxidase have involved the demonstration of reciprocity: not only do lipids have a dramatic influence on events at the catalytic site, but the ligands involved in catalysis have a strong influence on the affinity of the protein for lipids. Experiments have shown an energetic and conformational coupling between the catalytic active site and the lipid-binding site on pyruvate oxidase (Cunningham and Hager, 1971b; Russell *et al.*, 1977b; Schrock and Gennis, 1980). Figure 4 shows that in the absence of pyruvate and thiamin pyrophosphate, pyruvate oxidase interacts very weakly with dipalmitoylphosphatidylcholine vesicles, even at relatively high concentrations. In contrast, when the substrate and cofactor are present, most of the protein and phospholipid migrates as a single species in a sucrose gradient.

Other experiments have demonstrated that the affinity of the protein for detergents such as dodecyl sulfate is similarly enhanced in the presence of thiamin pyrophosphate and pyruvate (Schrock and Gennis, 1977). Under these circumstances the flavin moiety is reduced, as no electron acceptor is present. Reduction of the flavoprotein appears to be accompanied by a conformational change in the polypeptide that significantly alters the solution behavior of the protein. The reduced form of the flavoprotein has a tendency to self-aggregate and is highly surface labile, in addition to manifesting a higher affinity for both detergents and phospholipid vesicles. Thus, the reduced flavoprotein has the solution behavior typically associated with a hydrophobic membrane protein, whereas the oxidized form of the flavoprotein behaves like a typical water-soluble protein. Spectroscopic studies indicate that the conformational change is probably relatively subtle and local in nature (T. A. O'Brien, in preparation). No major alterations in the secondary structure of the protein, as determined by circular dichroism, accompany flavin reduction.

The most important point demonstrated by these studies is that protein–lipid interactions can be allosterically controlled in a manner similar to that already well documented for protein–protein interactions and interactions between proteins and small molecules. Conceivably, pyruvate oxidase is bound to the *E. coli* inner membrane only when the substrate and cofactor are present at sufficiently high concentrations. The rate of dissociation from the membrane might then be sufficiently slow so that the enzyme would remain bound to the membrane for the duration of the catalytic cycle.

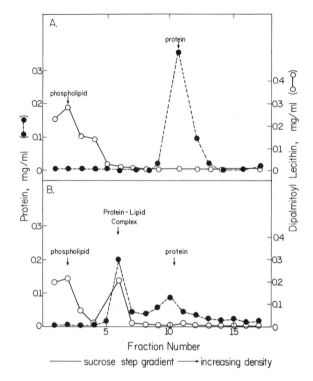

Figure 4. Sucrose step-gradient centrifugation of pyruvate oxidase in the presence and absence of the catalytic ligands. The samples analyzed contained 0.3 mg of dipalmitoylphosphatidylcholine and 0.3 mg of pyruvate oxidase in 0.3 ml of 0.1 M sodium phosphate, pH 5.7. (A) No catalytic ligands present. (B) 100 μM TPP, 100 mM sodium pyruvate, and 10 mM magnesium chloride present. (From Schrock and Gennis, 1980.)

2.4. Affinity Labeling of the Lipid Activation Site

The use of fatty acids as affinity labels for the lipid activation site of pyruvate oxidase is currently being explored (J. S. White and L. P. Hager, unpublished observations). Carboxylic acids react with carbodiimide reagents to form unstable derivatives that are susceptible to nucleophilic substitution by amines. Therefore, pyruvate oxidase, which contains high-affinity binding sites for fatty acids, can be specifically labeled as long as amino groups are available in the proximity of the high-affinity sites to complete amide-bond formation. Using lauric acid, a 12-carbon fatty acid, and the water-soluble carbodiimide reagent 1-ethyl-3-(3-dimethylaminopropyl)carbodiimide, pyruvate oxidase can be labeled at the lipid activation site with a stoichiometry of 1 lauric acid molecule per monomer of enzyme. The amino acid environment of the lipid activation site is presently being investigated using this affinity probe.

3. PROTEOLYTIC ACTIVATION OF PYRUVATE OXIDASE

In the presence of thiamin pyrophosphate, oxidized pyruvate oxidase is not modified by endoproteases such as trypsin and α-chymotrypsin (Hager, 1957). However, reduction of the flavoprotein by the substrate, pyruvate, renders a small region of the polypeptide accessible to attack by proteases (Hager, 1957). The subunit molecular weight of the reduced enzyme is altered from 60,000 to 56,000 upon modification by trypsin, α-chymotrypsin, and other endoproteases (Russell *et al.*, 1977a). Because the NH_2-terminal amino acid sequence of the 56,000-dalton subunit is the same as the sequence found for the 60,000-dalton subunit, it is presumed that the site of cleavage by the endoproteases is near the COOH-terminus (Russell, 1979).

The properties of this protease-modified form of the enzyme are noteworthy. Proteolytic cleavage results in the loss of the phospholipid- and detergent-binding properties characteristic of the native enzyme. Furthermore, the enzyme no longer self-aggregates even though the flavin is reduced, and remains a tetramer even at high protein concentrations (Raj *et al.*, 1977).

The simplest interpretation of these observations is that reduction of the flavoprotein exposes a hydrophobic region on the polypeptide that is a site of proteolytic cleavage. This conclusion is supported by the fact that the presence of detergents and phospholipids protects the enzyme against proteolytic cleavage, presumably by binding to the site of proteolysis (Russell *et al.*, 1977a). Finally, cleavage by proteases at this site has the same effect on enzyme activity as the binding of lipids: proteolysis activates the enzyme approximately 25-fold (Hager, 1957), in addition to destroying the lipid-binding properties of the protein (Russell *et al.*, 1977b). This strongly suggests a common site for proteolytic and amphiphilic activation of pyruvate oxidase.

Further insight into the mechanism of activation of pyruvate oxidase has been obtained through the use of carboxypeptidase Y. This exoprotease removes amino acids sequentially from the COOH terminus. Preliminary results have shown that carboxypeptidase Y is also capable of fully activating the reduced form of the enzyme (Russell, 1979); however, the subunit molecular weight of the modified form of the enzyme is not measurably different when examined on SDS-polyacrylamide gels. Apparently, this full activation of the enzyme is a result of the removal of only two or perhaps three amino acids from the COOH terminus. (Both arginine and leucine have been identified.) It appears that the low activity of the unactivated form of the enzyme is due to a self-inhibition resulting from what may be a direct interaction between the COOH-terminal amino acids and the active site or some other locus on the enzyme. The binding of the protein to the membrane or to amphiphiles alters this self-inhibiting interaction and activates the enzyme. Evidently, both endoproteases and exoproteases are able to effect the same alteration in the enzyme. The lipid-binding properties of the carboxypeptidase Y-modified form of pyruvate oxidase have not yet been examined.

4. THE CATALYTIC SITE OF PYRUVATE OXIDASE

The nature of the changes that occur at the enzyme active site in response to lipid binding is not known. However, chemical modification studies have been useful in identifying several amino acid residues that are apparently located at the active site. The reaction of a single arginine residue with phenylglyoxal is sufficient to prevent the binding of thiamin pyrophosphate to the enzyme (J. G. Koland and R. B. Gennis, unpublished observations). Employing the reagent *N*-bromosuccinimide, a single tryptophan residue has also been implicated in the binding of thiamin pyrophosphate (O'Brien and Gennis, 1980). These residues may be involved in the hydrophobic and electrostatic components of the binding of this cofactor to the enzyme. In addition, there appears to be one sulfhydryl group located at or near the active site (Houghton, 1979; J. G. Koland and R. B. Gennis, unpublished observations). Reaction of *N*-ethylmaleimide with this sulfhydryl group also prevents binding of thiamin pyrophosphate to the enzyme.

Because of the critical role played by the oxidation–reduction state of the FAD coenzyme in determining the conformation of the enzyme and its affinity for the membrane, effort has also been directed toward characterizing the flavin. The absorption spectrum of the flavin is highly structured and unusual for a flavoprotein (Williams and Hager, 1966; Palmer and Massey, 1968). The spectrum suggests that the flavin is in a very hydrophobic environment and perhaps deeply embedded in the enzyme. Efforts to measure the oxidation–reduction midpoint potential have been frustrated by the fact that redox mediators do not rapidly equilibrate with the enzyme. This may be an indication that the flavin is for the most part inaccessible to these molecules. Preliminary results, however, indicate that the midpoint potential is approximately −100 mV for the cooperative two-electron reduction of the flavin (M. W. Mather and R. B. Gennis, unpublished observations). This is reasonable in view of the fact that all of the cytochromes in *E. coli* have midpoint potentials that are more positive than this value (Reid and Ingledew, 1979; Pudek and Bragg, 1976; Hendler and Shrager, 1979). It has also been discovered that photochemical reduction of pyruvate oxidase in the presence of 5-deazaflavin results in the formation of a radical anion form of the flavin (M. W. Mather and R. B. Gennis, unpublished observations). Formation of a radical anion (as opposed to a neutral form of the radical) by one-electron reduction of the flavoprotein is unusual for dehydrogenases of this class (Massey and Hemmerich, 1980). This observation may be of significance both in terms of the mechanism of reduction of the flavin by hydroxylethyl-TPP (see Fig. 3) and in terms of the charge distribution about the flavin in the binding site.

5. PYRUVATE OXIDASE AND ELECTRON TRANSPORT

The *E. coli* electron transport chain is itself poorly characterized biochemically (Haddock and Jones, 1977). Figure 5 is a schematic of some of the generally accepted features. There are several *b*-type cytochromes involved in respiration, and the respiratory chain is branched both on the dehydrogenase side and on the oxidase side. Little is known about the mode of coupling of pyruvate oxidase and the other dehydrogenases to the electron transport chain.

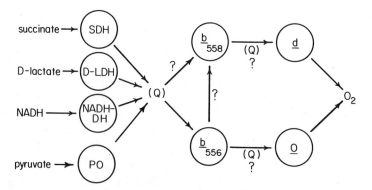

Figure 5. Scheme for aerobic electron transport in *E. coli*. Electrons from the various substrates flow through the appropriate dehydrogenases, through quinones, cytochromes, and finally to molecular oxygen. (Adapted from Haddock and Jones, 1977.)

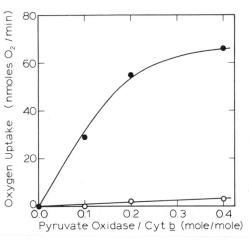

Figure 6. Pyruvate-supported oxygen consumption in a reconstituted pyruvate oxidase/*E. coli* inner membrane electron transport system. The indicated quantities of pyruvate oxidase were added to isolated inner membranes in 0.1 M sodium phosphate, 10 mM magnesium chloride, 100 μM TPP, and 200 mM sodium pyruvate, pH 6.0. Sufficient inner membrane was present so that the 3-ml volume contained 0.34 nmole of *b*-type cytochrome. Oxygen consumption was monitored polarographically. Results are shown for native ($\bullet$) and protease-activated ($\circ$) pyruvate oxidase. The protease-activated enzyme showed full activity when ferricyanide was used as an electron acceptor. Inner membranes were prepared according to Yamato *et al.* (1975). (J. G. Koland and R. B. Gennis, unpublished observations.)

Although the flavin coenzyme of pyruvate oxidase is reduced by pyruvate in a two-electron step, it is most likely reoxidized by a one-electron oxidant in the membrane. The natural electron acceptor in the *E. coli* membrane is not known, although it may be ubiquinone (Cunningham and Hager, 1975). When purified pyruvate oxidase is mixed with *E. coli* inner membranes that are devoid of pyruvate oxidase, pyruvate-supported oxygen consumption can be reconstituted (Williams and Hager, 1960; J. G. Koland and R. B. Gennis, unpublished observations) as shown in Fig. 6. In the absence of oxygen, all cytochromes in this reconstituted system are rapidly reduced via the substrate pyruvate (J. G. Koland and R. B. Gennis, unpublished observations). Reconstitution experiments with lipid-depleted *E. coli* membranes (Cunningham and Hager, 1975) have shown that ubiquinone is a functional element of this reconstituted system. Recent results (K. Kiuchi and L. P. Hager, unpublished observations) have identified palmitic acid as a second essential element. Finally, pyruvate oxidase that has been activated by α-chymotrypsin cannot be reconstituted with membrane preparations (Fig. 6) (J. G. Koland and R. B. Gennis, unpublished observations). Although proteolytic cleavage does not hamper the reoxidation of the reduced flavoprotein by a water-soluble oxidant such as ferricyanide, the same cleavage completely eliminates the ability of the enzyme to be reoxidized by its membrane-bound electron acceptor. These results suggest that the enzyme must be physically attached to the membrane to function as part of the reconstituted electron transport chain.

Future research will be directed primarily at determining which portions of the polypeptide are directly involved in membrane binding and determining the nature of the coupling between pyruvate oxidase and the *E. coli* respiratory chain.

REFERENCES

Blake, R. (1977). Ph.D. dissertation, University of Illinois.
Blake, R. Hager, L. P., and Gennis, R. B. (1978). *J. Biol. Chem.* **253**, 1963–1971.
Breslow, R. (1958). *J. Am. Chem. Soc.* **80**, 3719–3726.
Cunningham, C. C., and Hager, L. P. (1971a). *J. Biol. Chem.* **246**, 1575–1582.
Cunningham, C. C., and Hager, L. P. (1971b). *J. Biol. Chem.* **246**, 1583–1589.
Cunningham, C. C., and Hager, L. P. (1975). *J. Biol. Chem.* **250**, 7139–7146.
Haddock, B. A., and Jones, C. W. (1977). *Bacteriol. Rev.* **41**, 47–99.
Hager, L. P. (1957). *J. Biol. Chem.* **229**, 251–263.
Hager, L. P., and Krampitz, L. O. (1963). *Fed. Proc.* **22**, 536.

Hendler, R. W., and Shrager, R. I. (1979). *J. Biol. Chem.* **254,** 11288–11299.

Houghton, R. L. (1979). *Int. J. Biochem.* **10,** 205–208.

Krampitz, L. O. (1969). *Annu. Rev. Biochem.* **38,** 213–239.

Massey, V., and Hemmerich, P. (1980). *Biochem. Soc. Trans.* **8,** 246–257.

O'Brien, T. A., and Gennis, R. B. (1980). *J. Biol. Chem.* **255,** 3302–3307.

O'Brien, T. A., Schrock, H. L., Russell, P., Blake, R., and Gennis, R. B. (1976). *Biochim. Biophys. Acta* **452,** 13–29.

O'Brien, T. A., Blake, R., and Gennis, R. B. (1977). *Biochemistry* **16,** 3105–3109.

Palmer, G., and Massey, V. (1968). In *Biological Oxidations* (T. P. Singer, ed.), pp. 263–300, Interscience, New York.

Pudek, M. R., and Bragg, P. D. (1976). *Arch. Biochem. Biophys.* **174,** 546–552.

Raj. T., Russell, P., Flygare, W. H., and Gennis, R. B. (1977). *Biochem. Biophys. Acta* **481,** 42–49.

Reid, G. A., and Ingledew, W. J. (1979). *Biochem. J.* **182,** 465–472.

Russell, P. (1979). Ph.D. dissertation, University of Illinois.

Russell, P., Hager, L. P., and Gennis, R. B. (1977a). *J. Biol. Chem.* **252,** 7877–7882.

Russell, P., Schrock, H. L., and Gennis, R. B. (1977b). *J. Biol. Chem.* **252,** 7883–7887.

Schrock, H. L., and Gennis, R. B. (1977). *J. Biol. Chem.* **252,** 5990–5995.

Schrock, H. L., and Gennis, R. B. (1980). *Biochim. Biophys. Acta* **614,** 215–220.

Williams, F. R., and Hager, L. P. (1960). *Biochim. Biophys. Acta* **38,** 566–567.

Williams, F. R., and Hager, L. P. (1966). *Arch. Biochem. Biophys.* **116,** 168–176.

Yamato, I., Anraku, Y., and Hirosawa, K. (1975). *J. Biochem. (Tokyo)* **77,** 705–718.

11

Properties of Phospholipid Transfer Proteins

R. Akeroyd and K. W. A. Wirtz

1. INTRODUCTION

Owing to the extremely low critical micelle concentration of natural phospholipids, spontaneous transfer of monomer phospholipid molecules between membrane surfaces through an aqueous phase is very slow (Martin and MacDonald, 1976; Thilo, 1977). Movement of phospholipids would be restricted to the membrane bilayer if it were not for the ability of phospholipid transfer proteins to shuttle phospholipids between membranes (for reviews see Dawson, 1973; Wirtz, 1974; Zilversmit and Hughes, 1976; Kader, 1977). Since the original observation in 1968 that phospholipid transfer activity is present in the membrane-free cytosol of rat liver (Wirtz and Zilversmit, 1968), it has become clear that the cytosol of eukaryotic cells contains a number of phospholipid transfer proteins of different specificity. So far three distinct classes of transfer proteins have been purified to homogeneity: (1) the phosphatidylcholine transfer protein from bovine and rat liver (Kamp *et al.*, 1973; Poorthuis *et al.*, 1980); (2) the phosphatidylinositol transfer proteins from bovine brain and heart (Helmkamp *et al.*, 1974; DiCorleto *et al.*, 1979); (3) the nonspecific phospholipid transfer proteins from rat and bovine liver (Bloj and Zilversmit, 1977; Crain and Zilversmit, 1980a) and rat hepatoma (Dyatlovitskaya *et al.*, 1978). In this series, the phosphatidylcholine transfer proteins are specific for phosphatidylcholine whereas the phosphatidylinositol transfer proteins display a dual specificity with a preference for phosphatidylinositol and, to a lesser extent, phosphatidylcholine. The nonspecific proteins transfer all common diacyl phospholipids as well as cholesterol and are very likely identical to the sterol carrier protein 2 recently isolated from rat liver (Noland *et al.*, 1980). Less well characterized transfer proteins have also been found for other typical membrane components such as phosphatidylglycerol (Van Golde *et al.*, 1980), cholesterol (Erickson *et al.*, 1978), and glucosylceramide (Metz and Radin, 1980). This paper presents a summary on the characterization and properties of the transfer proteins purified to date.

R. Akeroyd and K. W. A. Wirtz • Laboratory of Biochemistry, State University of Utrecht, Transitorium III, NL-3584CH Utrecht, The Netherlands.

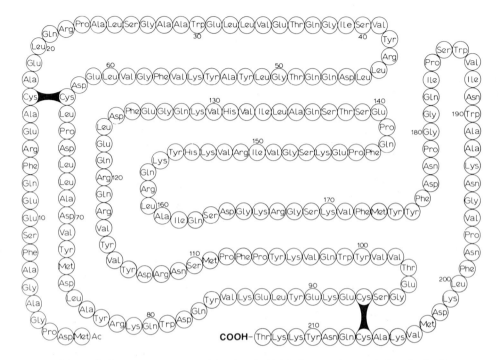

Figure 1. The primary structure of the phosphatidylcholine transfer protein from bovine liver.

2. PHOSPHATIDYLCHOLINE TRANSFER PROTEIN (PC-TP)

The first phospholipid transfer protein to be purified to homogeneity was PC-TP from bovine liver (Kamp *et al.*, 1973). Investigation of the primary structure has revealed PC-TP to consist of a single chain of 213 amino acids including two disulfide bonds (Akeroyd *et al.*, 1981). *N*-Acetylmethionine forms the blocked NH_2 terminus and threonine the COOH terminus (Fig. 1). The protein has a calculated molecular weight of 24,700 and a hydrophobicity index of 1356 calories per residue. This index is high relative to the molecular weight and may, in part, explain the tendency of PC-TP to aggregate (Kamp *et al.*, 1973). The protein has an isoelectric point of 5.8, which reflects an excess of two negative charges (12 Asp, 15 Glu, 16 Lys, 10 Arg, COOH-terminal Thr). Bovine PC-TP contains five Trp residues, which upon excitation have an emission spectrum with a maximum at 327 nm. This suggests that these residues are buried in the nonpolar region of the protein (Wirtz and Moonen, 1977). Recently, another phosphatidylcholine-specific transfer protein has been isolated from rat liver (Lumb *et al.*, 1976; Poorthuis *et al.*, 1980). This protein has an isoelectric point of 8.4 and a molecular weight of 28,000 as determined by sodium dodecyl sulfate polyacrylamide gel electrophoresis. Its amino acid composition is rather similar to that of bovine PC-TP; in spite of this similarity, antisera raised against either PC-TP were not cross-reactive. Rat PC-TP has at least one free sulfhydryl group, which is probably essential for its functioning (Illingworth and Portman, 1972).

As for the mode of action of phospholipid transfer proteins, studies on bovine PC-TP have provided the most extensive information to date. This protein acts as a carrier by forming a 1 : 1 molecular complex with phosphatidylcholine (Demel *et al.*, 1973). Upon formation of a collision complex with a membrane interface, PC-TP releases its bound phosphatidylcholine molecule into that interface (Kamp *et al.*, 1977). Upon disruption of the protein–interface complex, PC-TP extracts a phosphatidylcholine molecule from that interface, giving rise to a one-for-one molecular exchange process (Demel *et al.*, 1973; Machida and Ohnishi, 1978). If PC-TP forms a complex with an interface devoid of phosphatidylcholine, release of its bound phosphatidylcholine represents net transfer. Recently, bovine PC-TP has been demonstrated to facilitate the net transfer of phosphatidylcholine to vesicles containing a mixture of phosphatidylethanolamine and phosphatidic acid (Wirtz *et al.*, 1980) and to complexes of sphingomyelin and human apolipoprotein A-II (Wilson *et al.*, 1980). In agreement with the specificity of PC-TP, transfer of phosphatidylethanolamine and sphingomyelin in exchange for phosphatidylcholine was not observed (see also Kamp *et al.*, 1977).

A determining factor in the transfer process is the ability of PC-TP to form a complex with the interface. For example, PC-TP does not transfer phosphatidylcholine to vesicles of pure phosphatidylethanolamine or sphingomyelin (Kamp *et al.*, 1977; Wilson *et al.*, 1980). This lack of transfer may reflect a failure to form a complex as a result of the relatively high transition temperature of these phospholipids. Transfer is also inhibited when bovine PC-TP forms too strong a complex. This has been observed for vesicles that have a high content of acidic phospholipids such as phosphatidic acid and phosphatidylserine (Machida and Ohnishi, 1978, 1980; Wirtz *et al.*, 1979). Kinetic analysis has indicated that the dissociation constant of the protein–vesicle complex decreases with an increasing content of acidic phospholipids (van den Besselaar *et al.*, 1975). This effect on the dissociation constant runs parallel with a decreased rate of transfer. Apart from surface charge, the packing of phospholipids in the interface also appears to have a great effect on the transfer activity of PC-TP (DiCorleto and Zilversmit, 1977). It has been observed that the rate of phosphatidylcholine transfer to single bilayer vesicles is 100 times larger than to multilamellar liposomes (Machida and Ohnishi, 1980; Wirtz *et al.*, 1979).

At present, we have no evidence that PC-TP has to penetrate into the interface to either release or bind a phosphatidylcholine molecule. In view of its specificity, we assume that PC-TP has a recognition site for the phosphorylcholine moiety (Kamp *et al.*, 1977). In addition, upon interaction with the interface it may expose a hydrophobic peptide region that locally lowers the dielectric constant to the extent that a phosphatidylcholine molecule can flip from the membrane onto the protein. PC-TP has a number of hydrophobic peptide segments of which the most prominent are Val[98]-Val-Tyr-Trp-Gln-Val[103], Val[171]-Phe-Met-Tyr-Tyr-Phe[176], and Trp[186]-Val-Ile-Asn-Trp-Ala-Ala[192] (Akeroyd *et al.*, 1981). Application of photolabeled phosphatidylcholine has indicated that the hexapeptide Val[171]-Phe[176] forms part of the site that accommodates the phosphatidylcholine molecule (Moonen *et al.*, 1979). It remains to be established whether this peptide interacts directly with the interface.

3. PHOSPHATIDYLINOSITOL TRANSFER PROTEIN (PI-TP)

The prominence of phosphatidylinositol transfer activity in the 105,000*g* supernatant fractions from bovine brain prompted the isolation of PI-TP from this source (Helmkamp *et al.*, 1974; Demel *et al.*, 1977). This protein has also been isolated from bovine heart by a different procedure (DiCorleto *et al.*, 1979). From either tissue, two species of PI-TP (I and II) have been purified, which show striking similarities in molecular weight (approximately 33,000) and amino acid composition but differ in isoelectric point (I, pH 5.3; II, pH 5.6). In a variety of assays, PI-TP I and II were found to be equally active and to have a dual specificity with a distinct preference for phosphatidylinositol over phosphatidylcholine (Helmkamp *et al.*, 1976; DiCorleto *et al.*, 1979). Transfer activity for both phospholipids was completely inhibited by *N*-ethylmaleimide but only in the presence of phospholipid vesicles. This suggests that PI-TP exposes an essential sulfhydryl group upon interaction with the interface (DiCorleto *et al.*, 1979). At present, the significance of this particular combination of phospholipid transfer activities as displayed by PI-TP is not understood. However, PI-TP does transfer phosphatidylinositol from a donor membrane to phosphatidylcholine vesicles by exchanging phosphatidylinositol for phosphatidylcholine (Demel *et al.*, 1977; DiCorleto *et al.*, 1979). This may be part of the mechanism by which PI-TP maintains the levels of phosphatidylinositol in those membranes (e.g., plasma membranes) where a rapid turnover of phosphatidylinositol occurs but which lack the ability to synthesize phosphatidylinositol (Michell, 1975).

In contrast to bovine PC-TP, PI-TP appears to be much less sensitive to acidic phospholipids. By measuring the transfer of phosphatidylinositol from rat liver microsomes to phosphatidylcholine vesicles, Helmkamp (1980a) demonstrated an inhibition of transfer when phosphatidylinositol was incorporated into these vesicles or when, in addition, vesicles of pure phosphatidylinositol were added. Other acidic phospholipids, however, had virtually no effect. Determination of dissociation constants indicated that bovine PI-TP interacts most strongly with those interfaces that contained the phospholipid most actively transferred (Helmkamp, 1980a). The activity of PI-TP was also drastically affected by the fatty acid composition of phosphatidylcholine in the acceptor vesicles (Helmkamp, 1980b). Initial rates of phosphatidylinositol transfer decreased in the order egg phosphatidylcholine = dioleoyl- > dielaidoyl- > dimyristoylphosphatidylcholine. This was reflected in an increase of the apparent dissociation constant of the PI-TP–vesicle complex. In general, PI-TP appears to be a sensitive probe to changes in membrane fluidity.

Incubation with phospholipid monolayers and vesicles has shown that PI-TP binds phosphatidylinositol and phosphatidylcholine (Johnson and Zilversmit, 1975; Demel *et al.*, 1977). This strongly suggests that, in analogy with bovine PC-TP, PI-TP acts as a carrier between membranes.

4. NONSPECIFIC PHOSPHOLIPID TRANSFER PROTEIN (nsPL-TP)

In addition to PC-TP and PI-TP, the membrane-free cytosol of bovine liver contains phosphatidylethanolamine transfer activity. This activity is due to nsPL-TP, which recently has been purified to homogeneity from this source (Crain and Zilversmit, 1980a). Two species of nsPL-TP (I and II) have been identified, which have a similar molecular weight of 14,500 but differ in isoelectric point (I, pH 9.55; II, pH 9.75). This difference is probably due to the fact that nsPL-TP II has one Arg and one His residue, both of which are absent from nsPL-TP I. Both proteins lack Tyr and have one Trp, which explains the very low absorbance observed at 280 nm. Determination of the transfer activity indicated that these proteins are nonspecific in that they accelerate the transfer of phosphatidylcholine, phosphatidylethanolamine, phosphatidylinositol, phosphatidylserine, sphingomyelin, phosphatidylglycerol, phosphatidic acid, and cholesterol (Crain and Zilversmit, 1980a). Similar proteins (molecular weight, 13,000, isoelectric point pH 8.3–9.0) have been isolated from rat liver (Bloj and Zilversmit, 1977; Baranska and Grabarek, 1979). Interestingly, a low-molecular-weight nsPL-TP with an isoelectric point of 5.2 has been isolated from a fast-growing rat hepatoma and is absent from normal rat liver (Dyatlovitskaya *et al.*, 1978). On the other hand, phosphatidylethanolamine transfer activity was very low in a series of

Morris hepatomas, suggesting that nsPL-TP is virtually absent from these malignant tissues (Poorthuis *et al.*, 1980).

At present, it remains unresolved whether these proteins act as a carrier because the putative intermediate, i.e., the phospholipid–protein complex, has not been identified. However, under conditions where bovine PC-TP and PI-TP did not catalyze a net transfer of phospholipids, bovine nsPL-TP was shown to be very effective in catalyzing this process (Crain and Zilversmit, 1980b). This protein catalyzed the net transfer of phosphatidylcholine and phosphatidylinositol from multilamellar liposomes to either intact or delipidated human high-density lipoprotein and the net transfer of phosphatidylcholine from unilamellar vesicles to rat liver mitoplasts.

5. CONCLUSION

Mammalian tissues contain a number of proteins that mediate the transfer of specific classes of phospholipids between membranes. Recognition of the polar head group and the ability to act as a phospholipid carrier make the phospholipid transfer proteins interesting models in the study of lipid–protein interactions. Problems fundamental to the organization of membranes could possibly be answered by the investigation of these transfer proteins. At present, it still remains to be established in what manner these proteins fit in the total phospholipid metabolism of the cell. Their ability, however, to catalyze a net transfer of phospholipids argues in favor of these proteins being involved in membrane biosynthesis.

REFERENCES

Akeroyd, R., Moonen, P., Westerman, J., Puyk, C., and Wirtz, K. W. A. (1981). *Eur. J. Biochem.*, **114**, 385–391.

Baranska, J., and Grabarek, Z. (1979). *FEBS Lett.* **104**, 253–257.

Bloj, B., and Zilversmit, D. B. (1977). *J. Biol. Chem.* **252**, 1613–1619.

Crain, R. C., and Zilversmit, D. B. (1980a). *Biochemistry* **19**, 1433–1439.

Crain, R. C., and Zilversmit, D. B. (1980b). *Biochim. Biophys. Acta* **620**, 37–48.

Dawson, R. M. C. (1973). *Sub-Cell. Biochem.* **2**, 69–89.

Demel, R. A., Wirtz, K. W. A., Kamp, H. H., Geurts van Kessel, W. S. M., and van Deenen, L. L. M. (1973). *Nature New Biol.* **246**, 102–105.

Demel, R. A., Kalsbeek, R., Wirtz, K. W. A., and van Deenen, L. L. M. (1977). *Biochim. Biophys. Acta* **466**, 10–22.

DiCorleto, P. E., and Zilversmit, D. B. (1977). *Biochemistry* **16**, 2145–2150.

DiCorleto, P. E., Warach, J. B., and Zilversmit, D. B. (1979). *J. Biol. Chem.* **254**, 7795–7802.

Dyatlovitskaya, E. V., Timofeeva, N. G., and Bergelson, L. D. (1978). *Eur. J. Biochem.* **82**, 463–471.

Erickson, S. K., Meyer, D. J., and Gould, R. G. (1978). *J. Biol. Chem.* **253**, 1817–1826.

Helmkamp, G. M. (1980a). *Biochim. Biophys. Acta* **595**, 222–234.

Helmkamp, G. M. (1980b). *Biochemistry* **19**, 2050–2056.

Helmkamp, G. M., Harvey, M. S., Wirtz, K. W. A., and van Deenen, L. L. M. (1974). *J. Biol. Chem.* **249**, 6382–6389.

Helmkamp, G. M., Nelemans, S. A., and Wirtz, K. W. A. (1976). *Biochim. Biophys. Acta* **424**, 168–182.

Illingworth, D. R., and Portman, O. W. (1972). *Biochim. Biophys. Acta* **280**, 281–289.

Johnson, L. W., and Zilversmit, D. B. (1975). *Biochim. Biophys. Acta* **375**, 165–175.

Kader, J. C. (1977). In *Dynamic Aspects of Cell Surface Organization* (G. Poste and G. L. Nicolson, eds.), pp. 127–204, Elsevier/North-Holland, Amsterdam.

Kamp, H. H., Wirtz, K. W. A., and van Deenen, L. L. M. (1973). *Biochim. Biophys. Acta* **318**, 313–325.

Kamp. H. H., Wirtz, K. W. A., Baer, P. R., Slotboom, A. J., Rosenthal, A. F., Paltauf, F., and van Deenen, L. L. M. (1977). *Biochemistry* **16**, 1310–1316.

Lumb, R. H., Kloosterman, A. D., Wirtz, K. W. A., and van Deenen, L. L. M. (1976). *Eur. J. Biochem.* **69,** 15–22.

Machida, K., and Ohnishi, S. (1978). *Biochim. Biophys. Acta* **507,** 156–164.

Machida, K., and Ohnishi, S. (1980). *Biochim. Biophys. Acta* **596,** 201–209.

Martin, F. J., and MacDonald, R. C. (1976). *Biochemistry* **15,** 321–327.

Metz, R. J., and Radin, N. S. (1980). *J. Biol. Chem.* **255,** 4463–4467.

Michell, R. H. (1975). *Biochim. Biophys. Acta* **415,** 81–148.

Moonen, P., Haagsman, H. P., van Deenen, L. L. M., and Wirtz, K. W. A. (1979). *Eur. J. Biochem.* **99,** 439–445.

Noland, B. J., Arebalo, R. E., Hansbury, E., and Scallen, T. J. (1980). *J. Biol. Chem.* **255,** 4282–4289.

Poorthuis, B. J. H. M., van der Krift, T. P., Teerlink, T., Akeroyd, R., Hostetler, K. Y., and Wirtz, K. W. A. (1980). *Biochim. Biophys. Acta* **600,** 376–380.

Thilo, L. (1977). *Biochim. Biophys. Acta* **469,** 326–334.

van den Besselaar, A. M. H. P., Helmkamp, G. M., and Wirtz, K. W. A. (1975). *Biochemistry* **14,** 1852–1858.

Van Golde, L. M. G., Oldenborg, V., Post, M., Batenburg, J. J., Poorthuis, B. J. H. M., and Wirtz, K. W. A. (1980). *J. Biol. Chem.* **255,** 6011–6013.

Wilson, D. B., Ellsworth, J. L., and Jackson, R. L. (1980). *Biochim. Biophys. Acta* **620,** 550–561.

Wirtz, K. W. A. (1974). *Biochim. Biophys. Acta* **344,** 95–117.

Wirtz, K. W. A., and Zilversmit, D. B. (1968). *J. Biol. Chem.* **243,** 3596–3602.

Wirtz, K. W. A., and Moonen, P. (1977). *Eur. J. Biochem.* **77,** 437–443.

Wirtz, K. W. A., Vriend, G., and Westerman, J. (1979). *Eur. J. Biochem.* **94,** 215–221.

Wirtz, K. W. A., Devaux, P. F., and Bienvenue, A. (1980). *Biochemistry* **19,** 3395–3399.

Zilversmit, D. B., and Hughes, M. E. (1976). In *Methods in Membrane Biology* (E. D. Korn, ed.), Vol. 7, pp. 211–259, Plenum Press, New York.

II

The Physical Properties of Biological and Artificial Membranes

12

Freeze-Fracture Techniques Applied to Biological Membranes

Kurt Mühlethaler

1. INTRODUCTION

The development and application of the freeze-fracture technique to membrane research (Moor and Mühlethaler, 1963; Mühlethaler *et al.*, 1965) led to a better understanding of the complex structural arrangement of lipids and proteins. It is now used worldwide for topographical studies, being the only available method for producing a direct high-resolution image of large areas of membrane surfaces and internal fracture faces. At first, however, it was thought that freezing would cause phase transformation and structural rearrangement of the original membrane structure. This has been disproved with the help of indirect methods such as X-ray diffraction. As shown by Gulik-Krzywicki and Costello (1977), a perturbation of the molecular organization of hydrocarbon chains does not occur provided freeze quenching is carried out rapidly enough to avoid the formation of large ice crystals between lipid lamellae. Based on these results, much work has been carried out to improve the main preparative steps such as freezing, fracturing, and replication. In addition to these procedures, new methods for labeling membrane components, special devices for rapid quenching, handling procedures for split membranes, and elaborate image-processing procedures became available. Some of these achievements will be discussed in the following chapters.

2. LABELING OF MEMBRANE COMPONENTS

The most widely used immunochemical marker for the demonstration of antigens on cell surfaces is the IgG–ferritin conjugate introduced by Pinto da Silva and Branton (1970). The size of the conjugate is about 20 nm and allows positive identification of membrane

Kurt Mühlethaler • Swiss Federal Institute of Technology, Institute for Cell Biology, CH-8093 Zürich, Switzerland.

surface components spaced more than 25–30 nm apart. Distinction between the labeled molecules and the IgG–ferritin complexes becomes difficult if the size of peripheral membrane proteins is in the same order of magnitude. As shown by Berzborn *et al.* (1974), one way to overcome this handicap is to incubate the membrane with the antiserum alone. In this case, the antibody first reacts by only one of its binding sites with the particle. Subsequently, this complex moves laterally within the surface of the membrane until it meets another particle. The antibody then reacts with its second binding site, resulting in the formation of aggregates that can be clearly distinguished from the remaining particles. In this way Berzborn *et al.* (1974) were able to identify the coupling factor in thylakoids. For labeling glycoproteins the use of lectins (Figs. 1 and 2) can be recommended (Maurer, 1980). Because antibodies and lectins must react with the surface prior to quenching, it is not possible to localize these markers on a fractured membrane face. A new technique that overcomes this difficulty is the specific decoration by condensing substances, as introduced by Gross *et al.* (1978). The necessary requirement for this specific condensation is a clean membrane face. Under standard vacuum conditions (10^{-6} Torr) and at a temperature of $-100°C$, the freshly cleaved membrane face is contaminated with water within a second. If the vacuum pressure is lowered to 10^{-9} Torr, surface contamination is reduced to a factor of 1000. The buildup of a contaminating monolayer takes 15 min and allows metal evaporation under clean conditions. Under ultrahigh vacuum conditions, the specimens can be cleaved at considerably lower temperatures, which results in a better preservation of the membrane topography. At very low specimen temperatures ($-196°C$), water does not condense uniformly as a homogeneous layer. Instead, discrete ice crystals are formed by condensation on regions with higher binding energy for water molecules. Generally, the condensate crystals can be used as labels for specific regions on fracture faces. This decoration technique may be useful in obtaining more chemical information on membrane face.

3. SPLIT-MEMBRANE PREPARATION

As first observed by Branton (1966), frozen membranes fracture along the hydrophobic zone of the bilayer. In order to handle these "half" membranes, new methods for attaching the membranes to a substrate have been developed. Optically flat glass slides provide a convenient and appropriately smooth surface for attachment. Because both the membrane surface and the glass possess a net negative charge, the glass surface must be positively charged by polycations or other ligands. As first shown by Fisher (1975), polylysine is an appropriate "glue" for membrane attachment. An alternative method for ca-

---

Figure 1. Freeze-fracture micrograph of yeast plasma membrane. The P-fracture face shows paracrystalline arrays of particles. (From Maurer, 1980.)

Figure 2. Labeling of protoplasts with ferritin-conjugated concanavalin A, directed against glycoproteins. The distribution of the intramembranous particles is well indicated by the distribution of the ferritin molecules on the exoplasmic surface (ES). (From Maurer, 1980.)

Figure 3. Visualization of the cytoplasmic surface (PS) of the yeast plasma membrane showing the paracrystalline arrays of invertase particles. (From Maurer, 1980.)

Figures 4 and 5. Isolated photosynthetic membrane from *Rhodopseudomonas viridis* before and after digital image processing. The center-to-center distance of the repeating units in the hexagonal lattice is 130 Å. (From Wehrli and Kübler, 1980.)

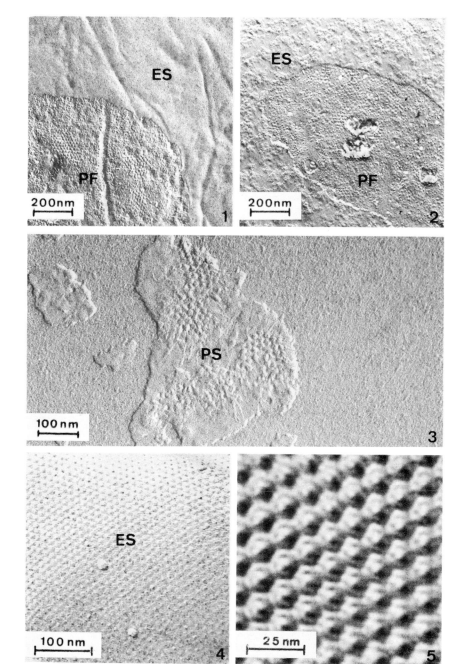

tionizing glass has been introduced by Sommer (1977). Cells can be firmly attached to glass by using Alcian blue as a ligand. This allows biochemical analysis of membrane halves or the preparation of the plasmatic surface (PS) after cell disruption and removal of most of the cytoplasmic contents (Fig. 13). A third technique for such studies has been developed by Büechi and Bächi (1979). A coverslip is treated with 3-aminopropyltrie-thoxysilane for the attachment of erythrocytes. In a second step the membrane is covalently linked to this ligand with glutaraldehyde. The visualization and labeling of the PS extends the application of the freeze-etching technique and allows morphological studies of transport processes through the plasmalemma.

4. IMAGE PROCESSING

Due to plastic deformation and other structural distortions during freezing and splintering or due to crystallite formation during metal evaporation, fine details of the specimen may become blurred and ill-defined. When the specimen has a periodic (crystalline) structure (Fig. 4), this "noise" can be reduced by lattice averaging (De Rosier and Klug, 1972). Recently, methods for averaging macromolecular structures not available in a regular form have also been introduced by Frank *et al.* (1978). Here, the randomly positioned particles are correlated with a single "reference" image. In Fig. 4 the highly ordered thylakoid membrane of *Rhodopseudomonas viridis* is shown after freeze fracturing and subsequent image enhancement (Fig. 5). The light-harvesting and reaction-center pigment—protein complexes and their associated electron-transport chains form rosettes of small subunits. These multienzyme complexes are organized into a hexagonal lattice with a periodicity of 13 nm (Wehrli and Kübler, 1980).

5. SUMMARY

The freeze-etching and freeze-fracturing technique widely used for membrane studies has been improved in several ways. Faster quenching and better splintering methods allow detailed studies of membrane surfaces and internal fracture planes. With special ligands attached to glass, the handling of "half" membranes and plasmatic surfaces has become easier. For localization studies, various labeling methods with antibodies, lectins, or decoration by condensation are available. For membranes with regular latticelike structures, it is advantageous to use image processing to average out minor structural defects caused by preparation procedures.

REFERENCES

Berzborn, R. J., Kopp, F., and Mühlethaler, K. (1974). *Z. Naturforsch.* **29c,** 694–699.
Branton, D. (1966). *Proc. Natl. Acad. Sci. USA* **55,** 1048–1056.
Büechi, M., and Bächi, Th. (1979). *J. Cell Biol.* **83,** 338–347.
De Rosier, D. J., and Klug, A. (1972). *J. Mol. Biol.* **65,** 469–488.
Fisher, K. A. (1975). *Science* **190,** 983–985.
Frank, J., Goldfarb, W., Eisenberg, D., and Baker, T. S. (1978). *Ultramicroscopy* **3,** 283–290.
Gross, H., Kübler, O., Bas, E., and Moor, H. (1978). *J. Cell Biol.* **79,** 646–656.
Gulik-Krzywicki, T., and Costello, M. J. (1977). *J. Microsc. (Oxford)* **112,** 103–113.
Maurer, A. (1980). Ph.D. thesis, Swiss Federal Institute of Technology, Zurich.

Moor, H., and Mühlethaler, K. (1963). *J. Cell. Biol.* **17,** 609–628.

Mühlethaler, K., Moor, H., and Szarkowski, J. W. (1965). *Planta* **67,** 305–323.

Pinto da Silva, P., and Branton, D. (1970). *J. Cell Biol.* **45,** 598–605.

Sommer, J. R. (1977). *J. Cell Biol.* **75,** 245a.

Wehrli, E., and Kübler, O. (1980). In *Electron Microscopy at Molecular Dimensions* (W. Baumeister and W. Vogell, eds.), pp. 48–56, Springer-Verlag, Berlin.

13

Electron Microscopy and Computer Image Reconstruction of Membrane Crystals

K. R. Leonard and H. Weiss

1. INTRODUCTION

Membrane proteins exist partially in a phospholipid bilayer and partially in the adjacent aqueous medium. For structural stability they therefore need as their environment a narrow band of nonpolar hydrocarbon bounded by polar groups. A pure aqueous solution is not satisfactory; likewise, a simple organic solvent would be unsuitable. The native environment can, however, be simulated successfully by detergent micelles (Tanford and Reynolds, 1976; Helenius *et al.*, 1979). This partially hydrophobic and partially hydrophilic property of membrane proteins renders it difficult to prepare three-dimensional crystals that are suitable for high-resolution single-crystal X-ray analysis. So far, only a small number of membrane proteins have been reported to crystallize (Tanaka *et al.*, 1980; Michel and Oesterhelt, 1980; Garavito and Rosenbusch, 1980).

The requirement for a membrane protein to exist at least partially within a phospholipid bilayer can, however, be of advantage for electron microscope specimen preparation. An increase in the local concentration of protein may bring about crystallization in a two-dimensional lattice defined spatially by the thickness of the lipid bilayer. The crystals are usually only one or two protein molecules thick and are thus very suitable for study by electron microscopy. Using crystalline material, electron microscope image analysis by Fourier methods can be carried out and the damaging effects of the electron beam can be reduced. A further advantage of using electron microscope images over diffraction methods in this type of structural study is that there is no phase problem. Phase information can be calculated directly from the images. For these reasons, only the analysis of two-dimensional membrane crystals will be discussed here, although it is also possible to carry out image analysis of single membrane proteins (Zingsheim *et al.*, 1980).

In a few cases, membrane crystals may exist naturally, such as the purple membrane of *Halobacter halobium* containing highly ordered crystalline patches of bacteriorhodopsin

---

K. R. Leonard and H. Weiss • European Molecular Biology Laboratory, 6900 Heidelberg, West Germany.

(Blaurock and Stoeckenius, 1971). Membrane crystals may also be produced artificially by delipidating membranes to increase the protein to lipid ratio such as for cytochrome oxidase (Vanderkooi *et al.*, 1972), or by reincorporating isolated protein–detergent complexes into phospholipid bilayers such as for cytochrome reductase (Wingfield *et al.*, 1979).

2. METHODS

Two methods have been used for obtaining three-dimensional information from two-dimensional crystals.

2.1. Low-Dose Microscopy of Unstained Images

This was first carried out by Unwin and Henderson (1975) in their study of the structure of the purple membrane. The authors used unstained membranes embedded in glucose to reduce the effect of dehydration in the vacuum of the electron microscope. By keeping the electron dose below 1 electron/$Å^2$, they were able to obtain electron diffraction intensities extending to 3.5-Å resolution. Digitization of the images and computer calculation of the Fourier transforms gave amplitudes and phases for reflections extending to 7 Å. Phase contrast was increased by underfocusing strongly. This, however, results in phase reversal of the reflections by 180° with successive maxima of the microscope contrast transfer function (Erickson and Klug, 1971). Data are also lost at the positions of minima in the transfer function. The reversed phases can be corrected, after taking a second high-dose image to determine the positions of the maxima (Thon rings). Missing data were filled in by taking a second image at a second defocus value where the minima in the transfer function occur in different positions. Because the limit to image formation at these very low exposures is the Poisson noise of the electron beam itself, averaging over thousands of unit cells was carried out in order to obtain a sufficiently high signal-to-noise ratio. This important, and so far unique, work demonstrated that in the purple membrane, structures 10 Å in diameter could be resolved, which could be interpreted as α helices seen in projection spanning the bilayer.

In order to obtain a three-dimensional structure for bacteriorhodopsin, they combined low-dose imaging with Fourier reconstruction methods (DeRosier and Klug, 1968; Hoppe *et al.*, 1968). Micrographs were taken at tilt intervals up to the maximum possible angle (normally 60°) and in different azimuthal directions. Because the electron exposure was kept below about 1 $e^-/Å^2$ to avoid damage, a different crystal was used for every micrograph.

Each tilted image, after digitization and Fourier transformation, defines a central section through the three-dimensional receprocal lattice of the structure. As the crystal is only one unit cell thick, the transform is continuous in the Z^* direction, i.e. normal to the membrane bilayer. It is thus made up of lattice lines parallel to the Z^* direction. The phases for reflections in each tilted view are refined to a common origin and the amplitudes scaled to each other. This then gives data for the variation in amplitude and phase along the lattice lines in the Z^* direction. In addition to the defocus correction described above, an additional correction is needed to allow for the difference in specimen position (i.e., focus) across a large crystal, tilted in the electron microscope. Smooth curves drawn through the data are interpolated at a suitable Z^* interval to give data in h, k, and l, which are used for Fourier synthesis and calculation of a three-dimensional electron scattering density map (Henderson and Unwin, 1975). The three-dimensional image reconstruction showed clearly the presence of seven α-helical segments arranged approximately at right angles to the plane of the membrane. Because the interhelical connecting regions could not be resolved, the exact extent of one molecule could not be established.

A second crystal form has since been prepared from isolated bacteriorhodopsin (Michel *et al.*, 1980). This crystallized in a different space group and enabled the boundary of one rhodopsin molecule to be defined unambiguously.

2.2. Negatively Stained Images

Image reconstruction of negatively stained membrane crystals makes use of the same principles as described above. However, because the stain provides higher contrast and is more beam resistant, the technical difficulties are decreased. It is not necessary to under-focus in order to obtain contrast and, as smaller image areas can be used, no correction is needed for the smaller change of focus across a tilted crystal. Furthermore, because the specimens are less radiation sensitive, a series of tilts can be carried out on a single crystal. The total dose for a series of tilts should, however, be kept as low as possible. Off-specimen focus and astigmatism correction should be used to avoid the possibility of stain redistribution (Unwin, 1974). The disadvantage of using negatively stained membrane crystals is that the resolution is limited to about 2 nm, so that only the overall molecular outline can be seen.

The requirements for low-dose imaging, namely large crystals with good long-range order, well distributed on the support so that images can be taken "blindly," and a small unit cell to increase the amount of averaging, are so severe that it is worthwhile to start with negatively stained material prior to attempting low-dose work.

3. RECONSTRUCTED MEMBRANE PROTEINS

The following membrane proteins have now been reconstructed to low resolution.

1. *Beef heart cytochrome oxidase* (EC 1.9.3.1). Two crystalline forms of cytochrome oxidase from beef heart mitochondria have been studied. One form, prepared by using Triton X-100, consists of two layers of the enzyme (Henderson *et al.*, 1977). The second form, prepared using deoxycholate, is only one enzyme molecule thick and may be a single-layer protein sheet, the presence of lipid being uncertain (Fuller *et al.*, 1979). Three-dimensional reconstruction carried out on the two forms indicated that the cytochrome oxidase molecule is a Y-shaped three-domain structure.

2. *Cytochrome reductase* (EC 1.10.2.2). The enzyme was isolated from *Neurospora* mitochondria as a protein–detergent complex. Single-layer membrane crystals or double-layer tubes were obtained by mixing the enzyme–detergent complex with phospholipid–detergent micelles and subsequently removing the detergent (Wingfield *et al.*, 1979). Three-dimensional reconstruction of the negatively stained single layers showed the enzyme to be an elongated dimer with asymmetric distribution of protein mass across the lipid bilayer (Leonard *et al.*, 1981). A smaller subunit complex also crystallized as single- and double-layer sheets (Hovmöller *et al.*, 1981), and three dimensional reconstruction is at present being carried out.

3. *Gap junction structures.* Unwin and Zampighi (1980) have carried out tilt reconstructions of negatively stained gap junction isolated from rat hepatic tissue, where it occurs naturally as hexagonally packed membrane crystals. The structure shows that one unit of the junction is a cylinder with a central channel composed of six elongated subunits that

are tilted slightly at an angle to the cylinder axis. After extended dialysis of the membrane crystals, the structure showed a change in the skewing of the subunits around the cylinder axis, so that the central channel was closed. It was suggested that *in vivo,* a similar angular shift of the subunits could result in closure of the communicating channels between cells.

4. PROSPECTS

The recent successful attempts of several laboratories to crystallize membrane proteins within phospholipid bilayers (Wingfield *et al.,* 1979; Fuller *et al.,* 1979; Michel *et al.,* 1980; Hovmöller *et al.,* 1981) encourage us to believe that similar two-dimensional membrane crystals can be produced with many other membrane proteins. Low-resolution structural information can then readily be obtained as described above. Higher resolution has, however, so far only been possible for bacteriorhodopsin, a small protein that is almost entirely located within the phospholipid bilayer and with relatively high scattering contrast between protein and lipid. Most membrane proteins have bigger unit cells so that a larger, well ordered, crystal is required for averaging. It is also likely that more of the protein mass will project into the aqueous phase, reducing the effective contrast between protein and environment.

Despite these problems, however, we are confident that it will be possible to solve the structure of other membrane proteins to the level of resolution found for bacteriorhodopsin.

REFERENCES

Blaurock, A. E. and Stoeckenius, W. (1971). *Nature New Biol.* **223,** 152–154.
DeRosier, D. J., and Klug, A. (1968). *Nature (London)* **217,** 130–134.
Erickson, H. P., and Klug, A. (1971). *Philos. Trans. Roy. Soc. London Ser. B* **261,** 105–118.
Fuller, S. D., Capaldi, R. A., and Henderson, R. (1979). *J. Mol. Biol.* **134,** 305–327.
Garavito, R. M., and Rosenbusch, J. P. (1980). *J. Cell Biol.* **86,** 327–329.
Helenius, A., Caslin, D. R. M., Fries, E., and Tanford, C. (1979). *Methods Enzymol.* **56,** 734–749.
Henderson, R., and Unwin, P. N. T. (1975). *Nature (London)* **257,** 28–32.
Henderson, R., Capaldi, R. A., and Leigh, J. S. (1977). *J. Mol. Biol.* **112,** 631–648.
Hoppe, W., Langer, R., Knesch, G., and Poppe, Ch. (1968). *Naturwissenschaften* **55,** 33–½⅛.
Hovmöller, S., Leonard, K. R., and Weiss, H. (1981). *FEBS Lett.* **123,** 118–122.
Leonard, K. R., Arad, T., Wingfield, P., and Weiss, H. (1981). *J. Mol. Biol,* **149,** 259–274.
Michel, H., and Oesterhelt, D. (1980). *Proc. Natl. Acad. Sci. USA* **77,** 1283–1285.
Michel, H., Oesterhelt, D., and Henderson, R. (1980). *Proc. Natl. Acad. Sci. USA* **77,** 338–342.
Tanaka, M., Suzuki, H., and Ozawa, T. (1980). *Biochim. Biophys. Acta* **612,** 295–298.
Tanford, C., and Reynolds, J. A. (1976). *Biochim. Biophys. Acta* **457,** 113–170.
Unwin, P. N. T. (1974). *J. Mol. Biol.* **87,** 657–670.
Unwin, P. N. T., and Henderson, R. (1975). *J. Mol. Biol.* **94,** 425–440.
Unwin, P. N. T., and Zampighi, G. (1980). *Nature (London)* **283,** 545–549.
Vanderkooi, T., Senior, A. E., Capaldi, R. A., and Hayashi, H. (1972). *Biochim. Biophys. Acta* **274,** 38–48.
Wingfield, P., Arad, T., Leonard, K. R., and Weiss, H. (1979). *Nature (London)* **280,** 696–697.
Zingsheim, H. P., Neugebauer, D.Ch., Barrantes, F. J., and Frank, J. (1980). *Proc. Natl. Acad. Sci. USA* **77,** 952–956.

14

NMR in the Study of Membranes

Oleg Jardetzky

The use of NMR for the study of membranes was introduced by Chapman and Salsbury (1966). Over the past decade several hundred reports of observations of ^1H, ^2H, ^{13}C, ^{19}F, and ^{31}P spectra on phospholipid vesicles and several natural membranes have appeared. The majority of these have addressed various aspects of the problems of order, orientation, mobility, and lateral diffusion in phospholipid bilayers and multilayers. Others have dealt with the conformations of the polar head groups of phospholipids, transmembrane exchange (flip-flop) between the two layers, interactions of small molecules—notably cholesterol—and local anesthetics with both model and natural membranes, and function of ion channels in reconstituted membranes, the fusion of phospholipid bilayers catalyzed by peptides and proteins, and the nature of protein–lipid interactions.

The qualitative observations of Chapman (1968) and subsequent ^1H NMR studies of Chan *et al.* (1971), Lee *et al.* (1972), Horwitz *et al.* (1973), and McLaughlin *et al.* (1973) have confirmed the conclusions reached by other methods (Luzzati, 1968): (1) Organized lipid structures may exist in either a quasi-crystalline state or a quasi-liquid state at higher temperatures, with a more or less well-defined phase transition in between characterized by an abrupt change in the resonance linewidth. (2) There is appreciable mobility of the lipid side chains, reflected in the narrower lines, in the quasi-liquid state. (3) There is a gradient of mobility, at least in simpler protein-free phospholipid bilayers, evident from the fact that the methyl resonances of the fatty acids are narrower than the methylene resonances.

A semiquantitative characterization of the gradient of mobility was possible from ^{13}C measurements (Metcalfe *et al.*, 1971; Levine *et al.*, 1972; Lee *et al.*, 1976), and detailed models of fatty acid side-chain motions were proposed, although strictly speaking relaxation data do not permit a clear distinction between the effects of packing on the amplitudes and on the rates of motion at each point in the chain, i.e., between *order* and *mobility*. Such a distinction can to some extent be made in the studies of quadrupole splittings in ^2H spectra. For this reason ^2H NMR experiments on selectively deuterated lipids have made

Oleg Jardetzky • Stanford Magnetic Resonance Laboratory, Stanford University, Stanford, California 94305.

the greatest impact in the study of membranes (Charvolin *et al.*, 1973; Seelig and Seelig, 1974; for a detailed review see Seelig, 1977). From the observed quadrupole splitting it is possible to define an average order parameter S for each methylene group of a fatty acid side chain in a bilayer. A plot of S as a function of chain length clearly shows a progressive decrease in order from the surface to the interior of a bilayer. While the trend is similar to that previously shown by spin labeling (Hubbell and McConnell, 1971), the two curves were not found to be identical, provoking a heated controversy. The explanation for the discrepancy favored by NMR spectroscopists is that the bulky spin label perturbs the order and ^{2}H quadrupole splitting gives a more accurate picture. However, it is impossible to exclude the possibility that the ESR and NMR parameters are averaged in a different manner by the same motion. Important in this context is the interpretation of the magic angle spinning experiment on oriented phospholipid multilayers (Seeling, 1978), which shows that the axis of motional averaging is identical with the bilayer normal. If one assumes that all averaging occurs about a single energy minimum, it follows that the fatty acids are arranged with their long axis along the bilayer normal. However, if, as generally believed, averaging results from *trans–gauche* isomerizations, this conclusion does not follow, and still another explanation for the discrepancy between ESR and NMR order parameters— the tilted arrangement of side chains proposed by McFarland and McConnell (1971)— cannot be ruled out.

Rates of lateral diffusion and transmembrane exchange, originally measured by Kornberg and McConnell (1971) using spin labels, have also been estimated by NMR (Lee *et al.*, 1973; Brulet and McConnell, 1976; Wennerstrom and Lindblom, 1977; Kuo and Wade, 1979) and found to be in the range of $1–5 \times 10^{-8}$ cm^2/sec for lateral diffusion and minutes to hours for transmembrane flip-flop.

The asymmetric distribution of lipids between the two layers in a phospholipid vesicle is detectable by NMR. Separate signals are often seen in the ^{31}P spectrum, corresponding to the inner and the outer layer (Berden *et al.*, 1974; McLaughlin *et al.*, 1975). The resonance of the outer layer can be selectively broadened or shifted by the addition of paramagnetic ions to which the vesicles are relatively impermeable (Bergelson and Barsukov, 1977, and references therein), and signals from the two layers can be studied separately. There has not been a sufficient number of studies of this type to permit any generalizations concerning the rules governing asymmetry, but the importance of such factors as the composition of the phospholipid mixture, vesicle size, and distribution of charges is apparent.

Two questions of phospholipid head-group conformation have been investigated by NMR: (1) conformer distribution in the choline or ethanolamine moiety (e.g., Batchelor *et al.*, 1972; Horwitz *et al.*, 1973) and (2) orientation of the glycerol backbone and the phosphamine moiety with respect to the bilayer plane. Firm conclusions on these points do not exist. Different interpretations can be given to the existing sets of data, depending on the assumptions made, especially concerning the nature of the averaging process. The ^{2}H and ^{31}P data (Griffin *et al.*, 1978; Seelig, 1978) are most consistent with the P–N axis of the head group lying parallel to the plane of the bilayer and rotating about the bilayer normal. On the other hand, the early findings (Hauser *et al.*, 1976) based on the lanthanide method suggested an orientation at an angle to the bilayer normal. Neither interpretation is compelling, given the nature of the data.

Numerous observations of the effects of cholesterol, local anesthetics, and other small molecules on the NMR spectra of phospholipids have been proposed (e.g., Yeagle *et al.*, 1975; Chatterjee and Brockerhoff, 1978), although the careful work of Oldfield *et al.*, (1978) and Jacobs and Oldfield (1979) indicates that a simple steric restriction of motion, which is different for different molecules, is sufficient to account for the findings. For the present, it is not possible to draw more specific conclusions from experiments of this type. First, in fluid medium, rapid averaging precludes a unique structural interpretation (Jardetzky and Roberts, 1981). Second, equations customarily used for the analysis of NMR data rest on the assumption that foreign molecules will be uniformly distributed in the phospholipid, forming a single phase. In the "solid" phase this is not a safe assumption, as shown by recent studies of Copeland and McConnell (1980) and Owicki and McConnell (1980) in which phospholipid–cholesterol mixtures were found to form alternating bands of cholesterol–depleted "solid" and cholesterol-rich "fluid" phases below the transition temperature. Thus, conventional theoretical analysis cannot be applied to such heterogeneous systems. Conclusions reported to date regarding the details of the interactions of various small molecules with phospholipids must therefore be regarded as highly uncertain. The same considerations apply to the study of hydration (e.g., Taylor, 1976; Finch and Schneider, 1975) and ion binding (James and Noggle, 1972; Hauser et al., 1975).

Protein–lipid interactions have also been extensively studied, as NMR in principle offers the possibility of detecting protein-bound lipid, provided the exchange between the bound and the free lipid is slow and the bound lipid is sufficiently immobilized to be distinguished. Studies of lipoproteins, using ^{1}H (Finer et al., 1975; Hauser, 1975), ^{31}C (e.g., Assman et al., 1974; Stoffel and Bister, 1975; Hamilton and Cordes, 1978), and ^{31}P (Yeagle et al., 1978), have proved disappointing in this regard. It can be safely concluded from the reported data that (1) all of the phospholipid is detectable, i.e., very little if any is immobilized, (2) the environment of individual groups and (3) their mobility do not differ much from pure phospholipid, and (4) the choline moieties are accessible to solvent. There has been considerable controversy concerning the extent of immobilization and exposure to solvent, but it must be observed that analysis of the data in terms of theories developed for systems undergoing isotropic motions is not very meaningful in this case. Anisotropy of motions cannot be taken into account without more detailed knowledge of the structure, and because this is lacking, few of the more detailed conclusions are on firm ground. In reconstituted protein-containing phospholipid vesicles, no firm evidence for the existence of a "boundary lipid" postulated on the basis of ESR measurements (Jost et al., 1973) could be found by NMR (Oldfield et al., 1977). A previous report (Dahlquist et al., 1977) confirming the existence of "boundary lipid" by NMR resulted from an experimental error (Dahlquist, 1979, personal communication), and a more recent report (Utsumi et al., 1980) is based on an unwarranted interpretation of the data. An effect of proteins on the ^{2}H quadrupole splitting in the spectra of bilayer lipids is nevertheless observable. This can be interpreted *either* as a decrease in order on protein incorporation as proposed (Kang et al., 1979) *or* as an as yet undetermined change in the pattern of averaging, which may include averaging between a "bound" and a "free" phase. If, as the data obtained on lipoproteins suggest, the lipid bound to the protein by thermodynamic criteria has an environment and mobility similar to those of the free lipid and exchange is rapid, very little definite information can be derived from an NMR spectroscopic measurement concerning the nature and the stoichiometry of the interaction. ^{19}F-labeled tyrosines of the M13 phage incorporated into cardiolipin vesicles have been shown to be inaccessible to solvent (Hagen et al., 1979).

NMR studies of natural membranes are still in their infancy, despite the fact that the first reports appeared about 10 years ago (^{1}H, Dea et al., 1971; ^{31}P, Davis and Inesi, 1972; ^{13}C, Robinson et al., 1972). From a qualitative interpretation of the spectra it is possible to infer the existence of a mobile phospholipid component in these membranes. More recent and sophisticated experiments (e.g., Nicolau et al., 1975; Brown et al., 1977; Smith et al., 1978; Stier et al., 1978; Gent and Ho, 1978; Davis et al., 1979; Kang et al., 1979) agree on this point, but permit few reliable additional conclusions.

NMR has been useful for demonstrating qualitatively that there is extensive mobility in phospholipid bilayers and natural membranes. Beyond this the interpretation of NMR findings on membranes and model membrane systems presents formidable theoretical difficulties, traceable in part to the anisotropy of the structure, but largely to extensive mobility and rapid exchange between different environments. In rapidly exchanging systems the observed NMR parameters represent ensemble and time averages, which give no clues concerning the states being averaged and do not permit a simple structural interpretation. Neglect of the averaging or incorrect assumptions about its nature can easily lead to physically meaningless interpretations of NMR data (Jardetzky, 1980; Jardetzky and Roberts, 1981). Yet in the application of the method to the study of membranes, very little attention has been paid to this problem. As a result, only qualitative conclusions can be accepted with confidence, and even the most sophisticated reports of quantitative analysis must be taken with a large grain of salt. The question of whether NMR can be more than a semiquantitative indicator of mobility and can make an important novel contribution to our understanding of membranes, as contrasted with our understanding of NMR phenomena observable in membranes, must still be regarded as open.

REFERENCES

Assman, G., Highet, R. J., Sokoloski, E. A., and Brewer, H. B. (1974). *Proc. Natl. Acad. Sci. USA* **71,** 3701–3705.

Batchelor, J. G., Prestegard, J. H., Cushley, R. J., and Lipsky, S. R. (1972). *Biochem. Biophys. Res. Commun.* **48,** 70–75.

Berden, J. A., Cullis, P. R., Hoult, D. I., McLaughlin, A. C., Radda, G. K., and Richards, R. E. (1974). *FEBS Lett.* **46,** 55–58.

Bergelson, L. D., and Barsukov, L. I. (1977). *Science* **197,** 224–230.

Brown, T. R., Ugurbil, K., and Shulman, R. G. (1977). *Proc. Natl. Acad. Sci. USA* **74,** 551–553.

Brulet, P., and McConnell, H. M. (1976). *J. Am. Chem. Soc.* **98,** 1314–1318.

Chan, S. I., Feigenson, G. W., and Seiter, C. H. (1971). *Nature* **231,** 110–112.

Chapman, D. (1968). In *Biological Membranes* (D. Chapman, ed.), pp. 189–202, Academic Press, New York.

Chapman, D., and Salsbury, N. J. (1966). *Trans. Faraday Soc.* **62,** 2607–2621.

Charvolin, J., Manneville, P., and Deloche, B. (1973). *Chem. Phys. Lett.* **23,** 345–348.

Chatterjee, N., and Brockerhoff, H. (1978). *Biochim. Biophys. Acta* **511,** 116–119.

Copeland, B. R., and McConnell, H. M. (1980). *Biochim. Biophys. Acta* **599,** 95–109.

Dahlquist, F. W., Muchmore, D. C., Davis, H. H., and Bloom, M. (1977). *Proc. Natl. Acad. Sci. USA* **74,** 5435–5439.

Davis, D. G., and Inesi, G. (1972). *Biochim. Biophys. Acta* **282,** 180–186.

Davis, J. H., Nichol, C. P., Weeks, G., and Bloom, M. (1979). *Biochemistry* **18,** 2103–2112.

Dea, P., Chan, S. I., and Dea, F. J. (1972). *Science* **175,** 206–209.

Finch, E. D., and Schneider, A. S. (1975). *Biochim. Biophys. Acta* **406,** 146–154.

Finer, E. G., Henry, R., Leslie, R. B., and Robertson, R. S. (1975). *Biochim. Biophys. Acta* **380,** 320–337.

Gent, M. P. N., and Ho, C. (1978). *Biochemistry* **17,** 3023–3038.

Griffin, R. G., Powers, L., and Pershan, P. S. (1978). *Biochemistry* **17,** 2718–2722.

Hagen, D. S., Weiner, J. H., and Sykes, B. D. (1979). *Biochemistry* **18,** 2007–2012.

Hamilton, J. A., and Cordes, E. H. (1978). *J. Biol. Chem.* **253,** 5193–5198.

Hauser, H. (1975). *FEBS Lett.* **60,** 71–75.

Hauser, H., Phillips, M. C., and Barratt, M. D. (1975). *Biochim. Biophys. Acta* **413,** 341–353.

Hauser, H., Phillips, M. C., Levine, B. A., and Williams, R. J. P. (1976). *Nature (London)* **261,** 390–394.

Horwitz, A. F., Michaelson, D., and Klein, M. P. (1973). *Biochim. Biophys, Acta* **298,** 1–7.

Hubbell, W. L., and McConnell, H. M. (1971). *J. Am. Chem. Soc.* **93,** 314–326.

Jacobs, R., and Oldfield, E. (1979). *Biochemistry* **18,** 3280–3285.

James, T. L., and Noggle, J. H. (1972). *Anal. Biochem.* **49,** 208–217.

Jardetzky, O. (1980). *Biochim. Biophys. Acta* **621,** 227–232.

Jardetzky, O., and Roberts, G. C. K. (1981). *NMR in Molecular Biology,* Academic Press, New York.

Jost, P. C., Griffiths, O. H., Capaldi, R. A., and Vanderkooi, G. (1973). *Biochim. Biophys. Acta* **311,** 141–152.

Kang, S. Y., Gutowsky, H. S., Hsung, J. C., Jacobs, R., King, T. E., Rice, D., and Oldfield, E. (1979). *Biochemistry* **18,** 3257–3267.

Kornberg, R. D., and McConnell, H. M. (1971). *Proc. Natl. Acad. Sci. USA* **68,** 2564–2568.

Kuo, A. L., and Wade, C. G. (1979). *Biochemistry* **18,** 2300–2308.

Lee, A. G., Birdsall, N. J. M., Levine, Y. K., and Metcalfe, J. C. (1972). *Biochim. Biophys. Acta* **225,** 43–56.

Lee, A. G., Birdsall, N. J. M., and Metcalfe, J. C. (1973). *Biochemistry* **12,** 1650–1659.

Lee, A. G., Birdsall, N. J. M., Metcalfe, J. C., Warren, G. B., and Roberts, G. C. K. (1976). *Proc. Roy. Soc. London Ser. B* **193,** 253–274.

Levine, Y. K., Birdsall, N. J. M., Lee, A. G., and Metcalfe, J. C. (1972)). *Biochemistry* **11,** 1416–1421.

Luzzati, V. (1968). In *Biological Membranes* (D. Chapman, ed.), pp. 71–123, Academic Press, New York.

McFarland, B. G., and McConnell, H. M. (1971). *Proc. Natl. Acad. Sci. USA* **68,** 1274–1278.

McLaughlin, A. C., Podo, F., and Blasie, J. K. (1973). *Biochim. Biophys. Acta* **330,** 109–121.

McLaughlin, A. C., Cullis, P. R., Hemminga, M. A., Hoult, D. I., Radda, G. K., Ritchie, G. A., Seeley, P. J., and Richards, R. E. (1975). *FEBS Lett.* **57,** 213–218.

Metcalfe, Metcalfe, J. C., Birdsall, N. J. M., Feeney, J., Lee, A. G., Levine, Y. K., and Partington, P. (1971). *Nature (London)* **233,** 199–201.

Nicolau, C., Dietrich, W., Steiner, M. R., Steiner, S., and Melnick, J. L. (1975). *Biochim. Biophys. Acta* **382,** 311–321.

Oldfield, E., Meadows, M., Rice, D., and Jacobs, R. (1978). *Biochemistry* **17,** 2727–2740.

Owicki, J. C., and McConnell, H. M. (1980). *Biophys. J.* **30,** 383–398.

Robinson, J. D., Birdsall, N. J. M., Lee, A. G., and Metcalfe, J. C. (1972). *Biochemistry* **11,** 2903–2902.

Seelig, A., and Seelig, J. (1974). *Biochemistry* **13,** 4839–4845.

Seelig, J. (1977). *Q. Rev. Biophys.* **10,** 353–418.

Seelig, J. (1978). *Biochim. Biophys. Acta* **515,** 105–140.

Smith, I. C. P., Tullock, A. P., Stockton, G. W., Schreier, S., Joyce, A., Butler, K. W., Boulanger, Y., Blackwell, B., and Bennet, L. G. (1978). *Ann. N.Y. Acad. Sci.* **308,** 8–31.

Stier, A., Finch, A. E., and Bosterling, B. (1978). *FEBS Lett.* **91,** 109–112.

Stoffel, W., and Bister, K. (1975). *Biochemistry* **14,** 2841–2847.

Taylor, R. P. (1976). *Arch. Biochem. Biophys.* **173,** 596–602.

Utsumi, H., Tunggal, B. D., and Stoffel, W. (1980). *Biochemistry* **19,** 2385–2390.

Wennerstrom, H., and Lindblom, G. (1977). *Q. Rev. Biophys.* **10,** 67–96.

Yeagle, P. L., Hutton, W. C., Huang, C.-H., and Martin, R. B. (1975). *Proc. Natl. Acad. Sci. USA* **72,** 3477–3481.

Yeagle, P. L., Martin, R. B., Pottenger, L., and Langdon, R. G. (1978). *Biochemistry* **17,** 2727–2710.

15

NMR of Protein–Lipid Interactions in Model and Biological Membrane Systems

Eric Oldfield

1. INTRODUCTION

Membranes are composed predominantly of lipids, proteins, and sterol molecules and are responsible at least in part for a wide variety of biochemical processes such as respiration, vision, photosynthesis, cell–cell recognition, and nerve impulse transmission. Not surprisingly then, there have been considerable efforts spent in attempting to characterize the molecular structure of, and intermolecular interactions between, individual membrane components, in an attempt to relate the structures of membranes to their function. In this short review we discuss recent developments in our understanding of the structure of membranes obtained by means of NMR spectroscopic techniques. We show that protein–lipid interactions in both model and intact biological membranes are characterized by a dynamic disordering of boundary-lipid hydrocarbon chains, as viewed by high-field deuterium NMR spectroscopy, while ^{31}P spectra indicate significant disordering and/or immobilization within the phospholipid polar head group region due to association with protein. The effects are very different from those seen with cholesterol, and are in marked contrast to the old ideas of rigid, ordered "boundary-lipid" surrounding membrane proteins. We also present results suggesting that it will soon be possible to directly monitor the effects of lipid on protein dynamics by means of NMR spectroscopy.

2. DEUTERIUM NMR OF LIPIDS IN MODEL SYSTEMS

The idea that membrane proteins are solvated by a layer of immobilized boundary lipid was first put forward by Jost *et al.* (1973a–c, 1977) originally for the cytochrome oxidase system, and subsequently for cytochrome b_5 by Dehlinger *et al.* (1974) and sarcoplasmic reticulum ATPase by Jost and Griffith (1978), and similar concepts have been

Eric Oldfield • School of Chemical Sciences, University of Illinois, Urbana, Illinois 61801.

presented by many other workers (Caron *et al*. 1974; Grant and McConnell, 1974; Hesketh *et al*., 1976; Marsh *et al*., 1978; Warren *et al*., 1974, 1975). In almost all instances, the evidence for rigid or immobilized "boundary lipid" has come from the "rigid-glass" (long correlation time) ESR spectrum of a nitroxide free radical introduced into the system, and it has been generally concluded that proteins increase order greatly for the first layer of lipid adjacent to protein, and that significant perturbations exist for the second and perhaps the third layers of lipid. Because it seemed to us plausible that nitroxides might not faithfully reproduce these interactions, due to their polar nature, we have used the nonperturbing techniques of ^{2}H and ^{31}P NMR spectroscopy to investigate these problems in more detail.

The theoretical background appropriate for consideration of the ^{2}H NMR spectra of lipid membranes is discussed in detail by Oldfield *et al*. (1978a,b, 1981) and by Seelig (1977) so we shall be content to briefly quote the principal result that for the ^{2}H nucleus (with spin $I = 1$ and in C–D bonds an asymmetry parameter $\eta = 0$) the allowed transitions correspond to $+1 \leftrightarrow 0$ and $0 \leftrightarrow -1$, and give rise to a "quadrupole splitting" of the NMR absorption line with separation between peak maxima of

$$\Delta \nu_Q = (3/2)(e^2 qQ/h)[(3 \cos^2 \theta - 1)/2]$$

$e^2 qQ/h$ is the deuterium quadrupole coupling constant, which has been found to be about 170 kHz for C–D bonds by Derbyshire *et al*. (1969), and θ is the angle between the magnetic field, H_0, and the principal axis of the electric field gradient tensor at the deuterium nucleus. For a rigid polycrystalline solid, all values of θ are possible and one obtains a so-called "powder pattern" lineshape (see Fig. 1A) in which the separation between peak maxima is about 127.5 kHz, and the separation between the distribution steps is twice this value. In biological membranes, however, there is considerable motion of the C–D vector, due for example to *gauche–trans* isomerization along the hydrocarbon chain, to chain tilt, to diffusion, and to chain rotation; thus, we must take an average in time of $(3 \cos^2\theta\text{-}1)$ for motions that are faster than 170 kHz. The result, in brief, is that the size of the quadrupole splitting decreases as the degree of order or "order parameter" (Seelig, 1977) of the system decreases.

Early studies demonstrated that addition of cholesterol to a deuterium-labeled dimyristoylphosphatidylcholine (DMPC) bilayer at a 1 : 1 mole ratio caused an almost twofold increase in order parameter of the lipid hydrocarbon chain, corresponding to an increase in ^{2}H quadrupole splitting from about 27 kHz to about 49 kHz (at 30°C; Oldfield *et al*., 1971). Below the phase transition temperature of the pure lipid, the hydrocarbon chains were prevented by cholesterol from crystallizing into the rigid α-crystalline gel state, and the observed quadrupole splitting remained at about 50 kHz even at 10°C, some 13°C below the gel-to-liquid-crystal phase transition temperature. These effects seen with ^{2}H NMR were precisely those predicted on the basis of earlier calorimetric studies by Ladbrooke *et al*. (1968), and ^{1}H NMR of sonicated vesicles by Chapman and Penkett (1966).

The effects of proteins on the ^{2}H NMR spectra of hydrocarbon-chain-labeled phospholipids were predicted, using one simple version of the boundary-lipid idea, to be an increase in ^{2}H NMR quadrupole splittings (or order parameters) of lipid associated with protein (Dahlquist *et al*., 1977; Hong and Hubbell, 1972; Jost and Griffith, 1978; Kleemann and McConnell, 1976; Marcelja, 1976; Marsh *et al*., 1978; Owicki *et al*., 1978; Schroder, 1977; Scott and Cherng, 1978). We show in Fig. 1 results we have obtained using cytochrome oxidase (EC 1.9.3.1) and sarcoplasmic reticulum ATPase (EC 3.6.1.3), the systems most frequently studied using physical techniques (Dahlquist *et al*., 1977;

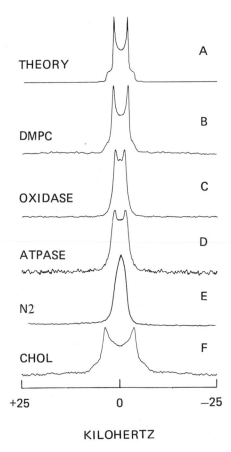

THEORY · A

DMPC · B

OXIDASE · C

ATPASE · D

N2 · E

CHOL · F

+25   0   −25

KILOHERTZ

Figure 1. Theoretical and experimental deuterium NMR spectra of ^{2}H-labeled lipids showing the effects of proteins and cholesterol on hydrocarbon chain order. (A) Theoretical ^{2}H powder pattern; the splitting is arbitrary. (B) 1-Myristoyl-2-(14,14,14-trideutero)myristoyl-sn-glycero-3-phosphocholine (DMPC-d_3) in excess water at 30°C. (C) As in (B) but lipid contains 67 wt% cytochrome oxidase (cytochrome c: oxygen oxidoreductase, EC 1.9.3.1). (D) As in (B) but lipid contains 65 wt% sarcoplasmic reticulum ATPase (EC 3.6.1.3). (E) As in (B) but lipid contains 67 wt% beef brain myelin proteolipid apoprotein. (F) As in (B) but lipid contains 33 wt% cholesterol. Deuterium NMR spectra were obtained at 34 MHz.

Hesketh *et al.*, 1976; Jost *et al.*, 1973a,b,c, 1977; Longmuir *et al.*, 1977; Marsh *et al.*, 1978; Moore *et al.*, 1978; Oldfield *et al.*, 1978a; Warren *et al.*, 1974, 1975), together with the much smaller model system beef brain myelin proteolipid apoprotein (N2) used by Curatolo *et al.* (1977) and by Papahadjopoulos *et al.* (1975). All systems were "complexed" or "reconstituted" with DMPC bilayers specifically deuterated at the terminal methyl position of the *sn*-2 chain, using standard techniques used by Curatolo *et al.* (1977), Oldfield *et al.* (1978a), and Warren *et al.* (1974). For comparison, also included is a ^{2}H NMR spectrum showing the effect of cholesterol, the sample again being prepared with standard techniques used by Oldfield *et al.* (1978b).

As may be seen from the spectra in Fig. 1, the effect of incorporating protein into the lipid bilayer is to cause a *decrease* in order parameter (quadrupole splitting) of the ^{2}H-labeled methyl group, a *disordering* rather than an ordering effect. Similar results shown by Oldfield *et al.* (1978a) are obtained with a wide range of other lipids and proteins. Cholesterol, on the other hand, causes a large *increase* in order parameter, consistent with its well-known "condensing" or ordering effect (Fig. 1F). Similar ordering effects are seen with cholesterol using lipids labeled at other chain positions, although the disordering effect of protein is by far the most pronounced at the methyl end of the hydrocarbon chain. Above the lipid gel-to-liquid-crystal phase transition temperature, therefore, it appears that proteins may tend to disorder the packing of lipid hydrocarbon chains somewhat, while cholesterol orders their packing. Notably, ESR studies of such protein–lipid samples in our laboratories have confirmed that *highly immobilized* nitroxide spin-label spectra are nevertheless obtained, suggesting that differences in time scale between the two resonance experiments, or specific interactions between the nitroxide probes and the protein surfaces, may be involved. In any case, because no two-component spectra are seen above the pure lipid phase transition temperature over a wide concentration range, exchange between "bound" and "free" lipid must be very fast ($\gtrsim 10^3$–10^4 sec^{-1}), suggesting that boundary lipid does not stay bound very long (Kang *et al.*, 1979a).

At temperatures below that of the gel-to-liquid-crystal phase transition, we find that both cholesterol *and* proteins prevent lipid hydrocarbon chains from crystallizing; however, cholesterol-containing samples are far more ordered than those containing protein.

The effects described above must clearly be related to the structures of the perturbing molecules—the proteins and cholesterol. It seems most likely that the rough, irregular surfaces of proteins (with their ~ 20 different amino acid side chains) may tend to cause lipid hydrocarbon chains to pack in a slightly disordered way into "vacancies" created by the side chains, and perhaps created in a time-dependent way by rotation and internal motions of the protein. Cholesterol, on the other hand, is a rigid tetracyclic structure, and inspection of molecular models reveals that it has rather planar sides, at least on the scale of a C–C bond length. Cholesterol thus acts as the rigid boundary assumed in the theoretical calculations of protein–lipid interaction discussed by Kleemann and McConnell (1976), Marcelja (1976), and Scott and Cherng (1978). Overall, our results suggest that boundary lipid is slightly more disordered than pure lipid, but that the rates of motion are probably slowed down due to protein–lipid steric interactions.

3. DEUTERIUM NMR OF LIPIDS IN BIOLOGICAL SYSTEMS

Specifically deuterated fatty acids have been incorporated into a variety of cell membrane systems over the past 10 years (e.g., by Davis *et al.*, 1979; Kang *et al.*, 1979b; Oldfield *et al.*, 1972; Stockton *et al.*, 1977), and ^{2}H NMR spectra have been obtained as a function of temperature for various membrane fractions and for isolated lipids dispersed in water. A general observation that may be made is that there is as yet no evidence for a "condensing" or "ordering" effect of protein on membrane lipid in any of the "intact" biological membranes examined. Cholesterol, on the other hand, again exhibits an ordering effect on lipid, even in the presence of membrane protein as shown by Stockton *et al.* (1977). Kang *et al.* (1980) have shown that in the *Acholeplasma laidlawii B* membrane system, protein has little effect on lipid order, due perhaps in part to the low protein-to-lipid ratio. For example, ^{2}H quadrupole splittings and linewidths observed for 4-, 6-, 8-, and 14-^{2}H-labeled tetradecanoic acid-enriched *A. laidlawii* cell membranes are within experimental error the same as those observed with the corresponding lipid extracts, and are also the same as those seen previously by Oldfield *et al.* (1978a) with bilayers of pure 1,2-dimyristoyl-*sn*-glycero-3-phosphocholine (DMPC) when examined immediately above the end of the solid-to-fluid phase transition.

In contrast, ^{2}H NMR spectra of hexadecan-1-oic acid-enriched *Escherichia coli* L48-2 cell membranes have in our laboratory shown considerable line broadening compared with spectra of their lipid extracts, and $\Delta\nu_Q$ values were slightly decreased. Results with these *E. coli* cell membranes showed essentially the same NMR lineshapes as those seen previously by Rice and Oldfield (1979) with the DMPC–gramicidin A' system: these include collapsed terminal methyl group quadrupole splittings and large (4–6 kHz) linewidths of methylene-segment chain resonances, emphasizing that a wide range of types of protein–

lipid interactions may be observed in microbial membrane systems. The results with intact *E. coli* membranes suggest that large-amplitude fluctuations of the methylene chain segments may be caused by protein–lipid interaction in some systems.

4. PHOSPHORUS NMR OF LIPIDS IN MODEL AND BIOLOGICAL MEMBRANES

^{31}P has a spin $I = \frac{1}{2}$ and a 100% natural abundance, so that it is a particularly attractive species for studying the dynamic structures of lipid membranes. ^{31}P NMR studies of the gel and liquid-crystalline states of a variety of phospholipids, together with the effects of cholesterol on head-group molecular motion, have already been conducted by a number of investigators (Barker *et al.*, 1972; Brown and Seelig, 1978; Cullis and deKruyff, 1976; Griffin, 1976; Griffin *et al.*, 1978; Kohler and Klein, 1976; Niederberger and Seelig, 1976; Seelig, 1978; Yeagle *et al.*, 1975), so it is therefore of some interest to compare these results with ^{31}P NMR studies of protein–lipid interactions.

We show in Fig. 2 selected results from a series of ^{31}P NMR experiments recently carried out in our laboratory revealing that addition of protein to a lecithin bilayer causes in all cases a decrease in $\Delta\sigma$ of the lipid phosphate head group. This is accompanied by increased T_1 and T_2 relaxation rates shown by Rajan *et al.* (1980), which suggests that in many cases lipid polar head groups may interact directly with membrane proteins, and that

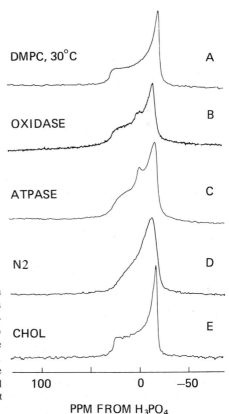

Figure 2. ^{31}P NMR spectra of DMPC in the presence of a variety of proteins, and of cholesterol. (A) DMPC in excess water, 30°C. (B) As (A) except lipid contains 67 wt% cytochrome oxidase. (C) As (A) except lipid contains 65 wt% sarcoplasmic reticulum ATPase. (D) As (A) except sample contains 67 wt% beef brain myelin proteolipid apoprotein. (E) As (A) except sample contains 33 wt% cholesterol. The weak features at 0 ppm in (B) and (C) arise from small phosphate molecule impurities. Spectra were obtained at 60.7 MHz under conditions of proton-decoupling.

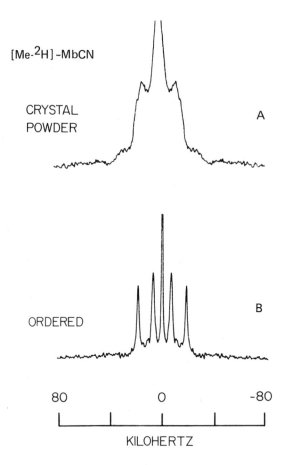

Figure 3. Deuterium NMR spectra of labeled protein crystals and membrane proteins obtained at 55.3 MHz. (A) Solid powder sample of methionine-labeled sperm whale cyanoferrimyoglobin hydrated with 90% saturated $(NH_4)_2SO_4$. (B) Magnetically ordered cyanoferrimyoglobin suspended in ~90% saturated $(NH_4)_2SO_4$. (C) [β-$^2H_1$]valine, 23°C. (D) [β-$^2H_1$]valine-enriched *Halobacterium halobium* purple membranes, in H_2O at 37°C. The sharp central peaks in (A), (B), and (D) are due to HO^2H.

as a result the head-group phosphate motion may become slower and more isotropic. Cholesterol also decreases $\Delta\sigma$ as shown by Brown and Seelig (1978) and Rajan *et al.* (1980), but in this case T_1 remains the same and there are slight increases in T_2 relaxation times from the values found in pure lipid bilayers, suggesting that cholesterol simply acts as an inert spacer molecule.

With intact biological membranes, spectral signal-to-noise ratios have been lower than those obtained with model systems and no differences between ^{31}P NMR spectra of membranes and their lipid extracts supporting the above ideas have yet been reported. Work in the area of intact biomembrane ^{31}P NMR has concentrated on observation and characterization of "isotropic" signal components, sometimes generated by the presence of protein or polypeptide as shown by Burnell *et al.* (1980), Rajan *et al.* (1980), and Stier *et al.* (1978). These results may be linked to the presence of "lipidic particles" in the lipid bilayer as discussed by Burnell *et al.* (1980) and Miller (1980).

At present, a plausible model of protein–lipid interaction is, therefore, one in which proteins "immobilize" lipid head groups, while simultaneously disordering somewhat the lipid hydrocarbon chains, due to the rough nature of the protein surface.

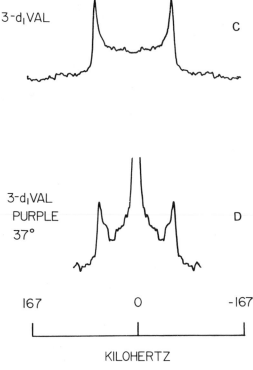

Figure 3. continued

5. DEUTERIUM NMR OF PROTEINS IN MODEL AND BIOLOGICAL SYSTEMS

The ideal way to monitor protein–lipid interactions would be to observe the effects of various membrane lipid constituents on protein static and dynamic structure directly, via observation of the NMR signals originating from the protein itself. Recently, considerable progress has been made in exploring this possibility.

First, it has been shown by Oldfield and Rothgeb (1980) and Rothgeb and Oldfield (1981) that it is possible to detect, resolve, and assign individual groups in (paramagnetic) protein crystals, a new type of "model system," using high-field ^{2}H NMR, as shown in Figs. 3A and B. Shown in Fig. 3A is a ^{2}H NMR spectrum of a random powder distribution of [*methyl*-^{2}H] methionine-labeled sperm whale cyanometmyoglobin microcrystals, while in Fig. 3B we show the spectrum obtained after suspension of the microcrystals in a strong magnetic field, which causes sample ordering as shown by Rothgeb and Oldfield (1981). Both ^{2}H-labeled sites are resolved in Fig. 3B, and have recently been assigned as shown by Rothgeb and Oldfield (1981). In principle, a variety of methods may be used to orient, or order, diamagnetic biological membranes, and thus resolve individual sites, if adequate spectrometer sensitivity is available to detect them. Again, the first steps in this area have been taken by Oldfield *et al.* (1981). ^{2}H NMR spectra of the pure amino acid [β-^{2}H$_1$]valine, in the solid state, and of [β-^{2}H$_1$]valine-labeled purple membranes from the extremely halophilic organism *Halobacterium halobium* (Figs. 3C and D) show the very rigid nature of the purple membrane protein, bacteriorhodopsin, and indicate that it should now be possible to study the dynamics of amino acid residues in many different biomembrane systems. It seems certain, therefore, that in the next 5 years a much more detailed picture of the effects of membrane lipid constituents on protein structure (and function) will be acquired using highfield NMR and isotopic labeling methods.

ACKNOWLEDGMENTS. Part of this work was supported by the National Institutes of Health (Grants CA-00595, HL-19481), by the National Science Foundation (Grants PCM 78-23021, PCM 79-23170), by the American Heart Association (Grant 80-867), by the Alfred P. Sloan Foundation, and by the Los Alamos Scientific Laboratory Stable Isotope Resource, which is jointly supported by the United States Department of Energy and the National Institutes of Health (Grant RR-99962), and has benefited from the use of facilities made available through the University of Illinois NSF Regional Instrumentation Facility (Grant CHE 79-16100). The author is an Alfred P. Sloan Foundation Fellow, 1978–1980, and a USPHS Research Career Development Awardee, 1979–1984.

REFERENCES

Barker, R. W., Bell, J. D., Radda, G. K., and Richards, R. E. (1972). *Biochim. Biophys. Acta* **260**, 161–163.
Brown, M. F., and Seelig, J. (1978). *Biochemistry* **17**, 381–384.
Burnell, E., Van Alphen, L., Verkleij, A., and DeKruijff, B. (1980). *Biochim. Biophys. Acta* **597**, 492–501.
Caron, F., Mateu, L., Rigny, P., and Azerad, R. (1974). *J. Mol. Biol.* **85**, 279–300.
Chapman, D., and Penkett, S. A. (1966). *Nature (London)* **211**, 1304–1305.
Cullis, P. R., and deKruyff, B. (1976). *Biochim. Biophys. Acta* **436**, 523–540.
Curatolo, W., Sakura, J. D., Small, D. M., and Shipley, G. G. (1977). *Biochemistry* **16**, 2313–2319.
Dahlquist, F. W., Muchmore, D. C., Davis, J. H., and Bloom, M. (1977). *Proc. Natl. Acad. Sci. USA* **74**, 5435–5439.
Davis, J. H., Nichol, C. P., Weeks, G., and Bloom, M. (1979). *Biochemistry* **18**, 2103–2112.
Dehlinger, P. J., Jost, P. C., and Griffith, O. H. (1974). *Proc. Natl. Acad. Sci. USA* **71**, 2280–2284.
Derbyshire, W., Gorvin, T. C., and Warner, D. (1969). *Mol. Phys.* **17**, 401–407.
Grant, C. W. M., and McConnell, H. M. (1974). *Proc. Natl. Acad. Sci. USA* **71**, 4653–4657.
Griffin, R. G. (1976). *J. Am. Chem. Soc.* **98**, 851–853.
Griffin, R. G., Powers, L., and Pershan, P. S. (1978). *Biochemistry* **17**, 2718–2722.
Hesketh, T. R., Smith, G. A., Houslay, M. D., McGill, K. A., Birdsall, N. J. M., Metcalfe, J. C., and Warren, G. B. (1976). *Biochemistry* **15**, 4145–4151.
Hong, K., and Hubbell, W. L. (1972). *Proc Natl. Acad. Sci. USA* **69**, 2617–2621.
Jost, P. C., and Griffith, O. H. (1978). In *Cellular Function and Molecular Structure: A Symposium on Biophysical Approaches to Biological Problems* (P. F. Agris, R. N. Loeppky, and B. D. Sykes, eds.), pp. 25–54, Academic Press, New York.
Jost, P. C., Capaldi, R. A., Vanderkooi, G., and Griffith, O. H. (1973a). *J. Supramol. Struct.* **1**, 269–280.
Jost, P., Griffith, O. H., Capaldi, R. A., and Vanderkooi, G. (1973b). *Biochim. Biophys. Acta* **311**, 141–152.
Jost, P. C., Griffith, O. H., Capaldi, R. A., and Vanderkooi, G. (1973c). *Proc. Natl. Acad. Sci. USA* **70**, 480–484.
Jost, P. C., Nadakavukaren, K. K., and Griffith, O. H. (1977). *Biochemistry* **16**, 3110–3114.
Kang, S. Y., Gutowsky, H. S., Hshung, J. C., Jacobs, R., King, T. E., Rice, D., and Oldfield, E. (1979a). *Biochemistry* **18**, 3257–3267.
Kang, S. Y., Gutowsky, H. S., and Oldfield, E. (1979b). *Biochemistry* **18**, 3268–3271.
Kang, S. Y., Kinsey, R., Rajan, S., Gutowsky, H. S., Gabridge, M. G., and Oldfield, E. (1980). *J. Biol. Chem.* **256**, 1155–1159.
Kleemann, W., and McConnell, H. M. (1976). *Biochim. Biophys. Acta* **419**, 206–222.
Kohler, S. J., and Klein, M. P. (1976). *Biochemistry* **15**, 967–973.
Ladbrooke, B. D., Williams, R. M., and Chapman, D. (1968). *Biochim. Biophys. Acta* **150**, 333–340.
Longmuir, K. J., Capaldi, R. A., and Dahlquist, F. W. (1977). *Biochemistry* **16**, 5746–5755.
Marčelja, S. (1976). *Biochim. Biophys. Acta* **455**, 1–7.
Marsh, D., Watts, A., Maschke, W., and Knowles, P. F. (1978). *Biochem. Biophys. Res. Commun.* **81**, 397–402.
Miller, R. G. (1980). *Nature (London)* **287**, 166–167.

Moore, B. M., Lentz, B. R., and Meissner, G. (1978). *Biochemistry* **17**, 5248–5255.

Niederberger, W., and Seelig, J. (1976). *J. Am. Chem. Soc.* **98**, 3704–3706.

Oldfield, E., and Rothgeb, T. M. (1980). *J. Am. Chem. Soc.* **102**, 3635–3637.

Oldfield, E., Chapman, D., and Derbyshire, W. (1971). *FEBS Lett.* **16**, 102–104.

Oldfield, E., Chapman, D., and Derbyshire, W. (1972). *Chem. Phys. Lipids* **9**, 69–81.

Oldfield, E., Gilmore, R., Glaser, M., Gutowsky, H. S., Hshung, J. C., Kang, S. Y., King, T. E., Meadows, M., and Rice, D. (1978a). *Proc. Natl. Acad. Sci. USA* **75**, 4657–4660.

Oldfield, E., Meadows, M., Rice, D., and Jacobs, R. (1978b). *Biochemistry* **17**, 2727–2740.

Oldfield, E., Janes, N., Kinsey, R., Kintanar, A., Lee, R. W. K., Rothgeb, T. M., Schramm, S., Skarjune, R., Smith, R., and Tsai, M.-D. (1981). *Biochem. Soc. Symp.* **46**, 155–181.

Owicki, J. C., Springgate, M. W., and McConnell, H. M. (1978). *Proc. Natl. Acad. Sci. USA* **75**, 1616–1619.

Papahadjopoulos, D., Moscarello, M., Eylar, E. H., and Isac, T. (1975). *Biochim. Biophys. Acta* **401**, 317–335.

Rajan, S., Kang, S. Y., Gutowsky, H. S., and Oldfield, E. (1980). *J. Biol. Chem.* **256**, 1160–1166.

Rice, D., and Oldfield, E. (1979). *Biochemistry* **18**, 3272–3279.

Rothgeb, T. M., and Oldfield, E. (1980). *J. Biol. Chem.* **256**, 1432–1446.

Schröder, H. (1977). *J. Chem. Phys.* **67**, 1617–1619.

Scott, H. L., and Cherng, S. L. (1978). *Biochim. Biophys. Acta* **510**, 209–215.

Seelig, J. (1977). *Q. Rev. Biophys.* **10**, 353–418.

Seelig, J. (1978). *Biochim. Biophys. Acta* **515**, 105–140.

Stier, A., Finch, S. A. E., and Bosterling, B. (1978). *FEBS Lett.* **91**, 109–112.

Stockton, G. W., Johnson, K. G., Butler, K. W., Tulloch, A. P., Boulanger, Y., Smith, I. C. P., Davis, J. H., and Bloom, M. (1977). *Nature (London)* **269**, 267–268.

Warren, G. B., Toon, P. A., Birdsall, N. J. M., Lee, A. G., and Metcalfe, J. C. (1974). *Biochemistry* **13**, 5501–5507.

Warren, G. B., Houslay, M. D., Metcalfe, J. C., and Birdsall, N. J. M. (1975). *Nature (London)* **255**, 684–687.

Yeagle, P. L., Hutton, W. C., Huang, C., and Martin, R. B. (1975). *Proc. Natl. Acad. Sci. USA* **72**, 3477–3481.

16

Current Views on Boundary Lipids Deduced from Electron Spin Resonance Studies

Philippe F. Devaux and Jean Davoust

1. PROBING OF THE LIPID–PROTEIN INTERFACE WITH FREELY DIFFUSIBLE SPIN LABELS

In 1973 Jost and co-workers showed that a spin-labeled fatty acid, with the nitroxide ring near the ω-2 acyl terminal (spin label I), gives rise to a composite spectrum when incorporated into vesicles made with cytochrome oxidase and various amounts of phospholipids. In order to detect the broad component corresponding to strongly immobilized probes, it was necessary to prepare vesicles with a low lipid-to-protein ratio. The immobilized component sensitive to the presence of proteins was attributed to spin labels located at the boundary of cytochrome oxidase. A tentative determination of the amount of "rigid" lipids led to the proposal that one layer of immobilized lipids surrounds each protein. This experiment was later repeated with spin-labeled phospholipids (spin label II) (Jost *et al.*, 1977). Other investigators have reproduced this experiment using cytochrome oxidase (Knowles *et al.*, 1979; Denes and Stanacev, 1978) or other intrinsic proteins such as rhodopsin (Pontus and Delmelle, 1975), Ca^{2+}-ATPase (Hesketh *et al.*, 1976), and lipophilin (Boggs *et al.*, 1976).

Chapman and collaborators were the first to point out (Chapman *et al.*, 1977), on the basis of experiments with gramicidin A incorporated into lipid bilayers, that perhaps the immobilized spin-labeled lipids seen at a low lipid-to-protein ratio were lipids trapped between protein aggregates. At physiological lipid-to-protein ratios the problem is more ambiguous; the existence of an immobilized component is spectroscopically difficult to prove, and it is equally difficult to prove that it does not exist.

Philippe F. Devaux and Jean Davoust • Institut de Biologie Physico-Chimique, 75005 Paris, France.

2. LIPID SPIN LABELS LINKED TO HYDROPHOBIC PROTEINS: MONITORING PROTEIN AGGREGATION

A simple way to enhance the spectral contribution of boundary lipids without changing the lipid-to-protein ratio is to link a spin-labeled lipid to a hydrophobic protein. A very high ratio of lipid to protein is not an obstacle anymore to the lineshape determination of the broad component because the contribution of the spin labels in the bulk lipids is reduced in principle to zero. Nevertheless, in 1975 we showed that a spin-labeled fatty acid forced to stay next to the ADP carrier of mitochondria gave rise to narrow lines when the probe was near the ω-2 terminal (Devaux *et al.*, 1975; Lauquin *et al.*, 1977) (Table I). The anchoring of the probe to this intrinsic protein was made with spin labels III and IV, which are derivatives of very specific inhibitors of the ADP carriers. Similarly, a spin-labeled long-chain acyl choline (spin label V) gave relatively narrow lines at the boundary of the acetylcholine receptor in *Torpedo marmorata* membrane fragments (Bienvenüe *et al.*, 1977). Thus, it is possible for a lipid chain to experience a high mobility even when it is in the direct vicinity of hydrophobic proteins included in native membranes.

The same results were obtained with spin-labeled fatty acid chains covalently attached to rhodopsin in disc membranes (Favre *et al.*, 1979) and in artificial rhodopsin/egg lecithin recombinants (Davoust *et al.*, 1979, 1980). A maleimide residue (spin label VI) or an isocyanate residue (spin label VII) anchored the labeled chain to the protein via a sulfhydryl group. The nitroxide was buried in the lipid core because it was totally inaccessible to a reducing agent that was added to the aqueous phase (Davoust *et al.*, 1980). The spectra at 37°C were slightly broader than those recorded with freely diffusible fatty acids, but showed no evidence of heterogeneity as long as the lipid-to-protein ratio was maintained at a high level (see Fig. 1b). An effective correlation time of the order of a few nanoseconds was calculated. However, if the lipid-to-protein ratio was decreased significantly below the physiological ratio (i.e., below 70/1 mole/mole), a component with a splitting of about 60 G appeared superimposed upon the narrow component (see Figs. 1d and e). A broad component can be generated not only by lipid depletion but also by protein cross-linking with glutaraldehyde or by prolonged illumination at 37°C. This latter treatment exposes new sulfhydryl groups on the rhodopsin molecule, which are then irreversibly cross-linked by intermolecular disulfide bridges. Finally, a broad component can be generated if rhodopsin is incorporated into phospholipids undergoing a liquid–solid transition. This process usually provokes lipid–protein segregation.

In conclusion, a broad component observed with covalently attached spin-labeled lipids always accompanies protein aggregation. Thus, the immobilization observed by ESR

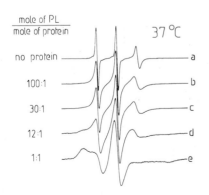

Figure 1. Influence of the lipid-to-protein ratio on the ESR spectra of spin label VI covalently bound to rhodopsin in rhodopsin–lecithin complexes (37°C). Spectrum e corresponds to lipid-free rhodopsin; it can be considered as a reference spectrum and used to quantify the amount of immobilized component present in other spectra. (Reproduced from Davoust *et al.*, 1979.)

Table I. Lipid Spin Labels

I $CH_3 - CH_2 - \underset{\underset{\text{(ring)}}{O \quad NO}}{C} - (CH_2)_{14} - COOH$

II $CH_3 - (CH_2)_{16} - COO - CH_2$

$CH_3 - CH_2 - \underset{O \quad NO}{C} - (CH_2)_{14} - COO - CH$

$CH_2 - O - \overset{O}{\underset{\underset{\overline{O}}{\|}}{P}} - O - (CH_2)_2 - \overset{+}{N}(CH_3)_3$

III $CH_3 - CH_2 - \underset{O \quad NO}{C} - (CH_2)_{14} - COS\ CoA$

IV $CH_3 - CH_2 - \underset{O \quad NO}{C} - (CH_2)_{14} - COO - Atractyloside$

V $CH_3 - CH_2 - \underset{O \quad NO}{C} - (CH_2)_{14} - COO(CH_2)\overset{+}{N}(CH_3)_3$

VI $CH_3 - CH_2 - \underset{O \quad NO}{C} - (CH_2)_{14} - COO(CH_2)N$ (maleimide ring with two =O)

VII $CH_3 - CH_2 - \underset{O \quad NO}{C} - (CH_2)_{14} - N=C=O$

VIII $CH_3 - CH_2 - \underset{O \quad NO}{C} - (CH_2)_{14} - COO(CH_2)_2NH$ (benzene ring with NO_2) $- N_3$

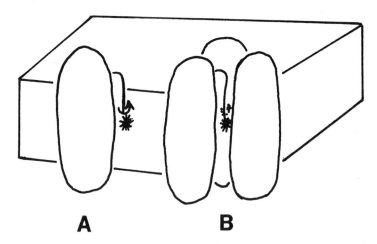

Figure 2. A spin-labeled lipid covalently attached to an intrinsic protein can be strongly immobilized only in the case of aggregated proteins (B).

is likely to be due to chain trapping between proteins (Fig. 2). In experiments with freely diffusible spin-labeled lipids and a low lipid-to-protein ratio, the immobilized component is also probably mainly indicative of protein aggregation.

Protein self-associations are not always artifacts, and spin-labeled lipids can be used to detect functional protein-oligomers. Spin label VI and spin label VIII have been used to probe Ca^{2+}-ATPase in sarcoplasmic reticulum (Fellmann *et al.*, 1980; Andersen *et al.*, 1981) and band 3 protein in erythrocytes (Andersen *et al.*, 1980). The spectra are remarkably different from the spectrum recorded with similar spin labels attached to rhodopsin in intact disc membranes. With the two former proteins under physiological conditions, a very large fraction of the spectrum ($\sim$ 80%) shows a strong immobilization (splitting of 60–63 G at 37°C). Saturation transfer spectroscopy demonstrates that the lipid probe is almost as immobilized as a spin label reporting on the overall mobility of the protein. The spectral observations with Ca^{2+}-ATPase and band 3 certainly reflect the oligomeric nature of these proteins (Vanderkooi *et al.*, 1977; Dorst and Schubert, 1979). The oligomericity of Ca^{2+}-ATPase was not taken into account in former studies of lipid–protein interaction (Metcalfe and Warren, 1977). It might be necessary also to reconsider ESR studies with cytochrome oxidase, in view of the fact demonstrated by Swanson *et al.* (1980) that cytochrome oxidase can form oligomers with full activity even at a high lipid-to-protein ratio.

In conclusion, broad components are indicative of protein–protein contacts, and covalently attached spin-labeled lipids can be used to quantitatively study protein solubility in lipids. However, the results of temperature studies will be partially obscured by the fact that, at low temperature, a restricted motion of spin-labeled lipids also can be observed around monomeric proteins dispersed in the lipids.

3. RESTRICTED MOTION AND STRONG IMMOBILIZATION: TWO DISTINCT STATES OF LIPID MOBILITY CAN BE DETECTED BY ESR AT LOW TEMPERATURES

The spectrum of a spin-labeled fatty acid chain covalently attached to rhodopsin has the appearance at 37°C of a homogeneous spectrum composed of three relatively narrow lines (see Fig. 1). A calculation performed in our laboratory (J. Davoust, M. Seigneuret, and P. F. Devaux, 1980, unpublished) assuming rapid exchange between a state of low mobility ($\tau_C \sim 10^{-7}$ S) and a state of high mobility ($\tau_C \sim 10^{-9}$ S) (Fig. 3) shows that an apparently homogeneous spectrum can be produced. An exchange rate was chosen of the order of the phospholipid diffusion rate at 37°C ($v > 10^7$ S^{-1}) Devaux *et al.*, 1973); this is justified by the fact that collisions with the phospholipids approaching the protein interface are responsible for the spin label exchanges between state 1 and state 2 (Fig. 3). If the exchange rate is reduced, a broad component appears, but with a splitting smaller than that of the original immobilized component. The conclusion is that an immobilized component with a splitting that is strongly temperature dependent can be generated, without implying chain trapping. This "motionally restricted" component disappears at high temperature and therefore is not to be confused with the "strongly immobilized component" seen with spin labels VI and VIII attached to Ca^{2+}-ATPase, band 3 or aggregated rhodopsin. At low temperature, however, it might be difficult to decide whether a composite spectrum is due to slow exchange or no exchange; hence, it might be difficult to differentiate between boundary effects and chain trapping. Only saturation transfer can permit resolution of such an ambiguity.

Watts *et al.*, (1979) were the first to stress that a spectral component corresponding to restricted motion can be detected in native biological membranes where protein self-

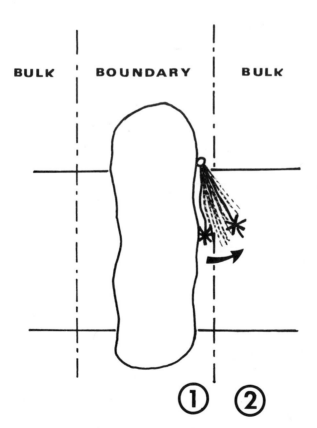

Figure 3. The motion of a probe can be different at the protein–lipid interface (boundary) and in the bulk lipids. If the exchange rate between state 1 and state 2 is rapid enough at the time scale of ESR, only an average mobility is detected.

association is unlikely to take place. They have shown that in intact disc membranes from bovine retina, a spin-labeled fatty acid or phospholipid gives rise to a component (associated with about 30% of the signal) with an outer splitting of 59 G at 0°C and 54 G at 24°C. This motionally restricted component seems to merge with the mobile component at high temperature. An effective rotational correlation time of a few tens of nanoseconds was attributed to the broad component at 24°C (Watts *et al.*, 1980), which indicates that boundary lipids are certainly not tightly bound to the surface of the proteins. Apparently, it is possible to detect this component with freely diffusible spin-labeled lipids in the disc membranes because the physiological lipid-to-protein ratio is relatively low. The same laboratory working with chromaffin granules could not detect this motionally restricted component (Fretten *et al.*, 1980). Using spin labels covalently attached to rhodopsin in reconstituted systems containing a high ratio of lipids, we have observed at −5 and 5°C a composite spectrum with a broad component associated with an outer splitting of about 60 G (Davoust *et al.*, 1979). This component was later attributed by us to protein aggregation at low temperature (Davoust *et al.*, 1980). It is more likely to correspond to the motionally restricted component, which disappears at high temperature due to rapid exchange. However, both phenomena may be involved.

4. THE QUANTITATION OF THE IMMOBILIZED COMPONENT

Quantitation requires the decomposition of spectra into the various components. This is a difficult task, for in most cases there is no unambiguous determination of the lineshape of each of the components. Briefly, three different approaches have been used.

1. The lineshape of the broad component is chosen a priori only by matching the extreme splitting. Computer subtraction permits one to deduce the lineshape and the percentage of the narrow component (Davoust *et al.*, 1979, 1980; Jost and Griffith, 1980).
2. Alternatively, one can try to fit the narrow component by using a set of reference spectra obtained with spin-labeled lipids in liposomes at different temperatures. The lineshape and percentage of the broad component can be deduced (Knowles *et al.*, 1979; Watts *et al.*, 1979).
3. A third method is based on a variational principle. By changing the conditions slightly, it is sometimes possible to modulate the ratio of the two components. Therefore, a linear combination of the experimental spectra makes it possible to decompose the spectra into the two "pure spectra" (Andersen *et al.*, 1981; Brotherus *et al.*, 1980).

In all cases it is necessary to choose somewhat arbitrary criteria to select the proper ratios. This means that appreciable error may be introduced. Definite conclusions concerning the amount of the two components are only meaningful if the two components are clearly separated in the experimental spectra.

5. SPECIFICITY OF LIPID–PROTEIN INTERACTIONS

Different approaches have been used with spin labels to investigate the specificity of lipid–protein interactions. One method relies on an estimation of the amount of motionally restricted lipids and/or trapped lipids. One pays particular attention to the broad component, which is generally easier to detect when the label is close to the binding site (supposedly in a polar region). Coupling of a fatty acid chain to various specific ligands was achieved in our laboratory (Devaux *et al.*, 1975; Brisson *et al.*, 1975; Lauquin *et al.*, 1977; Bienvenüe *et al.*, 1977) and in other laboratories (Birrell *et al.*, 1978; Brotherus *et al.*, 1980). The binding of these amphiphilic molecules was demonstrated by ESR. However, with the exception of a long-chain acyl CoA (Devaux *et al.*, 1975), these molecules are very artificial and may not give much information about the binding specificity of natural phospholipids. On the other hand, comparative studies of the amount of the motionally restricted component obtained with spin-labeled phosphatidylcholine, phosphatidylethanolamine, phosphatidylserine, or phosphatidic acid in the presence of cytochrome oxidase (Denes and Stanacev, 1978; Knowles *et al.*, 1980) or rhodopsin (Watts *et al.*, 1979) have shown that for interaction with these proteins there is a slight preference for charged phospholipids rather than zwitterionic phospholipids. The binding of natural phospholipids to intrinsic membrane proteins seems to be less specific than that to extrinsic proteins as shown by Birrell and Griffith (1976) and Stollery *et al.* (1980).

In our laboratory we have utilized a different method to measure the specificity of lipid–protein interactions with spin labels (Bienvenüe *et al.*, 1978; Davoust and Devaux, 1980) This method can be applied at 37°C, a temperature at which the motionally restricted component is practically undetectable. The method relies on a double labeling: the protein is labeled covalently with a [^{15}N] nitroxide, while a specific phospholipid is labeled with the ordinary [^{14}N]nitroxide (Fig. 4). The amount of spin–spin interaction is measured. Comparative studies should allow detection of preferential interactions, if any; however, with rhodopsin we have not been able to observe a significant difference between phosphatidylcholine, phosphatidylethanolamine, and phosphatidylserine. This confirms the prediction that the specificity is weak. The method nevertheless should be extended to other systems.

6. CONCLUSIONS

The spin-label method provides unique opportunities for the study of lipid–protein interactions in membranes for the following reasons:

1. The sensitivity of ESR permits the detection of one labeled lipid molecule anchored per protein molecule, and the mobility of the attached chain can be measured. Unambiguous differentiation between boundary and bulk lipids can therefore be achieved. A minor fraction of trapped lipids in an oligomeric protein can be detected that would not be seen by NMR.
2. The time scale of ESR seems to match exactly the domain of collision rates between lipids and protein. Thus, it is possible under appropriate conditions to distinguish between two components, whereas NMR can only average them.

Of course, refined information on the actual motion of lipid chains in contact with a protein can only be gathered with nonperturbing probes of the type used in NMR. This is meaningful only if parallel studies are undertaken on the structure, activity, and internal mobility of membrane proteins. Indeed, it should be kept in mind that the present discussion of lipid–protein interactions is based on very crude information about the proteins. Therefore, the "first-order" type of information provided by ESR is extremely valuable.

ACKNOWLEDGMENTS. This investigation was supported by research grants from the Délégation Générale à la Recherche Scientifique et Technique, the Centre National de la Recherche Scientifique (E.R.A. 690), and the University Paris VII.

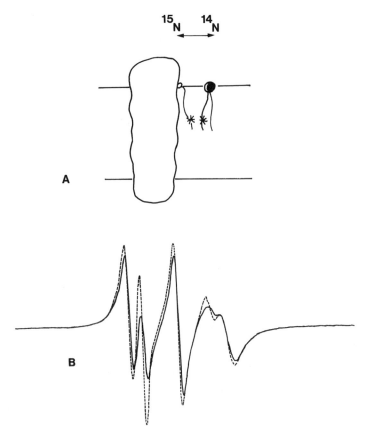

Figure 4. (A) Schematic representation of an experiment of double labeling, designed to measure the specificity of lipid–protein interactions. (B) Result of an experiment of double labeling with rhodopsin in phosphatidylcholine. Lipid-to-protein ratio, 80/1; percentage of labeled lipids, 4%; one ^{15}N label per rhodopsin. The dashed curve corresponds to computer addition of the spectra of the lipids and of the protein in the absence of interaction (Davoust and Devaux, 1980).

REFERENCES

Andersen, J. P., Fellmann, P., Møller, J. V., and Devaux, P. F. (1981). *Biochemistry* **20**, 4928–4936.

Andersen, J. P., Bienvenüe, A., Devaux, P. F., Fellmann P., Le Maire, M., Ohnishi, S. I., Sakaki, T., and Tsuji, A. (1980). 9th International Conference on Magnetic Resonance in Biological Systems (Bendor), Abstract 126.

Bienvenüe, A., Rousselet, A., Kato, G., and Devaux, P. F. (1977). *Biochemistry* **16**, 841–850.

Bienvenüe, A., Hervé, P., and Devaux, P. F. (1978). *C. R. Acad. Sci.* **287**, 1247–1250.

Birrell, G. B., and Griffith, O. H. (1976). *Biochemistry* **15**, 2925–2929.

Birrell, G. B., Sistrom, W. R., and Griffith, O. H. (1978). *Biochemistry* **17**, 3768–3773.

Boggs, J. M., Vail, W. M., and Moscarello, M. A. (1976). *Biochim. Biophys. Acta* **448**, 517–530.

Brisson, A. D., Scandella, C. J., Bienvenüe, A., Devaux, P. F., Cohen, J. B., and Changeux, J. P. (1975). *Proc. Natl. Acad. Sci. USA* **72**, 1087–1091.

Brotherus, J. R., Jost, P. C., Griffith, O. H., Keana, J. F., and Hokin, L. E. (1980). *Proc. Natl. Acad. Sci. USA* **77**, 272–276.

Chapman, D., Cornell, B. A., Eliasz, A. W., and Perry, A. (1977). *J. Mol. Biol.* **113**, 517–538.

Davoust, J., and Devaux, P. F. (1980). 9th International Conference on Magnetic Resonance in Biological Systems (Bendor), Abstract 133.

Davoust, J., Schoot, B., and Devaux, P. F. (1979). *Proc. Natl. Acad. Sci. USA* **76**, 2755–2759.

Davoust, J., Bienvenüe, A., Fellmann, P., and Devaux, P. F. (1980). *Biochim Biophys. Acta* **596,** 28–42.

Denes, A. S., and Stanacev, N. Z. (1978). *Can. J. Biochem.* **56,** 905–915.

Devaux, P. F., Scandella, C. J., and Mc Connell, H. M. (1973). *J. Magn. Reson.* **9,** 474–485.

Devaux, P. F., Bienvenüe, A., Lauquin, G. J. M., Brisson, A. D., Vignais, P. M., and Vignais, P. V. (1975). *Biochemistry* **14,** 1272–1280.

Dorst, H. J., and Schubert, D. (1979). *Hoppe-Seylers Z. Physiol. Chem.* **360,** 1605–1618.

Favre, E., Baroin, A., Bienvenüe, A., and Devaux, P. F. (1979). *Biochemistry* **18,** 1156–1162.

Fellmann, P., Andersen, J. P., Devaux, P. F., Le Maire, M., and Bienvenüe, A. (1980). *Biochem. Biophys. Res. Commun.* **95,** 289–295.

Fretten, P., Morris, S. J., Watts, A., and Marsh, D. (1980). *Biochim. Biophys. Acta* **598,** 247–259.

Hesketh, T. R., Smith, G. A., Houslay, M. D., McGill, K. A., Birdsall, N. S. H., Metcalfe, J. C., and Warren, G. B. (1976). *Biochemistry* **15,** 4145–4151.

Jost, P. C., and Griffith, O. H. (1980). *Ann. N.Y. Acad. Sci.* **348,** 391–407.

Jost, P. C., Griffith, O. H., Capaldi, R. A., and Van der Kooi, G. (1973). *Proc. Natl. Acad. Sci. USA* **70,** 480–484.

Jost, P. C., Nadakavukaren, K. K., and Griffith, O. H. (1977). *Biochemistry* **16,** 3110–3114.

Knowles, P. F., Watts, A., and Marsh, D. (1979).*Biochemistry* **18,** 4480–4487.

Knowles, P. F., Marsh, D., and Watts, A. (1980). 9th International Conference on Magnetic Resonance in Biological Systems (Bendor), Abstract 132.

Lauquin, G. J. M., Devaux, P. F., Bienvenüe, A., Villiers, C., and Vignais, P. V. (1977). *Biochemistry* **16,** 1202–1208.

Metcalfe, J. C., and Warren, G. B. (1977). *International Cell Biology* 1976–77 (Brinkley and K. Porter, eds.), pp. 15–23, Rockefeller University Press, New York.

Pontus, M., and Delmelle, M. (1975). *Biochim. Biophys. Acta* **401,** 221–230.

Stollery, J. G., Boggs, J. M., and Mascarello, M. A. (1980). *Biochemistry* **19,** 1219–1226.

Swanson, M. S., Thomas, D. D., and Quintanilha, A. T. (1980). *J. Biol. Chem.* **225,** 7494–7502.

Vanderkooi, J. M., Ierokomas, A., Nakamura, H., and Martonosi, A. (1977). *Biochemistry* **16,** 1262–1267.

Watts, A., Volstovski, I. D., and Marsh, D. (1979). *Biochemistry* **18,** 5006–5013.

Watts, A., Volstovski, I., and Marsh, D. (1980). 9th International Conference on Magnetic Resonance in Biological Systems (Bendor), Abstract 130.

17

Saturation Transfer EPR Studies of Rotational Dynamics in Membranes

David D. Thomas

1. INTRODUCTION

Beginning about 10 years ago, electron paramagnetic resonance (EPR) studies on spin-labeled lipids, pioneered by McConnell and co-workers, have helped elucidate the dynamic properties of the lipid components of both synthetic and biological membranes. While the conventional EPR technique is ideally suited to monitor the nanosecond rotational motions of fluid lipid hydrocarbon chains or small protein segments, that technique is insensitive to the slower motions (often in the microsecond range) that are typical for gel-phase lipid and entire membrane proteins. In contrast, saturation transfer EPR (ST-EPR), a technique introduced by Hyde and Dalton (1972), has its optimum sensitivity in the microsecond region (recent reviews: Thomas, 1978; Hyde and Dalton, 1979; Hyde and Thomas, 1980). Since the introduction of the ST-EPR capability in commercial spectrometers about 4 years ago, this technique has proven to be a powerful tool in the study of membrane dynamics, particularly the rotational motions of integral membrane proteins.

A major application of ST-EPR has been in the study of lipid–protein interactions. While the conventional EPR method has been used to characterize the effects of proteins on lipid motions, information on the effects of lipids on protein motions, provided by ST-EPR, is also essential. In several different membrane systems, EPR studies of both lipid and protein dynamics, accompanied by the manipulation of lipid and protein compositions, have provided new insights into lipid–protein interactions. On one level, these studies contribute to our general understanding of the physical chemistry of membranes. In addition, in studies of functional enzyme systems, correlations of EPR measurements with enzyme activity help to clarify the role of molecular dynamics in membrane function.

ST-EPR experiments involve the same types of nitroxide spin labels used in conventional EPR experiments, with nitroxide groups usually attached either to protein SH groups

David D. Thomas • Department of Biochemistry, University of Minnesota Medical School, Minneapolis, Minnesota 55455.

or to lipid hydrocarbon chains. Rotational motion is characterized by the rotational correlation time (τ_2), the characteristic time required for substantial reorientation of the spin label (Thomas, 1978). This time is determined by comparing the shape of the EPR spectrum with those obtained from computer simulations or experiments on model systems (Thomas *et al.*, 1976).

2. CALCIUM PUMP

Ca^{2+}-ATPase, the enzyme that pumps Ca^{2+} across the sarcoplasmic reticulum membrane in the presence of Mg^{2+} ATP, has been labeled with a maleimide spin label and studied by ST-EPR (Thomas and Hidalgo, 1978; Hidalgo *et al.*, 1978; Kirino *et al.*, 1978; Kaizu *et al.*, 1980). The spectrum of the spin-labeled enzyme in its endogenous lipid environment at 4°C (Thomas and Hidalgo, 1978) yielded an effective τ_2 value of 60 μsec. This result has been confirmed recently by optical techniques, as discussed elsewhere in this volume. Because this motion was stopped by cross-linking the proteins but not by immobilizing the membrane vesicles, the observed τ_2 was assigned to the overall motion of the enzyme within the membrane. This rotational motion is interesting for two reasons. It is consistent with models in which Ca^{2+} transport requires enzyme mobility, and it can be used to assess the effect of the lipid environment and other factors (such as protein aggregation) on the physical state of the enzyme. The protein mobility in sarcoplasmic reticulum has been found to increase with temperature, having a temperature dependence similar to that of enzymatic activity (Thomas and Hidalgo, 1978; Kirino *et al.*, 1978; Kaizu *et al.*, 1980). In addition, when endogenous lipid was replaced with the saturated lipid dipalmitoyl lecithin (which is much less fluid at 4°C), the protein's rotational mobility and enzymatic activity were both strongly inhibited (Hidalgo *et al.*, 1978). Similarly, both mobility and activity were inhibited by the addition of 10 mM Ca^{2+} or by partial delipidation. This correlation suggests that some degree of protein mobility is required for function, or at least that conditions restricting protein motion (e.g., rigid boundary lipid or protein aggregation) also inhibit the enzyme.

3. RHODOPSIN

Rhodopsin, the photoreceptor protein of retinal rod disc membranes, has been studied in the laboratories of Devaux and Ohnishi. In the first study, by Devaux's group, Baroin *et al.* (1977) attached a maleimide spin label to a fast-reacting SH group on rhodopsin in disc membranes and performed ST-EPR experiments. At 20°C, considerable rotational motion was observed, corresponding to an effective correlation time of 20 μsec. This motion was stopped when the proteins were cross-linked with glutaraldehyde, indicating that the observed motion was probably that of the protein as a whole. These results were in agreement with previous optical measurements, and were thus important in establishing the utility of ST-EPR in studying membrane protein motions. The EPR experiments went beyond the optical experiments, because they could be performed in both the presence and the absence of light. A brief exposure to light had no effect on the motion of rhodopsin, implying that light-induced protein aggregation (which would be expected to decrease rotational mobility) is not involved in visual transduction. Ohnishi's group (Kusumi *et al.*,

1978) obtained similar results with ST-EPR, except that they observed a slight decrease in mobility after prolonged exposure to light, as later confirmed by Devaux's group (Baroin *et al.*, 1979).

More recently, Devaux and co-workers (Baroin *et al.*, 1979; Favre *et al.*, 1979; Davoust *et al.*, 1980) have carried out studies to correlate the rotational mobility of the protein (probed with the standard maleimide spin label) with that of adjacent ("boundary") lipid chains (probed with a long-chain maleimide spin label). In disc membranes of normal composition, in which the protein is quite mobile (presumably, therefore, not aggregated), they found that the boundary-lipid probe was nearly as mobile as a probe in the bulk of the lipid. However, when the lipid content was reduced by phospholipase treatment, the protein and boundary-lipid probes were both immobilized. They concluded that the concept of a strongly immobilized boundary layer of lipid coating an intrinsic membrane protein may not be valid, at least for rhodopsin, and that the appearance of an immobilized lipid component in the presence of integral membrane proteins may, in general, be an indication that lipid has been trapped by protein aggregation.

Kusumi *et al.* (1980) carried out ST-EPR studies of MSL-rhodopsin in recombinant membrane vesicles formed from purified rhodopsin and synthetic lipid (dimyristoylphosphatidylcholine). They varied the protein-to-lipid ratio and correlated the observed protein mobility with that of spin-labeled fatty acids in the lipid phase.

4. CYTOCHROME OXIDASE

The terminal component of the electron transport chain in the mitochondrial inner membrane, cytochrome oxidase, has been the system most often cited in support of the hypothesis of an immobilized boundary layer of lipid surrounding each integral membrane protein. Swanson *et al.* (1980) carried out a study on cytochrome oxidase analogous to that performed on rhodopsin by Devaux and co-workers. In one reconstituted membrane preparation in which cytochrome oxidase was mobile (effective $\tau_2 = 40$ μsec), the boundary-lipid probe was also found to be nearly as mobile as a probe in the bulk lipid. In another preparation, with the same high enzymatic activity, both the protein and the boundary-lipid probes were substantially less mobile. These workers concluded that, for cytochrome oxidase, (1) as in the case of rhodopsin (Baroin *et al.*, 1979; Favre *et al.*, 1979), strong immobilization of lipid by protein may be due to protein aggregation, and (2) in contrast to the case of Ca^{2+}-ATPase (Hidalgo *et al.*, 1978), protein mobility does not appear to be required for enzymatic activity.

5. OTHER MEMBRANE PROTEINS

In an ST-EPR study of the spin-labeled acetylcholine receptor in endogenous membrane fragments, Rousselet and Devaux (1977) found that the protein was quite immobile, suggesting strong protein–protein interactions in these membranes. Devaux and co-workers (1978) have performed ST-EPR studies of the ADP carrier in rat heart mitochondria. Spectra were analyzed to assess the possible interaction of the ADP carriers with other proteins. Fung *et al.* (1979) have used ST-EPR to study the interactions of spectrin with actin and with membranes.

6. *SLOW LIPID MOTIONS*

In several of the above-mentioned saturation transfer studies on membrane proteins, spin labels have also been used to probe lipid chain motions. In most cases, it has been found that lipid motions are sufficiently rapid (in the nanosecond range) that conventional EPR suffices to detect them. However, ST-EPR was needed when lipids were in the gel phase (Hidalgo *et al.*, 1978) and when the membranes were so highly delipidated that the remaining lipid probes were immobilized by protein (Favre *et al.*, 1979). It is interesting to note that, in both of these cases, the ST-EPR spectra indicated that the lipid probes, although immobile on the nanosecond time scale, were more mobile than the protein probes.

In two recent studies (Marsh, 1980; Delmelle *et al.*, 1980), ST-EPR was used to study lipid model membranes in the gel phase, i.e., at temperatures below the gel-to-liquid-crystal phase transition. Both studies reported substantial hydrocarbon chain mobility in gel-phase lipid. Spectra of a dipalmitoylphosphatidylcholine (DPPC) spin label in DPPC dispersions showed a marked increase in rotational motion as the temperature increased above 25°C, a temperature that is well below the main phase transition (around 40°C) and that probably corresponds to the calorimetric "pretransition" (Marsh, 1980). Due to the well-known orientation of the nitroxide group relative to the membrane, Marsh was able to assign the spectral changes to long-axis rotation of the lipid probe. Delmelle *et al.* (1980) also studied DPPC in the gel phase, including studies on the effect of cholesterol, and recording spectra on oriented membrane as well as dispersions.

7. *THEORETICAL ANALYSIS*

As pointed out in many of the papers cited above, quantitative analysis of the spectra in terms of detailed motional models has been limited by the lack of theoretical simulations of the effects of anisotropic motion on ST-EPR spectra. These theoretical studies are in progress (Robinson and Dalton, 1980). Spectra are particularly difficult to analyze when both conventional and saturation transfer spectra show evidence of motion, as is almost always the case with lipid probes. In these cases, although ST-EPR does not usually yield an unambiguous description of molecular behavior, it can be used as a sensitive probe of changes or differences in molecular dynamics. For example, ST-EPR has been used to detect differences between the cell membranes of normal and dystrophic humans, under conditions where conventional EPR was insensitive to these changes (Wilkerson *et al.*, 1978; Swift *et al.*, 1980).

REFERENCES

Baroin, A., Thomas, D. D., Osborne, R., and Devaux, P. F. (1977). *Biochem. Biophys. Res. Commun.* **78**, 442–449.

Baroin, A., Bienvenüe, A., and Devaux, P. F. (1979). *Biochemistry* **18**, 1151–1155.

Davoust, J., Bienvenüe, A., Fellmann, P., and Devaux, P. F. (1980). *Biochim. Biophys. Acta* **596**, 28–42.

Delmelle, M., Butler, K. W., and Smith, I. C. P. (1980). *Biochemistry* **19**, 698–704.

Devaux, P. F., Baroin, A., Bienvenüe, A., Favre, E., Rousselet, A., and Thomas, D. D. (1978). In *Bioenergetics of Membranes* (L. Packer, ed.), pp. 47–54, Academic Press, New York.

Favre, E., Baroin, A., Bienvenüe, A., and Devaux, P. F. (1979). *Biochemistry* **18**, 1156–1162.

Fung, L. W.-M., Soo Hoo, M. J., and Meena, W. A. (1979). *FEBS Lett.* **105**, 379–383.

Hidalgo, C., Thomas, D. D., and Ikemoto, N. (1978). *J. Biol. Chem.* **253**, 6879–6887.

Hyde, J. S., and Dalton, L. R. (1972). *Chem. Phys. Lett.* **16**, 568–572.

Hyde, J. S., and Dalton, L. R. (1979). In *Spin Labeling II* (L. Berlina, ed.), pp. 1–70, Academic Press, New York.

Hyde, J. S., and Thomas, D. D. (1980). *Annu. Rev. Phys. Chem.* **31**, 293–317.

Kaizu, T., Kirino, Y., and Shimizu, H. (1980). *J. Biochem. (Tokyo)* **88**, 1837–1843.

Kirino, Y., Ohkuma, T., and Shimizu, H. (1978). *J. Biochem. (Tokyo)* **84**, 111–115.

Kusumi, A., Ohnishi, S., Ito, T., and Yoshizawa, T. (1978). *Biochim. Biophys. Acta* **507**, 539–543.

Kusimi, A., Sakaki, T., Yoshizawa, T., and Ohnishi, S. (1980). *J. Biochem. (Tokyo)* **88**, 1103–1111.

Marsh, D. (1980). *Biochemistry* **19**, 1632–1637.

Robinson, B. H., and Dalton, L. R. (1980). *J. Chem. Phys.* **72**, 1312–1324.

Rousselet, A., and Devaux, P. (1977). *Biochem. Biophys. Res. Commun.* **78**, 448–454.

Swanson, M. S., Quintanilha, A. T., and Thomas, D. D. (1980). *J. Biol. Chem.* **255**, 7494–7502.

Swift, L. L., Atkinson, J. B., Perkins, R. C., Jr., Dalton, L. R., and Le Quire, V. S. (1980). *J. Membr. Biol.* **52**, 165–172.

Thomas, D. D. (1978). *Biophys. J.* **24**, 211–223.

Thomas, D. D., and Hidalgo, C. (1978). *Proc. Natl. Acad. Sci. USA* **75**, 5488–5492.

Thomas, D. D., Dalton, L. R., and Hyde, J. S. (1976). *J. Chem. Phys.* **65**, 3006–3024.

Wilkerson, L. S., Perkins, R. C., Jr., Roelofs, R., Swift, L., Dalton, L. R., and Park, J. H. (1978). *Proc. Natl. Acad. Sci. USA* **75**, 838–841.

18

Lateral Diffusion of Membrane Proteins

Michael Edidin

Lateral diffusion of membrane proteins has been determined for many proteins (for reviews see Edidin, 1974; Cherry, 1979; Edidin, 1981), both of the plasma membrane and of internal membranes. The initial demonstration of diffusion, by following intermixing of surface antigens in newly formed heterokaryons, was essentially qualitative (Frye and Edidin, 1970). Solution of the diffusion equation for interdiffusion of components on two hemispheres (Huang, 1973) allowed quantitative estimations on cultured fibroblasts (Edidin and Wei, 1977a) and on labeled proteins of erythrocytes (Fowler and Branton, 1977).

Determination of lateral diffusion in heterokaryons was never widely used and has been superceded in recent years by the technique of fluorescence photobleaching and recovery (FPR) (Edidin et al., 1976; Jacobson et al., 1976; Schlessinger et al., 1976). In contrast to the heterokaryon approach, which is a cell population method for estimating diffusion, the FPR technique is a single-cell method. In it, a spot is bleached out of a uniformly labeled cell surface, and diffusion is monitored in terms of the time required for return of label to the bleached spot (Axelrod et al., 1976). FPR is a generalization of a technique first used by Poo and Cone (1974) to demonstrate lateral diffusion of rhodopsin in disc membranes of rod outer segments; measuring the return of unbleached rhodopsin by changes in absorbance of the bleached region. The method was generalized and made more sensitive by Peters and co-workers (1974) who used an extrinsic fluorophore rather than an intrinsic chromophore as the surface label. A further generalization was then developed by the three groups cited above, using labeled antibody fragments and labeled lectins.

Two main issues arose with the initial demonstration of protein lateral diffusion and have persisted to the present time. The first issue is the role of lateral diffusion in membrane function. Though the topic will not be further pursued here, we should note that lateral diffusion, effectively in two dimensions, serves to concentrate reactants and hence to increase the frequency of their interaction (Adam and Delbrück, 1969). Rate of lateral diffusion appears to be important in the reaction of cytochrome b_5 reductase and of monooxygenase in endoplasmic reticulum (Strittmatter and Rogers, 1975; Yang, 1977), and in the

Michael Edidin • Department of Biology, The Johns Hopkins University, Baltimore, Maryland 21218.

activation of adenylate cyclase by hormone–receptor complexes (Hanski *et al.*, 1979). Control of such diffusion-coupled reactions could be effected, at least in part, by control of the rate of lateral diffusion.

The control of diffusion rates is the second main issue raised by the values of diffusion coefficients. Though three-dimensional diffusion theory may not be strictly applicable to lateral diffusion in membranes (Saffman and Delbrück, 1975), it seems a good first approximation, and it indicates that a protein of 10–100,000 daltons should diffuse, given the apparent viscosity of membrane lipids (around 1 poise), at a rate $D \sim 3 \times 10^{-9}$ cm^2 sec^{-1}. Indeed, this value was found for rhodopsin by Poo and Cone (1974). However, most other values of D_{lat} for proteins have been at least an order of magnitude lower than this, ranging from 3×10^{-10} cm^2 sec^{-1} for peptide hormone and IgE receptors to $< 3 \times 10^{-12}$ cm^2 sec^{-1} for acetylcholine receptors in patches on cultured myotubes. As the FPR method uses a brief pulse of intense laser light to bleach the spot of interest, it has been suggested (Lepock *et al.*, 1978; Sheetz and Koppel, 1979; Bretscher, 1980) that the unexpectedly low values obtained by this method are due to photodamage, either heating or cross-linking of labeled proteins. Control experiments appear to rule out these possibilities. Axelrod (1977) calculated that the laser powers used raise the local temperature in the spot less than 1°C. Wolf *et al.* (1980) showed that the same values for D were found before and after half of all the label on a cell surface was bleached by continuous high levels of light. Finally, Wey *et al.* (1981) repeated the work of Poo and Cone, but using rhodamine-labeled rhodopsin, and obtained $D = 3.0 \pm 1.2 \times 10^{-9}$ cm^2 sec^{-1} by FPR as compared with $D = 3.5 \pm 1.5 \times 10^{-9}$ cm^2 sec^{-1} in the original work.

If the unexpectedly low coefficients of diffusion for proteins are not artifacts of the method of measurement, then it appears that most cells restrict free lateral diffusion in their surfaces. This restriction may be modulated (see for example Edelman, 1976; Edidin and Wei, 1977b). Such modulation would affect the rates of diffusion-mediated reactions and could serve to control these reactions. Furthermore, extremes of restriction on lateral diffusion appear to be found in epithelial cells and many other tissues in the body, for example, muscle, liver, and sperm. Cells of all of these tissues have polarized distributions of surface enzymes and receptors (references in Ziomek *et al.*, 1980). Though it is possible that the mechanism of restriction of diffusion differs in these tissue cells from the mechanisms effective in cultured cells, I believe this is unlikely.

Restriction of lateral diffusion could be due to interaction of membrane integral proteins with each other, or with membrane lipids, with extracellular coats, or with elements of the cytoplasm. The first mechanism offers some possibility of restriction. Fluid lipid viscosity may change by a factor of two or three when acyl chain composition or temperature is manipulated; transient associations of membrane proteins with one another probably are similarly effective in slowing diffusion. Lateral diffusion of bacteriorhodopsin or steroyl dextran depends on the concentration of the diffusing species incorporated into a bilayer (Cherry *et al.*, 1977; Wolf *et al.*, 1977). Neither effect can account for the 1000-fold range of measured diffusion coefficients.

Cell–cell associations appear to restrict lateral motion of acetylcholine receptors of muscle cells (Anderson and Cohen, 1977), and in general one might expect localization of ligand and receptor molecules at points where two cells adhere, or where cells adhere to surfaces. [See Michl *et al.* (1979) for behavior of macrophage F$_c$ receptors when the cells are plated on a surface of antibody–antigen complexes.] In a single experiment a protein of extracellular matrix, fibronectin, had no effect on diffusion of an integral membrane problem (Schlessinger *et al.*, 1977).

Several approaches indicate that interaction of membrane integral proteins with the cytoskeleton regulates their diffusion. Treatment of cells with cytochalasin B, which, among other effects, blocks polymerization of actin filaments, inhibits lateral diffusion measured by FPR (Schlessinger *et al.*, 1976) and in heterokaryons (Edidin and Wei, 1977a). The effect, though not expected from the studies cited, at least may indicate that cytochalasin-sensitive structures interact with membrane proteins. On the other hand, cytochalasin B interferes with glucose transport. Its effect on lateral diffusion could be exerted on cell metabolism, rather than specifically on microfilaments. [A brief review of the pharmacology of cytochalasins is given by Bray (1979).]

Direct isolation of complexes between cytoplasmic filaments and other elements of the cytoskeleton has been achieved for surface immunoglobulin of lymphocytes and for H-2 antigens of tumor cells (Flannagan and Koch, 1978; Koch and Smith, 1978). H-2/actin complexes were shed from unperturbed cells; immunoglobulin/actin complexes formed only after cross-linking of the surface Ig. Koch (1980) has developed these observations to suggest that integral membrane proteins interact weakly with microfilaments, and that cross-linking of these proteins creates complexes of higher affinities that can then anchor to the cytoplasm. This model, suggested earlier by dePetris and Raff (1973) and developed as well by Gershon (1979), is consistent with the observation that capped membrane proteins and protein models, stearoyl dextrans, do not freely diffuse in the time scale of FPR experiments (Dragsten *et al.*, 1979; Wolf *et al.*, 1977).

The model also parallels recent models of the association of spectrin and integral membrane proteins in erythrocytes. Bennett and Stenbuck (1979a) found that about 50% of the erythrocyte cytoskeletal protein spectrin associates specifically with integral protein, 2.1, of the red cell membrane. In turn, 15% of band 3, the major transmembrane glycoprotein of the erythrocyte, is associated with 2.1 (Bennett and Stenbuck, 1979b); thus, at least some of the transmembrane proteins are linked, indirectly, to the red cell cytoskeleton. Bennett and Stenbuck (1979a) are careful to point out that their method only isolates high-affinity complexes. They calculate that while they cannot detect associations with $K_D > 10^{-6}$ M, the local concentration of spectrin beneath the erythrocyte membrane is around 10^{-5} M. Thus, further restraints on lateral diffusion of band 3 could be achieved by direct interaction with spectrin. Lateral diffusion of fluorescein-labeled band 3 gives $D = 3 \times 10^{-11}$ cm^2 sec^{-1} at 30°C for freshly prepared ghosts (Fowler and Branton, 1977). If ghosts, while leaky, are treated with a fragment of 2.1 that dissociates spectrin from the membrane, diffusion of band 3 increases about 1.5-fold (Fowler and Bennett, 1978). This small effect is rather less than expected from the amount of spectrin removed, but is consistent with the observation that removal of spectrin has no effect on rotational diffusion of band 3 (Cherry *et al.*, 1976). Recently, diffusion of labeled band 3 was compared in normal and spherocytic mouse erythrocytes (Sheetz *et al.*, 1980). The spherocytes lack a cytoskeleton. Diffusion of band 3 in these cells was approximately 100-fold faster than in normal cells or reticulocytes from normal animals. Thus, a lesion of the red cell cytoskeleton massively affects lateral diffusion rates.

The interaction of cytoskeletal proteins and membrane proteins may require cell metabolism, continued synthesis of ATP. The clusters of acetylcholine receptors that form on cultured myotubes are dispersed when cells are treated with uncouplers and other metabolic poisons (Bloch, 1979), and treatment of isolated mouse intestinal epithelial cells with similar agents enhances the rate at which membrane enzymes redistribute in these cells (Ziomek *et al.*, 1980). As yet we cannot specify where and how ATP would be required in the regulation of lateral diffusion. One suspects that the answers will not be long in coming.

REFERENCES

Adam, G. M., and Delbrück, M. (1969). In *Structural Chemistry and Molecular Biology* (A. Rich and E. N. Davidson, eds.), pp. 198–215, Freeman, San Francisco.

Anderson, M. J., and Cohen, M. W. (1977). *J. Physiol. (London)* **268**, 757–773.

Axelrod, D. (1977). *Biophys. J.* **18**, 129–131.

Axelrod, D., Koppel, D. E., Schlessinger, J., Elson, E., and Webb, W. W. (1976). *Biophys. J.* **16**, 1055–1069.

Bennett, V., and Stenbuck, P. J. (1979a). *J. Biol. Chem.* **254**, 2533–2541.

Bennett, V., and Stenbuck, P. J. (1979b). *Nature (London)* **280**, 468–473.

Bloch, R. J. (1979). *J. Cell Biol.* **82**, 626–643.

Bray, D. (1979). *Nature (London)* **282**, 671.

Bretscher, M. S. (1980). *Trend Biochem. Sci.* October, VI–VII.

Cherry, R. J. (1979). *Biochim. Biophys. Acta* **559**, 289–327.

Cherry, R. J., Bürkli, A., Busslinger, M., Schneider, G., and Parrish, G. (1976). *Nature (London)* **263**, 389–393.

Cherry, R. J., Müller, U., and Schneider, G. (1977). *FEBS Lett.* **80**, 465–469.

dePetris, S., and Raff, M. C. (1973). In *Locomotion of Tissue Cells* (Ciba Foundation Symposium, new series 14), pp. 27–41, Elsevier, Amsterdam.

Dragsten, P., Henkart, P., Blumenthal, R., Weinstein, J., and Schlessinger, J. (1979). *Proc. Natl. Acad. Sci. USA* **76**, 5163.

Edelman, G. M. (1976). *Science* **192**, 218–226.

Edidin, M. (1974). *Annu. Rev. Biophys. Bioeng.* **3**, 179–201.

Edidin, M. (1981). In *New Comprehensive Biochemistry* (J. B. Finean and R. H. Michell, eds.), pp. 37–82, Elsevier/North-Holland, Amsterdam.

Edidin, M., and Wei, T. (1977a). *J. Cell Biol.* **75**, 475–482.

Edidin, M., and Wei, T. (1977b). *J. Cell Biol.* **75**, 483–489.

Edidin, M., Zagyansky, Y., and Lardner, T. J. (1976). *Science* **191**, 466–468.

Flannagan, J., and Koch, G. L. E. (1978). *Nature (London)* **273**, 278–281.

Fowler, V., and Bennet, V. (1978). *J. Supramol. Struct.* **8**, 215–221.

Fowler, V., and Branton, D. (1977). *Nature (London)* **268**, 23–26.

Frye, L. D., and Edidin, M. (1970). *J. Cell Sci.* **7**, 319–335.

Gershon, N. D. (1979). In *Physical Chemical Aspects of Cell Surface Events in Cellular Regulation* (C. DeLisi and R. Blumenthal, eds.), pp. 163–165, Elsevier/North-Holland, Amsterdam.

Hanski, E., Rimon, G., and Levitski, A. (1979). *Biochemistry* **18**, 846–853.

Huang, H. W. (1973). *J. Theor. Biol.* **40**, 11–17.

Jacobson, K., Wu, E. S., and Poste, G. (1976). *Biochim. Biophys. Acta* **433**, 215–222.

Koch, G. L. E. (1980). In *Cell Adhesion and Motility* (A. S. G. Curtis and J. D. Pits, eds.), pp. 425–444, Cambridge University Press, London.

Koch, G. L. E., and Smith, M. J. (1978). *Nature (London)* **273**, 273–278.

Lepock, J. R., Thompson, J. E., Kruuv, J., and Wallach, D. F. H. (1978). *Biochem. Biophys. Res. Commun.* **85**, 344–350.

Michl, J., Unkleless, J. C., Pieczonka, M. M., and Silverstein, S. C. (1979). *J. Cell Biol.* **83**, 295a.

Peters, R., Peters, J., Tews, R. H., and Bahr, W. (1974). *Biochim. Biophys. Acta* **367**, 282–294.

Poo, M.-M., and Cone, R. (1974). *Nature (London)* **247**, 438–441.

Saffman, P. G., and Delbrück, M. (1975). *Proc. Natl. Acad. Sci. USA* **72**, 3111–3113.

Schlessinger, J., Koppel, D. E., Axelrod, D., Jacobson, R., and Webb, W. W. (1976). *Proc. Natl. Acad. Sci. USA* **73**, 2409–2413.

Schlessinger, J., Barak, L. S., Hammes, G. G., Yamada, R. M., Pastan, I., Webb, W. W., and Elson, E. L. (1977). *Proc. Natl. Acad. Sci. USA* **75**, 5353–5357.

Sheetz, M. P., and Koppel, D. E. (1979). *Proc. Natl. Acad. Sci. USA* **76**, 3314–3317.

Sheetz, M. P., Schnidler, M., and Koppel, D. E. (1980). *Nature (London)* **285**, 510–512.

Strittmatter, P., and Rogers, M. (1975). *Proc. Natl. Acad. Sci. USA* **72**, 2650–2661.

Wey, C. L., Cone, R., and Edidin, M. (1981). *Biophys. J.* **33**, 225–232.

Wolf, D. E., Schlessinger, J., Elson, E. L., Webb, W. W., Blumenthal, R., and Henhart, P. (1977). *Biochemistry* **16**, 3476–3483.

Wolf, D. E., Edidin, M., and Dragsten, P. R. (1980). *Proc. Natl. Acad. Sci. USA* **77**, 2043–2045.

Yang, C. S. (1977). *Life Sci.* **21**, 1047–1057.

Ziomek, C. A., Schulman, S., and Edidin, M. (1980). *J. Cell Biol.* **86**, 849–857.

19

Rotational Diffusion of Membrane Proteins

Time-Resolved Optical Spectroscopic Measurements

Richard J. Cherry

In recent years, much effort has been directed toward measuring the mobility of proteins in membranes. These measurements have received considerable impetus from the growing awareness that the movements of membrane proteins are likely to have functional significance. In addition, diffusion of an individual protein is sensitive to interactions with other membrane components. Thus, mobility measurements provide a powerful method for directly elucidating structural interactions.

Techniques that have been developed for measuring both lateral and rotational motion of membrane proteins have been reviewed in detail (Cherry, 1979). The present article is concerned specifically with the rotational mobility of membrane proteins as measured by optical spectroscopy. Emphasis is placed on the latest developments in this area; the above review should be consulted for a more detailed description of earlier work and of methodology.

1. PRINCIPLES OF ROTATIONAL DIFFUSION MEASUREMENTS

Spectroscopic methods of measuring rotational motion depend on photoselection of an oriented population of excited molecules by linearly polarized light. Following flash excitation, the initial anisotropic distribution decays as molecules become randomized by Brownian rotation. The anisotropy decay may be measured by detecting either dichroism of absorption signals or polarization of emission signals.

Richard J. Cherry • Eidgenössische Technische Hochschule, Laboratorium für Biochemie, ETH-Zentrum, CH-8092 Zurich, Switzerland.

In order to measure the relatively slow rotation of membrane proteins, it is necessary to detect signals of long lifetime. Suitable signals can be obtained from a few intrinsic chromophores or, more generally, from the triplet state of a probe molecule. Triplet probes currently in use include various derivatives of eosin (Cherry and Schneider, 1976) and erythrosin (Moore *et al.*, 1979). Most measurements have so far been performed by measuring transient dichroism of absorption signals. In the case of triplet probes, the possibility of obtaining increased sensitivity by measuring phosphorescence depolarization (Moore *et al.*, 1979; Austin *et al.*, 1979) or delayed fluorescence depolarization (Greinert *et al.*, 1979) has recently been explored.

A flash photolysis apparatus for measuring transient dichroism has been described in detail elsewhere (Cherry, 1978). Experimental methods for measuring phosphorescence depolarization are discussed by Garland (this volume). In either case the data are analyzed by calculating the anisotropy $r(t)$, which for absorption signals is given by

$$r(t) = \frac{A_{\parallel}(t) - A_{\perp}(t)}{A_{\parallel}(t) + 2A_{\perp}(t)} \tag{1}$$

where $A_{\parallel}(t)$, $A_{\perp}(t)$ are the absorption changes at time t after excitation for light polarized parallel and perpendicular with respect to the polarization of the exciting flash.

The form of $r(t)$ depends on the type of motion present. A possible model for interpreting data obtained with membrane proteins is one in which the protein can only rotate around the membrane normal. The anisotropy is then predicted to decay according to the equation

$$r(t) = \left[\frac{r_0}{A_1 + A_2 + A_3} \right] [A_1 \exp(-t/\phi_{\parallel}) + A_2 \exp(-4t/\phi_{\parallel}) + A_3] \tag{2}$$

where $A_1 = (6/5)\sin^2\theta \cos^2\theta$; $A_2 = (3/10)\sin^4\theta$; $A_3 = (1/10)(3 \cos^2\theta - 1)^2$. θ is the angle between the transition dipole moment and the membrane normal, $\phi_{\parallel}$ the relaxation time about the membrane normal, and r_0 the experimental anisotropy at time zero. It is assumed that the transition dipole moments for excitation and measurement are parallel; otherwise, both orientations appear in the coefficients A_1, A_2, and A_3. Other aspects of the analysis and qualifications to the use of equation (2) are discussed in more detail elsewhere (Cherry, 1979).

2. BACTERIORHODOPSIN

Bacteriorhodopsin (BR) contains an intrinsic chromophore, retinal, which in the light-adapted state is in the all-*trans* form. Excitation of BR produces a complex cycle of spectroscopic intermediates (for review see Stoeckenius *et al.*, 1979). Rotation is usually measured by observing transient dichroism of ground-state depletion signals at 570 nm.

In the native purple membrane, BR is immobilized (Razi Naqvi *et al.*, 1973), as expected from the existence of a crystalline lattice. However, rotation of BR is observed when the protein is reconstituted into lipid vesicles (Cherry *et al.*, 1978). These BR–lipid vesicles provide a simple system that is of great value for testing theoretical models.

Figure 1 shows a typical anisotropy decay curve for BR in reconstituted dimyristoylphosphatidylcholine vesicles above the temperature of the lipid gel-to-liquid-crystal phase transition (T_c). The ability of equation (2) to fit the experimental data has recently been

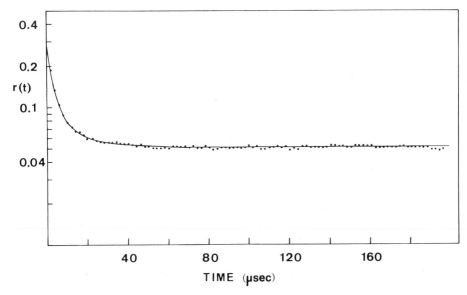

Figure 1. Anisotropy decay curve for BR in dimyristoylphosphatidylcholine vesicles (L/P = 220) at 28°C. The solid line is the best fit of equation (2) to the experimental points with the parameters $r_0 = 0.28$, $\phi_\parallel = 12.6$ μsec, and $\theta = 77°$.

tested by curve fitting with $\phi_\parallel$, θ, and r_0 as adjustable parameters (Cherry and Godfrey, 1981). It was found that a good fit is obtained provided the L/P (phospholipid/protein mole ratio) is greater than about 100 and the temperature is sufficiently above the T_c. The relaxation time $\phi_\parallel$ appeared to exhibit a weak dependence on temperature and on L/P. However, values of $\phi_\parallel$ measured between 25 and 37°C over the L/P range 140–250 all fell within the range of 15 ± 5 μsec.

The angle θ obtained from the curve fitting was $78 \pm 2°$. In an earlier study (Heyn *et al.*, 1977), θ was obtained only from the ratio $A_3/(A_1 + A_2 + A_3)$. In this case, in addition to the 78° solution, an angle of about 38° is also consistent with the experimental data. Recently, Hoffmann *et al.* (1980a) have obtained a somewhat higher value of the ratio $A_3/(A_1 + A_2 + A_3)$, occurring in the range where only a single solution of about 30° is possible. Although the reason for this discrepancy is not clear, it is more probable that 78° is the correct angle, for a better fit to the whole decay curve is obtained with this angle (Cherry and Godfrey, 1981). Moreover, it is much closer to the value of $\sim 70°$ independently determined from polarized absorption spectra of oriented purple membranes (Heyn *et al.*, 1977; Stoeckenius *et al.*, 1979; Acuna and Gonzalez-Rodriguez, 1979).

At sufficiently high L/P and temperature, the ability of equation (2) to fit the experimental data implies the existence of a single rotating species. From circular dichroism measurements, it is known that this species is the BR monomer (Heyn *et al.*, 1975). On cooling below the T_c, BR aggregates into a lattice and becomes immobilized (Cherry *et al.*, 1978).

3. BAND 3 PROTEINS

Rotational mobility of band 3, the anion transport protein of the human erythrocyte membrane, has been measured using eosin triplet probes (Cherry *et al.*, 1976; Austin *et*

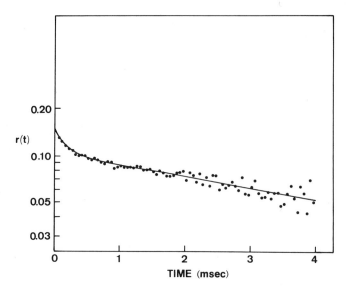

Figure 2. Anisotropy decay curve for eosin maleimide-labeled band 3 in human erythrocyte membranes at 37°C. Ghosts suspended in 5 mM phosphate buffer, pH 7.4. (Reproduced from Nigg and Cherry, 1979a.)

al., 1979). Figure 2 shows the anisotropy decay curve obtained with eosin maleimide-labeled band 3 in erythrocyte ghosts at 37°C. It is clear that the decay does not have the form predicted by equation (2) and found experimentally with BR. The difference can be explained by the existence of different populations of band 3 with different rotational mobilities (Nigg and Cherry, 1979a). Recently, evidence has been obtained that about 40% of band 3 has a restricted rotational mobility due to interaction with the erythrocyte cytoskeleton (Nigg and Cherry, 1980).

The above restriction can be removed by cleaving off the 40,000-dalton cytoplasmic segment of band 3 with low concentrations of trypsin. In these trypsinized membranes, the anisotropy decay curve at 45°C is qualitatively similar to that obtained with BR in lipid vesicles. As there is some evidence for a specific eosin-binding site on band 3 (Nigg *et al.*, 1979), it may be valid to fit the data by equation (2). This procedure yields a relaxation time of 130 μsec at 45°C.

Reducing the temperature of both trypsin-treated and untreated membranes produces a progressive loss of band 3 rotational mobility (Nigg and Cherry, 1979a, 1980). Below 20°C, most of band 3 is relatively immobile. The simplest explanation of this phenomenon is a temperature-dependent self-aggregation of band 3.

Self-association of band 3 into dimers was investigated by measuring band 3 rotation before and after cross-linking. In this way it was established that dimers preexist in the membrane and are not a product of the cross-linking reaction (Nigg and Cherry, 1979b).

Band 3 is immobilized when antibodies specifically directed against glycophorin A are bound to the membrane (Nigg *et al.*, 1980a). A similar effect was found with influenza virus particles, which also bind specifically to glycophorin (Nigg *et al.*, 1980b). Hence, it was deduced that band 3 forms a complex with glycophorin A in the erythrocyte membrane. However, glycophorin A has little influence on band 3 mobility, for very similar rotational motion was observed in En(a−) erythrocyte membranes, which completely lack glycophorin A (Nigg *et al.*, 1980c).

4. Ca²⁺,Mg²⁺–DEPENDENT ATPase IN SARCOPLASMIC RETICULUM

Measurements of the transient dichroism of iodoacetamidoeosin-labeled Ca^{2+}-ATPase in sarcoplasmic reticulum membranes reveal some interesting differences compared with results obtained with other membrane proteins. The anisotropy measured shortly after flash excitation is remarkably low at physiological temperature but increases upon reducing the temperature or fixing with glutaraldehyde (Bürkli and Cherry, 1981). These results indicate the existence of a motion that is faster than the resolution time of the experiments (20 μsec). It was argued that this fast motion corresponds in part to independent segmental motion of part of the enzyme containing the probe. Garland and Boxer (personal communication) have directly observed this fast motion by measuring, with improved time resolution, phosphorescence depolarization of an erythrosin probe.

Hoffmann *et al.* (1979) have also investigated rotational motion of the Ca^{2+}-ATPase using eosin isothiocyanate as the probe. Their data indicate a change in the temperature dependence of rotational motion at about 15°C, where Arrhenius plots of the ATPase activity also exhibit a change of slope. Such a correlation has also been observed in saturation transfer ESR experiments (Thomas and Hidalgo, 1978; Kirino *et al.*, 1978). However, the existence of internal flexibility of the Ca^{2+}-ATPase would considerably complicate the interpretation of these experiments.

Hoffmann *et al.* (1980b) have also studied the rotation of Ca^{2+}-ATPase reconstituted into dipalmitoylphosphatidylcholine vesicles. They found that the temperature at which the protein becomes immobilized is about 10°C below the T_c. This is comparable with results obtained with BR, which also becomes immobilized well below the T_c in dimyristoylphosphatidylcholine vesicles (Heyn *et al.*, 1981). In both cases, immobilization appears to result from a lipid–protein phase segregation followed by self-aggregation of the membrane proteins.

5. CYTOCHROME OXIDASE

Photodissociation of the heme a_3–CO complex in cytochrome oxidase provides the necessary long-lived spectroscopic change for measuring protein rotation. The $r(t)$ curve measured with cytochrome oxidase reconstituted into lipid vesicles is qualitatively similar to that obtained with BR (Kawato *et al.*, 1981). The theoretical expression for $r(t)$ for a circularly symmetric chromophore rigidly bound to a protein rotating only around the membrane normal has been derived by Kawato and Kinosita (1981). As with a single transition dipole moment, the ratio of the constant residual anisotropy r_∞ to r_0 is a function of the chromophore orientation. ESR and optical studies with oriented membranes indicate that the plane of heme a_3 is approximately perpendicular to the plane of the membrane (Blasie *et al.*, 1978). For this orientation, r_∞/r_0 is predicted to be 0.25. The experimental value of 0.28 ± 0.02 obtained by Kawato *et al.*, (1981) at high L/P confirms the perpendicular orientation of heme a_3 and demonstrates that in the reconstituted system all the cytochrome oxidase molecules are free to rotate over a time of 2 msec. Analysis of the decaying part of the $r(t)$ curve suggested the existence of different-sized small aggregates of cytochrome oxidase rotating with a mean relaxation time of about 500 μsec. Decreasing L/P resulted in higher values of r_∞/r_0, presumably due to self-association of a fraction of cytochrome oxidase into relatively immobile aggregates. Co-reconstitution of cytochrome bc_1 had no observable effect on the rotation of cytochrome oxidase, either in the presence or in the absence of cytochrome c.

Initial experiments with mitochondrial membranes were performed by Junge and De-vault (1975). They failed to detect any decay in dichroism for cytochrome oxidase in either pigeon heart mitochondria or beef heart submitochondrial particles. However, in further studies, rotational motion of cytochrome oxidase in both beef heart mitochondria and sub-mitochrondrial particles has been detected (Kawato *et al.* 1980). Although the shape of the $r(t)$ curve is qualitatively similar to that measured in reconstituted vesicles, an important difference is that r_∞/r_0 is about 0.5–0.6 in mitochondria. This indicates that about half of the cytochrome oxidase is relatively immobile, presumably due to protein–protein interactions. The remaining cytochrome oxidase rotates with a relaxation time in the order of 300 μsec. Further elucidation of interactions involving cytochrome oxidase will clearly be of interest in relation to mechanisms of electron transport.

6. OTHER CYTOCHROMES

Rotation of other cytochromes can also be studied by photodissociation of heme–CO. Richter *et al.* (1979) have observed rotational motion of cytochrome P-450 in rat liver microsomal membranes. Further experiments with rabbit liver microsomes suggest that the fraction of mobile cytochrome P-450 may vary according to species and the method of induction (McIntosh *et al.* 1980). Garland *et al.* (1979) found no decay of flash-induced transient dicroism for cytochrome a_1 in *Thiobacillus ferro-oxidans*, indicating immobilization. The same authors were unable to detect dichroism of cytochrome o in *E. coli*, suggesting that a rather rapid motion may be present.

Vaz *et al.* (1979) replaced the native heme group of cytochrome b_5 by rhodium(III)-protoporphyrin IX, which has a measurable triplet yield. From transient dichroism studies, they were able to investigate rotation of this modified cytochrome in dimyristoylphosphatidylcholine vesicles. The rather rapid anisotropy decays that they observed (0.4 μsec at 35°C, 9 μsec at 10°C) were interpreted in terms of a "wobbling in cone" model (Kinosita *et al.* 1977).

7. OTHER PROTEINS

The first measurement of rotation of a membrane protein was made with rhodopsin in isolated rod outer segments (Cone, 1972). From the decay of dichroism of absorption changes arising from excitation of the intrinsic retinal chromophore, a relaxation time of 20 μsec at 20°C was determined.

Lo *et al.* (1980) bound erythrosin-labeled α-bungarotoxin to acetylcholine receptors of *Torpedo* electric organ membranes. From phosphorescence depolarization measurements, they concluded that the receptor is immobilized, presumably due to the extensive protein–protein interactions that are thought to occur in this membrane.

The motion of eosin-labeled concanavalin A bound to Friend cells has been studied by Austin *et al.* (1979). They found no decay of phosphorescence emission anisotropy over the time range 1 μsec–4 msec, although the low values of anisotropy obtained indicated a rapid motion to be present.

Wagner and Junge (1980) have labeled the coupling factor CF_1 with eosin isothio-cyanate and reconstituted it into CF_1-depleted chloroplasts. From transient dichroism measurements, they obtained evidence for an increased rotational mobility of CF_1 upon simultaneous illumination and addition of ADP and P_i. A marked decrease in eosin triplet lifetime, interpreted as an "opening" of CF_1, was observed upon energization with a light-generated electrochemical potential difference.

Strambini and Galley (1980) have observed time-resolved phosphorescence depolarization from tryptophan residues of water-soluble proteins in viscous solution. This approach could find application to selected membrane proteins in reconstituted systems.

8. CONCLUDING REMARKS

The structural information that can be obtained from protein rotation measurements is well illustrated by the experiments with band 3 in the human erythrocyte membrane. These studies demonstrate the power of rotational diffusion measurements for investigating self-association of integral proteins, interactions between different integral proteins, and interactions between integral proteins and cytoskeletal structures. Functional significance of protein diffusion is most likely to be found in lateral motion. Here it is important to distinguish between lateral displacements over large distance ($> 1\mu m$) and local lateral diffusion. Long-range lateral diffusion may be measured by fluorescent photobleaching recovery (Axelrod *et al.*, 1976), while a local lateral diffusion coefficient can, with certain assumptions, be calculated from the rotational relaxation time. The validity of these assumptions is currently being tested. The local lateral diffusion coefficient is likely to be particularly important in reactions requiring collisions between membrane components.

Finally, the advantages of combining measurements of rotational motion using triplet probes with time-resolved fluorescence depolarization have recently been stressed (Cherry *et al.*, 1980). In this way, it is possible to measure rotational motion over the enormous time range of picoseconds to milliseconds, and hence to build up a detailed picture of the different motions occurring in a given system.

ACKNOWLEDGMENTS. I thank Professor P. B. Garland for communicating data prior to publication and the Swiss National Science Foundation for financial support.

REFERENCES

Acuna, A. U., and Gonzalez-Rodriguez, J. (1979). *An. Quim.* **75**, 630–635.
Austin, R. H., Chan, S. S., and Jovin, T. M. (1979). *Proc. Natl. Acad. Sci. USA* **76**, 5650–5654.
Axelrod, D., Koppel, D. E., Schlessinger, J., Elson, E., and Webb, W. W. (1976). *Biophys. J.* **16**, 1055–1069.
Blasie, J. K., Ericińska, M., Samuels, S., and Leigh, J. S. (1978). *Biochim. Biophys. Acta* **501**, 33–52.
Bürkli, A., and Cherry, R. J. (1981). *Biochemistry* **20**, 138–145.
Cherry, R. J. (1978). *Methods Enzymol.* **54**, 47–61.
Cherry, R. J. (1979). *Biochim. Biophys. Acta* **559**, 289–327.
Cherry, R. J., and Godfrey, R. E. (1981). *Biophys. J.* **36**, 257–276.
Cherry, R. J., and Schneider, G. (1976). *Biochemistry* **15**, 3657–3661.
Cherry, R. J., Bürkli, A., Busslinger, M., Schneider, G., and Parish, G. (1976). *Nature (London)* **263**, 389–393.
Cherry, R. J., Müller, U., Henderson, R., and Heyn, M. P. (1978). *J. Mol. Biol.* **121**, 283–298.
Cherry, R. J., Nigg, E. A., and Beddard, G. S. (1980). *Proc. Natl. Acad. Sci. USA* **77**, 5899–5903.
Cone, R. A. (1972). *Nature New Biol.* **236**, 39–43.
Garland, P. B., Davison, M. T., and Moore, C. M. (1979). *Biochem. Soc. Trans.* **7**, 1112–1114.
Greinert, R., Stärk, H., Stier, A., and Weller, A. (1979). *J. Biochem. Biophys. Methods* **1**, 77–83.
Heyn, M. P., Bauer, P. J., and Dencher, N. A. (1975). *Biochem. Biophys. Res. Commun.* **67**, 897–903.
Heyn, M. P., Cherry, R. J., and Müller, U. (1977). *J. Mol. Biol.* **117**, 607–620.
Heyn, M. P., Cherry, R. J., and Dencher, N. A. (1981). *Biochemistry* **20**, 840–849.

Hoffman, W., Sarzala, M. G., and Chapman, D. (1979). *Proc. Natl. Acad. Sci. USA* **76**, 3860–3864.

Hoffmann, W., Restall, C. J., Hyla, R., and Chapman, D. (1980a). *Biochim. Biophys. Acta* **602**, 531–538.

Hoffmann, W., Sarzala, M. G., Gomez-Fernandez, J. C., Goni, F. M., Restall, C. J., Chapman, D., Heppeler, G., and Kreutz, W. (1980b). *J. Mol. Biol.* **141**, 119–132.

Junge, W., and Devault, D. (1975). *Biochim. Biophys. Acta* **408**, 200–214.

Kawato, S., and Kinosita, K., Jr. (1981). *Biophys. J.* **36**, 277–296.

Kawato, S., Sigel, E., Carafoli, E., and Cherry, R. J. (1980). *J. Biol. Chem.* **255**, 5508–5510.

Kawato, S., Sigel, E., Carafoli, E., and Cherry, R. J. (1981). Submitted for publication.

Kinosita, K., Jr., Kawato, S., and Ikegami, A. (1977). *Biophys. J.* **20**, 289–305.

Kirino, Y., Ohkuma, T., and Shimizu, H. (1978). *J. Biochem.* **84**, 111–115.

Lo, M. M. S., Garland, P. B., Lamprecht, J., and Barnard, E. A. (1980). *FEBS Lett.* **111**, 407–412.

McIntosh, P. R., Kawato, S., Freedman, R. B. and Cherry, R. J. (1980). *FEBS Lett.* **122**, 54–58.

Moore, C. M., Boxer, D., and Garland, P. B. (1979). *FEBS Lett.* **108**, 161–166.

Nigg, E. A., and Cherry, R. J. (1979a). *Biochemistry* **18**, 3457–3465.

Nigg, E. A., and Cherry, R. J. (1979b). *Nature (London)* **277**, 493–494.

Nigg, E. A., and Cherry, R. J. (1980). *Proc. Natl. Acad. Sci. USA* **77**, 4702–4706.

Nigg, E. A., Kessler, M., and Cherry, R. J. (1979). *Biochim. Biophys. Acta* **550**, 328–340.

Nigg, E. A., Bron, C., Girardet, M., and Cherry, R. J. (1980a). *Biochemistry* **19**, 1887–1893.

Nigg, E. A., Cherry, R. J., and Báchi, T. (1980a). *Virology* **107**, 552–556.

Nigg, E. A., Gahmberg, C. G., and Cherry, R. J. (1980c). *Biochim. Biophys. Acta* **600**, 636–642.

Razi Naqvi, K., Gonzalez-Rodriguez, J., Cherry, R. J., and Chapman, D. (1973). *Nature New Biol.* **245**, 249–251.

Richter, C., Winterhalter, K. H., and Cherry, R. J. (1979). *FEBS Lett.* **102**, 151–154.

Stoeckenius, W., Lozier, R., and Bogolmoni, R. A. (1979). *Biochim. Biophys. Acta* **505**, 215–278.

Strambini, G. B., and Galley, W. C. (1980). *Biopolymers* **19**, 383–394.

Thomas, D. D., and Hidalgo, C. (1978). *Proc. Natl. Acad. Sci. USA* **75**, 5488–5492.

Vaz, W. L. C., Austin, R. H., and Vogel, H. (1979). *Biophys. J.* **26**, 415–426.

Wagner, T., and Junge, W. (1980). *FEBS Lett.* **114**, 327–333.

20

Rotational Diffusion of Membrane Proteins

Measurements by Light Emission

Peter B. Garland

The preceding chapter has described the use of triplet probes and their time-resolved optical spectroscopic detection in the measurement of rotational diffusion coefficients of membrane proteins. The present chapter is concerned with the detection of triplet probes by light emission measurements, and I shall describe three methods: phosphorescence (Austin *et al.*, 1979; Moore *et al.*, 1979), delayed fluorescence (Greinert *et al.*, 1979), and fluorescence depletion (Garland and Johnson, 1981). The aim of these alternative methods is to improve sensitivity, so that membrane proteins of low abundance (e.g., insulin receptors) can be studied. Because these methods all involve light emission, they can in principle be applied with a modified fluorescence microscope. The ambition then is to make measurements on the surfaces of intact cells, thereby bringing the sensitivity of rotational diffusion measurements to the same level as that obtainable for lateral diffusion using the fluorescence photobleaching recovery (FPR) method (Axelrod *et al.*, 1976; Koppel *et al.*, 1976), which can make measurements on a 1-um^2 spot of membrane containing as few as 10^3 fluorescently labeled target proteins. To achieve this performance for rotational diffusion requires an improvement in sensitivity of triplet detection of some 10 orders of magnitude over the best reported so far (Moore *et al.*, 1979), and this has now been achieved (Garland and Johnson, 1981).

A further important advantage that would arise from the development of a microscopic method of measuring triplet probes by emission is that the geometry of the sample (a single cell or tissue surface) can be so arranged that the propagation axes of the photoselecting light beam and of the measured emission are parallel to each other and also to the axis, normal to the plane of the membrane, around which membrane proteins are considered to

Peter B. Garland • Biochemistry Department, Dundee University, Dundee DD1 4HN, Scotland.

rotate. Under such conditions, the anisotropy set up by the original photoselecting triplet-exciting flash decays according to equation (1), developed from Cone (1972):

$$P_t = P_0 e^{-4Dt} \tag{1}$$

where D is the rotational diffusion coefficient of the protein under study, P_t is the polarization parameter measured at time t after the photoselection flash, and P_0 is the zero-time value. P is defined in terms of the emission intensities I measured either parallel ($\parallel$) or perpendicular ($\perp$) to the plane of a polarization of the photoselecting flash, and is given by

$$P = \frac{I(\parallel) - I(\perp)}{I(\parallel) + I(\perp)} \tag{2}$$

It is assumed that the transition dipole moments for absorption at the photoselecting wavelength and for the light emission process are parallel to each other. Equation (1) gives a much simpler analysis of data than the corresponding equation for a random suspension of membranes. In the latter case [see equation (2) of the preceding chapter], the decay of anisotropy is not a simple exponential but is characterized by two exponentials of unknown relative amplitudes plus a further time-independent term. Only in special cases, where there is a single and known angle between the transition dipole moment of the chromophore and the axis of rotation, can an unequivocal estimate of D, the rotational diffusion coefficient, be made. In general, extrinsic chromophores seem unlikely to give unique angles, although this is not so for the intrinsic chromophore of bacteriorhodopsin (Heyn *et al.*, 1977).

1. PHOTOLUMINESCENCE METHODS

A simplified electronic energy-level diagram for a triplet probe such as eosin is shown in Fig. 1. The points to note are that (1) excitation from the ground state to the excited singlet state S_1 is fast, photon absorption and excitation occurring in some 10^{-15}s sec; (2) radiative decay (fluorescence) from S_1 to the ground state occurs within a few nanoseconds

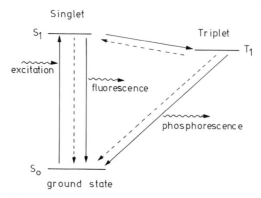

Figure 1. Photochemistry of eosin. The energy levels of the ground state (S_0), first excited singlet state (S_1), and triplet state (T_1) are indicated by horizontal bars. Interconversions are (1) excitation from S_0 to S_1 by photon absorption; (2) nonradiative decay from S_1 to S_0; (3) radiative decay from S_1 to S_0, or fluorescence; (4) intersystem crossing from S_1 to T_1; (5) thermally activated reversal from T_1 to S_1; (6) nonradiative decay of T_1 to S_0, including quenching; (7) radiative decay from T_1 to S_0, or phosphorescence. Thermally activated delayed fluorescence involves transitions from T_1 to S_1 and then to S_0.

after excitation; (3) intersystem crossing to the triplet state T_1 competes with direct decay of S_1 to the ground state; and (4) the lifetime of the triplet state at room temperature is at most a few milliseconds, there being decay to the ground state variously by a direct non-radiative transition (emitting phosphorescence), by quenching with other molecules (especially O_2, which must be excluded), and by return to S_1 (if energy levels permit) and then to the ground state. If this latter S_1-to-ground-state transition is radiative, it has the same wavelength characteristics as that of the short-lived or prompt fluorescence associated with the lifetime of S_1, but has a much longer lifetime paralleling that of the triplet state. For this reason the fluorescence arising from triplet return via S_1 is called delayed fluorescence. A full review of this topic is given by Parker (1968).

The quantum yields for phosphorescence of organic molecules are low at room temperature and in aqueous solution. Nevertheless, eosin (tetrabromofluorescein) and erythrosin (tetraiodofluorescein) have sufficient phosphorescence to permit detection of their triplet states under physiological conditions. Eosin is available as its isothiocyanate, maleimide, or iodoacetamide derivatives. Moore and Garland (1979) described the synthesis of erythrosin isothiocyanate. Of these two probes, erythrosin is the more sensitive by a factor of 10 or so, and because its relatively weak prompt fluorescence (550 nm) can be adequately removed by a high-pass filter from the phosphorescence (660 nm), the latter can be measured without need to gate the photomultiplier tube for protection against overwhelming and potentially damaging prompt fluorescence during the photoselecting laser flash. The lifetime of erythrosin is not more than a few hundred microseconds, and this probe is not as well suited for measuring rotations extending into the millisecond range (Garland and Moore, 1979). Eosin is less phosphorescent but more fluorescent, and Austin *et al.* (1979) used photomultiplier gating. However, eosin gives a longer triplet lifetime than erythrosin. In both cases, delayed fluorescence increases with increasing temperature (Parker, 1968), but the coincidence of the wavelengths for prompt and delayed fluorescence makes it essential to protect the photomultiplier during the photoselecting laser flash. Greinert *et al.* (1979) employed a mechanical shutter with a dead time of 70 μsec in their study of eosin delayed fluorescence. This dead time would be too long for many membrane studies, and photomultiplier gating would be preferable.

There is wide scope in the choice of laser for photoselection: N_2-pumped dye laser (Austin *et al.*, 1979); flash lamp-pumped dye laser, either slow repetition rate (Moore *et al.*, 1979) or fast (Lo *et al.* 1980); and frequency-doubled Nd-YAG (Chapman and Restall, 1981). In all cases, the fractional conversion of probe to triplet should not exceed 0.2–0.3 to preserve photoselection, and from that point onwards increased sensitivity must come from averaging repetitive traces. Yet another laser variant, a CW argon-ion laser can be acoustooptically modulated to give 20-μsec pulses focused onto a sample through the epi-illuminator and objective of a fluorescence microscope (Garland, 1981). With this method a sensitivity of about 10^8 molecules of erythrosin can be achieved. This is not sufficient to achieve the level of 10^3 molecules needed to study many interesting membrane proteins in single-cell surfaces. Because it seemed highly improbable that a probe with a very much higher phosphorescence quantum yield would exist, an alternative route for the achievement of higher sensitivity was taken, and is described below.

2. FLUORESCENCE DEPLETION METHODS

Optical spectroscopic measurements of the triplet state have used either the weak absorption band of the triplet itself (Cherry and Schneider, 1976; Austin *et al.*, 1979) or the stronger absorption of the ground state, which becomes depleted (Cherry and Schnei-

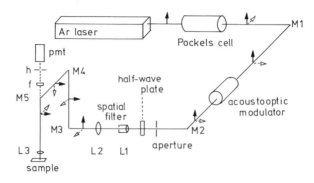

Figure 2. Apparatus for measurement of fluorescence depletion. The design is adapted from the fluorescence photobleaching recovery instrument described by Garland (1981): additional features are the transverse Pockels cell used to rotate the plane of laser beam polarization for alternate bleaches through 90°, and a half-wave plate to rotate all polarization planes as needed to neutralize polarization effects in the dichroic mirror (M5) of the fluorescence microscope epi-illuminator. M1–M4 are plane mirrors, f is a barrier filter (KV550), h is a pinhole in the image plane of the microscope objective, and pmt is the photomultiplier tube. Orthogonal planes of polarization are indicated with light and dark arrows. Lens L1 is the input of the spatial filter. Lens L2 focuses the laser beam at the image plane of the microscope objective L3.

der, 1976). In principle, ground-state depletion of a triplet probe that is fluorescent could as well be measured by a fluorescence depletion method, in which the absorption of light by the ground state is not measured by attenuation of a measuring light beam but by the prompt fluorescence that such a measuring light beam excites. An apparatus for making polarized fluorescence depletion measurements is shown in Fig. 2. The way in which it works is best considered by first explaining the fluorescence depletion measurements, then the polarization. The apparatus is based on a laser–microscope combination used for making FPR measurements. In fact, one advantage of this approach is that the one apparatus can be used to measure either rotational or lateral diffusion. The same extrinsic probe can be used for both methods.

The apparatus uses a 1-W CW argon-ion laser, acoustooptically modulated (Garland, 1981). The laser beam, either 514.5 nm (rhodamine or eosin) or 488 nm (fluorescein), passes through a spatial filter, is focused in the back image plane of the objective of a fluorescence microscope, and then is focused via the epi-illuminator and objective onto the sample, which may be a microscope slide or thin cuvette. Oxygen is removed from microscope slide samples by preparation in an anaerobic glove box, inclusion of an oxygen scavenging system in the sample (glucose, glucose oxidase, catalase), and sealing. Fluorescent emission is detected with a photomultiplier via a barrier filter and a pinhole in the objective focal plane at the top of the microscope (Koppel *et al.*, 1976). A laser probe of some 20-μsec duration and 1- to 5-mW intensity causes transient depletion (bleaching) of fluorescence due to triplet formation. The kinetics of fluorescence depletion decay are monitored with a measuring pulse of laser light, too brief (1 μsec) to cause significant further bleaching but of otherwise unattenuated intensity. The cycle of bleach and measure is reiterated with the timing of the measuring pulse swept across the period of interest (e.g., 2–20 msec), as with a boxcar detector. With appropriate sample and hold electronics, the complete curve for decay of fluorescence depletion is reconstructed in about 0.5–5 sec. The most sensitive probes are those with a high quantum yield for fluorescence and low quantum yield for triplet formation. Thus, the best for sensitivity is rhodamine, followed by fluorescein, eosin, and erythrosin. Calculation of the laser intensities and local

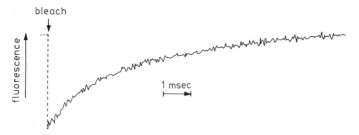

Figure 3. Triplet lifetime of rhodamine isothiocyanate conjugated with bovine serum albumin, measured by fluorescence depletion. Laser intensity into microscope, 2 mW at 514.5 nm. Fluorescence measured at 550 nm upwards. Bleach time, 30 μsec; 100× oil immersion microscope objective. Sample in 90% w/v glycerol at 20°C, prepared as a thin film. A total of 16 sweeps were averaged over 80 sec. The number of molecules of rhodamine at the focused laser spot of 1-μm diameter was about 4×10^3.

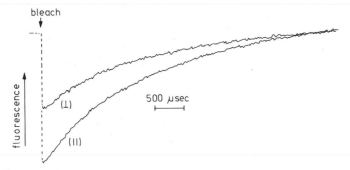

Figure 4. Time-independent anisotropy of fluorescence depletion signals excited in a solid solution of eosin. Laser beam at 514 nm, 1 mW, 5× microscope objective, eosin 3 μM in poly(methyl)methacrylate block prepared anaerobically. Fluorescence was measured at 550 m upwards. The measuring beam was polarized either parallel ($\parallel$) or perpendicular $\perp$) to the bleaching beam. Sixteen sweeps were averaged.

temperature rise at the focused spot on the sample can be made according to Axelrod (1977). In the worst case (rhodamine, 100× objective, 1-μm-diameter spot), the temperature rise at the spot is <0.1°C. Figure 3 shows the fluorescence depletion curve measured on about 4000 molecules of rhodamine-labeled serum albumin. It is clear that a detection sensitivity comparable to the FPR method has been achieved.

If the laser beam used for the bleaching pulse is plane polarized, then the ground-state depletion will be anisotropic. Thus, subsequent excitation of fluorescence with an *un*polarized laser beam will result in polarized emission as detected by detectors and analyzers parallel and perpendicular to the plane of polarization of the bleaching pulse. Alternatively, if the measuring pulse is alternately polarized either parallel or perpendicular to the bleaching pulse and emission is detected without an analyzer, then the same effect is achieved. This latter method has been used, and a Pockels cell (Fig. 2) controls the polarization planes. Figure 4 shows the time-independent anisotropy of the fluorescence depletion signal measured from an anaerobic solid solution of eosin, excited through a 5× objective.

3. CONCLUSIONS

This chapter has presented a state-of-the-art account of emission methods for measuring rotational diffusion of membrane proteins. The desirability of a highly sensitive microscopic method is described, as is its achievement by the fluorescence depletion method. The increase of sensitivity achieved over previously reported procedures is about 10 orders of magnitude. Photoluminescence methods of phosphorescence and delayed fluorescence are limited in their range of probes, and with their lower sensitivity may give way to the fluorescence depletion method. This is especially so for studies with an emphasis on intact cells or tissues, where parallel measurements of diffusion are also of interest.

ACKNOWLEDGMENTS. I thank the Medical Research Council and the Royal Society for financial support, Professor Dennis Chapman for a preprint of work in press, and Pauline Johnson (supported by a Science Research Council studentship) for the data of Figs. 3 and 4.

REFERENCES

Austin, R. H., Chan, S. S., and Jovin, T. M. (1979). *Proc. Natl. Acad. Sci. USA* **76,** 5650–5654.

Axelrod, D. (1977). *Biophys. J.* **18,** 129–131.

Axelrod, D., Koppel, D. E., Schlessinger, J., Elson, E., and Webb, W. W. (1976). *Biophys. J.* **16,** 1055–1069.

Chapman, D., and Restall, C. J. (1981). *Biochem. Soc. Symp.* **46,** 182–206.

Cherry, R. J., and Schneider, G. (1976). *Biochemistry* **15,** 3657–3661.

Cone, R. A. (1972). *Nature New Biol.* **236,** 39–43.

Garland, P. B. (1981). *Biophys. J.* **33,** 481–482.

Garland, P. B., and Moore, C. H. (1979). *Biochem. J.* **183,** 561–572.

Greinert, R., Stärk, H., Stier, A., and Weller, A. (1979). *J. Biochem. Biophys. Methods* **1,** 77–83.

Heyn, M. P., Cherry, R. J., and Müller, U. (1977). *J. Mol. Biol.* **117,** 607–620.

Johnson, P., and Garland, P. B. (1981). *FEBS Lett.* **132,** 252–256.

Koppel, D. E., Axelrod, D., Schlessinger, J., Elson, E. L., and Webb, W. W. (1976). *Biophys. J.* **16,** 1315–1329.

Lo, M. M. S., Garland, P. B., Lamprecht, J., and Barnard, E. A. (1980). *FEBS Lett.* **111,** 407–412.

Moore, C. H., and Garland, P. B. (1979). *Biochem. Soc. Trans.* **7,** 945–946.

Moore, C. H., Boxer, D. H., and Garland, P. B. (1979). *FEBS Lett.* **108,** 161–166.

Parker, C. A. (1968). In *Photoluminescence of Solutions,* pp. 392–394, Elsevier, Amsterdam.

21

Lipid–Protein Interactions in Monomolecular Layers

R. A. Demel

1. INTRODUCTION

Monolayers have proved to be a valuable model system for studying the physical properties of membrane constituents and the dynamic properties of biological membranes.

Monomolecular layers offer the advantages of a well-controlled system; the molecular packing of the surface components can be changed and the release from the interface or transfer to the interface of lipid or protein can be monitored continuously. As a further advantage, only small amounts of the material (2–10 nmoles of lipid and 1–20 μg of protein) are required.

2. DESCRIPTION OF THE METHOD

The trough containing the water phase is made of Teflon and contains 5–15 ml of buffer. Convenient dimensions are 5 cm long and wide and 0.5 cm deep, with an injection hole to add material to the subphase without disturbing the monolayer and a depression of 0.5-cm depth in the middle for a bar to stir the subphase (Fig. 1). The water used for the subphase must be free of organic contaminants. The salts used for the buffer solution should not appreciably change the surface tension of water. The surface tension of pure water at 20°C is 72.75 mN m^{-1}. The interfacial tension can be measured by the Wilhelmy plate method using a recording electrobalance and a sandblasted platinum plate 2 cm long and 1 cm high (Gaines, 1966; Demel, 1974). To obtain an ideal wetting of the platinum plate, so that the contact angle of liquid to plate is zero, the plate is cleaned with chromic acid and rinsed with distilled water. The surface pressure, which is the difference between the surface tension of clean water and the surface film, can reach for most phospholipids a

R. A. Demel • Laboratory of Biochemistry, State University of Utrecht, Transitorium III, NL-3584CH Utrecht, The Netherlands.

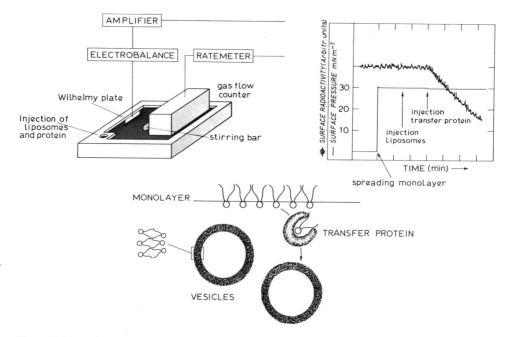

Figure 1. Measuring of the surface pressure and surface radioactivity, demonstrating the exchange of phosphatidylcholine between monolayer and liposomes.

value of up to 45 mN m^{-1}. The lipid material is carefully released onto the air–water interface from a capillary tube in small drops until the desired surface pressure is reached. As spreading solvent, chloroform (containing up to 20% methanol) or petrolether can be used. The molecular area at any surface pressure can be determined from the pressure–area curve of the lipid.

The surface radioactivity can be measured using a gas-flow detector with a Micromil window 4.3 × 1.3 cm (Demel, 1974). Because the intention is to measure the label only at the interface, the label should be quenched in the subphase. This condition is fulfilled for ^{14}C, ^{131}I, and partially for ^{32}P. A high specific activity is required due to the small amounts of material at the interface. Phosphatidyl[*methyl*-^{14}C]choline is obtained by the method of Stoffel *et al.* (1971) with a specific activity of 40 Ci mole^{-1}. Phosphatidyl[^{14}C]inositol with a specific activity of 1 Ci mole^{-1} can be prepared by inositol exchange with microsomal phosphatidylinositol and incorporation of myoinositol in the *de novo*-synthesized phosphatidylinositol (Strunecká and Zborowski, 1975). Proteins may be labeled with ^{131}I using lactoperoxidase. In case the radioactivity at the interface is too low to be detected directly (e.g., with ^{3}H, ^{125}I, or after transfer from liposomes in the subphase to the monomolecular film), the monomolecular film can be collected and measured as described below.

3. EXAMPLES OF APPLICATION

3.1 Phospholipases

A special case of lipid–protein interaction is the enzymatic hydrolysis of phospholipids. The degradation of the polar part of phospholipids by lipolytic enzymes can be followed by surface radioactivity. Bangham and Dawson (1960) used [^{32}P]phosphatidylcholine and [^{32}P]phosphatidylethanolamine to study the enzymatic hydrolysis by phospholipase B from *Penicillium notatum* and phospholipase C from *Clostridium perfringens* (Bangham and Dawson, 1962), and the effects of surface charge and divalent ions on the reaction. The importance of the surface pressure is indicated by the fact that [^{32}P]phosphatidylcholine monolayers of more than 40 mN m^{-1} are hydrolyzed by phospholipase C only when Ca^{2+} or UO_2^{2+} is present. Hydrolysis of low-pressure films (less than 30 mN m^{-1}) is observed in the absence of divalent cations.

The study of phospholipase A_2 activity at the air–water interface is more complicated (Eibl *et al.*, 1969; Schulman and Shah, 1967), for the products formed are not completely water soluble unless high pressures or short-chain phosphatidylcholines (e.g., dinonanoyl-phosphatidylcholine) are used. Detailed kinetic studies of pancreatic phospholipase A_2 on dinonanoylphosphatidylcholine under barostatic conditions have been described by Verger and de Haas (1973, 1976). The loss of hydrolysis products from the interface is compensated by a decrease in surface area. Eighty percent hydrolysis of unimolecular films of [^{32}P]phosphatidylcholine, [^{32}P]phosphatidylethanolamine, and [^{32}P]phosphatidylinositol was observed with phospholipase A_2 from *Naja naja* at 30 mN m^{-1}, but no activity was found at 42 mN m^{-1} (Dawson, 1966).

There are considerable differences in the lipolytic activity of phospholipases from different sources. Purified phospholipase A_2 from pig pancreas and *Crotalus adamanteus*, phospholipase D from cabbage, and phospholipase C from *Bascillus cereus* cannot hydrolyze the phospholipids of intact erythrocyte membranes, whereas phospholipase A_2 from *Naja naja*, bee venom, phospholipase C from *Clostridium welchii*, and the sphingomyeli-

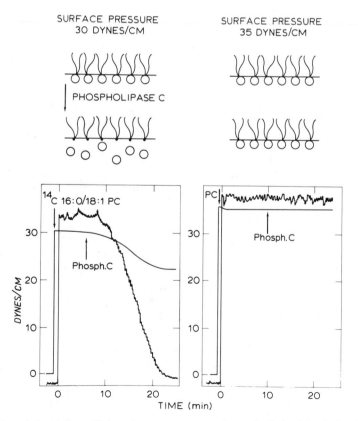

Figure 2. Action of phospholipase C from *B. cereus* on a monolayer of palmitoyloleoyl phosphatidyl[*methyl*-^{14}C]choline (PC).

nase from *Staphylococcus aureus* are highly active (Demel *et al.*, 1975). Using phosphatidyl[*methyl*-^{14}C]choline, it could be shown by following the decrease in surface radioactivity that there is a defined surface pressure at which phosphatidylcholine at the air–water interface can be hydrolyzed (Fig. 2). Phospholipases which hydrolyze the phospholipids of the intact erythrocyte membrane also hydrolyze phosphatidylcholine monolayers at a pressure above 31 mN m^{-1}. Negatively charged lipids decrease the molecular packing and lower the pressure at which hydrolysis can occur.

3.2. Phospholipid Exchange Proteins

The transfer of *intact* phospholipids from the interface to receptor membranes (e.g., liposomes or vesicles) is catalyzed by specific transfer proteins. Transfer proteins have been isolated from the 105,000g supernatant fraction of bovine liver (Kamp *et al.*, 1973; Crain and Zilversmit, 1980), heart (Ehnholm and Zilversmit, 1973), and brain (Helmkamp *et al.*, 1974). One protein from bovine liver (molecular weight 24,681) transfers only phosphatidylcholine. Another protein (molecular weight 14,500) is nonspecific and transfers not only phosphatidylcholine and phosphatidylethanolamine but also cholesterol. The two proteins from brain (molecular weights 32,300 and 32,800) preferentially transfer phosphatidylinositol, but also phosphatidylcholine. The properties and mode of action of these proteins were studied by monitoring the surface radioactivity and pressure. When the phosphatidylcholine transfer protein from bovine liver is injected under a monolayer of phosphatidyl[^{14}C]choline, a decrease of the surface radioactivity is observed while the surface pressure remains constant (Demel *et al.*, 1973). The decrease continues until an equilibrium value is reached. This value is governed by the ratio of phosphatidylcholine molecules in the monolayer to protein molecules in the subphase, demonstrating that upon interaction with the monolayer the protein exchanges its endogenous phosphatidylcholine for a molecule from the interface. By this exchange mechanism the protein functions as a carrier of phosphatidylcholine between separate membrane interfaces. This has been demonstrated by injection of phosphatidylcholine transfer protein into the subphase that connects a phosphatidyl[^{14}C]choline monolayer and an unlabeled phosphatidylcholine monolayer. The radioactivity decreases in the first monolayer and increases in the latter. After injection of unlabeled phosphatidylcholine vesicles under a monolayer of phosphatidyl[^{14}C]choline, a rapid decrease of the surface radioactivity is observed (Fig. 1). On the other hand, when the phosphatidylcholine in the vesicles is labeled, the appearance of radioactivity at the interface can be demonstrated. In this case the surface radioactivity will be too low to be detected by the gas-flow counter. The monomolecular film can be suctioned into a counting vial (Rietsch *et al.*, 1977) and its radioactivety assayed.

As a rule, the transfer of phosphatidylcholine to membranes lacking this phospholipid does not occur. Recently, however, a net transfer of phosphatidylcholine has been demonstrated to membranes composed of sphingomyelin and apoprotein A$_1$ (Wilson *et al.*, 1980).

The phosphatidylinositol transfer protein from bovine brain catalyzes the transfer of phosphatidyl[^{14}C]inositol from the monolayer to phosphatidylcholine vesicles in the subphase (Demel *et al.*, 1977) (Fig. 3). The transport of this negatively charged phospholipid seems to be coupled with the simultaneous transport of phosphatidylcholine in the opposite direction. Sphingomyelin inhibits the transfer reaction (Zborowski, 1979).

3.3. Serum Lipoproteins

Besides the possible need for lipid transfer in the cytoplasm for membrane biogenesis, specific transfer complexes also exist in the serum between serum apoproteins and lipids.

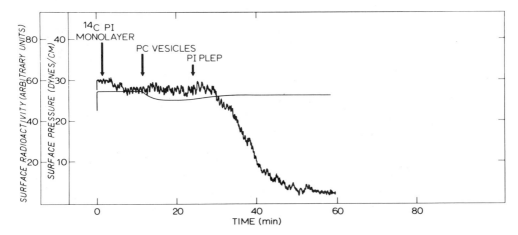

Figure 3. Transfer of phosphatidylinositol (PI) from a monolayer to phosphatidylcholine (PC) vesicles.

Chylomicrons, very-low-density lipoproteins, low-density lipoproteins, and high-density lipoproteins are the major carriers of lipid in blood (Jackson *et al.*, 1976). Each lipoproteins class contains specific proteins or apoproteins. Using bulk systems it has been difficult to determine any differences in the affinity of apoproteins for lipids. Study of the interaction of lipids with plasma apolipoproteins by the monolayer technique has shown that similar increases in surface pressure occur following injection of apolipoprotein A-I from human high-density lipoproteins and injection of arginine-rich protein from swine very-low-density lipoproteins underneath a monolayer of egg phosphatidyl[^{14}C]choline (Jackson *et al.*, 1979). However, with apolipoprotein C-II and apolipoprotein C-III there is a decrease in surface radioactivity, indicating that the apoproteins remove phosphatidylcholine from the interface. The removal is specific for phosphatidylcholine because cholesterol and phosphatidylinositol are not removed from the interface. The addition of phospholipid liposomes to the subphase greatly facilitates the apolipoprotein C-II-mediated removal of phosphatidylcholine from the interface.

Very-low-density lipoproteins of human plasma are degraded in the capillary endothelium by lipoprotein lipase. Upon hydrolysis of the triglycerides there is a loss of choles-

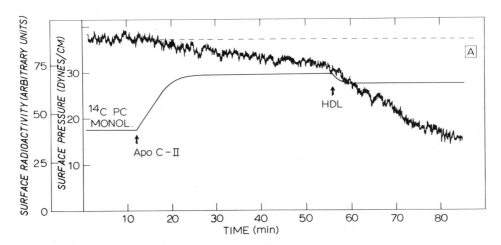

Figure 4. Effect of high-density lipoproteins (HDL) on the interaction of apolipoprotein C-II with egg phosphatidylcholine monolayers.

terol, phospholipids, and proteins from the lipoprotein particle. The proteins, mainly type C apoproteins, are found associated with the high-density lipoproteins. The addition of human plasma high-density lipoproteins or very-low-density lipoproteins to the subphase has been shown to increase three- to fourfold the apolipoprotein C-mediated removal of phosphatidyl[^{14}C]choline from the interface (Fig. 4). Low-density lipoproteins do not affect the rate of decrease. Cholesteryl esters comprising 70% of the total cholesterol of plasma from fasted individuals reside primarily in the low-density lipoproteins, with much of the remainder in high-density lipoproteins. Cholesteryl esters have practically no water solubility, nor can they be solubilized by phospholipids. A protein-mediated transfer of cholesteryl ester has been demonstrated recently (Harmony et al., 1982). Phosphatidylcholine monolayers containing up to 5 mole% [^{14}C]cholesteryl oleate were spread at the air–water interface; a lipid transfer complex from human plasma was able to remove the cholesteryl ester from the interface.

4. PERSPECTIVES

The examples given in this chapter demonstrate some of the possible uses of monomolecular films.

It is likely that similar studies on other enzymes that interact with membrane lipids or proteins will provide useful information on the conditions governing their activity.

REFERENCES

Bangham, A. D., and Dawson, R. M. C. (1960). Biochem. J. 75, 133–141.
Bangham, A. D., and Dawson, R. M. C. (1962). Biochim. Biophys. Acta 59, 103–115.
Crain, R. C., and Zilversmit, D. B. (1980). Biochim. Biophys. Acta 620, 37–48.
Dawson, R. M. C. (1966). Biochem. J. 98, 35C–37C.
Demel, R. A. (1974). Methods Enzymol. 32, 539–545.
Demel, R. A., Wirtz, K. W. A., Kamp, H. H., Geurts van Kessel, W. S. M., and van Deenen, L. L. M. (1973). Nature New Biol. 246, 102–105.
Demel, R. A., Geurts van Kessel, W. S. M., Zwaal, R. F. A., Roelofsen, B., and van Deenen, L. L. M. (1975). Biochim. Biophys. Acta 406, 97–107.
Demel, R. A., Kalsbeek, R., Wirtz, K. W. A., and van Deenen, L. L. M. (1977). Biochim. Biophys. Acta 466, 10–22.
Ehnholm, C., and Zilversmit, D. B. (1973). J. Biol. Chem. 248, 1719–1724.
Eibl, H., Demel, R. A., and van Deenen, L. L. M. (1969). J. Colloid Interface Sci. 29, 381–387.
Gaines, G. L. (1966). Insoluble Monolayers at Liquid–Gas Interfaces, Interscience, New York.
Harmony, J. A. K., Jackson, R. L., Ihm, J., Ellsworth, J. L., and Demel, R. A. (1981). FEBS Lett., submitted for publication.
Helmkamp, G. M., Harvey, M. S., Wirtz, K. W. A., and van Deenen, L. L. M. (1974). J. Biol. Chem. 249, 6382–6389.
Jackson, R. L., Morrisett, J. P., and Gotto, A. M. (1976). Physiol. Rev. 56, 259–316.
Jackson, R. L., Pattus, F., and Demel, R. A. (1979). Biochim. Biophys. Acta 556, 369–387.
Kamp, H. H., Wirtz, K. W. A., and van Deenen, L. L. M. (1973). Biochim. Biophys. Acta 318, 313–325.
Rietsch, J., Pattus, F., Desnuelle, P., and Verger, R. (1977). J. Biol. Chem. 252, 4313–4318.
Schulman, J. H., and Shah, D. O. (1967). J. Colloid Interface Sci. 25, 107–115.
Stoffel, W., Lekim, D., and Tschung, T. S. (1971). Hoppe-Seylers Z. Physiol. Chem. 352, 1058–1064.
Strunecká, A., and Zborowski, J. (1975). Comp. Biochem. Physiol. B. 50, 55–60.
Verger, R., and de Haas, G. H. (1973). Chem. Phys. Lipids 10, 127–136.
Verger, R., and de Haas, G. H. (1976). Annu. Rev. Biophys. Bioeng. 5, 77–117.
Wilson, D. B., Ellsworth, J. L., and Jackson, R. L. (1981). Biochim. Biophys. Acta 620, 550–561.
Zborowski, J. (1979). FEBS Lett. 107, 30–32.

22

Artificial Planar Bilayer Lipid Membranes

A Survey

H. Ti Tien

1. INTRODUCTION

About two decades ago, when a method for forming bilayer lipid membranes (BLMs) was discovered, the principal feature of the molecular architecture of biological membranes was being established as a bimolecular lipid leaflet (now called a lipid bilayer). First proposed by Gorter and Grendel in 1925, the existence of a lipid bilayer, which constitutes the key element of all biological membranes, has since been firmly established, largely through experiments with artificial BLMs (and liposomes).

Owing to space limitation, it is not possible to cite in this review all of the papers published in 1980 dealing with planar BLMs. Therefore, where possible, only the most recent references and a review article are given for the interested reader. For complete bibliographies and comprehensive reviews of work published before 1980, consult the following sources: Fendler (1980), Shamoo and Tivol (1980), Tien (1979), Antonov *et al.* (1979), and Andersen (1978).

2. NEW COMPOUNDS AND TECHNIQUES

Since their inception in 1960, BLMs have been formed from a variety of compounds, including ill-defined tissue extracts (Tien, 1974). Today, synthetic compounds tailor-made for bilayer formation are available. Kunitake and Okahata (1980) have synthesized ammonium amphiphiles of single-chain (hexadecyl), double-chain (didodecyl and dioctadecyl), and triple-chain (trioctyl) structure, as well as those containing a diphenylazomethine segment. They have demonstrated that some of these compounds possess esterolytic reac-

H. Ti Tien • Department of Biophysics, Michigan State University, East Lansing, Michigan 48824.

tivities. Alkyltrimethyl ammonium bromides in conjunction with cholesterol have been reported by Bagaveev *et al.* (1979). Pure lipids such as monoglycerides, diolein, and triglycerides form BLMs in the absence of alkane solvents. These BLMs are characterized by a large capacitance (0.59–0.96 $\mu F/cm^2$) and fluidity. Oxidized cholesterol, although not a new compound, has been systematically investigated by Robinson and Strickholm (1978). BLMs formed from oxidized cholesterol solutions are stable up to 3 hr even after 1 month's storage. Long-lived BLMs and "solvent-free" BLMs can also be formed. Both Mountz and Tien (1979) and Sanderman (1979) have prepared membranes on commercially available polycarbonate filters (Nucleopore membranes). Thus far, even lipid bilayers formed from monolayers according to the Tagaki technique (Tien, 1974) are not absolutely solvent-free, for petroleum jelly (Vaseline) or silicone grease must be used to coat the membrane support (White, 1980; Ebihara *et al.*, 1979).

Instruments and apparatus for BLM studies remain essentially unchanged with two exceptions. Huebner (1979) reported an apparatus for recording flash-induced voltage transient with 10-nsec resolution. Lopez and Tien (1980) have extended the range of the photoelectrospectrometry of BLMs into the ultraviolet region.

3. PHYSICAL STUDIES

White (1980), in addition to developing quantitative methods, has studied the physical chemistry of the BLM. He has reported that electric fields modify *n*-alkane solubility in the bilayer, causing it to shift from the bilayer to the Plateau–Gibbs border and microlenses. The solubility of *n*-alkanes in BLMs has been investigated by Gruen and Haydon (1980). They have reported that *n*-alkanes dissolve primarily into the central core of a BLM and alter significantly the BLM's thickness. *n*-Alkanes of increasing chain length have decreasing solubility, contrary to expectation. This phenomenon has also been considered at length by McIntosh *et al.* (1980) based on results obtained by the combined use of differential scanning calorimetry, X-ray diffraction, and monolayer techniques.

One problem associated with the BLM system is its stability. Snyder *et al.* (1980) have examined this problem experimentally, whereas Suezaki (1980) has analyzed it from a theoretical viewpoint. Due to the strong intermolecular interaction of lipids, the BLM can be in a true equilibrium state. The problem of the instability of BLMs has been considered in a series of papers by Chizmadzev and colleagues (1979). These experimental and theoretical studies are aimed at the elucidation of the mechanism of BLM breakdown in an electric field, which, it is hoped, may provide an important tool for studying the electric breakdown of biological membranes and possibly the nature of background conductivity of membranes. Using a charge-pulse relaxation technique Benz *et al.* (1979) found that the breakdown occurs in less than 10 nsec and proposed an electromechanical model to explain the results. Related experimental studies have also been reported by Shchipunov and Drachev (1980). Besides causing instability, an applied voltage can result in changes in reflectivity, bifacial tension, and capacitance of the BLM (Antonov *et al.*, 1979). Alteration of BLM capacitance can also be induced by temperature (Szekely and Morash, 1980) during mechanical oscillation. This method of perturbing the BLM causes a decrease in elasticity with decreasing temperature. A temperature gradient across the BLM can also perturb the system; thermopotentials have been observed in pigmented and nonpigmented BLMs (Tien, 1974; Antonov *et al.*, 1979). Another effective physical agent is ultrasound. Rohr and Rooney (1978) reported that ultrasound in the range of 0.5–1.4 W/cm^2 sped up the thinning of the BLM and ruptured it at 1.5 W/cm^2. Rohr and Rooney state that the use of acoustic energy on BLMs might provide insight in determining the specific effects of ultrasound on biological membranes.

4. DIFFUSION AND PERMEABILITY

Overton's rule states that $P_d = DK_{hc}$, where P_d is the permeability coefficient of a solute through the BLM, D the diffusion coefficient in hydrocarbon phase, and K_{hc} the partition coefficient; it has been found to be applicable to the BLM system for nonelectrolyte permeability (Orbach and Finkelstein, 1980). These authors challenge the recent emphasis on theoretically calculated diffusion coefficients, citing their lack of physiological relevance. Water, being the most biologically significant nonelectrolyte, is examined anew for its permeability in BLMs (Fettiplace and Haydon, 1980). The effects of benzyl alcohol have been studied by Ebihara *et al.* (1979).

5. ION TRANSPORT

The permeability of BLMs to ions is by far the most active area of research (Antonov *et al.*, 1979; Andersen, 1978). The compounds tested can be roughly divided into two categories: simple and complex.

5.1. Simple Compounds

These include tributyltin, plant hormone abscisic acid, heat-treated amino acids (Grote *et al.*, 1978), iron bathophenanthroline (Shchipunov and Drachev, 1980), and hashish compounds (Antonov *et al.*, 1979). The most dramatic effect of these compounds appears to be reduction of the BLM resistance.

5.2. Complex Compounds

Compounds in this category range from polypeptides and proteins to tissue extracts. Among the polypeptides are gramicidin A and alamethicin (Antonov *et al.*, 1979; Andersen, 1978). These compounds form channels and give rise to discrete conductance (Yoshida *et al.*, 1980). The matrix protein (porin) of *E. coli* can also form ion-selective channels in BLMs (Benz *et al.*, 1979). Other proteins such as cytochrome oxidase, proteolipid apoprotein (Ting-Beall *et al.*, 1979), dopamine-B-hydroxylase (Kafka *et al.*, 1978), plasma lipoproteins (Martsenjsky *et al.*, 1980, Repke *et al.*, 1980), and calcium-binding glycoprotein (Antonov *et al.*, 1979) greatly enhance the conductivity of the BLM. Stepwise increase in BLM conductance by membrane vesicles prepared from the electric organ of *Torpedo californica* has been reported by White and Miller (1979). Another excellent channel former, keyhold limpet hemocyanin, has been investigated by several workers (Antolini and Menestrina, 1979; McIntosh *et al.*, 1980). Currents through the hemocyanin channel show saturation at increasing concentrations of KC1, which in some aspects are similar to those obtained with alamethicin "excitability-inducing material" (Hoffman *et al.*, 1976; Andersen, 1978). Finally, an ion transport system in a BLM has been reconstituted using fragments of Na^+/K^+-ATPase (Shamoo and Tivol, 1980). Shamoo and Tivol have reviewed this work and provided excellent criteria for the reconstitution of ion transport systems.

6. BLM–LIPOSOME INTERACTIONS

Liposomes are lipid microvesicles or in single-layer form are BLMs in spherical configuration and have been extensively studied along with the BLM system (for reviews see Tien, 1974; Shamoo and Tivol, 1980; Fendler, 1980). To make these systems even more versatile, fusion of BLMs with liposomes has been carried out (Antonov *et al.*, 1979; Cohen *et al.*, 1980). This novel approach facilitates the introduction of ionophores, dyes, proteins, bacteriorhodopsin, or other compounds of interest from liposomes into the BLM (Lopez and Tien, 1980). Interactions between liposomes and the BLM in most cases are increased in the presence of Ca^{2+}. One ubiquitous biological phenomenon, exocytosis, is being studied using the combined liposome–BLM system (Cohen *et al.*, 1980). Suezaki (1980), on theoretical grounds, has attempted to clarify the thermodynamic relationship between the stability of BLMs and liposomes.

7. PHOTOEFFECTS IN BLMs

Photoelectric effects in BLMs discovered in 1968 have since been extensively investigated (Tien, 1974, 1979; Mauzerall, 1979). A variety of photoactive compounds such as chlorophylls, retinals, fluorescent probes, rhodopsin, and bacteriorhodopsin have been incorporated into BLMs (Lopez and Tien, 1980). The pigmented BLMs have been used to mimic the thylakoid membrane of the chloroplast and the visual receptor of the eye (Gambale *et al.*, 1979). Photoactive BLMs have been utilized in the field of photochemical conversion and storage of solar energy (Bolton and Hall, 1979; Lichtin, 1980). For example, with suitable electron acceptors and donors in the bathing solution separated by a pigmented BLM, light-induced redox reactions can occur at opposite interfaces. It might be possible to couple BLMs of this type in such a way as to effect photolysis of water into hydrogen and oxygen.

Of more immediate interest to biology are experiments using fluorescent probes in BLMs (Loew *et al.*, 1979). Dragsten and Webb (1978) reported that BLM-containing Merocyanine 540 responds to a potential step in two distinct time constants: one in less than 6 μsec and the other 0.1 sec. Both response amplitudes are proportional to applied voltages. The relationship between optical properties and the applied field, known as electrochromism, has been observed by Loew *et al.* (1979). They have presented evidence for a charge-shift electrochromatic mechanism and suggested that it might be possible to develop a universal set of probes for monitoring membrane potentials. Because the fluorescence response to a potential step of a BLM is much less than 6 μsec, the apparatus developed by Huebner (1979) for pigmented BLM studies should be of interest.

8. APPLICATIONS

Exploitation of the BLM system for clinical and analytical purposes goes back to the experiments of del Castillo *et al.*, in 1966 (Mountz and Tien, 1979). Recently, Thompson *et al.* (1980) reported the use of the BLM as an electrochemical sensor and reported the limits of detection of amphotericin B and valinomycin to be 10^{-9} and 10^{-11} M, respectively. In a study of the complex processes of drug absorption, Inui *et al.* (1977) have used the BLM system as a model for the intestinal lipid membrane. Although these methods are useful as laboratory tools, one severe drawback is the lack of long-term stability of the

BLM. Perhaps the use of Nucleopore filters (Mountz and Tien, 1979) as a BLM support might prolong the life expectancy of conventional BLMs.

9. CONCLUSIONS

The BLM system has furnished and will continue to furnish much important information directly pertinent to the problems posed by membrane biology and biophysics in areas such as active transport, phosphorylation, photosynthesis, vision, immunology, nerve conduction, and energy transduction. The BLM system should also be useful in understanding membrane biogenesis and other membrane-mediated life processes.

REFERENCES

Andersen, O. S. (1978). *Membr. Transp. Biol*.**1**, 369–446.

Antolini, R., and Menestrina, G. (1979). *FEBS Lett*. **100**, 377–381.

Antonov, V. F., Rovin, Y. G., and Trifmov, L. T. (1979). *A Bibliography of Bilayer Lipid Membranes: 1962–1975*, All Union Institute for Scientific Information, Moscow.

Bagaveev, I. A., Rovin, Y. G., and Starichkov, N. V. (1979). *Biofizika* **24**, 936–937.

Benz, R., Janko, K., and Lauger, P. (1979). *Biochim. Biophys. Acta* **551**, 236–247.

Bolton, J. R., and Hall, D. O., (1979). *Annu. Rev. Energy* **4**, 353–401.

Chizmadzev, Y. A., Abidor, I. G., Pastushev, V. F., and Arakelya, V. B. (1979). *Bioelectrochem. Bioenerg*. **6**, 37–87.

Cohen, F. S., Zimmerbe, J., and Finkelstein, A. (1980). *J. Gen. Physiol*. **75**, 251–270.

Dragsten, P. R., and Webb, W. W. (1978). *Biochemistry* **17**, 5228–5240.

Ebihara, L., Hall, J. E., MacDonald, R. C., and McIntosh, T. J. (1970). *Biophys. J.* **28**, 185–196.

Fendler, J. H. (1980). *J. Phys. Chem.* **84**, 1485–1496.

Fettiplace, R., and Haydon, D. A. (1980). *J. Physiol. Rev*. **60**, 510–550.

Gambale, F., Gliozzi, A., Pepe, I. M., Robello, M., and Rolandi, R. (1979). *Gazz. Chim. Ital*. **109**, 441–447.

Grote, J. R., Syren, R. M., and Fox, S. W. (1978). *Biosystems* **10**, 287–292.

Gruen, D. W. R., and Haydon, D. A. (1980). *Biophys. J.* **30**, 129–136.

Hoffman, R. A., Long, D. D., Arndt, R. A., and Roper, L. D. (1976). *Biochim. Biophys. Acta* **455**, 780–795.

Huebner, J. S. (1979). *Photochem. Photobiol*. **30**, 233–241.

Inui, K. I., Tabara, K., Hori, R., Kaneda, A., Muranish, S., and Sezaki, H. (1977). *J. Pharm. Pharmacol*. **29**, 22–26.

Kafka, M. S., Blumenthal, R., Walker, G. A., and Pollard, H. B. (1978). *Membr. Biochem*. **1**, 279–295.

Kunitake, T., and Okahata, Y. (1980). *J. Am. Chem. Soc*. **102**, 549–553.

Lichtin, N. N. (1980). *Chemtech* **10**, 254–262.

Loew, L. M., Scully, S., Simpson, L., and Waggoner, A. S. (1979). *Nature (London)* **281**, 497–499.

Lopez, J., and Tien, H. T. (1980). *Biochim. Biophys. Acta* **597**, 433–444.

McIntosh, T. J., Simon, S. A., and MacDonald, R. C. (1980). *Biochim. Biophys. Acta* **597**, 445–463.

Martsenjsky, O. V., Perova, N. V., Gerasimov, E. N., Tverdislov, V. A., and Elkarada, S. (1980). *Vopr. Med. Khim*. **26**, 484–489.

Mauzerall, D. (1979). In *Light-Induced Charge Separation in Biology and Chemistry* (H. Gerischer, ed.), pp. 241–254, Verlay-Chimie, Berlin.

Mountz, J., and Tien, H. T. (1979). *J. Bioenerg. Biomembr*., **10**, 139–151.

Orbach, E., and Finkelstein, A. (1980). *J. Gen. Physiol*. **75**, 427–436.

Repke, H., Berezi, A., and Matthies, H. (1980). *Acta Biol. Med*. **39**, 657–663.

Robinson, R. L., and Strickholm, A. (1978). *Biochim. Biophys. Acta* **509**, 9–20.

Rohr, K. R., and Rooney, J. A. (1978). *Biophys. J.* **23**, 33–40.

Sanderman, H. (1979). *Biochim. Biophys. Acta* **87**, 789–794.

Shamoo, A. E., and Tivol, W. F. (1980). *Curr. Top. Membr. Transp*. **14**, 57–126.

Shchipunov, Y. A., and Drachev, G. Y. (1980). *Biofizika* **25**, 921–923.

Snyder, T. S., Chaing, S. H., and Klinzing, G. E. (1980). *Sol. Energy Mat* **2**, 254–264.

Suezaki, Y. (1980). *J. Colloid Interface Sci*. **73**, 529–538.

Szekely, J. G., and Morash, B. D. (1980). *Biochim. Biophys. Acta* **599**, 73–80.
Thompson, M., Krull, J. J., and Worsfold, P. J. (1980). *Anal. Chem.* **117**, 121–145.
Tien, H. T. (1974). *Bilayer Lipid Membranes (BLM): Theory and Practice,* Dekker, New York.
Tien, H. T. (1979). In *Photosynthesis in Relation to Model Systems* (J. Barber, ed.), pp. 116–173, Elsevier/North Holland, Amsterdam.
Ting-Beall, H. P., Lees, M. B., and Robertson, J. D. (1979). *J. Membr. Biol.* **51**, 33–46.
White, S. H. (1980). *Science* **207**, 1075–1077.
Yoshida, M., Clark, A. F., and Swanson, P. D. (1980). *J. Membr. Sci.* **7**, 101–108.

ADDENDUM

Since the completion of this review a number of pertinent papers on planar BLMs have come to the author's attention. The incorporation of this new information into the text is not feasible. Therefore, a bibliography under seven group headings as reviewed is added so that the literature references are up to date.

New Compounds and Techniques

Hub, H. H., Hupfer, B., Koch, H., and Ringsdorf, H. (1980). *Angew. Chem.* **19**, 938–940.
Hub, H. H., Hupfer, B., Koch, H., and Ringsdorf, H. (1981). *J. Macromol. Sci. Chem.* **15**, 701–715.
Kossi, C. N., and Leblanc, R. M. (1981). *J. Colloid Interface Sci.* **80**, 426–436.
Mishra, J. P., and Barsainy, R. K. (1981). *Indian J. Chem. A.* **20**, 175–176.

Physical Studies

Ashcroft, R. G., Coster, H. G. L., and Smith, J. R. (1981). *Biochim. Biophys. Acta* **643**, 191–204.
Bivas, I., and Petrov, A. G. (1981). *J. Theor. Biol.* **88**, 459–483.
Brennan, B. W. (1981). *J. Acoust. Soc. Am.* **69**, 1217.
Derzhanski, A., Petrov, A. G., and Pavloff, Y. V. (1981). *J.Phys. (Paris) Lett.* **42**, L119–L122.
Donovan, J. J., Simon, M. I., Draper, R. K., and Montal, M. (1981). *Proc. Natl. Acad. Sci. USA* **78**, 172–176.
Kashchie, D., and Exerowa, D. (1980). *J. Colloid Interface Sci.* **77**, 501–511.
Kasumov, K. M. (1981). *Antibiotiki* **26**, 143–155.
Lakshminarayanaiah, N. (1980). **7**, 40–156.
Lin, G. S. B. (1980). *B. Math. Biol.* **42**, 601–625.
Passechnik, V. I., Perova, N. V., Hianik, T., and Gadzhalo, S. I. (1980). *Vopr. Med.* **26**, 493–497.
Passechnik, V. I., Hianik, T., and Artemova, L. G. (1981). *Stud. Biophys.* **83**, 139–146.
Schoch, P., and Sargent, D. F. (1980). *Biochim. Biophys. Acta* **602**, 234–247.
Sidorowi, A. (1980). *Adv. Mol. Int.* **16**, 275–280.
Trissl, H. W. (1981). *Biophys. J.* **33**, 233–242.

Diffusion and Permeability

Kaiser, G., Lambotte, P., Falmagne, P., Capiau, C., Zanen, J., and Ruysscha, J. M. (1981). *Int. Physiol.* **89**, B22–B23.
Matsumoto, S., Inoue, T., Kohda, M., and Ikura, K. (1980). *J. Colloid Interface Sci.* **77**, 555–563.
Sporanzi, N., and Cavallotti, C. (1980). *Experientia* **36**, 956.
Sugar, I. P. (1980). *Acta Biochim. Biophys. Acad. Sci. Hung* **15**, 73–75.
Tverdislov, V. A., El Karadagi, S., Martsenyuk, O. V., and Gerasimova, E. N. (1980). *Biofizika* **25**, 841–847.
Varanda, W., and Finkelst A. (1980). *J. Membr. Biol.* **55**, 203–211.

Ion Transport

Antonov, V. F., Petrov, V. V., Molnar, A. A., Predvodi, D. A., and Ivanov, A. S. (1980). *Nature (London)* **283**, 585–586.

Benz, R., and Zimmermann, U. (1980). *Bioelectrochemistry* **7**, 723–739.
Benz, R., Ishii, J., and Nakae, T. (1980). *J. Membr. Biol.* **56**, 19–29.
Blumenthal, R., Klausner, R. D., and Weinstei, J. N. (1980). *Nature (London)* **288**, 333–338.
Chernomordik, L. V., and Abidor, I. G. (1980). *Bioelectrochemistry* **7**, 617–623.
Derzhanski, A., Petrov, A. G., and Pavloff, Y. V. (1981). *J. Phys. (Paris) Lett.* **42**, L119–L122.
Fleischman, M., Gabriell, C., Labram, M. T. G., McMullen, A. I. and Wilmshur, T. H. (1980). *J. Membr. Biol.* **55**, 9–27.
Jordan, P. C. (1980). *Biophys. Chem.* **12**, 1–11.
Kasumov, K. M., Borisova, M. P., Ermishkin, L. N., Potseluyev, V. M., Silberstein, A. Y., and Vainshtein, V. A. (1979). *Biochim. Biophys. Acta.* **55**, 229–237.
Kolb, H.-A., and Frehland, E. (1980). *Biophys. Chem.* **12**, 21–34.
Latorre, R., Donovan, J. J., Koroshet, W., Tosteson, D. C., and Gisin, B. F. (1981). *J. Gen. Physiol.* **77**, 387–417.
Lukyanen, A. I., Berestov, G. N., and Evtodien, Y. V. (1980). *Biofizika* **25**, 82–86.
McLaughlin, S., Mulrine, N., Gresalfi, T., Vaio, G., and McLaughlin, A. (1981). *J. Gen. Physiol.* **77**, 445–473.
Pryashevskaya, J. S., and Shchipun, Y. A. (1980). *Biofizika* **25**, 848–849.
Tverdislov, V. A., Elkarada, S., Martseny, O. V., and Gerasimo, E. N. (1980). *Biofizika* **25**, 841–847.

BLM–Liposome Interactions

Duzgunes, N., and Ohki, S. (1981). *Biochim. Biophys. Acta* **640**, 734–747.
Sokolov, Y. V., and Lishko, V. K. (1980). *Ukr. Biokhim. Zh.* **52**, 700–705.
Zimmerberg, J., Cohen, F. S., and Finkelst. A. (1980). *Science* **210**, 906–908.

Photoeffects in BLMs

Berns, D. S., and Alexandrowica, G. (1980). *Photobiochemistry* **1**, 353–360.
Feldberg, S. W., Armen, G. H., Bell, J. A., Chang, C. K., and Wang, C. B. (1981). *Biophys. J.* **34**, 149–163.
Hong, F. T. (1980). *Adv. Chem. Ser.* **198**, 211–237.
Lojewska, Z., and Kutnik, J. (1981). *Stud. Biophys.* **82**, 127–135.

23

Hydrophobic Labeling and Cross-Linking of Membrane Proteins

Hans Sigrist and Peter Zahler

1. INTRODUCTION

In recent years increasing effort has been put into the selective modification of membrane proteins (Hubbard and Cohn, 1976; Ji, 1979; Peters and Richards, 1977). The initially cumbersome task of modifying membrane proteins from within the lipid bilayer was stimulated by the expected answers to several basic questions in membrane biochemistry.

1. Which parts of and to what extent are membrane proteins inserted in the lipid bilayer?
2. What are the structures of vectorially acting membrane proteins in the crucial hydrophobic domain?
3. What is the nature of hydrophobic interactions between membrane-integrated domains of a single protein, among neighboring proteins, and between proteins and lipids?

To attempt the elucidation of these problems, membrane proteins were hydrophobically labeled and the modified proteins structurally characterized to ascertain the site(s) of modification. The resulting effect on the proteins' function was investigated. Topological interrelationships occurring in the apolar membrane phase were studied by hydrophobic inter- and intramolecular cross-linking.

Classical labeling procedures, in general, make use of water-soluble reagents. These probes interact predominantly with those parts of membrane proteins exposed to the aqueous exterior of the bilayer, thereby providing information on the topological disposition of membrane proteins. Information on additional structural and topological features is to be gained by hydrophobic labeling of the intramembranous segments of membrane proteins. However, the techniques available for apolar modification of membrane proteins are limited in number. Characterization of the so-labeled hydrophobic peptides is difficult to

Hans Sigrist and Peter Zahler • Institute of Biochemistry, University of Bern, CH-3012 Bern, Switzerland.

achieve. Exploration of the intramembrane domain has therefore lagged far behind that of the aqueous-exposed membrane milieu.

The properties of membrane proteins depend on the defined characteristics of a given membrane system. For modification reactions, membrane proteins have to be considered as individual, invariable reactants. The reagents, however, used for membrane labeling can be selected from an abundant variety of modifiers, differing in size, polarity, reactivity, and their reaction mechanism. To specifically modify the apolar domain of membrane proteins the labels have to fulfill certain requirements. (1) The reagent must be hydrophobic and partition in favor of the apolar phase. (2) The functional groups of the membrane protein have to be available in the reactive form. (3) The reaction, the label, and the reaction products are required to be nonionogenic and noncharged. (4) Water should be involved neither as reactant nor as product. These criteria were selected for minimal perturbation of the membrane system (Sigrist and Zahler, 1978).

2. HYDROPHOBIC LABELING REAGENTS

2.1. Photoreagents

Hydrophobic azides and diazirines fulfill the listed requirements for apolar modification of membrane proteins. The common characteristic of these labels is the presence of a chemical group that can be specifically activated by light to form a highly reactive nitrene or carbene, respectively. The reactive molecular species insert into otherwise chemically inert bonds, preferentially C–H bonds. A critical comparative description of the reactivity of these reagents and their utility for membrane labeling studies is described elsewhere in this volume (Chapter 24). Some representative structures of azides and diazirines are summarized in the first section of Table I. Synthetic procedures and applications of the listed reagents for probing integral domains of biological membranes are described in the corresponding references. The selectivity of the labels for the apolar membrane phase is implied by their hydrophobicity, resulting in preferential partitioning into the lipid phase of the bilayer. Favorable distribution into the apolar domain is thus additionally enforced by the nature of the radioactive isotope used ($^{125}I \gg {}^3H/{}^{14}C$) and a high specific radioactivity of the photoreagents (Gitler and Bercovici, 1980; Bayley and Knowles, 1980).

Whereas aryl azides, aryl diazirines, and adamantane diazirines are forced to stay in the apolar lipid phase by their apolarity, azides bearing fatty acids, either singly (Wisnieski and Bramhall, 1979) or as integrated parts of phospholipids, make use of the molecules' amphipathic character to trap the reactive center in the apolar phase of the bilayer. The use of photoactive diacylphospholipids to modify membrane components from within the lipid bilayer has been initiated by Chakrabarti and Khorana (1975) and by Stoffel *et al.* (1976). In recent investigations the incorporation of various photoreactive fatty acids into phospholipids has been reported: ω-(*m*-azidophenoxy) undecanoic acid, ω-(diazirinophenoxy) undecanoic acid, ω-(trifluorodiazopropionoxy)lauric acid (Gupta *et al.*, 1979), 12-amino-(4-*N*-3-nitro-l-azidophenyl)lauric acid (Bisson *et al.*, 1979; Montecucco *et al.*, 1979), 12-(*p*-azidophenyl)-9,10-dithio-dodecanoic acid (Brunner and Richards, 1980).

The mentioned photoreagents have frequently been employed to investigate the disposition of integral membrane proteins from within the lipid bilayer. The use of nitrene- or carbene-generating reagents is advantageous for topological investigations but restricted for structural studies due to the unselective and nonstoichiometric interaction of these probes with membrane proteins and lipids. In addition, the broad reactivity of photogenerated

Table I. Monofunctional Hydrophobic Labeling Reagents

Reagent	Structural formula	References
Photoreagents		
Phenylazide	(phenyl)—N_3	Abu-Salah and Findlay (1977), Bayley and Knowles (1978a)
1-Azido-4-iodobenzene	I—(phenyl)—N_3	Klip and Gitler (1974), Wells and Findlay (1979a, b, 1980)
5-Iodonaphthyl-1-azide (INA)	(naphthyl with N_3 and I)	Klip et al. (1976), Bercovici and Gitler (1978), Sigrist-Nelson et al. (1977), Karlish et al. (1977), Kahane and Gitler (1978), Gitler and Bercovici (1980), Cerletti and Schatz (1979)
Hexanoyldiiodo-N-(4-azido-2-nitrophenyl) tyramine	N_3—(phenyl with NO_2)—NH—$(CH_2)_2$—(phenyl with two I)—O—$C(=O)$—$(CH_2)_4$—CH_3	Owen et al. (1980)

Table I. (Continued)

Reagent	Structural formula	References
3-Phenyl diazirine		Bayley and Knowles (1978b)
1-Spiro (adamantane-4,4′-diazirine)		Goldman et al. (1979), Bayley and Knowles (1980)
3-Trifluoromethyl-3-phenyl diazirine		Brunner et al. (1980)
12- (4-azido-2-nitrophenoxy) stearoylglucosamine	$CH_3 - (CH_2)_5 - CH - (CH_2)_{10} - C - NH - Glc$ 	Wisnieski and Bramhall (1979)
Group-specific reagents		
Phenylisothiocyanate		Sigrist and Zahler (1978, 1981a), Sigrist et al. (1980a,b; 1981), Kempf et al. (1981a)

2-Naphthylisothiocyanate
(NITC)

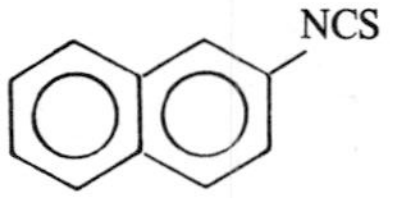

Sigrist and Allegrini (unpublished)

4-*N,N*-dimethylaminoazobenzene-
4'-isothiocyanate
(DABITC)

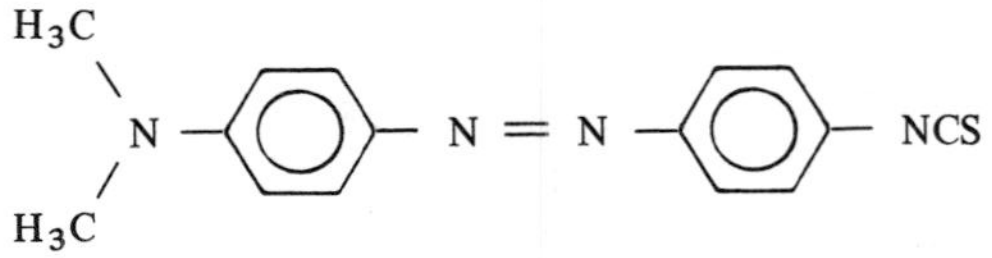

Kempf *et al.* (1981b)

N,N'-dicyclohexylcarbodiimide
(DCCD)

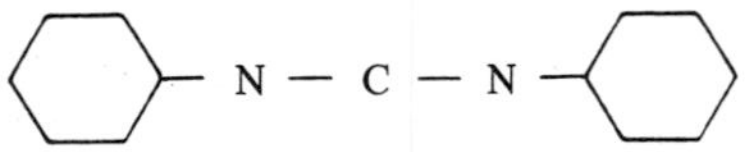

Catell *et al.* (1971), Sebald *et al.*
(1979), Sigrist-Nelson *et al.* (1978)

N-(2,2,6,6-tetramethyl-piperidyl-
1-oxyl)-N'-(cyclohexyl)carbodiimide
(NCCD)

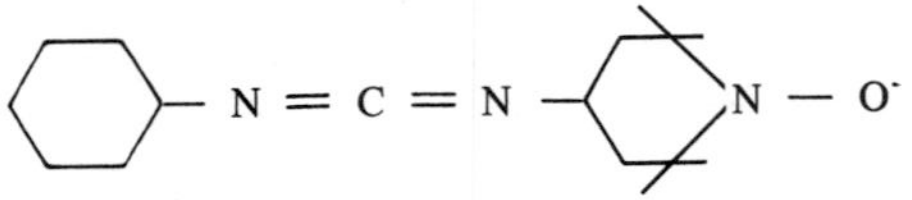

Azzi *et al.* (1973), Sigrist-Nelson
and Azzi (1979)

N-Ethylmaleimide
(NEM)

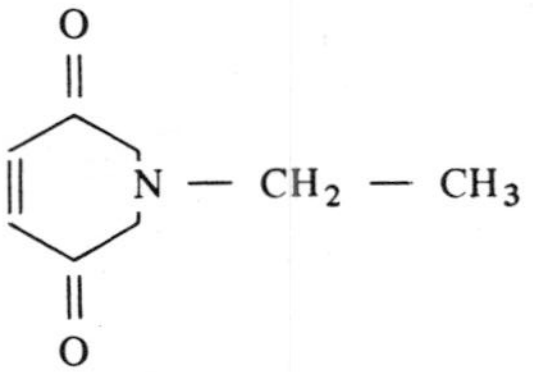

Alexander (1958), Thorley-Lawson
and Green (1977), Haest *et al.* (1979)

probes implies nonselective modification of various amino acid side chains (Chowdhry and Westheimer, 1979; Matheson *et al.*, 1977; Brunner and Richards, 1980).

2.2. Group-Specific Reagents

The apolar reagents listed in Table Ib demonstrate preference for proteinaceous functional groups positioned within the hydrophobic domain. They interact with either amino, carboxyl, or thiol groups. The amino acid sequences of membrane-integrated proteins strongly suggest the presence of such functional groups within intramembranous segments. Protein functional groups should thus be available and accessible to hydrophobic modifiers.

2.2.1. Arylisothiocyanates

Chemical labeling using arylisothiocyanates has been introduced to achieve selective modification of defined amino acid side chains of membrane-integrated protein segments (Sigrist and Zahler, 1978). For modification of proteins, the pH-controlled reactivity of protein nucleophilic groups can be used to obtain selective labeling by these reagents. Arylisothiocyanates are known to form covalent bonds with nucleophilic groups solely in their nonprotonated form ($RS^- \gg RO^- > RNH_2$; Drobnika *et al.*, 1977). As reported in Table II, ϵ-amino groups of lysine residues exposed to the aqueous phase are, at neutral pH, protonated and accordingly not reactive with arylisothiocyanates. In contrast, the buried pH-independent amino functions may be in a reactive (deprotonated) state, making modification with hydrophobic arylisothiocyanates feasible (Table II). Cysteine thiols, histidine imidazole NH and α-amino groups exposed to the aqueous phase are, if present, partially deprotonated at neutral pH and are as a result reactive with arylisothiocyanates. Topological selectivity is obtained by addition of the hydrophobic arylisothiocyanate in the presence of its polar structural analog (e.g., p-sulfophenylisothiocyanate).

Group-directed hydrophobic modification of membrane-integrated segments by arylisothiocyanates has been applied to bacteriorhodopsin. Labeling of purple membranes with phenylisothiocyanate resulted in covalent modification of a lysine ϵ-amino group of bacteriorhodopsin (Sigrist *et al.*, 1980b, 1981). The phenylisothiocyanate binding site is not

Table II. Interaction of Arylisothiocyanates with Amino Acid Side-Chain Nucleophilic Groups

Amino acid nucleophilic group		Intrinsic dissociation constant	Reactive form of the nucleophile	Nucleophilic form present at neutral aqueous pH	
				Polar domain	Apolar domain
Cysteine	–SH	8.3	R–S⁻	R–S⁻/R–SH	R–SH
Serine/ threonine	–OH	>14	R–O⁻	R–OH	R–OH
Tyrosine	–OH	9.6[a]	R–O⁻	R–OH	R–OH
Lysine	–NH₂	10.2[a]	R–NH₂	R–NH₃⁺	R–NH₂
Imidazole (histidine)	⟩NH	7.0[a]	⟩N⁻	⟩N⁻/ NH	⟩NH

[a] Data from Glazer (1976).

accessible for the water-soluble analog *p*-sulfophenylisothiocyanate, indicating the intramembranous disposition of the protein segment. Covalent binding of phenylisothiocyanate to the erythrocyte anion transport protein, band 3, has been reported (Sigrist *et al.*, 1980a). With regard to this membrane-integrated protein, 4–5 moles of phenylisothiocyanate was covalently bound per mole of protein. The label binds preferentially to a transmembrane COOH-terminal fragment whose apparent molecular weight is 10,000 daltons (Kempf *et al.*, 1981a). The sarcoplasmic Ca^{2+}-stimulated ATPase has topologically been investigated for nucleophile accessibility by water-soluble or hydrophobic arylisothiocyanates. The presence of hydrophobically located nucleophiles has been demonstrated (Sigrist and Zahler, 1981a). Hydrophobic modification of membrane proteins can also be obtained using the fluorescent probe 2-naphthylisothiocyanate or the colored 4-*N,N'*-dimethylaminoazobenzene-4'-isothiocyanate as group-specific modifier. Labeled proteins will thus become fluorescent (Sigrist and Allegrini, unpublished) or highly chromophoric, respectively (Kempf *et al.*, 1981b).

2.2.2. Carbodiimides

Carbodiimides are probably the most widely used reagents for forming amide bonds. The reaction of a carboxyl group with a carbodiimide yields first an *O*-acylurea that, if an amine is present, reacts to form an amide. Competing with this reaction is the rearrangement to an *N*-acylurea, which is stable and no longer reacts with nucleophiles. The latter reaction, first described by Hoare and Koshland (1967), favorably occurs in the absence of nucleophiles (e.g., amino groups) and water (Previero *et al.*, 1973). The reaction of free acetic acid with di-*p*-tolylcarbodiimide in carbon tetrachloride results in the corresponding *N*-acetylurea (Smith *et al.*, 1958). It is therefore not surprising that apolar carbodiimides (e.g., *N,N'*-dicyclohexyl-carbodiimide) were frequently applied for hydrophobic modification of membrane-buried carboxyl groups.

The most outstanding example for dicyclohexylcarbodiimide modification are the dicyclohexylcarbodiimide-binding proteins, which are membrane-integrated parts of energy-transducing ATPase complexes. The sequences of at least six of these proteins from various sources have been established. Dicyclohexylcarbodiimide binds covalently to γ-(β)carboxyl groups of glutamic (aspartic) acid (Sebald *et al.*, 1979). The apolar disposition of carbodiimide binding sites has been documented by inaccessibility either to the polar analog 1-ethyl-3-(3-dimethylaminopropyl)carbodiimide (Sigrist-Nelson and Azzi, 1980; Pick and Racker, 1979), or to ascorbic acid when spin-labeled proteins were investigated (Azzi *et al.*, 1973; Sigirst-Nelson and Azzi, 1979).

2.2.3. Maleimides

Maleimides are relatively stable reagents that form covalent bonds stoichiometrically with sulfhydryl groups (Alexander, 1958). The existence of *N*-ethylmaleimide-reactive thiol groups within the membrane has been documented in the study of Abbott and Schachter (1976), which compared the accessibility of membrane-impermeant and membrane-permeant maleimides. *N*-Ethylmaleimide and its analogs have been widely applied for membrane protein modification, including topographical studies (Thorley-Lawson and Green, 1977; Rao, 1979; Reithmeier and Rao, 1979; Haest *et al.*, 1979).

Table III. Hydrophobic Cross-Linking Reagents

Reagent	Structural formula	References
Homobifunctional reagents		
1,5-Diazidonaphthalene (DAN)		Mikkelsen and Wallach (1976)
4,4'-Diazidobiphenyl (DAB)		Mikkelsen and Wallach (1976)
4,4'-Dithiobisphenylazide (DPA)		Mikkelsen and Wallach (1976), Kiehm and Ji (1977)
N,N'-Phenylene dimaleimide (PDM)		Wold (1972), Weiss and McCarty (1977)
Heterobifunctional reagents		
p-Azidophenylisothiocyanate (APITC)		Sigrist and Zahler (1980a, 1981b) Sigrist et al. (in preparation)

5-Isothiocyanato-1-naphthaleneazide (ITCNA)

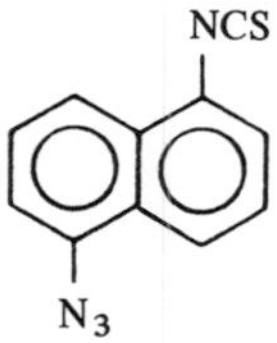

Sigrist and Zahler (1981b),
Sigrist et al. (in preparation)

N-p-azidophenyl-N'-cyclohexyl carbodiimide (ACCD)

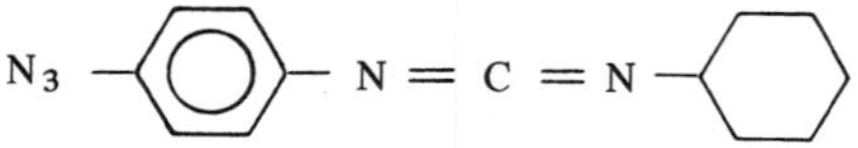

Sigrist et al. (in preparation)

4-Azidophenyl maleimide (APM)

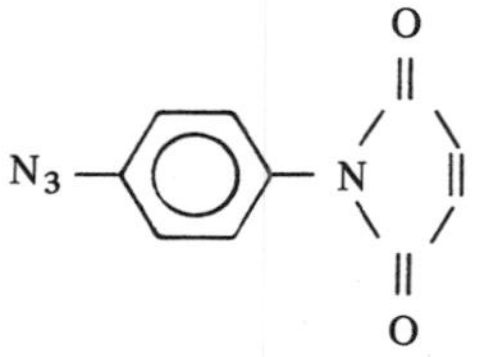

Trommer et al. (1977)

N-(4-azidophenylthio) phthalimide (APTP)

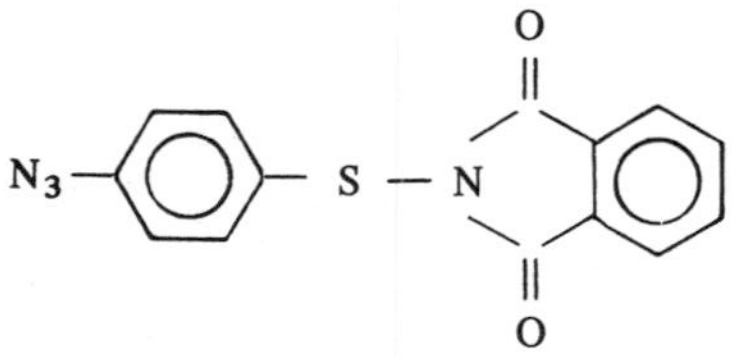

Kiehm and Ji (1977)

Di-N-(2-nitro-4-azidophenyl) cystamine-S,S-dioxide (DNCO)

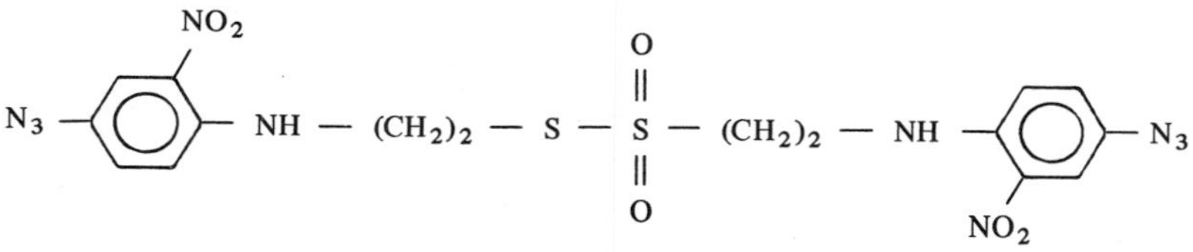

Huang and Richards (1977),
Peters and Richards (1977)

3. HYDROPHOBIC CROSS-LINKING REAGENTS

Cross-linkers are the major tool for predicting molecular associations in biological membranes. They allow structural studies in the dimension range of 5 to 20 Å. Nearest-neighbor analysis can therewith be performed *in situ*. Intramolecular cross-linking studies then become feasible, resulting in structural information. By cross-linking membrane components from within the lipid phase, changes in the surface charge distribution due to reagent binding are avoided.

In recent years, various hydrophobic cross-linking probes have been designed to specifically investigate the apolar membrane domain (Table III). These reagents possess a high potential for the study of membrane topography with regard to hydrophobic inter- and intramolecular relationships.

3.1. Homobifunctional Cross-Linking Reagents

The application of hydrophobic homobifunctional reagents for intramembranous cross-linking was first described by Mikkelsen and Wallach (1976). In this study, 1,5-diazido-naphthalene, 4,4'-diazidobiphenyl, and 4,4'-dithiobisphenylazide were synthesized and their application characterized in the erythrocyte membrane system. In addition to high-molecular-weight products, dimerization of erythrocyte band 3 was observed upon cross-linking with the cleavable 4,4'-dithiobisphenylazide (see also Kiehm and Ji, 1977).

The bifunctional maleimide, N,N'-phenylene dimaleimide, is insoluble in aqueous media. However, when added as a solid to solutions of bovine serum albumin, intermolecular cross-links result (Wold, 1972). O-Phenylene dimaleimide induces intramolecular cross-links in the subunit of the chloroplast coupling factor CF_1 (Weiss and McCarty, 1977). To date, dimaleimides have generally been applied for –SH cross-linking studies. However, because most of the reagents used are hydrophobic, they may be utilized as well for specific cross-link formation in the apolar membrane phase, if aqueous-exposed thiol functions have previously been protected.

3.2. Heterobifunctional Cross-Linking Reagents

Heterobifunctionality in hydrophobic cross-linking reagents offers the advantage of forming the cross-link in stepwise reactions. Initial attachment of the cross-linker optionally follows the specificity that has been described for group-selective hydrophobic probes. The primary site of interaction is thus topologically defined and can be identified in the primary protein sequence of modified proteins. The heterofunction is intentionally chosen to be highly and unspecifically reactive. The time necessary for covalent interaction should be as short as possible to guarantee random insertion into mobile membrane components. Photogenerated nitrenes and carbenes optimally meet these requirements. In addition, they offer the advantage of selective activation of the heterofunction by exposure to light.

Table III lists some representative heterobifunctional reagents that have been, or might be, utilized for group-specific hydrophobic cross-linking. The target functions for primary interactions are thiol, amino, and carboxyl groups of membrane-integrated protein fragments. The synthesis of 4-azidophenyl maleimide (Trommer *et al.*, 1977) and N-(4-azidophenylthio) phthalimide (Kiehm and Ji, 1977) have been described. The latter compound can be transferred to un-ionized sulfhydryl groups in erythrocyte membranes (Ji, 1979). Huang and Richards (1977) successfully applied the lipid-soluble, cleavable cross-linking reagent di-N-(2-nitro-4-azidophenyl)cystamine-S,S-dioxide for the cross-linking of erythrocyte membrane proteins. The cross-linked complexes formed were similar to but not identical with those produced by water-soluble bisimidates or Cu^{2+}-orthophenanthroline.

The azidoarylisothiocyanates, *p*-azidophenylisothiocyanate and 5-isothiocyanato-1-naphthalene azide, have been utilized for hydrophobic cross-linking of bacteriorhodopsin in purple membranes (Sigrist and Zahler, 1980, 1981b). Both reagents compete for phenylisothiocyanate binding. Upon photoactivation, homopolymers are recovered demonstrating covalent interaction of both the arylisothiocyanate and the azido function. Cross-linking of *p*-azidophenylisothiocyanate-labeled band 3 in a vesicular system resulted in dimerization of the anion transport protein. N-(*p*-azidophenyl)-N'-cyclohexyl carbodiimide has been applied to investigate the environment of hydrophobically located carboxyl groups. The interaction of this reagent with the dicyclohexylcarbodiimide-binding proteolipid has been investigated. (Sigrist *et al.*, in preparation).

4. CONCLUSIONS

The presented overview has documented the versatile pioneering attempts to explore the apolar membrane phase by chemical modification. The current list of hydrophobic labels and cross-linking reagents is diverse and growing rapidly. Depending on the structural and chemical characteristics of the reagents used, chemical probes can be directed towards topologically distinguishable areas of integrated membrane proteins. If properly applied, hydrophobic labeling studies will render important information on both structural and functional features of the intrinsic membrane domain, where crucial steps in transmembrane processes occur.

ACKNOWLEDGMENTS. This work was supported by the Swiss National Science Foundation (Grant 3.674-0.80) and by the Central Laboratories of the Swiss Blood Transfusion Service SRK, Bern. The authors thank Dr. K. Sigrist-Nelson for valuable discussions.

REFERENCES

Abbott, R. E., and Schachter, D. (1976). *J. Biol. Chem.* **251**, 7176–7183.
Abu-Salah, K. M., and Findlay, J. B. C. (1977). *Biochem. J.* **161**, 223–228.
Alexander, N. M. (1958). *Anal. Chem.* **30**, 1292–1294.
Azzi, A., Bragadin, M. A., Tamburro, M. A., and Santato, M. (1973). *J. Biol. Chem.* **248**, 5520–5526.
Bayley, H., and Knowles, J. R. (1978a). *Biochemistry* **17**, 2414–2419.
Bayley, H., and Knowles, J. R. (1978b). *Biochemistry* **17**, 2420–2423.
Bayley, H., and Knowles, J. R. (1980). *Biochemistry* **19**, 3887–3892.
Bercovici, F., and Gitler, C. (1978). *Biochemistry* **17**, 1484–1489.
Bisson, R., Montecucco, C., Gutweniger, H., and Azzi, A. (1979). *J. Biol. Chem.* **254**, 9962–9965.
Brunner, J., and Richards, F. M. (1980). *J. Biol. Chem.* **255**, 3319–3329.
Brunner, J., Senn, H., and Richards, F. M. (1980). *J. Biol. Chem.* **255**, 3313–3318.
Catell, K., Lindop, C., Knight, I., and Beechey, R. (1971). *Biochem. J.* **125**, 169–177.
Cerletti, N., and Schatz, G. (1979). *J. Biol. Chem.* **254**, 7746–7751.
Chakrabarti, P., and Khorana, H. G. (1975). *Biochemistry* **14**, 5021–5033.
Chowdhry, V., and Westheimer, F. H. (1979). *Annu. Rev. Biochem.* **48**, 293–325.
Drobnika, L., Kristian, P., and Augustin, J. (1977). In *The Chemistry of Cyanates and Their Thioderivatives* (S. Patai, ed.), Part 2, pp. 1002–1222, Wiley, New York.
Gitler, C., and Bercovici, T. (1980). *Ann. N.Y. Acad. Sci.* **346**, 199–211.
Glazer, A. (1976). In *The Proteins* (H. Neurath and R. L. Hill, eds.), 3rd ed., pp. 1–103, Academic Press, New York.

Goldman, D. N., Pober, J. S. White, J., and Bayley, H. (1979). *Nature (London)* **280,** 481–483.
Gupta, C. M., Radhakrishnan, R., Gerber, G. E., Olsen, W. L., Quay, S. C., and Khorana, H. G. (1979). *Proc. Natl. Acad. Sci. USA* **76,** 2595–2599.
Haest, C. W. M., Kamp, D., and Deutike, B. (1979). *Biochim. Biophys. Acta* **557,** 363–371.
Hoare, D. C., and Koshland, D. E. (1967). *J. Biol. Chem.* **242,** 2447–2453.
Huang, C., and Richards, F. M. (1977). *J. Biol. Chem.* **252,** 5514–5521.
Hubbard, A., and Cohn, Z. (1976). In *Biochemical Analysis of Membranes* (A. H. Maddy, ed.), pp. 427–501, Chapman & Hall, London.
Ji, H. (1979). *Biochem. Biophys. Acta* **559,** 39–69.
Kahane, J., and Gitler, C. (1978). *Science* **201,** 351–352.
Karlish, S. J. D., Jorgensen, P. L., and Gitler, C. (1977). *Nature (London)* **269,** 715–717.
Kempf, C., Brock, C., Sigrist, H., Tanner, M. J. A., and Zahler, P. (1981a). *Biochim. Biophys. Acta,* **641,** 88–98.
Kempf, C., Sigrist, H., and Zahler, P. (1981b). *FEBS Lett.* **124,** 225–228.
Kiehm, D. J., and Ji, T. H. (1977). *J. Biol. Chem.* **252,** 8524–8531.
Klip, A., and Gitler, C. (1974). *Biochim. Biophys. Res. Commun.* **60,** 1155–1162.
Klip, A., Darszon, A., and Montal, M. (1976). *Biochim. Biophys. Res. Commun.* **72,** 1350–1358.
Matheson, R. R., van Went, H. E., Burgess, A. W., Weinstein, L. I., and Sheraga, H. A. (1977). *Biochemistry* **16,** 396–403.
Mikkelsen, R. B., and Wallach, D. F. H. (1976). *J. Biol. Chem.* **251,** 7413–7416.
Montecucco, C., Bisson, R., Pitotti, A., Dabbeni-Sala, F., and Gutweniger, H. (1979). *Biochem. Soc. Trans.* **7,** 954–955.
Owen, M. J., Knott, J. C. A., and Crumpton, M. J. (1980). *Biochemistry* **19,** 3092–3099.
Peters, K., and Richards, F. M. (1977). *Annu. Rev. Biochem.* **46,** 523–551.
Pick, U., and Racker, E. (1979). *Biochemistry* **18,** 108–113.
Previero, A., Derancourt, J., Coletti-Previero, M. A., and Laursen, R. A. (1973). *FEBS Lett.* **33,** 135–138.
Rao, A. (1979). *J. Biol. Chem.* **254,** 5303–3511.
Reithmeier, R. A. F., and Rao, A. (1979). *J. Biol. Chem.* **254,** 6151–6155.
Sebald, W., Hoppe, J., and Wachter, E. (1979). In *Function and Molecular Aspects of Biomembrane Transport* (E. Quagliariello, F. Palmieri, S. Papa, and M. Klingenberg, eds.), pp. 63–74, Elsevier/North-Holland, Amsterdam.
Sigrist, H., and Zahler, P. (1978). *FEBS Lett.* **95,** 116–120.
Sigirst, H., and Zahler, P. (1980). *FEBS Lett.,* **113,** 307–311.
Sigrist, H., and Zahler, P. (1981a). *J. Bioenerg. Biomembr.* **13,** 89–101.
Sigrist, H., and Zahler, P. (1981b). *Methods Enzymol.* Submitted for publication.
Sigrist, H., Kempf, C., and Zahler, P. (1980a). *Biochim. Biophys. Acta* **597,** 137–144.
Sigrist, H., Allegrini, P. R., Strasser, R. J., and Zahler, P. (1980b). In *The Blue Light Syndrome* (H. Senger, ed.), pp. 30–37, Springer-Verlag, Berlin.
Sigrist, H., Allegrini, P. R., Zahler, P., Abdulaev, N. G., Feigina, M. Yu., and Ovchinnikov, Yu. (1981). Submitted for publication.
Sigrist-Nelson, K., and Azzi, A. (1979). *J. Biol. Chem.* **254,** 4470–4474.
Sigrist-Nelson, K., and Azzi, A. (1980). *J. Biol. Chem.* **255,** 10638–10643.
Sigrist-Nelson, K., Sigrist, H., Bercovici, T., and Gitler, C. (1977). *Biochim. Biophys. Acta* **468,** 163–176.
Sigrist-Nelson, K., Sigrist, H., and Azzi, A. (1978). *Eur. J. Biochem.* **92,** 9–14.
Smith, M., Moffatt, J. G., and Khorana, H. G. (1958). *J. Am. Chem. Soc.* **80,** 6204–6212.
Stoffel, W., Salem, K., and Köchemeier, U. (1976). *Hoppe-Seylers Z. Physiol. Chem.* **357,** 917–924.
Thorley-Lawson, D. A., and Green, M. N. (1977). *Biochem. J.* **167,** 739–748.
Trommer, W. E., Friebel, K., Kiltz, H., and Kolkenbrock, H. (1977). In *Protein Crosslinking* (M. Friedman, ed.), pp. 187–195, Plenum Press, New York.
Weiss, M. A., and McCarty, R. E. (1977). *J. Biol. Chem.* **252,** 8007–8012.
Wells, E., and Findlay, J. B. C. (1979a). *Biochem. J.* **179,** 257–264.
Wells, E., and Findlay, J. B. C. (1979b). *Biochem. J.* **179,** 265–272.
Wells, E., and Findlay, J. B. C. (1980). *Biochem. J.* **187,** 719–725.
Wisnieski, B. J., and Bramhall, J. S. (1979). *Biochem. Biophys. Res. Commun.* **87,** 308–313.
Wold, F. (1972). *Methods Enzymol.* **25,** 623–651.

24

Photoactivated Hydrophobic Reagents for Integral Membrane Proteins

A Critical Discussion

Hagan Bayley

1. INTRODUCTION

Photoactivated hydrophobic reagents are being developed to label those regions of integral membrane proteins that lie within the hydrocarbon core of the lipid bilayer. There are three goals. First, we wish to have a collection of reagents that react entirely from within the bilayer and to a negligible extent from the aqueous phase. Second, we would like to label, without discrimination, all the functional groups exposed to the hydrocarbon core of the membrane. A third goal is to make reagents that will label integral proteins at points that are at defined distances from the surface of the membrane.

The photoactivated reagents that have been developed to carry out these tasks may be divided into three groups; namely, in order of increasing molecular complexity: simple hydrophobic reagents (Table I), amphipathic reagents (Fig. 1), and phospholipid analogs (Fig. 2). The hydrophobic and amphipathic reagents can be introduced directly into native biological membranes; the lipids can be introduced by vesicle fusion or with phospholipid exchange proteins, synthesized *in vivo* in cells auxotrophic for fatty acids, or used in reconstituted proteoliposomes. Photoactivated reagents are used because there is no danger of them reacting in undesired regions of the membrane during the binding or reconstitution processes. Besides, only they have the potential of yielding the highly reactive species that are required to achieve the second and perhaps the third of the three goals.

Although this review focuses on the simple hydrophobic reagents, many of the points that are made apply also to the amphipathic reagents and the phospholipid analogs. To

Hagan Bayley • Department of Biochemistry, College of Physicians & Surgeons, Columbia University, New York, New York 10032.

Table I. Comparison of Three Well-Characterized Hydrophobic Reagents [a]

Reagent	Structure	λ_{max}/ϵ	P_{rbc} [b]	Reactive species	References
Iodonaphthylazide	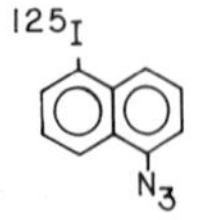	310 nm/21,400	163,000	Nitrene and/or azacyclo-heptatetraene	Bercovici *et al.* (1978), Kahane and Gitler (1978), Karlish *et al.* (1977), Tarrab-Hazdai *et al.* (1980)
Adamantane diazirine	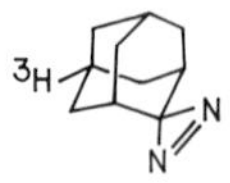	372 nm/245	1,750	Carbene and diazoada-mantane	Bayley and Knowles (1978a,b, 1980a,b), Goldman *et al.* (1979), Farley *et al.* (1980)
3-Trifluoromethyl-3-(*m*-iodophenyl)diazirine	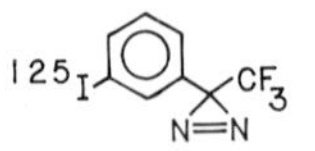	353 nm/266	24,000	Carbene (diazo com-pound is inert)	Brunner (1981), Brunner and Semenza (1981)

[a] Other simple hydrophobic reagents have been used. For aryl azides see Cerletti and Schatz (1979), Wells and Findlay (1980), Owen *et al.* (1980). Pyrene sulfonyl azide has also been used (Sator *et al.*, 1979).
[b] P_{rbc}, the partition coefficient of the reagent into red blood cell membranes, is defined as: (ligand bound/mg membrane protein)/(free ligand/μl buffer).

obtain a wider perspective the reader should also consult the recent articles of Brunner (1981), Robson *et al.*, (1981), and Sigrist and Zahler (this volume).

1.1. Goal 1: Reaction within the Bilayer

Three lines of evidence have been presented to show that simple hydrophobic reagents can be made that label from within the lipid bilayer. First, the fatty acyl chains of single-bilayer vesicles suspended in aqueous media can be derivatized by the more reactive car-

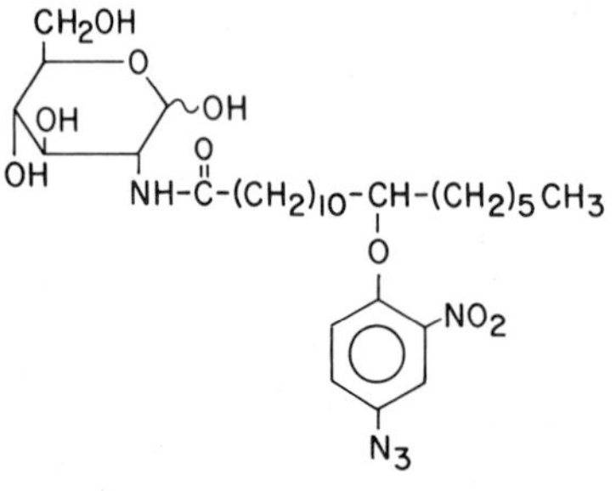

Figure 1. An amphipathic reagent. See Wisnieski and Bramhall (1981) and references therein for details.

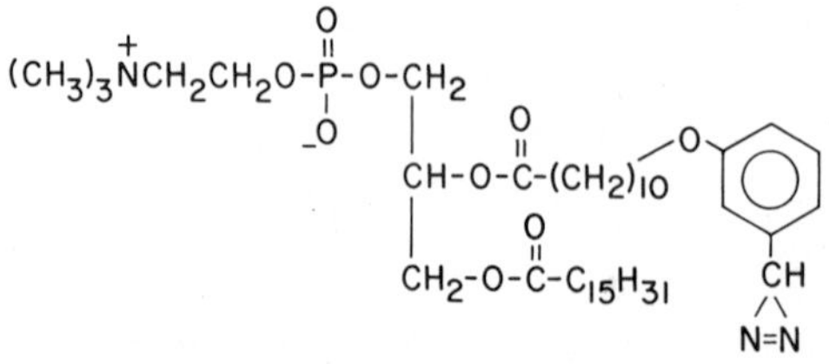

Figure 2. A photolabile phospholipid analog (Radhakrishnan *et al.*, 1980). A variety of lipid analogs containing diazirine, trifluorodiazopropionyl, aryl azide, and alkyl azide functionalities have been made (see Radhakrishnan *et al.*, 1980, 1981; Brunner and Richards, 1980; Bisson *et al.*, 1979; Stoffel *et al.*, 1978).

bene reagents. Carbon–hydrogen bond insertion may be demonstrated by methanolysis of the derivatized lipids and examination of the fatty acid methyl esters by combined gas chromatography and mass spectrometry (Bayley and Knowles, 1978b). Second, with certain reagents (for examples see Table I), in which are combined sufficient hydrophobicity and intrinsic reactivity, only integral membrane proteins and not peripheral proteins are heavily labeled. This may be demonstrated by labeling membranes such as the plasma membrane of the human red blood cell for which the disposition of the proteins has been deduced from independent evidence (Bercovici *et al.*, 1978; Bayley and Knowles, 1980b; Brunner and Semenza, 1981). Third, it has been demonstrated that only the membrane-bound portions of integral membrane proteins are labeled (Kahane and Gitler, 1978; Goldman *et al.*, 1979; Brunner, 1981). Analogous demonstrations have been made with amphipathic reagents (Hu and Wisnieski, 1979) and with phospholipid analogs (see Radhakrishnan *et al.*, 1980; Robson *et al.*, 1981).

1.2. Goal 2: Reactivity toward All Functional Groups within the Lipid Bilayer

Chemical reagents are generally unreactive toward hydrophobic amino acids, which are expected to be the predominant exposed residues within the lipid bilayer. Some membrane-associated polypeptide segments are indeed extraordinarily rich in amino acids with hydrocarbon side chains; e.g., the COOH-terminal region of the HLA antigens (Ploegh *et al.*, 1980) and the NH₂-terminal region of isomaltase (Frank *et al.*, 1978). In other cases, such as bacteriorhodopsin, regions of the protein that are believed to lie within the bilayer contain polar or charged groups; but here it is likely that these residues are not in contact with the hydrocarbon core of the bilayer but are involved in intermolecular and intramolecular interactions (Engelman *et al.*, 1980). Again, the amide linkages of the peptide backbone are likely to be protected from the hydrocarbon phase and to be hydrogen-bonded within such structures as α helices. It is apparent that an ideal reagent should be capable of reacting with the most inert functionalities, namely carbon–hydrogen bonds. A second and related desirable property of a hydrophobic reagent would be the ability to insert into a carbon–hydrogen bond in the presence of nucleophilic functional groups such as $-NH_2$ and $-OH$. A reagent possessing this property would label all regions of a particular polypeptide that were accessible to it; further, it would not discriminate between a region comprising (for example) only leucine and valine, and one that also contained a nucleophilic serine residue.

The known photochemistry of aryl azides suggests that the nitrenes derived from them will not undergo efficient intermolecular reaction with hydrocarbons even under favorable circumstances. Accordingly, when phenylnitrene is generated photochemically in phospholipid vesicles suspended in aqueous medium, it does not react efficiently with saturated fatty acyl chains, whereas the isosteric, isoelectronic phenyl carbene does insert into carbon–hydrogen bonds under the same conditions (Bayley and Knowles, 1978a,b). These experiments suggest that carbenes may be used to label chemically inert polypeptides within the lipid bilayer in the absence of reactive groups that might compete for the reagent.

The second criterion, reaction with such unreactive groups as carbon–hydrogen bonds in the presence of more reactive functionalities, has not yet been achieved, and several lines of evidence support this statement. Even carbenes are selective in their reactions: photogenerated singlet methylene, a highly indiscriminate species, reacts with the hydroxyl of *t*-butanol 11 times more rapidly than with a carbon–hydrogen bond (Kerr *et al.*, 1967). Furthermore, besides the desired carbenes or nitrenes, the reagents that have been used until recently yield, when irradiated, unwanted electrophilic and highly selective reactive intermediates that are relatively long-lived. Aryl azides yield azacycloheptatetraenes (Chapman, 1979), and such species or derivatives of them may be responsible for much of the observed labeling from these reagents (Nielsen *et al.*, 1978; Mas *et al.*, 1980; for a discussion see Staros, 1980). Nitrenes derived from alkyl azides (e.g., Stoffel *et al.*, 1978) are expected to rearrange to imines, and the mechanism of the photochemical reaction between azidoalkyl lipids and membrane proteins is unclear. Diazirines usually rearrange in part to diazo compounds (Smith and Knowles, 1975; Bayley and Knowles, 1978b), which are expected to react with nucleophiles, in particular carboxylic acids (e.g., Asp and Glu). Because the formation of electrophilic ketenes was a problem with the original diazoester photoaffinity reagents devised by Westheimer, the improved trifluorodiazopropionyl functionality (Chowdhry *et al.*, 1976) was chosen for attachment to phospholipids (see Radhakrishnan *et al.*, 1980, and references therein). A most important recent development has been the introduction of trifluoromethylphenyldiazirines, the diazo rearrangement products of which are quite inert (Brunner *et al.*, 1980; Brunner and Richards, 1980; Brunner, 1981; Brunner and Semenza, 1981) (Table I).

Evidence for selectivity and the existence of long-lived species has been obtained in experiments with biological membranes. (1) The protein-to-lipid labeling ratio is far higher than would be expected on statistical grounds and it varies with the reagent (compare Bercovici *et al.*, 1978; Bayley and Knowles, 1980b; Brunner and Semenza, 1981). (2) When different reagents are used on the same membranes, different labeling patterns are often obtained, both at the level of which polypeptides are labeled (e.g., for acetylcholine receptor compare Sator *et al.*, 1979; Tarrab-Hazdai *et al.*, 1980), and even at the level of which fragments of a polypeptide are labeled (for Na^+/K^+-ATPase, see below). (3) Recently, Ross and Khorana (1982) have shown by sequence analysis that a 1H-aryl diazirino lipid reacts mainly with a single residue (Glu 70) of glycophorin A in reconstituted membranes, and it is likely that this is a reaction of the diazo isomer of the diazirine.

1.3. Goal 3: Reaction at a Fixed Depth within the Bilayer

Amphipathic and phospholipid reagents have been designed with the view of labeling proteins from different depths within a membrane. The success of such reagents will depend on the product of two factors: the reactivity and the molecular rigidity of the reagent. In the case of photochemical coupling to neighboring fatty acyl chains, where most of the positions within the bilayer are of equal reactivity, flexible molecules (e.g., Fig. 2) have been shown to react in a region broadly centered around the expected depth of the reactive group linked to the extended chain of the reagent (Gupta *et al.*, 1979; Czarniecki and Breslow, 1979; Radhakrishnan *et al.*, 1980). However, it is obvious that if the photogenerated reactive group spends just a small fraction of its lifetime in a region that is many times more reactive than its most frequently occupied dwelling place, it will react in an undesired region of the bilayer (or even outside the bilayer if a simple hydrophobic reagent is used). Indeed, in the case of the weakly bound, comparatively unreactive photolysis products of phenyl azide, the labeling of red cell membrane proteins was greatly reduced in the presence of the water-soluble scavenger glutathione (Bayley and Knowles, 1980b). More recently, photoactivated fatty acid chains of lipids have been observed to loop around and react close to the membrane surface (Brunner and Richards, 1980; Ross and Khorana, 1982). There is a clear requirement for more rigid molecules in such experiments.

2. THE USE OF HYDROPHOBIC REAGENTS

The present view then is that hydrophobic reagents are useful for distinguishing integral from peripheral proteins and for detecting those regions of integral proteins that lie within or very close to the bilayer. For the time being, it would seem prudent to use several reagents with different reactive groups and of varying molecular shape for each membrane protein investigated. At present, hydrophobic reagents cannot be used to label indiscriminately all the polypeptide segments associated with the hydrocarbon region of a membrane, nor have molecules yet been developed that will label at given depths within the bilayer, except in model systems.

Full experimental details for the synthesis and use of hydrophobic reagents may be found in the references cited (see especially: Bercovici *et al.*, 1978; Bayley and Knowles, 1980b; Brunner and Semenza, 1981; Hu and Wisnieski, 1979; Wisnieski and Bramhall, 1981; Robson *et al.*, 1981; Radhakrishnan *et al.*, 1981; Brunner and Richards, 1980; Bisson *et al.*, 1979; Bisson and Montecucco, 1981). The precautions that must be taken when using simple hydrophobic reagents are briefly restated here. Analogous caveats often apply to the amphipathic molecules, but the phospholipids are free of some of the potential problems of the simpler molecules as they are unlikely to dissociate from the lipid bilayer or bind to hydrophobic pockets in proteins.

Clearly the extent of labeling of peripheral proteins should always be low or negligible. Endogenous peripheral proteins (e.g., actin), proteins that bind nonspecifically to the surface of the membrane (e.g., lysozyme to negatively charged membranes), or IgG against specific surface antigens (Prochaska *et al.*, 1980; this would seem to be a useful and generally applicable control) may be monitored for the incorporation of radioactivity. Another useful test is to conduct a labeling experiment in the presence of an impermeant scavenger (Bayley and Knowles, 1978a,b). The addition of glutathione led to a massive reduction in the extent of labeling of red cell membrane proteins by phenyl azide, a weakly bound reagent (Bayley and Knowles, 1980b). Finally, it should be ensured that there are no tight-binding sites for a reagent. This can be done by monitoring the protein-to-lipid labeling ratio over a wide range of concentrations (Bayley and Knowles, 1980b; Farley *et al.*, 1980) or by looking at the effects of prelabeling with a nonradioactive reagent and then labeling with the radioactive compound (Bayley and Knowles, 1980b).

3. SOME RECENT APPLICATIONS OF HYDROPHOBIC REAGENTS

To illustrate the points made above, three examples of hydrophobic labeling are examined in more detail.

3.1. Structure of Na$^+$/K$^+$-ATPase

The Na$^+$/K$^+$-ATPase (or "sodium pump") of eukaryotic cells serves to couple the free energy of hydrolysis of ATP to the translocation of Na$^+$ and K$^+$ ions across the plasma membrane against their concentration gradients. Both the large catalytic subunit of the protein (α, 100,000 daltons) and the smaller glycoprotein subunit (β, 50,000 daltons) are firmly attached to the cell membrane. Two attempts have been made to determine the nature of this association by using hydrophobic reagents. In the first [5-^{125}I]iodonaphthylazide was used (Karlish *et al.*, 1977), and in the second the reagent was [^{3}H]adamantane diazirine (Farley *et al.*, 1980).

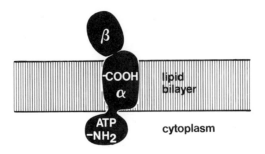

Figure 3. A tentative model for the disposition of the α and β subunits of $Na^+/^+K$-ATPase with respect to the lipid bilayer. The model is based in part on labeling studies with adamantane diazirine (Farley *et al.*, 1980).

When the relative extents of labeling of the two subunits were examined, both reagents gave similar patterns: the α subunit was strongly labeled and the β subunit hardly at all. The most agreeable interpretation of this result is that the β subunit is a peripheral protein tightly bound to the α subunit, an integral protein (Fig. 3). Of course, as indicated in the preceding discussion, there remains the possibility that the β subunit is simply unreactive toward hydrophobic reagents (but the two used in these studies have quite different properties) and the more intriguing possibility that the β subunits do penetrate the bilayer (indeed, the β subunits are quite resistant to proteolysis; Castro and Farley, 1979) but are shielded from the hydrocarbon domain by neighboring α subunits.

The agreement between the two studies ends when the distribution of label within the α subunit is considered. Label from adamantane diazirine was located in at least two broadly defined regions in the COOH-terminal half of the α chain (Fig. 3), whereas iodonaphthylazide appeared to label the polypeptide almost exclusively in a 12,000-dalton region at the NH$_2$-terminus. Such discordant results could arise for two reasons. First, one of the reagents might bind to a particular site on the α chain, not necessarily within the bilayer, and affinity label the polypeptide. There is some evidence for this in the case of iodonaphthylazide and Na^+/K^+-ATPase. When used at higher concentrations, proportionally less of this reagent is incorporated into the α chain on irradiation, suggesting that the labeling site can be saturated, which should not be the case with an indiscriminate hydrophobic reagent. Second, there arises the problem of the relative reactivity of the two reagents toward the various functional groups in the hydrophobic segments of the α chain. If a reactive group with a predilection for reaction with a hydrophobic reagent lay within the bilayer, a saturation phenomenon might also be observed.

Na^+/K^+-ATPase is known to undergo conformational changes during the catalytic cycle. Simply put, there is a form with K^+ bound (E_K) and a form with Na^+ bound (E_{Na}). It would be interesting to define the two forms in terms of their dispositions in the lipid bilayer and in principle this could be done using hydrophobic reagents. If one form were labeled more strongly than another, a reasonable conclusion would be that a larger part of its surface was exposed to the hydrocarbon domain. Such a situation might arise if a segment of the polypeptide in question moved into the bilayer or if two subunits within the bilayer dissociated. By now, the reader will, of course, be aware that the movement of an occluded reactive group into the hydrophobic region might produce the same result. The lack of agreement between labeling with iodonaphthylazide and adamantane diazirine extends to the two conformers. In the case of the azide, E_K incorporated 10–25% more label than E_{Na}, but when the diazirine was used, E_K was labeled 30–40% less heavily than E_{Na}. The sites of attachment of the two labels are different (see above), and it might be argued that in potassium-containing media the NH$_2$-terminal region becomes more accessible to the hydrocarbon core of the bilayer while the opposite is true for the COOH-terminal region. But, considering the problems discussed here and the need for more detailed peptide maps of both labeled E_K and E_{Na}, such a conclusion would be clearly premature.

3.2. Structure of the HLA-DR Antigen

The HLA-DR antigen is a product of the major histocompatibility complex, which is involved in several classes of cell–cell interaction in the immune system. The antigen is composed of two associated glycoprotein subunits (34,000 and 29,000 daltons) that are tightly bound to the plasma membranes of lymphocytes. In two recent studies the antigen was labeled with the hydrophobic reagents [^{3}H]adamantane diazirine (Kaufman and Strominger, 1979) and [^{125}I]hexanoyldiiodo-N-(4-azido-2-nitrophenyl)tyramine (Owen *et al.*, 1980).

The azide labeled the 34,000-dalton chain far more strongly than the 29,000-dalton chain and as both chains were reactive after dissociation by heating in Triton X-100, the authors favored the interpretation that the tight association of the heavy chain with the light chain protected the latter, which was known to span the membrane, from the hydrophobic reagent. Kaufman and Strominger (1979) found, however, that both chains were labeled with adamantane diazirine, suggesting that the light chain is exposed to the bilayer but in this state it is relatively unreactive toward the photolysis products of the aryl azide. In addition, the distribution of label after two sequential proteolytic cleavages near the COOH terminus of each chain suggested that they both have a structure similar to glycophorin, with a short COOH-terminal hydrophilic domain preceded by a short hydrophobic domain that lies within the lipid bilayer. The study of Owen and co-workers is noteworthy for its interesting data on the labeling of both soluble and membrane proteins with the hydrophobic azide in the presence of membranes, liposomes, detergents, or none of these. For instance, serum IgM was labeled in solution but was protected when Triton X-100 was present, whereas integral proteins (e.g., the heavy chains of HLA-A, -B, and -DR) were not protected by the detergent.

3.3. The Kinetics of Penetration of a Membrane by Cholera Toxin

Cholera toxin, which indirectly activates adenylate cyclase by ADP-ribosylation of the GTP-binding protein, is believed to penetrate the membranes of cells with which it interacts. Wisnieski and Bramhall (1981), in a futuristic study, have investigated the penetration of toxin into membranes of Newcastle disease virus that had been preincubated with the amphipathic reagent 12-(4-azido-2-nitrophenoxy)stearoyl glucosamine (Fig. 1). It was previously postulated that a pentamer of B subunits [the subunit structure of cholera toxin is $B_5(A_1-A_2)_1$] first binds to the membrane receptor and then penetrates the bilayer to form a channel through which the disulfide-linked dimer A_1-A_2 enters, A_1 being released into the cytoplasm after reduction of the disulfide linkage. In the study considered here, only the A_1 subunit was labeled by the amphipathic reagent, and it could be demonstrated, using a 15-sec photolysis time, that the extent of labeling increased to a maximum approximately 1 min after mixing toxin with virus and then decreased to approximately half the maximal value over 15 min. The authors believe that the reagent is confined to the outer half of the bilayer and propose that the slow second phase in which labeling of the A_1 subunit decreases may represent translocation of A_1 to the inner half of the bilayer (Fig. 4). Because the extent of labeling does not fall to zero, it was further suggested that A_1 finally distributes between both halves of the leaflet in a dynamic equilibrium. While this is an attractive possibility, the arguments presented elsewhere in this review suggest several alternative

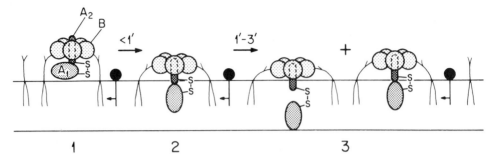

Figure 4. The interaction of cholera toxin with the membrane of Newcastle disease virus. Step 1: B subunits bind to ganglioside receptors. Step 2: Within 1 min of binding at 37°C, A_1 inserts into the outer monolayer of the membrane. Step 3: Within 2 min of insertion, A_1 is moving between the outer and the inner monolayers in a dynamic equilibrium. The small arrows represent the photoreactive group of the labeling reagent (see Fig. 1). (From Wisnieski and Bramhall, 1981.)

conformational changes that might explain the time dependence of the extent of labeling of A_1 (see Section 3.1). A further problem in determining the time course of an interaction using an aryl azide is the lifetime of the photogenerate species. The work of Mas *et al.* (1980) suggests that unless suitable precautions are taken, 30–50% of the observed labeling might occur after photolysis over a period of more than 30 min. Despite these possible imperfections, the potential strength of the approach taken by Wisnieski and Bramhall is clear.

4. PROSPECTS

The development of new reagents such as the trifluoromethylphenyldiazirines is of considerable importance, and future achievements should include further progress on the reactivity problem. It may not be possible to produce a single reagent that will randomly label all functional groups within a membrane, but it should be possible to find a set of reagents that together will do so. For example, free radicals might be used for reaction with carbon–hydrogen bonds (Galardy *et al.*, 1973). A notable attempt (Erni and Khorana, 1980) to decrease the bulk of the photolabile group by using $\alpha,\alpha,\alpha',\alpha'$-tetrafluorodialkyl diazirines failed because of intramolecular reactions, and more work in this area is desirable.

The problem of achieving reaction at a defined depth within the membrane should be solved by producing reagents so rigid that the different reactivities of various groups within the bilayer are of no consequence, unless the proteins themselves have appreciable transverse mobility. It seems likely though that the first reagents to be proved to react in a more localized region than the entire hydrocarbon core of the bilayer will be those where the reactive group is confined to one side of the membrane, or to the bilayer surface. The former might be achieved by using a phospholipid exchange protein to place a lipid analog in the outer half of the bilayer of sealed vesicles (Brunner and Richards, 1980) or by the addition of an amphipathic reagent to sealed vesicles (Simon and Wisnieski, 1980), the latter by using a phospholipid analog or an amphipathic reagent with a reactive group in the polar region of the molecule (Radhakrishnan *et al.*, 1980; Bisson and Montecucco, 1981).

Another area under active investigation is the testing of photochemical homo- and heterobifunctional hydrophobic cross-linking reagents. The review of Sigrist and Zahler (this volume) should be consulted on this topic.

Finally, the potential for time-dependent studies has been demonstrated by the investigation of membrane penetration by cholera toxin discussed above and by an investigation of complement binding to membranes (Hu *et al.*, 1981). Such experiments will be placed on a surer foundation if the possible presence of long-lived photogenerated intermediates is ruled out.

5. RECENT DEVELOPMENTS

Girdlestone *et al.* (1981) have recently demonstrated that a subunit of succinate ubiquinone reductase (SD_1) is labeled by a phospholipid containing a photoreactive group designed to lie close to the membrane surface. The same subunit is only lightly labeled by a lipid analog with a deeply buried reactive group; therefore, looping around to the surface by the buried group is only a minor problem in this system.

Montecucco *et al.* (1981) have used four different azide reagents to label Na^+/K^+-ATPase, and in the case of a phospholipid analog with a deeply buried photoactivatable group quite extensive labeling of the β subunit was obtained.

Jorgensen *et al.* (1982) have examined the labeling of Na^+/K^+-ATPase with iodonaphthylazide in more detail and found that the COOH-terminal region of the α subunit is labeled at a higher reagent concentration than that used previously.

Tomasi and Montecucco (1981) have obtained new results on the interaction of cholera toxin with membranes using photoactivatable phospholipid analogs.

For a recent description of the biosynthetic incorporation of photoactivatable fatty acids into lipids the work of Quay *et al.* (1981) should be consulted.

REFERENCES

Bayley, H., and Knowles, J. R. (1978a). *Biochemistry* **17**, 2414–2419.
Bayley, H., and Knowles, J. R. (1978b). *Biochemistry* **17**, 2420–2423.
Bayley, H., and Knowles, J. R. (1980a). *Ann. N.Y. Acad. Sci.* **346**, 45–58.
Bayley, H., and Knowles, J. R. (1980b). *Biochemistry* **19**, 3883–3892.
Bercovici, T., Gitler, C., and Bromberg, A. (1978). *Biochemistry* **17**, 1484–1489.
Bisson, R., and Montecucco, C. (1981). *Biochem. J.* **193**, 757–763.
Bisson, R., Montecucco, C., Gutweniger, H., and Azzi, A. (1979). *J. Biol. Chem.* **254**, 9962–9965.
Brunner, J. (1981). *Trends Biochem. Sci.* **6**, 44–46.
Brunner, J., and Richards, F. M. (1980). *J. Biol. Chem.* **255**, 3319–3329.
Brunner, J., and Semenza, G. (1981). *Biochemistry*, in press.
Brunner, J., Senn, H., and Richards, F. M. (1980). *J. Biol. Chem.* **255**, 3313–3318.
Castro, J., and Farley, R. A. (1979). *J. Biol. Chem.* **254**, 2221–2228.
Cerletti, N., and Schatz, G. (1979). *J. Biol. Chem.* **254**, 7746–7751.
Chapman, O. L. (1979). *Pure Appl. Chem.* **51**, 331–339.
Chowdhry, V., Vaughan, R., and Westheimer, F. H. (1976). *Proc. Natl. Acad. Sci. USA* **73**, 1406–1408.
Czarniecki, M. F., and Breslow, R. (1979). *J. Am. Chem. Soc.* **101**, 3675–3676.
Engelman, D. M., Henderson, R., McLachlan, A. D., and Wallace, B. A. (1980). *Proc. Natl. Acad. Sci. USA* **77**, 2023–2027.
Erni, B., and Khorana, H. G. (1980). *J. Am. Chem. Soc.* **102**, 3888–3896.
Farley, R. A., Goldman, D. W., and Bayley, H. (1980). *J. Biol. Chem.* **255**, 860–864.
Frank, G., Brunner, J., Hauser, H., Wacker, H., Semenza, G., and Zuber, H. (1978). *FEBS Lett.* **96**, 183–188.
Galardy, R. E., Craig, L. C., and Printz, M. P. (1973). *Nature New Biol.* **242**, 127–128.

Girdlestone, J., Bisson, R., and Capaldi, R. A. (1981). *Biochemistry* **20,** 152–156.

Goldman, D. W., Pober, J. S., White, J., and Bayley, H. (1979). *Nature (London)* **280,** 841–843.

Gupta, C. M., Costello, C. E., and Khorana, H. G. (1979). *Proc. Natl. Acad. Sci. USA* **76,** 3139–3143.

Hu, V. W., and Wisnieski, B. J. (1979). *Proc. Natl. Acad. Sci. USA* **76,** 5460–5464.

Hu, V. W., Esser, A. F., Podack, E. R., and Wisnieski, B. (1981). *J. Immunol.,* in press.

Jorgenson, P. L., Karlish, S. J. D., and Gitler, C. (1982). *J. Biol. Chem.,* in press.

Kahane, I., and Gitler, C. (1978). *Science* **201,** 351–352.

Karlish, S. J. D., Jorgensen, P. L., and Gitler, C. (1977). *Nature (London)* **269,** 715–717.

Kaufman, J. F., and Strominger, J. L. (1979). *Proc. Natl. Acad. Sci. USA* **76,** 6304–6308.

Kerr, J. A., O'Grady, B. V., and Trotman-Dickenson, A. F. (1967). *J. Chem. Soc. A,* 897.

Mas, M. T., Wang, J. K., and Hargrave, P. A. (1980). *Biochemistry* **19,** 684–692.

Montecucco, C., Bisson, R., Gache, C., and Johannsson, A. (1981). *FEBS Let.* **128,** 17–21.

Nielsen, P. E., Leick, V., and Buchardt, O. (1978). *FEBS Lett.* **94,** 287–290.

Owen, M. J., Knott, J. C. A., and Crumpton, M. J. (1980). *Biochemistry* **19,** 3092–3099.

Ploegh, H. L., Orr, H. T., and Strominger, J. L. (1980). *Proc. Natl. Acad. Sci. USA* **77,** 6081–6085.

Prochaska, L., Bisson, R., and Capaldi, R. A. (1980). *Biochemistry* **19,** 3174–3177.

Quay, S. C., Radhakrishnan, R., and Khorana, H. G. (1981). *J. Biol. Chem.* **256,** 4444–4449.

Radhakrishnan, R., Gupta, C. M., Erni, B., Robson, R. J., Curatolo, W., Majumdar, A., Ross, A. H., Takagaki, Y., and Khorana, H. G. (1980). *Ann. N.Y. Acad. Sci.* **346,** 165–198.

Radhakrishnan, R., Robson, R. J., Takagaki, Y., and Khorana, H. G. (1981). *Methods Enzymol.,* **72,** 408–433.

Robson, R. J., Radhakrishnan, R., Ross, A. H., Takagaki, Y., and Khorana, H. G. (1981). In *Lipid–Protein Interactions* (O. H. Griffiths and P. Jost, eds.), Wiley, New York, in press.

Ross, A. H., and Khorana, H. G. (1982). Submitted for publication.

Sator, V., Gonzalez-Ros, J. M., Calvo-Fernandez, P., and Martinez-Carrion, M. (1979). *Biochemistry* **18,** 1200–1206.

Simon, P., and Wisnieski, B. J. (1980). *Fed. Proc.* **39,** 2190.

Smith, R. A. G., and Knowles, J. R. (1975). *J. Chem. Soc. Perkin Trans. 2,* 686–694.

Staros, J. V. (1980). *Trends Biochem. Sci.* **5,** 320–322.

Stoffel, W., Schreiber, C., and Scheefers, H. (1978). *Hoppe-Seyler's Z. Physiol. Chem.* **359,** 923–931.

Tarrab-Hazdai, R., Bercovici, T., Goldfarb, V., and Gitler, C. (1980). *J. Biol. Chem.* **255,** 1204–1209.

Tomasi, M., and Montecucco, C. (1981). *J. Biol. Chem.* **256,** 11177–11181.

Wells, E., and Findlay, J. B. C. (1980). *Biochem. J.* **187,** 719–725.

Wisnieski, B. J., and Bramhall, J. S. (1981). *Nature (London)* **289,** 319–323.

25

Optical Probes of the Potential Difference across Membranes

Alan S. Waggoner

1. INTRODUCTION

The membrane potential is an important property of cells. For example, it is well known that the nerve impulse is a transient depolarization of the axon membrane potential. As well, the transmembrane electrical potential difference is critical to the proper function of all nonexcitable cells and organelles. The importance becomes clear when one counts the number of ion, sugar, amino acid, neurotransmitter, and metabolite transport systems that are driven directly or indirectly by the membrane potential.

Because many interesting cells and organelles are too small to be punctured by microelectrodes, a number of other probing techniques have been developed for determining the membrane potential. Most of these techniques involve measurement of the distribution of labeled ions between the compartments on the two sides of the membrane. Radioactively labeled permeant ions like [³H]triphenylmethylphosphonium⁺, [¹⁴C]thiocyanate⁻, and ⁸⁶Rb⁺-valinomycin have frequently been used (Rottenberg, 1979). Hydrogen ion distribution can be followed with pH meters (Macey *et al.*, 1978), K^+ distribution can be determined (Schummer *et al.*, 1980), spin-label distribution can be monitored by electron spin resonance (Cafiso and Hubbell, 1978), and distribution of permeant dyes can be determined with absorbance or fluorescence instruments.

This short review will focus only on the use of dye molecules (Fig. 1) and endogenous chromophores as probes of membrane potential. It will be general and will cover only a small number of recent applications of this new technique to problems in bioenergetics and neurophysiology. Several recent comprehensive reviews are available (Freedman and Laris, 1981; Waggoner, 1979a; Rottenberg, 1979; Bashford and Smith, 1979; Cohen and Salzberg, 1978).

Alan S. Waggoner • Department of Chemistry, Amherst College, Amherst, Massachusetts 01002.

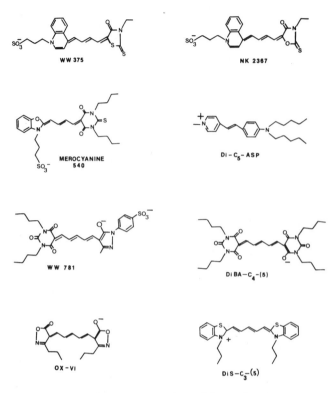

Figure 1. Structures of eight potential-sensitive dyes. Three of the top four dyes are merocyanines, diS-C₃-(5) and di-C₅-ASP are cyanine dyes, and the three others are oxonol dyes.

2. MEMBRANE POTENTIALS IN NONEXCITABLE CELLS

In 1974 Hoffman and Laris first demonstrated the usefulness of cyanine dyes for measuring membrane potentials in suspensions of cells. In the same year Sims and colleagues showed that the cyanine dyes behave as membrane permeant cations that distribute across membranes according to the potential difference. In hyperpolarized red cells, the fluorescence of dye that has been accumulated into the cells is quenched. The quenching occurs because of a combination of dye aggregation, of dye binding to hemoglobin, and of concentration quenching (Sims *et al.*, 1974; Tsien and Hladky, 1978; Freedman and Hoffman, 1979; Krasne, 1980a,b). An example of the magnitude of the fluorescence changes that can be seen with the cyanine dye diS-C₃-(5) is shown in Fig. 2. At very low concentrations, certain cationic cyanine dyes and anionic oxonol dyes that do not readily aggregate show fluorescence increases rather than decreases when cells are hyperpolarized. It is also possible to determine membrane potential changes by measuring dye absorption. In order to translate fluorescence or absorbance changes from an experiment into membrane potential units, a calibration curve is usually generated by setting the membrane potential at known values while monitoring the optical signal (Waggoner, 1979b; Freedman and Laris, 1981).

Freedman and Laris (1981) have reviewed more than 150 publications that appeared between 1974 and 1980 in which cyanine, oxonol, and merocyanine dyes were used to study nonexcitable cells, organelles, and vesicles. In the next few paragraphs I shall mention just a few examples of recent work with these dyes.

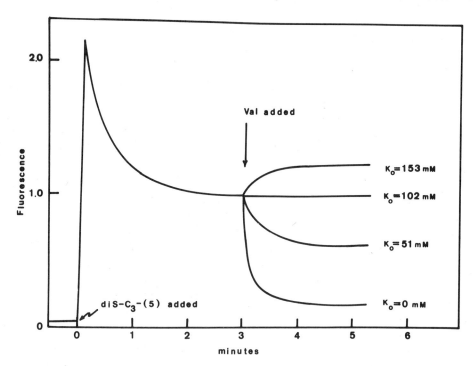

Figure 2. Characteristic changes in fluorescence intensity with time of diS-C_3-(5) in a 0.17% suspension of normal human red cells in NaC1–Tris medium (pH 7.4) and in mixtures where KC1–Tris was substituted for NaC1–Tris to give the $[K_0]$ values listed in the figure. The cellular potassium concentration, $[K_c]$, = 152 mmoles/liter of cell H_2O. Dye and valinomycin were added where indicated. Fluorescence was recorded at 670 nm with excitation at 622 nm. (From Sims et al., 1974.)

Two papers have been published in which cyanine dyes were used to monitor the kinetics of membrane potential changes of host cells during early stages in infection. Labedan and Letellier (1981) interpreted fluorescent changes of diI-C_1-(5) in terms of a transient depolarization in *E. coli* following T_4 or T_5 infection. After repolarization begins, additional adsorption of phage produces no further depolarizing effects. Bayer and Bayer (1981) monitored ANS and diO-C_5-(3) fluorescence of *E. coli* and *Salmonella* strains during infections with a variety of normal and defective phages. They suggest that the delayed but strong fluorescence responses indicate a surface "walk," which is terminated upon encounter of an injection site and depolarization of the cell.

Hacking and Eddy (1981) used diS-C_3-(5) to study mechanisms of amino acid transport in ascites tumor cells. The cells were shown to absorb 2-amino isobutyrate, glycine, L-leucine, L-isoleucine, and certain other amino acids electrogenically. The fluorescence studies permitted determination of K_m and V_{max} values for these amino acids. As a result of another series of experiments with diS-C_3-(5), Wright *et al*. (1981) were able to conclude that the major tricarboxylic acid cycle intermediates are transported via a common Na^+-dependent transport system in renal brush border membranes.

Bacterial bioenergetics continues to be studied with potential-sensitive dyes. For example, Laszlo and Taylor (1981) have used diS-C_3-(5) fluorescence to study the role of electron transport in the aerotaxis of *Salmonella typhimurium*. In cases like this one, the interpretation of optical changes involves few problems. Zaritsky *et al*. (1981), on the other hand, have shown that distribution techniques for measuring membrane potential under certain experimental conditions can produce opposite results. In experiments with

Bacillus subtilis, it was found that fluorescence changes of diS-C_3-(5) and distribution changes of [^{3}H]triphenylmethylphosphonium$^+$ were artifactual, and occurred without membrane potential changes, as judged by ^{86}Rb$^+$ uptake experiments in the presence of valinomycin.

It is acknowledged that much caution is needed in carrying out and interpreting experiments with optical membrane potential probes (Waggoner, 1979a,b). In a paper concerned with the mechanism of voltage-sensitive dye responses on sarcoplasmic reticulum, Beeler *et al.* (1981) point out the difficulties in analyzing absorbance and fluorescence changes of cyanine and oxonol dyes when divalent ion concentrations (e.g., Ca^{2+}) vary widely. The magnitude of membrane surface potentials, which depend upon the ionic makeup and strength of the bathing medium, strongly affects the membrane binding constants of charged hydrophobic dyes.

A serious problem arises when cyanine dyes and other permeant ions are used to determine the plasma membrane potentials of cells containing organelles that may also be electrically polarized. For example, a substantial fraction of cationic probe molecules taken into the cell from the bathing medium actually may be sequestered within the mitochondria. In this situation the plasma membrane potential will be overestimated if the estimate is based solely on the amount of dye taken up by the cells. While poisoning the mitochondria would eliminate dye accumulation by these structures, the cellular phosphate potential will be altered as well, so that transport properties and the electrical potential of the plasma membrane will be different.

The potential-dependent distribution of dyes into intracellular compartments has a positive side. It may be possible to estimate the membrane potentials of the organelles of individual tissue culture cells. For example, Johnson *et al.* (1981) found that diO-C_5-(3), rhodamine 123, and several other ionic, permeant dyes strongly stain the mitochondria of cultured fibroblasts. The staining is reduced by electron transport inhibitors. Cohen *et al.* (1981) found that the staining of fibroblasts increases over fivefold in less than 15 min when the cells move from the G_0 to the G_1 state of the cell cycle. Again, most of the staining is in the mitochondria. If indeed the increased fluorescence of the mitochondria of cycling fibroblasts is due to increased energization of the mitochondria, the result is surprising because microelectrode studies of fibroblasts (Cone, 1975) indicate that cycling cells have a much lower plasma membrane potential (-10 mV) and confluent cells (-60 mV).

3. OPTICAL MONITORING OF ELECTRICAL ACTIVITY IN EXCITABLE CELLS

Just before 1970 Tasaki *et al.* (1968) and Cohen *et al.* (1970) and their colleagues began looking at changes in fluorescence of ANS and biological stains on excited nerve fibers. Much progress has been made since this early work, in which extensive signal averaging was needed to obtain clean optical signals from the excited neurons. With the potential-sensitive dyes available today, it is possible to monitor single calcium action potentials from growth cones of cultured neurons using a mini He–Ne laser microbeam (Grinvald and Farber, 1981). By using 10×10 arrays of photodiodes, Grinvald *et al.* (1981a) have simultaneously determined the electrical activity of most of the large neurons in the brain of the barnacle (Fig. 3). Cohen and his collaborators hope to use this optical approach to determine the wiring diagrams of simple nervous systems and to gain insights into mechanisms of learning and behavior. In other experiments, Grinvald *et al.* (1981b)

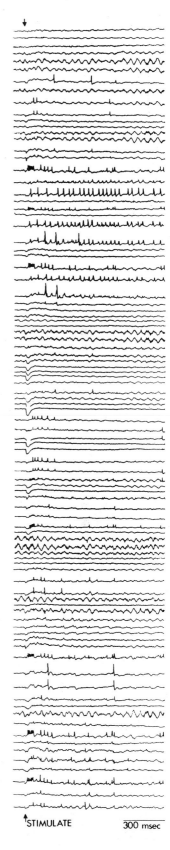

Figure 3. When the image of a dye-treated barnacle ganglion is focused on a 10 × 10 array of photodetectors, electrical activity of individual nerve cells in the ganglion can be inferred from the photodetectors' outputs. Here each trace shows the output of one detector. Changes in the absorbance of light by a single nerve cell can cause as many as 6 of the 100 detectors to respond simultaneously if the cell's image happens to impinge on that many. In this record, about 16 or 17 cells are active. The arrows indicate the time of an electrical shock given to nerve fibers leading into the ganglion. (Courtesy of Lawrence B. Cohen.)

STIMULATE   300 msec

used fluorescent oxonol dyes for simultaneous optical measurements of electrical activity from multiple sites on processes of cultured neurons, and Grinvald and Segal (1981) monitored spatial spread of electrical activity in pre- and postsynaptic elements in rat hippocampal slices using optical probes and a matrix of photodetectors. A nonfluorescent merocyanine dye was used by Salzberg *et al.* (1981) to monitor the spread of excitation in a salivary gland, while Morad and Salama (1979) characterized absorption and fluorescence changes of merocyanine dyes in frog hearts. Most optical studies of excitable tissue before 1980 have been reviewed elsewhere (Waggoner, 1979a; Freedman and Laris, 1981).

Such sophisticated optical approaches to biological problems would not be possible without significant progress in the development of sensitive dye molecules that do not alter synaptic activity or produce photodynamic damage in the stained preparations. During the past 10 years, Cohen, Waggoner, and their collaborators have screened over 1000 dyes in an effort to find optical probes for detecting optical signals in excitable cells (Gupta *et al.*, 1981). Hundreds of new optical probes were designed and synthesized as part of these efforts because it was found that dye molecules that gave good optical signal-to-noise ratios in squid axon experiments could often be considerably improved by appropriate synthetic modification of their chemical structures.

In recent years, Loew has used molecular orbital theory to design a new series of potential-sensitive dyes that work by an electrochromic mechanism (Lowe and Simpson, 1981). Recent screening experiments with these new dyes applied to squid axons (Cohen, Loew, and Grinvald, unpublished) indicate that they may be very useful, especially for monitoring very rapid voltage changes across membranes.

Photosynthetic organisms have natural electrochromic membrane potential indicators. The wavelength shifts of carotenoid absorption bands continue to be exploited in bioenergetics experiments with chloroplasts and photosynthetic bacteria (Junge, 1977).

Considerable progress has also been made in elucidating the mechanisms of the rapidly responding dyes that are used with excitable tissues (Waggoner, 1979a). Loew and Simpson (1981) provide evidence supporting an electrochromic mechanism for the diASP dyes, which undergo large dipole moment changes or charge shifts in going from the ground to the excited state. Most of the merocyanine, cyanine, and oxonol dyes show absorption and fluorescence changes that result from physical movement of the charged or dipolar dye within the membrane. It is not surprising that the large electric-field changes ($> 10^5$ V/cm) within excitable membranes can cause translation of charged dyes to new solvent environments in the membrane or rotate dye molecules with large permanent dipole moments. Furthermore, most of these potential-sensitive dyes have highly solvent-dependent absorption and emission properties.

In summary, the development and application of sensitive optical probes of membrane potential has grown rapidly in the last 6 years. While development of more sensitive probes in the future would be extremely valuable, only a few laboratories are active in this effort. The mechanisms of only a few dyes are known, and more probes need to be developed that are less toxic to biological preparations.

REFERENCES

Bashford, C. L., and Smith, J. C. (1979). *Methods Enzymol.* **55**, part F, 569–586.
Bayer, M. E., and Bayer, M. H. (1981). *Proc. Natl. Acad. Sci. USA*, **78**, 5618–5622.
Beeler, T. J., Farmen, R. H., and Martonosi, A. N. (1981). *J. Membr. Biol.*, **62**, 113–137.
Cafiso, D. S., and Hubbell, W. L. (1978). *Biochemistry* **17**, 187–195.
Cohen, L. B., and Salzberg, B. M. (1978). *Rev. Physiol. Biochem. Pharmacol.* **83**, 35–88.

Cohen, L. B., Landown, D., Shrivastav, B. B., and Ritchie, J. M. (1970). *Biol. Bull. (Woods Hole, Mass.)* **139,** 418–419.

Cohen, R. L., Muirhead, K. A., Gill, J. E., Waggoner, A. S., and Horan, P. K. (1981). *Nature (London)*, **290,** 593–595.

Cone, D. C. (1975). *Ann. N.Y. Acad. Sci.* **238,** 420–435.

Freedman, J. C., and Hoffman, J. F. (1979). *J. Gen. Physiol.* **74,** 187–212.

Freedman, J. C., and Laris, P. C. (1981). *Intl. Rev. Cytol.* Suppl. 12, 177–246.

Grinvald, A., and Farber, I. C. (1981). *Science*, **212,** 1164–1167.

Grinvald, A., and Segal, M. (1981). *Soc. Neurosci,* **7,** 889.

Grinvald, A., Cohen, L. B., Lesher, S., and Boyle, M. B. (1981a). *J. Neurophysiol.*, **45,** 829–840.

Grinvald, A., Ross, W. N., and Farber, I. C. (1981b). *Proc. Natl. Acad. Sci. USA*, **78,** 3245–3249.

Gupta, R. K., Salzberg, B. M., Grinvald, A., Cohen, L. B., Kamino, K., Lesher, S., Boyle, M. B., Waggoner, A. S., and Wang, C. H. (1981). *J. Membr. Biol.* **58,** 123–137.

Hacking, C., and Eddy, A. A. (1981). *Biochem. J.*, **194,** 415–426.

Hoffman, J. F., and Laris, P. C. (1974). *J. Physiol. (London)* **239,** 519–552.

Johnson, L. V., Walsh, M. L., Bockus, B. J., and Chen, L. B. (1981). *J. Cell Biol.* **88,** 526–535.

Junge, W. (1977). *Annu. Rev. Plant Physiol.* **28,** 503–536.

Krasne, S. (1980a). *Biophys. J.* **30,** 415–440.

Krasne, S. (1980b). *Biophys. J.* **30,** 441–462.

Labedan, B., and Letellier, L. (1981). *Proc. Natl. Acad. Sci. USA* **78,** 215–219.

Laszlo, D. J., and Taylor, B. L. (1981). *J. Bacteriol.* **145,** 990–1001.

Loew, L. M., and Simpson, L. (1981). *Biophys. J.*, **34,** 353–365.

Macey, R. I., Adorante, J. S., and Orme, F. W. (1978). *Biochim. Biophys. Acta* **512,** 284–295.

Morad, M. and Salama, G. (1979). *J. Physiol. (London)* **292,** 267–295.

Rottenberg, H. (1979). *Methods, Enzymol.* **55,** part F, 547–569.

Salzberg, B. M., Senseman, D. M., and Salama, G. (1981). *Biophys. J.* **33,** 90a.

Schummer, U., Schieffer, H.-G., and Gerhardt, U. (1980). *Biochim. Biophys. Acta* **600,** 998–1006.

Sims, P. J. Waggoner, A. S., Wang, C., and Hoffman, J. F. (1974). *Biochemistry* **13,** 3315–3329.

Tasaki, I., Watanabe, A., Sandlin, R., and Carnay, L. (1968). *Proc. Natl. Acad. Sci. USA* **61,** 883–888.

Tsien, R. Y., and Hladky, S. B. (1978). *J. Membr. Biol.* **38,** 73–97.

Waggoner, A. S. (1979a). *Annu. Rev. Biophys. Bioeng.* **8,** 47–68.

Waggoner, A. S. (1979b). *Methods Enzymol.* **55,** part F, 689–695.

Wright, S. H., Krasne, S., Kippen, I., and Wright, E. M. (1981). *Biochim. Biophys. Acta* **640,** 767–778.

Zaritsky, A., Kihara, M., and Macnab, R. M. (1981). *J. Membr. Biol.*, **63,** 215–231.

26

The Use of Detergents for the Isolation of Intact Carrier Proteins, Exemplified by the ADP, ATP Carrier of Mitochondria

M. Klingenberg

1. INTRODUCTION

In recent years, progress in handling detergents for the solubilization and isolation of integral membrane protein has been outstanding. Beginning in 1969, when our group embarked on the task of isolating and purifying the ADP, ATP carrier, no success was reached in the first 3 years because we used the then-popular methods for isolation of membrane proteins such as solubilization with cholate and deoxycholate combined with ammonium sulfate fractionation, organic solvents such as butanol, chloroethanol, lysolecithin, and even SDS (Klingenberg *et al.*, 1974). Our objective was to isolate the membrane protein not only in a pure but also in the native state. Nonionic detergents were hardly used at that time, although they were already known to be excellent solubilizers of membranes. However, their use was not very popular, because they seemed difficult to handle during the purification; it seemed impossible to remove the large excess of detergent required for solubilization. Moreover, in several cases there was no clear assay for characterizing the protein in the native state.

By applying the nonionic detergent Triton X-100 for solubilization and purification (Riccio *et al.*, 1975a; Klingenberg *et al.*, 1979a), it has been possible to obtain the ADP, ATP carrier in the native state at a high yield, virtually free of other proteins (Riccio *et al.*, 1975b). Subsequently, Triton X-100 has been used to isolate other proteins from mitochondria, such as the cytochrome bc_1 complex (Riccio *et al.*, 1977), and therefrom cytochromes *b* (von Jagow *et al.*, 1978a) and c_1 (Klingenberg, unpublished), the uncoupling protein from brown fat adipose tissue (Lin and Klingenberg, 1980), and the P_1 carrier (Wohlrab, 1980; Kolbe *et al.*, 1981). For comparing various detergents, the ADP, ATP

M. Klingenberg • Institute for Physical Biochemistry, University of Munich, 8000 Munich 2, West Germany.

carrier has turned out to provide particularly simple criteria for following the success of the solubilization and preservation of the native structure of the membrane protein.

2. SOLUBILIZATION

The solubilization of membrane proteins by detergents is now the method of choice, replacing earlier methods such as organic solvents or phospholipases. The investigation of the solubilization of the ADP, ATP carrier by detergents was undertaken, using binding with the highly specific inhibitor carboxyatractylate (CAT) as an assay that, at the same time, serves as indicator for the intactness of the solubilized protein (Klingenberg *et al.*, 1979a; Riccio *et al.*, 1975a,b). The hitherto most widely applied detergents for mitochondria, cholate and deoxycholate, irreversibly abolished the binding capacity of CAT on solubilization (Klingenberg *et al.*, 1974; Riccio *et al.*, 1975a). We therefore abandoned the popular cholate/ammonium sulfate fractionation procedure. Nevertheless, some years later, a purification of the ADP, ATP carrier by this procedure was claimed (Shertzer and Racker, 1976), however, the published results actually contradicted this claim. Although the strongly ionic detergent SDS completely denatures the ADP, ATP carrier, purification of the protein with full binding capability not only for CAT but also for ADP, ATP has nevertheless been stated (Bojanovski *et al.*, 1976). Again this report has not been substantiated.

The first attempts to solubilize with cholate and with nonionic detergents such as Triton, indicated that the ADP, ATP carrier is quite labile as assayed by CAT binding. A breakthrough was achieved when the ADP, ATP carrier was first loaded with CAT in the mitochondria and subsequently treated by detergents (Riccio *et al.*, 1975a). In this case, the CAT–protein complex could be fully recovered in the solubilized extract using Triton X-100. Its stability permitted further purification (Riccio *et al.*, 1975b). On the other hand, solubilization of the unloaded carrier gave only a diminished capacity for CAT binding, which rapidly disappeared within a few hours. Obviously, loading with the tightly bound CAT protected the ADP, ATP carrier from Triton-induced denaturation.

The solubilizing power of various detergents on the mitochondrial membrane, and on the ADP, ATP carrier in particular, could be determined; the following sequence was established (Klingenberg *et al.*, 1979a,b), using the detergent-to-protein weight ratio of beef heart mitochondria as parameter: SDS, aminoxide (dodecyldimethylamine oxide, 3-lauramido-*N*,*N*-dimethylpropylamine oxide), lauroylbetaine, Triton X-100, deoxycholate, cholate, octylglucoside, Brij 58, Lubrol WX, the latter two failing to cause solubilization. Among these, the nonionic (Triton X-100), the zwitterionic (lauroylbetaine), and the dipolar (aminoxide) detergents preserve the CAT–protein complex in solution, whereas the anionic detergents (SDS and cholate) destroy the complex. If one correlates the solubilizing power in terms of detergent-to-carrier-protein mole ratios, the large weight differences between the dipolar-type and the polyoxyethylene-type detergents disappear.

An important factor facilitating solubilization of mitochondria is high ionic strength. The solubilizing power is increased nearly twofold by an ionic strength of about 0.4 M (Klingenberg, 1979a). The strong assistance of the salt concentration in the solubilization of the mitochondrial membrane may be explained by the high content of acidic phospholipids. Another peculiarity is the inability of linear alkylpolyoxyethylene detergents to disintegrate the mitochondrial membrane (Klingenberg *et al.*, 1979a). Only the branched isotridecyl-polyoxyethylene (Emulphogen) solubilizes with moderate efficiency (Klingenberg *et al.*, 1979a; Brandolin *et al.*, 1974) and the isooctylphenyl derivative (Triton) with high efficiency. However, the cetyl derivates (Brij and Lubrol) are good solubilizers of the Na^+/K^+ (Hokin *et al.*, 1973; Esmann *et al.*, 1979) and Ca^{2+}-ATPases (Le Maire *et al.*, 1976). The reason for the difference between the mitochondrial and other membranes is not clear. It probably resides in the different lipid composition. Plasma membranes have a relatively high content of cholesterol and little acidic phospholipids.

With the ADP, ATP carrier system the question was also approached whether the solubilization of an integral membrane protein is achieved first by disintegrating the lipid matrix and then enveloping the protein with detergent, or by a more direct attack of detergents on the protein, before the membrane is fully solubilized (Klingenberg *et al.*, 1979b). By following with increasing detergent concentration the release of total protein, the CAT–protein complex, and of phospholipid, it became clear that the ADP, ATP carrier is solubilized parallel with the phospholipid rather than with the bulk protein. In other words, the disintegration of the phospholipid bilayer is a prerequisite for the solubilization of the integral membrane protein. Then, at the hydrophobic surfaces of the protein, the phospholipid is replaced by a detergent envelope, forming a soluble mixed protein–detergent micelle (Tanford, 1980; Helenius and Simons, 1975).

In line with its function, it can be expected that the ADP, ATP carrier penetrates through the membrane. Therefore, transport proteins should be released by detergents only on complete disintegration of the membrane. Other partially membrane-anchored proteins are cleaved out of the membrane at an earlier stage of membrane disintegration.

The ADP, ATP carrier also offers unique possibilities to study the mildness of detergents toward the native structure of membrane proteins. The carrier can be brought into different functional states in which the tendency to denaturation strongly varies. For example, with the highly specific inhibitor bongkrekate (BKA), the ADP, ATP carrier can be brought into the "m" state, in which the binding center is open toward the matrix side. With CAT the carrier is in the "c" state, open to the "c" side. By loading the carrier first with BKA in the membrane, the m state of the ADP, ATP carrier can be solubilized and isolated as BKA–protein complex (Aquila *et al.*, 1976). However, this form is much more labile than the CAT–protein complex. Still less stable is the unloaded carrier, and further instability results from adding the substrate ADP or ATP to the solubilized protein (Klingenberg *et al.*, 1979a). More labile states of the carrier therefore require milder detergents and render a more critical assay for this quality. For example, among all the detergents tested, only with Triton X-100 was it possible to isolate the intact BKA–protein complex. In other nonionic detergents such as Emulphogen BC, and much more so in octylglucoside, the BKA–protein complex disintegrates within a few hours, and in "dipole" detergents the BKA–protein complex is even more rapidly destroyed.

On solubilization of the unloaded protein, the native state can be tested by the capability to accept CAT or BKA (Krämer *et al.*, 1977). In Triton, binding capacity for CAT and also for BKA is retained for a few hours (Klingenberg *et al.*, 1979b). With dipole detergents, binding of CAT but not of BKA is found (Krämer *et al.*, 1977). From these and other assays the following sequence of mildness for the various detergents was established: Triton X-100, Emulphogen, 3-lauramido-*N,N*-dimethylpropylamine oxide, dodecyldimethylamine oxide, octylglucoside, cholate, deoxycholate. This refers to commercial-grade detergents, and it cannot be excluded that occasionally some contaminants are responsible for the damaging effect.

The mechanism of the labilization appears to be complex. One reason for the different stability of the ADP, ATP carrier in the c vs. m state is an apparently more closed structure of the c state. The carrier protein is unusually sensitive to endogenous proteases, which tend to attack the m state, e.g., the BKA–protein complex (Aquila *et al.*, 1976). This open conformation appears also to be more accessible to the denaturing action of detergents. Still more labile is the m state not liganded with BKA. This can be deduced from the strong destabilization of the unloaded protein in detergents by addition of substrate. The solubilized unloaded carrier is at first largely in the c-state, as known from immunological studies, and addition of ADP and ATP rapidly converts the protein into the m state from which it rapidly disintegrates.

3. PURIFICATION OF DETERGENT-SOLUBILIZED MEMBRANE PROTEIN

The relatively high concentration of nonionic detergents is often a deterrent to further purification. Their removal by dialysis is difficult because of the low critical micelle concentration. It has been found that for the purification of the ADP, ATP carrier, removal of excess detergent can be delayed to the last purification step. An astonishingly efficient first purification step was found to be the application of the crude extract to hydroxyapatite. Under appropriate conditions, at moderate salt concentrations, the ADP, ATP carrier passes through hydroxyapatite whereas most other proteins are retained (Klingenberg et al., 1979a; Riccio et al., 1975b; Aquila et al., 1976; von Jagow et al., 1978b). Thus, an approximately eightfold purification is obtained. The procedure is extremely fast and therefore very useful for the labile forms. Using hydroxyapatite in a batch procedure, an 80% pure preparation can be obtained within 5 min. including centrifugation, (Krämer and Klingenberg, 1977). The nonadsorptivity is interpreted to reflect shielding of the protein by extensive coverage with the detergent. Only a protein with a large hydrophobic surface should behave this way. The denatured ADP, ATP carrier, obtained after several hours in crude Triton extract without CAT addition, does not pass through the hydroxyapatite, presumably because the protein is partially unfolded and a larger hydrophilic surface is exposed (Engel et al., 1980; Aquila et al., 1976). Thus, hydroxyapatite also aids in the separation of denatured from native integral membrane protein.

Other integral membrane proteins—the uncoupling protein from brown adipose tissue (Lin and Klingenberg, 1980) and the P_i carrier for mitochondria (Wohlrab, 1980; Kolbe et al., 1981)—have also been found to not adsorb to hydroxyapatite in a Triton extract. Even adsorbed membrane proteins can be purified from hydroxyapatite by elution with increasing concentration of P_i, in good yield and without denaturation, e.g., the cytochrome bc_1 complex (Riccio et al., 1977; Engel et al., 1980; Klingenberg et al., 1979a). Following this example, also in other laboratories the Triton-solubilized preparation of bc_1 complex has now replaced (Weiss and Wingfield, 1979) the earlier cholate-solubilized preparation (Yu et al., 1974). Cytochrome b (von Jagow et al., 1978b) is another case that shows the unusually high yield and simplicity of the combined application of Triton and hydroxyapatite as a purification method.

The enriched extract can be applied to get filtration with clear separation from other proteins. The ADP, ATP carrier is eluted at a surprisingly low K_m value, indicative of an unusually large Stokes radius, which has been determined by careful calibration to about 63 Å, in agreement with sedimentation–diffusion measurements (Hackenberg and Klingenberg, 1980). Because the protein elutes close to the bulk of the free Triton, separation from excess Triton and solubilized phospholipids is only partial. Nevertheless, the yield is high, there is no denaturation, and if one selects only the peak fractions, there is also extensive purification.

Although using these procedures the ADP, ATP carrier appears to be 95% pure, there is still a large excess of Triton and phospholipid present in the form of mixed micelles. A separation is highly desirable for most biochemical or physicochemical studies. The difference in the specific volume between pure Triton and the Triton–protein mixed micelle can be utilized to separate most of the free Triton from the protein in a sucrose gradient. This mild procedure also removes the bulk of the phospholipids and other highly hydrophobic proteins (Hackenberg and Klingenberg, 1980). The sucrose gradient is also useful for substituting one detergent with another. For example, in a sucrose gradient containing octyl-glucoside, nearly all Triton can be removed (Hackenberg, unpublished). Thus, in the mixed protein–detergent micelles, one detergent can fully replace another.

In summary, the total purification of the Triton X-100-solubilized protein requires essentially three steps, none of which includes absorption. The yields are high and denaturation negligible. It avoids ion-exchange and ligand affinity chromatography, which denature the membrane protein by breaking up the sensitive protein–detergent micelle structure. By nearly the same procedure, the uncoupling protein from brown fat mitochondria has also been purified with a high yield and in the native state (Lin and Klingenberg, 1980).

4. STRUCTURE OF CARRIER–DETERGENT MICELLE

Any discussion on detergent–protein interaction has to consider the tendency of nonionic detergents to form relatively large micelles of a hundred or more molecules even at a rather low detergent concentration. The thermodynamic stability of these micelles is inversely related to the critical micelle concentration. After solubilization of membranes, mixed micelles are formed between detergent, protein, and phospholipid. The mixed detergent–protein micelle should be monodisperse and reflect the protein entity in the membrane originally surrounded by a phospholipid envelope. The micelle size is influenced by the content of protein and phospholipid. Therefore, the number of detergent molecules present in mixed micelles may deviate considerably from that in pure detergent micelles. This point is sometimes misunderstood in the literature, and the number of Triton molecules is assumed to be fixed with or without protein (Weiss and Wingfield, 1979). There are also good thermodynamic reasons to suggest that the Triton content varies with the protein-to-phospholipid ratio.

Hydrodynamic studies, in particular ultracentrifugation, have been useful for characterizing the micelles. However, several problems must be overcome, such as obtaining accurate data for the specific densities of the three micelle components and the peculiar hydrodynamic properties of the polyoxyethylene detergents. These problems have now been largely solved (Tanford and Reynolds, 1976; Tanford, 1980) and more reliable ultracentrifuge data are available. A specific technical problem with Triton X-100, the contribution to the protein ultraviolet absorbance, has also been overcome (Hackenberg and Klingenberg, 1980), and the aromatic absorbance of Triton is now utilized for following the mixed protein–Triton micelle. This requires careful removal of excess free Triton by sucrose gradient centrifugation. By ultraviolet scanning of the boundaries of the mixed protein micelle, the contributions of the residual free Triton micelles and of the monomeric Triton can be differentiated, and diffusion constants for the mixed Triton–protein micelle determined.

According to these data, the ADP, ATP carrier micelle consists of 150 molecules Triton X-100, 18 molecules phospholipids and 2 molecules 30,000-dalton protein (Hackenberg and Klingenberg, 1980). Total molecular mass according to sedimentation equilibrium is 178,000 daltons, out of which the ADP, ATP carrier represents 60,000 daltons, the rest being Triton and phospholipids. The dimeric structure of the ADP, ATP carrier agrees with the effective molecular mass calculated from the binding site for CAT (i.e., 58,000 daltons) (Riccio *et al.*, 1975b).

In the micelle the small protein molecule is visualized to be encased by a large number of detergent molecules that cover the extensive hydrophobic surface formerly embedded in the lipid bilayer. The relatively large Stokes radius and the high frictional ratios can be best explained by assuming an oblate ellipsoid micelle in which the protein axis coincides with the broad axis of the ellipsoid surface of 52 Å (Hackenberg and Klingenberg, 1980). A large detergent ring forms a Stokes diameter of up to 130 Å on the protein. The considerable hydration of the polyoxyethylene chain contributes also to the large Stokes radius. Very similar data on the Triton–protein micelle were obtained for the isolated uncoupling protein from brown fat, which has a similar molecular mass of $2 \times 32,000$ daltons (Lin *et al.*, 1980). We believe that this micelle structure is typical for deeply embedded integral membrane proteins, which may have a molecular size just penetrating the membrane. Besides the two mitochondrial proteins mentioned, bacteriorhodopsin and the P_1 carrier have similar molecular mass.

5. REINCORPORATION OF THE PROTEIN–DETERGENT MICELLE INTO PHOSPHOLIPIDS

The reconstitution of the transport function of the ADP, ATP carrier and other transport proteins requires reinsertion into the membrane. Vesicles generated by various methods from phospholipid mixtures, e.g., egg lecithin, avidly incorporate the ADP, ATP carrier from the Triton micelle. The Triton mixes with phospholipids and does not destroy the phospholipid membrane if the Triton content is below 5%. The fragile unloaded ADP, ATP carrier protein is stabilized to a high degree after incorporation into phospholipids (Krämer and Klingenberg, 1977). The binding capacity for BKA and CAT is not only increased but remains stable for several days. This and other evidence shows that phospholipids are essential for preserving the native protein. The removal of detergents of low critical micelle concentration from the phospholipid mixture by dialysis is virtually impossible. As all detergents destabilize the isolated protein to varying extents, the beneficial effect of phospholipids is considered to be largely structural, no specific phospholipid–protein interaction having been observed.

For reconstituting transport activity, the liposomes generated by mixing the vesicles with the protein–Triton micelles must be resonicated in order to form unilamellar vesicles (Krämer and Klingenberg, 1979). The phospholipid composition strongly influences transport activity. In particular, phosphatidylethanolamine and cardiolipin enhance the exchange activity of the ADP, ATP carrier (Krämer and Klingenberg, 1980).

REFERENCES

Aquila, H., Eiermann, W., and Klingenberg, M. (1976). *Abstracts of the 10th International Congress of Biochemistry*, p. 345.

Bojanovski, D., Schlimme, E., Wang, C. S., and Alaupovic, P. (1976). *Eur. J. Biochem.* **71**, 539–548.

Brandolin, G., Meyer, C., Defaye, G., Vignais, P. M., and Vignais, P. V. (1974). *FEBS Lett.* **46**, 149–153.

Engel, W. D., Schägger, H., and von Jagow, G. (1980). *Biochim. Biophys. Acta* **592**, 211–222.

Esmann, M., Skou, J. C., and Christansen, C. (1979). *Biochim. Biophys. Acta* **567**, 410–420.

Hackenberg, H., and Klingenberg, M. (1980). *Biochemistry* **19**, 548–555.

Helenius, A., and Simons, K. (1975). *Biochim. Biophys. Acta* **415,** 29–79.

Hokin, L. E., Dahl, J. L., Deupree, J. D., Dixon, J. F., Hackney, J. F., and Perdue, J. F. (1973). *J. Biol. Chem.* **248,** 2593–2605.

Klingenberg, M., Riccio, P., Aquila, H., Schmiedt, B., Grebe, K., and Topitsch, P. (1974). In *Membrane Proteins in Transport and Phosphorylation* (G. F. Azzone, M. Klingenberg, E. Quagliariello, and N. Siliprandi, eds.), pp. 229–243, North–Holland, Amsterdam.

Klingenberg, M., Aquila, H., Krämer, R., Babel, W., and Feckl, J. (1977). In *Biochemistry of Membrane Transport* (G. Semenza and E. Carafoli, eds.), pp. 567–579, Springer-Verlag, Berlin.

Klingenberg, M., Aquila, H., and Riccio, P. (1979a). *Methods Enzymol.* **56,** 229–233.

Klingenberg, M., Hackenberg, H., Eisenreich, G., and Mayer, I. (1979b). In *Function and Molecular Aspects of Biomembrane Transport* (E. Quagliariello, F. Palmieri, S. Papa and, M. Klingenberg, eds.), pp. 291–303, Elsevier/North-Holland, Amsterdam.

Kolbe, H. V. J., Böttrich, J., Genchi, G., Palmieri, F., and Kadenbach, B. (1981). *FEBS Lett.* **124,** 265–269.

Krämer, R., and Klingenberg, M. (1977). *Biochemistry* **16,** 4954–4961.

Krämer, R., and Klingenberg, M. (1979). *Biochemistry* **18,** 4209–4215.

Krämer, R., and Klingenberg, M. (1980). *FEBS Lett.* **119,** 257–260.

Krämer, R., Aquila, H., and Klingenberg, M. (1977). *Biochemistry* **16,** 4949–4953.

Le Maire, M., Møller, J. V., and Tanford, C. (1976). *Biochemistry* **15,** 2336–2342.

Lin, C. S., and Klingenberg, M. (1980). *FEBS Lett.* **113,** 299–303.

Lin, C. S., Hackenberg, H., and Klingenberg, M. (1980). *FEBS Lett.* **113,** 304–306.

Riccio, P., Aquila, H., and Klingenberg, M. (1975a). *FEBS Lett.* **56,** 129–132.

Riccio, P., Aquila, H., and Klingenberg, M. (1975b). *FEBS Lett.* **56,** 133–138.

Riccio, P., Schägger, H., Engel, W. D., and von Jagow, G. (1977). *Biochim. Biophys. Acta* **459,** 250–262.

Shertzer, H. G., and Racker, E. (1976). *J. Biol. Chem.* **251,** 2446–2452.

Tanford, C. (1980). *The Hydrophobic Effect: Formation of Micelles and Biological Membrane,* Wiley,New York.

Tanford, C., and Reynolds, J. A. (1976). *Biochim. Biophys. Acta* **457,** 133–170.

von Jagow, G., Schägger, H., Engel, W. D., Machleidt, W., Machleidt, I., and Kolb, H. J. (1978a). *FEBS Lett.* **91,** 121–125.

von Jagow, G., Schägger, H., Engel, W. D., Riccio, P., Kolb, H. J., and Klingenberg, M. (1978b). *Methods Enzymol.* **53,** 92–98.

Weiss, H., and Wingfield, P. (1979). *Eur. J. Biochem.* **99,** 151–160.

Wohlrab, H. (1980). *J. Biol. Chem.* **255,** 8170–8173.

Yu, C. A., Yu, L., and King, T. E. (1974). *J. Biol. Chem.* **249,** 4905–4910.

III

Biosynthesis of Cell Membranes: Selected Membrane-Bound Metabolic Systems

27

Protein Translocation across the Membrane of the Endoplasmic Reticulum

Bernhard Dobberstein and David I. Meyer

1. INTRODUCTION

The translocation of proteins across membranes is a feature common to all cells. In eukaryotes several distinct membranes are endowed with this capacity. Certain proteins (largely secretory) are translocated across the membrane of the endoplasmic reticulum (ER), while others are translocated across the membrane of mitochondria, chloroplasts, or peroxisomes. An understanding of the phenomenon of protein translocation depends upon answering three major questions: (1) What determines whether a protein will remain in the cytoplasm or will be translocated across a membrane? (2) How are the proteins selected to cross a specific membrane (endoplasmic reticulum, mitochondrial, chloroplast, or peroxisomal)? (3) What is the mechanism by which the actual physical translocation across the membrane occurs? This latter aspect requires the transfer of large hydrophilic moieties through a hydrophobic membrane and is certainly the most difficult to understand in molecular terms.

Two modes of translocation have been distinguished. In the first, referred to as co-translational, proteins are transferred across the membrane *during* their synthesis on membrane-bound ribosomes. In the second, proteins cross the membrane *after* their synthesis, and thus the process is called posttranslational translocation. The latter mode has been demonstrated for proteins synthesized in the cytoplasm and then imported into chloroplasts (reviewed by Chua and Schmidt, 1979) or mitochondria (reviewed by Schatz, 1979) and might also occur in the case of peroxisomal proteins (Lazarow, 1980). In contrast, translocation across the membrane of the endoplasmic reticulum is tightly coupled to translation (Kreibich *et al.*, 1978). This seems also to be the case for bacterial proteins that cross the inner bacterial membrane, although for one phage protein a posttranslational mechanism has been proposed (recent reviews: Davis and Tai, 1980; Inouye and Halegoua, 1980; Wickner, 1980).

Bernhard Dobberstein and David I. Meyer • European Molecular Biology Laboratory, 6900 Heidelberg, West Germany.

In this review only the cotranslational pathway will be discussed focusing on protein translocation across the membrane of the endoplasmic reticulum. Special consideration is given to the components involved in this translocation.

2. SIGNAL SEQUENCES

Nascent secretory proteins usually contain an NH_2-terminal peptide extension called a "signal" or a "leader" sequence. It has been proposed that this sequence directs the initiated ribosomal complex onto the membrane and facilitates the translocation of the remainder of the polypeptide chain across the membrane (Blobel and Dobberstein, 1975). This sequence is thought to be crucial not only for selecting the specific translocation site on the membrane, but also for the translocation process itself. After part of the nascent polypeptide has traversed the membrane, the signal sequence is cleaved by an enzyme called signal or leader peptidase. The following evidence in support of the "signal hypothesis" has been accumulated using eukaryotic as well as prokaryotic systems: (1) Signal sequences have been found and characterized for a large number of secretory proteins (Blobel, 1980). (2) Signal sequences are a feature of only those proteins that are translocated (either partially or completely) across the membrane and are absent from those that remain in the cytoplasm. (3) A synthetic signal peptide was found to compete with the membrane translocation of nascent hormones (Majzoub *et al.*, 1980). (4) Mutations that impair translocation of a mutated protein all map within the signal sequence region (Emr *et al.*, 1980). (5) Secretory proteins in which the signal sequence has been removed by gene manipulation techniques are no longer translocated across a membrane (Talmadge *et al.*, 1980a,b).

Is the signal sequence the only part of a secretory protein that is required for translocation? In β-lactamase, a bacterial periplasmic protein, a mutation near the COOH-terminal end has been described, which results in the cytoplasmic accumulation of this protein (Koshland and Botstein, 1980). Unfortunately, no data on the cotranslational transport of β-lactamase are available. For large secretory proteins, which are cotranslationally translocated across the membrane, an involvement of the COOH-terminal part in the translocation process can be excluded. Here the NH_2-terminal end has already crossed the membrane before the COOH-terminal part is synthesized. It is conceivable, however, that some feature of the COOH terminus plays a role in making the translocation process irreversible.

Because signal sequences are necessary for protein translocations, the question arises whether they are also sufficient to translocate *any* protein across the membrane. The genetically engineered addition of a signal sequence to the NH_2 terminus of the cytoplasmic protein β-galactosidase did not result in its translocation (Moreno *et al.*, 1980). Hydrophobic regions might, however, prevent such a cytoplasmic protein from being translocated.

From the available sequence data of more than 40 signal sequences some general features become apparent: (1) Signal sequences are usually 15–30 residues in length. (2) Their primary sequences are largely different from one another. (3) Secondary structure predictions and anti-signal-peptide antibodies, however, reveal a similar conformation (Austen, 1979; Baty and Lazdunski, 1979). (4) They contain a stretch of nonpolar residues in the middle region, and at either end polar or hydrophilic residues (reviewed by Inouye and Halegoua, 1980). (5) Amino acid residues with short side chains—such as Ala, Cys, Gly, and Ser—are found at the cleavage site.

The site in a signal sequence involved in protein translocation and that specifying its cleavage by the peptidase are different. Certain modifications of a signal sequence can interfere with its cleavage, without affecting the translocation process (Inouye and Halegoua, 1980; Hortin and Boime, 1980). There are also proteins that do not contain a cleavable signal sequence. It has been proposed that ovalbumin contains the functional equivalent of a signal sequence located internally between amino acid residues 234 and 253 (Lingappa *et al.*, 1979). This segment shows homology to signal sequences of other secretory proteins and, if used at very high concentrations, blocks the translocation of other secretory proteins. As ovalbumin can be expressed and secreted by bacteria, it will be possible by genetic manipulation to unequivocally locate the region(s) in the protein necessary for protein translocation (Fraser and Bruce, 1978). The idea of internal uncleaved signal sequences is appealing, as it could easily explain how proteins that span the membrane more than once are inserted. In such a case, protein loops might be sequentially inserted (Blobel, 1980). No evidence for such a mechanism exists up to now.

When the signal sequence interacts with the membrane of the endoplasmic reticulum, what does it react with? As the major common feature of signal sequences is their hydrophobic character, an interaction with the lipid bilayer would be suggested. Such an interaction alone, however, could not explain the restriction of the translocation process to a particular membrane (such as the rough endoplasmic reticulum). The existence of a protein receptor for signal sequences is suggested by the observation that a synthetic signal peptide competes with nascent prehormones for a limited number of specific binding sites (Majzoub *et al.*, 1980).

Does one receptor in the membrane recognize all possible signal sequences or is each signal sequence individually recognized by its own receptor? Secretory proteins from very different classes such as mammals, fish, and even insects could all be translocated across the microsomal membrane from dog pancreas in a cell-free system (Blobel *et al.*, 1979). A hybrid protein containing a eukaryotic signal sequence was recognized by *E. coli* membranes and efficiently translocated across the bacterial membrane (Talmadge *et al.*, 1980a). As it is unlikely that bacterial or dog membranes contain individual receptors for precursor proteins from such vastly different sources, the above-described data would suggest that there is only one or a few signal receptors present in the membrane of the endoplasmic reticulum.

3. SIGNAL PEPTIDASE

During the translocation of nascent polypeptide chains across the membrane, the signal sequence is cleaved by an enzyme called a signal peptidase. Such a cotranslational cleavage was suggested by the finding that the signal sequence is already removed from nascent polypeptide chains found on membrane-bound ribosomes (Blobel and Dobberstein, 1975). A proteolytic cleavage of the nascent chain would also explain why the intact precursor chains are rarely found *in vivo* in the tissue (Blobel and Dobberstein, 1975; Patzelt *et al.*, 1978).

The signal peptidase has been solubilized from membranes of the rough ER (Jackson and Blobel, 1977). It retains its specificity whether a nascent or a completed presecretory protein is used as a substrate. As it is not inactivated by proteolytic enzymes added to intact membrane vesicles, it has been suggested that the enzyme is localized on the cisternal side of the ER (Jackson and Blobel, 1980). Attempts to purify signal peptidase have been made in bacterial and eukaryotic systems (Jackson and Blobel, 1980; Zwizinski and

Wickner, 1980). Whether there is only one or more signal peptidases is not known. Because presecretory proteins in all heterologous translocation systems tested so far have been cleaved with fidelity, it could be argued that only one signal peptidase is present.

Regarding the specificity of signal peptidase, a conformational rather than a sequence requirement can be assumed. Amino acids with short side chains at the site of cleavage might play a role in this respect. What is the fate of the signal peptide after it is removed from the nascent chain? As it contains a stretch of hydrophobic residues in its middle portion, it most likely remains associated with the membrane. One can only speculate about its disposition in the membrane (Inouye and Halegoua, 1980).

4. MEMBRANE COMPONENTS INVOLVED IN TRANSLOCATION

The translocation of proteins across the membrane of the rough ER can be reconstituted in a cell-free system supplemented with microsomal vesicles (Blobel and Dobberstein, 1975; Szcesna and Boime, 1976). When proteins have been translocated into the vesicles, they should be inaccessible to exogenously added protease (Blobel and Dobberstein, 1975). Such an *in vitro* assay was prerequisite for identifying and characterizing membrane components involved in translocation. It allowed the dissection of the microsomal membrane into inactive components that could be reassembled subsequently into a functional entity.

Membrane proteins dissected from microsomal membranes with high salt, trypsin, or elastase in conjunction with high salt were shown to be required for translocation (Warren and Dobberstein, 1978; Walter *et al.*, 1979; Meyer and Dobberstein, 1980). For their translocation-restoring function an unblocked sulfhydryl group is essential (Jackson *et al.*, 1980). The active fragment obtained after elastolytic cleavage has an apparent molecular weight of 60,000, is basic in character, and has such an accessible sulfhydryl group (Meyer and Dobberstein, 1980).

Its location on the cytoplasmic side of the membrane would suggest that it is involved in the initial interaction of the ribosome or the nascent polypeptide chain with the membrane. The precise function of this molecule remains to be determined. The relationship between the salt-extracted and protease–salt-extracted proteins is unclear. They might represent different parts of the same parent molecule or separate protein entities. Further characterization of the active membrane-derived components will be required to assess their relationship.

In contrast to these proteins, which were characterized on the basis of their functioning in translocation, two membrane proteins of the rough ER, called ribophorins I and II, were characterized by their ability to physically interact with ribosomes (Kreibich *et al.*, 1978). Their functional involvement in protein translocation is unclear.

5. MEMBRANE PROTEINS

Certain membrane proteins also have to cross a membrane, at least partially. A small hydrophobic segment anchors these proteins in the membrane, leaving the COOH-terminal portion on the cytoplasmic side. Their integration into a membrane can be assumed at least in its initial phase to be identical to that of secretory proteins. That this assumption is correct was demonstrated for several viral [G protein of vesicular stomatitis virus, Semliki Forest virus, and sindbis virus glycoproteins (Katz *et a*., 1977; Garoff *et al.*, 1978; Bonatti *et al.*, 1979)] and plasma membrane proteins [H-2 and HLA antigens (Dobberstein *et al.*, 1979; Krangel *et al.*, 1979)]. These proteins are synthesized on ribosomes bound to the endoplasmic reticulum, and their nascent chains (which contain a signal sequence) are translocated across the membrane of the rough ER. Translocation stops when the hydrophobic segment anchors the protein in the membrane. Do secretory and the aforementioned membrane proteins require the same sites in the ER for their translocation or insertion? It was found that precursors for the G protein of vesicular stomatitus virus competed for translocation with presecretory proteins (Lingappa *et al.*, 1978). This would suggest that signal peptides of secretory and certain membrane proteins are structurally and functionally equivalent and require the same sites for their translocation or insertion.

6. OUTLOOK

Progress made in recent years in understanding protein translocation across membranes resulted mainly from two approaches: (1) use of cell-free systems in which components effecting translocation can be experimentally tested and characterized (reviewed by Blobel *et al.*, 1979) and (2) manipulation and mutation of genes coding for secretory proteins and determination of their localization within the cell compartments (reviewed by Emr *et al.*, 1980). We are, however, far away from understanding protein translocation across membranes in molecular terms. Using the above approaches or a combination of them, further answers to the following central questions in protein translocation should become available: (1) What features of the signal sequence are essential for membrane interaction and signal peptidase action? (2) What are the membrane receptors interacting with the nascent polypeptide chain or the ribosome? (3) What is the driving force for protein translocation? (4) What are the characteristics of the intramembranous environment during translocation of the hydrophilic portion of a protein across the membrane?

REFERENCES

Austen, B. M. (1979). *FEBS Lett.* **103**, 308–313.
Baty, D., and Lazdunski, C. (1979). *Eur. J. Biochem.* **102**, 503–507.
Blobel, G. (1980). *Proc. Natl. Acad. Sci. USA* **77**, 1496–1500.
Blobel, G., and Dobberstein, B. (1975). *J. Cell. Biol.* **67**, 835–851, 852–862.
Blobel, G., Walter, P., Chang, C. N., Goldman, B., Erickson, A. H., and Lingappa, V. R. (1979). In *Symposia of the Society for Experimental Biology*, Symposium XXXIII, pp. 9–36, Cambridge University Press, London.
Bonatti, S., Canadda, R., and Blobel, G. (1979). *J. Cell Biol.* **80**, 219–224.
Chua, N.-H., and Schmidt, G. W. (1979). *J. Cell Biol.* **81**, 461–483.
Davis, B. D., and Tai, P.-C. (1980). *Nature (London)* **283**, 433–438.
Dobberstein, B., Garoff, H., Warren, G., and Robinson, P. J. (1979). *Cell* **17**, 759–769.
Emr, S. D., Hall, M. N., and Silhavy, T. J. (1980). *J. Cell Biol.* **86**, 701–711.
Fraser, T., and Bruce, B. J. (1978). *Proc. Natl. Acad. Sci. USA* **75**, 5936–5940.
Garoff, H., Simons, K., and Dobberstein, B. (1978). *J. Mol. Biol.* **124**, 587–600.
Hortin, G., and Boime, I. (1980). *J. Biol. Chem.* **255**, 8007–8010.
Inouye, M., and Halegoua, S. (1980). *CRC Crit. Rev. Biochem.* **7**, 339–371.
Jackson, R. C., and Blobel, G. (1977). *Proc. Natl. Acad. Sci. USA* **74**, 5598–5602.
Jackson, R. C., and Blobel, G. (1980). *Ann. N.Y. Acad. Sci.* **343**, 391–404.
Jackson, R. C., Walter, P., and Blobel, G. (1980). *Nature (London)* **286**, 174–176.
Katz, F. N., Rothman, J. E., Lingappa, V. R., Blobel, G., and Lodish, H. F. (1977). *Proc. Natl. Acad. Sci. USA* **74**, 3278–3282.
Koshland, D., and Botstein, D. (1980). *Cell* **20**, 719–760.

Krangel, M. S., Orr, H. T., and Strominger, J. L. (1979). *Cell* **18,** 979–991.

Kreibich, G., Czako-Graham, M., Grebenau, R., Mok, W., Rodriguez-Boulan, E., and Sabatini, D. (1978). *J. Supramol. Structure* **8,** 279–302.

Lazarow, P. B. (1980). *Ann. N.Y. Acad. Sci.* **343,** 293–301.

Lingappa, V. R., Katz, F. N., Lodish, H. F., and Blobel, G. (1978). *J. Biol. Chem.* **253,** 8667–8670.

Lingappa, V. R., Lingappa, J. R., and Blobel, G. (1979). *Nature (London)* **281,** 117–121.

Majzoub, J. A., Rosenblatt, M., Fennick, B., Mannus, R., Kronenberg, H. M., Potts, J. T., Jr., and Habener, J. F. (1980). *J. Biol. Chem.* **255,** 11478–11483.

Meyer, D. I., and Dobberstein, B. (1980). *J. Cell Biol.* **87,** 498–502, 503–508.

Moreno, F., Fowler, A., Hall, M., Silhavy, T., Zabin, I., and Silhavy, M. (1980). *Nature (London)* **286,** 356–360.

Patzelt, C., Labrecque, A. D., Duguid, J. R., Carroll, R. J., Keim, P. S., Heinrikson, R. L., and Steiner, D. F. (1978). *Proc. Natl. Acad. Sci. USA* **75,** 1260–1264.

Schatz, G. (1979). *FEBS Lett.* **103,** 203–211.

Szcesna, E., and Boime, I. (1976). *Proc. Natl. Acad. Sci. USA* **73,** 1179–1183.

Talmadge, K., Stahl, S., and Gilbert, W. (1980a). *Proc. Natl. Acad. Sci. USA* **77,** 3369–3373.

Talmadge, K., Kaufman, J., and Gilbert, W. (1980b). *Proc. Natl. Acad. Sci. USA* **77,** 3988–3992.

Walter, P., Jackson, R. C., Marcus, M. M., Lingappa, V. R., and Blobel, G. (1979). *Proc. Natl. Acad. Sci. USA* **76,** 1795–1799.

Warren, G., and Dobberstein, B. (1978). *Nature (London)* **273,** 569–571.

Wickner, W. (1980). *Science* **210,** 861–868.

Zwizinski, C., and Wickner, W. (1980). *J. Biol. Chem.* **255,** 7973–7977.

28

Synthesis and Assembly of the Vesicular Stomatitis Virus Glycoprotein and Related Glycoproteins

Asher Zilberstein and Harvey F. Lodish

1. INTRODUCTION

Several cell surface proteins, of quite diverse origin, have in common a number of key structural characteristics: the surface glycoproteins of vesicular stomatitis virus (VSV) and sindbis virus; influenza HA glycoprotein; and possibly surface glycoproteins of other viruses; glycophorin, a major erythrocyte surface protein; and the heavy chain of the HLA-A and H-2 major histocompatibility antigens. All of these are structural proteins, and appear to be involved in cell–cell or virus–cell recognition. All of these polypeptides span the phospholipid bilayer only once (Fig. 1). In all of these proteins a region of 10–30 amino acids at the COOH terminus faces the cytoplasmic surface. This region contains a number of hydrophilic amino acids. Adjacent to this region is a sequence of 20–25 very hydrophobic amino acids believed to be the segment that spans the lipid membrane. Adjacent to either the NH_2 or the COOH end of this hydrophobic stretch are often a series of lysine and arginine residues; possibly, these interact with the phosphate residues of the phospholipid bilayer. Recent work indicates that fatty acids are covalently bound to serine or threonine residues near the COOH terminus of at least some of these proteins. The remainder of the polypeptide chain (which varies considerably in size among these polypeptide species) is on the extracytoplasmic surface, as are all of the attached carbohydrate chains.

These proteins are glycoproteins. The carbohydrate residues may be either *N*-linked (on asparagine) or *O*-linked (to serine or threonine). The structures of most *N*-linked oligosaccharides on membrane and secreted glycoproteins are similar (Kornfeld and Kornfeld, 1976, 1980); an example—the structure of the two identical "complex" *N-linked oligosac-*

Asher Zilberstein and Harvey F. Lodish • Department of Biology, Massachusetts Institute of Technology, Cambridge, Massachusetts 02139.

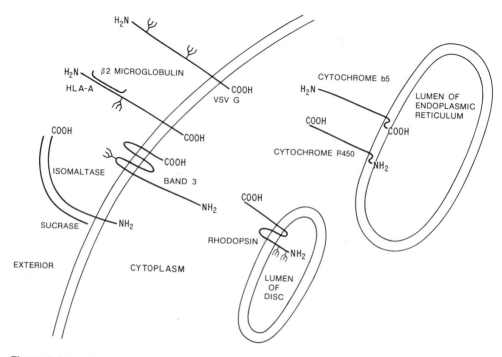

Figure 1. Schematic structure of a number of integral membrane glycoproteins. The symbol Ψ denotes an aspar-agine-linked oligosaccharide.

charides on VSV G protein (Reading *et al.*, 1978)—is shown in Fig. 2a. Differences among different proteins reflect primarily the number of branches (two or three) and the number of sialic acid residues (Kornfeld and Kornfeld, 1976, 1980). One member of this class of proteins—the HLA-A heavy chain—forms on its extracytoplasmic domain a tight nonco-valent interaction with a molecule of a soluble glycoprotein, β2 microglobulin (Barnstable *et al.*, 1978; Strominger *et al.*, 1980). *O*-linked sugars are prevalent on glycophorin (Gahmberg *et al.*, 1980).

Much more is known about the biosynthesis of these viral glycoproteins, and the structurally related proteins HLA-A and glycophorin, than about any other class of mem-brane proteins. Their biogenesis is unusually complex, involving covalent addition of both carbohydrate and fatty acid residues to the polypeptide backbone, and also processing of both the polypeptide and carbohydrate chains. We shall focus on several aspects of this process. First, how is transmembrane asymmetry achieved and how are these polypeptides inserted into membranes? Second, how and where does glycosylation—in particular, the addition of the common asparagine-linked oligosaccharides—take place? Finally, how and when do these glycoproteins move from their site of synthesis in the rough endoplasmic reticulum (ER), and how do they achieve their final distribution in the cell surface mem-brane? What modifications take place during maturation? The discussion will focus on the viral glycoproteins but, unless there is an explicit statement to the contrary, the results are consistent with the more limited data available on the synthesis of glycophorin and HLA-A.

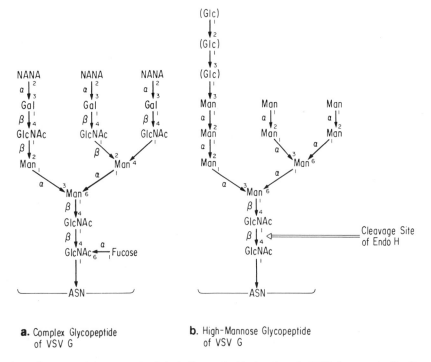

a. Complex Glycopeptide of VSV G

b. High-Mannose Glycopeptide of VSV G

Figure 2. (a) Structure of the asparagine-linked oligosaccharide found on the VSV glycoprotein (Reading *et al.*, 1978). (b) Structure of the two high-mannose carbohydrate chains found on the microsomal form of VSV G protein (taken from Li *et al.*, 1978), including the site of cleavage by Endo H.

2. TRANSMEMBRANE BIOSYNTHESIS OF THE VSV G PROTEIN

VSV encodes five polypeptides, all of which are structural components of the virions, and directs the synthesis of five corresponding mRNA species (Wagner, 1975; Pringle, 1977; Katz *et al.*, 1977a,b). mRNAs for four of them (N, NS, M, and L) are translated on free polyribosomes, and the newly made polypeptides are soluble in the cell cytoplasm. In particular, M protein is synthesized on free polysomes, and becomes localized to the inner surface of the plasma membrane during budding and assembly of VSV (Knipe *et al.*, 1977b). G mRNA, in contrast, is exclusively bound to the ER, and at all stages of its maturation G itself is bound to membranes (Morrison and Lodish, 1975) (Figs. 3, 4).

Immediately after its synthesis, G spans the ER membrane (Fig. 4). About 30 amino acids at the COOH terminus remain exposed to the cytoplasm. The balance of the poly-peptide, including the NH_2 terminus and the two asparagine-linked carbohydrate chains, face the lumen of the ER, and are protected from extravesicular protease digestion by the permeability barrier of the ER membrane (Katz *et al.*, 1977a,b; Lingappa *et al.*, 1978; Katz and Lodish, 1979). Note that the orientation of microsomal G with respect to the membrane is the same as that of G on the surface of infected cells and in virions (Fig. 1). The same 30 COOH-terminal amino acids are exposed on the cytoplasmic surface, while the carbohydrate chains and the bulk of the polypeptide, including the NH_2 terminus, re-main extracytoplasmic (Lingappa *et al.*, 1978; Katz and Lodish, 1979; Rose *et al.*, 1980). Thus, the overall membrane topology of G is preserved as the protein matures from the ER to the cell surface.

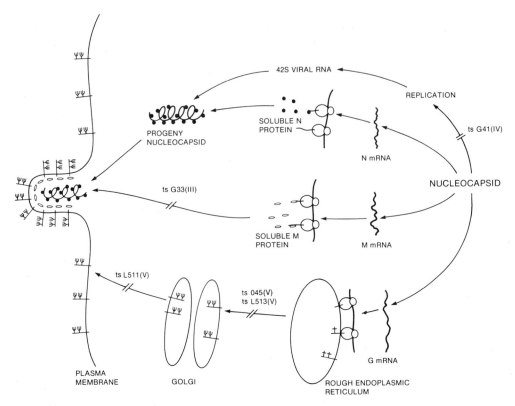

Figure 3. Schematic diagram illustrating the pathways of maturation of the major structural proteins of VSV and the proposed site of block in virion assembly for certain temperature-sensitive mutants (redrawn after Knipe *et al.*, 1977c). The symbol † denotes a high-mannose oligosaccharide, and Ψ denotes a complex oligosaccharide in G.

3. INTERACTION OF G mRNA AND NASCENT GLYCOPROTEIN WITH MICROSOMAL MEMBRANES

The G mNRA appears to be bound to the ER membrane via the nascent G chain, for treatment with puromycin, which causes premature termination of the growing polypeptide, causes release of the G mRNA from the ER. Unlike the case of mRNAs encoding secretory proteins, there is apparently no ionic linkage between the ribosomes and membranes, for dislodging of the mRNA does not require treatment with solutions of high salt concentration (Lodish and Froshauer, 1977).

Several recent experiments established that the growing G chain is extruded across the ER membrane into the lumen, similar to the way in which a nascent secretory protein is processed:

1. G. protein synthesized *in vitro* in the absence of membranes (G_0) contains 16 amino acids at the NH_2 terminus that are absent from the form of G made either in the presence of ER membranes or made by the cell (G_1) (Lingappa *et al.*, 1978). Most of these residues are highly hydrophobic, and resemble in structure and function "signal" peptides found at the NH_2 terminus of presecretory proteins (Devillers-Thiery *et al.*, 1975).

2. If a preparation of rough ER, from which the endogenous ribosomes have been removed, is added to a wheat germ or reticulocyte cell-free extract that is translating VSV G mRNA, the resultant G protein (1) has lost the NH_2-terminal 16 amino acids, (2) contains two "high-mannose" carbohydrate chains attached to asparagine residues, and (3) is inserted as a transmembrane protein in the ER with the same orientation as in the infected cell (Katz *et al.*, 1977a,b; Lingappa *et al.*, 1978). During synchronized *in vitro* protein synthesis in wheat germ extracts, the ER vesicles must be added to the protein synthesis reaction before the nascent chain is about 80 amino acids in length in order for the growing molecule to be subsequentgly inserted into the ER bilayer and to be glycosylated (Rothman and Lodish, 1977). Because about 30–40 of these 80 NH_2-terminal residues would, at this key time, still be embedded in the large ribosome subunit, this result establishes that it is the 40–50 NH_2-terminal residues, containing the 16 amino acids of the signal sequence, that are crucial in directing proper interaction of the nascent chain with the membrane. Presumably, if the nascent chain is of greater length when the membranes are added, the NH_2 terminus is folded in such a fashion that it cannot interact with receptors on the ER membrane that recognize this region of the nascent polypeptide.

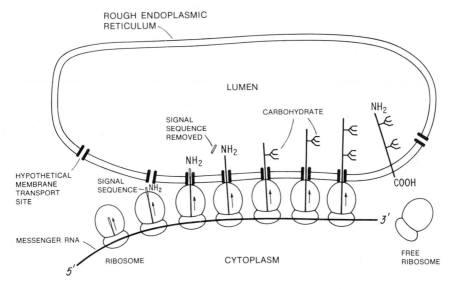

Figure 4. Model for synthesis, glycosylation, and transmembrane synthesis of VSV G protein. Among the first 50 amino acid residues of G is a "signal sequence" that identifies the protein as one destined to be inserted into the membrane of the rough ER. Because some 40 amino acids remain buried in the ribosome, the signal sequence does not emerge until the polypeptide is about 70 amino acid units long. At that time the signal sequence is recognized by some molecule, presumably a protein, in the membrane of the ER. This hypothetical protein is thought to facilitate the passage of the polypeptide through the lipid bilayer. Once in the lumen of the reticulum, at least part—16 amino acids—of the signal sequence is removed. The protein continues to elongate, and as it grows it is extruded through the membrane and folds up in the lumen. As it enters, two identical, preformed carbohydrate side chains (Fig. 2b) are transferred to it from a lipid carrier. Proteins secreted by the cell pass all the way through the membrane in this manner, but for reasons that are discussed in the text, G becomes stuck at about the time translation is completed, with some 30 amino acids remaining in the cytoplasm. Thus, the completed glycoprotein has its NH_2 terminus, most of its bulk, and all of its carbohydrates in the lumen of the endoplasmic reticulum, and a short stub that includes the COOH terminus on the cytoplasmic side. Once the protein has folded, it cannot be pulled out of the membrane, nor can it execute a transmembrane flip-flop; it is anchored in an asymmetric orientation.

4. TERMINATION OF GLYCOPROTEIN SYNTHESIS AND INSERTION INTO THE MEMBRANE

It is thus clear that insertion of the nascent glycoprotein into the ER membrane is a cotranslational event; it requires the presence of appropriate membrane vesicles during synthesis of the bulk of the polypeptide chain. We assume that, as is the case for nascent secretory proteins, the NH_2 terminus is extruded into the lumen of the ER (Palade, 1975), although there is no direct evidence for this point. It is also not known whether the growing peptide chain passes directly through the phospholipid bilayer, or whether it traverses the membrane in a proteinaceous channel (Von Heijne, 1980). Nor is it clear whether energy, in addition to that expanded in forming the peptide bond, is essential for this process. Nonetheless, it appears that at least the initial stages of insertion of the G protein are similar to those of a secretory protein (for which the above unanswered questions also apply). Why, then, is the nascent G protein not completely extruded into the ER lumen as is a secretory protein; why does it remain embedded in the ER as a transmembrane protein?

The following simple consideration shows that this is not the relevant question; the real problem is to determine why a secreted protein is completely extruded into the lumen of the ER! Consider the situation when synthesis of a molecular of either a secretory protein or VSV G protein has just been completed. The 30 COOH-terminal amino acids will be on the cytoplasmic side of the ER, embedded in the large ribosome subunit. The next 20–30 proximal residues will be spanning the ER membrane (whether they are passing through a proteinaceous channel or through the lipid matrix itself cannot at present be determined). At this point the two ribosome subunits separate from each other and from the mRNA. Presumably, the nascent chain is released from the ribosome. Whether or not the motive force of peptide bond synthesis, or ribosome translocation, provides the energy for extrusion of the *nascent* chain, it is clear that at this stage this motive force is no longer available to push the 50–60 COOH-terminal residues into and across the ER membrane. One likely possibility is that the continued extrusion of the COOH terminus of a secretory protein across the ER membrane is driven by the folding of the NH_2-terminal part of the protein molecule within the lumen; such a polypeptide could not exist stably as a transmembrane protein. On the other hand, the just-completed VSV G protein can achieve its final configuration without any movement with respect to the bilayer! It simply remains as a transmembrane polypeptide, anchored by the hydrophobic segment near the COOH terminus, which is localized within the phospholipid bilayer when polypeptide chain elongation is completed.

An interesting feature of this model is that it may explain why a number of diverse polypeptides (VSV G HLA-A, glycophorin) all have less than 30 amino acid residues at the COOH terminus exposed on the cytoplasmic surface of the membrane; it is precisely these residues that were embedded in the 60 S ribosome subunit at the instant the polypeptide chain was terminated. A prediction of this model is that all species of this class of transmembrane proteins that are inserted cotranslationally never have substantially more than 30 amino acids on the cytoplasmic side of the ER membrane.

5. BIOGENESIS OF TWO SINDBIS VIRUS GLYCOPROTEINS

Sindbis provides a rather different and very important system with which to study biogenesis of transmembrane glycoproteins. Sindbis, like VSV, is a lipid-enveloped virus, but contains only three structural proteins. Two are envelope glycoproteins that are integral membrane proteins (E_1 and E_2); one (E_2), like VSV G, spans the lipid membrane with a few amino acids, at the COOH terminus, exposed to the cytoplasmic surface. E_1 is also embedded in the lipid membrane near the COOH terminus, but may not be transmembrane (Wirth *et al.*, 1977; Garoff and Soderland, 1978). The third protein, core (C), is internal to the membrane and, like VSV N, is complexed to the viral RNA genome. In marked contrast to the case of VSV-infected cells, all three of these proteins are synthesized from one polyadenylated mRNA 26 S, which contains the nucleotide sequences found at the 3' end of the virion 42 S RNA. Both in cell-free systems and in infected cells, a single initiation site is used for the synthesis of all three proteins encoded by the 26 S RNA (Clegg and Kennedy, 1975; Cancedda *et al.*, 1975). The core protein is synthesized first, followed by the two envelope glycoproteins (E_1 and pE_2, a precursor to E_2). C, E_1, and pE_2 are derived by proteolytic cleavage of the nascent chain (Fig. 5). The gene order in Semliki Forest virus, a close relative of sindbis, and presumably also in Sindbis, is core–pE_2–E_1 (Lachmi and Kaariainen, 1976) as has been confirmed by the complete nucleotide sequence of 26 S RNA (Garoff *et al.*, 1980). Thus, one RNA encodes two very different types of proteins: a soluble cytoplasmic protein (C) and two integral membrane proteins.

During infection the 26 S RNA is found mainly in membrane-bound polysomes that synthesize all three virion proteins (Wirth *et al.*, 1977). Vesicles containing sindbis 26 S RNA in polysomes will direct cell-free synthesis of all three sindbis structural proteins, C_1, pE_2, and E_1. All of the newly made C protein is on the outside (cytoplasmic side) of the vesicles, while E_1 and pE_2 are sequestered in the vesicles. About 30 amino acids of E_2 remain exposed to the cytoplasm, as is the case with synthesis of VSV G (Wirth *et al* 1977; Garoff *et al.*, 1978).

A model (Wirth *et al.*, 1977; Garoff *et al.*, 1978, 1980) explaining these and other results on the translation of 26 S RNA is shown in Fig. 5. A ribosome begins translating the core protein at the 5' end of a free 26 S mRNA. As soon as the core protein segment is finished, it is removed by a protease, thus exposing the NH_2 terminus of the pE_2 protein. This initiates an interaction with the membrane, and is subsequently transferred into and across the membrane, presumably through a protein channel. The NH_2 segment of pE_2 contains a hydrophobic region that probably functions as a signal peptide, but it is not cleaved from the polypeptide (Bonatti and Blobel, 1979; Bonatti *et al.*, 1979). As in the case of VSV G protein, this interaction leads to the binding of the polysome to the ER. As the ribosome continues to traverse the mRNA, completed pE_2 is cleaved and remains embedded in or sequestered by the membrane. E_1 is then translated and inserted into the membrane, again attaching the polysome to the membrane (see Fig. 5).

Because the same ribosomes synthesize both soluble (C) and membrane (E_1 and pE_2) proteins, it is clear that the specificity of binding the 26 S RNA to membranes in such a way as to transfer only the glycoproteins into the membrane cannot reside in the ribosomes alone. The 60 S ribosome subunit may interact directly with membrane proteins, but this is neither the primary interaction nor is this sufficient to result in specific insertion of the membrane proteins. Similarly, binding of the mRNA directly to the membrane cannot alone account for the specificity of this interaction. We conclude that the nascent chain of the two glycoproteins—the only other possibility—determines the specific insertion of the membrane proteins and has a major role in the binding of polysomal 26 S RNA to membranes.

How E_1 is translocated across the ER membrane is not clear. In Semliki Forest virus the NH_2 terminus of E_1 is separated from the COOH terminus of pE_2 by a 60-residue peptide (Garoff *et al.*, 1980). This peptide is cleaved from the polyprotein during translation, and may function as a cleaved (albeit long) leader peptide of E_1. Evidence from a

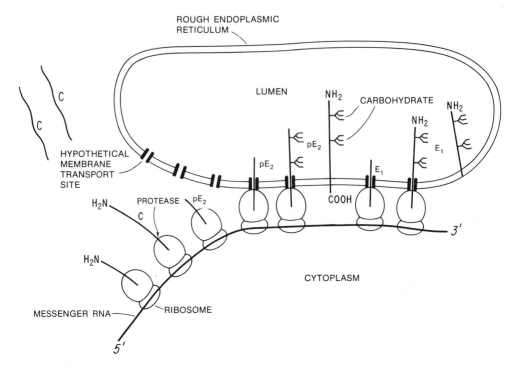

Figure 5. Model for synthesis of sindbis virus proteins, subsequent nascent cleavage, and sequestration of the envelope proteins pE_2 and E_1 by the ER (Wirth *et al.*, 1977; Garoff *et al.*, 1978, 1980).

Semliki Forest virus mutant indicates that E_1 possesses a leader peptide independent of that used for pE_2 (Hashimoto *et al.*, 1981).

6. INTERACTION OF NASCENT SINDBIS GLYCOPROTEIN AND THE ER MEMBRANE

Strong support for this model has come recently from two sources. First, Garoff *et al.* (1978) have done *in vitro* translation and ER vesicle addition experiments with the 26 S RNA similar to those done with VSV G mRNA (Rothman and Lodish, 1977). In synchronized *in vitro* translation of 26 S mRNA, ER membranes can be added as late as the time when the soluble C protein has just been completed and cleaved from the growing chain, and still allow subsequent normal membrane insertion of the E_1 and pE_2 proteins. If, however, membrane addition is delayed beyond this point, after synthesis of an appreciable portion of pE_2, the E_1, and pE_2 proteins subsequently made are not inserted into the ER phospholipid bilayer, nor is E_1 cleaved from pE_2. C, however, is properly cleaved. This is consistent with the notion that it is the NH_2 terminus of nascent pE_2 that is crucial in directing interaction of the complex of ribosomes, mRNA, and the growing polypeptide to ER membranes. If the NH_2 segment of nascent pE_2 is too long, it cannot interact productively with the ER receptors.

Second, we have investigated the properties of a temperature-sensitive mutant of sindbis virus, ts2 (Wirth *et al.*, 1979). This mutant fails to cleave the structural proteins at the nonpermissive temperature, resulting in the production of a polyprotein of 130,000 molecular weight (referred to as the ts2 protein). The order of the proteins in this polypeptide is presumed to reflect the gene order, NH_2-core–pE_2-E_1-COOH. Therefore, in the ts2 protein, the NH_2-terminal sequence of each glycoprotein, E_1 and pE_2, is internal. Although both ts2 protein and ts2 26 S mRNA are bound to microsomal membranes, no part of the ts2 protein is localized to the luminal side of the ER. Although the envelope protein sequences are present, they are not inserted into the membrane and are not protected from proteolysis. This result is consistent with the notion that the proteolytic cleavage between core and pE_2 exposes a sequence, presumably at the NH_2 terminus of pE_2, that is essential for proper insertion of pE_2 into the ER membrane. At least one other proteolytic cleavage does not occur in ts2-infected cells: that between pE_2 and E_1. Presumably, the principal defect of the ts2 mutation is the inhibition of C–pE_2 cleavage; the cleavage between E_1 and pE_2 probably does not occur because pE_2 is not properly inserted into the membrane.

7. TRANSPORT OF TRANSMEMBRANE VIRAL GLYCOPROTEINS FROM THE ROUGH ER TO THE PLASMA MEMBRANE

This remains one of the least understood processes in membrane biogenesis. These proteins do not appear on the cell surface until 20–45 min after the synthesis of the polypeptide chain, depending on the temperature and strains of cells and virus employed. The polypeptides move first to an intracellular smooth membrane component, presumably the Golgi (Bergmann *et al.*, 1981) where, as discussed below, glycosylation is completed (Knipe *et al.*, 1977a,b; Robbins *et al.*, 1977; Kornfeld *et al.*, 1978; Hunt *et al.*, 1978). During a late stage in maturation, possibly also in the Golgi, several molecules of fatty acid are covalently linked to the glycoprotein (Schmidt and Schlesinger, 1979). In what manner, and by what force, these integral membrane proteins are channeled, first to the Golgi and then to the surface, is obscure. Recent evidence suggests that clathrin-containing "coated vesicles" are involved in both stages of transport of G (Rothman and Fine, 1980). A related problem is elucidation of the roles, if any, glycosylation and lipid addition might play in these processes.

8. PROCESSING OF THE ASPARAGINE-LINKED OLIGOSACCHARIDE CHAINS OF VSV

The two aspragaine-linked complex oligosaccharides on each molecule of VSV G (Fig. 2) have a structure (Li *et al.*, 1978; Reading *et al.*, 1978) very similar, if not identical, to that found on a number of typical secreted glycoproteins, such as immunoglobins (Kornfeld and Kornfeld, 1976, 1980).

The biosynthesis of these complex-type asparagine-linked oligosaccharides is a multistep process that involves both addition and removal of specific saccharide residues (Fig. 6). A branched oligosaccharide precursor containing three glucose, nine mannose, and two *N*-acetylglucosamine residues is preformed on a lipid carrier molecule, localized in the rough ER (Fig. 2b). This oligosaccharide chain is transferred, en bloc to the nascent G polypeptide. Synchronized *in vitro* translation studies established that one of the chains is added when the nascent G is about one-third completed, the other when it is about 70% complete (Rothman and Lodish, 1977). The *N*-linked oligosaccharides are invariably found in the tripeptide sequences Asn-X-Ser/Thr. Recent work suggests that this tripeptide is the minimal substrate for the oligosaccharide-protein transferase (Hanover and Lennarz, 1980).

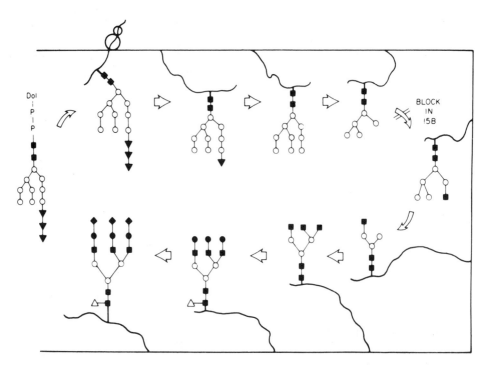

Figure 6. Proposed sequence for the synthesis of complex-type oligosaccharides (Kornfeld *et al.*, 1978). Dol = dolichol, the lipid carrier of the oligosaccharide. ■, *N*-acetylglucosamine; ○, mannose; ▼, glucose; ●, galactose; ◆, sialic acid; △, fucose.

The oligosaccharide-protein transferase (Hanover and Lennarz, 1980). The oligosaccharides invariably face the luminal side of the ER, and it is believed that the enzyme that transfers the saccharide to the polypeptide is localized on the luminal surface of the ER (Fig. 4).

Immediately after transfer to the polypeptide, one or two of the glucose residues are removed. Further processing of the oligosaccharide occurs only 10–20 min after synthesis of G, presumably at the time the protein is transferred to the Golgi complex. In a stepwise, concerted set of reactions, the remaining glucose residues and six of the nine mannose residues are removed from the oligosaccharide, and the "peripheral" sugar residues *N*-acetylglucosamine (three residues per chain), galactose (three residues), sialic acid (one to three residues per chain), and fucose (one residue) are added (Fig. 6). Oligosaccharide processing is completed about 10 min before the G protein reaches the cell surface (Knipe *et al.*, 1977a).

Although Golgi membranes have not been purified, or even studied to any great extent, in the cell lines used for HLA-A synthesis or virus infections, it is thought that this organelle is the site of these late carbohydrate-processing events. In liver cells the Golgi fraction is greatly enriched for the two α-mannosidases, and several of the monosaccharide transferases that can function in cell-free systems in these reactions (Turco and Robbins, 1979; Tabas and Kornfeld, 1979; Bretz *et al.*, 1980; Schachter and Roseman, 1980).

The structure of the *N*-linked high-mannose oligosaccharide on certain glycoproteins is different from that of complex oligosaccharides (Fig. 2), in that they lack fucose, sialic acid, galactose, and the three peripheral GlcNAc residues. Rather, they contain six to nine mannose residues, as well as two GlcNAc residues adjacent to the Asn residue. In some cases, the complex and high-mannose Asn-linked carbohydrates are found on the same glycoprotein molecule (Kornfeld and Kornfeld, 1976, 1980).

Despite this difference in structure, it is believed that both complex and high-mannose oligosaccharides are derived from a common precursor. The high-mannose oligosaccharides found on the mature form of certain glycoproteins resemble an intermediate ($Man_{6-9}GlcNAc_2$) in processing of the complex oligosaccharides (Fig. 6). It is believed that both classes of N-linked oligosaccharides do derive from this same intermediate, the difference being the action of an α-mannosidase on the $Man_{6-9}GlcNAc_2$ oligosaccharides. As both complex and high-mannose oligosaccharides are occasionally found on the same protein molecule, the differences in processing may reflect the accessibility of a particular sugar chain to a Golgi mannosidase.

9. MODIFICATION AND MATURATION OF TRANSMEMBRANE GLYCOPROTEINS

What, if any, is the function of these oligosaccharides in maturation of G? Tunicamycin is an antibiotic that blocks the first stage in formation of the oligosaccharide–lipid donor; in its presence the G polypeptide is synthesized but contains no sugar residues. Nonglycosylated G is inserted normally as a transmembrane protein in the ER (Rothman *et al.*, 1978), but does not normally mature to the cell surface at any temperature (Leavitt *et al.*, 1977a,b). [There is one interesting exception: nonglycosylated G synthesized by one specific strain of VSV will mature to the cell surface at 32°C but not at 37°C (Gibson *et al.*, 1978, 1979).] This indicates that at least the initial, high-mannose, oligosaccharides are generally essential for some step in intracellular transport.

Studies on the solubilization of G proteins by detergents suggest that the failure to glycosylate the nascent G protein can affect the folding of the molecule and, as a consequence, produce an increased sensitivity to temperature during interactions of the protein with the phospholipid membrane (Gibson *et al.*, 1979). A similar abnormal G is produced in cells (Thy-1⁻) in which a $Glc_3Man_5GlcNAc_2$ oligosaccharide is added to nascent G (Gibson *et al.*, 1981). Clearly, the size of the oligosaccharide precursor is important for the proper folding of the nascent protein. Moreover, the physicochemical properties of the G molecule also depend on the size and structure of the mature carbohydrate side chains (Gibson *et al.*, 1981). For instance, cell lines such as 15B (Fig. 6) are defective in some of the later carbohydrate-maturation steps. VSV infection of these cells results in a near-normal yield of virus particles. G protein moves normally to the cell surface and is incorporated normally into virions, even though G lacks most of the terminal sugar residues galactose, fucose, and sialic acid (Schlesinger *et al.*, 1976). Apparently, these peripheral sugars do not play a role in intracellular movement of G although, in some cases, they can affect the aggregation state of G protein in lipids (Gibson *et al.*, 1981).

Many secretory proteins contain the same asparagine-linked oligosaccharide chains as found in VSV G and sindbis virus glycoproteins. Secretion of some, but not all, of these proteins will take place even in the presence of tunicamycin. For instance, the rate and extent of secretion of glycosylated and unglycosylated fibronectin by fiberoblasts (Olden *et al.*, 1978) or transferrin by a cultured line of rat hepatoma cells are the same (Strous and Lodish, 1980). Secretion of some classes of immunoglobulins, but not others, will occur if addition of the asparagine-linked oligosaccharide is blocked (Hickman and Kornfeld, 1978). Unglycosylated HLA-A protein matures to the cell surface at the same rate, and to the same extent, as does the normally glycosylated form (Owen *et al.*, 1980). A similar result has been obtained with glycophorin (Gahmberg *et al.*, 1980). Clearly, the requirement for glycosylation is an idiosyncratic property of the individual protein species. It has been suggested that addition of the high-mannose oligosaccharide chain modifies the conformation of the polypeptide with respect to the membrane phospholipid (Gibson *et al.*, 1979); in some cases this alteration may be essential for subsequent movement.

Each molecule of HLA-A and H-2 heavy chain is normally covalently bound, on the extracytoplasmic surface, to a molecule of the soluble glycoprotein $\beta 2$ microglobulin (Fig. 1). This attachment normally takes place either while the HLA-A chain is nascent or shortly after it is completed (Dobberstein *et al.*, 1979; Krangel *et al.*, 1979). Proper transmembrane insertion of HLA-A is not dependent on attachment of $\beta 2$ microglobulin (Owen *et al.*, 1980). Interestingly, in a cell line (Daudi) unable to produce $\beta 2$ microblobulin, HLA-A remains in the ER and will not mature to the cell surface (Ploegh *et al.*, 1979). Possibly, the addition of this polypeptide modifies the conformation of HLA-A in the membrane, so that it can be moved to the Golgi and thence to the cell surface.

10. ANALYSIS OF MATURATION-DEFECTIVE MUTANTS OF VSV AND SINDBIS VIRUS

Animal cells infected with enveloped viruses, such as VSV or sindbis virus, provide very attractive model systems for studying the biosynthesis, processing, and intracellular transport of membrane glycoproteins. One of the major advantages is that temperature-sensitive maturation-defective mutants, each with a specific alteration in the structural gene encoding a viral glycoprotein, are available. These mutants can be used to dissect the steps involved in the maturation pathway of a glycoprotein and to study the roles of several posttranslational modification reactions and the interaction with other viral components.

10.1. VSV Mutants

Temperature-sensitive (ts) mutants in complementation group (V) of VSV, which corresponds to the structural gene of G, are available and can be selectively isolated (Lafay, 1974; Lodish and Weiss, 1979). All the ts (V) mutants studied so far share several common properties. Cells infected with ts (V) mutants produce at the nonpermissive temperature markedly reduced yields of viruslike particles, which are noninfectious and specifically deficient in G. The G polypeptide is synthesized normally and is metabolically stable; however, it fails to mature to the cell surface at the nonpermissive temperature (Knipe *et al.*, 1977b,c; Lodish and Weiss, 1979; Zilberstein *et al.*, 1980).

We have analyzed the structure of the G molecules that are synthesized and accumulate in ts (V)-infected cells, with respect to the known biochemical parameters of posttranslational processing and intracellular transport of G. The results of these studies demonstrate that mutations in the gene encoding for G result in blocks at very different stages in the maturation pathway (Zilberstein *et al.*, 1980).

A major analytical tool utilized in these studies, to probe for the functional transport of G from the ER to the Golgi system, is the enzyme endo-β-N-acetylglucosaminidase H (Endo H). This is based on the substrate specificity of the enzyme, which cleaves between the two proximal N-acetylglucosamine residues of the large high-mannose-type oligosaccharides, as indicated in Fig. 2. Complex-type carbohydrate side chains are resistant to this enzyme, and oligosaccharides that are intermediates in processing show intermediate sensitivities, according to their structure (Tarentino and Maley, 1974; Tai *et al.*, 1977; Robbins *et al.*, 1977). Conversion of a carbohydrate side chain of a glycoprotein from the Endo H-sensitive to the Endo H-resistant form is, therefore, a consequence of its transport from the ER to the Golgi and the subsequent oligosaccharide-processing reactions.

This analysis has shown that the ts (V) mutants of VSV can be subdivided into (at least) two distinct subclasses, with respect to the stage of posttranslational processing at which the block occurs (Zilberstein *et al.*, 1980):

1. The mutants ts L513(V), ts M501(V), and ts 045(V) encode G molecules that are blocked at the nonpermissive temperature, at an early pre-Golgi step in G maturation. In cells infected with these mutants, the G polypeptide chain is synthesized normally and is inserted into and spans the ER membrane, at both the permissive and the restrictive temperatures, in a manner that is indistinguishable from that of wild-type G. The newly synthesized G is also normally glycosylated and has lost the NH_2-terminal hydrophobic signal sequence. However, processing of the carbohydrate side chains from the Endo H-sensitive to the Endo H-resistant form is blocked, and no fatty acid is attached to the G molecule, at the nonpermissive temperature.

A more detailed carbohydrate analysis of the ts L513(V) G molecule that accumulates at the nonpermissive temperature has shown that the very first oligosaccharide-processing reactions, which are localized to the rough ER, do occur—removal of the glucose residues from one of the nonreducing termini. All the subsequent Golgi-mediated carbohydrate-processing reactions are effectively blocked (Zilberstein *et al.*, 1980). These results are consistent with the previous suggestion, made on the basis of subcellular fractionation experiments, that the G molecule in ts M501(V)- or ts 045(V)-infected cells is unable to be transported from the rough ER to Golgi system at the nonpermissive temperature (Knipe *et al.*, 1977c). The structure of the ts L513(V) G protein synthesized at the *permissive* temperature is very similar to that of wild-type G. The only difference, observed so far, between these two G molecules is that VSV particles produced by ts (V)-infected cells at the permissive temperature are thermolabile (Lodish and Weiss, 1979; Weiss and Bennett, 1980).

2. ts L511(V) represents a novel class among the maturation-defective mutants of animal viruses that have been characterized so far. G maturation in cells infected with this ts (V) mutant proceeds normally through most of the Golgi-mediated functions involved in processing of the carbohydrate side chains, including terminal sialylation. However, this G molecule does not undergo two posttranslational modification reactions that take place with wild-type G.

a. One hexose residue, fucose, is not added to the oligosaccharide side chains of ts L511(V) G. Interestingly, this defect is expressed at both the permissive and the nonpermissive temperatures (Zilberstein *et al.*, 1980). Clearly, addition of fucose to G is not an obligatory requirement for proper maturation of G and virion assembly.

b. Schlesinger and his co-workers have shown that the VSV G molecule is also modified by the covalent attachment of one or two molecules of fatty acid directly to the polypeptide chain (Schmidt and Schlesinger, 1979; Schmidt *et al.*, 1979). This reaction is selectively blocked at the nonpermissive temperature in cells infected with any of the ts (V) mutants analyzed (Zilberstein *et al.*, 1980).

This analysis tentatively indicates the step where the G-bound fatty acid might be required. Fatty acid attachment normally occurs shortly before G is converted from the Endo H-sensitive to the Endo H-resistant form (Schmidt and Schlesinger, 1980). Despite the fact that fatty acid is not added to ts L511(V) G, this molecule undergoes most of the carbohydrate-processing reactions in the Golgi and acquires most of the terminal sugars (except fucose) with normal or nearly normal kinetics. Clearly, covalent addition of lipid is not required for most of the terminal stages of oligosaccharide modification. If it does have a function, it may be in the subsequent movement of glycoproteins to the cell surface. It appears that maturation of ts L511(V) G to the cell surface, at the nonpermissive temperature, is specifically blocked at this step.

The mechanisms by which the ts mutation prevents these two late processing reactions of ts L511(V) G are not clear; at least two alternative explanations are possible. The mutation in G could result in a change in the conformation of the molecule, so that it is not transported to the specific sites where the modification reactions take place, or its substrate activity for the modifying enzymes could be markedly impaired. The observations that fucosylation of ts L511(v) G is defective even at the permissive temperature and that fatty acid attachment is a relatively early Golgi function favor the second possibility. The amino acid sequence of G might be altered, such that the conformation of the glycoprotein at 32°C is compatible with intracellular transport, insertion into the plasma membrane, and budding, but the conformation of this molecule at 40°C is incompatible with these processes.

One critical question remains to be answered: whether the defects in these posttranslational processing reactions in the maturation of ts L511(V) and ts L513(V) glycoproteins are primary or secondary effects of the amino acid changes that underlie the temperature-sensitive phenotypes. Detailed analysis of temperature-resistant revertants of these mutants will be of great interest, as they might clarify the roles of addition of fucose and fatty acid in the maturation of G. It is our hope that comparative sequence analysis of the G proteins of wild-type VSV, ts (V) mutants, and their revertants will shed light on the "domains" in the G molecule that are important for the intracellular transport of G from the rough ER to the Golgi system, and from the Golgi to the cell surface.

10.2. Sindbis Virus Mutants

That the VSV-infected cell is the simplest model system for studying glycoprotein biosynthesis and maturation is evident from the following:

1. VSV has only one glycoprotein species—G.
2. VSV G has two identical carbohydrate side chains.
3. Transport of VSV G to the cell surface is independent of that of the other viral proteins (Knipe *et al.*, 1977b,c).

Many other enveloped viruses contain more than one glycoprotein, each of which may have several different oligosaccharides, and in a few cases their maturation is interdependent. Characterization of maturation-defective ts mutants of such viruses has been useful in assigning biological activities to each glycoprotein and in demonstrating a functional association between them during intracellular transport and virion assembly.

As mentioned earlier, sindbis virus has two envelope glycoproteins, E_1 and E_2, while the closely related Semliki Forest virus has three, E_1, E_2, and E_3. pE_2 and E_1 are sequestered cotranslationally within the lumen of the ER membrane system and are transported to the plasma membrane as a relatively stable pE_2–E_1 complex (Bracha and Schlesinger, 1976; Jones *et al.*, 1977; Ziemiecki and Garoff, 1978; Garoff *et al.*, 1980). pE_2 is cleaved to E_2 at a late stage of development, just before E_1 and E_2 appear on the cell surface (Smith and Brown, 1977). Although experimental evidence suggested that pE_2, like E_1 and E_2, may be present on the surface of infected cells, the presence of pE_2 on the cell surface has never been demonstrated directly by any procedure (Erwin and Brown, 1980; Garoff *et al.*, 1980). Each of the mature glycoproteins E_1 and E_2 has at least one high-mannose and one complex-type carbohydrate side chain (Keegstra *et al.*, 1975; Hakimi and Atkinson, 1980). [E_1 of Semiliki Forest virus has only one potential glycosylation site (Garoff *et al.*, 1980).]

Using a combination of genetic and immunological approaches, the ts maturation-defective mutants of sindbis virus were divided into three complementation groups:

1. Mutants in complementation group (C)—such as ts 2(C) and ts 9(C)—are altered in the gene encoding the core (C) protein. Cells infected with these mutants accumulate, at the nonpermissive temperature, the 130,000-dalton precursor for the three structural proteins of sindbis virus (see Section 6). Genetic analysis of ts (C) mutants and *in vitro* translation of the sindbis virus and Semliki Forest virus 26 S RNAs provide indirect evidence that the C protein itself is responsible for the proteolytic cleavage at the C/pE_2 junction (Cancedda *et al.*, 1975; Garoff *et al.*, 1978).

2. Sindbis virus mutants in complementation group (D) encode an altered E_1 glycoprotein. In cells infected at the nonpermissive temperature with these mutants, no functional glycoprotein reaches the cell surface and virus assembly is effectively blocked (Brown and Smith, 1975; Bell and Waite, 1977). Interestingly, the cleavage of pE_2 to E_2 is also blocked under these conditions. This finding, together with the observation that cleavage of pE_2 to E_2 is also blocked in cells infected with ts (E) mutants (see next), first suggested that pE_2 and E_1 might exist as a complex during viral replication. This was confirmed, first, by the ability of antisera specific for E_1 (as well as anti-E_2 antisera) to inhibit the cleavage of pE_2 to E_2 in the ER membrane very soon after their synthesis (Ziemiecki and Garoff, 1978).

Biochemical analysis of the pE_2 and E_1 molecules that are synthesized and accumulate at the nonpermissive temperature in cells infected with ts 23(D) or ts 10(D) has shown that all the Golgi-mediated carbohydrate-processing reactions are blocked: neither E_1 nor pE_2 is converted from the Endo H-sensitive to the Endo H-resistant form, although the glucose residues are removed in the ER (D. F. Wirth, personal communication). In addition, fatty acid attachment to pE_2 (and E_1) is selectively blocked at the nonpermissive temperature (Schmidt *et al.*, 1979).

From these studies it seems that the block in maturation of the viral envelope glycoproteins in cells infected with ts (D) mutants of sindbis virus is very similar to that observed in cells infected with the "early" mutants of VSV [such as ts L513(V)]. In contrast, results from subcellular fractionation experiments suggest that in ts 23(D)-infected cells at the nonpermissive temperature, at least a part of the pE_2 and E_1 molecules are transported from the rough ER into a smooth ER/Golgi fraction, and the major block in their maturation to the cell surface is in the transport from the Golgi system to the plasma membrane (Erwin and Brown, 1980). If this is so, then these forms of pE_2 and E_1 are not recognizable by any of the known posttranslational modification systems in the Golgi. In any case, no quantitative estimate of the proportion of the pE_2 and E_1 molecules in the cell that are actually transported into the Golgi can be obtained from fractionating cells into three relatively crude and significantly cross-contaminated membrane fractions (Erwin and Brown, 1980).

3. Maturation-defective mutants in complementation group (E) of sindbis virus are altered in the structural gene encoding pE_2. No mature or budding virons can be detected in cells infected with ts (E) mutants at the nonpermissive temperature (Smith and Brown,

1977; Bell and Waite, 1977). The biochemical lesion(s) in the maturation of the viral glycoproteins in cells infected with ts (E) mutants of sindbis is very different from that in cells infected with the VSV ts L511(V) mutant. At the nonpermissive temperature, E_1 matures to the cell surface in ts 20(E)-infected cells (Smith and Brown, 1977; Bell and Waite, 1977). The cleavage of pE_2 to E_2 is blocked, but unlike the situation in ts 23(D)-infected cells, the pE_2 form becomes associated with the plasma membrane (Smith and Brown, 1977). Both pE_2 and E_1 have fatty acid attached to them (Schmidt et al., 1979), and large numbers of nucleocapsids are attached to the inner surface of the plasma membrane, possibly by direct association of the C protein with the viral glycoproteins in the virus-modified plasma membrane (Brown and Smith, 1975).

We do not know the nature of the primary ts biochemical lesion in the maturation of the viral envelope glycoproteins in cells infected with ts (E) mutants. Possibly, the association of pE_2 with E_1 and the nucleocapsids at the inner surface of the plasma membrane is nonfunctional (as noted earlier, no form of pE_2 has as yet been demonstrated on the cell surface, by any technique), and thus virus assembly and budding are blocked. Alternatively, the primary defect may be that the conformation of this pE_2 molecule is not recognized by the enzyme that cleaves normal pE_2 to E_2, and in the absence of E_2 no virus assembly can take place. We need to know more about the exact sequence of events during Sindbis virus assembly and budding in order to discriminate between these possibilities.

In either case, the pE_2 molecule synthesized in ts (E)-infected cells has a temperature-dependent conformation that is clearly different from that of the pE_2 form synthesized in wild-type-infected cells. This is also demonstrated by two other structural abnormalities of this ts (E) pE_2 molecule. First, a block in processing of the carbohydrate side chains of pE_2 and E_1, which is very similar to that observed in cells infected with ts (D) mutants, is also observed in ts (E)-infected cells (D. F. Wirth, personal communication). Clearly complete processing of any of the carbohydrate side chains on pE_2 and E_1, from the high mannose to the mature complex type, is not absolutely required for migration of the glycoproteins to the plasma membrane and for the functional insertion of E_1 on the cell surface. However, glycosylation per se is required for a late step in sindbis virus maturation and assembly, for in tunicamycin-treated cells cleavage of pE_2 to E_2, maturation of the pE_2 and E_1 polypeptides to the cell surface, and virion assembly are blocked (Schwarz et al., 1976; Leavitt et al., 1977a,b). Second, the pE_2 molecule synthesized at the nonpermissive temperature in ts (E)-infected cells is not recognized by anti-E_2 antisera, which react with pE_2 from wild-type or ts (D)-infected cells (Bell and Waite, 1977).

This analysis of the maturation-defective mutants of sindbis virus demonstrates that the proteolytic cleavage of pE_2 to E_2 is essential for the functional assembly and budding of sindbis virus. This is dependent on both the structure of the pE_2 molecule itself, which can be modified by ts (E) mutations, by blocking glycosylation with tunicamycin, or by abnormal completion of the complex-type carbohydrate side chains (Gottlieb et al., 1979), and its proper interaction with E_1, which is modified by ts (D) mutations or by anti-E_1 antisera.

ACKNOWLEDGMENTS. Research in the authors' laboratory was supported by Grants AI 08814 and AM 15322 from the National Institutes of Health.

We thank Dr. D. Wirth for communicating her results on the carbohydrate analysis of the sindbis virus mutants prior to publication, and Miriam D. Boucher for the patience, understanding, and skill involved in the preparation of the manuscript.

REFERENCES

Barnstable, C. J., Jones, E. A., and Crumpton, M. J. (1978). *Br. Med. Bull.* **34,** 241–246.

Bell, J. W., and Waite, M. R. F. (1977). *J. Virol.* **21,** 788–791.

Bergmann, J. E., Takuyasu, K. T., and Singer, S. J. (1981). *Proc. Natl. Acad. Sci. USA* **78,** 1476–1750.

Bonatti, S., and Blobel, G. (1979). *J. Biol. Chem.* **254,** 12261–12264.

Bonatti, S., Cancedda, R., and Blobel, G. (1979). *J. Cell Biol.* **80,** 219–224.

Bracha, M., and Schlesinger, M. J. (1976). *Virology* **74,** 441–449.

Bretz, R., Bretz, H., and Palade, G. E. (1980). *J. Cell Biol.* **84,** 87–101.

Brown, D. T., and Smith, J. F. (1975). *J. Virol.* **15,** 1262–1266.

Cancedda, R., Villa-Komaroff, L., Lodish, H. F., and Schlesinger, M. J. (1975). *Cell* **6,** 215–222.

Clegg, J. C. S., and Kennedy, S. I. T. (1975). *J. Mol. Biol.* **97,** 401–411.

Devillers-Thiery, A., Kindt, T., Scheele, G., and Blobel, G. (1975). *Proc. Natl. Acad. Sci. USA* **72,** 5016–5020.

Dobberstein, B., Garoff, H., and Warren, G. (1979). *Cell* **17,** 759–769.

Erwin, C., and Brown, D. T. (1980). *J. Virol.* **36,** 775–786.

Gahmberg, C. G., Jokinen, J., Karhi, K. K., and Anderson, L. C. (1980). *J. Biol. Chem.* **255,** 2169–2175.

Garoff, H., and Schwarz, R. L. (1978). *Nature (London)* **274,** 487–490.

Garoff, H., and Soderland, H. (1978). *J. Mol. Biol.* **124,** 535–549.

Garoff, H., Simons, K., and Dobberstein, B. (1978). *J. Mol. Biol.* **124,** 587–600.

Garoff, H., Frischauf, A. M., Simons, K., Lehrach, H., and Delius, H. (1980). *Nature (London)* **228,** 236–241.

Gibson, R., Leavitt, R., Kornfeld, S., and Schlesinger, S. (1978). *Cell* **13,** 671–679.

Gibson, R., Schlesinger, S., and Kornfeld, S. (1979). *J. Biol. Chem.* **254,** 3600–3607.

Gibson, R., Kornfeld, S., and Schlesinger, S. (1981). *J. Biol. Chem.* **256,** 456–462.

Gottlieb, C., Kornfeld, S., and Schlesinger, S. (1979). *J. Virol.* **29,** 344–351.

Hakimi, J., and Atkinson, P. H. (1980). *Biochemistry* **19,** 5619–5624.

Hanover, J. A., and Lennarz, W. J. (1980). *J. Biol. Chem.* **255,** 3600–3604.

Hashimoto, K., Erdei, S., Keranen, S., Saraste, J., and Kaariainen, L. (1981). *J. Virol.,* in press.

Hickman, S., and Kornfeld, S. (1978). *J. Immunol.* **121,** 990–996.

Hunt, L. A., Etchison, J. R., and Summers, D. F. (1978). *Proc. Natl. Acad. Sci. USA* **75,** 754–758.

Jones, K. J., Scupham, R. K., Pfeil, J. A., Wan, K., Sagik, B. P., and Bose, H. R. (1977). *J. Virol.* **21,** 778–787.

Katz, F. N., and Lodish, H. F. (1979). *J. Cell Biol.* **80,** 416–426.

Katz, F. N., Rothman, J. E., Knipe, D. M., and Lodish, H. F. (1977a). *J. Supramol. Struct.* **7,** 353–370.

Katz, F. N., Rothman, J. E., Lingappa, V. R., Blobel, G., and Lodish, H. F. (1977b). *Proc. Natl. Acad. Sci. USA* **74,** 3278–3282.

Keegstra, K., Sefton, B., and Burke, D. (1975). *J. Virol.* **16,** 613–620.

Knipe, D., Lodish, H. F., and Baltimore, D. (1977a). *J. Virol.* **21,** 1121–1127.

Knipe, D. M., Baltimore, D., and Lodish, H. F. (1977b). *J. Virol.* **21,** 1128–1139.

Knipe, D. M., Baltimore, D., and Lodish, H. F. (1977c). *J. Virol.* **21,** 1149–1158.

Kornfeld, R., and Kornfeld, S. (1976). *Annu. Rev. Biochem.* **45,** 217–237.

Kornfeld, R., and Kornfeld, S. (1980). In *Biochemistry of Glycoproteins and Proteoglycans* (W. J. Lennarz, ed.), pp. 1–34, Plenum Press, New York.

Kornfeld, S., Li, E., and Tabas, I. (1978). *J. Biol. Chem.* **253,** 7771–7778.

Krangel, M. S., Orr, H. T., and Strominger, J. L. (1979). *Cell* **18,** 979–991.

Lachmi, B. E., and Kaariainen, L. (1976). *Proc. Natl. Acad. Sci. USA* **73,** 1936–1940.

Lafay, F. (1974). *J. Virol.* **14,** 1220–1228.

Leavitt, R., Schlesinger, S., and Kornfeld, S. (1977a). *J. Biol. Chem.* **252,** 9018–9023.

Leavitt, R., Schlesinger, S., and Kornfeld, S. (1977b). *J. Virol.* **21,** 375–385.

Li, E., Tabas, I., and Kornfeld, S. (1978). *J. Biol. Chem.* **253,** 7762–7770.

Lingappa, V. R. Katz, F. N., Lodish, H. F., and Blobel, G. (1978). *J. Biol. Chem.* **253,** 8667–8670.

Lodish, H. F., and Froshauer, S. (1977). *J. Cell Biol.* **74,** 358–364.

Lodish, H. F., and Weiss, R. A. (1979). *J. Virol.* **30,** 177–189.

Morrison, T., and Lodish, H. F. (1975). *J. Biol. Chem.* **250,** 6955–6962.

Olden, K., Pratt, R. M., and Yamada, K. (1978). *Cell* **13,** 461–473.

Owen, M. J., Kissonerghis, A. M., and Lodish, H. F. (1980). *J. Biol. Chem.* **255,** 9678–9684.

Palade, G. (1975). *Science* **189,** 347–358.

Ploegh, H., Cannon, L. E., and Strominger, J. L. (1979). *Proc. Natl. Acad. Sci. USA* **76,** 2273–2277.

Pringle, C. R. (1977). In *Comprehensive Virology* (H. Fraenkel-Conrat and R. R. Wagner, eds.), Vol. IX, pp. 239–290, Plenum Press, New York.

Reading, C. L., Penhoet, E. E., and Ballou, C. E. (1978). *J. Biol. Chem.* **253**, 5600–5612.

Robbins, P. W., Hubbard, S. C., Turco, S. J., and Wirth, D. F. (1977). *Cell* **12**, 893–900.

Rose, J. K., Welch, W. J., Sefton, B. M., Esch, F. S., and Ling, N. C. (1980). *Proc. Natl. Acad. Sci. USA* **77**, 3884–3888.

Rothman, J. E., and Fine, R. E. (1980). *Proc. Natl. Acad. Sci. USA* **77**, 780–784.

Rothman, J. E., and Lodish, H. F. (1977). *Nature (London)* **269**, 775–780.

Rothman, J. E., Katz, F. N., and Lodish, H. F. (1978). *Cell* **15**, 1447–1454.

Schachter, H., and Roseman, S. (1980). In *The Biochemistry of Glycoproteins and Proteoglycans* (W. J. Lennarz, ed.), pp. 85–160, Plenum Press, New York.

Schachter, H., Jabal, I., Hudgin, R. L., Pineteric, L. McGuire, E. J., and Roseman, S. (1970). *J. Biol. Chem.* **245**, 1090–1100.

Schlesinger, S., Gottlieb, C., Feil, P., Gelb, N., and Kornfeld, S. (1976). *J. Virol.* **17**, 239–246.

Schmidt, M. F. G., and Schlesinger, M. (1979). *Cell* **17**, 813–819.

Schmidt, M. F. G., and Schlesinger, M. J., (1980). *J. Biol. Chem.* **255**, 3334–3339.

Schmidt, M. F. G., Bracha, M., and Schlesinger, M. J. (1979). *Proc. Natl. Acad. Sci. USA* **76**, 1687–1691.

Schwarz, R. T., Rohrschneider, J. M., and Schmidt, M. J. (1976). *J. Virol.* **19**, 782–791.

Smith, J. F., and Brown, D. T. (1977). *J. Virol.* **22**, 662–678.

Strominger, J. L., Englehard, V. H., Fuks, A., Guild, B. C., Hyafil, F., Kaufman, J. F., Korman, A. J., Kostyk, T. G., Krangel, M. S., Lancet, D., de Castro, L. J. A., Mann, D. L., Orr, H. T., Parham, P., Parker, K. C., Ploegh, H. L., Pober, J. S., Robb, R. J., and Shackelford, D. A. (1980). In *The Role of the Major Histocompatibility Complex in Immunobiology* (F. Benacerraf and M. E. Dorf, eds.), Garland Press, in press.

Strous, G. J. A. M., and Lodish, H. F. (1980). *Cell* **22**, 709–717.

Tabas, I, and Kornfeld, S. (1979). *J. Biol. Chem.* **254**, 11655–11663.

Tai, T., Yamashita, K., and Kobata, A. (1977). *Biochem. Biophys. Res. Commun.* **78**, 434–441.

Tarentino, A. L., and Maley, F. (1974). *J. Biol. Chem.* **249**, 811–817.

Turco, S. J., and Robbins, P. W. (1979). *J. Biol. Chem.* **254**, 4560–4567.

Von Heijne, G. (1980). *Eur. J. Biochem.* **103**, 431–438.

Wagner, R. R. (1975). In *Comprehensive Virology* (H. Fraenkel-Conrat and R. R. Wagner, eds.), Vol. IV pp. 1–94, Plenum Press, New York.

Weiss, R. A., and Bennett, P. L. P. (1980). *Virology* **100**, 252–274.

Wirth, D. F., Katz, F., Small, B., and Lodish, H. F. (1977). *Cell* **10**, 253–263.

Wirth, D. F., Lodish, H. F., and Robbins, P. W. (1979). *J. Cell Biol.* **81**, 154–162.

Yamashita, K., Tachibana, Y., and Kobata, A. (1978). *J. Biol. Chem.* **253**, 3862–3869.

Ziemiecki, A., and Garoff, H. (1978). *J. Mol. Biol.* **122**, 259–269.

Zilberstein, A., Snider, M. D., Porter, M., and Lodish, H. F. (1980). *Cell* **21**, 417–427.

29

Biosynthesis of Mitochondrial Membrane Proteins

Robert O. Poyton, Gary Bellus, and Ann-Louise Kerner

1. INTRODUCTION

Current interest in membrane biogenesis is focused on a series of questions that pertain to: the structure and regulation of genes for membrane proteins; the routes taken by membrane proteins and lipids from their sites of synthesis in the cell to their sites of insertion in the membrane; the coordination between the synthesis of membrane proteins and lipids; and the mechanisms by which proteins and lipids are inserted into the membrane bilayer so as to ensure its continued asymmetry. In considering the biogenesis of mitochondrial membranes, these questions must be extended to include, in addition, the respective roles played by mitochondrial and nuclear genomes in the assembly process. In this review we will first briefly summarize the coordination between nuclear and mitochondrial genomes in mitochondrial biogenesis and then discuss the pathways followed by nuclear and mitochondrial gene products into the genetic mosaic of the inner mitochondrial membrane.

2. ROLES OF MITOCHONDRIAL AND NUCLEAR GENETIC SYSTEMS IN MITOCHONDRIAL MEMBRANE ASSEMBLY

It is now clear that the assembly of a respiratory-competent mitochondrion in all eukaryotic cells is the end result of the fruitful collaboration between two distinct intracellular genetic systems—that localized within the mitochondrial matrix and that located in the nucleus and cytosol. With the biosynthetic inventory of mitochondrial constituents now nearing completion (recent review: Poyton, 1980), it is clear that all but a handful of those proteins that reside in the mitochondrion are nuclear gene products. These proteins are translated into the cytoplasmic compartment of eukaryotic cells and are subsequently trans-

Robert O. Poyton and Gary Bellus • Department of Molecular, Cellular, and Developmental Biology, University of Colorado, Boulder, Colorado 80309.   ***Ann-Louise Kerner*** • Department of Biochemistry, University of Connecticut Health Center, Farmington, Connecticut 06032.

ported into the mitochondrion (recent review: Neupert and Schatz, 1981). They play key roles in mitochondrial transcription and translation, in mitochondrial lipid and heme synthesis, in substrate oxidation by the TCA cycle, and in mitochondrial electron transport and oxidative phosphorylation (reviews: Schatz and Mason, 1974; Tzagoloff *et al.*, 1979; Poyton, 1980). In short, they are major components of all four mitochondrial compartments (i.e., inner and outer mitochondrial membranes, matrix and intermembrane space).

Those few polypeptides encoded on the mitochondrial genome are, on the other hand, restricted in their distribution to the inner mitochondrial membrane where they are subunits of oligomeric complexes that are genetic mosaics composed of nuclear-encoded polypeptides as well. Three of these mosaic complexes (cytochrome *c* oxidase, coenzyme QH_2-cytochrome *c* reductase, and oligomycin-sensitive ATPase) are membrane proteins (Henderson *et al.*, 1977; von Jagow and Sebald, 1980; Sebald, 1977), while a fourth is the small subunit of the mitochondrial ribosome (Terpstra *et al.*, 1979; Groot *et al.*, 1979).

When considered together, the above findings have made clear that while the biogenesis of the outer mitochondrial membrane is entirely under nuclear control, the biogenesis of the inner mitochondrial membrane is under the control of both mitochondrial and nuclear genetic systems. This dual genetic origin of the protein components of the inner mitochondrial membrane has raised a number of questions that are now under active investigation. First, it has focused interest on the nature and function of those few mitochondrial proteins that are encoded by the mitochondrial genome. Second, it has posed questions concerning the mechanism(s) by which the cell coordinates the expression of both mitochondrial and nuclear sets of proteins, especially those (e.g., the subunits of cytochrome *c* oxidase) that are present in equimolar amounts in the inner mitochondrial membrane. And third, it has made clear that the inner mitochondrial membrane grows *bitropically* by the insertion of new proteins into both of its surfaces (i.e., cytoplasmic and matrix). This mode of membrane growth is unique among all other biological membranes, which grow *monotropically* by the insertion of new proteins into their cytoplasmic surface only (Blobel, 1980). Although it is not clear if these two modes of membrane growth are fundamentally different and if, for instance, the *bitropic* growth of the inner mitochondrial membrane is responsible for its convoluted morphology (Munn, 1974), it has become of some interest to compare the mechanism(s) of insertion of mitochondrial gene products (into the matrix face of the inner membrane) with the mechanisms of insertion of nuclear gene products into the cytoplasmic face of the membrane.

3. MITOCHONDRIAL GENE PRODUCTS ARE INTEGRAL PROTEINS OF THE INNER MEMBRANE

Although the inner mitochondrial membrane contains approximately 100 different integral polypeptides, only seven of these are mitochondrial gene products (Fig. 1). It has been known for some time that these polypeptides have a high content of hydrophobic amino acids and are insoluble in the absence of added detergents. During recent years a great deal of effort has been directed at trying to establish their topographical disposition with respect to the two surfaces of the inner membrane (Racker, 1976; De Pierre and Ernster, 1977; Bell *et al.*, 1979; Ludwig *et al.*, 1979, 1980). While these studies are still incomplete, it is now clear that all mitochondrial gene products (with the possible exception of the var 1 protein associated with the small mitochondrial ribosomal subunit) are either partially or completely buried in the hydrophobic milieu of the inner membrane (Fig. 2). These studies have also made clear that nuclear gene product may be either integral or

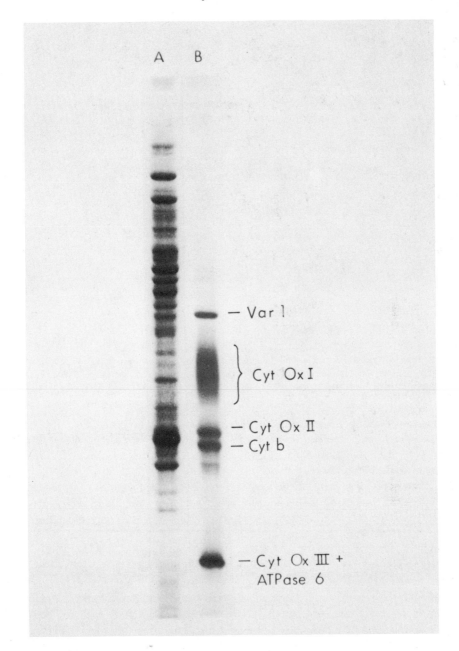

Figure 1. Identification of those integral inner mitochondrial membrane proteins in *Saccharomyces cerevisiae* that are mitochondrial gene products. Total mitochondrial inner-membrane proteins have been separated by SDS-polyacrylamide gel electrophoresis and stained with Coomassie blue (lane A). Mitochondrial gene products, labeled *in vitro* with [^{35}S]methionine as described by McKee *et al.* (1981), separated by SDS-polyacrylamide gel electrophoresis and autoradiographed are shown in lane B. Subunit 9 of ATPase is not shown.

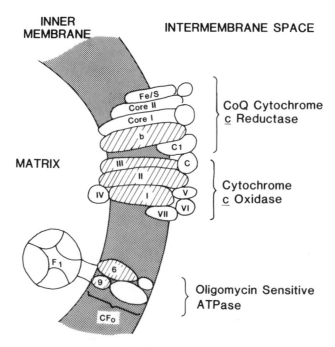

INNER MEMBRANE

INTERMEMBRANE SPACE

MATRIX

CoQ Cytochrome $\underline{c}$ Reductase

Cytochrome $\underline{c}$ Oxidase

Oligomycin Sensitive ATPase

Figure 2. Topographical disposition of mitochondrial gene products across the inner mitochondrial membrane. Cross-hatched subunits are mitochondrial gene products and clear subunits are nuclear gene products. (Adopted from Racker, 1976; De Pierre and Ernster, 1977; Bell *et al.*, 1979; Ludwig *et al.*, 1980; Poyton, 1980.)

peripheral inner-membrane proteins. For example, cytochrome c and subunits IV and VI of yeast cytochrome c oxidase are peripheral polypeptides whereas subunits V and VII of yeast cytochrome c oxidase are integral polypeptides (De Pierre and Ernster, 1977; George-Nascimento and Poyton, 1981).

A more exact placement of the various regions of mitochondrial gene products with respect to the two sides of the inner membrane should be possible in the near future as their primary sequences become available. Recently, in a series of elegant studies, Tzagoloff and his co-workers have determined the DNA sequences for cytochrome c oxidase subunits I, II, and III (Bonitz *et al.*, 1980; Coruzzi and Tzagoloff, 1979; Thalenfeld and Tzagoloff, 1980), cytochrome b (Nobrega and Tzagoloff, 1980), and oligomycin-sensitive ATPase subunit 9 (Macino and Tzagoloff, 1979) from *Saccharomyces cerevisiae*. Upon application of the parameters of Chou and Fasman (1978) and Segrest and Feldman (1974) to these sequences, it has been possible to identify those regions of each polypeptide that have the correct length, conformation, and hydrophobicity to span the inner membrane (for example, see Fig. 3). By use of lipid-soluble photoaffinity labels, surface (water-soluble) covalent probes, and refined peptide maps of mitochondrial gene produts, it will be possible to experimentally verify, or dispute, these predictions.

4. ORGANIZATION AND REGULATION OF MITOCHONDRIAL STRUCTURAL GENES

The organization of the mitochondrial genome and the control of its expression by nuclear gene products have been studied extensively in a variety of eukaryotes (recent

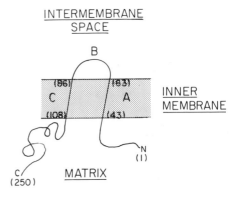

Figure 3. Possible orientation of subunit II of yeast cytochrome *c* oxidase in the inner mitochondrial membrane. The DNA sequence of subunit II (Coruzzi and Tzagoloff, 1979) reveals stretches of hydrophobic amino acids (residues 43–63 and 86–108) that are long enough to span the inner membrane. Both stretches have a high probability of forming α helices as determined by the rules of Chou and Fasman (1978).

reviews: Mahler and Perlman, 1979; Attardi *et al.*, 1979), especially human (HeLa) and yeast (*Saccharomyces*). These studies have revealed that the size and organization of mitochondrial genomes vary dramatically from organism to organism, that the mitochondrial genetic code is, for some amino acids, different than that used for either prokaryotes or eukaryotes (Barrell *et al.*, 1979; Macino *et al.*, 1979), and that fewer tRNAs seem to be employed for the translation of mitochondrial mRNAs than for the translation of either prokaryotic or eukaryotic nuclear mRNAs (review: Attardi, 1981). While a detailed discussion of the structure and expression of mitochondrial genomes is beyond the scope of this review, a few points that are especially relevant to the assembly of the inner mitochondrial membrane merit some consideration. First, the structural genes for mitochondrially encoded membrane polypeptides are *dispersed* around the genome, interrupted by rRNA and tRNA genes (Fig. 4). Transcripts for these genes are derived from larger transcripts that overlap tRNA, rRNA, and other mRNA transcripts. This is especially striking for the heavy strand of HeLa mitochondrial DNA where the tRNA genes seem to punctuate (and separate) genes coding for the rRNAs and protein (Attardi *et al.*, 1979; Chomyn *et al.*, 1981). The separate disposition of the structural genes for mitochondrially encoded membrane proteins makes it clear that they constitute separate cistrons rather than a common operon. Whether these separate genes are regulated coordinately or independently of one another remains to be determined. Second, some mitochondrially encoded membrane proteins in yeast are products of complex split genes containing introns and exons. These include the structural gene for cytochrome *b* and cytochrome *c* oxidase subunit I (Church *et al.*, 1979; Nobrega and Tzagoloff, 1980; Bonitz *et al.*, 1980). An interesting, yet still poorly understood, interaction between these two genes has been observed in mutants within the noncoding regions (introns) of the cytochrome *b* gene. Such mutants effect the expression of the cytochrome oxidase subunit I gene (Slonimski *et al.*, 1978). Grivell and his collaborators (Van Ommen *et al.*, 1980) have proposed recently that this interaction is affected at the level of RNA processing. Third, mitochondrial transcripts are translated inside of the organelle on ribosomes tightly bound to the inner membrane (Kuriyama and Luck, 1973; Spithill *et al.*., 1978). And fourth, the synthesis of mitochondrially encoded membrane proteins is regulated extensively by nuclear gene products. While some of these proteins (e.g., initiation and elongation factors) regulate the synthesis of all mitochondrial gene products, others appear from both genetic (Ono *et al.*, 1975; Tzagoloff *et al.*, 1975) and biochemical (Poyton and Kavanagh, 1976) studies to be specific regulators of individual mitochondrially encoded membrane proteins. The identification of these regulatory proteins and the elucidation of their mode of action promises to be a fascinating chapter in understanding the assembly of the inner mitochondrial membrane.

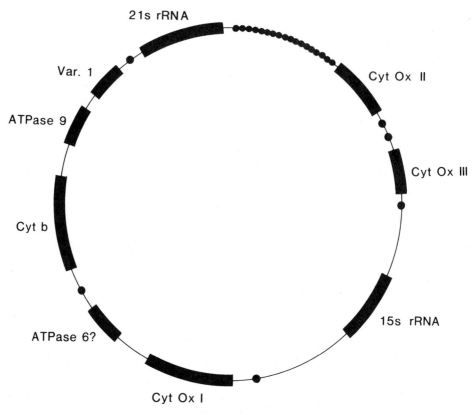

Figure 4. The mitochondrial genome of *Saccharomyces cerevisiae*. Structural genes for rRNA and proteins are represented by bars and those for tRNA by circles. Genes for proteins are designated by the name of the protein they encode. The only mitochondrially encoded polypeptide that is not an integral membrane protein is var 1. This protein is localized in the small subunit of mitochondrial ribosomes (Groot *et al.*, 1979; Terpstra *et al.*, 1979).

5. PATHWAYS OF PROTEIN TRANSPORT INTO THE INNER MEMBRANE

One of the most unusual aspects of the assembly of the inner mitochondrial membrane is the bitropic insertion of newly synthesized integral membrane polypeptides into both of its surfaces (Fig. 5). This aspect becomes especially interesting when one considers that the pathways followed by the mitochondrially encoded polypeptides, which are inserted into the matrix face of the inner membrane, are quite different than those followed by the nuclear-encoded polypeptides, which are inserted into the cytoplasmic face of the membrane.

Mitochondrially encoded polypeptides are translated on mitochondrial ribosomes that are tightly bound to the matrix face of the membrane. Like secretory proteins, these polypeptides are most likely inserted into the inner mitochondrial membrane *cotranslationally* as they are being elongated on the ribosome (Sevarino and Poyton, 1980; Poyton *et al.*, 1980). Recent studies have revealed that some of these polypeptides have transient "leader peptides" at their NH₂ termini, while others do not. In *S. cerevisiae*, it has been shown from *in vitro* synthesis experiments (Sevarino and Poyton, 1980; De Ronde *et al.*, 1980;

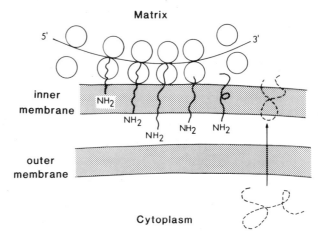

Figure 5. Bitropic insertion of polypeptides into the mitochondrial inner membrane.

McKee *et al.*, 1981) and DNA sequence data (Fox, 1979; Coruzzi and Tzagoloff, 1979) that subunit II of cytochrome *c* oxidase is synthesized as a precursor with a 15-amino-acid NH_2-terminal leader peptide. In *Neurospora crassa,* there is evidence for similar precursors to subunits I and II of cytochrome *c* oxidase (Werner *et al.*, 1980). However, in human cytochrome *c* oxidase, it appears likely that neither of these polypeptides is initially translated as precursors with transient leader peptides (Chomyn *et al.*, 1981).

It is not yet clear if some mitochondrial gene products have uncleaved NH_2-terminal leader peptides like those proposed recently for sindbis virus glycoprotein pE_2 (Bonatti and Blobel, 1979). However, a comparison of the sequence of the leader peptide on the precursor to cytochrome *c* oxidase subunit II with the NH_2-terminal sequence of mature ATPase subunit 9, a polypeptide synthesized without a leader peptide (Tzagoloff *et al.*, 1979; Macino and Tzagoloff, 1979), makes this possibility unlikely. As seen from Fig 6, there is little, if any, sequence homology between these two amino acid sequences. Upon application of the Chou and Fasman (1978) rules to these sequences, it has become clear that the leader peptide to subunit II of cytochrome *c* oxidase possesses some properties common to other leader peptides. Indeed, the region from residues −9 to −2 is rich in hydrophobic amino acids (hydrophobicity index 2.44; Segrest and Feldman, 1974) and has a high probability of forming a β-pleated sheet, a structure that is very common in leader peptides (Austen, 1979). Other common properties are a basic residue (position −10) within five

Cytochrome c Oxidase Leader Peptide

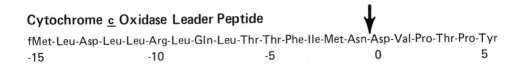

fMet-Leu-Asp-Leu-Leu-Arg-Leu-Gln-Leu-Thr-Thr-Phe-Ile-Met-Asn-Asp-Val-Pro-Thr-Pro-Tyr

-15   -10   -5   0   5

ATPase subunit 9

fMet-Gln-Leu-Val-Leu-Ala-Ala-Lys-Tyr-Ile-Gly-Ala-Gly-Ile-Ser-Thr-Ile-Gly-Leu-Leu-Gly

5   10   15   20

Figure 6. Comparison of the NH_2-terminal sequences of the structural gene for subunit II of yeast cytochrome *c* oxidase and subunit 9 of yeast oligomycin-sensitive ATPase. The first 15 amino acids of the subunit II structural gene are part of a transient leader peptide (McKee *et al.*, 1981). The mature subunit II polypeptide starts at residue 16 as indicated by the arrow.

residues of the NH_2 terminus and proline residues (positions +3 and +5) within six residues of the cleavage site. In contrast to these structural features of the leader peptide to cytochrome *c* oxidase subunit II, the NH_2 terminus of ATPase subunit 9, although high in hydrophobic amino acids, has a high probability of forming an α helix, not a β-pleated sheet. While these results clearly argue against an uncleaved leader peptide on ATPase subunit 9, they do not rule out the possibility that ATPase subunit 9 (or other mitochondrial gene products that lack transient leader peptides) has internal peptide segments that perform leader peptide functions. In view of the overall hydrophobicities of mitochondrial gene products, this seems a likely possibility.

Unlike mitochondrial gene products, nuclear-encoded polypeptides that are inserted into or through the inner mitochondrial membrane are translated on cytoplasmic ribosomes not bound to the inner membrane. Although the intracellular sites of synthesis of most cytoplasmically translated proteins remain to be determined, three different cytoplasmic polysome populations have been implicated so far. Polysomes bound to the surface of the endoplasmic reticulum have been proposed as the site of synthesis of glutamate dehydrogenase (Godinot and Lardy, 1973; Kawajiri *et al.*, 1977a,b). malate dehydrogenase (Kawajiri *et al.*, 1977a), and an unspecified mixture of mitochondrial proteins from rat liver (Shore and Tata, 1977). "Free" or "soluble" cytoplasmic polysomes have been implicated as the sites of synthesis of cytochrome *c* from rat liver (Saez De Cordova *et al.*, 1977) and *N. crassa* (Korb and Neupert, 1978), carbamyl phosphate synthetase from rat liver (Raymond and Shore, 1979), and the ADP–ATP carrier from *N. crassa* (Zimmerman and Neupert, 1980). And polysomes bound to the outer mitochondrial membrane have been proposed as the sites of synthesis of the α, β, and γ subunits of yeast F_1-ATPase (Ades and Butow, 1980). Irrespective of their sites of synthesis, it is now clear from a number of studies (Hallermayer *et al.*, 1977; Korb and Neupert, 1978; Poyton and McKemmie, 1979; Maccechini *et al.*, 1979; Zimmerman and Neupert, 1980; Ades and Butow, 1980) that most, if not all, cytoplasmically translated mitochondrial proteins are inserted into the mitochondrion *posttranslationally* after their polypeptide chains have been completely elongated (review: Neupert and Schatz, 1981).

At present, little is known about the molecular mechanisms underlying the *posttranslational* sequestration of cytoplasmically translated proteins by mitochondria. For convenience, it is useful to view the import of these proteins by mitochondria as a two-event process: (1) the targeting of the protein to the mitochondrial surface and (2) the internalization of the protein and its subsequent sequestration into either matrix, inner membrane, intermembrane space, or outer-membrane compartment. It seems likely that the first event involves the interaction between a *membrane-specific selection sequence*, on the protein to be targeted, and a receptor, on the mitochondrial surface. Possible candidates for membrane-specific selection sequences include leader peptides present on precursors to these proteins, internal peptide regions present in the mature peptides themselves, or ancillary peptides with which these proteins associate in the cytoplasm. A number of recent studies have demonstrated that some cytoplasmically made proteins are translated as larger precursors, while others are not (Neupert and Schatz, 1981). Although it is likely that those synthesized as larger precursors have NH_2-terminal leader peptides (Mihara and Blobel, 1980), there is no direct evidence that the leader peptide plays a role in targeting the protein to the mitochondrial surface. This is clearly not the case for those proteins synthesized without leader peptides. Future studies can be expected to examine the role, if any, of the leader peptide in the targeting step, and the functional relationship between leader peptides on proteins synthesized as precursors and internal peptide regions on proteins synthesized in their mature form. These studies may reveal the presence of internal signal sequences. It will also be of some interest to determine if either type of protein becomes associated with ancillary polypeptides during their journey to the mitochondrion. Such ancillary polypeptides could serve two possible functions. On the one hand, they may serve to keep hydrophobic membrane proteins, such as the ADP–ATP carrier protein (Zimmerman and Neupert, 1980) "soluble" in the cytosol. On the other hand, they may serve to "pilot" cytoplasmic proteins (Poyton *et al.*, 1980) to their mitochondrial surface receptor in much the same way as the B subunit of cholera toxin recognizes plasma membrane surface receptors for its A subunit effectomer (Bennett and Cuatrecasas, 1977).

The second event in the internalization of cytoplasmically made proteins is also poorly understood at present. It will be of interest to determine how inner-membrane proteins are sequestered into the inner membrane, rather than one of the other three mitochondrial compartments. It will also be of interest to study the role of an energized inner membrane on the insertion process (Nelson and Schatz, 1979; Zimmerman and Neupert, 1980), and to determine what role, if any, the concurrent cotranslational incorporation of mitochondrially translated polypeptides into the matrix face of the membrane plays in the insertion of these cytoplasmically translated polypeptides into the cytoplasmic face of the membrane.

6. SUMMARY AND FUTURE PROSPECTS

From the studies discussed above it is clear that of the two mitochondrial membranes only the inner membrane is a genetic mosaic composed of proteins synthesized both inside and outside of the mitochondrion. The mitochondrially made polypeptides are translated on mitochondrial ribosomes bound to the inner membrane and are inserted into the membrane cotranslationally. On the other hand, those nuclear gene products that are integral polypeptides of the inner membrane are translated on cytoplasmic ribosomes and are inserted into the membrane posttranslationally. As both classes of proteins reside within the inner membrane, it seems likely that both contain membrane insertion sequences. It also seems likely that nuclear gene products contain, in addition, membrane-specific selection sequences, which assure that they are targeted correctly to the mitochondrial surface, rather than incorrectly to the surface of some other intracellular organelle. The major challenges for the future are: (1) the identification of both of these sequences; (2) a better understanding of the physiological conditions necessary for the interaction between mitochondrial and nuclear genomes and for the insertion of the protein products of both genomes into the membrane; and (3) an analysis of the sequential events underlying the assembly of the oligomeric enzyme complexes of the inner membrane after their subunits have been inserted into the membrane proper.

REFERENCES

Ades, I. Z., and Butow, R. A. (1980). *J. Biol. Chem.* **255**, 9918–9924.

Attardi, G. (1981). *Trends Biochem. Sci.* **6**, 86–89.

Attardi, G., Cantatore, P., Ching, E., Crews, S., Gelfand, R., Merkel, C., and Ojala, C. (1979). In *Extrachromosomal DNA* (D. J. Cummings, P. Borst, I. B. Dawid, S. M. Weissman, and C. F. Fox, eds.), pp. 443–469, Academic Press, New York.

Austen, B. M. (1979). *FEBS Lett.* **103**, 308–313.

Barrell, B. G., Bankier, A. T., and Drouin, J. (1979). *Nature (London)* **282**, 189–194.

Bell, R. L., Sweetland, J., Ludwig, B., and Capaldi, R. A. (1979). *Proc. Natl. Acad. Sci. USA* **76,** 741–745.

Bennett, V., and Cuatrecasas, P. (1977). In *The Specificity and Action of Animal, Bacterial, and Plant Toxins* (P. Cuatrecasas, ed.), pp. 3–66, Chapman & Hall, London.

Blobel, G. (1980). *Proc. Natl. Acad. Sci. USA* **77,** 1496–1500.

Bonatti, S., and Blobel, G. (1979). *J. Biol. Chem.* **254,** 12261–12264.

Bonitz, S. G., Coruzzi, G., Thalenfeld, B. E., Tzagoloff, A., and Macino, G. (1980). *J. Biol. Chem.* **255,** 11927–11941.

Chomyn, A., Hunkapiller, M. W., and Attardi, G. (1981). *Nucleic Acid Res.,* **9,** 867–877.

Chou, P. Y., and Fasman, G. D. (1978). *Annu. Rev. Biochem.* **47,** 251–276.

Church, G. M., Slonimski, P. P., and Gilbert, W. (1979). *Cell* **18,** 1209–1215.

Coruzzi, G., and Tzagoloff, A. (1979). *J. Biol. Chem.* **254,** 9324–9330.

De Pierre, J. W., and Ernster, L. (1977). *Annu. Rev. Biochem.* **46,** 201–262.

De Ronde, A., Van Loon, A. P. G. M., Grivelli, L. A., and Kohli, J. (1980). *Nature (London)* **287,** 361–363.

Fox, T. D. (1979). *Proc. Natl. Acad. Sci. USA* **76,** 6534–6538.

George-Nascimento, C., and Poyton, R. O. (1981). *J. Biol. Chem.,* **256,** 9363–9370.

Godinot, C., and Lardy, H. A. (1973). *Biochemistry* **12,** 2051–2060.

Groot, G. S. P., Mason, T. L., and Van Harten-Loosbroek, N. (1979). *Mol. Gen. Genet.* **174,** 339–342.

Hallermayer, G., Zimmerman, R., and Neupert, W. (1977). *Eur. J. Biochem.* **81,** 523–532.

Henderson, R., Capaldi, R. A., and Leigh, J. S. (1977). *J. Mol. Biol.* **112,** 631–648.

Kawajiri, K., Harano, T., and Omura, T. (1977a). *J. Biochem.* **82,** 1403–1416.

Kawajiri, K., Harano, T., and Omura, T. (1977b). *J. Biochem.* **82,** 1417–1423.

Korb, H., and Neupert, W. (1978). *Eur. J. Biochem.* **91,** 609–620.

Kuriyama, Y., and Luck, D. J. L. (1973). *J. Cell Biol.* **59,** 776–784.

Ludwig, B., Downer, N. W., and Capaldi, R. A. (1979). *Biochemistry* **18,** 1401–1407.

Ludwig, B., Prochaska, L., and Capaldi, R. A. (1980). *Biochemistry* **19,** 1516–1523.

Maccechini, M.-L., Rudin, Y., Blobel, G., and Schatz, G. (1979). *Proc. Natl. Acad. Sci. USA* **76,** 343–347.

Macino, G., and Tzagoloff, A. (1979). *J. Biol. Chem.* **254,** 4617–4623.

Macino, G., Coruzzi, G., Nobrega, F. G., Li, M., and Tzagoloff, A. (1979). *Proc. Natl. Acad. Sci. USA* **76,** 3784–3785.

McKee, E. E., Sevarino, K. A., Bellus, G., and Poyton, R. O. (1981). In *The Biochemistry of Yeasts* (G. G. Stewart, C. Robinow, B. Johnson, E. R. Tustanoff, M. A. La Chance, and I. Russel, eds.), Pergamon Press, Elmsford, N.Y., in press.

Mahler, H. R., and Perlman, P. S. (1979). In *Extrachromosomal DNA* (D. J. Cummings, P. Borst, I. B. Dawid, S. M. Weissman, and C. F. Fox, eds.), pp. 11–33, Academic Press, New York.

Mihara, K., and Blobel, G. (1980). *Proc. Natl. Acad. Sci, USA* **77,** 4160–4164.

Munn, E. A. (1974). *The Structure of Mitochondria*, Academic Press, New York.

Nelson, N., and Schatz, G. (1979). *Proc. Natl. Acad. Sci. USA* **76,** 4365–4369.

Neupert, W., and Schatz, G. (1981). *Trends Biochem. Sci.* **6,** 1–4.

Nobrega, F. G., and Tzagoloff, A. (1980). *J. Biol. Chem.* **255,** 9828–9837.

Ono, B., Fink, G., and Schatz, G. (1975). *J. Biol. Chem.* **250,** 775–782.

Poyton, R. O. (1980). *Curr. Top. Cell. Regul.* **17,** 231–295.

Poyton, R. O., and Kavanagh, J. (1976). *Proc. Natl. Acad. Sci. USA* **73,** 3947–3951.

Poyton, R. O., and McKemmie, E. (1979). *J. Biol. Chem.* **254,** 6772–6780.

Poyton, R. O. Sevarino, K., George-Nascimento, C., and Power, S. D. (1980). *Ann. N.Y. Acad. Sci.* **343,** 275–292.

Racker, E. (1976). *A New Look at Mechanisms in Bioenergetics*, Academic Press, New York.

Raymond, Y., and Shore, G. C. (1979). *J. Biol. Chem.* **254,** 9335–9338.

Saez De Cordova, C., Cohen, R., and Gonzalez-Cadavid, N. F. (1977). *Biochem. J.* **166,** 305–313.

Schatz, G. ,and Mason, T. L. (1974). *Annu. Rev. Biochem.* **43,** 51–87.

Sebald, W. (1977). *Biochim. Biophys. Acta* **463,** 1–27.

Segrest, J. P., and Feldman, R. J. (1974). *J. Mol. Biol.* **87,** 853–858.

Sevarino, K. A., and Poyton, R. O. (1980). *Proc. Natl. Acad. Sci. USA* **77,** 142–146.

Shore, G., and Tata, J. R. (1977). *J. Cell Biol.* **72,** 726–743.

Slonimski, P. P., Claisse, M. L., Foucher, M., Jacq, C., Kochko, A., Lamouroux, A., Pajot, P., Perrodin, G., Spyridakis, A., and Wambier-Kluppel, M. (1978). In *Biochemistry and Genetics of Yeasts* (M. Bacila, B. L. Horecker, and A. D. M. Stoppani, eds.), pp. 391–401, Academic Press, New York.

Spithill, T. W., Trembath, M. K., Lukins, H. B., and Linnane, A. W. (1978). *Mol. Gen. Genet.* **164,** 155–162.

Terpstra, P., Zanders, E., and Butow, R. A. (1979). *J. Biol. Chem.* **254,** 12653–12661.

Thalenfeld, B. E., and Tzagoloff, A. (1980). *J. Biol. Chem.* **255,** 6173–6180.

Tzagoloff, A., Akai, A., and Needleman, R. B. (1975). *J. Biol. Chem.* **250,** 8228–8235.

Tzagoloff, A., Macino, G., and Sebald, W. (1979). *Annu. Rev. Biochem.* **48,** 419–441.

Van Ommen, G. J. B., Boer, P. H., Groot, G. S. P., De Haan, M., Roosendaal, E., Grivell, L. A., Haid, A., and Schweyen, R. (1980). *Cell* **20,** 173–183.

von Jagow, G., and Sebald, W. (1980). *Annu. Rev. Biochem.* **49,** 281–314.

Werner, S., Machleidt, W., Bertrand, H., and Wild, G. (1980). In *The Organization and Expression of The Mitochondrial Genome* (A. M. Kroon and C. Saccone, eds.), pp. 399–411, North-Holland, Amsterdam.

Zimmerman, R., and Neupert, W. (1980). *Eur. J. Biochem.* **190,** 217–229.

30

Membrane-Bound Enzymes of Cholesterol Biosynthesis from Lanosterol

James L. Gaylor

1. TYPE-REACTIONS CATALYZED BY MICROSOMAL ENZYMES

All of the enzymes required for the synthesis of cholesterol from lanosterol, the first steroid intermediate (Fig. 1), are bound to endoplasmic reticulum that is isolated from cell-free homogenates as microsomes. Water-insoluble sterol intermediates are generated in the membrane and metabolized by membrane-bound enzymes without diffusion of steroids from the membrane. A plausible 19-step sequence of enzymatic reactions can now be written as a result of studies that have progressed from organic synthesis of the presumed intermediate, generation of the intermediate in microsomes, further enzymatic conversion of the intermediate to cholesterol, solubilization and purification of the membrane-bound enzyme that acts on the intermediate, and finally reconstitution of groups of solubilized enzymes into artificial membranes. This brief synopsis outlines current knowledge, presents some generalizations, and updates a recent review that describes details on the enzymology (Gaylor, 1981).

Four types of enzymes are involved: mixed-function oxidases, lyases, reductases, and an isomerase. Although 19 transformations probably occur, the complexity of the membrane-bound, multienzymatic system is reduced substantially by several of the enzymes acting on more than one substrate.

Of the type-reactions the mixed-function oxidases are the most complex enzymatically because in addition to the terminal enzyme that activates molecular oxygen and catalyzes steroid conversion, the membrane also contains electron carriers for transfer of reducing equivalents (White and Coon, 1980). Liver microsomes contain two principal electron transport processes *both* of which are needed for conversion of lanosterol into cholesterol (Fig. 2). Although the NADPH-cytochrome P-450 transport chain is quantitatively more abundant in liver microsomes and is involved in oxidation of a wide variety of structurally

James L. Gaylor • Central Research and Development, Glenolden Laboratory, E. I. DuPont and Company, Glenolden, Pennsylvania 19707.

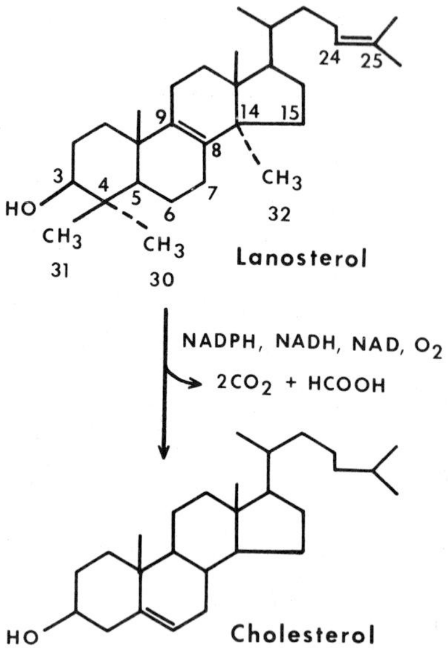

Figure 1. Structures of lanosterol and cholesterol. NADPH and NADH supply reducing equivalents to the mixed-function oxidases. In the same reactions oxygen is consumed. NAD is required for redox reactions. The stoichiometry indicated is lanosterol (C_{30})→cholesterol (C_{27}) + $2CO_2$ + HCOOH.

dissimilar compounds, only one part of cholesterol synthesis appears to be cytochrome P-450 dependent, the oxidation of the C-32 methyl group (Fig. 2). C-32 oxidative demethylation is sensitive to inhibition by CO (Gibbons *et al.*, 1979). Pascal *et al.* (1980) proposed a nonoxidative step for conversion of the C-32 hydroxymethyl group to the aldehyde, but their study of further metabolism of synthetic intermediates rather than formation of oxygenated compounds may not be conclusive. Although the cytochrome P-450 that catalyzes C-32 attack has been solubilized with Emulgen 913 (Fisher *et al.*, 1981), a hydroxylated sterol has not, as yet, been isolated. However, isolated yeast microsomal cytochrome P-450 has been purified. The yeast cytochrome P-450 apparently catalyzes three oxidative reactions in sequence (Aoyama and Yoshida, 1978; Ohba *et al.*, 1978).

Oxidation of the C-30 methyl group is much clearer. The same mixed-function oxidase catalyzes, in sequence, attack of the 4α-methyl group to yield an alcohol, aldehyde, and finally a 4α-carboxylic acid (Miller and Gaylor, 1970). The stoichiometry of the oxidase has been established (Gaylor *et al.*, 1975) and, quite recently, absolute dependence upon microsomal cytochrome b_5 has been shown for yeast and liver (Aoyama *et al.*, 1981;

ELECTRON TRANSPORT:

REACTIONS CATALYZED:

NADH⟶ cytochrome b_5 ⟶ cytochrome b_5
reductase

NADPH⟶ cytochrome P-450 ⟶ cytochrome P-450
reductase

$\left\{ \begin{array}{l} \text{C-30: } CH_3 \longrightarrow CH_2OH \longrightarrow CHO \longrightarrow COOH \\ \text{C-31: } CH_3 \longrightarrow CH_2OH \longrightarrow CHO \longrightarrow COOH \\ \qquad \Delta^7 \longrightarrow \Delta^{5,7} \end{array} \right.$

C-32: $CH_3 \longrightarrow CH_2OH \overset{?}{\longrightarrow} CHO \overset{?}{\longrightarrow} \left\{ \begin{array}{l} CHO \\ 15\alpha\text{-OH} \end{array} \right.$

Figure 2. Microsomal electron transport and the associated mixed-function oxidases of cholesterol synthesis from lanosterol.

Fukushima *et al.*, 1981), respectively. The C-30 oxidase has been solubilized with Renex 690 (Gaylor and co-workers), but significant enrichment has not been achieved.

The final oxygenase of the pathway (also cytochrome b_5 dependent) catalyzes insertion of the 5-double bond into cholesta-7,24-dien-3β-ol to yield cholesta-5,7,24-trien-3β-ol (Reddy *et al.*, 1977). Brady and co-workers (1980) have shown that the oxygenases of cholesterol biosynthesis are supported more effectively by endogenously rather than exogenously reduced pyridine nucleotide.

After oxidation, membrane-bound lyases catalyze cleavage of the leaving group. In C-32 oxidation, formic acid is formed with generation of either an 8(14)-double bond (Pascal *et al.*, 1980) or, if additional oxidation had occurred, an 8,14-diene (Gibbons *et al.*, 1979; Ohba *et al.*, 1978). From oxidation at C-30 to a carboxylic acid, an NAD-dependent decarboxylase catalyzes the release of CO_2 concomitant with dehydrogenation of the 3β-hydroxyl group to a 3-ketone. The decarboxylase has been solubilized with deoxycholate and partially purified (Rahimtula and Gaylor, 1972).

Five NADPH-dependent reductions appear to occur. Reduction of 3-ketosteroids to 3β-hydroxy alcohols is catalyzed by an enzyme that has been solubilized with Lubrol WX and partially purified (Gaylor, 1972). Reduction occurs after each of two C-30 decarboxylations. Three double bonds must be reduced: Δ^{14} after oxidative elimination of C-32; Δ^5 after introduction of Δ^5; and Δ^{24} after transformations of the steroid nucleus. No isolation and purification of NADPH-dependent double bond reductases has been reported; some reductases have been studied while membrane bound (Dempsey, 1974).

Rat liver microsomes contain an isomerase that catalyzes the conversion of Δ^8- to Δ^7-sterols (Yamaga and Gaylor, 1978). The enzyme has now been solubilized with octylglucoside (Y. K. Paik and J. L. Gaylor, unpublished observation), and purification is under way.

2. EFFECTS OF THE MEMBRANE

Clearly, the most obvious problem related to the conduct of these studies is that membrane-bound enzymes catalyze transformations without natural interruption of the process. Best evidence for precursor–product relationships is obtained by artificially interrupting the overall process. Several reactions require oxygen; they can be interrupted by anaerobiosis. Similarly, all but two reactions require either oxidized or reduced pyridine nucleotide. Thus, by removing endogenous pyridine nucleotides (Miller *et al.*, 1967; Brady *et al.*, 1980) and introducing only the desired nucleotide (with an appropriate redox generating system), the process can be interrupted experimentally. Without purifying any of the membrane-bound enzymes, the complete 10-step oxidative demethylation of the 4-gem-dimethyl group was elucidated by interruption (Gaylor, 1981):

3β—OH		3β—OH		3=O		3β—OH
R–$\begin{cases}4\alpha CH_3\\4\beta CH_3\end{cases}$	$\xrightarrow[\text{NADH}+O_2]{\rightarrow\rightarrow\rightarrow}$	R–$\begin{cases}4\alpha COOH\\4\beta CH_3\end{cases}$	$\xrightarrow[CO_2]{\text{NAD}}$	R–$\begin{cases}4\alpha CH_3\\4\beta H\end{cases}$	$\xrightarrow{\text{NADPH}}$	R–$\begin{cases}4\alpha CH_3\\4\beta H\end{cases}$
R–$\begin{cases}4\alpha CH_3\\4\beta H\end{cases}$	$\rightarrow\rightarrow\rightarrow$	R–$\begin{cases}4\alpha COOH\\4\beta H\end{cases}$	$\xrightarrow[CO_2]{}$	R–$\begin{cases}4\alpha H\\4\beta H\end{cases}$	$\longrightarrow$	R–$\begin{cases}4\alpha H\\4\beta H\end{cases}$

The enzymes may be organized into a multienzyme system in the membrane. Perhaps the striking overall efficiency of the rate of the process suggests that an organized system does exist. Studies on similar enzymes, e.g., acyl-CoA : cholesterol acyltransferase (Lichtenstein and Brecher, 1980) and squalene epoxidase (Friedlander *et al.*, 1980), as well as electron transfer from exogenously reduced pyridine nucleotides indicate that sterol-processing enzymes may be on the cytoplasmic surface but deeply buried in the membrane. Translocation of the sterol substrates may be facilitated by noncatalytic proteins (Dempsey, 1974; Friedlander *et al.*, 1980; Nakamura and Sato, 1979). However, the earlier suggestion that sterol intermediates are "carried" *to* the endoplasmic reticulum is unlikely because the intermediate sterols are generated *in* the membrane.

Because the enzymes are bound to the same membrane, regulation of the rate of each step may be coordinated through common mechanisms. As 3-hydroxy-3-methylglutaryl-CoA reductase, the initial rate-limiting reaction of cholesterol biosynthesis, is also bound to the same membrane, it is attractive to speculate that the activities of all of the membrane-bound enzymes may be regulated in concert (Gaylor, 1981). For example, inhibition of sterol synthesis by oxygenated sterols affects both membrane-bound 3-hydroxy-3-methylglutaryl-CoA reductase and a terminal step, the 14α-demethylation of lanosterol (Gibbons *et al.*, 1980). Diurnal rhythms and effects of dietary cholesterol are also similar. As organization of the enzymes in membranes is better understood, both the common mechanisms and the possible synchrony of activity regulation should become apparent.

Cytosolic proteins that affect activities of the membrane-bound enzymes of sterol biosynthesis are under close study. The soluble protein that stimulates squalene epoxidase and 2,3-oxidosqualene-lanosterol cyclase (Friedlander *et al.*, 1980) binds to microsomal vesicles (Caras *et al.*, 1980) as does cytosolic Z protein, which affects activities of several microsomal enzymes (Billheimer and Gaylor, 1980). The former appears to modulate activities via translocation of squalene and water-insoluble substrates, whereas the latter facilitates vesicular uptake of water-soluble cofactors. An intact microsomal membrane is needed to observe changes in activity. A third cytosolic protein catalyzes transfer of cholesterol, cholesterol precursors, and other lipids between membranes without binding of the protein to membranes (Noland *et al.*, 1980; Trzaskos and Gaylor, 1981). Oxygenated sterols bind to a fourth cytosolic protein (Kandutsch and Thompson, 1980). Because the various cytosolic proteins exhibit different functions, the earlier suggestion of a single cytosolic protein and a single "carrier" function has been far too restrictive.

The ultimate test of functionality of enzymes isolated from membranes is to reconstitute purified enzymes in artificial membranes. For example, functionality and substrate specificity for a purified form of cytochrome P-450 have been shown by incorporation into synthetic vesicles (Hall *et al.*, 1979). Such studies with the microsomal sterol biosynthetic enzymes will be under way soon.

Finally, what is the effect of the membrane on the enzymes? For maximal membrane function, the α face of cholesterol and other product sterols appears to be unencumbered via the microsomal enzymes (Rohmer *et al.*, 1979; Dahl *et al.*, 1980; Nes *et al.*, 1978; Yeagle *et al.*, 1977). Thus, it should not be surprising to observe that enzymatically only the angular methyl groups on the α face are oxidized from lanosterol. Indeed, it appears that all of the 19 steps in the conversion of lanosterol to cholesterol can be written as being initiated via α-face attack. Thus, membrane-bound substrates, enzymes, and noncatalytic proteins may function spatially as dictated by the membranous location of the catalytic process.

REFERENCES

Aoyama, Y., and Yoshida, Y. (1978). *Biochem. Biophys. Res. Commun.* **85,** 28–34.

Aoyama, Y., Yoshida, Y., Sato, R., Susani, M., and Ruis, H. (1981). *Biochim. Biophys. Acta* **663,** 194–202.

Billheimer, J. T., and Gaylor, J. L. (1980). *J. Biol. Chem.* **255,** 8128–8135.

Brady, D. R., Crowder, R. D., and Hayes, W. J. (1980). *J. Biol. Chem.* **255,** 10624–10629.

Caras, I. W., Friedlander, E. J., and Bloch, K. (1980). *J. Biol. Chem.* **255,** 3575–3580.

Dahl, C. E., Dahl, J. S., and Bloch, K. (1980). *Biochemistry* **19,** 1462–1467.

Dempsey, M. D. (1974). *Annu. Rev. Biochem.* **43,** 967–990.

Fisher, G. F., Fukishima, H., and Gaylor, J. L. (1981). *J. Biol. Chem.* **256,** 4388–4394.

Friedlander, E. J., Caras, I. W., Fen, L., Lin, H., and Bloch, K. (1980). *J. Biol. Chem.* **255,** 8042–8045.

Fukishima, H., Grinstead, G. F., and Gaylor, J. L. (1981). *J. Biol. Chem.* **256,** 4822–4826.

Gaylor, J. L. (1972). *Adv. Lipid Res.* **10,** 89–141.

Gaylor, J. L. (1981). In *Biosynthesis of Isoprenoid Compounds* (J. W. Porter, ed.), Vol. I, Chap. 10, Wiley, New York.

Gaylor, J. L., Miyake, Y., and Yamano, T. (1975). *J. Biol. Chem.* **250,** 7159–7167.

Gibbons, G. F., Pullinger, C. R., and Mitropoulos, K. A. (1979). *Biochem. J.* **183,** 309–315.

Gibbons, G. F., Pullinger, C. R., Chen, H. W., Cavenee, W. K., and Kandutsch, A. A. (1980). *J. Biol. Chem.* **255,** 395–400.

Hall, P. F., Watanuki, M., and Hamkalo, B. A. (1979). *J. Biol. Chem.* **255,** 547–552.

Kandutsch, A. A., and Thompson, E. B. (1980). *J. Biol. Chem.* **255,** 10813–10826.

Lichtenstein, A. H., and Brecher, P. (1980). *J. Biol. Chem.* **255,** 9098–9104.

Miller, W. L., and Gaylor, J. L. (1970). *J. Biol. Chem.* **245,** 5369–5374.

Miller, W. L., Kalafer, M. E., Gaylor, J. L., and Delwiche, C. V. (1967). *Biochemistry* **6,** 2673–2678.

Nakamura, N., and Sato, R. (1979). *Biochem. Biophys. Res. Commun.* **89,** 900–906.

Nes, W. R., Sekula, B. C., Nes, W. D., and Adler, J. H. (1978). *J. Biol. Chem.* **253,** 6218–6225.

Noland, B. J., Arebalo, R. E., Hansbury, E., and Scallen, T. J. (1980). *J. Biol. Chem.* **255,** 4282–4289.

Ohba, M., Sato, R., Yoshida, Y., Nishino, T., and Katsuki, H. (1978). *Biochem. Biophys. Res. Commun.* **85,** 21–27.

Pascal, R. A., Chang, P., and Schroepfer, G. J. (1980). *J. Am. Chem. Soc.* **102,** 6599–6601.

Rahimtula, A. D., and Gaylor, J. L. (1972). *J. Biol. Chem.* **247,** 9–15.

Reddy, V. V. R., Kupfer, D., and Caspi, E. (1977). *J. Biol. Chem.* **252,** 2797–2801.

Rohmer, M., Bouvier, P., and Ourisson, G. (1979). *Proc. Natl. Acad. Sci. USA* **76,** 847–851.

Trzaskos, J. M., and Gaylor, J. L. (1982). *J. Biol. Chem.* **257,** in press.

White, R. E., and Coon, M. J. (1980). *Annu. Rev. Biochem.* **49,** 315–356.

Yamaga, N., and Gaylor, J. L. (1978). *J. Lipid Res.* **19,** 375–382.

Yeagle, P. L., Martin, R. B., Lala, A. K., Lin, H.-K., and Bloch, K. (1977). *Proc. Natl. Acad. Sci. USA* **74,** 4924–4926.

31

Oligosaccharide Conformation and the Control of Oligosaccharide Assembly

A Model for Carbohydrate-Mediated Information Transfer

Harry Schachter, Saroja Narasimhan, Noam Harpaz, and Gregory D. Longmore

1. INTRODUCTION

It is now generally accepted that glycoproteins and glycolipids, members of a large group of macromolecules called complex carbohydrates or glycoconjugates, are important constituents of the mammalian cell membrane. It has also been suggested that the oligosaccharide moieties of cell surface complex carbohydrates serve as probes with which the cell interacts with its environment and through which the environment delivers signals to the interior of the cell (Hughes, 1976; Sharon, 1979; Atkinson and Hakimi, 1980). Cell surface sugars have, in fact, been shown to be present on receptors for viruses, hormones, toxins, interferon, bacteria, and mitogenic lectins; however, in only a limited number of cases has the oligosaccharide moiety itself been proven to be the basis for the recognition signal. A series of phenomena in which sugars obviously must determine recognition are the various biological effects that lectins have on certain cells (Sharon, 1979; Lis and Sharon, 1977), e.g., mitogenic stimulation, agglutination, toxicity, etc. An analogous area of recent research activity has been in the field of "mammalian lectins," cell surface and intracellular membrane receptors that recognize specific sugar moieties; for example, mammalian hepatocytes have been shown to carry on their surface and on intracellular membranes a receptor for molecules with terminal galactose residues, and human fibroblast lysosomal

Harry Schachter, Saroja Narasimhan, Noam Harpaz, and Gregory D. Longmore • Research Institute, Hospital for Sick Children, and Department of Biochemistry, University of Toronto, Toronto, Ontario M5G 1X8, Canada.

membranes are believed to carry a receptor that recognizes and binds mannose-6-phosphate-containing macromolecules thereby determining the routing of lysosomal hydrolases to the lysosome (Neufeld and Ashwell, 1980).

If cell membrane oligosaccharides are to serve as recognition signals in such receptor functions and in the even more complex processes associated with intercellular recognition, cell growth and differentiation, malignant transformation and metastasis, and so on, it is clear that they must be capable of forming and maintaining relatively stable three-dimensional conformations capable of specific interaction with a complementary molecule. This interaction is a form of information transfer in which the information resides not in a nucleic acid or polypeptide sequence but rather in an oligosaccharide structure. The oligosaccharide may in theory interact with a complementary molecule that is itself an oligosaccharide or with a molecule such as a protein.

During the course of our work on the elongation of N-glycosidically linked oligosaccharides within the mammalian Golgi apparatus (Schachter and Roseman, 1980), we observed a series of phenomena that together provide an excellent model system for the interaction of a specific three-dimensional oligosaccharide structure with various protein receptors. These observations will be briefly reviewed in this chapter and the conclusions that will be drawn are that oligosaccharides can indeed take up specific three-dimensional shapes in solution, that relatively minor changes in the primary monosaccharide sequence of the oligosaccharide structure can have important effects on the three-dimensional shape of the molecule, and that the three-dimensional shape of the oligosaccharide can determine binding to protein.

2. OLIGOSACCHARIDE SYNTHESIS AND PROCESSING

A great deal of information has recently been obtained on the synthesis of N-glycosidically linked oligosaccharides. There is a preassembly step in which a large oligosaccharide becomes attached by a pyrophosphate link to the polyisoprenol lipid dolichol. The resulting dolichol pyrophosphate oligosaccharide has in several systems been shown to have the structure $Glc_3Man_9(GlcNAc)_2$-pyrophosphate-dolichol; the oligosaccharide is transferred from the lipid intermediate to an asparagine residue of nascent polypeptide in the sequence -Asn-X-Ser(Thr) (see Struck and Lennarz, 1980, for a detailed discussion). Oligosaccharide processing then occurs in the endoplasmic reticulum and Golgi apparatus such that all three Glc residues and zero to six Man residues are removed by specific glycosidases (see Kornfeld and Kornfeld, 1980; Schachter and Roseman, 1980). If less than four Man residues are processed, the oligosaccharide remains in the "simple" or "high-mannose" form containing six to nine Man residues. However, in the synthesis of the "complex" or "N-acetyllactosamine" type of oligosaccharide, four Man residues are removed to form the $Man_5(GlcNAc)_2Asn-X$ structure shown in Fig. 1. This structure then undergoes a complex set of reactions to form a variety of possible N-acetyllactosamine structures. The present discussion will be restricted to "biantennary" structures in which each terminal Man of the $Man_3(GlcNAc)_2Asn-X$ core carries only a single antenna with a GlcNAc residue attached in $\beta2$ linkage. However, structures have been described containing three or four antennae (Kornfeld and Kornfeld, 1980) and even five or six antennae (Takasaki *et al.*, 1980). It is as yet unknown what signals the synthesis of a simple versus a complex oligosaccharide; in contrast, the degree of branching of a complex oligosaccharide probably depends on competition between several glycosyltransferases for common substrates. The latter type of control is illustrated at several points in the scheme of Fig. 1.

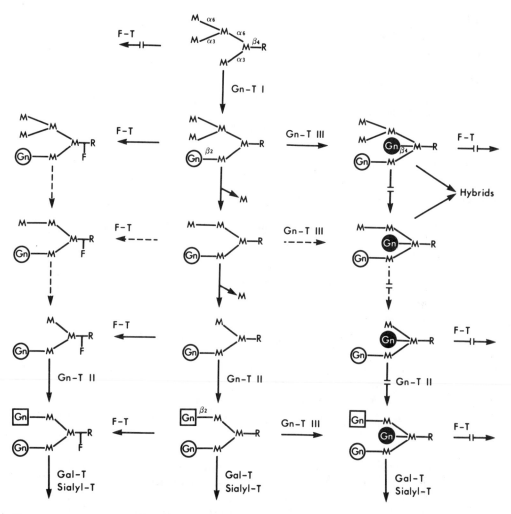

Figure 1. Pathways for the elongation within the Golgi apparatus of N-glycosidically linked oligosaccharides of the "complex" or "N-acetyllactosamine" type. Abbreviations: M, D-Man; Gn, D-GlcNAc; R, GlcNAcβ4-GlcNAc-Asn-X; F, L-Fuc attached α6 to the Asn-linked GlcNAc; Gn-T, N-acetylglucosaminyltransferase; F-T, fucosyltransferase. Discontinuous arrows indicate paths that have not been definitively established.

2.1. UDP-GlcNAc: α-D-mannoside β2-N-acetylglucosaminyltransferase I

An interesting lectin-resistant mutant cell line was characterized several years ago independently by three groups, Kornfeld in St. Louis, our own group in collaboration with Stanley and Siminovitch in Toronto, and Hughes in London (see Stanley, 1980, for a review). This line was resistant to the toxic actions of several lectins and was shown to lack the enzyme UDP-GlcNAc: α-D-mannoside β2-N-acetylglucosaminyltransferase I (Gn-T I), which attaches GlcNAc in β2 linkage to the Manα3 terminus of the Man$_5$(GlcNAc)$_2$Asn-X structure (Fig. 1). Gn-T I-deficient cells accumulate this Man$_5$-containing oligosaccharide on their glycoproteins and are incapable of carrying out the further elongation steps shown in Fig. 1, i.e., they cannot add Fuc to the core, nor remove two

Man residues to produce the Man_3-containing core, nor begin a second antenna, nor add sialic acid or galactose residues. These cells lack Gn-T I but have normal levels of a very similar enzyme, Gn-T II, which initiates the second antenna (Fig. 1). Recent work in our laboratory (S. Narasimhan, D. Tsai, and H. Schachter, unpublished) has shown the presence of a third GlcNAc transferase, which incorporates a GlcNAc residue in β linkage to the β-linked Man of the core (Gn-T III, Fig. 1); this GlcNAc residue has been termed an interantennary GlcNAc residue because it is inserted between the first two antennae. It is believed to cause major disruptions of the three-dimensional oligosaccharide structure with resultant profound effects on the biosynthetic pathways. These effects will now be discussed in some detail.

2.2. The Interantennary GlcNAc Alters Oligosaccharide Conformation

We have obtained direct evidence that the insertion of the interantennary GlcNAc may cause dramatic changes in the conformation of the oligosaccharide. This evidence derives from high-resolution proton NMR spectroscopy studies on glycopeptides carried out in collaboration with Drs. J. P. Carver and A. A. Grey using the 360-MHz facility at Toronto; conditions for carrying out such studies have been described (Narasimhan *et al.*, 1980). Table I lists some typical NMR data illustrating the effect of the interantennary GlcNAc residue. Comparing the spectra for glycopeptides MM and MGn, it is evident that the addition of a GlcNAc residue in $\beta 2$ linkage to the $Man\alpha 3$ terminus of the core causes a significant shift in the signal of the C-2 hydrogen of this Man residue (from 4.068 to 4.190 ppm) and a minor shift in the signal of the C-1 hydrogen (from 5.128 to 5.138 ppm); however, no other signals are affected, indicating that the attachment of GlcNAc at C-2 of the $\alpha 3$-linked Man residue causes only a limited and localized alteration in the conformation of this residue. Comparison of the spectra for MGn and MGn(Gn), on the

Table I. Chemical Shifts (in ppm) of Glycopeptide Hydrogens Derived from High-Resolution Proton NMR Spectra

	Glycopeptide structure[a]		
Hydrogens	MM	MGn	MGn(Gn)
Carbon-1			
$\beta 4$-Man	4.773	4.773	4.747
$\alpha 3$-Man	5.128	5.138	5.059
$\alpha 6$-Man	4.923	4.922	4.961
Carbon-2			
$\beta 4$-Man	4.261	4.261	4.178
$\alpha 3$-Man	4.068	4.190	4.242
$\alpha 6$-Man	3.972	3.976	4.108

[a] M, D-Man; Gn, D-GlcNAc; R, GlcNAc$\beta 4$(Fuc$\alpha 6$)GlcNAc-Asn-X. Glycopeptide isolation and nomenclature are described in Narasimhan *et al.* (1980). Spectra for the C-1 hydrogens were run at 70–82°C and for the C-2 hydrogens at 20–30°C.

Table II. Glycopeptide Elution Patterns on Concanavalin A–Sepharose Chromatography

Glycopeptide structure[a]	Abbreviation[b]	Elution pattern[c]
M α6 / Gnβ2M α3 \ Mβ4R	MGn	Binds to column, elutes as sharp peak with α-methylmannoside
M α6 / Gnβ2M α3 \ Gnβ4Mβ4R	MGn(Gn)	Does not bind to column, elutes in a retarded manner
Gnβ2M α6 / Gnβ2M α3 \ Mβ4R	GnGn	Binds to column, elutes as broad peak with α-methylmannoside
Gnβ2M α6 / Gnβ2M α3 \ Gnβ4Mβ4R	GnGn(Gn)	Does not bind to column, elutes in a retarded manner

[a] M, D-Man; Gn, D-GlcNAc; R, GlcNAcβ4(±Fucα6)GlcNAc-Asn-X.
[b] See Narasimhan *et al.* (1980).
[c] See Narasimhan *et al.* (1979) and Harpaz and Schachter (1980b).

other hand, indicates major shifts in both the C-1 and the C-2 hydrogens of not only the β-linked Man residue, to which the interantennary GlcNAc is attached, but also of the α3- and α6-linked Man residues. These data indicate that the interantennary GlcNAc changes the conformation not only of the Man residue to which it is attached but also of the two α-linked Man residues of the core. It is suggested that the interantennary GlcNAc forces the two branches of the core further apart. Other pairs of glycopeptides differing only in the absence or presence of the interantennary GlcNAc have also been studied, and in all cases, signals similar to those shown in Table I were obtained.

2.3. The Interantennary GlcNAc Alters Interaction with Concanavalin A

The three-dimensional perturbations suggested by the NMR data are supported by the effect of the interantennary GlcNAc on several oligosaccharide–protein interaction systems. Narasimhan *et al.* (1979) showed that under certain conditions, glycopeptides could be classified into four groups depending on their elution patterns from concanavalin A–Sepharose columns. Table II shows that the interantennary GlcNAc weakens the interaction with concanavalin A–Sepharose. It is clear from studying a large series of glycopeptides in this system (Narasimhan *et al.*, 1979; Harpaz and Schachter, 1980b) that the two terminal Man residues present in the Man$_3$(GlcNAc)$_2$Asn-X core are essential for strong bind-

Table III. Substrate Specificity of Bovine Colostrum GlcNAc Transferases I and II

| Glycopeptide[a] | Gn-T I | | Gn-T II |
	K_m (mM)	V_{max} (μmoles/min per mg)	K_m (mM)
MM	0.20	0.083	Inactive
MGn	10	0.110	0.1
MGn(Gn)		Inactive	Inactive

[a] See Table I for structures.

ing to concanavalin A–Sepharose. It is suggested that the interantennary GlcNAc forces these two Man residues apart and thereby interferes with the fit on the concanavalin A molecule.

2.4. The Interantennary GlcNAc Prevents Gn-T I and II Action

Gn-T I has been highly purified and Gn-T II has been partially purified from bovine colostrum (Harpaz and Schachter, 1980a). The substrate specificities of these two enzymes are summarized in Table III. The physiological role of Gn-T I is indicated in Fig. 1; this role is almost certainly not addition of GlcNAc to MGn, a very poor substrate (Table III). However, it is of interest that MGn(Gn) is not a substrate for Gn-T I under conditions where Gn-T I action on MGn is readily detected; this implies that the interantennary GlcNAc prevents Gn-T I action although ideally one should test a glycopeptide such as

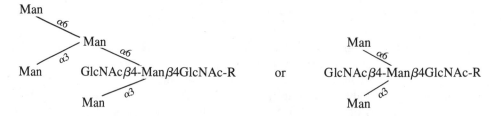

to be sure of this point. These glycopeptides are, however, not available for testing.

The situation for Gn-T II is far more conclusive; MGn is an excellent substrate whereas MGn(Gn) is totally inactive (Table III and Fig. 1). Thus, it appears that the interantennary GlcNAc so distorts the shape of the oligosaccharide as to prevent proper binding to these two enzymes.

2.5. The Interantennary GlcNAc Prevents the Action of Gn-T I-Dependent α-Mannosidase

As shown in Fig. 1, based on the data of Harpaz and Schachter (1980b), the presence of the interantennary GlcNAc prevents conversion of the Man_5-containing structure to the Man_4 structure by the action of Gn-T I-dependent α-mannosidase. This has an important effect on the biosynthetic pathway. In hen oviduct, for example, the action of Gn-T III seems to be more effective than α-mannosidase action, leading to the formation of a large amount of "hybrid" structures in ovalbumin; these hybrid oligosaccharides contain *N*-acetyllactosamine arms on the Manα3 terminus of the core and high-mannose structures on the Manα6 terminus of the core. They also invariably contain an interantennary GlcNAc, and it is suggested that it is the incorporation of this residue that shunts synthesis toward the hybrid mode by turning off mannosidase action. We tested a Man₄-containing hybrid structure and showed it also to be resistant to α-mannosidase, as expected from the fact that ovalbumin contains hybrid structures with both four and five Man residues. The effect of the interantennary GlcNAc on removal of Man residues is another example indicating interference with oligosaccharide–protein interaction.

2.6. The Interantennary GlcNAc Inhibits L-Fucose Incorporation

We have shown (Wilson *et al.*, 1976; Longmore and Schachter, 1980) that L-fucose is incorporated in α6 linkage to the Asn-linked GlcNAc residue only after Gn-T I has acted (Fig. 1). Recent studies have compared four pairs of glycopeptides differing only in the absence or presence of the interantennary GlcNAc (Table IV). It is evident that in all four cases, the interantennary GlcNAc has completely prevented Fuc incorporation. This dependence of Fuc incorporation on Gn-T I action and its inhibition by Gn-T III action (Fig. 1) probably explain why high-mannose and hybrid structures do not contain Fucα6 linked to the core GlcNAc.

Table IV. Substrate Specificity of Pork Liver Golgi GDP-L-Fuc: β-N-acetylglucosaminide(Fuc→Asn-linked GlcNAc)α6-fucosyltransferase

Glycopeptide structure	R' = H-		R' = GlcNAcβ4-
	K_m (mM)	V_{max} (μmoles/hr per mg)	

M
  α6
    R'-Mβ4R
  α3
Gnβ2M
	0.6	0.19	Inactive

Gnβ2M
  α6
    R'-Mβ4R
  α3
Gnβ2M
	0.08	0.11	Inactive

Gnβ2M
  α6
    R'-Mβ4R
  α3
Gnβ2M
  β4|
  Gn
	0.3	0.18	Inactive

M
  α6
    M
  α3 α6
M   R'-Mβ4R
      α3
    Gnβ2M
	—	>0.09	Inactive

3. SUMMARY

Incorporation of an interantennary GlcNAc into *N*-glycosidically linked oligosaccharides causes significant perturbation of the three-dimensional structure of the oligosaccharide, probably by pushing apart the two β2-linked antennae, and thereby alters and weakens the interaction of the oligosaccharide with at least five proteins, i.e., concanavalin A, Gn-T I, Gn-T II, Gn-T I-dependent α-mannosidase, and GDP-L-Fuc: β-*N*-acetylglucosaminide(Fuc→Asn-linked GlcNAc)α6-fucosyltransferase.

ACKNOWLEDGMENT. This work was supported by the Medical Research Council of Canada.

REFERENCES

Atkinson, P. H., and Hakimi, J. (1980). In *The Biochemistry of Glycoproteins and Proteoglycans* (W. J. Lennarz, ed.), pp. 191–239, Plenum Press, New York.

Harpaz, N., and Schachter, H. (1980a). *J. Biol. Chem.* **255,** 4885–4893.

Harpaz, N., and Schachter, H. (1980b). *J. Biol. Chem.* **255,** 4894–4902.

Hughes, R. C. (1976). *Membrane Glycoproteins.* Butterworths, London.

Kornfeld, R., and Kornfeld, S. (1980). In *The Biochemistry of Glycoproteins and Proteoglycans* (W. J. Lennarz, ed.), pp. 1–34, Plenum Press, New York.

Lis, H., and Sharon, N. (1977). In *The Antigens* (M. Sela, ed.), Vol. IV, pp. 429–529, Academic Press, New York.

Longmore, G., and Schachter, H. (1980). *Fed. Proc.* **39,** 2002.

Narasimhan, S., Wilson, J. R., Martin, E., and Schachter, H. (1979). *Can. J. Biochem.* **57,** 83–96.

Narasimhan, S., Harpaz, N., Longmore, G., Carver, J. P., Grey, A. A., and Schachter, H. (1980). *J. Biol. Chem.* **255,** 4876–4884.

Neufeld, E. F., and Ashwell, G. (1980). In *The Biochemistry of Glycoproteins and Proteoglycans* (W. J. Lennarz, ed.), pp. 241–266, Plenum Press, New York.

Schachter, H., and Roseman, S. (1980). In *The Biochemistry of Glycoproteins and Proteoglycans* (W. J. Lennarz, ed.), pp. 85–160, Plenum Press, New York.

Sharon, N. (1979). In *Structure and Function of Biomembranes* (K. Yagi, ed.), pp. 63–82, Japan Scientific Societies Press, Tokyo.

Stanley, P. (1980). In *The Biochemistry of Glycoproteins and Proteoglycans* (W. J. Lennarz, ed.), pp. 161–189, Plenum Press, New York.

Struck, D. K., and Lennarz, W. J. (1980). In *The Biochemistry of Glycoproteins and Proteoglycans* (W. J. Lennarz, ed.), pp. 35–83, Plenum Press, New York.

Takasaki, S. Ikehira, H., and Kobata, A. (1980). *Biochem. Biophys. Res. Commun.* **92,** 735–742.

Wilson, J. R., Williams, D., and Schachter, H. (1976). *Biochem. Biophys. Res. Commun.* **72,** 909–916.

32

Multifunctional Glucose-6-phosphatase of Endoplasmic Reticulum and Nuclear Membrane

Robert C. Nordlie

1. INTRODUCTION

Work beginning in 1963 in our laboratory (Nordlie and Arion, 1964) and that of Stetten (Stetten and Taft, 1964) has shown glucose-6-phosphatase to be a multifunctional enzyme capable of synthesis of glucose-6-P via potent phosphotransferase activities [reaction (1)] as well as glucose-6-P hydrolysis [reaction (2)]. Since that time a major emphasis has been placed upon PP_i and carbamyl-P as phosphoryl donors (RP) and glucose as acceptor, although other compounds also will function as substrates.

$$RP + \text{glucose} \rightarrow \text{glucose-6-}P + R \qquad (1)$$
$$\text{glucose-6-}P + H_2O \rightarrow \text{glucose} + P_i \qquad (2)$$

The broad distribution, multifunctional nature, specificity, methods of assay, kinetics, and physiological functions of glucose-6-phosphatase phosphotransferase have been considered in detail in a number of recent reviews (Nordlie, 1971, 1974, 1976, 1979). Purification of the brain enzyme (Anchors and Karnovsky, 1975) has been achieved, but not that of the hepatic enzyme. Within the past 7 or 8 years, the major emphasis in this area has focused on (1) interrelationships between the membranous nature of the enzyme and its catalytic behavior, and (2) possible physiological functions for the transferase activities.

In a chapter appearing in 1976 in *The Enzymes of Biological Membranes* (Nordlie and Jorgenson, 1976), we concisely reviewed interrelationships between the catalytic behavior of this enzyme and the membranous structures of which it is a part. With this chapter as a point of departure, we will emphasize here recent developments regarding the relationships

Robert C. Nordlie • Department of Biochemistry, University of North Dakota School of Medicine, Grand Forks, North Dakota 58202.

of the membranous nature of glucose-6-phosphatase to catalysis, and physiological implications related thereto.

2. GLUCOSE-6-PHOSPHATASE

2.1. Latency

Since the pioneering work of Beaufay and de Duve (1954), glucose-6-phosphatase has been known to be intimately associated with cellular membranes, to be sensitive to modifications in membrane phospholipids, and to display a partial latency* (i.e., detergent sensitivity) when assayed in isolated microsomal preparations (see Nordlie and Jorgenson, 1976). The latency is more pronounced with phosphate substrates such as PP_i, carbamyl-P, and mannose-6-P than with glucose-6-P. Further, the degree of latency of all microsomal activities may be altered through hormonal manipulations *in vivo* (see Nordlie and Jorgenson, 1976, for primary references). An increase in latency accompanying the increase in enzyme activity in response to insulin deprivation has been noted: a significant decrease in latency accompanies the elevation in enzymatic activity in response to glucocorticoids. This glucocorticoid effect may be superimposed upon the response to diabetes. Interesting differences in the degree of latency also have been noted between microsomes derived from rough and smooth endoplasmic reticulum, between microsomes prepared from livers of rats of various ages (Goldsmith and Stetten, 1979), and in isolated nuclei and nuclear membranes as compared with microsomes (Gunderson and Nordlie, 1975).

2.2. Physiological Significance of Latency

Recent studies from four independent groups have substantiated the physiological relevance of latency of glucose-6-phosphatase (Bialek *et al.*, 1977; Narisawa *et al.*, 1978; Sann *et al.*, 1980; Lange *et al.*, 1980). All of these studies were concerned with what is now termed type IB glycogenesis. Each case involved young patients displaying the classic symptoms of von Gierke's disease (type I glycogenosis), including pronounced postabsorptive hypoglycemia. Surprisingly, examination of liver biopsy specimens indicated a marked increase in the fraction of total glucose-6-phosphatase activity that was latent, and presumably nonfunctional, as compared to normal control tissue. For example, in the first of these studies reported (Bialek *et al.*, 1977), we noted an increase from a normal value of 40% to 75% in the latency of glucose-6-P phosphohydrolase in type IB liver samples. Also of interest, the latency of phosphotransferase activities was unchanged or actually decreased (Nordlie, 1981; Lange *et al.*, 1980).

2.3. Mechanisms Underlying Latency of Microsomal Glucose-6-phosphatase

There are two schools of thought on the molecular mechanism underlying the differential latencies of activities of microsomal glucose-6-phosphatase. Both the author (Nordlie, 1971, 1974; Gunderson and Nordlie, 1975) and Stetten (Stetten and Burnett, 1967) have hypothesized, based on their own observations and the earlier suggestions of Ernster

---

* "Latency" refers to that proportion of total intrinsic enzyme present in membranous preparations that is manifest in the assay procedure only following total disruption of the biomembrane by detergent supplementation or other means.

et al. (1962), that changes in the lipid environment, enzyme conformation within that environment, and/or enzyme orientation within the biomembrane may serve to constrain and thus regulate differentially various hydrolytic and synthetic capabilities of glucose-6-phosphatase. Recent studies in this laboratory on the effects of polyamines (Nordlie *et al.*, 1979) and the thermotropic effects seen with the microsomal preparations in the absence and presence of polyamines (Johnson and Nordlie, 1980) support this concept. Earlier studies of Pollak *et al.* (1971) and Snoke and Nordlie (1972) indicate selective constraints imposable on the activated enzyme by phospholipid supplementation.

In contrast, Arion and his group (see Arion *et al.*, 1975) have carried out a series of sophisticated kinetic analyses suggesting the involvement of a multicomponent system consisting of a fundamental catalytic unit of broad specificity located on the luminal surface of the endoplasmic reticulum, working in conjunction with a phosphate-substrate-specific translocase system serving to carry phosphate substrate(s) from the cytosol of the hepatocyte to the catalytic unit. This latter concept has received additional support from some recent studies in which (1) diazobenzene sulfonate and proteases were found to affect glucose-6-P hydrolysis but not pyrophosphatase of untreated microsomes (Nilsson *et al.*, 1978); (2) a selective inhibition of glucose-6-P hydrolysis by pyridoxal phosphate was noted with intact microsomes (Gold and Widnell, 1976); (3) selective inhibition by 4-acetamido-4'-isothiocyanostilbene-2,2'-disulfonic acid and 4,4'-diisothiocyanostilbene-2,2'-disulfonic acid of glucose-6-phosphatase of intact microsomes was investigated (Zoccoli and Karnovsky, 1980); (4) the effects of controlled proteolysis and phospholipase C action, antibodies to microsomes, and effects of temperature were considered (Schulze and Speth, 1980); (5) selective thermal inactivation of the catalytic unit (Arion *et al.*, 1976a), and interactions of microsomes with mannose-6-P, mannose, and EDTA (Arion *et al.*, 1976b) were studied; and (6) selective inhibition by phlorizin in the absence and presence of detergent was considered (Arion *et al.*, 1980a; see also Lygre and Nordlie, 1969). Arion *et al.* (1980b) have presented kinetic evidence in support of the concept that permeability of microsomes to P_i, PP_i, and carbamyl-P is mediated by a second translocase distinct from that for glucose-6-P. In their current view, this second transporter functions primarily in the direction from the catalytic unit to the cytosol, in contrast to that for glucose-6-P.

In terms of the transporter/catalytic unit concept, type IB glycogenosis is envisaged as a defect in the translocase system for glucose-6-P (Lange *et al.*, 1980). Increase in the degree of latency of glucose-6-phosphatase in diabetes or fasting (see above) is explained as due to an increase in the number of functional catalytic units relative to transporter units in insulin deprivation. In contrast, the responses to glucocorticoid administration are rationalized in terms of an increase in translocases relative to catalytic units (Arion *et al.*, 1976a).

Nilsson *et al.* (1978), recognizing the limitations of the kinetic approach, have pointed out that ". . . the definitive evidence [for the translocase concept] must come either from the isolation and reconstitution of the glucose-6-P carrier or actual measurement of the pool of glucose-6-P within the microsomal vesicles." Ballas and Arion (1977) have achieved limited success with the latter approach.

We (Nordlie, 1981) have suggested that features of both the transporter hypothesis and the enzyme conformation/membrane morphology concept may well be relevant within the cell. Possibly, membrane-related conformational changes may serve to modify and thus control transporters (as well as modulate the catalytic unit) selectively and hence discriminately limit access of substrates to the microsomal glucose-6-phosphatase, per se. As Arion and Nordlie (1967) previously have noted the manifestation of the glucocorticoid effect in the presence as well as absence of the protein synthesis inhibitor actinomycin D, it follows that the increase in transporter, if it occurs, must involve an *activation* of existing translocase rather than an increased synthesis of same. This could be effected, we would suggest, either by perturbation of the biomembrane triggered by hormone action (Nordlie, 1974; Band and Jones, 1980), or possibly by covalent modification. In this latter regard, it has been suggested that glucose-6-phosphatase might be regulated by a phosphorylation/dephosphorylation mechanism (Burchell and Burchell, 1980). These ideas are supported also by Dyatlovitskaya *et al.* (1979), who reinvestigated the phospholipid dependence of glucose-6-phosphatase by employing lipid-exchange proteins to modify selectively the phospholipid microenvironment of isolated microsomal preparations; and by Schulze and Speth (1980), whose work with controlled proteolysis and phospholipase C action and the use of antibodies to microsomes has indicated that glucose-6-phosphatase is buried within the microsomal membrane rather than being exposed on either side, and that phospholipids may be involved in glucose-6-*P* transport as well as hydrolysis. Other supportive evidence has emerged from our work with nuclei and hepatocytes, which is reviewed briefly below.

2.4. Studies with Isolated Nuclei and Nuclear Membrane Preparations

Gunderson and Nordlie (1975) have shown that avian nuclei contain 16–19% of total hepatic glucose-6-phosphatase phosphotransferase. In marked contrast to microsomal preparations, various hydrolytic and synthetic activities of glucose-6-phosphatase of intact, isolated nuclei exhibited little or no latency. Catalytic characteristics of the enzyme of nuclei compared favorably with those of detergent-treated microsomes from the same liver. Fragmentation and isolation of nuclear membrane led to the development of approximately 50% latency of all activities, indicating convincingly interrelationships between membrane morphology and catalytic behavior of glucose-6-phosphatase within that membrane. Responses of the enzyme of nuclei to hormonal manipulations differed dramatically from those of microsomes from the same livers (Nordlie *et al.*, 1979). A three-to fourfold increase was seen in diabetes or in response to glucocorticoids, whether or not detergent was added before assay. Latency was unaltered in diabetes, and was *increased* in response to hydrocortisone.

2.5. Studies with Isolated Hepatic Parenchymal Cells

Jorgenson and Nordlie (1980) have developed a technique whereby synthetic and hydrolytic activities may be studied *in situ* within the endoplasmic reticulum and nuclei of the filipin-treated hepatocyte. With such preparations, patterns of latency were observed that were quite different from those noted with microsomes isolated from these same cells. Latencies of 27, 55, 54, and 18% were noted at pH 7.4 with cellular glucose-6-*P* phosphohydrolase, carbamyl-*P* : glucose phosphotransferase, mannose-6-*P* phosphohydrolase, and PP_i-glucose phosphotransferase, respectively. Corresponding values determined under identical conditions with microsomes isolated from these same hepatocytes were 40, 74, 95, and 73%. In marked contrast to isolated microsomes, hepatocyte PP_i-glucose phosphotransferase was quite active at pH 7.4. K_m values were in generally good agreement with those noted with the detergent-dispersed microsomal preparations. With hepatocyte preparations, the degree of latency of both hydrolase and phosphotransferase increased in fasting and diabetes. For example, an increase from 27% to 38% in the latency of glucose-6-*P* phosphohydrolase was observed after a 48-hr fast (Nordlie and Jorgenson, 1981). From these studies it was concluded than when provided with adequate levels of substrates, a significant proportion of total phosphotransferase as well as glucose-6-*P* phosphohydrolase activity may be manifested by glucose-6-phosphatase within the cell, and that hormonally alterable latency of these activities *in situ* may be a physiological regulatory mechanism for control of these activities. These observations are consistent with either the conformational concept or the translocase hypothesis. If the latter is to prevail, however, it would appear that transporters for PP_i, carbamyl-*P*, mannose-6-*P* and glucose-6-*P* must normally be functional *in situ,* and that a somewhat selective physical or functional loss of these transporters must accompany the isolation of microsomes from hepatocytes (Jorgenson and Nordlie, 1980).

2.6. Physiological Considerations

With isolated perfused liver preparations from both fed and 48-hr-fasted rats, Alvares and Nordlie (1977) observed rates of glucose uptake far in excess of that explainable on the basis of assayed hepatic glucokinase plus hexokinase. Glucose uptake was inhibited by 3-*O*-methyl-D-glucose, an inhibitor of glucose-6-phosphatase phosphotransferase that does not affect glucokinase or hexokinase, and by ornithine, which may compete via ornithine transcarbamylase with glucose-6-phosphatase phosphotransferase for carbamyl-*P* (Alvares and Nordlie, 1977). Nordlie *et al*. (1980) have studied the effects of varied glucose loads on the level of glucose-6-*P* and rate of glycogenesis in perfused rat livers. Results support a step involving glucose phosphorylation via a high-K_m enzyme supplemental to glucokinase (i.e., a metabolic "push") as the primary site involved in the well-documented glucose concentration dependence of metabolic flux from glucose to hepatic glycogen. Recent studies (Nordlie and Sukalski, 1980) involving quantitative considerations of metabolic cycling and its control at the glucose $\rightleftarrows$ glucose-6-*P* site in liver indicate the need not only for constraints of glucose-6-*P* phosphohydrolase by latency, and inhibition by glucose, P_i, and bicarbonate, but in addition the requirement for hepatic glucose phosphorylative capacity above and beyond that of glucokinase. Johnson and Nordlie (1977) have found that micromolar levels of Cu^{2+} may activate phosphotransferase activity of the enzyme while inhibiting glucose-6-*P* phosphohydrolase. Veech and colleagues have implicated a PP_i-glucose phosphotransferase in hepatic metabolism (Lawson and Veech, 1979) and have observed an increase in hepatic PP_i levels to the millimolar range following injection of fatty acids (Veech *et al*., 1980), while Cohen *et al*. (1980) have demonstrated movement of carbamyl-*P* from mitochondria to cytosol.

3. CONCLUSIONS

All of these observations suggest that controllable latency, by whatever mechanism, may be a key means for regulation of activities of glucose-6-phosphatase *in vivo*, and that synthetic as well as hydrolytic functions of this multifunctional enzyme may play significant roles in maintenance of glucose homeostasis under diabetic and perhaps other conditions (Nordlie, 1981).

ACKNOWLEDGMENT. This work was supported by Grant AM 07141 from the National Institutes of Health.

REFERENCES

Alvares, F. L., and Nordlie, R. C. (1977). *J. Biol. Chem.* **252**, 8404–8414.
Anchors, J. M., and Karnovsky, M. L. (1975). *J. Biol. Chem.* **250**, 6408–6416.
Arion, W. J., and Nordlie, R. C. (1967). *J. Biol. Chem.* **242**, 2207–2210.
Arion, W. J., Wallin, B. K., Lange, A. J., and Ballas, L. M. (1975). *Mol. Cell. Biochem.* **6**, 75–83.
Arion, W. J., Lange, A. J., and Ballas, L. M. (1976a). *J. Biol. Chem.* **251**, 6784–6790.
Arion, W. J., Ballas, L. M., Lange, A. J., and Wallin, B. K. (1976b). *J. Biol. Chem.* **251**, 4901–4907.
Arion, W. J., Lange, A. J., and Walls, H. E. (1980a). *J. Biol. Chem.* **255**, 10387–10395.
Arion, W. J., Lange, A. J., Walls, H. E., and Ballas, L. M. (1980b). *J. Biol. Chem.* **255**, 10396–10406.
Ballas, L. M., and Arion, W. J. (1977). *J. Biol. Chem.* **252**, 8512–8518.
Band, G. C., and Jones, C. T. (1980). *FEBS Lett.* **119**, 190–194.
Beaufay, H., and de Duve, C. (1954). *Bull. Soc. Chim. Biol.* **36**, 1551–1568.
Bialek, D. S., Sharp, H. L., Kane, W. J., Elders, J., and Nordlie, R. C. (1977). *J. Pediatr.* **91**, 838.
Burchell, A., and Burchell, B. (1980). *FEBS Lett.* **118**, 180–184.
Cohen, N. S., Cheung, C.-W., and Raijman, L. (1980). *J. Biol. Chem.* **255**, 10248–10255.
Dyatlovitskaya, E. V., Lemenovskaya, A. F., and Bergelson, L. D. (1979). *Eur. J. Biochem.* **99**, 605–612.
Ernster, L., Siekevitz, P., and Palade, G. (1962). *J. Cell Biol.* **15**, 541–578.
Gold, G., and Widnell, C. C. (1976). *J. Biol. Chem.* **251**, 1035–1041.
Goldsmith, P. K., and Stetten, M. R. (1979). *Biochim. Biophys. Acta* **583**, 133–147.
Gunderson, H. M., and Nordlie, R. C. (1975). *J. Biol. Chem.* **250**, 3552–3559.
Johnson, W. T., and Nordlie, R. C. (1977). *Biochemistry* **16**, 2458–2466.
Johnson, W. T., and Nordlie, R. C. (1980). *Life Sci.* **26**, 297–302.
Jorgenson, R. A., and Nordlie, R. C. (1980). *J. Biol. Chem.* **255**, 5907–5915.
Lange, A. J., Arion, W. J., and Beaudet, A. L. (1980). *J. Biol. Chem.* **255**, 8381–8383.
Lawson, J. W. R., and Veech, R. L. (1979). *J. Biol. Chem.* **254**, 6528–6537.
Lygre, D. G., and Nordlie, R. C. (1969). *Biochim. Biophys. Acta* **185**, 360–366.
Narisawa, K., Igarashi, Y., Otomo, H., and Tada, K. (1978). *Biochem. Biophys. Res. Commun.* **83**, 1360–1364.
Nilsson, O. S., Arion, W. J., Depierre, J. W., Dallner, G., and Ernster, L. (1978). *Eur. J. Biochem.* **82**, 627–634.
Nordlie, R. C. (1971). In *The Enzymes* (P. D. Boyer, ed.), 3rd ed., Vol. 4, pp. 543–609, Academic Press, New York.
Nordlie, R. C. (1974). In *Current Topics in Cellular Regulation* (B. L. Horecker and E. R. Stadtman, eds.), Vol. 8, pp. 33–117, Academic Press, New York.
Nordlie, R. C. (1976). In *Gluconeogenesis* (M. A. Mehlman and R. W. Hanson, eds.), pp. 93–152, Wiley, New York.
Nordlie, R. C. (1979). *Life Sci.* **24**, 2397–2404.
Nordlie, R. C. (1981). In *Glucose Formation and Utilization in Mammals* (C. Veneziale, ed.), pp. 291–314. University Park Press, Baltimore.
Nordlie, R. C., and Arion, W. J. (1964). *J. Biol. Chem.* **239**, 1680–1685.
Nordlie, R. C., and Jorgenson, R. A. (1976). In *The Enzymes of Biological Membranes* (A. Martonosi, ed.), Vol. 2, pp. 465–491, Plenum Press, New York.
Nordlie, R. C., and Jorgenson, R. A. (1981). *J. Biol. Chem.* **256**, 4768–4771.
Nordlie, R. C., and Sukalski, K. A. (1980). *Fed. Proc.* **39**, 2146.
Nordlie, R. C., Johnson, W. T., Cornatzer, W. E., Jr., and Twedell, G. W. (1979). *Biochim. Biophys. Acta* **585**, 12–23.
Nordlie, R. C., Sukalski, K. A., and Alvares, F. L. (1980). *J. Biol. Chem.* **255**, 1834–1838.
Pollak, J. K., Malor, R., Morton, M., and Ward, K. A. (1971). In *Autonomy and Biogenesis of Mitochondria and Chloroplasts* (N. K. Boardman, A. W. Linnane, and R. M. Smilie, eds.), pp. 27–41, North-Holland, Amsterdam.
Sann, L., Mathieu, M., Bourgeois, J., Bienvenue, J., and Bethenod, M. (1980). *J. Pediatr.* **96**, 691–694.
Schulze, H.-U., and Speth, M. (1980). *Eur. J. Biochem.* **106**, 505–514.
Snoke, R. E., and Nordlie, R. C. (1972). *Biochim. Biophys. Acta* **258**, 188–205.
Stetten, M. R., and Burnett, F. F. (1967). *Biochim. Biophys. Acta* **139**, 138–147.
Stetten, M. R., and Taft, H. L. (1964). *J. Biol. Chem.* **239**, 4041–4046.
Veech, R. L., Cook, G. A., and King, M. T. (1980). *FEBS Lett.* **117**, K65–K72.
Zoccoli, M. A., and Karnovsky, M. L. (1980). *J. Biol. Chem.* **255**, 1113–1119.

33

The Role of the Microsomal Membrane in Modulating the Activity of UDP-Glucuronyltransferase

David Zakim and Donald A. Vessey

1. INTRODUCTION

Hepatic microsomal UDP-glucuronyltransferase activities (EC 2.4.1.17) constitute a family of enzymes (Bock *et al.*, 1979; Dutton and Greig, 1957; Tukey *et al.*, 1978; Zakim *et al.*, 1973a) that detoxify several endogenously synthesized toxins, therapeutic agents, and environmental pollutants (Dutton, 1966). All forms of the enzyme are integral components of the microsomal membrane. There is considerable evidence that the phospholipids of this membrane are important for efficient function of UDP-glucuronyltransferase. For example, phospholipids within the microsomal membrane appear to regulate the kinetic function of UDP-glucuronyltransferase via direct interactions with the enzyme (Erickson *et al.*, 1978; Zakim and Vessey, 1976). Moreover, the microsomal lipids have significance for the delivery of water-insoluble aglycones to the active site of at least some of these enzymes (Zakim and Vessey, 1977; Boyer *et al.*, 1980). This review is limited to a discussion of the role of the microsomal phospholipids as modulators of the function of UDP-glucuronyltransferases. Other literature should be consulted for recent advances in the purification of different forms of UDP-glucuronyltransferases (Bock *et al.*, 1979; Burchell, 1978, 1980; Gorski and Kasper, 1977; Yuasa, 1977), induction of these activities (Bock *et al.*, 1979; Lamartiniere *et al.*, 1979; Owens, 1977), and their patterns of fetal and neonatal development (Dutton *et al.*, 1976; Dutton, 1978; Goldstein *et al.*, 1980; Wishart and Dutton, 1977).

David Zakim and Donald A. Vessey • Liver Studies Unit, Veterans Administration Medical Center, San Francisco, California 94121, and Departments of Medicine and Pharmacology, University of California, San Francisco, California 94143.

2. DIRECT EFFECTS OF PHOSPHOLIPIDS ON THE PROPERTIES OF UDP-GLUCURONYLTRANSFERASE

Perturbations of the lipid portion of the microsomal membrane by treatments with detergents, phospholipases, and phosphatides (Graham and Wood, 1969; Lueders and Kuff, 1967; Zakim and Vessey, 1975a) alter the kinetic properties of UDP-glucuronyltransferase assayed with a variety of aglycones. For example, the activity of the enzyme increases after treatment of microsomes with detergents. There are two divergent interpretations of the significance of intact membrane phospholipids for regulating the enzyme. The first, referred to as the compartmentation model, proposes that the active site of UDP-glucuronyltransferase is on the inside of the microsomal vesicle (Berry *et al.*, 1975; Hallinan, 1978). The intact microsomal membrane is believed in this view to limit the activity of UDP-glucuronyltransferase, functioning as a barrier to the free access of the water-soluble UDP-glucuronic acid to the active site. The alternate proposal to explain regulation of UDP-glucuronyltransferase by microsomal phospholipids posits that (1) the enzyme can exist as several different conformational isomers; (2) different isomers have variable kinetic properties (Vessey and Zakim, 1971; Zakim and Vessey, 1976); and (3) the stabilities of different isomers are determined by the enzyme's lipid environment.

2.1. Examination of the Compartmentation Model

The compartmentation model is attractive because it accounts readily for activations of UDP-glucuronyltransferase in response to rupturing microsomal vesicles. On the other hand, it does not explain several experimental observations. Thus, (1) activations of UDP-glucuronyltransferase measured at V_{max} are different after treatment of microsomes with cholate as compared with pure phospholipase A_2 or lysolecithin (Zakim and Vessey, 1975b). (2) Each of these treatments activates the forward and reverse reactions catalyzed by UDP-glucuronyltransferase to a different extent (Zakim and Vessey, 1975b). (3) Recent experiments indicate that UDP-glucuronic acid does not penetrate to the inside of microsomal vesicles (Finch *et al.*, 1979). These studies utilized pH-induced splitting of the NMR spectrum of ^{31}P as an analytical tool. The technique is exceedingly sensitive, and at the same time it is straightforward in its application (4) Treatment of microsomes with phospholipase A_2 alters the specificity of the UDP-glucuronic binding site of UDP-glucuronyltransferase (Zakim *et al.*, 1973b). (5) Treatments with detergents and phospholipase A_2 have differential effects on the affinity of UDP-glucuronyltransferase for UDP and UDP-sugars other than UDP-glucuronic acid (Zakim *et al.*, 1973b; Zakim and Vessey, 1975b). (6) The affinity of untreated UDP-glucuronyltransferase for UDP-glucuronic acid is modulated by UDP-*N*-acetylglucosamine (Vessey *et al.*, 1973). This allosteric regulation is destroyed by treatments with detergent and phospholipase A_2 (Zakim *et al.*, 1973b). To explain the observations under (4), (5), and (6), the proponents of the compartmentation model have proposed that access of UDP-glucuronic acid to the active site of UDP-glucuronyltransferase in intact microsomes is effected via a specific microsomal transport system (Berry and Hallinan, 1978). There is no independent evidence for such a transporter. No modifications of the compartmentation model can explain the data under (1) and (2). Hence, it seems that the compartmentation model for regulation of UDP-glucuronyltransferase by microsomal phospholipids explains only a limited amount of the relevant kinetic data.

2.2. Examination of Data Relating to Direct Regulation of UDP-Glucuronyltransferase by Phospholipids

All of the data bearing on the regulation of the kinetic properties of UDP-glucuro-nyltransferase in intact and modified microsomes are compatible with the idea of dynamic regulation by protein–lipid interactions (cf. Vessey and Zakim, 1978; Zakim and Vessey, 1976). The true test of the idea that phospholipids can regulate the activity of UDP-glu-curonyltransferases, must come, however, from studies with delipidated, reconstituted systems. It has been found, in this regard, that phospholipids alter the binding affinity of substrates for UDP-glucuronyltransferase as well as activity at V_{max} when added to partially purified, delipidated forms of the enzyme (Erickson *et al.*, 1978; Hochman *et al.*, 1981). This is true, as well, for purified, delipidated enzyme (Hochman *et al.*, 1981). Phospho-lipids also modify the kinetic function of pure forms of UDP-glucuronyltransferase that are not completely delipidated (Gorski and Kasper, 1978).

Delipidated UDP-glucuronyltransferase has residual activity, but activity is enhanced as much as 1000-fold on adding phospholipids to delipidated enzyme (Hochman and Zakim, 1981). Activity appears to be stimulated only by phospholipids containing a phosphoryl-choline group (Erickson *et al.*, 1978). Also, the extent of activation and the detailed kinetic properties of reconstituted forms of UDP-glucuronyltransferase depend on the number, chain length, and degree of unsaturation of the acyl groups of the added phospholipid. An important observation with regard to reconstitution experiments is that detergents, especially nonionic detergents, inhibit phospholipid-induced reactivation of delipidated enzyme (Er-ickson *et al.*, 1978). In addition, residual phospholipids can prevent activation of purified forms of UDP-glucuronyltransferase (D. Zakim and Y. Hochman, unpublished observations). Because phospholipids, depending on their composition, have variable effects on the activity of UDP-glucuronyltransferase, conclusions about the phospholipid dependence of this enzyme depend on studies with completely delipidated enzyme.

Data from at least two laboratories suggest that the activity of a purified form of UDP-glucuronyltransferase does not depend on addition of phospholipids (Bock *et al.*, 1979; Burchell and Hallinan, 1978). Interpretation of these results is uncertain, however, for two reasons. First, no chemical evidence for complete delipidation of enzyme is presented. Second, enzyme was purified in the presence of nonionic detergents. As noted above, these inhibit phospholipid-induced reconstitution of activity (Erickson *et al.*, 1978). Also, it is more difficult to remove nonionic detergents from membrane-bound proteins than is generally acknowledged (Allen *et al.*, 1980). The best evidence indicates, therefore, that lipid–protein interactions modulate the substrate binding and catalytic properties of UDP-glucu-ronyltransferases. We believe the evidence with delipidated enzyme establishes that UDP-glucuronyltransferase can exist as different conformational isomers with different kinetic properties, and that the relative stability of these isomers depends on the structure of lipids in their environment.

Possibly, the functional significance of interactions between phospholipids and UDP-glucuronyltransferases will vary for different substrate-specific forms of the enzyme. Experimental testing of this idea requires precision in determining the amounts of detergent and phospholipid present in purified preparations of enzyme prior to addition of phospho-lipids.

3. BIOLOGICAL SIGNIFICANCE OF THE MEMBRANE LOCATION OF UDP-GLUCURONYLTRANSFERASE

UDP-glucuronyltransferase activity is measured most often with water-soluble substrates. However, most of the substrates for these enzymes in intact animals are water insoluble (for example, steroids, bilirubin, and a number of therapeutic agents). These water-insoluble compounds partition readily into cell membranes (Boyer *et al.*, 1980; Tipping *et al.*, 1979; Zakim and Vessey, 1977; Zakim *et al.*, 1981). We see, therefore, that UDP-glucuronyltransferase and its substrates that are water-insoluble could be concentrated in the same two-dimensional compartment of the cell. This concurrence of events has functional significance (Zakim and Vessey, 1977; Zakim *et al.*, 1981). Studies of the conjugation of estrone show that the pool of estrone sequestered by the microsomal membrane has direct access to the active site of UDP-glucuronyltransferase in that estrone reaches the active site via lateral diffusion in the plane of the membrane (Zakim and Vessey, 1977). It does not reach the active site through the aqueous phase. In fact, the rate of hydration of estrone bound to a lipid bilayer is too slow ($t_{1/2} = 24$ min) to support observed rates of glucuronidation in a suspension of microsomes in water. The assembly of UDP-glucuronyltransferase in an extended bilayer membrane thus is an important structural feature for the function of the enzyme independent of the influence of lipid–protein interactions on the kinetic properties of these enzymes.

In addition to sequestering water-insoluble substrates for UDP-glucuronyltransferase within the same limited compartment of the cell that contains this enzyme, the lipids of the membrane may contribute to catalysis of these substrates by diminishing the entropy term of the free energy of activation. Although there is no direct evidence on this aspect of membrane-enzyme function, the limited data available indicate that steroid molecules have specific orientations in lipid bilayers (Munck, 1957). This may be so for other classes of water-insoluble substrates.

ACKNOWLEDGMENTS. Work from the authors' laboratory was supported by grants from the National Science Foundation, the National Institutes of Health, and basic institutional support from the Veterans Administration.

REFERENCES

Allen, T. M., Romans, A. Y., Kercret, H., and Segrest, J. P. (1980). *Biochim. Biophys. Acta* **601**, 328–342.
Berry, C., and Hallinan, T. (1978). *Biochem. Soc. Trans.* **6**, 178–180.
Berry, C., Stellon, A., and Hallinan, T. (1975). *Biochim. Biophys. Acta* **403**, 335–344.
Bock, K. W., Josting, D., Lilienblum, W., and Pfeil, H. (1979). *Eur. J. Biochem.* **98**, 19–26.
Boyer, T. D., Zakim, D., and Vessey, D. A. (1980). *J. Biol. Chem.* **255**, 627–631.
Burchell, B. (1978). *Biochem. J.* **173**, 749–757.
Burchell, B. (1980). *FEBS Lett.* **111**, 131–135.
Burchell, B., and Hallinan, T. (1978). *Biochem. J.* **171**, 821–824.
Dutton, G. J., (ed.) (1966). In *Glucuronic Acid Free and Combined*, pp. 186–299, Academic Press, New York.
Dutton, G. J. (1978). *Annu. Rev. Pharmacol. Toxicol.* **18**, 17–35.
Dutton, G. J., and Greig, C. G. (1957). *Biochem. J.* **66**, 52P.
Dutton, G. J., Wishart, G. J., Leakey, J. E. A., and Goheer, M. A. (1976). In *Drug Metabolism: From Microbe to Man* (D. V. Parke and R. L. Smith, eds.), pp. 71–90, Taylor & Francis, London.
Erickson, R. H., Zakim, D., and Vessey, D. A. (1978). *Biochemistry* 17, 3706–3711.
Finch, S. A. E., Slater, T. F., and Steir, A. (1979). *Biochem. J.* **177**, 925–930.

Goldstein, R. B., Vessey, D. A., Zakim, D., Mock, N., and Thaler, M. (1980). *Biochem. J.* **186,** 841–845.

Gorski, J. P., and Kasper, C. B. (1977). *J. Biol. Chem.* **252,** 1336–1343.

Gorski, J. P., and Kasper, C. B. (1978). *Biochemistry* **17,** 4600–4605.

Graham, A. B., and Wood, G. C. (1969). *Biochem. Biophys. Res. Commun.* **37,** 567–575.

Hallinan, T. (1978). In *Conjugation Reactions in Drug Biotransformation* (A. Aitio, ed.), pp. 257–267, Elsevier/North-Holland, Amsterdam.

Hochman, Y., and Zakim, D. (1981). Submitted for publication.

Hochman, Y., Zakim, D., and Vessey, D. A. (1981). *J. Biol. Chem.,* **256,** 4783–4788.

Lamartiniere, C. A., Dieringer, C. S., Kita, E., and Lucier, G. W. (1979). *Biochem. J.* **180,** 313–318.

Leuders, K. K., and Kuff, E. L. (1967). *Arch. Biochem. Biophys.* **120,** 198–203.

Munck, A. (1957). *Biochim. Biophys. Acta* **24,** 507–514.

Owens, I. S. (1977). *J. Biol. Chem.* **252,** 2827–2833.

Tipping, E., Ketterer, B., and Christodoulides, L. (1979). *Biochem. J.* **180,** 319–326.

Tukey, R. H., Billings, R. E., and Tephly, T. R. (1978). *Biochem. J.* **171,** 659–663.

Vessey, D. A., and Zakim, D. (1971). *J. Biol. Chem.* **246,** 4649–4656.

Vessey, D. A., and Zakim, D. (1978). In *Conjugation Reactions in Drug Biotransformation* (A. Aitio, ed.), pp. 247–255, Elsevier/North-Holland, Amsterdam.

Vessey, D. A., Goldenberg, J., and Zakim, D. (1973). *Biochim. Biophys. Acta* **309,** 58–66.

Wishart, G. J., and Dutton, G. J. (1977). *Biochem. J.* **168,** 507–511.

Yuasa, A. (1977). *J. Coll. Dairying* **7** (Suppl.), 103–156.

Zakim, D., and Vessey, D. A. (1975a). *J. Biol. Chem.* **250,** 342–343.

Zakim, D., and Vessey, D. A. (1975b). *Biochim. Biophys. Acta* **410,** 61–73.

Zakim, D., and Vessey, D. A. (1976). In *The Enzymes of Biological Membranes* (A. Martonosi, ed.), Vol. 2, pp. 443–461, Plenum Press, New York.

Zakim, D., and Vessey, D. A. (1977). *J. Biol. Chem.* **252,** 7534–7537.

Zakim, D., Goldenberg, J., and Vessey, D. A. (1973a). *Biochim. Biophys. Acta* **309,** 67–74.

Zakim, D., Goldenberg, J., and Vessey, D. A. (1973b). *Eur. J. Biochem.* **38,** 59–63.

Zakim, D., Vessey, D. A., and Boyer, T. D. (1981). In *Molecular Basis of Drug Action* (T. P. Singer and R. Ondarza, eds.), pp. 205–221, Elsevier/North-Holland, Amsterdam.

IV

The Structure, Composition, and Biosynthesis of Membranes in Microorganisms

34

Structural and Functional Asymmetry of Bacterial Membranes

Milton R. J. Salton

1. INTRODUCTION

Plasma membranes of both gram-positive and gram-negative bacteria perform a variety of functions including transport of ions and metabolites, secretion of proteins, energization, biosynthetic stages in cell wall peptidoglycan, lipopolysaccharide, and capsular polysaccharide formation, and the biosynthesis of membrane lipids. All of these functions are organized in a single membrane system, in contrast to the functional compartmentalization characteristic of the membranous organelles (e.g., mitochondria, endoplasmic reticulum, Golgi, and lysosomes) of eukaryotic cells (Stanier, 1970; Carlile, 1980). In addition to plasma membranes, gram-negative bacteria possess a distinctive outer membrane (Inouye, 1979), and certain groups of bacteria possess specialized intracellular membranes such as chromatophores of photosynthetic bacteria, cytomembranes of nitrifying organisms, and spore membranes of sporulating bacteria. The only other membrane structure seen in many bacteria, especially in gram-positive organisms, is the mesosome. The origins and functions of mesosomes have been widely reviewed and there is still much speculation and some evidence as to their role and significance in the bacterial cell and cell cycle (Ghosh, 1974; Salton and Owen, 1976; Higgins *et al.*, 1981). Central to an understanding of the structure–function relationships of these various bacterial membrane systems are (1) resolution of the complexity of the components in plasma membranes, outer membranes of gram-negative bacteria, and specialized membranes such as chromatophores, and (2) establishment of the molecular architecture and sidedness of the membranes and derived vesicles. Much progress has been made in recent years by resolving the variety of polypeptides in membrane structures by sodium dodecyl sulfate (SDS)-polyacrylamide gel electrophoresis (SDS-PAGE) and by the application of two-dimensional (''crossed'') immunoelectrophoresis (CIE) in the identification of membrane antigens and/or enzymes under essentially

Milton R. J. Salton • Department of Microbiology, New York University School of Medicine, New York, New York 10016.

nondenaturing solubilization conditions with nonionic detergents such as Triton X-100 (Owen and Salton, 1975a; Owen and Smyth, 1977; Smyth *et al.*, 1978). Moreover, the CIE analysis of bacterial membranes has been used in combination with conventional antibody-absorption techniques to provide a powerful approach to the problem of establishing the asymmetric structure of bacterial membranes, vesicles, and chromatophores (Owen and Salton, 1975a; Salton and Owen, 1976; Owen and Kaback, 1978, 1979; Perille-Collins *et al.*, 1980).

2. IMMUNOELECTROPHORETIC ANALYSIS OF BACTERIAL MEMBRANE ANTIGENS

Although SDS-PAGE has provided much useful information about the variety of proteins in membranes, very few enzymes survive SDS treatment and assigning functions to the components is often based on circumstantial evidence. A procedure in which retention of biological activity is combined with high resolution of complex mixtures of components (antigens/enzymes), such as those found in bacterial membranes, would thus have many advantages over SDS-PAGE. The CIE method together with enzyme (zymogram) staining (Uriel, 1971; Owen and Smyth, 1977) provides such a procedure and has facilitated the analysis of the variety of antigens and certain enzymes in plasma membranes of the gram-positive organisms *Micrococcus lysodeikticus* (Owen and Salton, 1975a, 1977) and *Bacillus subtilis* (Rutberg *et al.*, 1978) inner and outer membranes of *Escherichia coli* (Smyth *et al.*, 1978) and *Neisseria gonorrhoeae* (Perille-Collins and Salton, 1980), membrane vesicles of *E. coli* (Owen and Kaback, 1978, 1979), chromatophores of *Rhodopseudomonas sphaerioides* (Perille-Collins *et al.*, 1979; Elferink *et al.*, 1979), and antigens of *Acholeplasma laidlawii* membranes (Johansson and Hjértén, 1974).

Analysis of Triton X-100-solubilized membrane antigens of *M. lysodeikticus* with antibodies generated to isolated plasma membranes revealed the presence of 27 discrete immunoprecipitates; of these, five major antigens were identified by zymogram staining as ATPase, succinate dehydrogenase, malate dehydrogenase, and two immunologically distinct NADH dehydrogenases (Owen and Salton, 1975a, 1977). Succinylated lipomannan, the membrane amphiphile of *M. lysodeikticus* (Owen and Salton, 1975b), was identified as one of the major membrane antigens; it was considerably enriched in isolated mesosome fractions, which in contrast to the plasma membranes contained extremely small amounts of ATPase and dehydrogenases (Salton and Owen, 1976). The immunoprecipitate pattern for *B. subtilis* membranes was simpler than that of *M. lysodeikticus* and could be related to the lower solubility of its membranes in Triton X-100 (Rutberg *et al.*, 1978). This latter problem presents one of the limitations of the CIE analysis (Owen and Smyth, 1977).

Antibodies to *E. coli* K12 envelopes were used by Smyth *et al.* (1978) in resolving over 46 distinct antigenic components of the inner (plasma) membrane; of these, 10 immunoprecipitates were identified as membrane enzymes by zymogram staining: ATPase, succinate-, NADH-, D-lactate-, 6-phosphogluconate-, glutamate- (two separate components), dihydro-orotate-, and glycero-3-phosophate dehydrogenases, and a protease. A similar CIE analysis (Owen and Kaback, 1978) has been performed on the *E. coli* ML308-225 strain used so extensively in the membrane vesicle studies of active transport by Kaback (1972). Solubilized *E. coli* ML308-225 membrane vesicles as prepared for transport studies (Kaback, 1971) gave over 50 immunoprecipitates when reacted with antivesicle antibodies, and seven of the immunoprecipitates possessed similar enzyme activities to those found in *E. coli* K12 (Owen and Kaback, 1978, 1979). Although some antigens usually found in the cytoplasm or in the outer membrane (e.g., lipopolysaccharide, lipoprotein) were detected in these transport vesicles, it was concluded by Owen and Kaback (1978) that the degree of contamination was minimal. Thus, CIE analysis of the vesicles has made a valuable contribution to a fuller characterization of a membrane transport model of considered biochemical importance. In addition, Owen *et al.* (1980) have detected seven discrete iron-containing antigens in *E. coli* ML308-225 membrane vesicles by growing the organism in the presence of ^{59}Fe. The iron seemed to be largely in the form of nonheme derivatives, and three of the labeled immunoprecipitates corresponded to NADH dehydrogenase, NADPH dehydrogenase, and glutamate dehydrogenase and one labeled precipitate was identifiable as Braun's lipoprotein (Owen *et al.*, 1980).

Isolated outer membranes of *E. coli* K12 yielded some 25 immunoprecipitates on CIE analysis, but in contrast to the inner membranes, the major immunoprecipitates of the outer membrane formed a slow-moving cluster of seven antigens, two of which were more prominent and identified as Braun's lipoprotein and lipopolysaccharide (Smyth *et al.*, 1978). Similar general differences between inner- and outer-membrane antigen patterns of *N. gonorrhoeae* were also seen recently in the study of Perille-Collins and Salton (1980).

The chromatophore membrane of the facultative photosynthetic bacterium *R. sphaeroides* represents a differentiated intracytoplasmic membrane formed in response to reduced oxygen tension, and it is suggested that it is continuous with the peripheral cytoplasmic membrane and that the isolated chromatophores have an inside-out orientation, i.e., opposite that of the *in situ* plasma membrane. Because of the specialized nature of the chromatophore membrane in photosynthesis and its apparent origin by differentiation of the plasma membrane, it provides an attractive and comparatively rare instance of a readily accessible, naturally occurring inside-out vesicle and differentiated bacterial membrane system. Two groups of investigators have independently studied the chromatophore membranes of *R. sphaeroides* by CIE analysis. Perille-Collins *et al.* (1979) detected 31 antigens in Triton X-100-solubilized chromatophore membranes and identified the light-harvesting bacteriochlorophyll *a*–protein complex as a major immunoprecipitate in addition to the photochemical reaction center, NADH dehydrogenase, and L-lactate dehydrogenase. Elferink *et al.* (1979) found a similar array of some 50 immunoprecipitates by CIE analysis of both the plasma membranes and the chromatophores, but neither the light-harvesting protein nor the photochemical reaction-center protein was identified. Although there were some quantitative differences between the two types of membranes, malate dehydrogenase and four distinct immunoprecipitates staining for NADH dehydrogenase were identical for both. Perille-Collins *et al.* (1979) were unable to detect ATPase and succinate dehydrogenase in their preparations, whereas both were identifiable in fractions studied by Elferink *et al.* (1979).

These studies serve to illustrate how the complexity of components in the multifunctional bacterial plasma membrane structures and specialized membranes such as chromatophores can be resolved both immunologically as antigens and biochemically as enzymes or other functional proteins (e.g., light-harvesting bacteriochlorophyll *a*–protein complex). It is of course apparent that the functions of many of the precipitated antigens seen in the CIE analyses are unknown, and for a multitude of reasons it may be impossible to approach complete enzymatic identification by these procedures. Thus, antibody inhibition, lack of suitable enzyme assays applicable to the CIE system, inactivation by solubilization procedures, and loss of components in detergent-insoluble fractions may all account for identification failures. Nonetheless, the resolution and specificity achievable by CIE both for antigens and for enzymes are impressive. None of the immunoprecipitates showed multiple enzyme staining, i.e., more than one activity in a given precipitate, and where more than one immunoprecipitate stained for a particular activity (e.g., NADH dehydrogenase), they were almost invariably immunologically distinct. Moreover, CIE analysis has the potential or capability of detection and quantitation of nanogram levels of a given component (e.g., lipomannan, Owen and Salton, 1976) and has been used to determine the amounts of molecular species (e.g., ATPase, two immunologically distinct NADH dehydrogenases, and lipomannan antigen) in extracts of *M. lysodeikticus* membranes (Perille-Collins and Salton, 1979).

3. MEMBRANE ASYMMETRY

As membranes constitute the barrier between the external environment or compartment and the inside of cells, it is not surprising that they are believed to be asymmetric and that the outside and inside of the membrane "do different things" (Singer, 1975). Thus, membrane components possessing carbohydrate residues have been shown to exist almost exclusively on the outer or external face of the membrane where they can function in cell recognition and cell–cell contact processes. The inner, cytoplasmic face of erythrocyte membranes has been shown to be the site of certain enzymes (Singer, 1975). It would be expected, therefore, that the bacterial plasma membrane with its multiplicity of biosynthetic, transport, secretory, and mitochondrial-type functions must rely on a highly asymmetric structure to perform and conserve these various functions. Indeed, it was shown by ferritin-conjugated antibody labeling that the F_1-ATPase is asymmetrically located only on one face (cytoplasmic) of the membrane of *M. lysodeikticus* (Oppenheim and Salton, 1973). Such a location for the F_1-ATPase was an inherent requirement for its role in the chemiosmotic hypothesis of Mitchell (1968). The asymmetric disposition of the F_1-ATPase of *M. lysodeikticus* membranes was confirmed by the combined methology of antibody absorption with intact protoplasts and CIE, and moreover, the asymmetry was extended to other enzymes including malate, succinate, and two antigenically distinct NADH dehydrogenases (Owen and Salton, 1975a, 1977), which were found exclusively on the cytoplasmic face of the plasma membrane. Antibody-absorption CIE studies also established that the succinylated lipomannan (a major antigen) and several antigens (glycoproteins?) reacting with concanavalin A were exposed on the outer face of the protoplast membrane. These results suggested that the bacterial plasma membrane presented an external surface rich in carbohydrate residues, a feature in common with mammalian membranes. So far no outer-membrane enzyme marker has been reported, but it seems possible that some of the enzymes involved in peptidoglycan assembly and metabolism may be part of or loosely associated with the outer face. The release and recovery of the penicillin-sensitive, D-alanine carboxypeptidase of *M. lysodeikticus* in the protoplasting fluid (Linder and Salton, 1975) adds support to this suggestion.

Determination of membrane asymmetry by the methodology first used with *M. lysodeikticus* (Owen and Salton, 1975a, 1977) has been extended by Owen and Kaback (1978, 1979) to resolve the question of the orientation of *E. coli* ML308-225 membrane vesicles. Although it was believed that these membrane vesicles had the same right-side-out topography of the plasma membrane in intact cells of *E. coli* ML308-225 (Kaback, 1971), conflicting reports appeared suggesting mixed populations of normal and inverted vesicles to almost completely inverted membrane vesicles (see Hare *et al.*, 1974; Futai, 1974; Adler and Rosen, 1977). Immunoadsorption experiments with osmotically prepared membrane vesicles of *E. coli* ML308-225 revealed that over 95% of the membranes were in the form of sealed vesicles with the same orientation as that of the membrane in intact cells (Owen and Kaback, 1979). They concluded that dislocation of components from the inner to the outer surface did not exceed 10% during membrane vesicle preparation. The highly asymmetric nature of the vesicle membrane was similar to that established for *M. lysodeikticus* protoplast membranes, in that many of the identifiable enzymes (e.g., NADH dehydrogenase, D-lactate dehydrogenase, 6-phosphogluconate dehydrogenase, ATPase) were minimally exposed for reaction with antibodies unless the vesicles were disrupted (Owen and Kaback, 1978, 1979).

Similar approaches have been used by Perille-Collins *et al.* (1980) and Elferink *et al.* (1979) in assessing the asymmetry of *R. sphaeroides* chromatophores and membrane vesicles. Perille-Collines *et al.* (1980) found that the photochemical reaction center, the light-harvesting bacteriochlorophyll *a*–protein complex, L-lactate dehydrogenase, and NADH dehydrogenase were all exposed on the outer surface of the isolated chromatophores, thus demonstrating their opposite (inside-out) orientation to the plasma membrane (as determined with osmotically protected spheroplasts). These results are in general agreement with those of Elferink *et al.* (1979) who estimated that about 25 and 95% of the ATPase, succinate dehydrogenase, and NADH dehydrogenase was accessible for reaction with antibodies in membrane vesicle and chromatophore preparations, respectively. These enzymes were thus almost completely exposed on the outer face of the chromatophore fractions.

The highly asymmetric character of bacterial plasma membranes, sealed membrane vesicles prepared by the method of Kaback (1971) for use in transport studies, and chromatophores has been established by antibody-absorption CIE techniques. Thus, membrane-bound dehydrogenases and the energy-transducing F_1-ATPases are found on the cytoplasmic face of the plasma membranes and the intracellular chromatophores as they exist in the intact cell. The immunological approaches used in these studies of bacterial membrane structure and function are sufficiently encouraging in their sensitivity and specificity to suggest that they could be exploited in probing the molecular architecture of membrane transport systems.

REFERENCES

Adler, L. W., and Rosen, B. P. (1977). *J. Bacteriol.* **129,** 959–966.

Carlile, M. J. (1980). In *The Eukaryotic Microbial Cell* (G. W. Gooday, D. Lloyd, and A. P. J. Trinci, eds.), pp. 1–40, Cambridge University Press, London.

Elferink, M. G. L., Hellingwerf, K. J., Michels, P. A. M., Seyen, H. G., and Konings, W. N. (1979). *FEBS Lett.* **107,** 300–307.

Futai, M. (1974). *J. Membr. Biol.* **15,** 15–28.

Ghosh, B. K. (1974). *Sub-Cell. Biochem.* **3,** 311–367.

Hare, J. F., Olden, K., and Kennedy, E. P. (1974). *Proc. Natl. Acad. Sci. USA* **71,** 4843–4846.

Higgins, M. L., Parks L. C., and Daneo-Moore, L. (1981). In *Organization of Prokaryotic Cell Membranes* (B. K. Ghosh, ed.), Vol. 2, pp. 78–84, CRC Press, Boca Raton, Fla.

Inouye, M., (ed.) (1979). In *Bacterial Outer Membranes*, pp. 1–12, Wiley, New York.

Johansson, K.-E., and Hjértén, S. (1974). *J. Mol. Biol.* **86,** 341–348.

Kaback, H. R. (1971). *Methods Enzymol.* **22,** 99–120.

Kaback, H. R. (1972). *Biochim. Biophys. Acta* **265,** 367–416.

Linder, R. and Salton, M. R. J. (1975). *Eur. J. Biochem.* **55,** 291–297.

Mitchell, P. (1968). *Chemiosmotic Coupling and Energy Transduction*, Glynn Research, Bodmin.

Oppenheim, J. D., and Salton, M. R. J. (1973). *Biochim. Biophys. Acta* **298,** 297–322.

Owen, P., and Kaback, H. R. (1978). *Proc. Natl. Acad. Sci. USA* **75,** 3148–3152.

Owen, P., and Kaback, H. R. (1979). *Biochemistry* **18,** 1422–1426.

Owen, P., and Salton, M. R. J. (1975a). *Proc. Natl. Acad. Sci. USA* **72,** 3711–3715.

Owen, P., and Salton, M. R. J. (1975b). *Biochim. Biophys. Acta* **406,** 214–234.

Owen, P., and Salton, M. R. J. (1976). *Anal. Biochem.* **73,** 20–26.

Owen, P., and Salton, M. R. J. (1977). *J. Bacteriol.* **132,** 974–985.

Owen, P., and Smyth, C. J. (1977). In *Immunochemistry of Enzymes and Antibodies* (M. R. J. Salton, ed.), pp. 147–202, Wiley, New York.

Owen, P., Kaczorowski, G., and Kaback, H. R. (1980). *Biochemistry* **19,** 596–600.

Perille-Collins, M. L., and Salton, M. R. J. (1979). *Biochim. Biophys. Acta* **553,** 40–53.

Perille-Collins, M. L., and Salton, M. R. J. (1980). *Infect. Immun.* **30,** 281–288.

Perille-Collins, M. L., Mallon, D. E., and Niederman, R. A. (1979). *J. Bacteriol.* **139,** 1089–1092.

Perille-Collins, M. L., Mallon, D. E., and Niederman, R. A. (1980). *J. Bacteriol.* **143,** 221–230.

Rutberg, B., Hederstedt, L., Holmgren, E., and Rutberg, L. (1978). *J. Bacteriol.* **136,** 304–311.

Salton, M. R. J., and Owen, P. (1976). *Annu. Rev. Microbiol.* **30,** 451–482.

Singer, S. J. (1975). In *Cell Membranes, Biochemistry, Cell Biology and Pathology* (G. Weissmann and R. Claiborne, eds.), pp. 35–44, HP Publishing, New York.

Smyth, C. J., Siegel, J., Salton, M. R. J., and Owen, P. (1978). *J. Bacteriol.* **133,** 306–319.

Stanier, R. Y. (1970). In *Organization and Control in Prokaryotic and Eukaryotic Cells* (H. P. Charles and B. C. J. G. Knight, eds.), pp. 1–38, Cambridge University Press, London.

Uriel, J. (1971). *Methods Immunol. Immunochem.* **3,** 294–321.

35

The Cell Membrane of Mycoplasmas

Shmuel Razin

Most mycoplasmologists support the proposal of Gibbons and Murray (1978) endowing the mycoplasmas with the lofty status of one of the four major divisions in the kindgom Procaryotae, the division named Mollicutes (*mollis* soft + *cutes* skin) to denote the lack of cell walls in these organisms. The recent report that penicillin-binding proteins and enzymes of peptidoglycan synthesis are absent from mycoplasmas as against their presence in the plasma membrane of the wall-less bacterial L-forms (Martin *et al.*, 1980) supports the idea of a separate division for mycoplasmas. A different view has recently been expressed by Fox *et al.* (1980) who, on the basis of nucleotide sequences in 16 S rRNA, argue that mycoplasmas are genealogically wall-less descendents of clostridia, and consequently should not be given a separate high taxonomic status. Notwithstanding this somewhat philosophical controversy, the fact is that mycoplasmas differ from other prokaryotes in several unique properties, most useful in membrane studies. Thus, the mycoplasmas are unique in being the only self-replicating organisms with a single membranous structure— the plasma membrane. This is, perhaps, their greatest advantage in membrane studies, for it facilitates the isolation of pure plasma membranes uncontaminated by other membrane types. Moreover, the mycoplasmas' lack of cell walls enables the application of gentle and simple techniques, such as osmotic lysis for membrane isolation (Razin, 1978, 1981).

In addition, mycoplasmas have the smallest genome size among self-replicating organisms. The limited genetic information dictates limited biosynthetic abilities. Thus, the mycoplasmas are partially or totally incapable of fatty acid synthesis, and depend on the growth medium for their supply. This dependence has been exploited most effectively to introduce controlled alterations in the fatty acid composition of mycoplasma membranes (reviewed in Razin, 1978, 1981). The flexibility of membrane fatty acid composition is so great that it is possible to obtain membranes with essentially one fatty acid ("fatty acid-homogeneous" membranes). This is done by growing the fatty acid auxotroph *Mycoplasma mycoides* subsp. *mycoides* with elaidate as the only fatty acid in a semidefined medium (Rodwell, 1968), or by inhibiting fatty acid synthesis in *Acholeplasma laidlawii* by avidin

Shmuel Razin • Department of Membrane and Ultrastructure Research, The Hebrew University-Hadassah Medical School, Jerusalem, Israel.

(Silvius *et al.*, 1980). The fatty acid-homogeneous membranes exhibit sharp gel-to-liquid-crystal phase transitions, indicating that the broadness of the lipid phase transition peak in "normal" membranes is mainly due to fatty acid heterogeneity rather than to diversity of the lipid head groups (Silvius *et al.*, 1980).

It is conceivable that for the membrane to function, the lipid bilayer must be in the liquid-crystal state, at least partially. According to McElhaney (1974), up to about one-half of the membrane lipid in *A. laidlawii* may be transformed into the gel state without apparent effects on cell growth, and the existence of less than one-tenth of the membrane lipid in a fluid state is sufficient to support some cell growth, albeit at greatly reduced rates. The broad range of lipid phase transition midpoint temperatures observable in various fatty acid-homogeneous membranes strongly suggests that membrane lipid fluidity can vary appreciably without impairing essential membrane functions. Silvius *et al.* (1980) conclude that membrane fluidity need not be tightly regulated in order to preserve proper membrane function. Accordingly, the major function of "homeoviscous adaptation" in prokaryotes is to adjust the lipid phase transition temperature range for optimal membrane functioning at the growth temperature rather than to maintain an absolutely constant membrane fluidity.

The mycoplasmas appear to employ two major mechanisms for regulating membrane fluidity: one is based on alterations in membrane polar lipid composition, and the other on sterol incorporation. The extensive work of McElhaney's group published in 1977–1978 (reviewed by Razin, 1978) revealed that the sterol-nonrequiring *Acholeplasma* species adjust their fatty acid composition at various levels of *de novo* synthesis of saturated fatty acids, by incorporation and elongation of exogenous fatty acids and their utilization for complex membrane lipid synthesis. However, it appears that changes in the head groups of polar membrane lipids also take place during adjustment of membrane fluidity. The lipid species synthesized by *A. laidlawii* include the typical prokaryotic acidic phospholipids, phosphatidylglycerol and diphosphatidylglycerol (about 30% of membrane lipids), the glycolipids monoglucosyldiglyceride (MGDG) and diglucosyldiglyceride (DGDG) and phosphoglucolipids. Wieslander *et al.* (1980) showed that the MGDG-to-DGDG ratio in *A. laidlawii* membranes significantly increases in response to conditions leading to higher membrane viscosity, such as low temperature, excess of saturated over unsaturated fatty acids in the medium, and the presence of cholesterol. MGDG and DGDG have wedgelike and rodlike molecular shapes, respectively. Consequently, MGDG forms a reversed hexagonal-phase structure, while DGDG forms a lamellar phase. Because the lamellar phase is the only lipid structure compatible with a functional biological membrane, the balance between lipids forming lamellar and other mesophase structures must be kept within certain limits. As the membrane becomes more fluid, the balance should favor the lamellar DGDG to keep the membrane stable.

Another function of glycolipids in membranes has been recently proposed by Smith (1980). Accordingly, the neutral glycolipids in *Acholeplasma* membranes act as "spacer" molecules, separating the acidic phospholipid molecules from each other and in this way decrease the charge density of the membrane, and stabilize its structure. In this respect glycolipids act synergistically with divalent cations. The zwitterionic phospholipids, e.g., phosphatidylethanolamine and phosphatidylcholine, act as spacer molecules in membranes of gram-negative bacteria and in eukaryotic cells.

Membrane fluidity in most mycoplasmas—those requiring sterols for growth—is apparently regulated by sterols. For this function, the sterol should have a planar ring system, a free 3β-OH group, and an aliphatic side chain. Cholesterol fulfills all three structural requirements and is usually superior to other sterols in growth promotion of mycoplasmas (Razin, 1978). The recent finding by Bloch's group (Odriozola *et al.*, 1978; Dahl *et al.*,

1980) that *M. capricolum* can grow with sterols that do not fulfill all three structural requirements indicates that sterols may have additional functions to that of regulating bulk lipid fluidity. Thus, the finding that lanosterol, a sterol incompetent of modulating membrane fluidity, supports suboptimal growth of *M. capricolum* has been taken to suggest that sterols may also act as spacer molecules, like glycolipids in *Acholeplasma* membranes (Dahl *et al.*, 1980). Moreover, Dahl *et al.* (1981) have recently shown that small amounts of cholesterol considerably enhance uptake of unsaturated fatty acids in lanosterol-grown *M. capricolum*. This finding is of great interest as it suggests that cholesterol may interact specifically with an enzyme(s) involved in fatty acid activation, transfer, or incorporation into membrane phospholipids.

The sterol-requiring mycoplasmas were also shown to differ from *Acholeplasma* in the positional distribution of fatty acids in membrane lipids (Rottem and Markowitz, 1979). In lipids of *Mycoplasma* species the unsaturated fatty acids are preferentially bound to position 1 of the glycerol and the saturated acids to position 2, the reverse order of that found in membrane lipids in nature. This is another of the peculiar properties of mycoplasmas, but whether it has anything to do with sterol requirement has to be clarified.

Mycoplasmas serve as excellent tools for studying the mechanism of cholesterol transfer from serum lipoproteins to membranes, for the plasma membrane of the mycoplasmas interacts directly with the cholesterol donor, and the cholesterol taken up is not esterified or modified. Furthermore, results of uptake experiments are not blurred by endogenous sterol synthesis as in eukaryotic systems. Among serum lipoproteins, those having the highest cholesterol-to-phospholipid ratio, such as human low-density lipoproteins, are the best cholesterol donors (Slutzky *et al.*, 1977). Transfer of free cholesterol to *Acholeplasma* membranes occurs during a transient contact of the lipoprotein particle with the membrane. In sterol-requiring mycoplasmas, the contact appears to be tighter, or more prolonged, as it suffices for transfer of some of the lipoprotein phospholipid and esterified cholesterol to the membrane (Razin *et al.*, 1980). It is proposed that the sterol-requiring mycoplasmas possess receptors on their surface, probably of a protein nature, responsible for better contact of the lipoprotein particle with the membrane (Efrati *et al.*, 1981).

The uptake of significant amounts of cholesteryl esters by mycoplasmas (Razin *et al.*, 1980) poses an enigma, as cholesteryl esters do not commonly form a significant component of biological membranes. The evidence available speaks against the possibility of the esters being part of lipoprotein particles adsorbed to the cell surface. Recent calorimetric data suggest that the esters form large droplets or pockets within the mycoplasma membrane (Melchior and Rottem, 1981).

The lack of a cell wall in mycoplasmas facilitates studies on the disposition of membrane components, as these studies depend primarily on the use of macromolecular labeling agents, such as enzymes and antibodies, which may be unable to penetrate the wall barrier. In mycoplasma membranes, as in other biological membranes, the transbilayer distribution of phospholipids and glycolipids is asymmetrical, as determined by phospholipase A_2 treatment (Bevers *et al.*, 1977) or by lactoperoxidase-mediated iodination (Gross and Rottem, 1979). The fact that all of the mycoplasma cholesterol is of exogenous origin has been utilized to show that cholesterol taken up by growing cells flip-flops rapidly from the outer to the inner half of the lipid bilayer. The final transbilayer distribution of the sterol depends on growth conditions and on the structure of the sterol molecule (Clejan *et al.*, 1981).

The first strides toward the molecular characterization of mycoplasma membrane proteins have already been made following development of methods for their solubilization and fractionation (reviewed by Razin, 1978, 1981). Amino acid analysis of only a few of these proteins revealed a preponderance of hydrophobic acids, while in other proteins the amino acid composition resembled that of hydrophilic ones (Johansson *et al.*, 1979). The finding of a glycoprotein in membranes of *M. pneumoniae* (Kahane and Brunner, 1977) is of particular interest, considering the rarity of glycoproteins in prokaryotes.

A very strong case can be built in support of the existence of a contractile protein system associated with the mycoplasma membrane, as a variety of contractile processes, including chemotaxis, are observable in mycoplasmas (Daniels *et al.*, 1980; Razin, 1981). Several reports suggesting the presence of actinlike proteins in mycoplasmas are available (Neimark, 1977; Williamson *et al.*, 1979), and a fibrillar cytoskeleton has recently been observed in Triton-treated *M. pneumoniae* (Meng and Pfister, 1980). Fibrils made of a protein resembling tubulin in molecular weight, but not in molecular properties, were isolated from *S. citri* (Townsend *et al.*, 1980). Clearly, further studies in this direction will be rewarding in view of the scarcity of data on contractile elements in prokaryotes.

Another interesting development concerns the effects of membrane potential on exposure of proteins on the mycoplasma cell surface. The degree of exposure of proteins on the *A. laidlawii* cell surface decreased by short-circuiting the membrane potential and proton gradient by valinomycin and carbonylcyanide *m*-chlorophenyl hydrazone, respectively (Amar *et al.*, 1978), and increased on energization of *M. capricolum* cells (Le Grimellec *et al.*, 1981). These findings may be relevant to studies of interactions among cells that depend on the availability of protein receptors on their surface. The first clue leading in this direction is the finding that *M. pneumoniae* attachment to inert surfaces depends on the energized state of the cells (Feldner *et al.*, 1981). The recent successful application of probes for measuring membrane potential and ΔpH in mycoplasmas (Schummer *et al.*, 1980; Leblanc and Le Grimellec, 1979; Rottem *et al.*, 1981) assures that well-controlled experiments can now be carried out on the effect of the energized state of the membrane on the disposition of its components.

The Mg^{2+}-ATPase of mycoplasmas differs from similar ATPases of other prokaryotes in being tightly associated with the membrane and in depending on membrane lipids for activity (Silvius and McElhaney, 1980). The tight association of the enzyme with membrane lipids and its high sensitivity to detergents have thwarted all efforts to isolate and characterize it. The crucial question is, obviously, whether the mycoplasmal enzyme can translocate protons. The answer appears to be positive, as the ATPase activity was shown to be instrumental in sugar transport in *A. laidlawii* and in K^+ transport in *M. mycoides* subsp. *capri*, processes driven by the protonmotive force (Tarshis and Kapitanov, 1978; Leblanc and Le Grimellec, 1979), and in ATP production resulting from a ΔpH created in *Ureaplasma urealyticum* cells during urea hydrolysis (Romano *et al.*, 1980). Moreover, inhibition by dicyclohexylcarbodiimide of mycoplasma motility (Daniels *et al.*, 1980) supports the proton-translocating nature of the mycoplasmal ATPase.

Fermentative *Mycoplasma* and *Spiroplasma* species utilize the efficient phosphoenolpyruvate-dependent sugar phosphotransferase system for sugar transport (Cirillo, 1979). Although one would expect the preparation of membrane vesicles active in transport to be easy with the wall-less mycoplasmas, results have usually been disappointing (Razin, 1981). The reasons for the difficulties in preparing sealed mycoplasma membrane vesicles are yet to be clarified.

The role of membrane components in mycoplasma pathogenicity has become one of the important issues in mycoplasma research. Animal mycoplasmas are surface parasites firmly adhering to and colonizing the epithelial lining of the respiratory and genital tracts of infected animals (Razin, 1978, 1981). In this case no wall separates between the plasma membrane of the parasite and that of its host. Although evidence supporting fusion of the two membranes is disputable, recent studies by Wise *et al.* (1978) indicate transfer of antigens between the membranes, an event that may trigger immunological responses of serious consequences to the host. The intimate association between mycoplasmas and host cell membranes is also reflected by the ''capping'' of mycoplasmas adhering to lymphocytes, followed by shedding of membrane vesicles of presumed host origin (Stanbridge and Weiss, 1978). This phenomenon is apparently related to the well-known induction of blast transformation by mycoplasmas.

Sialic acid moieties on the host cell surface serve as specific receptors for the respiratory pathogens *M. pneumoniae*, *M. gallisepticum*, and *M. synoviae*. There are sound indications that mycoplasma membrane proteins interact specifically with the sialic acid receptors (Hu *et al.*, 1977; Razin, 1981). The first strides have already been made to isolate and characterize these proteins (Banai *et al.*, 1980) with the final aim of developing a highly specific vaccine that will prevent disease by interfering with the initial step of attachment of the parasite to its host.

REFERENCES

Amar, A., Rottem, S., and Razin, S. (1978). *Biochem. Biophys. Res. Commun.* **84,** 306–312.
Banai, M., Razin, S., Bredt, W., and Kahane, I. (1980). *Infect. Immun.* **30,** 628–634.
Bevers, E. M., Singal, S. A., Op den Kamp, J. A. F., and van Deenen, L. L. M. (1977). *Biochemistry* **16,** 1290–1295.
Cirillo, V. P. (1979). In *The Mycoplasmas* (M. F. Barile and S. Razin, eds.), Vol. I, pp. 323–349, Academic Press, New York.
Clejan, S., Bittman, R., and Rottem, S. (1981). *Biochemistry,* **20,** 2200–2206.
Dahl, J. S., Dahl, C. E., and Bloch, K. (1980). *Biochemistry* **19,** 1467–1471.
Dahl, J. S., Dahl, C. E., and Bloch, K. (1981). *J. Biol. Chem.* **256.** 87–91.
Daniels, M. J., Longland, J. M., and Gilbert, J. (1980). *J. Gen. Microbiol.* **118,** 429–436.
Efrati, H., Rottem, S., and Razin, S. (1981). *Biochim. Biophys. Acta,* **641,** 386–394.
Feldner, J., Bredt, W., and Razin, S. (1981). *Infect. Immun.,* **31,** 107–113.
Fox, G. E., Stackebrandt, E., Hespell, R. B., Gibson, J., Maniloff, J., Dyer, T. A., Wolfe, R. S., Balch, W. E., Tanner, R. S., Magrum, L. J., Zablen, L. B., Blakemore, R., Gupta, R., Bonen, L., Lewis, B. J., Stahl, D. A., Luehersen, K. R., Chen, K. N., and Woese, C. R. (1980). *Science* **209,** 457–463.
Gibbons, N. E., and Murray, R. G. E. (1978). *Int. J. Syst. Bacteriol.* **28,** 1–6.
Gross, Z., and Rottem, S. (1979). *Biochim. Biophys. Acta* **555,** 547–552.
Hu, P. C., Collier, A. M., and Baseman, J. B. (1977). *J. Exp. Med.* **145,** 1328–1343.
Johansson, K.-E., Pertoft, H., and Hjerten, S. (1979). *Int. J. Biol. Macromol.* **1,** 111–118.
Kahane, I., and Brunner, H. (1977). *Infect. Immun.* **18,** 273–277.
Leblanc, G., and Le Grimellec, C. (1979). *Biochim. Biophys. Acta* **554,** 168–179.
Le Grimellec, C., Lajeunesse, D., and Rigaud, J. L. (1982). *Rev. Infect. Dis.,* in press.
McElhaney, R. N. (1974). *J. Mol. Biol.* **84,** 145–157.
Martin, H. H., Schilf, W., and Schiefer, H.-G. (1980). *Arch. Microbiol.* **127,** 297–299.
Melchior, D. L., and Rottem, S. (1981). *Eur. J. Biochem.* **117,** 147–153.
Meng, K. E., and Pfister, R. M. (1980). *J. Bacteriol.* **144,** 390–399.
Neimark, H. C. (1977). *Proc. Natl. Acad. Sci. USA* **74,** 4041–4045.
Odriozola, J. M., Waitzkin, E., Smith, T. L., and Bloch, K. (1978). *Proc. Natl. Acad. Sci. USA* **75,** 4107–4109.
Razin, S. (1978). *Microbiol. Rev.* **42,** 414–470.
Razin, S. (1981). In *Organization of Prokaryotic Cell Membranes* (B. K. Ghosh, ed.), Vol. I, pp. 180–273, CRC Press, Boca Raton, Fla.
Razin, S., Kutner, S., Efrati, H., and Rottem, S. (1980). *Biochim. Biophys. Acta* **598,** 628–640.
Rodwell, A. W. (1968). *Science* **160,** 1350–1351.
Romano, N., Tolone, G., Ajello, F., and Licata, R. L. (1980). *J. Bacteriol.* **144,** 830–832.
Rottem, S., and Markowitz, O. (1979). *FEBS Lett.* **107,** 379–382.
Rottem, S., Linker, C., and Wilson, T. H. (1981). *J. Bacteriol.,* **145,** 1299–1304.
Schummer, U., Schiefer, H.-G., and Gerhardt, U. (1980). *Biochim. Biophys. Acta* **600,** 998–1006.

Silvius, J. R., and McElhaney, R. N. (1980). *Proc. Natl. Acad. Sci. USA* **77**, 1255–1259.
Silvius, J. R., Mak, N., and McElhaney, R. N. (1980). *Biochim. Biophys. Acta* **597**, 199–215.
Slutzky, G. M., Razin, S., Kahane, I., and Eisenberg, S. (1977). *Biochemistry* **16**, 5158–5163.
Smith, M. W. (1980). Ph.D. thesis, Brown University, Providence.
Stanbridge, E. J., and Weiss, R. L. (1978). *Nature (London)* **276**, 583–587.
Tarshis, M. A., and Kapitanov, A. B. (1978). *FEBS Lett.* **89**, 73–77.
Townsend, R., Archer, D. B., and Plaskitt, K. A. (1980). *J. Bacteriol.* **142**, 694–700.
Wieslander, A., Christiansson, A., Rilfors, L., and Lindblom, G. (1980). *Biochemistry* **19**, 3650–3655.
Williamson, D. L., Blaustein, D. I., Levine, R. J. C., and Elfvin, M. J. (1979). *Curr. Microbiol.* **2**, 143–145.
Wise, K. S., Cassell, G. H., and Acton, R. T. (1978). *Proc. Natl. Acad. Sci. USA* **75**, 4479–4483.

36

Lipoproteins from the Bacterial Outer Membranes

Their Gene Structures and Assembly Mechanism

Masayori Inouye

1. INTRODUCTION

The envelope of gram-negative bacteria consists of two distinct membranes, the cytoplasmic membrane and the outer membrane. In contrast to the cytoplasmic membrane, the outer membrane contains a small number of major proteins, accounting for more than 90% of the total outer-membrane protein (see reviews by Inouye, 1979a; DiRienzo *et al.*, 1978). As these major outer-membrane proteins (lipoprotein, OmpA protein, and matrix proteins) are the most abundant of all cellular proteins, they are readily purified and characterized.

The lipoprotein is the most thoroughly investigated of the major outer-membrane proteins in *Escherichia coli* from its protein structure to its gene, and provides an excellent system to study membrane biogenesis. In this review, I will mainly focus on the recent progress in the lipoprotein research including the structure of the lipoprotein gene (*lpp*) and the *lpp* mRNA, and the secretory mechanism of the lipoprotein across the cytoplasmic membrane. The reader is referred to a review for other aspects of the lipoprotein (Inouye, 1979b).

2. PROTEIN STRUCTURE

The lipoprotein, which consists of 58 amino acid residues, lacks histidine, tryptophan, glutamic acid, glycine, proline, and phenylalanine as shown in Fig. 1 (Braun and Bosch,

Masayori Inouye • Department of Biochemistry, State University of New York, Stony Brook, New York 11794.

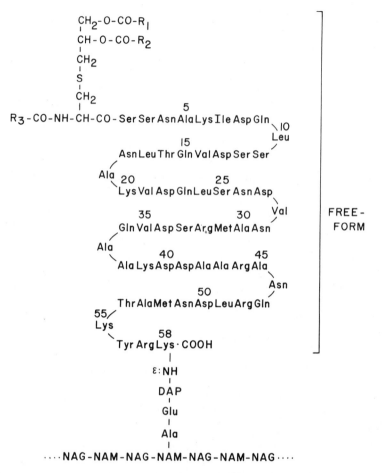

Figure 1. The complete chemical structure of the bound form of the *E. coli* lipoprotein (Braun and Bosch, 1972; Hantke and Braun, 1973; Nakamura *et al.*, 1980). DAP, diaminopimelic acid; NAM, *N*-acetylmuramic acid; NAG, *N*-acetylglucosamine. R_1, R_2, and R_3 represent hydrocarbon chains of fatty acids.

1972; Hantke and Braun, 1973; Nakamura *et al.*, 1980). The NH_2-terminal structure of the lipoprotein is unique, consisting of a glycerylcysteine to which two fatty acids are linked by two ester linkages, and one fatty acid by an amide linkage. There are estimates of 7.2×10^5 lipoprotein molecules per cell, which makes the lipoprotein the most abundant protein in terms of numbers of molecules (see review by Inouye, 1979b). One-third of the lipoprotein molecules are linked by the ϵ-amino group of their COOH-terminal lysine to the carboxyl group of every 10th to 12th *meso*-diaminopimelic acid residue of the peptidoglycan (Braun and Bosch, 1972; Inouye *et al.*, 1972). Although the exact function of the lipoprotein is still obscure, it is considered to play an important role in maintaining the integrity of the outer-membrane structure (see review by Inouye, 1979b).

3. GENE STRUCTURE

A point mutation in the *lpp* gene of *E. coli* causes an alteration in the structure of all lipoprotein molecules in the cell, indicating that there is only one *lpp* gene in the *E. coli* chromosome (Inouye *et al.*, 1977b). How, then, is the single *lpp* gene in the cell controlled to produce the lipoprotein as the most abundant protein in the cell? This question led us to study the structure of the *lpp* mRNA and the *lpp* gene. Because of the unusual stability of the *lpp* mRNA, it has been purified (Wang *et al.*, 1979) and its entire base sequence has been determined (Pirtle *et al.*, 1980; Nakamura *et al.*, 1980).

Using the purified *lpp* mRNA as a probe, the *lpp* gene of *E. coli* has been cloned and its entire DNA sequence has been determined (Nakamura and Inouye, 1979). Recently, the *lpp* genes from *Serratia marcescens* (Nakamura and Inouye, 1980) and *Erwinia amylovora* (Yamagata *et al.*, 1981), which among the Enterobacteriaceae are highly divergent from *E. coli*, have been cloned and their DNA sequences have also been determined. These genes are expressed in *E. coli*, and their lipoproteins are normally assembled in the *E. coli* outer membrane.

Figure 2 shows the DNA sequences of three DNA fragments encompassing the *lpp* genes from *E. coli*, *Er. amylovora*, and *S. marcescens*. The transcription initiation and termination sites are deduced from the complete nucleotide sequence of the *E. coli lpp* mRNA (Nakamura *et al.*, 1980). Surprisingly, in spite of the fact that *E. coli* is highly divergent from the other two bacteria, the DNA sequences of the *lpp* genes are strongly conserved: (1) The promoter region (base No. −45 to −1) is highly conserved (87% homology between *Er. amylovora* and *E. coli*, and 93% between *Er. amylovora* and *S. marcescens*). (2) The structural gene (base No. 1 to 319) is also highly conserved (89% between *Er. amylovora* and *E. coli*, and 88% between *Er. amylovora* and *S. marcescens*). (3) The sequences of the 5′ nontranslated regions of the mRNAs show especially close homology (97% between *Er. amylovora* and *E. coli*, and 92% between *Er. amylovora* and *S. marcescens*).

It should be noted that most of mismatchings in the translated regions occur at the third bases of codons so that they do not cause alterations in the amino acid sequences of the lipoproteins. On the other hand, extensive mismatchings are observed immediately after the transcription termination sites as well as in the regions upstream of the promoters (−45 to −260). There seems to be no consistent pattern in these mismatching regions.

The DNA sequences of the *lpp* promoters (−1 to −45) contain "-35 regions" as well as "Pribnow boxes" as found in all promoters of *E. coli* so far determined (Siebenlist *et al.*, 1980). However, the most striking feature of the *lpp* promoters is their extremely high AT contents: 80, 80, and 78% for *E. coli*, *Er. amylovora*, and *S. marcescens*, respectively. It is also interesting to note that the spacer regions between the *lpp* promoter and the unknown gene upstream to the *lpp* gene (−45 to −260) have rather high AT contents: 68% for *E. coli*, compared with a genome average of 49%; 56% for *Er. amylovora*, compared with a genome average of 50%; and 61% for *S. marcescens*, compared with a genome average of 42%. The AT richness of the promoter sequence is believed to destabilize the helix structure and facilitate RNA polymerase-mediated strand unwinding necessary for the initiation of transcription (Chamberlin, 1976). The extremely high content in the *lpp* promoter region is, therefore, likely to be essential for the efficient transcription of the *lpp* gene.

4. mRNA STRUCTURE

The *lpp* mRNA has been purified from *E. coli* (Wang *et al.*, 1979), and its complete nucleotide sequence has been determined (Nakamura *et al.*, 1980). This is the only mRNA ever purified and sequenced in prokaryotic systems. Its possible secondary structure is shown in Fig. 3. It consists of 322 nucleotides, and there are respectively 38 and 50

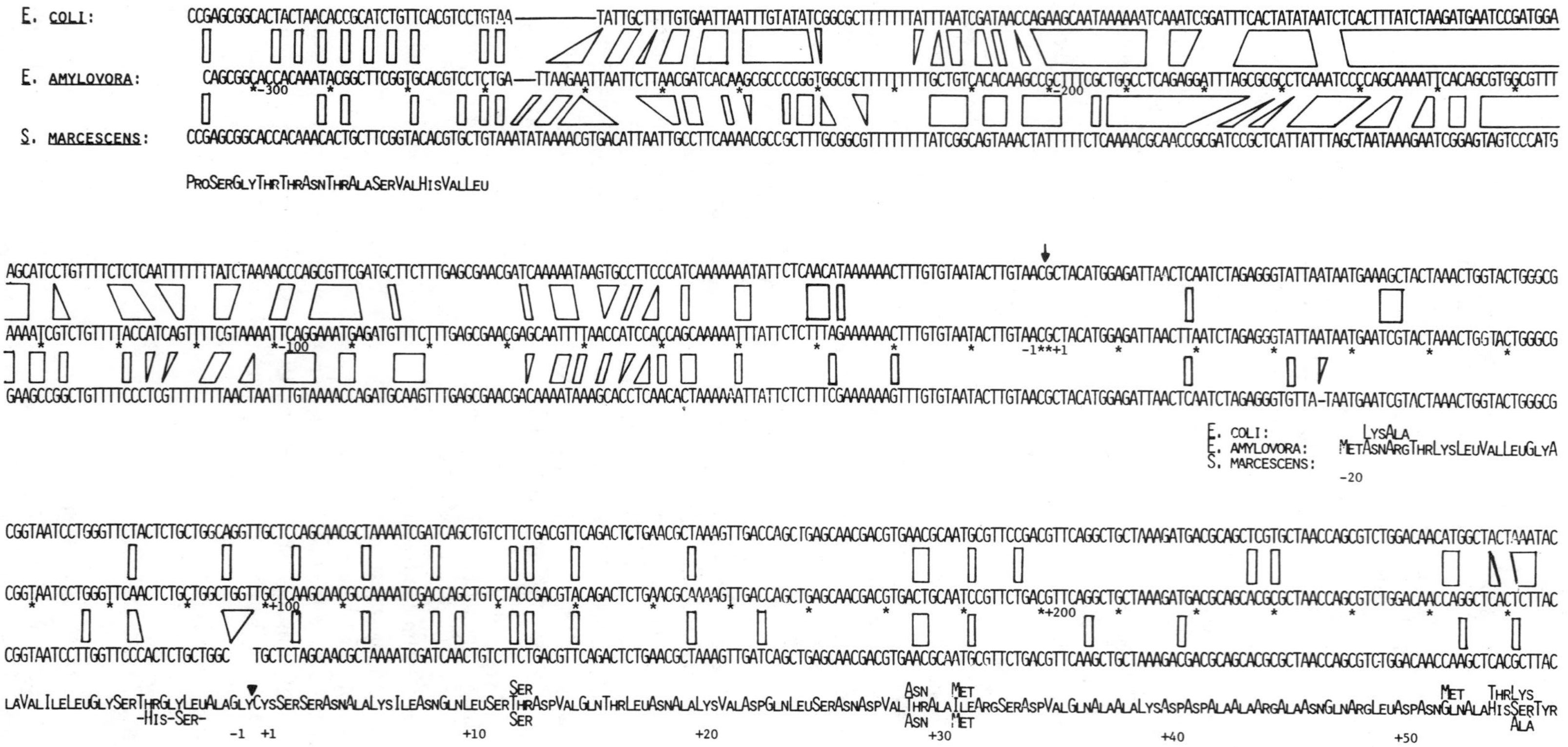

E. COLI:

E. AMYLOVORA:

S. MARCESCENS:

ProSerGlyThrThrAlaSerValHisValLeu

LysAla
MetAsnArgThrLysLeuValLeuGlyA
-20

E. COLI:
E. AMYLOVORA:
S. MARCESCENS:

-1***1

-1**+1

-300

-200

-100

+1

+100

+200

+30

+40

+50

+10

+20

-1   +1

LeuValIleLeuGlyLysThrGlyLeuValGlyCysSerSerAsnValAlaLysIleLeuAsnGlnLeuSer
-HisSer-

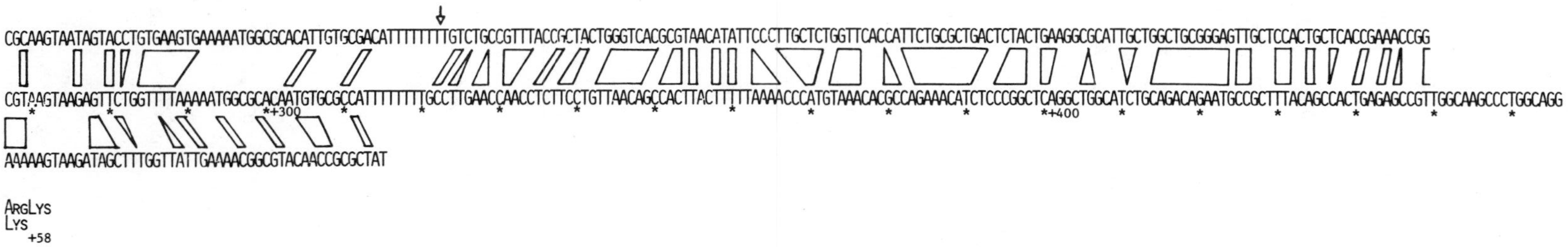
CGCAAGTAATAGTACCTGTGAAGTGAAAAATGGCGCACATTGTGCGACATTTTTTTTGTCTGCCGTTTACCGCTACTGGGTCACGCGTAACATATTCCCTTGCTCTGGTTCACCATTCTGCGCTGACTCTACTGAAGGCGCATTGCTGGCTGCGGGAGTTGCTCCACTGCTCACCGAAACCGG
CGTAAGTAAGAGTTCTGGTTTTAAAAATGGCGCACAATGTGCGCCATTTTTTTTGCCTTGAACCAACCTCTTCCTGTTAACAGCCACTTACTTTTTAAAACCCATGTAAACACGCCAGAAACATCTCCCGGCTCAGGCTGGCATCTGCAGACAGAATGCCGCTTTACAGCCACTGAGAGCCGTTGGCAAGCCCTGGCAGG
+300
+400
AAAAAGTAAGATAGCTTTGGTTATTGAAAACGGCGTACAACCGCGCTAT
ArgLys
Lys
+58

Figure 2. DNA sequences encompassing the *E. coli*, *Er. amylovora*, and *S. marcescens lpp* genes (Yamagata *et al*., 1981). Only sense strands are shown. The first nucleotide of the *lpp* mRNA deduced from *E. coli* mRNA is numbered as +1. *E. coli* mRNA start (↓) and stop (⇣) sites are indicated. Nonhomologous regions are connected by empty bars. Amino acid sequences deduced from the DNA sequences are shown. Cleavage site of the prolipoprotein is indicated by arrowhead (▼). The Cys residue of the prolipoprotein at the cleavage site is numbered as +1. Amino acid sequences of *E. coli* and *S. marcescens* are shown only where they differ from those of *Er. amylovora*. Straight lines in nucleotide sequences are only for the purpose of aligning amino acid sequences.

nucleotides in the 5' and 3' nontranslated regions. It has several unique features: (1) The amounts of AU in the 5' and 3' nontranslated regions are 66 and 62%, respectively, in contrast to 49% in the coding region. (2) Exactly the same sequence of 12 nucleotides, $\overset{30}{\text{GUAUUAAUAAUG}}\overset{41}{}$, is found as for the 80 S ribosome binding site in brome mosaic virus RNA 4, a eukaryotic mRNA. In both cases, AUG, the last three nucleotides, corresponds to the initiation codon. (3) There are two possible 70 S ribosome binding sites, $\overset{8}{\text{GGAG}}\overset{11}{}$ and $\overset{25}{\text{AGAGGGU}}\overset{31}{}$. (4) The first nucleotide after the initiation codon AUG is A, which has been shown to cause the most efficient binding of tRNA$^{\text{f-Met}}$ to Qβ-RNA (Taniguchi and Weissmann, 1978). (5) All three termination codons (UAA, UAG, and UGA) are used in place. (6) The codon usage in the *lpp* mRNA is unusual. In spite of the several possible codons for each amino acid due to the ambiguity at the third base of the genetic code, only a few codons are used. Out of 50 possible codons to code for the prolipoprotein (a precursor of the lipoprotein), only 25 codons are used (Nakamura *et al.*, 1980). For example, all of the nine leucine residues in the prolipoprotein are coded by only one codon, CUG, out of six possible degenerate leucine codons. All the codons used in the *lpp* mRNA appear to be translated by major isoaccepting species of tRNAs for their respective amino acids. This fact is considered to be important for efficient translation of the mRNA. (7) As shown in Fig. 3, the *lpp* mRNA can form nine stable stem-and-loop structures. Similar secondary structures can also be formed for the *S. marcescens* (Nakamura and Inouye, 1980) and *Er. amylovora* (Yamagata *et al.*, 1981) *lpp* mRNAs. These structures may be important for the unusual stability of the *lpp* mRNA. The hairpin structure I is the most stable structure ($\Delta G = 21.1$ kcal). Very similar structures II and I at the 3' end of the *lpp* mRNA are also found at the 3' end of the gene for the OmpA protein, another major outer-membrane protein (Movva *et al.*, 1980). (8) In spite of a large number of secondary structures in the *lpp* mRNA, there is no stable hairpin structure in the first 64 residues from the 5' end so that the ribosome binding site in the 5'-end nontranslated region may be fully exposed to ribosomes. (9) A nucleotide sequence of 40 residues from $\overset{131}{\text{U}}$ to $\overset{170}{\text{G}}$ seems to be duplicated twice in the *lpp* mRNA: the sequence of 21 residues from $\overset{108}{\text{A}}$ to $\overset{128}{\text{G}}$ with only three differences in the 3'-end half, and the sequence of 35 residues from $\overset{194}{\text{U}}$ to $\overset{228}{\text{C}}$ (Nakamura *et al.*, 1980).

5. SECRETION ACROSS THE CYTOPLASMIC MEMBRANE

The lipoprotein has been shown to be produced from a secretory precursor, the prolipoprotein, which has a peptide extension (signal peptide) at the NH$_2$ terminal (Inouye *et al.*, 1977a). As shown in Fig. 2, the prolipoproteins of *E. coli* and *Er. amylovora* have signal peptides of 20 amino acid residues, whereas the signal peptide of the *S. marcescens* prolipoprotein consists of 19 residues. Besides the lipoprotein, several other *E. coli* outer-membrane proteins as well as the periplasmic proteins have been shown to be produced from secretory precursors, and the amino acid sequences of their signal peptides have also been determined (see review by Inouye and Halegoua, 1980).

On the basis of the structural analysis of the amino acid sequences of these signal peptides, the loop model has been proposed to explain the function of the signal peptide in protein secretion across the membrane (Inouye and Halegoua, 1980; DiRienzo *et al.*, 1978). In this model, in contrast to the original signal hypothesis (Blobel and Dobberstein, 1975),

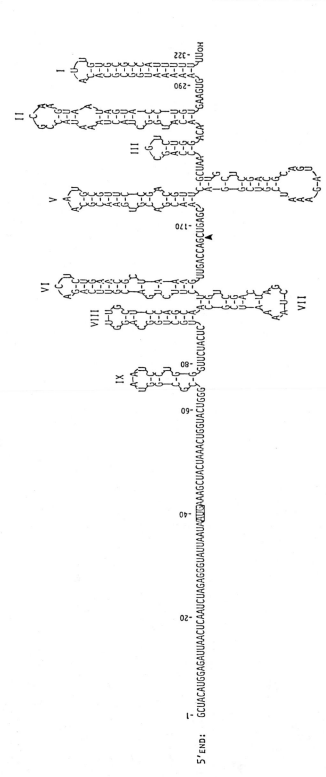

Figure 3. Possible secondary structure of the *E. coli lpp* mRNA (Nakamura *et al.*, 1980). AUG (38 to 41) is the initiation codon. The ΔG values for hairpin stem-and-loop structures I to IX are -21.1, -11.3, -1.0, -9.8, -13.8, -3.8, -1.1, -11.6, and -1.6 kcal/mole, respectively.

the signal peptide is not linearly translocated across the membrane. Instead, the NH$_2$-terminal section of the signal peptide, which is negatively charged at neutral pH, forms stable salt linkages with the positively charged inner surface of the cytoplasmic membrane. Subsequently, the hydrophobic section of the signal peptide is progressively inserted into the cytoplasmic membrane by forming a loop. As the peptide further elongates, the loop is further extended so that the cleavage site of the signal peptide is eventually exposed to the outside surface of the cytoplasmic membrane. The signal peptidase then cleaves off the signal peptide to detach the NH$_2$-terminal end of the secreted protein from the membrane.

One of the ways to examine this working hypothesis is to block the signal peptidase so that one can study the localization and the mode of the interaction of the precursor protein with the membrane. Several methods have been developed to inhibit signal peptidases: toluene treatment (Halegoua et al., 1977; Sekizawa et al., 1977), phenethyl alcohol treatment (Halegoua and Inouye, 1979), and reduction of membrane lipid fluidity (DiRienzo and Inouye, 1979). Besides these methods, specific inhibitors for signal peptidase have been developed. An antibiotic developed by Inukai et al. (1978a,b; Nakajima et al., 1978) has been shown to be a specific inhibitor of the prolipoprotein signal peptidase (Inukai et al., 1978c). This antibiotic is a cyclic peptide consisting of L-serine, L-allo-threonine, glycine, N-methylleucine, L-allo-isoleucine, and 3-hydroxy-2-methylnonaic acid (Nakajima et al., 1978) and provides an excellent means to characterize the prolipoprotein accumulated in the membrane. The prolipoprotein accumulated in the presence of globomycin has been shown to be already modified by glycerol and fatty acid(s) (Hussain et al., 1980). The drug can also be used for preferential selection of deletion mutants in the lpp gene (Zwiebel et al., 1981). Recently, it has been shown that benzyloxycarbonylalanine-chloromethylketone also serves as a specific inhibitor of the prolipoprotein signal peptidase (Maeda, Glass, and Inouye, unpublished).

REFERENCES

Blobel, G., and Dobberstein, B. (1975). *J. Cell Biol.* **67**, 835–851.

Braun, V., and Bosch, V. (1972). *Eur. J. Biochem.* **28**, 51–69.

Chamberlin, M. J. (1976). In *RNA Polymerase* (R. Lodish and M. Chamberlin, eds.), pp. 159–191, Cold Spring Harbor Laboratory, Cold Spring Harbor, N.Y.

DiRienzo, J. M., and Inouye, M. (1979). *Cell* **17**, 155–161.

DiRienzo, J. M., Nakamura, K., and Inouye, M. (1978). *Annu. Rev. Biochem.* **47**, 481–532.

Halegoua, S., and Inouye, M. (1979). *J. Mol. Biol.* **130**, 39–61.

Halegoua, S., Sekizawa, J., and Inouye, M. (1977). *J. Biol. Chem.* **252**, 2324–2330.

Hantke, K., and Braun, V. (1973). *Eur. J. Biochem.* **34**, 284–296.

Hussain, M., Ichihara, S., and Mizushima, S. (1980). *J. Biol. Chem.* **255**, 3707–3712.

Inouye, M., (ed.) (1979a). In *Bacterial Outer Membranes: Biogenesis and Functions,* pp. 1–12, Wiley, New York.

Inouye, M. (1979b). In *Biomembranes* (L.A. Manson, ed.), Vol. 10, pp. 141–208, Plenum Press, New York.

Inouye, M., and Halegoua, S. (1980). *CRC Crit. Rev. Biochem.* **7**, 339–371.

Inouye, M., Shaw, J., and Shen, C. (1972). *J. Biol. Chem.* **247**, 8154–8159.

Inouye, S., Wang, S., Sekizawa, J., Halegoua, S., and Inouye, M. (1977a). *Proc. Natl. Acad. Sci. USA* **74**, 1004–1008.

Inouye, S., Lee, N., Inouye, M., Wu, H. C., Suzuki, H., Nishimura, S., and Hirota, Y. (1977b). *J. Bacteriol.* **132**, 308–313.

Inukai, M., Enokita, R., Torikata, A., Nakajima, M., Iwada, S., and Arai, M. (1978a). *J. Antibiot.* **31**, 410–420.

Inukai, M., Nakajima, M., Osawa, M., Haneishi, T., and Arai, M. (1978b). *J. Antibiot.* **31**, 421–425.

Inukai, M., Takeuchi, M., Shimizu, K., and Arai, M. (1978c). *J. Antibiot.* **31**, 1203–1205.

Movva, N. R., Nakamura, K., and Inouye, M. (1980). *J. Biol. Chem.* **255**, 27–29.

Nakajima, M., Inukai, M., Haneishi, T., Terahara, A., Arai, M., Kinoshita, T., and Tamura, C. (1978). *J. Antibiot.* **31,** 426–432.

Nakamura, K., and Inouye, M. (1979). *Cell* **18,** 1109–1117.

Nakamura, K., and Inouye, M. (1980). *Proc. Natl. Acad. Sci. USA* **77,** 1369–1373.

Nakamura, K., Pirtle, R. M., Pirtle, I. L., Takeishi, K., and Inouye, M. (1980). *J. Biol. Chem.* **255,** 210–216.

Pirtle, R. M., Pirtle, I. L., and Inouye, M. (1980). *J. Biol. Chem.* **255,** 199–209.

Sekizawa, J., Inouye, S., Halegoua, S., and Inouye, M. (1977). *Biochem. Biophys. Res. Commun.* **77,** 1126–1133.

Siebenlist, U., Simpson, R. B., and Gilbert, W. (1980). *Cell* **20,** 269–281.

Taniguchi, T., and Weissmann, C. (1978). *J. Mol. Biol.* **118,** 533–565.

Wang, S., Pirtle, R., Pirtle, I., Small, M., and Inouye, M. (1979). *Biochemistry* **18,** 4270–4277.

Yamagata, H., Nakamura, K., and Inouye, M. (1981). *J. Biol. Chem.*, **256,** 2194–2198.

Zwiebel, L., Inukai, M., Nakamura, K., and Inouye, M. (1981). *J. Bacteriol.*, **145,** 654–656.

37

Biosynthesis and Assembly of Outer Membrane Lipoprotein in E. coli

Henry C. Wu

1. INTRODUCTION

The outer membrane of gram-negative bacteria contains abundant copies of a limited number of proteins, the so-called major outer membrane proteins. One of the major outer membrane proteins in *E. coli* and other gram-negative bacteria is the murein lipoprotein (Braun and Rehn, 1969). The NH_2 terminus of this protein consists of a novel amino acid, glycerylcysteine. Two moles of fatty acid are attached to the glyceryl moiety through ester linkage, and one mole of fatty acid is amide-linked to the amino group of the cysteine moiety (Hantke and Braun, 1973).

Lipoprotein was first isolated as the murein-bound form. The ϵ-NH_2 group of the COOH-terminal lysine of murein lipoprotein is amide-linked to the COOH group of *meso*-diaminopimelic acid (Braun and Wolff, 1970). Lipoprotein also exists as the so-called free form, i.e., not linked to peptidoglycan (Inouye *et al.*, 1972). Murein lipoprotein is distinguished by its unique structural features as well as by the unusual stability of its mRNA (Hirashima *et al.*, 1973) and is present as the most abundant protein in *E. coli*. In addition, it is probably one of the most extensively studied proteins in *E. coli* (Braun and Bosch, 1972; Nakamura *et al.*, 1980; Nakamura and Inouye, 1979). In this review, I shall attempt to summarize the current knowledge of biosynthesis and assembly of lipoprotein in *E. coli* and to define our ignorance as well. Several extensive reviews on this protein have appeared (Braun, 1975; Inouye, 1979; Osborn and Wu, 1980; Wu *et al.*, 1980).

Henry C. Wu • Department of Microbiology, Uniformed Services University of the Health Sciences, Bethesda, Maryland 20014.

2. BIOSYNTHESIS OF LIPOPROTEIN

2.1. Prolipoprotein as the Precursor for Lipoprotein

The primary translation product of lipoprotein mRNA is a precursor protein called prolipoprotein, which contains a peptide extension of 20 extra amino acids at the NH_2 terminus (Halegoua *et al.*, 1977; Inouye *et al.*, 1977b). Several lines of evidence support the contention that prolipoprotein is indeed a biosynthetic precursor for lipoprotein. A mutant lipoprotein has been identified that corresponds to uncleaved prolipoprotein with a single amino acid substitution of aspartic acid for glycine at the 14th position of the prolipoprotein (Lin *et al.*, 1978, 1980b). As will be described in a later section, prolipoprotein accumulates *in vivo* in *E. coli* cells treated with a novel antibiotic, globomycin (Inukai *et al.*, 1978c; Hussain *et al.*, 1980). Removal of globomycin allows the conversion of unprocessed prolipoprotein to lipoprotein both *in vitro* and *in vivo* (Inukai *et al.*, 1978c; Hussain *et al.*, 1980).

The structure of prolipoprotein differs from that of mature murein lipoprotein in a number of ways. Consequently, there must occur a series of posttranslational modifications and processing events prior to its translocation to the outer membrane. These modifications include (1) covalent attachment of glycerol to the sulfhydryl group of cysteine followed by *O*-acylation; (2) proteolytic removal of the signal peptide at the NH_2-terminus of prolipoprotein; (3) *N*-acylation of the new NH_2-terminal cysteine; and (4) the joining of free-form lipoprotein to the murein sacculus.

2.2. Biosynthesis of the Lipid Moiety Covalently Attached to Lipoprotein

Based on both *in vivo* pulse-chase experiments and fusion of labeled phospholipid vesicles with intact cells, we have postulated that the biosynthesis of the covalently linked lipid moieties in lipoprotein proceeds as follows (Chattopadhyay and Wu, 1977; Chattopadhyay *et al.*, 1979a,b; Lai *et al.*, 1980; Lai and Wu, 1980):

apoprolipoprotein + phosphatidylglycerol
→ glycerylcysteine-prolipoprotein + phosphatidic acid

glyceryl-cysteine-prolipoprotein + phospholipid
→ diglyceride-cysteine-prolipoprotein + lysophospholipid

diglyceride-cysteine-prolipoprotein
→ diglyceride-cysteine-lipoprotein + signal peptide

diglyceride-cysteine-lipoprotein + phospholipid
→ mature lipoprotein (*N*-acyl diglyceride-cysteine-lipoprotein) + lysophospholipid

2.3. Processing of Prolipoprotein

The processing of prolipoprotein is very rapid *in vivo*, for uncleaved prolipoprotein cannot be detected by pulse labeling. Prolipoprotein can be accumulated *in vivo* by any one of three means: use of a membrane perturbant such as toluene (Halegoua *et al.*, 1977); use of antibiotics such as globomycin (Inukai *et al.*, 1978c; Hussain *et al.*, 1980); or use of bacterial mutants (Lin *et al.*, 1978, 1980b). Globomycin is a cyclic peptide antibiotic produced by *Streptomyces*, and its chemical structure and mode of action have been elucidated by Arai and his co-workers (Inukai *et al.*, 1978a,b,c, 1979; Nakajima *et al.*, 1978). This drug was found to specifically inhibit the processing of prolipoprotein (Inukai *et al.*, 1978c; Hussain *et al.*, 1980), and its specificity of action has been attributed to a structural similarity with that of the signal sequence of prolipoprotein (Inukai *et al.*, 1978c). The inhibition of prolipoprotein processing is reversible both *in vivo* and *in vitro*. Mizushima and his colleagues, in addition to confirming these findings (Hussain *et al.*, 1980), have shown that prolipoprotein in globomycin-treated cells is already modified to contain both glycerol and ester-linked fatty acyl moieties. The glyceride-containing prolipoprotein was found to accumulate in the cytoplasmic membrane fraction of the globomycin-treated cells (Hussain *et al.*, 1980). Both of these observations may be important in the understanding of the mode of action of globomycin.

Two lipid-deficient (pro)lipoprotein mutants have been isolated that are more resistant to globomycin than wild-type parental strains (Lai *et al.*, 1981a). This observation provides the basis for an alternative hypothesis for the mode of action of globomycin. Instead of acting as a structural analog of the substrate (prolipoprotein) in the processing reaction, globomycin may bind to the substrate (glyceride-containing prolipoprotein), thereby interfering with processing by the signal peptidase. This interpretation would account for the apparent globomycin resistance of lipid-deficient lipoprotein mutants. Finally, there remains an enigma: while both the mature glyceride-containing lipoprotein and the lipid-deficient prolipoprotein containing signal peptide are translocated to the outer membrane extremely rapidly (Lin *et al.*, 1980c), prolipoprotein containing both glyceride and signal sequence accumulates in the cytoplasmic membrane of the globomycin-treated cells. It remains to be seen whether globomycin interferes directly or indirectly with the translocation of glyceride-prolipoprotein through the formation of a globomycin–prolipoprotein complex.

2.4. Attachment of Lipoprotein to Murein Sacculus

Although lipoprotein was first discovered in the form covalently attached to peptidoglycan, the assembly of lipoprotein into the murein sacculus has not been as extensively studied as the biosynthesis of the free-form lipoprotein. However, a few interesting observations have been made regarding the structural requirements of lipoprotein as a substrate for this attachment reaction (Fig. 1). A mutation near the COOH terminus of lipoprotein (*lpp*-1 mutant of Hirota, Inouye *et al.*, 1977a) does not affect the assembly of the mutant lipoprotein into the murein sacculus (Wu *et al.*, 1977) even though the mutant lipoprotein is deficient in ester-linked fatty acid (Inouye *et al.*, 1977a; Rotering and Braun, 1977). A mutant strain that synthesizes a prolipoprotein totally deficient in glycerol and both ester- and amide-linked fatty acids has also been described, and the prolipoprotein synthesized by this mutant is very poorly assembled into the peptidoglycan (Wu *et al.*, 1977). Although the amount of murein-bound lipoprotein found in this mutant is low (4% of wild type), the murein-bound lipoprotein contains both glyceride and amide-linked fatty acid (Lin *et al.*, 1980d). Finally, the glyceride-containing prolipoprotein that accumulates in globomycin-treated cells is assembled into the peptidoglycan, albeit less efficiently than the wild-type lipoprotein (Inukai *et al.*, 1979). These data taken together strongly suggest that removal of the signal peptide is not essential for incorporation into peptidoglycan, but that modification by glyceryl transferase and possibly partial *O*-acylation of glyceryl prolipoprotein are required for this reaction. It is not clear whether the low efficiency of assembly of

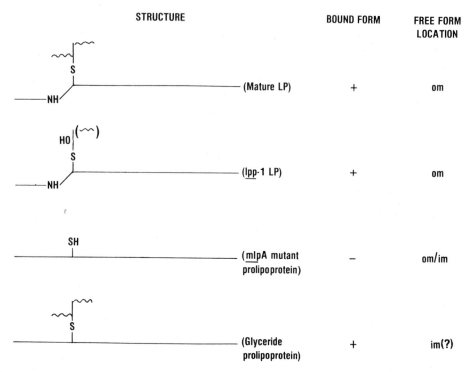

STRUCTURE · BOUND FORM · FREE FORM LOCATION

(Mature LP) · + · om

(lpp-1 LP) · + · om

(mlpA mutant prolipoprotein) · – · om/im

(Glyceride prolipoprotein) · + · im(?)

Figure 1. Effects of alterations in lipoprotein structures on the assembly of lipoproteins into the murein sacculus and outer membrane. om, outer membrane; im, inner membrane.

glyceride-containing prolipoprotein into the peptidoglycan in globomycin-treated cells is due to the presence of the signal peptide, to its localization in the cytoplasmic membrane, or to the presence of bound globomycin. It has been assumed that assembly of lipoprotein into the murein sacculus occurs subsequent to its translocation to the outer membrane. Thus, it is somewhat surprising that prolipoprotein is indeed assembled into the murein sacculus, for the bulk of prolipoprotein synthesized in globomycin-treated cells accumulates in the cytoplasmic membrane. It is conceivable that the murein-bound prolipoprotein in globomycin-treated cells is derived from the fraction of glyceride-containing prolipoprotein that has been translocated to the outer membrane.

3. ASSEMBLY OF LIPOPROTEIN INTO THE OUTER MEMBRANE

3.1. Signal Hypothesis as Applied to the Assembly of Lipoprotein

The assembly of lipoprotein into the outer membrane is consistent with many but not all features envisioned by the cotranslational mode of protein secretion according to the signal hypothesis. Thus, the precursor form of lipoprotein contains an NH_2-terminal extension enriched with hydrophobic amino acids characteristic of the signal sequences found in many of the precursors for secretory proteins (Habener et al., 1978). The kinetics of assembly of lipoprotein into the outer membrane is consistent with a cotranslational mode; the order of appearance of lipoprotein in the outer membrane appears to be the same as that of completion of synthesis of the nascent chains (Lin et al., 1980a). On the other hand, there has been no direct evidence for the synthesis of lipoprotein on membrane-bound ribosomes in *E. coli*. It is clear, however, that proteolytic cleavage of the signal peptide in the prolipoprotein is not required for the translocation and assembly of lipoprotein into the outer membrane (Lin *et al*., 1978, 1980c).

3.2. A Unique Signal Peptidase for Prolipoproteins

There have been a number of observations suggesting that there exist distinct signal peptidases for individual major outer-membrane proteins. Ito (1978) reported the selective inhibition of processing of OmpF and OmpC precursor proteins by protease inhibitor(s). In addition, the processing of prolipoprotein is inhibited by toluene at a concentration less than that required for the inhibition of processing of pro-OmpF (and pro-OmpC) protein(s) (Sekizawa *et al*., 1977). Both phenethyl alcohol (Halegoua and Inouye, 1979) and reduced membrane fluidity (DiRienzo and Inouye, 1979) have differential effects on the processing of precursors for lipoprotein, OmpA and OmpF (OmpC) proteins. Procaine and other local anesthetics preferentially inhibit processing of precursors of protein a(3b) (Gayda *et al*., 1979) and alkaline phosphatase (Lazdunski *et al*., 1979). All these observations can be interpreted to support the notion that each outer-membrane/periplasmic protein is processed by a sequence-specific protease. On the other hand, the selective inhibition of processing by certain agents may be due to the differential effects of these agents on other components required for processing or on the translocation process *per se* rather than affecting the signal peptidases directly. For example, these agents may affect the conformation of the precursor proteins or other components of the export machineries (signal peptide receptor), or the ability of individual precursor proteins to reach a common processing enzyme. Definitive resolution of this problem must await biochemical and genetic characterization of signal peptidases.

Evidence is rapidly accumulating that Braun's lipoprotein, the major lipoprotein in the *E. coli* cell envelope, is not the only protein containing covalently-linked glyceride. In fact, the number of newly recognized lipoprotein species is rapidly increasing (Mizuno, 1979; Ichihara *et al*., 1981). These lipoproteins have one feature in common, i.e., they appear to be synthesized as lipid-containing precursor proteins, and the processing of these prolipoproteins is inhibited by globomycin. Presumptive evidence also indicates that the amino acid sequences near the site of modification and cleavage for two of the lipoproteins (Braun's lipoprotein and Mizuno's peptidoglycan-associated lipoprotein) are similar if not identical. This raises the interesting possibility that the processing of these prolipoproteins may be catalyzed by a unique signal peptidase that recognizes the modified cysteine. The prevalence of lipoproteins in gram-negative bacteria suggests the modification of a number of structurally unrelated proteins by a common enzyme(s). Both immunological studies and DNA hybridization data suggest that the primary structure of Braun's lipoprotein is unique and distinct from the newly discovered lipoproteins described by Ichihara *et al*. (1981). The lack of sequence homology and the large variation in the apparent molecular weights of these various lipoproteins strongly suggest that the putative enzymes responsible for the formation of glyceride-cysteine recognize a relatively limited amino acid sequence near the site of modification and processing (Fig. 2). This speculation has recently been verified by the demonstration that penicillinase coded by the gene from *B. licheniformis* is modified to contain covalently linked glyceride when this gene is expressed in *E. coli* (Lai *et al*., 1981b). This seemingly surprising result is consistent with the above hypothesis inasmuch as the precursor form of this enzyme contains the tetrapeptide -Leu-Ala-Gly-Cys-, within the signal sequence, and this sequence is identical to the sequence in prolipoprotein at the

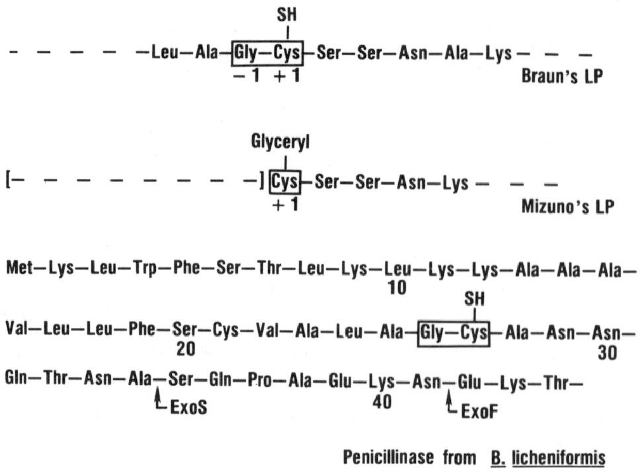

SH
– – – – –Leu–Ala–Gly–Cys–Ser–Ser–Asn–Ala–Lys – – –
–1 +1
Braun's LP
Glyceryl
[– – – – – – – – –] Cys–Ser–Ser–Asn–Lys – – –
+1
Mizuno's LP
Met–Lys–Leu–Trp–Phe–Ser–Thr–Leu–Lys–Leu–Lys–Lys–Ala–Ala–Ala–
10
SH
Val–Leu–Leu–Phe–Ser–Cys–Val–Ala–Leu–Ala–Gly–Cys–Ala–Asn–Asn–
20
30
Gln–Thr–Asn–Ala–Ser–Gln–Pro–Ala–Glu–Lys–Asn–Glu–Lys–Thr–
ExoS
40
ExoF
Penicillinase from B. licheniformis

Figure 2. NH_2-terminal sequences of Braun's lipoprotein, Mizuno's lipoprotein, and *B. licheniformis* penicillinase.

met Gly–Cys Lys
NH2
Lipid-free
Prolipoprotein
Phosphatidylglycerol
Glyceryl transferase
glyceryl
met Gly–Cys Lys
NH2
Glyceryl-Prolipoprotein
Phospholipid
Acyl Transferase
Glyceride
met Gly–Cys Lys
NH2
Glyceride-Prolipoprotein
signal peptidase
Glyceride
NH2–Cys Lys
NH2
Glyceride-Lipoprotein
Phospholipid
Acyl Transferase
Glyceride
Acyl–NH–Cys Lys
NH2
Free-form
Lipoprotein
Peptidoglycan
Glyceride
Acyl–NH–Cys Lys
NH
C=O
m–DAP–Peptidoglycan
Bound-form
Lipoprotein

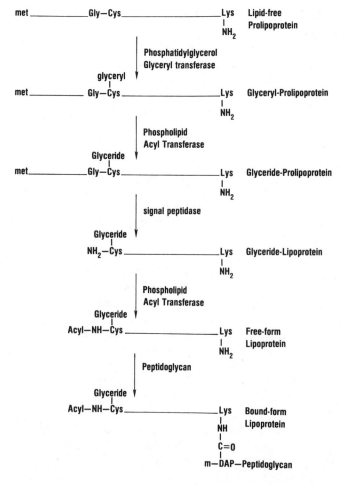

Figure 3. Pathway for the biosynthesis and assembly of outer-membrane Braun's lipoprotein in *E. coli*.

site of modification (Fig. 2). The identification of glycerylcysteine in penicillinase synthesized in *E. coli* (Lai *et al.*, 1981b) further supports this interpretation.

4. CONCLUSION: A POSTULATED PATHWAY FOR THE ASSEMBLY OF LIPOPROTEIN

Figure 3 summarizes the current model for the biosynthesis and assembly of lipoprotein in *E. coli*. The identification of the series of enzymes involved in this pathway remains a major challenge in the near future. The discovery of new lipoprotein species raises the question as to why the cell chooses to modify this class of proteins by covalently attaching glyceride and fatty acids to the NH_2 termini of these proteins. The functions of these lipoproteins and the role of the lipid moiety in their biogenesis or functions are just a few of many important unanswered questions.

ACKNOWLEDGMENTS. The work described in this article was supported by grants from the National Institutes of Health (GM-28811) and the American Heart Association (78-601).

REFERENCES

Braun, V. (1975). *Biochim. Biophys. Acta* **415,** 335–377.
Braun, V., and Bosch, V. (1972). *Proc. Natl. Acad. Sci. USA* **69,** 970–974.
Braun, V., and Rehn, K. (1969). *Eur. J. Biochem.* **10,** 426–438.
Braun, V., and Wolff, H. (1970). *Eur. J. Biochem.* **14,** 387–391.
Chattopadhyay, P. K., and Wu, H. C. (1977). *Proc. Natl. Acad. Sci. USA* **74,** 5318–5322.
Chattopadhyay, P. K., Lai, J. S., and Wu, H. C. (1979a). *J. Bacteriol.* **137,** 309–312.
Chattopadhyay, P. K., Engel, R., Tropp, B. E., and Wu, H. C. (1979b). *J. Bacteriol.* **138,** 944–948.
DiRienzo, J. M., and Inouye, M. (1979). *Cell* **17,** 155–161.
Gayda, R. C., Henderson, G. W., and Markovitz, A. (1979). *Proc. Natl. Acad. Sci. USA* **76,** 2138–2142.
Habener, J. F., Rosenblatt, M., Kemper, B., Kronenber, H. M., Rich, A., and Potts, J. T., Jr. (1978). *Proc. Natl. Acad. Sci. USA* **75,** 2616–2620.
Halegoua, S., and Inouye, M. (1979). *J. Mol. Biol.* **130,** 39–61.
Halegoua, S., Sekizawa, J., and Inouye, M. (1977). *J. Biol. Chem.* **252,** 2324–2330.
Hantke, K., and Braun, V. (1973). *Eur. J. Biochem.* **34,** 284–296.
Hirashima, A., Childs, G., and Inouye, M. (1973). *J. Mol. Biol.* **79,** 373–389.
Hussain, M., Ichihara, S., and Mizushima, S. (1980). *J. Biol. Chem.* **255,** 3707–3712.
Ichihara, S., Hussain, M., and Mizushima, S. (1981). *J. Biol. Chem.* **256,** 3125–3129.
Inouye, M. (1979). *Biomembranes* **10,** 141–208.
Inouye, M., Shaw, J., and Shen, C. (1972). *J. Biol. Chem.* **247,** 8154–8159.
Inouye, S., Lee, N., Inouye, M., Wu, H. C., Suzuki, H., Nishimura, Y., Iketani, H., and Hirota, Y. (1977a). *J. Bacteriol.* **132,** 308–313.
Inouye, S., Wang, S. S., Sekizawa, J., Halegoua, S., and Inouye, M. (1977b). *Proc. Natl. Acad. Sci. USA* **74,** 1004–1008.
Inukai, M., Enokita, H., Torikata, A., Nakahara, M., Iwado, S., and Arai, M. (1978a). *J. Antibiot.* **31,** 410–420.
Inukai, M., Nakajima, M., Ōsawa, M., Haneishi, T., and Arai, M. (1978b). *J. Antibiot.* **31,** 421–425.
Inukai, M., Takeuchi, M., Shimizu, K., and Arai, M. (1978c). *J. Antibiot.* **31,** 1203–1205.
Inukai, M., Takeuchi, M., Shimizu, K., and Arai, M. (1979). *J. Bacteriol.* **140,** 1098–1101.
Ito, K. (1978). *Biochem. Biophys. Res. Commun.* **82,** 99–107.
Lai, J. S., and Wu, H. C. (1980). *J. Bacteriol.* **144,** 451–453.
Lai, J. S., Philbrick, W. M., and Wu, H. C. (1980). *J. Biol. Chem.* **255,** 5384–5387.
Lai, J. S., Philbrick, W. M., Hayashi, S., Inukai, M., Arai, M., Hirota, Y., and Wu, H. C. (1981a). *J. Bacteriol.*, **145,** 657–660.

Lai, J. S., Sarvas, M., Brammar, W. J., Neugebauer, K., and Wu, H. C. (1981b). *Proc. Natl. Acad. Sci. USA* **78**, 3506–3510.

Lazdunski, C., Baty, D., and Pages, J. M. (1979). *Eur. J. Biochem.* **96**, 49–57.

Lin, J. J. C., Kanazawa, H., Ozols, J., and Wu, H. C. (1978). *Proc. Natl. Acad. Sci. USA* **75**, 4891–4895.

Lin, J. J. C., Giam, C. Z., and Wu, H. C. (1980a). *J. Biol. Chem.* **255**, 807–811.

Lin, J. J. C., Kanazawa, H., and Wu, H. C. (1980b). *J. Biol. Chem.* **255**, 1160–1163.

Lin, J. J. C., Kanazawa, H., and Wu, H. C. (1980c). *J. Bacteriol.* **141**, 550–557.

Lin, J. J. C., Lai, J. S., and Wu, H. C. (1980d). *FEBS Lett.* **109**, 50–54.

Mizuno, T. (1979). *J. Biochem.* **86**, 991–1000.

Nakajima, M., Inukai, M., Haneishi, T., Terahara, A., Arai, M., Kinoshita, T., and Tamura, C. (1978). *J. Antibiot.* **31**, 426–431.

Nakamura, K., and Inouye, M. (1979). *Cell* **18**, 1109–1117.

Nakamura, K., Pirtle, R. M., Pirtle, I. L., Takeishi, K., and Inouye, M. (1980). *J. Biol. Chem.* **255**, 210–216.

Osborn, M. J., and Wu, H. C. (1980). *Annu. Rev. Microbiol.* **34**, 369–422.

Rotering, H., and Braun, V. (1977). *FEBS Lett.* **83**, 41–44.

Sekizawa, J., Inouye, S., Halegoua, S., and Inouye, M. (1977). *Biochem. Biophys. Res. Commun.* **77**, 1126–1133.

Wu, H. C., Hou, C., Lin, J. J. C., and Yem, D. W. (1977). *Proc. Natl. Acad. Sci. USA* **74**, 1388–1392.

Wu, H. C., Lin, J. J. C., Chattopadhyay, P. K., and Kanazawa, H. (1980). *Ann. N.Y. Acad. Sci.* **343**, 368–383.

38

Direct Demonstration of Cotranslational Secretion of Proteins in Bacteria

Phang-C. Tai and Bernard D. Davis

1. INTRODUCTION

In animal cells a special role for the ribosomes bound to the endoplasmic reticulum was recognized early by Palade and co-workers (reviewed in Palade, 1975), who observed a parallel between the abundance of such ribosomes and the secretion of proteins. Various secretory proteins were shortly shown to be synthesized on the membrane-bound polysomes, and various cytoplasmic proteins on the free polysomes (reviewed in Rollerston, 1974). This distribution suggested that secreted proteins might cross the membrane as growing chains, rather than after completion. In support of this model, Redman and Sabatini (1966) found that puromycin released incomplete chains to the interior of the microsomes.

The existence of a leading, signal sequence on a polypeptide, serving to initiate its secretion, was postulated by Blobel and Sabatini (1971) and was independently demonstrated by Milstein *et al.* (1972). The demonstration depended on recognizing an additional property of the signal, which made it readily accessible to experimental study: cleavage during secretion. They found that when mRNA from myeloma cells was translated by an *in vitro* system lacking homologous membrane, the light IgG chain appeared as a larger precursor, containing an additional NH_2-terminal sequence that was absent from the final authentic product. Milstein and co-workers proposed that the extra sequence (molecular weight ~3000) provided a signal for attaching the growing chain of this secretory protein to the membrane. Cowan *et al.* (1973) further reported that during synthesis of the precursor, homologous membrane fragments could convert it to the mature protein, but after its completion the cleavage of the signal sequence was prevented. Furthermore, Schechter *et al.* (1975) found that the extra sequence of the IgG precursor has a marked preponderance of hydrophobic amino acids, and they suggested that its hydrophobicity might mediate the

Phang-C. Tai and Bernard D. Davis • Bacterial Physiology Unit, Harvard Medical School, Boston, Massachusetts 02115.

307

initial interaction of the chains with membrane. Moreover, a similarly hydrophobic sequence, following a very short, early basic region, was found in the signal sequence of several additional proteins (Devillers-Thiery *et al.*, 1975). An extensive series of studies by Blobel and co-workers (reviewed by Blobel *et al.*, 1979) led to increased interest in the "signal hypothesis" and to a more detailed model, in which the interactions of signal sequence and membrane are thought to induce aggregation of membrane proteins to form a tunnel around the secreted chain (Blobel and Dobberstein, 1975). More recently, precursors with a hydrophobic NH_2-terminal signal sequence have also been identified for several proteins that are secreted from bacteria (Inouye and Beckwith, 1977; Randall *et al.*, 1978; Sarvas *et al.*, 1978; Sutcliffe, 1978).

Nevertheless, the presence of a hydrophobic initial sequence on membrane-bound polysomes does not establish whether its binding is followed by cotranslational secretion or whether the elongating chain folds against the membrane surface and is subsequently transferred by engulfment by the membrane after release from the ribosome. (In fact, posttranslational secretion has since been found.) More direct tests for cotranslational secretion would therefore be desirable.

2. DIRECT EVIDENCE FOR COTRANSLATIONAL SECRETION

Cotranslational secretion of a protein would be demonstrated directly if one could label the end of the growing chain protruding from a membrane and show that the other end was still attached to a ribosome as peptidyl-tRNA. Such experiments seemed feasible in bacteria (although not in eukaryotic cells), as the far side of the bacterial cytoplasmic membrane should be accessible to external labeling.

We therefore undertook to label radioactively the protruding nascent chains on spheroplasts of *Escherichia coli* (Smith *et al.*, 1977). The polysomes of the growing cells were stabilized by rapid chilling plus addition of chloramphenicol; the cells were converted to spheroplasts to allow accessibility of the cytoplasmic membrane; and the spheroplasts were treated with ^{35}S-labeled acetylmethionyl methylphosphate sulfone, which reacts with free amino groups (of protein and phospholipid) on the cell surface but does not penetrate cell membranes (Bretscher, 1971). The spheroplasts were then washed and gently disrupted by osmotic lysis, and the membrane–polysome complexes were separated, by sedimentation in a sucrose gradient or by gel filtration, from the free membrane fragments and the free polysomes. The polysomes from the membrane–polysome complexes were freed of membrane by washing with detergent. They were found to retain about 6% of the initial label in the complexes (2% of the total incorporation). More recently, use of diazo[^{125}I]iodosulfanilic acid, a nonpenetrating reagent that labels only tyrosine and histidine residues and not phospholipid, has increased about fourfold the proportion of cellular label that is recovered on polysomes (Smith *et al.*, 1979).

The label on the polysomes (from either reagent) is evidently attached via peptidyl-tRNA: it was released by a low Mg^{2+} concentration; subsequent cleavage by dilute alkali decreased the average molecular weight by about that of tRNA; and the label was also released by puromycin or by chain completion *in vitro*. Moreover, the product of chain completion, as expected, had an average molecular weight about twice that of the material released by puromycin. (This growth of the labeled molecules is particularly cogent evidence, for it excludes the possibility that the small amount of label in the purified polysomes was merely contamination.) Finally, the demonstration of cotranslational secretion was completed by identifying a periplasmic enzyme, alkaline phosphatase, among the products of completion of the extracellularly labeled growing chains (Smith *et al.*, 1977).

Using the same procedure, we have also demonstrated cotranslational secretion of several extracellular proteins in gram-positive bacteria: α-amylase of *Bacillus subtilis* (Smith *et al.*, 1979), the toxin of *Corynebacterium diphtheriae* (Smith *et al.*, 1980), and penicillinase of *Bacillus licheniformis* (Smith *et. al.*, 1981). Cotranslational secretion has also been indicated by the demonstration that in protoplasts nascent chains, still attached to ribosomes, can be shortened by extracellular treatment with Pronase (Smith *et al.*, 1978b; see below).

Both in bacteria and in eukaryotic cells precursors with cleavable signal sequences have been demonstrated not only for proteins destined for secretion but also for proteins incorporated into a membrane (cf. reviews by Osborn and Wu, 1980; Wickner, 1980; Inouye, this volume). Moreover, some of these membrane proteins have been shown to enter the membrane during chain growth. Thus, incomplete protein chains have been recovered from the outer membrane of pulse-labeled *E. coli* cells (de Leij *et al.*, 1978). Moreover, during the insertion of a viral protein into the membrane of the endoplasmic reticulum, the protein can be glycosylated, on the intraluminal surface, only during the early part of chain growth and not later (Katz *et al.*, 1977; Rothman and Lodish, 1977).

3. SIGNIFICANCE OF MEMBRANE-ASSOCIATED RIBOSOMES IN BACTERIA

The demonstration of cotranslational secretion (and incorporation into membrane) has incidentally settled the long-standing question of the functional significance of the association of ribosomes with membrane in bacteria. In thin sections of these cells, unlike eukaryotic cells, it has not been possible to distinguish membrane-bound and free polysomes by electron microscopy, because the population of ribosomes is too dense. Hence, even though membrane fragments in bacterial lysates have long been observed to contain ribosomes, it has not been certain whether the attachment is functional or is due to artificial trapping. A functional role, like that observed earlier in eukaryotes, is now established.

A functional attachment was also suggested by the finding that the membrane-bound polysome fraction of *E. coli* produced a somewhat greater abundance of a secretory protein (alkaline phosphatase) than did the free fraction (Cancedda and Schlessinger, 1974). Similarly, Randall *et al.* (1977) prepared membrane–polysome fractions whose products were enriched in several secreted or incorporated proteins, while containing very little of a cytoplasmic protein (elongation factor EFTu). In our laboratory we have achieved a more complete separation, yielding virtually exclusive synthesis of alkaline phosphatase, α-amylase, penicillinase, and diphtheria toxin on membrane-bound polysomes of various organisms, while cytoplasmic proteins (elongation factors EFTu and EFG) were made exclusively on free polysomes. After removal of membrane from the complexes with detergent, chain completion often yields a product that is a precursor to the mature protein. These complexes thus appear to be a good source of precursor proteins.

4. MECHANISM OF CHAIN SECRETION AFTER INITIAL ATTACHMENT

With cotranslational secretion established, it is now a major challenge to determine how it occurs. Though the hydrophobicity of the signal segment might lead to its spontaneous insertion between lipid molecules in the membrane, the extrusion of the subsequent, largely hydrophilic chain through the membrane would require energy. This energy could conceivably be supplied by the ribosome or by the membrane. Hence, as a beginning of a study of the mechanism of secretion, we have inquired into the nature of the attachment of ribosomes to membrane, and also the length of the nascent chains embedded in the membrane.

4.1. Length of Nascent Chains Embedded in Membrane

When protoplasts of *B. subtilis* with pulse-labeled peptide chains were treated with Pronase, to remove any protruding segments, it was found that membrane (plus ribosomes) protected a length of 49 amino acid residues for about half of the chains recovered from membrane-bound polysomes (Smith *et al.*, 1978b). (The rest were mostly longer and of varying length; they were presumably being folded into the membrane.) The ribosomes protected a uniform length of 28 residues (see also Malkin and Rich, 1967), and there was no intervening sequence between ribosome and membrane that could be cleaved by Pronase. The difference of 21 residues may be taken as the approximate length of the chain surrounded by membrane. As the thickness of bacterial membrane is about 75 Å, the nascent chain spanning the membrane is probably in the extended form, in which a segment of 21 residues would have this length.

The protection of a segment of secreted chain from proteolysis incidentally eliminates one conceivable mechanism of secretion: embedding of the ribosome into the membrane so that the chain leaving it is already extracellular. Furthermore, we found that when the chains protruding from protoplasts were excised by treatment with Pronase, the polysomes remained complexed with the membrane even at 37°C for 30 min (Smith *et al.*, 1978b); hence, the ribosomes carrying peptides bind firmly to the membrane (either directly or via the nascent chain) and are not held passively by an extracellular, folded peptide "knot."

4.2. Attachment of Polysomes to Membrane

In purified rough microsomes from animal cells (Adelman *et al.*, 1973) and in chloroplasts (Chua *et al.*, 1976), release of the nascent chains from the ribosomes by puromycin did not release the ribosomes from the membrane unless the salt concentration was elevated or a cytoplasmic factor (Blobel, 1976) was provided. The ribosomes therefore appear to have two types of interaction with membrane: an indirect attachment through the nascent chain, and a direct attachment. Two proteins found only in the membrane from rough microsomes, and absent from smooth microsomes, have been suggested to be involved in the attachment and have been termed "ribophorins" (Kreibich *et al.*, 1978). However, with purified membrane–polysome complexes from bacteria, we found that puromycin (plus EFG and GTP) released the ribosomes even under ionic conditions (high Mg^{2+}, low K^+) that might be expected to stabilize a weak association (Smith *et al.*, 1978a, 1979). This finding suggests that ribosomes in bacteria are attached to the membrane solely by their nascent chains. If so, the ribosomes cannot push the chains through the membrane. However, the possibility cannot be excluded that the secreting ribosomes do have a direct attachment but it is lost on release of the embedded chain, because of the conformational effect of that loss on the ribosome or on a membrane protein.

Alternatively, an active secretory machinery in the membrane, surrounding the nascent chain, could transport it unidirectionally by transducing metabolic energy. As in the active transport of small molecules, the energy could be derived either from the proton-motive force or from intracellular high-energy phosphate.

In the membrane from the membrane–ribosome complexes of *B. subtilis*, we have observed several proteins that are absent from the free membrane fraction (D. Marty, P.-

C. Tai, and B. D. Davis, unpublished). These proteins could be part of either an active machinery of secretion or a passive channel (including a ribosome attachment site), and other kinds of experiments will be necessary to define their role.

5. CONCLUDING REMARKS

Since the discovery of the NH_2-terminal signal sequence of an immunoglobulin precursor in 1972, our knowledge of the mechanism of protein secretion has progressed a good deal. The general outline of the process is very similar in bacteria and in eukaryotic cells. The simultaneous translation–extrusion mode of protein secretion has now been directly demonstrated in bacteria, by extracellular labeling of growing chains. However, it is not the only mechanism. Posttranslational secretion (and incorporation into membrane), after synthesis in the cytosol, has also been demonstrated, in both prokaryotes and eukaryotes, with or without formation of a cleavable precursor (Koshland and Botstein, 1980; Nichols *et al.*, 1980; Dobberstein *et al.*, 1977; Highfield and Ellis, 1978; see also reviews by Davis and Tai, 1980; Wickner, 1980).

The central problem of cotranslational protein secretion is now the identification of the molecular mechanism that ensures unidirectional transfer of the growing chain. The available evidence seems to point toward a specific machinery for active transport of proteins across the membrane, but direct attachment of the ribosome, permitting it to supply the energy, is not rigorously excluded. Some other unsolved problems may also be noted: whether the signal sequence is inserted into lipid linearly (Blobel and Dobberstein, 1975) or as a loop (DiRienzo *et al.*, 1978); and whether the channel surrounding the chain exists preformed or is induced to assemble from separate molecules in the membrane (and possibly in the cytosol). The relation of the cleavage enzyme to the channel, and the fate of the signal segment after cleavage, also remain to be determined.

As there is no evident systematic difference between the signal sequences of secreted and of incorporated proteins, the information that directs an attached chain to either of these fates, and to different membranes, remains to be determined. In bacteria the presence of a multilayered rigid envelope presents a special problem for secretion of proteins into the medium (in contrast to secretion into the periplasm): it is not known whether these proteins pass through the successive layers or are transiently attached to enter regions of the cytoplasmic membrane that become exposed to the exterior. In addition, the segregation of different proteins into the inner and the outer membrane of gram-negative bacteria presents a problem of membrane morphogenesis analogous to that of eukaryotic cells.

The mechanism of posttranslational secretion is also not clear. Cytoplasmic synthesis preceding secretion seems to be incompatible with provision of energy by a membrane-attached ribosome; it would be compatible either with embedding of the folded protein in the membrane (followed, for secretory proteins, by release on the other side), or with insertion of the unfolded chain into an active machinery of transport. Energy has been found to be necessary for the posttranslational secretion of a chloroplast protein (Grossman *et al.*, 1980; see also Wickner, 1980; Nelson and Schatz, 1979).

The genetic manipulation of bacteria, to yield mutants or recombinants altered in various functions, provides a powerful tool for the study of the secretory process. Such studies (see Garwin and Beckwith, this volume) have led to several conclusions. (1) An intact signal sequence is essential but not sufficient (Silhavy *et al.*, 1976, 1977; Bedouelle *et al.*, 1980; Emr *et al.*, 1980). (2) Secretory proteins and membrane proteins may interact initially with membranes at the same sites: defective secretory proteins that stick in the membrane were found to interfere with export of several outer-membrane proteins, which were then accumulated in precursor form (Bassford *et al.*, 1979); and a single mutation can affect secretion of both kinds of protein (Wanner *et al.*, 1979). (3) The effect of a mutation in the signal sequence could be reversed by a suppressor mutation that may alter a ribosomal protein (Silhavy *et al.*, 1979). (4) A mutant precursor that could not be cleaved was still exported to the outer membrane (Lin *et al.*, 1978; Wu, this volume).

Studies with bacteria, which have advantages for genetic and biochemical manipulations, will no doubt continue to shed much light on the general mechanism of protein secretion. In addition, the bacterial apparatus can evidently recognize a eukaryotic signal, for bacteria containing the ovalbumin gene or the insulin gene secrete its product (Fraser and Bruce, 1978; Mercereau-Puijalon *et al.*, 1978; Talmadge *et al.*, 1980). Knowledge of the mechanism of secretion might therefore have implications for the practical use of recombinant bacteria.

REFERENCES

Adelman, M. R., Sabatini, D. D., and Blobel, G. (1973). *J. Cell Biol.* **56,** 206–229.

Bassford, P. J., Silhavy, T. J., and Beckwith, B. J. (1979). *J. Bacteriol.* **139,** 19–31.

Bedouelle, H., Bassford, P. J., Fowler, A. V., Zabin, I., Beckwith, J., and Hofnung, M. (1980). *Nature (London)* **285,** 78–81.

Blobel, G. (1976). *Biochem. Biophys. Res. Commun.* **68,** 1–7.

Blobel, G., and Dobberstein, B. (1975). *J. Cell Biol.* **67,** 835–851.

Blobel, G., and Sabatini, D. D. (1971). In *Biomembranes* (L. A. Manson, ed.), Vol. 2, pp. 193–195, Plenum Press, New York.

Blobel, G., Walter, P., Chang, C. N., Goldman, B. M., Erickson, A. H., and Lingappa, R. (1979). *Sym. Soc. Exp. Biol.* **33,** 9–36.

Bretscher, M. S. (1971). *J. Mol. Biol.* **58,** 775–781.

Cancedda, R., and Schlessinger, M. J. (1974). *J. Bacteriol.* **117,** 290–301.

Chua, N. H., Blobel, G., Siekevitz, P., And Palade, G. E. (1976). *J. Cell. Biol.* **71,** 497–514.

Cowan, N. J., Harrison, T. M., Brownlee, G. G., and Milstein, C. (1973). *Biochem. Soc. Trans.* **1,** 1247–1250.

Davis, B. D., and Tai, P.-C. (1980). *Nature (London)* **283,** 433–438.

de Leij, L., Kingma, J., and Witholt, B. (1978). *Biochim. Biophys. Acta* **512,** 365–376.

Devillers-Thiery, A., Kindt, T. Scheele, G. and Blobel, G. (1975). *Proc. Natl. Acad. Sci. USA* **72,** 5016–5020.

DiRienzo, J. M., Nakamura, K., and Inouye, M. (1978). *Annu. Rev. Biochem.* **47,** 481–532.

Dobberstein, B., Blobel, G., and Chua, N. H. (1977). *Proc. Natl. Acad. Sci. USA* **74,** 1082–1085.

Emr, S. D., Clement, J. M., Silhavy, T., Hedgpeth, J., and Hofnung, M. (1980). *Nature (London)* **285,** 82–85.

Fraser, T., and Bruce, B. J. (1978). *Proc. Natl. Acad. Sci. USA* **75,** 5936–5940.

Grossmman, A., Bartlett, S., and Chua, N.-H. (1980). *Nature (London)* **285,** 625–628.

Highfield, P. E., and Ellis, R. J. (1978). *Nature (London)* **271,** 420–424.

Inouye, H., and Beckwith, J. (1977). *Proc. Natl. Acad. Sci. USA* **74,** 1440–1444.

Katz, F. N., Rothman, J. E., Lingappa, V. R., Blobel, G., and Lodish, H. F. (1977). *Proc. Natl. Acad. Sci. USA* **74,** 3278–3282.

Koshland, D., and Botstein, D. (1980). *Cell* **20,** 749–760.

Kreibich, G., Ulrich, B. C., and Sabatini, D. D. (1978). *J. Cell Biol.* **77,** 464–487.

Lin, J. J. C., Kanazawa, H., Ozols, J., and Wu, H. C. (1978). *Proc. Natl. Acad. Sci. USA* **75,** 4891—4895.

Malkin, L. L., and Rich, A. (1967). *J. Mol. Biol.* **26,** 329–346.

Mercereau-Puijalon, O., Royal, A., Cami, B., Garapin, A., Krust, A., Gannon, F., and Kourilsky, P. (1978). *Nature (London)* **275,** 505–510.

Milstein, C., Brownlee, G. G., Harrison, T. M., and Mathews, M. B. (1972). *Nature New Biol.* **239,** 117–120.

Nelson, N., and Schatz, G. (1979). *Proc. Natl. Acad. Sci. USA* **76,** 4365–4369.

Nichols, J., Tai, P.-C., and Murphy, J. R. (1980). *J. Bacteriol.* **144,** 518–523.

Osborn, M. J., and Wu, H. C. (1980). *Annu. Rev. Microbiol.* **34,** 369–422.

Palade, G. (1975). *Science* **189,** 347–358.

Randall, L. L., Hardy, S.J.S., and Josefsson, L. G. (1977). *Eur. J. Biochem.* **75**, 43–53.

Randall, L. L., Hardy, S. J. S., and Josefsson, L. G. (1978). *Proc. Natl. Acad. Sci. USA* **75**, 1209–1212.

Redman, C. M., and Sabatini, D. D. (1966). *Proc. Natl. Acad. Sci. USA* **56**, 608–615.

Rollerston, F. S. (1974). *Sub-Cell. Biochem.* **3**, 91–117.

Rothman, J. E., and Lodish, H. F. (1977). *Nature (London)* **269**, 775–780.

Sarvas, M., Hirth, K. P., Fuchs, E., and Simons, K. (1978). *FEBS Lett.* **95**, 76–80.

Schechter, I., McKean, D. J., Guyer, R., and Terry, W. (1975). *Science* **188**, 160–162.

Silhavy, T. J., Casadaban, M. J., Shuman, H. A., and Beckwith, J. R. (1976). *Proc. Natl. Acad. Sci. USA* **73**, 3423–3427.

Silhavy, T. J., Shuman, H. A., Beckwith, J., and Schwartz, M. (1977). *Proc. Natl. Acad. Sci. USA* **74**, 5411–5415.

Silhavy, T. J., Bassford, P., and Beckwith, J. (1979). In *Bacterial Outer Membrane* (M. Inouye, ed.), pp. 203–254, Wiley, New York.

Smith, W. P., Tai, P.-C., Thompson, R. C., and Davis, B. D. (1977). *Proc. Natl. Acad. Sci. USA* **74**, 2830–2834.

Smith, W. P., Tai, P.-C., and Davis, B. D. (1978a). *Proc. Natl. Acad. Sci. USA* **75**, 814–817.

Smith, W. P., Tai, P.-C., and Davis, B. D. (1978b). *Proc. Natl. Acad. Sci. USA* **75**, 5922–5925.

Smith, W. P., Tai, P.-C., and Davis, B. D. (1979). *Biochemistry* **18**, 198–202.

Smith, W. P., Tai, P.-C., Murphy, J. R., and Davis, B. D. (1980). *J. Bacteriol.* **141**, 184–189.

Smith, W. P., Tai,'P.-C., and Davis, B. D. (1981). *Proc. Natl. Acad. Sci. USA* **78**, 3501–3508.

Sutcliffe, J. G. (1978). *Proc. Natl. Acad. Sci. USA* **75**, 3737–3741.

Talmadge, K., Stahl, S., and Gilbert, W. (1980). *Proc. Natl. Acad. Sci. USA* **77**, 3369–3373.

Wanner, B. L., Sarthy, A., and Beckwith, J. (1979). *J. Bacteriol.* **140**, 229–239.

Wickner, W. (1980). *Science* **210**, 861–868.

39

Genetic Approaches for Studying Protein Localization

Jeffrey L. Garwin and Jon Beckwith

1. INTRODUCTION

The process of protein compartmentalization is common to both prokaryotic and eukaryotic cells. In eukaryotic cells, proteins are apportioned specifically to the cytoplasm, to any of a variety of organelles, or to the surrounding medium. While less complex, gram-negative bacteria also specifically localize proteins to subcellular compartments: cytoplasm, inner membrane, outer membrane, or periplasm (the space between the inner and the outer membrane).

Biochemical and cytological approaches to the problem of protein localization (for review see Davis and Tai, 1980) have revealed that many proteins destined for noncytoplasmic locations are made on membrane-bound ribosomes. *In vitro*, in the absence of membranes, these proteins are synthesized as precursor forms that are larger than the mature form by 15–30 amino acid residues at the NH_2 terminus. Cell-free protein synthesis systems have been used to verify that the sequestration of proteins into vesicles is accompanied by the cleavage of the NH_2-terminal signal sequence. The signal hypothesis, codified in most detail by Blobel and Dobberstein (1975), postulates that the NH_2-terminal signal sequence plays a critical role in directing the nascent protein to a particular proteinaceous secretion site at the membrane. The crucial testable predictions of this model are twofold: there is information essential for the secretion process in particular domains of all secreted proteins, and there is an export apparatus and intracellular traffic control system that recognizes these signals and interacts with them.

For studying protein secretion (or any other biological process), genetic approaches facilitate testing whether models or *in vitro* systems accurately represent a physiological reality. Furthermore, genetic studies frequently reveal unsuspected aspects or components of the process. One advantage the geneticist has in approaching a complex problem such

Jeffrey L. Garwin and Jon Beckwith • Department of Microbiology and Molecular Genetics, Harvard Medical School, Boston, Massachusetts 02115.

as protein localization is that she/he need not dissect the system before starting the analysis. As initial experiments assay the function of whole organisms, simple and elegant experiments are compatible with a high level of ignorance. The first step of a genetic analysis is the isolation of mutant strains with functional lesions in the process under study. Further analysis may be biochemical, attempting to identify which specific component has been affected, and how. Or it may be genetic. Genetic mapping may reveal that the gene affected is one previously identified. Different mutant isolates can be compared to determine if the underlying lesions reside in the same or in separate functions (complementation analysis), and the order of separable functions can be ascertained. In addition, it is often possible to obtain second site mutations that suppress the original mutant phenotype.

Starting around 1976, geneticists began addressing two kinds of questions. First, what features of the secreted protein are specific determinants of protein localization? Second, do other, interacting factors exist that promote protein compartmentalization? Representative examples of these studies will be briefly discussed below, in order to clarify the genetic reasoning and techniques that have been useful.

2. DETERMINANTS OF SECRETION WITHIN THE SECRETED PROTEIN

2.1. Analyzing the NH₂ Terminus of Secreted Proteins Using Protein Fusions

By modifying the structure of the secreted proteins genetically and then analyzing biochemically the subsequent protein localization, genetic studies have verified that determinants of protein localization are found within the secreted protein itself. The first genetic modifications analyzed were protein fusions, hybrid *E. coli* proteins that contain peptide sequences from both a secreted protein and a cytoplasmic protein (reviewed extensively by Silhavy *et al.*, 1979). The NH₂ terminus of the hybrid protein was derived from the NH₂-terminal peptide of a secreted protein, while a large, enzymatically functional fragment of β-galactosidase (the *lacZ* gene product) comprised the remainder of the protein. The large size of the β-galactosidase moiety and the fact that the hybrid protein retains β-galactosidase activity facilitated subcellular localization studies. Thus, the question could be asked, can the NH₂ terminus of a secreted protein alter the localization of a cytoplasmic protein?

Hybrid proteins comprised of the maltose binding protein (*malE* gene product) and β-galactosidase were cytoplasmic if the *malE* portion was small, but membrane-associated if the NH₂-terminal *malE* portion was large. In no case was a *malE–lacZ* hybrid protein found secreted to the periplasm, the normal location of the maltose binding protein (Bassford *et al.*, 1979). Hybrid proteins between the phage lambda receptor (*lamB* gene product) and β-galactosidase were also studied, and fell into four classes by subcellular localization and β-galactosidase phenotype. Class I and class II protein fusions were localized to the cytoplasm. However, some of the class III and class IV *lamB–lacZ* hybrid protein was localized to the outer membrane, the normal location of the lambda receptor (for comprehensive review of *lamB–lacZ* fusions, see Emr *et al.*, 1980a).

Certain *malE* and *lamB* protein fusions conferred maltose sensitivity on the host cell (Bassford *et al.*, 1979; Silhavy *et al.*, 1977). That is, induction of synthesis of large amounts of the hybrid protein by addition of maltose to the culture medium resulted in cell death. Mutations were selected that resulted in the loss of the maltose-sensitive phenotype without loss of maltose-inducible β-galactosidase activity. These mutations were found to have altered the *malE* or *lamB* portion of the fused gene (Bassford and Beckwith, 1979; Emr and Silhavy, 1980). Such altered hybrid proteins were found to be localized to the cytoplasm. When the mutation was moved back into the unfused *malE* or *lamB* gene, the otherwise wild-type protein was localized to the cytoplasm, in precursor form, instead of to its normal compartment. DNA sequence analysis of these mutant genes revealed deletions or point mutations that altered the amino acid sequence in the central portion of the signal sequence, the NH_2-terminal peptide that is present in precursor forms but not mature forms of the secreted proteins (Bedouelle *et al.*, 1980; Emr *et al.*, 1980b).

The membrane localization of some hybrid proteins points to the important role of the NH_2 terminus in determining protein export. The abnormal localization of proteins with signal sequence mutations is strong evidence that the hydrophobic segment of the signal sequence plays an essential role in the secretion process. While an intact signal sequence clearly plays a central role in determining export, it is also clear that the nature of the remainder of the exported proteins is important for proper localization. This conclusion is based in part on the analysis of localization of different classes of *lamB–lacZ* hybrid proteins. Class II hybrid protein was not localized to the outer membrane, despite carrying a complete signal sequence plus 15 amino acid residues of the mature lambda receptor protein. Class III and class IV fusions carry more *lamB,* and do localize some hybrid protein to the outer membrane (Emr *et al.*, 1980b). In contrast, no *malE–lacZ* hybrid proteins are localized to the periplasm, despite substantial NH_2-terminal *malE* sequence (Bassford *et al.*, 1979). Whether the attached β-galactosidase moiety is actively interfering with secretion, or the missing COOH-terminal portion of the exported protein plays an active role in secretion is not clear from these studies.

2.2. Analyzing the COOH Terminus of Secreted Proteins Using Chain-Terminating Mutations

Other genetic modifications have been used to analyze the importance of other domains of secreted proteins. Koshland and Botstein (1980) analyzed the localization of mutant and wild-type β-lactamase carried on a P-22 phage of *Salmonella*. β-Lactamase is normally secreted to the periplasm. The mutations were all chain-terminating mutations, resulting in the loss of specific amounts of the COOH terminus of the protein. Analysis of the localization of these incomplete proteins in P-22-infected, UV-irradiated *Salmonella* suggested that the approximately 21 COOH-terminal residues of wild-type β-lactamase are necessary for proper localization. In contrast, for chain-terminating mutations in the *malE* gene of *E. coli*, Ito and Beckwith (1981) have found that the COOH terminus does not appear necessary for localization of mature protein fragments to the periplasm. It is possible that differences in experimental conditions can account for the different results, although it is also possible that the two secreted proteins differ in the location of their essential secretion domains.

2.3. Effect of Processing Determinants on Secretion

Model and co-workers have taken still another genetic approach in their studies of the localization and maturation of filamentous phage coat protein (Boeke *et al.*, 1980; Russel and Model, 1981). Using a mutant with a chain-terminating (amber) codon in the second position of the mature protein sequence, they determined the effect of substituting various amino acids at that position. They used either amber-suppressor host *E. coli* or else phage revertants of the chain-terminating locus. Certain substitutions were without effect on the rate of processing of precursor to mature size M13 coat protein, whereas other substitutions slowed processing from a half-time of 2 sec to a half-time of 60 sec. None of the substitutions altered the kinetics of insertion of precursor into the membrane. The authors inferred from these studies that the domain recognized by the enzyme(s) removing the signal sequence extends beyond the cleavage site into the sequence of the mature protein. They also observed membrane association and signal peptide cleavage as sequential events, under conditions that permit maturation of coat protein into mature phage particles. As even in the revertant strains no precursor form of coat protein is found in the mature virion, it seems likely that the signal sequence, although necessary for protein localization, is incompatible with normal mature protein function and localization to the phage coat. A potentially useful hypothesis is that proteins are addressed to specific subcellular compartments by transient covalent modifications that are normally removed in the final maturation process.

Wu and co-workers have isolated a mutant in the signal sequence of *E. coli* lipoprotein that affects the processing of the protein (Lin *et al.*, 1978). While substantial amounts of the protein are localized to the proper compartment, the outer membrane, the protein is not processed. Thus, this mutant allows the important conclusion that processing and export are dissociable steps: i.e., processing is *not necessary* for export.

3. MUTATIONS IN THE SECRETION APPARATUS

3.1. Mutants with an Abnormal Secretion Phenotype

Mutants in the secretion apparatus have been identified either by the loss of a normal secretion phenotype or by the alteration of a phenotype conferred by an abnormal protein.

Wanner *et al.* (1979) sought mutations that reduced secretion of the periplasmic enzyme alkaline phosphatase (*phoA*) of *E. coli*. Mutations genetically unlinked to that structural gene were obtained by mutagenizing a strain deleted for *phoA*. After mutagenesis, the bacteria were plated on indicator plates and tested for their ability to make alkaline phosphatase, the *phoA* gene being introduced by spraying the colonies with a specialized transducing phage. A novel class of mutants, *perA*, was detected by this approach. While it is not clear that *perA* mutations affect secretion as opposed to gene regulation, the general approach should be useful in detecting secretion mutants.

Novick and Schekman (1979) obtained and analyzed pleiotropic mutants of a mutagenized yeast strain using protocols that illustrate a number of important genetic principles. They reasoned that a secretion or protein localization defect may well be lethal to the cell if expressed continuously. They therefore screened temperature-sensitive colonies for simultaneous abolition of secretion of two unrelated secreted proteins, acid phosphatase and invertase. The colonies were able to grow normally at 24°C, but were killed after prolonged incubation at 37°C. Colloidal silica density gradients revealed that the mutant strains possessed an altered density, sufficiently different from the wild type to permit use of these gradients as an initial enrichment procedure (Novick *et al.*, 1980). The gradients were used, therefore, in a subsequent mutant hunt. The efficiency of enrichment was 10- to 25-fold, yielding 485 mutant clones.

The 188 mutants that were analyzed in detail accumulated both invertase and acid phosphatase at the nonpermissive temperature. The mutants could be placed in 23 complementation groups. The existence of 23 complementation groups implies the function of at least that many factors in the secretion of invertase and acid phosphatase in yeast. However, many of these factors may be necessary for assembly of organelles of the secretion pathway, and thus only indirectly affect the secretion process. When representatives of the 23 complementation groups were examined by electron microscopy, it was observed that three different types of membranous organelles accumulated after growth at 37°C. The order of secretion events was inferred from cytological analysis of strains with lesions in two different complementation groups. The rationale behind these experiments is that if a secretory pathway progresses from one type of organelle to another, and two different mutations are present in the same cell, then the organelles accumulating will be those characteristic of the mutation earlier in the pathway.

3.2. Mutants That Suppress a Phenotype Conferred by an Abnormal Protein

Mutations in the secretory apparatus of *E. coli* have been isolated by identifying strains that reverse, or suppress, a mutant phenotype. Emr *et al*. (1981) have been analyzing extragenic suppressors of a *lamB* signal sequence mutation. Presumably, these mutations have altered the cellular export machinery in such a way that the mutant signal sequence is now recognized and the protein is exported again. Many of these suppressor mutations map in a cluster of genes coding for ribosomal proteins. This result implies some important role of the ribosome in secretion and should permit the identification of the specific ribosomal protein affected.

In our laboratory, an approach has been developed by D. Oliver for the direct selection of mutants altering the cell's secretory apparatus. This selection is based on certain properties of a *malE–lacZ* gene fusion strain. The synthesis of hybrid protein, which is membrane-bound, is inducible by the addition of maltose to the growth medium, resulting in high β-galactosidase activity. However, under uninduced conditions, cells producing the *malE–lacZ* protein have unexpectedly low β-galactosidase activity and exhibit a Lac⁻ phenotype. It was previously inferred that the low enzymatic activity was due to the membrane localization of the hybrid β-galactosidase. Confirming this inference, mutant derivatives of this strain that could grow on lactose were shown to have localized at least some of the hybrid protein to the cytoplasm. The mutations characterized fall into two general classes: signal sequence mutations in the *malE* portion of the hybrid gene, and mutations genetically unlinked to the hybrid gene that have pleiotropic effects on secretion. Among the latter, a number of conditional lethal mutations were found (Oliver and Beckwith, 1981).

Conditional lethal mutations of this sort or other mutants that have pleiotropic effects on secretion can be used to determine the elements of the secretory apparatus. Genetic mapping of the mutations may identify them as affecting already-characterized *E. coli* genes. This would allow ready identification of the gene product. Otherwise, techniques exist for identifying protein products of genes once they have been cloned from the *E. coli* chromosome onto phage (Jaskunas *et al.*, 1975) or plasmids (Sancar *et al.*, 1979). Furthermore, extracts and subcellular fractions of such mutant strains can be used for *in vitro* studies to dissect the secretion process and identify the step of secretion that is affected. Finally, it should be possible to isolate mutations in genes unlinked to the original mutation site that suppress the conditional lethal phenotype. These mutations could facilitate identifying other gene products involved in secretion. Jarvik and Botstein (1973, 1975) used a series of temperature-sensitive and cold-sensitive suppressors to elucidate the order of events in another complex pathway, the morphogenesis of phage P-22.

4. CONCLUSION

From the brief descriptions above of work recently published and in progress, it should be apparent that genetic approaches have considerable utility for elucidating the process of protein localization, especially in bacterial and yeast systems. On the basis of the biochemical similarities already discovered, it appears likely that the more complex eukaryotic systems will exhibit many aspects of compartmentalization pathways that are analogous to those present in systems more amenable to genetic analysis. Most strikingly, it has recently been shown that *E. coli* can secrete and correctly process rat preproinsulin carrying the eukaryotic signal sequence (Talmadge *et al.*, 1980a,b).

A genetic approach is most useful for defining the scope and essential parameters of a process, and for analyzing the control systems and sequence of steps in a pathway. The time course of genetic analysis is such that laying the groundwork and finding the first interesting mutants are the rate-limiting steps. Once these first mutants are obtained, analysis of their phenotypes often permits more rapid isolation of different mutants affecting the process under investigation. We therefore anticipate that the pace of genetic analysis of protein localization will increase rapidly in the next few years.

Finally, mutants greatly facilitate biochemical analysis. *In vitro* systems can be analyzed using components from strains isogenic except for a single defect in a complex process. The identification *in vitro* of a genetically deleted essential component is much easier than analysis of the effect of a biochemical modification of the same complex system. The fidelity of *in vitro* systems can be assessed using the *in vivo* phenotype as a standard. Ultimately, cloning the loci involved in secretion should increase the availability of purified protein components for *in vitro* structure–function studies.

ACKNOWLEDGMENTS. We thank Dr. P. Model for a preprint of work in press, and Dr. M. Chow for critical review of the manuscript. This work was supported by grants to J.B. from the American Cancer Society and the National Science Foundation. J.G. was supported by a postdoctoral fellowship from the National Institutes of Health.

REFERENCES

Bassford, P., and Beckwith, J. (1979). *Nature (London)* **277**, 538–541.
Bassford, P. J., Jr., Silhavy, T. J., and Beckwith, J. R. (1979). *J. Bacteriol.* **139**, 19–31.
Bedouelle, H., Bassford, P. J., Jr., Fowler, A. V., Zabin, I., Beckwith, J., and Hofnung, M. (1980). *Nature (London)* **285**, 78–81.
Blobel, G., and Dobberstein, D. (1975). *J. Cell Biol.* **67**, 835–851.
Boeke, J. D., Russel, M., and Model, P. (1980). *J. Mol. Biol.* **144**, 103–116.
Davis, B. D., and Tai, P.-C. (1980). *Nature (London)* **283**, 433–438.
Emr, S. D., and Silhavy, T. J. (1980). *J. Mol. Biol.* **141**, 63–90.
Emr, S. D., Hall, M. N., and Silhavy, T. J. (1980a). *J. Cell Biol.* **86**, 701–711.
Emr, S. D., Hedgpeth, J., Clement, J.-M., Silhavy, T. J., and Hofnung, M. (1980b). *Nature (London)* **285**, 82–85.
Emr, S. D., Hanley-Way, S., and Silhavy, T. J. (1981). *Cell* **23**, 79–88.
Ito, K. and Beckwith, J. R. (1981). *Cell* **25**, 143–150.
Jarvik, J., and Botstein, D. (1973). *Proc. Natl. Acad. Sci. USA* **70**, 2046–2050.
Jarvik, J., and Botstein, D. (1975). *Proc. Natl. Acad. Sci. USA* **72**, 2738–2742.
Jaskunas, S. R., Lindahl, L., Nomura, M., and Burgess, R. R. (1975). *Nature (London)* **257**, 458–462.
Koshland, D., and Botstein, D. (1980). *Cell* **20**, 749–760.
Lin, J. J. C., Kanazawa, J., Ozols, J., and Wu, H. C. (1978). *Proc. Natl. Acad. Sci. USA* **75**, 4891–4895.
Novick, P., and Schekman, R. (1979). *Proc. Natl. Acad. Sci. USA* **76**, 1858–1862.

Novick, P., Field, C., and Schekman, R. (1980). *Cell* **21,** 205–215.

Oliver, D. B., and Beckwith, J. (1981). *Cell* **25,** 765–772.

Russel, M., and Model, P. (1981). *Proc. Natl. Acad. Sci. USA,* **78,** 1717–1721.

Sancar, A., Hack, A. M., and Rupp, W. D. (1979). *J. Bacteriol.* **137,** 692–693.

Silhavy, T. J., Shuman, H. A., Beckwith, J., and Schwartz, M. (1977). *Proc. Natl. Acad. Sci. USA* **74,** 5411–5415.

Silhavy, T. J., Bassford, P. J., Jr., and Beckwith, J. R. (1979). In *Bacterial Outer Membranes: Biogenesis and Functions* (M. Inouye, ed.), pp. 203–254, Wiley, New York.

Talmadge, K., Kaufman, J., and Gilbert, W. (1980a). *Proc. Natl. Acad. Sci. USA* **77,** 3988–3992.

Talmadge, K., Stahl, S., and Gilbert, W. (1980b). *Proc. Natl. Acad. Sci. USA* **77,** 3369–3373.

Wanner, B. L., Sarthy, A., and Beckwith, J. (1979). *J. Bacteriol.* **140,** 229–239.

40

The Biosynthesis of Bacterial Wall Teichoic Acids

I. C. Hancock and J. Baddiley

As bacteria grow and divide they must continuously renew and enlarge their cell walls. Peptidoglycan chains synthesized from intracellular precursors must be extruded and cross-linked to the wall, lying outside the cytoplasmic membrane, while in gram-positive bacteria polymers such as teichoic acids, teichuronic acids, and polysaccharides become covalently linked to the external peptidoglycan. The way these processes are catalyzed and controlled by membrane proteins is a fascinating field of study. This review concerns recent developments in our knowledge of how one class of wall polymer, teichoic acid, is synthesized and integrated into the wall. The details of teichoic acid structure and earlier work on its biosynthesis have been thoroughly reviewed by Baddiley (1972).

Teichoic acids occur in many species of gram-positive bacteria, and in a variety of structures, characterized by repeating units of alditol phosphates or sugar phosphates, linked by phosphodiesters. The types most frequently encountered are poly(glycerol phosphate) and poly(ribitol phosphate) or variants of these (see Fig. 1).

The teichoic acids are covalently linked to the peptidoglycan of the cell wall by means of a small "linkage unit" (LU) that differs from the main polymer chain. The structure of this unit has been determined in three species: *Staphylococcus aureus* H and *Bacillus subtilis* W23, which contain ribitol teichoic acids glycosylated at C-4 with *N*-acetylglucosamine and β-D-glucose, respectively, and *Micrococcus varians* ATCC 29750, which contains poly(*N*-acetylglucosamine phosphate) (Coley *et al.*, 1978). In all three the linkage unit consists of tri(glycerol phosphate)-*N*-acetylglucosamine 1-phosphate (Fig. 2). This unit intervenes between the terminal phosphate of the main teichoic acid chain and the C-6 hydroxyl of a muramic acid residue in the peptidoglycan, to which it is linked by the 1-phosphate of the LU *N*-acetylglucosamine (Heckels *et al.*, 1975). In the case of glycerol teichoic acids the linkage unit glycerol phosphate trimer would be indistinguishable from the polymer main chain, but the LU *N*-acetylglucosamine 1-phosphate can still be detected

I. C. Hancock and J. Baddiley • Microbiological Chemistry Research Laboratory, The University, Newcastle upon Tyne NE1 7RU, England.

(a) glycerol teichoic acids

Bacillus subtilis 168

R = α-glucosyl

Ala = D-alanyl

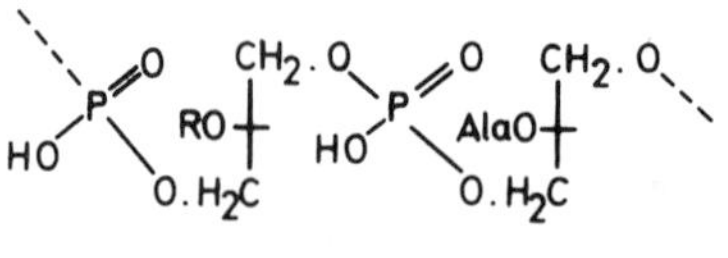

poly(glycerol phosphate)

Bacillus licheniformis

Lactobacillus plantarum C106

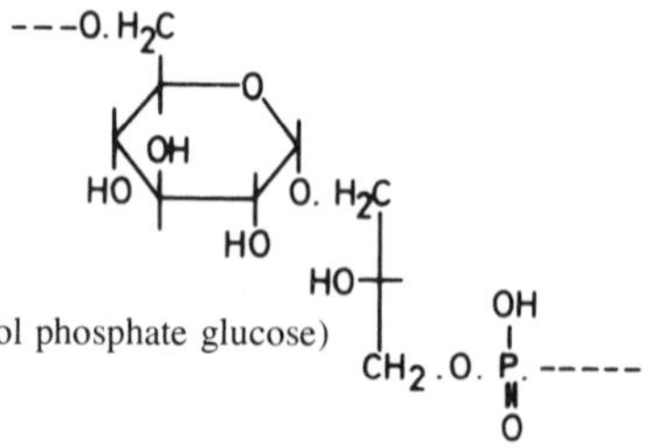

poly(glycerol phosphate glucose)

(b) ribitol teichoic acids

Staphylococcus aureus H

R = β-N-acetylglucosamine

Bacillus subtilis

R = β-glucosyl

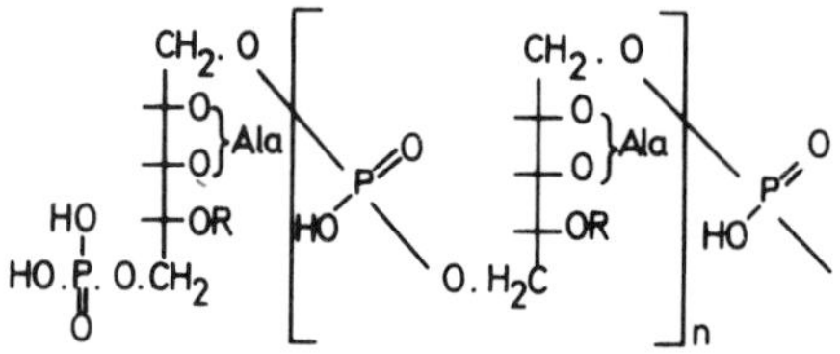

poly(ribitol phosphate)

(c) sugar phosphate polymers

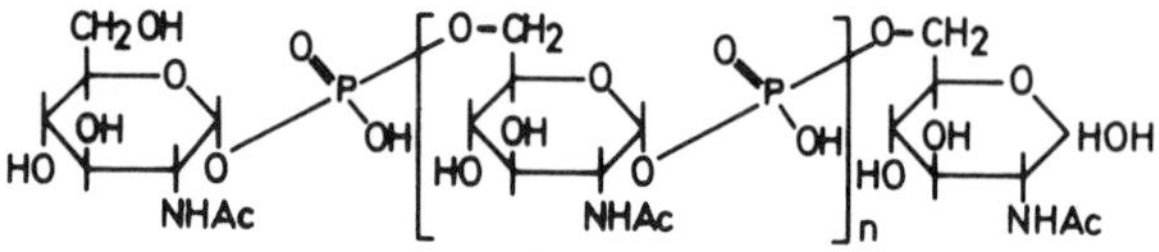

poly(N-acetylglucosamine phosphate) from

Micrococcus varians ATCC 29750

Figure 1. Structure of teichoic acids.

in the linkage region of *B. subtilis* NCIB 3610 (Coley *et al.*, 1978). Teichoic acids so linked to the cell wall comprise up to 60% of its dry weight and impart a large negative charge to the cell surface.

Following the identification by Baddiley and his co-workers of the alditol nucleotides CDP-glycerol and CDP-ribitol, and the subsequent identification of teichoic acids in a number of bacterial species, several groups of workers, notably Glaser and his colleagues, demonstrated that preparations of isolated cytoplasmic membrane catalyzed the synthesis of poly(glycerol phosphate) or poly(ribitol phosphate) from the appropriate nucleotide. Divalent cations, usually Mg^{2+}, were required at concentrations in the range 5–50 mM for maximum activity (Burger and Glaser, 1964; Ishimoto and Strominger, 1966). The polymeric product remained bound to the membrane under mild conditions, but treatment with aqueous phenol, dilute acid or alkali released it into solution. The nature of the membrane-

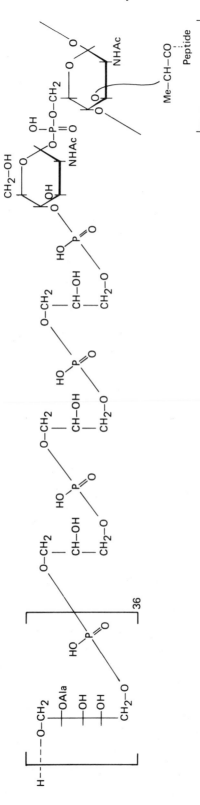

Figure 2. Scheme for the linkage of ribitol teichoic acid to peptidoglycan in a mutant of *S.aureus* H.

bound "primer" for the polymerization remained unknown until Glaser and his co-workers succeeded in identifying an amphiphilic molecule with acceptor activity in *B. subtilis* (Mauck and Glaser, 1972a) and *S. aureus* (Fiedler and Glaser, 1974). Treatment of membranes with Triton X-100 solubilized teichoic acid polymerases and the acceptor, which could be separated by subsequent chromatography in the presence of the surfactant. The solubilized enzyme from *S. aureus* was entirely dependent on added acceptor, and the absence of detectable incomplete polymer chains from the reaction mixture suggested that individual molecules of enzyme and acceptor remained tightly associated until chain growth was complete (Fiedler and Glaser, 1974). The acceptor closely resembled the membrane teichoic acids of the two strains and was named lipoteichoic acid carrier (LTC).

LTC probably is not identical with native membrane teichoic acid. Lambert *et al.* (1977) were able to separate acceptor activity from the bulk of the membrane teichoic acid extracted from *S. aureus* by Triton X-100, while in a recent study Fischer and Rösel (1980) have shown that native membrane teichoic acid only gains acceptor activity after most of the *O*-alanyl ester and glycosyl substituents have been removed from the poly(glycerol phosphate) chain. Alanyl esters are extremely labile at pH values above neutrality, and the latter authors believe that LTC is an artifactual breakdown product of membrane teichoic acid, which arises during preparation of membranes at slightly alkaline pH. The alanyl-LTA ester linkages have a half-life of about 4 hr at pH 8, 37°C (Childs and Neuhaus, 1980). Fischer and Rösel (1980) have found that the lipoteichoic acid of *S. aureus* exhibits a range of degrees of alanylation, and the possibility of a small amount of material with low enough substitution to permit LTC activity cannot be ruled out.

Despite uncertainty about its function *in vivo*, LTC is evidently the principal acceptor for polymerization in isolated membrane preparations when teichoic acid synthesis is studied only in the presence of the nucleotide precursor of the main chain and divalent cations (Hancock *et al.*, 1976). Under these conditions, the direction of chain growth during synthesis of poly(glycerol phosphate) (Kennedy and Shaw, 1968) and teichoic acids containing *N*-acetylglucosamine phosphate (Hussey *et al.*, 1969) has been determined. The chains grow by the transfer of new repeating units to the end of the chain distant from the acceptor in the membrane, that is, to the "nonreducing" end in polymers containing sugars in the chain or to the end bearing a free primary hydroxyl group in the case of alditol phosphate polymers. This is the reverse of the direction in which peptidoglycan chains grow (Ward and Perkins, 1973). There, new units are inserted at the reducing end of the growing glycan so that the newer end of the chain is attached to the membrane-bound lipid carrier. This has important implications for the arrangement of the biosynthetic enzymes in the membrane. Whereas the nongrowing end of the peptidoglycan chain may extend into, and be linked to, the wall during chain elongation, both ends of a growing teichoic acid chain must maintain contact with the membrane until chain elongation is complete.

The discovery of the linkage unit that attaches teichoic acid to peptidoglycan (Heckels *et al.*, 1975) has led to rapid growth of our knowledge of cell wall assembly. Hancock and Baddiley (1976) described the synthesis of LU from CDP-glycerol and UDP-*N*-acetylglucosamine in membrane preparations from *S. aureus*, *B. subtilis* W23, and *Micrococcus* sp. 2102 (*M. varians* ATCC 29750) and showed that it became attached to teichoic acid even in the absence of peptidoglycan, so that attachment of teichoic acid to LU must have preceded its attachment to peptidoglycan. They suggested that LU was first synthesized in a lipid-bound form, to which teichoic acid subsequently became attached, rendering the LU-lipid water soluble. Wyke and Ward (1975) had already found that wall–membrane preparations from *B. licheniformis* were capable of attaching teichoic acid to concomitantly synthesized peptidoglycan, but because CDP-glycerol was required for teichoic acid synthesis and UDP-*N*-acetylglucosamine for peptidoglycan synthesis in this strain, they did not detect a separate dependence on these two nucleotides for linkage of the two polymers. Simultaneously with the report of LU synthesis by Hancock and Baddiley (1976), Bracha and Glaser (1976a) described the dependence of linkage of poly(ribitol phosphate) to peptidoglycan on CDP-glycerol and UDP-*N*-acetylglucosamine in a wall–membrane preparation from *S. aureus*. Subsequent work with a number of bacterial strains (Bracha and Glaser, 1976b; McArthur *et al.*, 1978, 1980a) has confirmed that LU is synthesized as a membrane-bound lipid before teichoic acid is attached to it (Fig. 3). Wyke and Ward (1977a) have shown that the phosphate that attaches LU to muramic acid originates in UDP-*N*-acetylglucosamine in *B. licheniformis*.

The lipid intermediates I to IV (Fig. 3) have been separated and characterized (McArthur *et al.*, 1978, 1980a; Roberts *et al.*, 1979), and the antibiotic tunicamycin has been shown to inhibit potently the synthesis of the first lipid in the pathway (lipid I), which is believed to be prenyl pyrophosphate *N*-acetylglucosamine. A mutant of *S. aureus* lacking the ability to synthesize this lipid has no teichoic acid in its wall, although membrane preparations retain the ability to synthesize poly(ribitol phosphate) attached to LTC (Bracha *et al.*, 1978) and can synthesize additional teichoic acid on addition of a mixture of LU-lipid intermediates extracted from the wild-type strain. This result confirms the evidence of Roberts *et al.* (1979) that the membrane-bound enzyme that catalyzes the attachment of teichoic acid to LU lacks the specificity to distinguish between the complete LU-lipid (lipid IV) and the intermediate containing two glycerol phosphate residues (lipid III), which is the predominant component of the mixture of lipids that accumulates in membranes of *S. aureus* and *M. varians* incubated with CDP-glycerol and UDP-*N*-acetylglucosamine (McArthur *et al.*, 1978). The fact that LU in the wall contains three glycerol phosphate

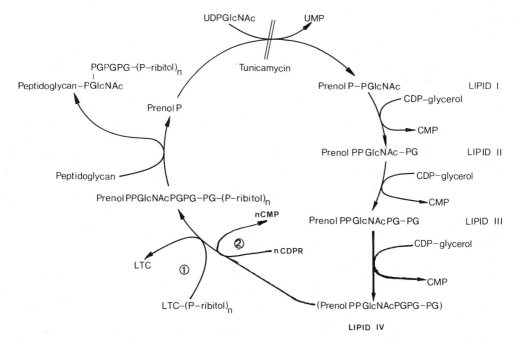

Figure 3. Pathway for the biosynthesis of linkage unit in *Staphylococcus aureus*, showing the alternative routes for attachment of teichoic acid to linkage unit: (1) from lipoteichoic acid carrier (LTC); (2) by direct transfer of ribitol phosphate residues from CDP-ribitol.

residues therefore suggests a considerable degree of spatial organization of the enzymes and lipid carriers responsible for linkage *in vivo*.

In much of the work on LU synthesis, it has been assumed (Bracha and Glaser, 1976a; McArthur *et al.*, 1980a; Wyke and Ward, 1977a) that the teichoic acid chain is assembled on LTC, then transferred intact to LU-lipid (see Fig. 3). The recent demonstration that teichoic acid synthesized *in vitro* on LTC in one batch of membrane can be transferred to LU-lipid synthesized in another batch of membrane when the two lots of membrane are mixed (McArthur *et al.*, 1981) supports this view. On the other hand, it has been reported that poly(ribitol phosphate) synthesized on LTC by the solubilized polymerase from *S. aureus* is not transferred to LU-lipid by *S. aureus* membranes, whereas the soluble polymerase can utilize LU-lipid directly as an acceptor in the absence of LTC (Bracha *et al.*, 1978). These uncertainties about the identity of the *in vivo* acceptor for polymerase and the superficial resemblance of the amphiphilic LU-lipid to a short-chain version of LTC led Fischer and Rösel (1980) to suggest that the true acceptor for the polymerase is LU-lipid and that LTC is an artifact absent from live bacteria. These workers have enzymatically shortened the poly(glycerol phosphate) chain of LTC and have shown that, while the optimum length for acceptor activity is 20 repeating units, it retains activity down to a mean chain length of about 4 units. It should be pointed out, however, that the terminal glycerol phosphate residues of LU and LTC must have, as a result of their different modes of synthesis, opposite stereochemical configurations. The glycerol phosphate arising from CDP-glycerol, in LU, is D-glycerol 1-phosphate, while that in LTC, arising from phosphatidylglycerol (Glaser and Lindsay, 1974), is D-glycerol 3-phosphate. It is not known whether stereochemical specificity for acceptor occurs in the polymerase.

Little is known about the glycosylation of wall teichoic acids. Glycosyl transferases that catalyze the glycosylation of teichoic acid from the appropriate sugar nucleotide have been detected in membrane preparations from many bacteria, and the UDP-glucose: poly(glycerol phosphate) glucosyl transferase of *B. subtilis* 168 has been studied by Brooks *et al.* (1971). Unlike the teichoic acid polymerases, this enzyme appears to be a peripheral membrane protein. Some becomes solubilized during normal membrane isolation procedures, and the membrane-bound form can be extracted into solutions of chaotropic agents such as sodium perchlorate, in an active form. The degree of association of the enzyme with the membrane depends on the stage of growth of the bacterial culture, and is particularly low at early exponential phase (Brooks *et al.*, 1971). The glycosyl transferases are specific for a particular sugar nucleotide and type of teichoic acid. The stage of synthesis of wall teichoic acid at which glycosylation takes place is unknown.

Both wall and membrane teichoic acids usually carry *O*-esterified D-alanine residues either on the alditol residues in the main chain or on the glycosyl substituents. The degree of alanylation varies, but may approach one alanine residue per polymer repeating unit. Nothing is known about the mechanism of alanylation of wall teichoic acid or the stage at which it occurs. Neuhaus *et al.* (1974) studied the process with membrane teichoic acid, where endogenous lipoteichoic acid in membranes of *Lactobacillus casei* accepted alanyl residues from an activated form of alanine synthesized by a soluble enzyme. They proposed the series of reactions:

$$\text{D-alanine} + \text{enzyme (E1)} + \text{ATP} \rightarrow \text{AMP-alanine-E1} + PP_i$$

(Baddiley and Neuhaus, 1960),

$$\text{AMP-alanine-E1} + \text{LTA} \rightarrow \text{alanyl-LTA} + \text{E1} + \text{AMP}$$
$$\text{E2}$$

where E1 is the soluble enzyme and E2 is membrane-bound. Childs and Neuhaus (1980) have recently characterized the membrane-bound acceptor in more detail and have confirmed that it is the membrane teichoic acid. In principle, the same type of mechanism could lead to the attachment of alanyl residues to wall teichoic acid.

In spite of many *in vitro* studies of the mechanism of attachment of teichoic acid to peptidoglycan, there remains considerable uncertainty about the stages in the biosynthesis of these two polymers at which they become linked in growing cells. Wall–membrane preparations of *S. aureus* (Bracha and Glaser, 1976a) and *B. subtilis* W23 (Wyke and Ward, 1977b), and toluenized cells of *B. subtilis* W23, *B. subtilis* 168 (Hancock, 1981), and *M. varians* (Hancock and Baddiley, 1976) synthesize teichoic acid and attach it to the cell wall in the absence of peptidoglycan synthesis, whereas linkage in a wall–membrane preparation of *B. licheniformis* required concomitant peptidoglycan synthesis (Wyke and Ward, 1975). The wall acceptor of the teichoic acid in these cases appears to be peptidoglycan chains already cross-linked to the wall by peptide bridges but still in the process of glycan-chain elongation at the time of cell disruption and therefore still bound to the membrane by their growing ends (Ward and Perkins, 1973). These systems do not elongate teichoic acid chains already linked to the wall; incorporation into the wall is entirely dependent on *de novo* synthesis of LU. Although LU can be incorporated slowly into the wall without concomitant teichoic acid synthesis, the units so incorporated do not act as acceptors for teichoic acid synthesized subsequently, in either wall–membrane preparations or toluenized cells of *B. subtilis* W23 (Wyke and Ward, 1977b; Hancock, 1981). This evidence implies that teichoic acid chains are complete and attached to LU before they are linked to peptidoglycan.

There is evidence that teichoic acid can be linked only to peptidoglycan that is already attached by peptide cross-bridges to the wall. Tynecka and Ward (1975) examined the peptidoglycan released into the growth medium by cultures of a poorly lytic strain of *B. licheniformis* treated with penicillin. This material, which was shown not to be a wall-turnover product but rather new peptidoglycan whose attachment to wall was inhibited by the antibiotic, had no teichoic acid attached to it. In contrast to this result, McArthur *et al.* (1980b) have demonstrated linkage of concomitantly synthesized teichoic acid and peptidoglycan by a membrane preparation from *M. varians* in which no cross-linking of peptidoglycan was possible. These results raise the possibility of regulatory mechanisms that are inoperable in cell-free systems.

Because teichoic acid LU becomes attached to peptidoglycan already attached to the cell wall in wall–membrane and toluenized cell preparations, it must be assumed that this reaction takes place on the outer surface of the cytoplasmic membrane. The precursors of teichoic acid and LU, the sugar nucleotides and alditol nucleotides, are synthesized in the cytoplasm by the action of pyrophosphorylases that are believed to be soluble enzymes (Anderson *et al.*, 1973). However, the locations of the membrane-bound enzymes that catalyze the intervening reactions in the synthesis of LU and teichoic acid are less clear and none of the enzymes has so far been purified to homogeneity. Despite this paucity of information there is some indirect evidence that the enzymes comprise a tightly organized transmembrane complex that may in turn work in a coordinated way with the enzymes of peptidoglycan synthesis. This concept was first put forward by Watkinson *et al.* (1971) who obtained evidence that teichoic acid synthesis and peptidoglycan synthesis shared access to a pool of prenyl phosphate carrier lipid in *S. lactis* I3. Bacitracin did not inhibit the synthesis of teichoic acid catalyzed by membrane preparations in the absence of peptidoglycan synthesis. If, however, peptidoglycan precursors were also present, bacitracin did reduce the rate of teichoic acid synthesis. The antibiotic is known to sequester prenyl pyrophosphate formed during peptidoglycan synthesis and thus prevent its recycling for further rounds of polymer synthesis. It was concluded that, as prenyl pyrophosphate was not formed during teichoic acid synthesis, the effect of bacitracin must have been to reduce the size of a prenyl monophosphate pool common to the two pathways. Similar results were later obtained with *B. licheniformis* and *B. subtilis* (Anderson *et al.*, 1972). The elegant work of Mauck and Glaser (1972b) clearly showed that molecules of teichoic acid and peptidoglycan found linked together in the wall of *B. subtilis* were synthesized concomitantly, and the same conclusion was reached by Tomasz *et al.* (1975) in a study of *Streptococcus pneumoniae*. Moreover, Ward (1981) has demonstrated that in walls of *B. licheniformis* and *B. subtilis* that contain covalently linked polymers in addition to teichoic acid, individual peptidoglycan chains always carry only one type of secondary polymer or teichoic acid. These observations support the view that sites of synthesis of peptidoglycan in the membrane are in close contact with the sites of synthesis of other wall polymers.

The first evidence for organization of the teichoic acid-synthesizing enzymes came from an investigation of teichoic acid synthesis during synchronous germination of spores of *B. subtilis* (Chiu *et al.*, 1968). The cells produced both fully glucosylated and unglucosylated polymer. The two polymers began to be synthesized at slightly different times, and only the acquisition of unglucosylated polymer was inhibited by chloramphenicol. Apparently, only a proportion of the biosynthetic sites acquired glucosyl transferase, and indeed, walls of vegetative cells of this strain contain a mixture of fully glucosylated and unglucosylated teichoic acids (Chiu *et al.*, 1966). In some strains of *S. aureus* each biosynthetic site apparently contains either an α- or a β-*N*-acetylglucosamine transferase but not both; the teichoic acid chains can be separated immunologically into those substituted only with the α anomer and those carrying only the β anomer (Tori *et al.*, 1964).

Because the teichoic acid LU is assembled on a prenyl phosphate lipid, it is attractive to suppose that this lipid forms a center around which the enzymes cluster. Leaver *et al.* (1981) have solubilized enzymes that synthesize teichoic acid and LU, from membranes of *M. varians* in Triton X-100. Fractionation of the extract on a sucrose density gradient yielded a protein fraction that catalyzed the synthesis of the main polymer chain in the absence of LU precursors and another much less dense fraction that catalyzed the synthesis and accumulation of LU-lipid intermediates I to III but not polymer synthesis. The latter fraction, which had a distinctive polypeptide composition, could not be fractionated further without loss of activity and therefore appeared to consist of a complex of the enzymes associated with prenol lipid acceptor. The first reaction in this complex is the transfer of *N*-acetylglucosamine 1-phosphate form UDP-*N*-acetylglucosamine to the prenyl phosphate. This reaction is analogous to the initial reaction of peptidoglycan polymerization, the transfer of *N*-acetylmuramylpeptide phosphate to prenyl phosphate, which Neuhaus and his colleagues have termed the translocase reaction. It should be emphasized, however, that there is no evidence that such reactions involve translocation of the sugar phosphate group across the membrane while attached to the lipid carrier. All the physical studies of prenyl phosphate carrier lipids in membranes have failed to detect any potential for transmembrane flip-flop motion (Weppner and Neuhaus, 1977; Hanover and Lennartz, 1979; McCloskey and Troy, 1980). The balance of probabilities favors a transmembrane protein complex as the mediator of group translocation in the synthesis of cell wall polymers. In this connection, the observation (Bertram *et al.*, unpublished) that *B. subtilis* W23 cells suspended in isotonic buffer rapidly acquire the ability to synthesize teichoic acid and peptidoglycan with remarkable efficiency from externally added nucleotide precursors, as their walls begin to autolyse, is particularly interesting. No membrane permeability to nucleotides can be detected in these protoplasting cells, or in free protoplasts, which are also highly active. The polymeric products are mainly extracellular and their synthesis can be blocked reversibly by extracellular trypsin; activity is regained on the addition of trypsin inhibitor. When the protoplasts are gently burst, the total synthetic activity approximately doubles. It therefore appears that either all or some of the biosynthetic sites are permanently exposed on the outer surface of the membrane, but are shielded from the external environment by close contact with the wall, or more probably, that the protein complex at each site is temporarily exposed to the outside of the membrane at some stage in the biosynthetic cycle. This implies some type of reorientation or rotation of the complex in the membrane during biosynthesis of the polymers. Lack of fluidity in the surrounding bilayer might hinder this movement and account for the inefficiency of membranes of *M. varians* in attaching main polymer chain to LU-lipid at temperatures below 35°C, where lipids are rapidly synthesized (Roberts *et al.*, 1979).

REFERENCES

Anderson, R. G., Hussey, H., and Baddiley, J. (1972). *Biochem. J.* **127**, 11–25.

Anderson, R. G., Douglas, J., Hussey, H., and Baddiley, J. (1973). *Biochem. J.* **136**, 871–876.

Baddiley, J. (1972). In *Essays in Biochemistry* (P. N. Campbell and F. Dickens, eds.), Vol. 8, pp. 35–77, Academic Press, New York.

Baddiley, J., and Neuhaus, F. C. (1960). *Biochem. J.* **75**, 579–587.

Bracha, R., and Glaser, L. (1976a). *J. Bacteriol.* **125**, 872–879.

Bracha, R., and Glaser, L. (1976b). *Biochem. Biophys. Res. Commun.* **72**, 1091–1094.

Bracha, R., Davidson, R., and Mirelman, D. (1978). *J. Bacteriol.* **134**, 412–418.

Brooks, D., Mays, L. L., Hatefi, Y., and Young, F. E. (1971). *J. Bacteriol.* **107**, 223–229.

Burger, M. M., and Glaser, L. (1964). *J. Biol. Chem.* **239**, 3168–3177.

Childs, W. C., and Neuhaus, F. C. (1980). *J. Bacteriol.* **143**, 293–301.

Chiu, T., Burger, M. M., and Glaser, L. (1966). *Arch. Biochem. Biophys.* **116**, 358–367.

Chiu, T., Younger, J., and Glaser, L. (1968). *J. Bacteriol.* **95**, 2044–2050.

Coley, J., Tarelli, E., Archibald, A. R., and Baddiley, J. (1978). *FEBS Lett.* **88**, 1–9.

Fiedler, F., and Glaser, L. (1974). *J. Biol. Chem.* **249**, 2684–2689.

Fischer, W., and Rösel, P. (1980). *FEBS Lett.* **119**, 224–226.

Glaser, L., and Lindsay, B. (1974). *Biochem. Biophys. Res. Commun.* **59**, 1137–1144.

Hancock, I. C. (1981). *European J. Biochem.* **119**, 85–90.

Hancock, I. C., and Baddiley, J. (1976). *J. Bacteriol.* **125**, 880–886.

Hancock, I. C., Wiseman, G., and Baddiley, J. (1976). *FEBS Lett.* **69**, 75–78.

Hanover, J. A., and Lennartz, W. J. (1979). *J. Biol. Chem.* **254**, 9237–9246.

Heckels, J. E., Archibald, A. R., and Baddiley, J. (1975). *Biochem. J.* **149**, 637–647.

Hussey, H., Brooks, D., and Baddiley, J. (1969). *Nature (London)* **221**, 665–666.

Ishimoto, N., and Strominger, J. L. (1966). *J. Biol. Chem.* **241**, 639–645.

Kennedy, L. D., and Shaw, D. R. D. (1968). *Biochem. Biophys. Res. Commun.* **32**, 861–865.

Lambert, P. A., Coley, J., and Baddiley, J. (1977). *FEBS Lett.* **79**, 327–330.

Leaver, J., Hancock, I. C., and Baddiley, J. (1981). *J. Bacteriol.* **146**, 847–852.

McArthur, H. A. I., Roberts, F. W., Hancock, I. C., and Baddiley, J. (1978). *FEBS Lett.* **86**, 193–200.

McArthur, H.A.I., Hancock, I. C., Roberts F. W., and Baddiley, J. (1980a). *FEBS Lett.* **111**, 317–323.

McArthur, H. A. I., Roberts, F. W., Hancock, I. C., and Baddiley, J. (1980b). *Bioorg. Chem.* **9**, 55–62.

McArthur, H. A. I., Hancock. I. C., and Baddiley, J. (1981). *J. Bacteriol.* **145**, 1222–1231.

McCloskey, M. A., and Troy, F. A. (1980). *Biochemistry* **19**, 2056–2060.

Mauck, J., and Glaser, L. (1972a). *Proc. Natl. Acad. Sci. USA* **69**, 2386–2390.

Mauck, J., and Glaser, L. (1972b). *J. Biol. Chem.* **247**, 1180–1187.

Neuhaus, F. C., Linzer, R., and Reusch, V. M. (1974). *Ann. N.Y. Acad. Sci.* **235**, 502–518.

Roberts, F. W., McArthur, H. A. I., Hancock, I. C., and Baddiley, J. (1979). *FEBS Lett.* **97**, 211–216.

Tomasz, A., McDonnell, M., Westphal, M., and Zanati, E. (1975). *J. Biol. Chem.* **250**, 337–341.

Tori, M., Kabat, E. A., and Bezer, A. E. (1964). *J. Exp. Med.* **120**, 13–15.

Tynecka, Z., and Ward, J. B. (1975). *Biochem. J.* **146,** 253–267.

Ward, J. B. (1981). *Microbiol. Rev.* **45,** 211–243.

Ward, J. B., and Perkins, B. (1973). *Biochem. J.* **135,** 721–728.

Watkinson, R. J., Hussey, H., and Baddiley, J. (1971). *Nature (London)* **229,** 57–59.

Weppner, L., and Neuhaus, F. C. (1977). *J. Biol. Chem.* **252,** 2296–2303.

Wyke, A. W., and Ward, J. B. (1975). *Biochem. Biphys. Res. Commun.* **65,** 877–885.

Wyke, A. W., and Ward, J. B. (1977a). *FEBS Lett.* **73,** 159–163.

Wyke, A. W., and Ward, J. B. (1977b). *J. Bacteriol.* **134,** 412–418.

41

Solution Structure of Acyl Carrier Protein

Charles O. Rock and John E. Cronan, Jr.

The role of acyl carrier protein (ACP) in the biosynthesis of membrane lipid precursors is well established (for review see Vagelos, 1971, 1973; Prescott and Vagelos, 1972). ACP and its thioesters interact specifically with at least a dozen enzymes of fatty acid biosynthesis in *Escherichia coli*, and acyl-ACPs are substrates for the *sn*-glycerol-3-phosphate acyltransferase. The complete primary structure of ACP is known (Vanaman *et al.*, 1968). The molecular weight is 8847, and the protein contains a 4'-phosphopantetheine moiety as a prosthetic group attached via a phosphodiester linkage to Ser 36. The acyl intermediates are attached as thioesters to the terminal sulfhydryl of the prosthetic group. Recent work on the regulation of membrane lipid biogenesis has focused attention on several enzyme systems that utilize acyl-ACP substrates. The involvement of ACP in these pathways has renewed interest in the solution structure of ACP and how it relates to the biochemical behavior of this protein. The purpose of this chapter is to summarize and review recent experimental findings on the structure of ACP particularly with regard to lipid–protein interactions.

Recent experimental evidence is at odds with the view of Takagi and Tanford (1968) that ACP is a typical globular protein, compactly folded and sparingly hydrated. Rock and Cronan (1979b) have reported that ACP has a Stokes radius of 19.6 Å as determined by gel filtration. This radius is close to that expected for a globular protein of molecular weight 20,000, but sedimentation equilibrium experiments demonstrated that ACP is a monomer of molecular weight 9000 in solution (Rock and Cronan, 1979b). Therefore, the anomalous elution position of ACP was concluded to be due to the protein moiety possessing an asymmetric shape, and a frictional ratio of 1.43 was calculated (Rock and Cronan, 1979b). Frictional ratios of this magnitude have been reported for both calmodulin and troponin c (Dedman *et al.*, 1977), and thus structural asymmetry may be a general property of acidic proteins. The highly acidic nature of ACP can account for many of its properties. These include an isoelectric point of pH 4.1, a pH of minimum solubility of 3.9, high α-

_____

Charles O. Rock • Department of Biochemistry, St. Jude Children's Research Hospital, Memphis, Tennessee 38101.   ***John E. Cronan, Jr.*** • Department of Microbiology, University of Illinois, Urbana, Illinois 61807.

helical content, and anomalous behavior in gel electrophoresis systems containing anionic detergents (Rock and Cronan, 1979b).

Enzymatic synthesis of ACPSH thioesters was an important first step in investigating the structure of acyl-ACP. Earlier chemical methods for making acyl-ACP substrates introduced protein modifications that have been shown to result in a drastic loss of ACP secondary structure (Schulz, 1975; Abita *et al.*, 1971) and biological activity (Schulz, 1975; Jaworski and Stumpf, 1974). Jaworski and Stumpf (1974) were the first to use an enzymatic approach and were able to prepare a mixture of saturated acyl-ACPs using chloroplast extracts. A more general method using the enzyme acyl-ACP synthetase from *E. coli* (Ray and Cronan, 1976; Rock and Cronan, 1979a, 1981a) has been developed to prepare a wide range of acyl-ACP chain lengths (Rock and Garwin, 1979; Rock *et al.*, 1981b). Acyl-ACP synthetase preparations possessing different degrees of purity have been used to prepare acyl-ACPs having chain lengths between 8 and 18 carbons (Spencer *et al.*, 1978, 1979; Rock and Garwin, 1979; Rock *et al.*, 1981b). Homogeneous samples of acyl-ACP are obtained using reversed-phase chromatography on octyl-Sepharose as the final step to resolve ACP from acyl-ACP (Rock and Garwin, 1979). In contrast to ACP, acyl-ACP does adsorb to this column and can be subsequently eluted from the column by gradient or step elution with aqueous solutions of 2-propanol. In addition to the physical characterization of these thioesters described below, acyl-ACPs prepared by this method are excellent substrates for enzymes of fatty acid (Garwin *et al.*, 1980) and phospholipid (Rock *et al.*, 1981c) synthesis. With this method, one is assured of obtaining unmodified acyl-ACP that can be used as a standard to evaluate other synthetic approaches.

The drawback to the acyl-ACP synthetase method is that it is time consuming and not generally applicable to all acyl-ACP chain lengths in high yield. The yield of acyl-ACP is highest for myristic and palmitic acids, less for their unsaturated counterparts, and decreases steadily with decreasing chain length. Acyl-ACPs having chain lengths less than 8 carbons have not been prepared using the acyl-ACP synthetase method. The yields of 8- and 10-carbon saturated acyl-ACPs are poor, and repeated attempts to prepare *cis*-3-decanoyl-ACP have been unsuccessful. Recently, a chemical method was developed based on the specific acylation of the prosthetic group sulfhydryl by *N*-acylimidazoles (Cronan and Klages, 1981). Under proper conditions, only the sulfhydryl group is acylated, and comparison of these acyl-ACPs to those prepared enzymatically has shown that the protein moiety retains its native structure (Cronan and Klages, 1981). This method has permitted the synthesis of short-chain acyl-ACP that cannot be prepared by acyl-ACP synthetase.

ACP is a highly charged, acidic water-soluble protein that functions as a carrier of a hydrophobic acyl moiety. One of the central questions being investigated is: Does noncovalent interaction occur between the fatty acid and the protein moieties, and if so, what are the conformational consequences? The ease of purification (Rock and Cronan, 1980), small size, high solubility, and simple amino acid composition make ACP an ideal subject for NMR spectroscopy, and this technique gave the first experimental suggestion of fatty acid–protein interaction. Gally *et al.* (1978) observed two fluorine resonances of equal height and width for 6,6-difluorotetradecanoyl-ACP, suggesting substantial fatty acid–protein interaction at this point in the acyl chain. In contrast, the analysis of 13,13-difluorotetradecanoyl-ACP gave a spectrum with one broad peak indicative of less, if any, interaction at this position. Further study (Spencer and Prestegard, unpublished) has shown that the splitting in the fluorine signal from 6,6-difluorotetradecanoyl-ACP is pH dependent, and they conclude that the splitting is due to the protonation of a protein functional group with a p*K* of 6.3. The protein spectrum of ACP is well resolved, and specific assignments have been made to all resonances in the aromatic region (Gally *et al.*, 1978). Proton resonances from histidine and one of the two phenylalanines are perturbed in palmitoyl-ACP as compared to ACP, but a glutamic acid residue has been suggested as the titratable group responsible for the observed splitting. As glutamic acid is the most abundant amino acid in ACP, the specific residue responsible is unknown.

ACP has been shown to undergo reversible conformational change between its native α-helical ($\approx 70\%$) structure and a random coil (Schulz, 1975). A steady loss of ACP secondary structure as judged by circular dichroism is observed with increasing pH (Schulz, 1975) or temperature (Schulz, 1977). The conformation change occurring in both instances can be reversed by addition of divalent cations and to a lesser extent by monovalent salts. ACP binds several molecules of CA^{2+} at pH 7.6 (Schulz, 1972), and Argos (1977) has pointed out the similarity between ACP and several Ca^{2+}-binding proteins. However, no specific requirement for Ca^{2+} or other ion has been reported for any ACP function. The cause of the pH-induced denaturation of ACP has been attributed to charge repulsion between adjacent carboxyl functions of this highly acidic protein (Schulz, 1975). Divalent cations, and salts in general, are considered to counter this effect by charge neutralization (Schulz, 1975).

Another manifestation of pH-induced denaturation is an increase in the Stokes radius of ACP as judged by gel filtration chromatography (Rock and Cronan, 1979b). A comparison of acyl-ACP to ACP was possible due to the abnormal stability of the thioester bond of acyl-ACP to base hydrolysis (Rock and Cronan, 1979b). Significantly, under denaturing conditions, acyl-ACP possesses a smaller Stokes radius than ACP, showing that the presence of the acyl group stabilizes the protein moiety to pH-induced hydrodynamic expansion (Rock and Cronan, 1979b). Acyl-ACPs having chain lengths from 8 to 18 carbons were found to stabilize the protein to the same extent. In an extension of this work, a native gel electrophoresis system (Rock and Cronan, 1981b) has been employed to analyze the conformational stability of a variety of ACP derivatives (Rock et al., 1981a). Attachments of hydrophilic ligands to the sulfhydryl of ACP resulted in less stable protein structures, whereas the presence of a hydrophobic thioester resulted in stabilization of the protein conformation (Rock et al., 1981a). All ACP derivatives resolved by this gel system were found to have unique ^{31}P NMR spectra at neutral pH and relatively high ionic strength (Rock et al., 1981a). Less stable ACP derivatives were found to have ^{31}P NMR chemical shifts displaced downfield from ACP, whereas acyl-ACPs were found to have a chemical shift upfield from ACP. As in the gel filtration experiments, acyl-ACP chain lengths from 14 to 18 carbon atoms were found to have identical ^{31}P NMR spectra and to stabilize the protein moiety to the same extent.

All acyl-ACP chain lengths examined in the above studies have identical effects on the protein, showing that they all possess the minimum structural features required to confer maximum stability to the protein moiety. Recently, short-chain acyl-ACPs have been prepared and the stability of these derivatives tested (Cronan, 1982, unpublished) using the gel electrophoresis system. The order of stability was found to be octyl > hexyl > butyl > acetyl > free ACP. The adsorption of these acyl-ACPs to octyl-Sepharose has also been examined. As expected for true reversed-phase chromatography, a linear relationship between the log of the elution position and the carbon number was established for acyl-ACP chain lengths 10 to 18 (Rock and Garwin, 1979). Cronan (1982, unpublished) found that octyl-ACP falls just below the calibration line, but lower homologs fall far below the line and acetyl-ACP and butyl-ACP do not bind at all. Taken together, these results strongly support the idea that the major lipid–protein interaction in acyl-ACP occurs between the first six carbons of the acyl chain and the protein.

ACP has not been successfully crystallized despite the efforts of several laboratories.

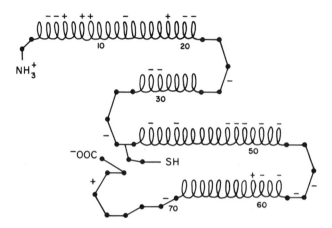

Figure 1. Schematic diagram of the predicted secondary structure of ACP. Residues are represented in their respective conformational state: ℓ, helical; •—•, coil; β turns are indicated by chain reversals.

In order to bring together the available data into a consistent picture in the absence of crystallographic data and suggest future avenues of research, a model for ACP secondary structure was constructed using a predictive algorithm (Rock and Cronan, 1979b). The method used has been shown to be quite accurate particularly with small, conformationally flexible proteins (Chou and Fasman, 1978). Four α-helical regions interrupted by three β turns were significantly predicted. The 4'-phosphopantetheine prosthetic group is attached to Ser 36, which was predicted to occupy the fourth position of the second β turn in the molecule (Fig. 1). The predicted ACP structure provides a rationale for most of the physical behavior of ACP discussed above. First, the α-helical content predicted is close to that observed experimentally. Second, ACP is predicted to be an asymmetric molecule with a long axis of 43 Å. This value is much closer to the observed Stokes radius than the 32-Å radius expected for a globular molecule. Third, the model offers an explanation for the increased stability of acyl-ACP to pH-induced denaturation. The only two hydrophobic regions of ACP are predicted to occur in adjacent helical segments in close proximity to the thiol group. The length of this domain was calculated to be 7.5 Å and would be sufficient to accommodate the first six carbon atoms of a fatty acid chain.

The results discussed above suggest that ACP may play a more complex role in the mechanism and/or regulation of membrane lipid synethsis than previously believed. Taken together, they show that ACP exists in specific conformational states depending on the structure of the group attached to the active-site sulfhydryl. The possibility that the acyl-chain specificity of enzymes that utilize ACP substrates may be due in large part to recognition of specific ACP conformers needs to be explored.

ACKNOWLEDGMENTS. The authors' research reported herein was supported by NIH Grants GM 29053 and GM 28035 (C.O.R.), GM 26165, and AI 15650 (J.E.C.), NSF Grant 79-25689 (J.E.C.), NCI Cancer Center (CORE) Grant CA 21765 (C.O.R.), and ALSAC (C.O.R.).

REFERENCES

Abita, J. P., Lazdunski, M., and Ailhaud, G. (1971). *Eur. J. Biochem.* **23,** 412–420.
Argos, P. (1977). *Biochemistry* **16,** 665–672.

Chou, P. Y., and Fasman, G. D. (1978). In *Advances in Enzymology* (Meister, A., ed.), Vol. 48, pp. 45–148, Wiley, New York.

Cronan, J. E., Jr., and Klages, A. L. (1981). *Proc. Natl. Acad. Sci. USA* **78**, 5440–5444.

Dedman, J. R., Potter, J. D., Jackson, R. L., Johnson, D., and Means, A. R. (1977). *J. Biol. Chem.* **252**, 8415–8422.

Gally, H. U., Spencer, A. K., Armitage, I. M., Prestegard, J. H., and Cronan, J. E., Jr. (1978). *Biochemistry* **17**, 5377–5382.

Garwin, J. L., Klages, A. L., and Cronan, J. E., Jr. (1980). *J. Biol. Chem.* **255**, 11949–11956.

Jaworski, J. G., and Stumpf, P. K. (1974). *Arch. Biochem. Biophys.* **162**, 166–173.

Prescott, D. J., and Vagelos, P. R. (1972). *Adv. Enzymol.* **36**, 269–311.

Ray, T. K., and Cronan, J. E., Jr. (1976). *Proc. Natl. Acad. Sci. USA* **73**, 4374–4378.

Rock, C. O., and Cronan, J. E., Jr. (1979a). *J. Biol. Chem.* **254**, 7116–7122.

Rock, C. O., and Cronan, J. E., Jr. (1979b). *J. Biol. Chem.* **254**, 9778–9785.

Rock, C. O., and Cronan, J. E., Jr. (1980). *Anal. Biochem.* **102**, 362–364.

Rock, C. O., and Cronan, J. E., Jr. (1981a). *Methods Enzymol.* **71**, 163–168.

Rock, C. O., and Cronan, J. E., Jr. (1981b). *Methods Enzymol.* **71**, 341–351.

Rock, C. O., and Garwin, J. L. (1979). *J. Biol. Chem.* **254**, 7123–7128.

Rock, C. O., Cronan, J. E., Jr., and Armitage, I. B. (1981a). *J. Biol. Chem.* **256**, 2669–2674.

Rock, C. O., Garwin, J. L., and Cronan, J. E., Jr. (1981b). *Methods Enzymol.* **72**, 397–403.

Rock, C. O., Goelz, S. E., and Cronan, J. E., Jr. (1981c). *J. Biol. Chem.* **256**, 736–742.

Schulz, H. (1972). *Biochem. Biophys. Res. Commun.* **46**, 1446–1453.

Schulz, H. (1975). *J. Biol. Chem.* **250**, 2299–2304.

Schulz, H. (1977). *FEBS Lett.* **78**, 303–306.

Spencer, A. K., Greenspan, A. D., and Cronan, J. E., Jr. (1978). *J. Biol. Chem.* **253**, 5922–5926.

Spencer, A. K., Greenspan, A. D., and Cronan, J. E., Jr. (1979). *FEBS Lett.* **101**, 253–256.

Takagi, T., and Tanford, C. (1968). *J. Biol. Chem.* **243**, 6432–6435.

Vagelos, P. R. (1971). *Curr. Top. Cell Regul.* **3**, 119–166.

Vagelos, P. R. (1973). In *The Enzymes* (P. D. Boyer, ed.), 3rd ed., Vol. 8, pp. 155–199, Academic Press, New York.

Vanaman, T. C., Wakil, S. J., and Hill, R. L. (1968). *J. Biol. Chem.* **243**, 6420–6431.

V

Bioenergetics of Electron and Proton Transport in Mitochondria

42

Thermodynamic Analysis of Biomembrane Energy Transduction

H. V. Westerhoff and K. van Dam

1. INTRODUCTION

The success (Boyer *et al.*, 1977) of the chemiosmotic hypothesis (Mitchell, 1961) has led to the general insight that metabolism (chemistry) and transport (physics) are connected by two rather than by one tie. Transport is no longer parallel to metabolism—it is intimately linked to it at the level of oxidative or photophosphorylation.

A possibly important reason why general acceptance of the chemiosmotic hypothesis as the primary working model (Boyer *et al.*, 1977) has taken so much time is the fact that it is the result of a confrontation of two paradigms (Kuhn, 1970): on the one hand, the view of metabolic enzymology, working within the framework of kinetics (concentrations, Michaelis constants, maximal velocities; Roberts, 1977), and on the other hand, electrophysiology, discussing experiments in terms of permeabilities, electromotive forces, and membrane potentials (Cole, 1968). Mitchell (1961) reconciled the two approaches by showing that theoretically the chemical Gibbs free energy stored in the terminal pyrophosphate bond of ATP is equivalent to and, at least theoretically, convertible into an electrochemical gradient of protons. Since then, this interconvertibility has experimentally been shown in oxidative and photophosphorylation (for a review see Boyer *et al.*, 1977).

The bridge between the metabolic and the transport approach is, however, not yet complete with this interconvertibility. The question still needs to be solved how the rates of metabolic reactions depend on transport processes and vice versa.

Enzyme kinetics (Roberts, 1977), though a powerful tool in the study of isolated soluble enzymes, is too complicated already to describe analytically a set of more than two enzyme-catalyzed reactions. Therefore, when it is applied to a system consisting of interlocked transport and metabolic reactions, either severe equilibrium assumptions are made with respect to partial processes (Wilson *et al.*, 1979), the transport per se is not considered

H. V. Westerhoff and K. van Dam • Laboratory of Biochemistry, University of Amsterdam, 1018TV Amsterdam, The Netherlands.

at all (Wilson *et al.*, 1979; Stoner and Sirak, 1979), or the treatment is confined to a single transport reaction (Hladky, 1980).

On the other hand, the electrophysiological way of describing transport (Cole, 1968) is not suited for inclusion of biochemical reactions at either side of the membrane, nor is it well equipped to describe saturation characteristics that are so common in all biological (enzyme-catalyzed) reactions, even at the level of transport itself.

Many authors have been aware of this gap in our knowledge and have suggested approaches to a solution. It is the purpose of this contribution to evaluate these approaches and to point out the problems that are still to be investigated. Moreover, we shall demonstrate how a most promising method of description can be applied to systems that include both transport and metabolism.

2. PHENOMENOLOGICAL THERMODYNAMICS OF PROCESSES NEAR EQUILIBRIUM

It is to classical thermodynamics that we owe the possibility of comparing heat to work (in equilibrium) by expressing both as energy. The nonequilibrium extension (see Katchalsky and Curran, 1967) makes it possible to describe any system that is close to equilibrium by expressing the flows (reaction rates, fluxes) as linear functions of all corresponding forces (ΔG's of the reactions, free-energy gradients). For our discussion it is especially relevant that each flow will in principle depend on all forces, reflecting the fact that this theory is able to describe the interaction between, for instance, heat flow and flow of matter, or between metabolism and transport. The phenomenon of "active transport" is the simplest example: the transmembrane flux of a solute may be directed against its electrochemical gradient because it also depends on the free-energy difference of a coupled chemical reaction.

Rottenberg *et al.* (1970) have applied these principles to oxidative phosphorylation. They defined the system as consisting of oxidation (rate J_o, force ΔG_o), ATP synthesis (rate J_p, force ΔG_p), and transmembrane proton flux (rate J_H, force $\Delta \tilde{\mu}_H$). Then they wrote down the resulting proportional equations, the so-called phenomenological equations:

$$J_o = L_{oo}\Delta G_o + L_{oH}\Delta \tilde{\mu}_H + L_{op}\Delta G_p$$

$$J_H = L_{oH}\Delta G_o + L_{HH}\Delta \tilde{\mu}_H + L_{pH}\Delta G_p$$

$$J_p = L_{op}\Delta G_o + L_{pH}\Delta \tilde{\mu}_H + L_{pp}\Delta G_p$$

It is important to note that, in addition to the general principles, these authors also applied another powerful theorem of nonequilibrium thermodynamics: Onsager's reciprocal relations. This theorem states that the cross coefficients relating the flows to each other's forces are equal (Katchalsky and Curran, 1967). This reciprocity was experimentally verified (Rottenberg, 1973), although some authors have found small but reproducible deviations (Stucki, 1980; Van den Berg, personal communication). In addition, it was shown that oxidation and phosphorylation rates do depend linearly on the Gibbs free energy of oxidation and phosphorylation (Rottenberg, 1973; Stucki, 1980). As a result, the values of the proportionality constants L could be calculated from the experimental data.

From the proportionality constants the magnitude of the so-called coupling constant,

q, which is defined as a combination of L's, could be calculated (Rottenberg, 1978). This coupling constant can vary between 1 for full coupling and 0 for no coupling at all.

Thus, an internally consistent black-box description of oxidative phosphorylation was accomplished. Several points, however, made this description either unacceptable or irrelevant in the eyes of some bioenergeticists and transport scientists:

1. The assumed proportional and reciprocal relations between flows and forces can only be generally proven to hold near equilibrium, whereas many metabolic and transport processes occur far from equilibrium.
2. The assumed proportional relations between flows and forces are evidently not universally valid, as they are in conflict with the observed saturation behavior of enzyme-catalyzed reactions.
3. The resulting relations are uninformative: they describe the system as a black box and do not indicate the influence of molecular parameters (such as the H^+/O stoichiometry, the specific activity of translocators, the proton permeability of the membrane, etc.) on the overall performance of the system.

3. "MECHANISTIC" THERMODYNAMICS OF PROCESSES NEAR EQUILIBRIUM

Especially the last point was taken seriously by a number of groups (Blumenthal *et al.*, 1967; Oster *et al.*, 1973; Lagarde, 1976; van Dam and Westerhoff, 1977; Hill, 1979; Westerhoff *et al.*, 1979; Mikulecky *et al.*, 1979; van Dam *et al.*, 1980). They came up with solutions that are basically similar. Rather than describing the complete system of fully coupled reactions as a whole (Rottenberg *et al.*, 1970; Rottenberg, 1978; Stucki, 1980), these authors first took stock of the system (or a hypothetical model of the system) in terms of the mutually independent processes that form its basis. For the chemiosmotic concept of oxidative phosphorylation, this would be (Lagarde, 1976; van Dam and Westerhoff, 1977): oxidative proton pump + ATPase proton pump + proton leak + substrate translocators, product translocators. Then for each of these processes a relation between the flow of the process and the force driving the process would be written down. Most often each of these relations was taken as proportional, as in the theory of near-equilibrium thermodynamics (see above). The most important reason for this is that with proportional relations the mathematical manipulations that follow are simplest. If, for simplicity's sake, we also use proportional relations here, we can easily show that basically only two types of processes exist:

1. An autonomous process, for instance the (facilitated) diffusion of a substance i across a membrane. The relation for this process will be $J_i = L_i^1 \, \Delta\bar{\mu}_i$, where the superscript 1 (from "leak") refers to the fact that the process is autonomous.
2. A process that consists of two strictly coupled processes. For a chemiosmotic reaction:

$$n_H^p H^+_{in} + ATP \rightleftharpoons n_H^p H^+_{out} + ADP + P_i$$

the relations will then be for the rate of ATP hydrolysis:

$$J_p = L_p(\Delta G_p + n_H^p \Delta\bar{\mu}_H)$$

and for the associated rate of H^+ movement:

$$J_H^p = n_H^p J_p = n_H^p L_p(\Delta G_p + n_H^p \Delta\tilde\mu_H).$$

Every other process can be reduced to a combination of autonomous and strictly coupled processes. As an example we may take an ATPase proton pump with a slip in it, so that the ATP can be partly split without movement of protons. One may describe such a slipping pump as:

$$J_p^H = L_p(\Delta G_p + n_H^p \Delta\tilde\mu_H)$$
$$J_H^p = n_H^p L_p(\Delta G_p + n_H^p \Delta\tilde\mu_H)$$
$$J_p^l = L_p^l \Delta G_p$$
$$J_p = J_p^H + J_p^l = (L_p + L_p^l)\Delta G_p + n_H^p L_p \Delta\tilde\mu_H$$

This can be compared with the most general description of a not fully coupled pump which follows from phenomenological irreversible thermodynamics (Rottenberg, 1978) as well as from a near-equilibrium kinetic description (Hill, 1977):

$$J_p = L_{pp}\Delta G_p + L_{pH}\Delta\tilde\mu_H$$
$$J_H = L_{pH}\Delta G_p + L_{HH}\Delta\tilde\mu_H$$

in which $q^2 = L_{pH}^2/(L_{pp} \cdot L_{HH}) < 1$.

Taking

$$L_{pp} = L_p^l + L_p$$
$$L_{pH} = n_H^p \cdot L_p$$
$$L_{HH} = (n_H^p)^2 \cdot L_p$$

the two descriptions are identical.

A system that consists of n autonomous, m fully coupled, and k partially coupled processes can be described by $(n + 2m + 3k)$ basic equations. Usually, only a limited number of parameters in a system are accessible to experimental determination, and thus the present description would not be suitable to work with if the number of equations could not be greatly diminished. This simplification is possible thanks to the occurrence of steady-state conditions. Some of these are relatively trivial. If one assumes rapid equilibration of the proton electrochemical potential within each aqueous phase, then the protonmotive force that affects the oxidation rate in oxidative phosphorylation is identical to the proton-motive force that drives phosphorylation. Moreover, the proton fluxes through the oxidative proton pump and through the ATPase cannot be experimentally discriminated; one can only measure the sum of the two. Other simplifications are less trival, but open to experimental checks. For instance, in steady-state oxidative phosphorylation the intramitochondrial concentrations of phosphate, adenine nucleotides, and oxidative substrate and product are time independent. The number of equations can now be greatly reduced, because the fluxes through the adenine nucleotide translocator and the phosphate translocator have to equal the rate of ATP synthesis.

For mitochondrial oxidative phosphorylation the following equations result:

$$J_o = (L_o + L_o^l)\Delta G_o + n_H^o L_o \Delta\tilde\mu_H$$
$$J_H = n_H^o L_o \Delta G_o + \{(n_H^o)^2 L_o + L_H^l + (n_H^p)^2 L_p\}\Delta\tilde\mu_H + n_H^p L_p \Delta G_p$$
$$J_p = n_H^p L_p \Delta\tilde\mu_H + (L_p + L_p^l)\Delta G_p$$

Comparison with the result of phenomenological irreversible thermodynamics shows that the third shortcoming of the latter has been overcome: the coefficients are no longer phenomenological only, they refer to the underlying processes. For instance, we can now see what the effect of adding a protonophore to the system will be: L_H^l (the proton conductance coefficient of the inner mitochondrial membrane) will increase and thereby L_{HH} increases. Moreover, all other coefficients are expected to remain constant.

The above set of equations is strictly connected with the mechanism of coupling between oxidation and phosphorylation that was assumed in the first place—the chemiosmotic one; for a chemical coupling mechanism each phenomenological coupling coefficient would be replaced by an expression different from the one derived above (Westerhoff, in preparation). Thus, the present description might be called mechanistic irreversible thermodynamics (van Dam and Westerhoff, 1980) in contrast to the earlier phenomenological irreversible thermodynamics.

An interesting property of the above mechanistic irreversible thermodynamic equations is (Oster *et al.*, 1973) that they obey Onsager's reciprocal relations. The reason for this is that the equations start from linear combinations of the basic equations that possess the same symmetry.

4. MOSAIC THERMODYNAMICS

Of the three drawbacks of phenomenological thermodynamics, only one has been eliminated up to this point. We still face the problem that the near-equilibrium proportional relations may not be applicable to the biological system under consideration. Moreover, the effect of saturable enzyme kinetics is still not accounted for in the present equations.

There are at least two possibilities to proceed further from this point. One (Mikulecky *et al.*, 1979) is to find first, either by experiment or by theoretical considerations, the actual exact relation between flow and force of each elementary process in the relevant region far from equilibrium. In detail, such relations are often much more complex than linear. The result is that a combination of elementary processes becomes a mathematically complex and analytically unsolvable mixture. The use of a computer may make it possible to simulate the system (Mikulecky *et al.*, 1979). Although it is then possible in principle to obtain an exact and experimentally testable description, the method has the disadvantage that is common to all computer-assisted studies: the spectator loses sight of the effects that changes in the underlying mechanisms may have on the overall result.

The second possibility aims at a close inspection of the basic relations, to see whether under certain conditions these relations may be approximated by simpler relations with which the mathematical manipulations are still analytically tractable. After the theoretical and experimental demonstration by Prigogine *et al.* (1948) that for chemical reactions close to equilibrium the rates are proportional to the free-energy differences, Rottenberg (1973) was the first to consider conditions far from equilibrium. He showed that for enzyme-catalyzed reactions, at constant substrate or product concentration, in addition to the near-equilibrium proportionality region there generally exists a second range of reaction rates that are linear with (but not proportional to) the Gibbs free-energy change of the reaction. Only recently this important work was extended by van der Meer *et al.* (1980) to include an often more relevant boundary condition, namely that where the sum of substrate and product remains constant. Later it was recognized that the occurrence of an additional linear, but not proportional, relation between the rate of a reaction and the Gibbs free-energy difference of that reaction for a large range of velocities, is not unique for enzyme-catalyzed reactions (van Dam and Westerhoff, 1980). Rothschild *et al.* (1980) have given a more general analysis of the origin of linear relations between flows and forces in biological processes. Also the considerations by Stucki (personal communication) that proportional flow–force relations correspond with higher efficiencies and that biological systems may have evolved to make use of this fact, suggest that approximation of flow–force relations by linear equations is a good starting point. Independent of these theoretical justifications, there exist a number of experimental verifications of linearity between flow and force.

van Dam *et al.* (1980) were the first to add the above approach to the mechanistic irreversible thermodynamic approach. In doing this, they also included two additional features of the far-from-equilibrium linear domain. The dependence of the rate of a process on the force of that process was taken as generally different from the dependence in the near-equilibrium domain. Moreover, the dependence may not be equal for the two forces that determine a given coupled process (Oster *et al.*, 1973; Arata and Nishimura, 1980; van Dam and Westerhoff, 1980). The following equations for oxidative phosphorylation resulted:

$$J_o = (L_o^1 + L_o)(\Delta G_o - \Delta G_b^{\ddagger}) + n_H^o \gamma_o L_o \Delta \bar{\mu}_H$$
$$J_H = n_H^o L_o(\Delta G_o - \Delta G_b^{\ddagger}) + \{(n_H^o)^2 \gamma_o L_o + L_H^1 + (n_H^p)^2 L_p\}\Delta \bar{\mu}_H + n_H^p \Delta G_p$$
$$J_p = n_H^p L_p \Delta \bar{\mu}_H + (n_H^p)^2 L_p \Delta G_p$$

The manner in which the different basic processes appear in these equations is different. Proton leakage is only present through an L_H^1 term, the oxidative proton pump is present through L_o, L_o^1, n_H^o, γ_o, and $\Delta G_b^{\ddagger}$, whereas the ATPase proton pump does not bring with it a γ_p or a $\Delta G_p^{\ddagger}$. Consequently, the resulting equations look like a mosaic, consisting of numerous elements with different colors. We will therefore term this type of approach "mosaic irreversible thermodynamics."

5. THE SATURATION BEHAVIOR OF ENZYME-CATALYZED REACTIONS

Rather than giving only the linear part of the flow–force relation of enzyme-catalyzed reactions, Rottenberg (1973) and van der Meer *et al.* (1980) calculated this relation throughout the whole force region. It appears that generally flow–force relations are S shaped; the linear region is limited by regions that reflect that the rate cannot exceed a certain value. Although this effect can be the single result of the chosen boundary condition, it shows that the straightforward contradiction between saturable enzyme kinetics and linear flow–force relations can be solved by always realizing that the linear approximation is valid for only about 75% of the velocity range. For high and low free-energy differences, a much better approximation may be that the rate of the reaction is independent of the free-energy difference. The best approximation for actual flow–force relations is probably the one given in Fig. 1.

6. APPLICATION OF MOSAIC IRREVERSIBLE THERMODYNAMICS

The physically best defined system to which mosaic irreversible thermodynamics has been applied is bacteriorhodopsin, incorporated into unilamellar liposomes. The theoretical

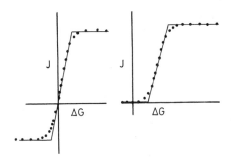

Figure 1. Proposed approximation (solid lines) for the general relation (dots) between rate (J) and free-energy difference (ΔG) of an (enzyme-catalyzed) reaction. The dots have been calculated from Michaelis–Menten kinetics for a reversible reaction with the boundary condition that the sum of substrate and product concentrations remains constant (van der Meer *et al.*, 1980). On the left a small value of $\Delta G°$ is chosen, on the right a large $\Delta G°$; this corresponds to an easily reversible or an irreversible reaction, respectively.

description of this system (Westerhoff *et al.*, 1979) was set up according to the above-mentioned principles and may well serve as a general example of how to develop such a description. Already in a semiquantitative experimental analysis (Hellingwerf *et al.*, 1979) the method was seen to be productive: it allowed us to understand the effects of nigericin and valinomycin on the pH gradient and the membrane potential. The initial membrane potential predicted by the mosaic irreversible thermodynamic analysis was experimentally observed. Moreover, the analysis inspired Hellingwerf *et al.* (1978) to investigate whether bacteriorhodopsin should be considered an ideal current source (proton flow independent of protonmotive force) or a nonideal voltage source (proton flow inhibited by an opposed protonmotive force). The experimental results supported the voltage source model, at that moment a finding certainly not expected. At relatively low membrane potential values both the rate of the light-driven proton pump and the rate of passive proton leakage depend linearly on the pH gradient (Arents *et al.* and Westerhoff *et al.*, both in preparation).

The fact that it is possible at all to treat a light-driven pump as a linear energy converter (see also Rottenberg, 1978) is unexpected in view of theoretical objections to this in the literature (Keizer, 1976; Hill, 1977). However, it may be that these theoretical objections are not stiff (Westerhoff, in preparation), and furthermore, the success of the description gives hope that linear relations can be used for other photochemical systems.

The mosaic irreversible thermodynamic approach has also been applied to more complicated systems, such as mitochondrial oxidative phosphorylation. It is self-evident that more experimental material is required to judge the complete consistency of the description in such systems. Nevertheless, the description itself certainly helps in devising critical experiments to test certain details of a proposed model. For instance, the mosaic approach led to a method to determine the ratio between the number of protons required to synthesize one molecule of ATP and the number of protons released per oxidation event (the "theoretical P/O ratio"; van Dam *et al.*, 1980). A unique aspect of this method is that it allows the determination of these important stoichiometries under conditions where there is true steady-state oxidative phosphorylation.

It can be expected that the mosaic irreversible thermodynamic approach will be applied with success to many other biological systems that otherwise are not easily analyzed in a quantitative way. This expectation is based on the fact that in all cases up to now the method has shown great heuristic value.

REFERENCES

Arata, H., and Nishimura, M. (1980). *Biophys. J.* **32**, 791–806.

Arents, J. C., Van Dekken, H., Hellingwerf, K. J., and Westerhoff, H. V. (1981). *Biochemistry* **20**, 5114–5123.

Blumenthal, R., Caplan, S. R., and Kedem, O. (1967). *Biophys. J.* **7**, 735–757.

Boyer, P. D., Chance, B., Ernster, L., Mitchell, P., Racker, E., and Slater, E. C. (1977). *Annu. Rev. Biochem.* **46**, 955–1026.

Cole, K. S. (1968). In *Membranes, Ions and Impulses*, University of California Press, Berkeley.

Hellingwerf, K. J., Schuurmans, J. J., and Westerhoff, H. V. (1978). *FEBS Lett.* **92**, 181–186.

Hellingwerf, K. J., Arents, J. C., Scholte, B. J., and Westerhoff, H. V. (1979). *Biochim. Biophys. Acta* **547**, 561–582.

Hill, T. L. (1977). *Free Energy Transduction in Biology*, Academic Press, New York.

Hill, T. L. (1979). *Proc. Natl. Acad. Sci. USA* **76**, 2236–2238.

Hladky, S. B. (1980). In *Current Topics in Membranes and Transport* (F. Bronner and A. Kleinzeller, eds.), Vol. 12, pp. 54–164, Academic Press, New York.

Katchalsky, A., and Curran, P. F. (1967). *Non-equilibrium Thermodynamics in Biophysics*, Harvard University Press, Cambridge, Mass.

Keizer, J. (1976). *J. Chem. Phys.* **64**, 4466–4474.

Kuhn, T. S. (1970). *The Structure of Scientific Revolutions*, 2nd ed., University of Chicago Press, Chicago.

Lagarde, A. E. (1976). *Biochim. Biophys. Acta* **426**, 198–217.

Mikulecky, D. C., Huf, E. G., and Thomas, S. R. (1979). *Biophys. J.* **25**, 87–105.

Mitchell, P. (1961). *Nature (London)* **191**, 144–148.

Oster, G. F., Perelson, A. S., and Katchalsky, A. (1973). *Q Rev. Biophys.* **6**, 1–134.

Prigogine, I., Outer, P., and Herbo, C. (1948). *J. Phys. and Colloid Chem.* **52**, 321–331.

Roberts, D. V. (1977). *Enzyme Kinetics*, Cambridge University Press, London.

Rothschild, K. J., Ellias, S. A., Essig, A., and Stanley, H. E. (1980). *Biophys. J.* **30**, 209–230.

Rottenberg, H. (1973). *Biophys. J.* **13**, 503–511.

Rottenberg, H. (1978). In *Progress in Surface and Membrane Science* (D. A. Cadenhead and J.F. Danielli, eds.), Vol. XII, pp. 245–325, Academic Press, New York.

Rottenberg, H., Caplan, S. R., and Essig, A. (1970). In *Membranes and Ion Transport* (E. E. Bittar, ed.), pp. 165–191, Interscience, New York.

Stoner, C. D., and Sirak, H. D. (1979). *J. Bioenerg. Biomembr.* **11**, 113–146.

Stucki, J. W. (1980). *Eur. J. Biochem.* **109**, 269–283.

van Dam, K., and Westerhoff, H. V. (1977). In *Structure and Function of Energy Transducing Membranes* (K. van Dam and B. F. van Gelder, eds.), pp. 157–167, Elsevier, Amsterdam.

van Dam, K., and Westerhoff, H. V. (1980). *Recl. Trav. Chim. Pays-Bas, J. R. Neth. Chem. Soc.* **99**, 329–333.

van Dam, K., Westerhoff, H. V., Krab, K., van der Meer, R., and Arents, J. C. (1980). *Biochim. Biophys. Acta* **591**, 240–250.

van der Meer, R., Westerhoff, H. V., and van Dam, K. (1980). *Biochim. Biophys. Acta* **591**, 488–493.

Westerhoff, H. V., Scholte, B. J., and Hellingwerf, K. J. (1979). *Biochim. Biophys. Acta* **547**, 544–560.

Westerhoff, H. V., Hellingwerf, K. J., Arents, J. C., Scholte, B. J., and van Dam, K. (1981). *Proc. Natl. Acad. Sci. USA* **78**, 3554–3558.

Wilson, D. F., Erecińska, M., Drown, C., and Silver, J. A. (1979). *Arch. Biochem. Biophys.* **195**, 485–493.

43

Regulation of in Vivo Mitochondrial Oxidative Phosphorylation

David F. Wilson

1. INTRODUCTION

The regulation of mitochondrial respiration *in vivo* is designed to maintain a supply of cellular ATP under conditions for which its hydrolysis is sufficiently energetic to do the required metabolic work. *In vivo*, various metabolic pathways in the mitochondrial matrix (citric acid cycle, fatty acid oxidation, etc.) contain dehydrogenases, which catalyze the transfer of reducing equivalents from substrates to NAD^+ and FAD to form NADH and $FADH_2$. These coenzymes are then reoxidized by the respiratory chain, serving as the source of the reducing equivalents that reduce molecular oxygen to water. The amounts of NADH and $FADH_2$ available to the respiratory chain are determined by the metabolic pathways being utilized, but in general, NADH provides most of the total reducing equivalents. Metabolic regulation ensures that these equivalents (NAD couple) are available at a relatively constant oxidation–reduction potential (from -240 mV to -280 mV depending on cell type and metabolic status). This differs markedly from the conditions normally utilized for evaluation of the regulation of mitochondrial functions in suspensions of isolated mitochondria. In the latter case, reducing substrate is added in excess and the potential of the NAD couple can be as negative as -350 mV. This is one of the reasons care must be exercised in extrapolating from *in vitro* measurements to *in vivo* function.

The energetics of mitochondrial oxidative phosphorylation have been extensively evaluated both in suspensions of isolated mitochondria and in intact cells and tissue and the reaction is remarkably efficient. The principal source of reducing equivalents is the intramitochondrial NAD couple, and the overall reaction is

$$H^+ + NADH + 3ADP + 3P_i + \frac{1}{2}O_2 \rightarrow NAD^+ + H_2O \tag{1}$$

David F. Wilson • Department of Biochemistry and Biophysics, University of Pennsylvania, Philadelphia, Pennsylvania 19104.

Measurement of the concentrations of the individual reactants allows calculation of the energetics for this reaction. In suspensions of isolated mitochondria oxidizing malate plus glutamate as substrates, of the approximately 217 kJ of free energy available on oxidation of each mole of NADH, approximately 182 kJ is conserved in the synthesis of 3 moles of ATP for an efficiency of 84% (Erecińska *et al.*, 1974). Comparable calculations for intact cells and tissue give an efficiency of approximately 67–76% (Wilson *et al.*, 1974a,b; Erecińska *et al.*, 1978). The reactions of oxidative phosphorylation, with the exception of the reactions involving molecular oxygen, have long been known to be reversible, and analysis of the energetics of the reactions from NAD to cytochrome *c* (two phosphorylation sites):

$$NADH + 2c^{3+} + 2ADP + 2P_i \rightleftharpoons NAD^+ + 2ATP + 2c^{2+} \tag{2}$$

show the reaction to be associated with a free-energy change of approximately zero (Erecińska *et al.*, 1974; Wilson *et al.*, 1974a,b). That is, this part of oxidative phosphorylation appears to be near equilibrium, in agreement with the observed reversibility of these reactions.

2. DEPENDENCE OF THE RESPIRATORY RATE ON [ATP], [ADP], AND [P_i]

Regulation of any metabolic pathway is designed to provide the necessary product (synthesis of ATP from ADP and P_i) in the required cellular concentration by utilizing metabolites (NADH or $FADH_2$ and O_2) at their physiological concentrations. Because cellular consumption of ATP occurs primarily in the cytoplasm, mitochondrial oxidative phosphorylation must utilize cytosolic ADP and P_i to provide cytosolic ATP. Actual phosphorylation occurs in the mitochondrial matrix space, making transport of these metabolites an essential intermediate of oxidative phosphorylation, but *in vivo* the rate of ATP synthesis must be dependent on the cytosolic levels of ATP, ADP, and P_i. Agreement exists that the respiratory rates of suspensions of isolated mitochondria are dependent on the [ATP]/[ADP] ratio rather than on the concentration of the individual reactants (Klingenberg and Schollmeyer, 1961; Owen and Wilson, 1974; Holian *et al.*, 1977) when [P_i] is held constant. In general, the respiratory rate at a given [ATP]/[ADP] is reported to be essentially independent of the total adenine nucleotide concentration ([ADP] + [ATP]) (for an exception see Davis and Lumeng, 1975). The role of inorganic phosphate concentration ([P_i]) is much more controversial. Some workers have reported that it is equally as important as [ADP] and [ATP] and that the respiratory rate may best be described as a function of [ATP]/[ADP][P_i] (Owen and Wilson, 1974; Holian *et al.*, 1977), while others report that [P_i] is less effective than [ATP] or [ADP] (Davis and Lumeng, 1975). It is interesting to note that workers reporting that [P_i] is an equal participant also observe that the rate is strictly dependent on [ATP]/[ADP], while those reporting [P_i] as a less than full participant observe a dependence on total adenine nucleotide concentration ([ATP] + [ADP]). This suggests a fundamental difference in the mitochondria used, presumably due to differences in isolation procedures.

In suspensions of intact cells and in tissues, dependence of mitochondrial respiration on [ATP]/[ADP][P_i] is clearly observed (Erecińska *et al.*, 1977, 1979; Hassinen and Hiltunen, 1975; Reed, 1976). Experimentally induced changes in intracellular [P_i] that result in no change in respiratory rate show that the [ATP]/[ADP][P_i] remains constant. Thus, a decrease in intracellular $[P_i]$ in liver cells from 3.4 mM to 0.8 mM was accompanied by a decrease in [ATP]/[ADP] from 4.2 to 1.6, leaving the $[ATP]/[ADP][P_i]$ essentially unchanged (Erecińska *et al.*, 1977).

3. DEPENDENCE OF THE RESPIRATORY RATE ON THE INTRAMITOCHONDRIAL [NADH] AND [NAD$^+$]

The reducing equivalents for reduction of oxygen are provided primarily by the citric acid cycle and fatty acid oxidation. The rate of respiration at given concentrations of [ATP], [ADP], and $[P_i]$ (or $[ATP]/[ADP][P_i]$) is dependent on the available reducing equivalents (NADH or FADH$_2$), with NADH, playing a dominant role *in vivo*. In general, the respiratory rate increases with increasing available NADH, but most hypotheses are not specific as to the nature of this dependence. It has been proposed (Owen and Wilson, 1974; Wilson *et al.*, 1977, 1979b; Erecińska *et al.*, 1978) that the respiratory rate increases as a direct function of the intramitochondrial [NADH]/[NAD$^+$]. The evidence for this hypothesis is provided by comparison of the behavior of suspensions of isolated mitochondria to which highly reducing substrates have been added with that of mitochondria in intact cells as well as the responses of *in vivo* respiration to changes in intramitochondrial [NADH]/[NAD$^+$] (Wilson *et al.*, 1977, 1979b; Erecińska *et al.*, 1978). A dependence on [NADH]/[NAD$^+$] would not be surprising as each regulated reaction of the citric acid cycle as well as the β-OH acyl-CoA dehydrogenase are directly or indirectly dependent on [NAD$^+$]/[NADH]. Emphasis on NADH as a substrate recognizes its role as quantitatively the principal electron donor for oxidative phosphorylation *in vivo*, but it is well known that *in vitro* other primary substrates can be utilized.

4. ON THE ROLE OF ADENINE NUCLEOTIDE TRANSLOCASE IN REGULATION OF MITOCHONDRIAL RESPIRATION

The adenine nucleotide translocase is responsible for catalyzing the exchange of adenine nucleotides (ATP and ADP) across the inner mitochondrial membrane. This activity is essential to oxidative phosphorylation because ADP must be transported into the mitochondrial matrix to be phosphorylated and ATP must be returned to the cytoplasm. It has been proposed (Heldt, 1967; Heldt and Pfaff, 1969; Kemp *et al.*, 1969; Davis and Lumeng, 1975; Akerboom *et al.*, 1977; Lemasters and Sowers, 1979) that the rate of exchange of ATP and ADP across the membrane (adenine nucleotide translocase) is responsible for determining the rate of mitochondrial respiration. Consistent with this concept are the following observations.

1. The adenine nucleotide exchange reaction is reported to be electrogenic and the transmembrane electrical potential (negative inside) favors transport of ATP out and ADP in. This would make the reaction potentially subject to control by the transmembrane electrical gradient (Klingenberg *et al.*, 1969; Klingenberg and Rottenberg, 1977; Villiers *et al.*, 1979).
2. The rate of adenine nucleotide translocation is dependent on the extramitochondrial [ATP]/[ADP] (Souverijn *et al.*, 1973; Pfaff and Klingenberg, 1968) as is the respiratory rate when $[P_i]$ is held constant (see below, however).

3. The effect of translocase inhibitors (atractyloside and carboxyatractyloside) on the metabolism of liver cells has been examined by Akerboom *et al.* (1977) and Stubbs *et al.* (1978). The former authors concluded that their data show that the translocase is rate limiting, while the latter argued that the observed inhibitory pattern does not necessarily indicate that the translocator is rate limiting. Similar studies on suspensions of isolated rat liver mitochondria (Lemasters and Sowers, 1979) have been interpreted as indicating the translocase is rate limiting.

Opposed to the concept of regulation of the respiratory rate by the adenine nucleotide translocase are the following lines of evidence.

1. The rate of translocation of adenine nucleotides in suspensions of isolated mitochondria is essentially independent of $[P_i]$ (Nohl and Klingenberg, 1978), while most workers observe a dependence of the respiratory rate on $[P_i]$.
2. The rate of adenine nucleotide translocation is dependent on [ATP]/[ADP], but in both suspensions of isolated mitochondria (Owen and Wilson, 1974; Holian *et al.*, 1977) and intact cells (Erecińska *et al.*, 1977) the [ATP]/[ADP] can be markedly increased or decreased with no change in respiratory rate ([ATP]/[ADP]$[P_i]$ held constant).
3. Analysis of the energetics of the reaction shown in equation (1) indicate that it is near equilibrium ($\Delta G \approx 0$), the accuracy of the analysis making it unlikely that any component part of the phosphorylation reaction is displaced from equilibrium by more than threefold (this would introduce a $-\Delta G$ of 2.9 kJ/mole ATP).
4. In the prokaryotic organism *Paracoccus denitrificans*, which does not require an adenine nucleotide translocase, the dependence of oxidative phosphorylation on [ATP], [ADP], and $[P_i]$ is very similar to that seen in mitochondrial oxidative phosphorylation in eukaryotic organisms (Erecińska *et al.*, 1978).
5. Decreasing the adenine nucleotide content of rat liver mitochondria causes a parallel decrease in the measured rate of translocation but has little effect on the respiratory rate (state 3 or 4) until more than 50% of the total is removed (Asimakis and Aprille, 1980).

The role of the adenine nucleotide translocase in oxidative phosphorylation necessarily involves the intramitochondrial adenine nucleotide pool, and the extent to which it serves as an intermediate in oxidative phosphorylation remains an open question. Klingenberg and co-workers (see, for example, Nohl and Klingenberg, 1978) have reported both kinetic and thermodynamic evidence that this pool is fully competent in oxidative phosphorylation, while Vignais *et al.* (1975) have suggested a special relationship between the ATP synthetase and the adenine nucleotide translocase. This special relationship is proposed to be a structural (spatial) relationship that gives newly synthesized ATP (intramitochondrial) preferential access to the translocase binding site and therefore preferential exchange for extramitochondrial ADP. Also anomalous is the report (Davis and Lumeng, 1975) that there is no change in mitochondrial [ATP]/[ADP] over a wide range of extramitochondrial [ATP]/[ADP] and respiratory rate values, a result incompatible with the intramitochondrial adenine nucleotides being an essential intermediate in oxidative phosphorylation. Brawand *et al.* (1980) observed that the intramitochondrial [ATP]/[ADP] changes in the same direction as the extramitochondrial values when the respiratory rate of suspensions of mitochondria was varied, but the changes were still much less than expected for a true intermediate in oxidative phosphorylation.

5. DEPENDENCE OF THE OXIDATIVE PHOSPHORYLATION ON OXYGEN CONCENTRATION

Oxygen is one of the essential substrates for oxidative phosphorylation, and clearly when it is present in limiting concentrations the capacity for ATP synthesis will be compromised. A critical question of physiological chemistry is the oxygen concentration dependence of the capacity for ATP synthesis and its influence on the specialized systems that have evolved to deliver oxygen to tissues. In suspensions of both intact cells (Longmuir, 1957; Wilson *et al.*, 1979a,b) and isolated mitochondria (Petersen *et al.*, 1974; Oshino *et al.*, 1974; Sugano *et al.*, 1974), the rate of respiration has been reported to be essentially independent of oxygen concentrations above 5 μM and to have an apparent K_m value of less than 0.5 μM. More recently, measurements in suspensions of intact cells have been extended to a general evaluation of energy metabolism including the cellular [ATP], [ADP], lactate and pyruvate production, and cytochrome c reduction. These extended measurements, although confirming the low apparent K_m for oxygen as observed from the respiration, showed that other parameters of cellular energy metabolism (cytochrome c reduction, [ATP]/[ADP], lactate/pyruvate) were all dependent on oxygen concentration to values greater than 100 μM (Wilson *et al.*, 1979a,b; Jones and Mason, 1978). This has been interpreted (Wilson *et al.*, 1979a,b) as evidence that oxidative phosphorylation is dependent on oxygen concentrations up to approximately 100 μM but this dependence is not observed in the respiratory rate because the rate is determined by the rate of ATP utilization. This may be visualized as follows: a decrease in oxygen concentration to a new steady state causes a decrease in the rate of ATP synthesis due to a change in the kinetics of cytochrome oxidase. The rate of ATP utilization remains unchanged, however, and a progressive fall in [ATP] and increase in [ADP] and [P_i] occur. This continues until the stimulating effect of decreased [ATP]/[ADP][P_i] on the respiratory rate results in a new steady state in which the rate of ATP synthesis (respiration) is again equal to the rate of ATP utilization. This "compensation" is primarily expressed as reduction of cytochrome c and activation of cytochrome c oxidase activity. These changes hold the respiratory rate essentially constant until the respiratory chain can no longer compensate for the decreased availability of oxygen or until the decreasing availability of ATP inhibits the rate of ATP utilization. A detailed model and its kinetic behavior have been presented (Wilson *et al.*, 1979b).

This model and conclusions are in marked contrast to the view (Chance, 1976) that the mitochondrial respiratory chain has a very low K_m for oxygen and oxygen is not limiting until its concentration is below approximately 1 μM ($K_m < 0.1$ μM). Comparison of the oxygen dependence of the reduction of mitochondrial respiratory chain components with the oxygenation of myoglobin in perfused heart (Chance, 1976; Tamura *et al.*, 1978) or enzyme activities such as urate oxidase in perfused livers (Sies, 1978) shows that the two decrease in parallel despite very different *in vitro* oxygen dependences. This has been interpreted (Chance, 1976) as evidence for very steep oxygen diffusion gradients that result in the oxygen concentration at the mitochondrion being much lower than that in the cytosol.

The oxygen gradients generated by mitochondrial oxygen consumption can be evaluated by comparison with oxygen electrodes where similar questions have been analyzed in detail. Oxygen electrodes have been constructed with dimensions comparable to those of mitochondria, and because they function as total oxygen sinks, the generated oxygen gradients can be accurately measured and calculated from a sound experimental and theoretical basis. A clean platinum oxygen cathode develops currents of approximately 2.6×10^{-12}

A/mm Hg per μm^2 surface area (Schneiderman and Goldstick, 1976) when appropriately polarized. The latter may represent the diffusion limit for oxygen. Mitochondria, however, (assuming a 1-μm sphere containing 0.5 nmole cytochrome a_3/mg protein, 2 μl volume /mg protein, and a respiratory rate of 12 moles o_2/mole cytochrome a_3 per sec), consume oxygen at a rate that is equivalent to an electrical current of 1.85×10^{-13} A/μm^2 surface area. Thus, even at an oxygen pressure of only 1 mm Hg, a micro oxygen electrode consumes 10 times more oxygen per unit surface area than does a mitochondrion at saturating oxygen concentrations. In these calculations, rather high rates of mitochondrial respiration were assumed, and for most cells lower values are observed. Liver mitochondria, for example, contain less than 0.35 nmole cytochrome a_3/mg protein and respire approximately 2 moles of O_2/sec per mole cytochrome a_3 for an equivalent current of 2.1×10^{-14} A/μm^2 surface area. This means that liver mitochondria consume oxygen at approximately 1% of the diffusion-limited rate even when the oxygen pressure is only 1 mm Hg. These considerations make it unlikely that there are steep oxygen diffusion gradients between the cytosol and the mitochondria. The data for perfused organs (Chance, 1976; Tamura *et al.*, 1978; Sies, 1978) were for saline-perfused organs, and the low oxygen capacity of the perfusate could have resulted in a large decrease in pO_2 between the arterial and the venous ends of the capillaries. This would give heterogenous "patches" of near normoxic and hypoxic tissue, consistent with the experimental observations. The *in vivo* and *in vitro* experimental data on the oxygen dependence of mitochondrial oxidative phosphorylation can be rationalized if the apparent K_m for oxygen is strongly dependent on the state of reduction of cytochrome c and the [ATP]/[ADP][P_i]. This would allow the *in vivo* oxygen dependence to be markedly different from that normally observed for mitochondria *in vitro*. Resolution of this question will probably not be complete until the details of the mechanism of oxygen reduction by cytochrome c oxidase and its regulation are known.

Although great progress has been made in our understanding of the regulation of mitochondrial respiration *in vivo,* there are several areas that require further evaluation. For example, two areas of vital importance to physiological chemistry are: (1) the oxygen concentration dependence of oxidative phosphorylation under cellular conditions and (2) the effect of adenine nucleotide translocation on the overall rate and energetics of oxidative phosphorylation. Current controversies have sharply delineated the problems involved, and well-defined hypotheses have been put forward. This is the basis for scientific progress and provides hope for early resolution of these controversies.

ACKNOWLEDGMENT. This work was supported by Grant GM 21524 from the National Institutes of Health.

REFERENCES

Akerboom, T. M., Bookelmann, H., and Tager, J. M. (1977). *FEBS Lett.* **74,** 50–54.
Asimakis, G. K., and Aprille, J. R. (1980). *Arch. Biochem. Biophys.* **203,** 307–316.
Brawand, F., Folly, G., and Walter, P. (1980). *Biochim. Biophys. Acta* **590,** 285–289.
Chance, B. (1976). *Circ. Res.* **38,** 131–138.
Davis, E. J., and Lumeng, L. (1975). *J. Biol. Chem.* **250,** 2275–2282.
Erecińska, M., Veech, R. L., and Wilson, D. F. (1974). *Arch. Biochem. Biophys.* **160,** 412–421.
Erecińska, M., Stubbs, M., Miyata, Y., Ditre, C. M., and Wilson, D. F. (1977). *Biochim. Biophys. Acta* **462,** 20–35.
Erecińska, M., Wilson, D. F., and Nishiki, K. (1978). *Am. J. Physiol.* **234**(3), C82–C89.
Erecińska, M., Davis, J. S., and Wilson, D. F. (1979). *Arch. Biochem. Biophys.* **197,** 463–469.
Hassinen, I. E., and Hiltunen, K. (1975). *Biochim. Biophys. Acta* **408,** 319–330.

Heldt, H. W. (1967). In *Mitochondrial Structure and Compartmentalization* (E. Quagliariello, S. Papa, E. C. Slater, and J. M. Tager, eds.), pp. 260–267, Adriatica Editrice, Bari.

Heldt, H. W., and Pfaff, E. (1969). *Eur. J. Biochem.* **10**, 494–500.

Holian, A., Owen, C. S., and Wilson, D. F. (1977). *Arch. Biochem. Biophys.* **181**, 164–171.

Jones, D. P., and Mason, H. S. (1978). *J. Biol. Chem.* **253**, 4874–4880.

Kemp, A., Jr., Groot, G. S. P., and Reitsma, H. J. (1969). *Biochim. Biophys. Acta* **180**, 28–34.

Klingenberg, M., and Rottenberg, H. (1977). *Eur. J. Biochem.* **73**, 125–130.

Klingenberg, M., and Schollmeyer, P. (1961). *Biochem. Z.* **335**, 243262.

Klingenberg, M., Heldt, H. W., and Pfaff, E. (1969). In *The Energy Level and Metabolic Control in Mitochondria,* (S. Papa, J. M. Tager, E. Quagliarello, and E. C. Slater, eds.) pp. 237–253, Adriatica Editrice, Bari.

Lemasters, J. J., and Sowers, A. E. (1979). *J. Biol. Chem.* **254**, 1248–1251.

Longmuir, I. S. (1957). *Biochem. J.* **57**, 378–382.

Nohl, H., and Klingenberg, M. (1978). *Biochim. Biophys. Acta.* **505**, 155–169.

Oshino, N., and Sugano, T., Oshino, R., and Chance, B. (1974). *Biochim. Biophys. Acta* **368**, 298–310.

Owen, C. S., and Wilson, D. F. (1974). *Arch. Biochem. Biophys.* **161**, 581–591.

Petersen, L. C., Nicholls, P., and Degn, H. (1974). *Biochem. J.* **142**, 247–252.

Pfaff, E., and Klingenberg, M. (1968). *Eur. J. Biochem.* **6**, 66–79.

Reed, E. B. (1976). *Life Sci.* **19**, 1307–1322.

Schneiderman, G., and Goldstick, T. K. (1976). *Adv. Exp. Med. Biol.* **75**, 9–16.

Souverijn, J. M., Huisman, L. A., Rosing, J., and Kemp, A., Jr. (1973). *Biochim. Biophys. Acta* **305**, 185–198.

Stubbs, M., Vignais, P. V., and Krebs, H. A. (1978). *Biochem. J.* **172**, 333–342.

Sugano, T., Oshino, N., and Chance, B. (1974). *Biochim. Biophys. Acta* **347**, 340–358.

Tamura, M., Oshino, N., Chance, B., and Silver, I. A. (1978). *Arch. Biochem. Biophys.* **191**, 8–22.

Vignais, P. V., Vignais, P. M., and Doussiere, J. (1975). *Biochim. Biophys. Acta* **376**, 219–230.

Villiers, C., Michejda, J. W., Block, M., Lauquin, G. J. M., and Vignais, P. V. (1979). *Biochim. Biophys. Acta* **546**, 157–170.

Wilson, D. F., Stubbs, M., Veech, R. L., Erecińska, M., and Krebs, H. A. (1974a). *Biochem. J.* **140**, 57–64.

Wilson, D. F., Stubbs, M., Oshino, N., and Erecińska, M. (1974b). *Biochemistry* **13**, 5305–5311.

Wilson, D. F., Owen, C. S., and Holian, A. (1977). *Arch. Biochem. Biophys.* **182**, 749–762.

Wilson, D. F., Erecińska, M., Drown, C., and Silver, I. A. (1979a). *Arch. Biochem. Biophys.* **195**, 485–493.

Wilson, D. F., Owen, C. S., and Erecińska, M. (1979b). *Arch. Biochem. Biophys.* **195**, 494–504.

44

Proton Translocation by Cytochrome Oxidase

Mårten Wikström

1. INTRODUCTION

Cytochrome oxidase (EC 1.9.3.1) is the terminal oxidoreduction complex of the respiratory chain. It is located in the inner mitochondrial membrane, and is also found in the cell membrane of some bacteria. Cytochrome oxidase catalyzes the reduction of O_2 to water by electrons donated by cytochrome c. It also conserves a large fraction of the free energy liberated in the redox reaction for subsequent synthesis of ATP. The cytochrome oxidase segment of the respiratory chain has indeed traditionally been called the "third site," or simply "Site 3," of oxidative phosphorylation.

It is generally agreed and presently well covered in textbooks of biochemistry (see e.g. Stryer, 1981) that electron transfer and ATP synthesis are linked by an electrochemical proton gradient ($\Delta\bar{\mu}_{H^+}$), as proposed in the chemiosmotic theory (Mitchell, 1966). Although this linkage concept is, no doubt, a major achievement in our understanding of oxidative phosphorylation, it does not by itself solve the mechanism. The latter would require knowledge of how $\Delta\bar{\mu}_{H^+}$ is generated by respiration and how it is utilized by the ATP synthase.

In this review I will briefly show why and how recent research on cytochrome oxidase has contributed to our understanding of the mechanism of oxidative phosphorylation. The results from this research have necessitated a fundamental departure from the o/r or redox loop principle of proton translocation postulated in the chemiosmotic theory (Mitchell, 1966, 1979). In the main part of the paper I hope to demonstrate that the most recent research on cytochrome oxidase has led to an unprecedented level of understanding of the principle and the molecular elements of primary energy conservation. Whether these concepts may be generalized to hold also for other coupling "sites" apart from the oxidase is presently only a matter of speculation.

Mårten Wikström • Department of Medical Chemistry, University of Helsinki, SF-00170 Helsinki 17, Finland.

2. THE PRINCIPLE OF GENERATION OF $\Delta\bar{\mu}_{H+}$ BY CYTOCHROME OXIDASE

It was discovered in 1977 that the cytochrome oxidase reaction is linked to true translocation of H^+ from the matrix (M) to the cytoplasmic (C) side of the inner mitochondrial membrane (Wikström, 1977; Wikström and Saari, 1977). This finding was subsequently confirmed (Sigel and Carafoli, 1978; Sorgato *et al.*, 1978; Krab and Wikström, 1979; see review by Wikström and Krab, 1979), including the important verification from experiments with isolated and purified cytochrome oxidase reincorporated ("reconstituted") into the membranes of liposomes (Wikström and Saari, 1977; Krab and Wikström, 1978; Casey *et al.*, 1979; Sigel and Carafoli, 1979, 1980; Coin and Hinkle, 1979). More recently, it was shown that the cytochrome oxidase in membranes of *Paracoccus denitrificans* also functions as a proton pump (Van Verseveld *et al.*, 1981).

The essence of these findings may be summarized as follows: For transfer of each electron from cytochrome c to O_2, there is an overall or net (i) release of $1H^+$ into the aqueous C phase, (ii) uptake of $2H^+$ from the aqueous M phase, and (iii) translocation of two electrical charge equivalents from the M to the C side of the membrane.

This contrasts greatly with the redox loop model (Mitchell, 1966, 1979), according to which (i) is zero, (ii) is $1H^+$, and (iii) is one charge equivalent. The difference is not merely one of stoichiometry (and thereby of the energetics), but has more far-reaching consequences. The redox loop model is, in fact, unable to explain net H^+ translocation by cytochrome oxidase for the reaction involves no classical redox center or carrier of the H-transfer type (like ubiquinone, for example). Consequently, the primary event of energy conservation at cytochrome oxidase must follow a different principle. We have called this principle a redox-linked proton pump (Wikström and Krab, 1979). It corresponds to what Mitchell has termed a completely indirect chemiosmotic mechanism (see Mitchell, 1979). This refers to the fact that the proton cannot be translocated on the reduced form of the redox-active center (as in a redox loop), but that proton/electron coupling must be more indirect and mediated, presumably, by the protein structure (cf. conformational changes). One of the fundamental differences between this type of mechanism and that of a redox loop is that the former does not require that oxidoreduction reactions be vectorially oriented across the membrane (or its electroosmotic barrier).

Against this background it is understandable that there are objections against this conclusion (Mitchell and Moyle, 1979; Lorusso *et al.*, 1979; Papa *et al.*, 1980). These, as well as some disagreement about the stoichiometry of H^+ translocation by cytochrome oxidase (see Alexandre and Lehninger, 1979; Azzone *et al.*, 1979), have been discussed in some detail elsewhere (Wikström and Krab, 1979, 1980; Wikström *et al.*, 1981a). The objections are based on experiments with intact mitochondria for which the system complexity is considerable. The results with the purified and liposome-reconstituted enzyme (see above) provide strong support to our conclusion. More recent functional and structural data provide further support, as will be shown below.

3. INVOLVEMENT OF THE REDOX CENTERS IN PROTON TRANSLOCATION

The minimum catalytic unit of cytochrome oxidase, i.e., the cytochrome aa_3 monomer, consists of four dissimilar redox centers, viz. two hemes (a and a_3) and two protein-bound coppers (Cu_A and Cu_B). Of these, heme a_3 and Cu_B are closely associated both physically and functionally, forming a binuclear a_3/Cu_B center that catalyzes the reduction of O_2 (see Malmström, 1979). The function of cytochrome a and of Cu_A has usually been regarded merely to be one of electron transfer catalysis between cytochrome c and the a_3/Cu_B center (but see below).

The discovery that the enzyme functions as a redox-linked proton pump initiated active search for the redox center the activity of which would be coupled to H^+ translocation (see Wikström and Krab, 1979). Cytochromes a and a_3 are apparently the best candidates, because their midpoint redox potentials (E_m) are functions of pH, which is not the case for the two coppers.

The large energy-dependent spectral shift observed in ferric heme a_3 (Erecińska *et al.*, 1972) suggested that cytochrome a_3 may be the component involved in proton translocation (Wikström, 1977; Wikström and Saari, 1977; Wikström and Krab, 1979). However, this would be difficult to reconcile with the involved function of cytochrome a_3 in the reduction of O_2 (Wikström, 1981a,b). This dilemma was recently resolved by the finding that the spectral shift is simply due to energy-dependent partial reversal of electron transfer between the a_3/Cu_B center and O_2. In this reaction water is apparently oxidized to peroxide, which remains bound to the center, the electrons being transferred backwards to cytochrome c (Wikström, 1981b). The associated change in the ligand field of heme a_3 provides a rational explanation of the spectral shift and thus removes the experimental basis for a direct role of cytochrome a_3 in proton translocation.

Attention was therefore directed to the possibility that cytochrome a may have this role (see below).

3.1. Cytochrome a is the Redox Element of the Proton Pump

If the activity of a redox center is linked to proton translocation, the center should exhibit certain specific properties (Wikström, 1981a; Wikström *et al.*, 1981b). One such property is its existence in two interconvertible states, comparable to input and output, respectively, in any energy converter. Because shuttling between these states during activity may be expected to be slower than electron transfer per se, there may be conditions where such a redox center behaves kinetically as if it consisted of *two distinct entities*. Transient oxidation and reduction as well as steady-state kinetics of cytochrome a are unique in showing exactly this property (Wikström, 1981a; Wikström *et al.*, 1981a,b).

A second expected property purports to the dependence of the redox center on H^+ activity, which is a function of the essential proton/electron coupling in the pump. In certain states of the transducer, there should be a unique *sidedness* in this property with respect to the two aqueous phases on each side of the membrane (see Wikström, 1981a). Again this is observed for cytochrome a, the E_m of which is uniquely a function of pH in the M phase in cyanide-inhibited mitochondria (Artzatbanov *et al.*, 1978).

A third point of evidence also relates to the pH dependence of the E_m of cytochrome a, and is perhaps the most striking. However, this is deferred to Section 4.

It is concluded that several independent findings suggest that cytochrome a is the redox element of the proton pump.

3.2. The Role of the a_3/Cu_B Center in Energy Conservation

Although the a_3/Cu_B center is probably not involved in the functioning of the proton pump, it must nevertheless, in a sense, be considered to contribute to energy conservation. As the most recent evidence (Wikström, 1981a; Section 4) suggests that the H^+ ions re-

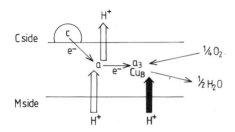

Figure 1. The principle of proton translocation by cytochrome oxidase. The electron and proton transfer reactions are depicted schematically for cytochrome oxidase in the mitochondrial membrane. Proton translocation linked to oxidoreduction of cytochrome a (thick unshaded arrows) constitutes the function of the proton pump. Note that $\Delta\tilde{\mu}_H+$ is also generated by the acceptance of the a_3/Cu$_B$ center of electrons from the C side and protons from the M side (thick shaded arrow) of the membrane.

quired in formation of water at this center are taken from the M side, and as the electrons derive from cytochrome c on the C side, the overall reaction is generation of $\Delta\tilde{\mu}_{H^+}$. In fact, this is the very function still maintained by Mitchell (1979) to be entirely responsible for energy conservation by cytochrome oxidase. Note, however, that it does not necessarily imply that it is the electron that is translocated across the electroosmotic barrier (Wikström, 1981a). This function is, in any case, energetically equivalent to complete translocation of $1H^+/e^-$ across the membrane, even though no H^+ is released on the C side (Wikström *et al.*, 1981a). Energetically, it is therefore *of equal significance* with the proton pump (in which $1H^+/e^-$ is translocated linked to oxidoreduction of cytochrome a) to which it is connected in series (see Fig. 1).

4. THE ROLE OF SUBUNIT III IN PROTON TRANSLOCATION

It is conceivable that a proton translocator may require specific H^+-conducting structures. The precedent of this is the H^+ channel in the F_0 segment of the H^+-translocating ATPase of mitochondria, bacteria, and chloroplasts. The H^+ conductance may be blocked by binding of the carboxyl-group reagent dicyclohexylcarbodiimide (DCCD) to this segment. The DCCD-binding protein contains two hydrophobic stretches of amino acids. In the middle of one of these there is a single glutamic acid residue, which is the DCCD-binding site (Sebald *et al.*, 1980; see review by Wikström *et al.*, 1981a).

Casey *et al.* (1980) demonstrated that DCCD also blocks the proton pump of cytochrome oxidase, and that this may be associated with binding of DCCD to one of the hydrophobic subunits of the enzyme that are synthesized by the mitochondrion, viz. subunit III (see review by Azzi, 1980). Moreover, DCCD seems to bind to a glutamic acid residue located in an otherwise hydrophobic segment of subunit III (Prochaska *et al.*, 1981; Azzi, 1980). The analogy to the DCCD-binding protein of the ATPase is striking and suggests that subunit III may be involved in proton translocation by the oxidase.

This notion is strongly supported by the finding that subunit III may be removed with retained electron transfer activity and optical spectra, but with apparent loss of net H^+ translocation (Saraste *et al.*, 1981). Respiratory control, however, is retained in liposomes reconstituted with the subunit III-free enzyme. In the present context, this can only be explained by assuming a "local uncoupling" (see below) of the proton pump, and that the overall reaction is still generating $\Delta\tilde{\mu}_{H^+}$ (at half the efficiency) due to the organization of the a_3/Cu$_B$ center (see Section 3.2). In this situation, there should only be translocation of one charge equivalent across the membrane per transferred electron (see section 2). Recent data have confirmed this prediction (Penttilä and Wikström, 1981).

"Local uncoupling" of the proton pump (the membranes must remain comparatively impermeable to H^+ because $\Delta\tilde{\mu}_{H^+}$ is sustained) is equivalent to a phenomenon of "molecular slipping" where the normally tight coupling between electron transfer and H^+ translocation is loosened so that electron transfer may take place independently.

"Local uncoupling" on removal of subunit III would suggest that this polypeptide may be much more intimately involved in the mechanism of proton translocation than merely by forming a passive H^+ conductance. This notion has recently gained strong support from experiments performed by Mr. Timo Penttilä in this laboratory (Penttilä and Wikström, 1981).

Penttilä first confirmed that the E_m of cytochrome a is a function of pH in the isolated enzyme. He then found that this pH dependence is lost almost completely upon removal of subunit III. This is a very important finding because it suggests that the major redox-linked acid/base group coupled to cytochrome a may be located in subunit III. There is little doubt that this group is of fundamental importance in the proton pump mechanism (Wikström and Krab, 1979; Wikström, 1981a; Wikström *et al.*, 1981a,b). Because cytochrome a is retained on removal of subunit III, this result suggests that the crucial proton/electron coupling of the proton pump takes place *between separate polypeptide chains*. These findings also nicely weave together the notions of cytochrome a as the redox element of the pump and the intimate involvement of subunit III in proton translocation, supporting both concepts simultaneously.

5. CONCLUDING REMARKS

As discussed previously in some detail (Wikström and Krab, 1979; Wikström, 1981a; Wikström *et al.*, 1981a,b), the minimal elements of a redox-linked proton pump are a redox center and an acid/base center. Both should possess certain specific properties of which the most crucial is that of their mutual linkage. It is likely that both these elements have now been identified in the cytochrome oxidase proton pump. If so, our understanding of primary energy conservation in respiration has finally reached the detailed molecular level. It should be obvious from this review that the structural information has been of crucial importance in this progress.

It is also of some interest to note that cytochrome oxidase shows, in all likelihood, a clear-cut case of *intersubunit* proton/electron coupling in primary energy conservation. Though this possibility has been repeatedly suggested in the literature, its existence is now for the first time supported experimentally. This should help considerably to remove the existing prejudice against so-called indirect chemiosmotic coupling, which, however, thus far has not meant much more than a "wiggly line" connecting redox and proton transfer events in reaction schemes (see Mitchell, 1979). The present data do not define this "wiggly line" in molecular terms, but they definitely indicate its existence and, more important, where to look for the answer in future experiments.

Finally, a note of caution may be appropriate. While the above conclusions seem reasonably well founded for cytochrome oxidase, it is not certain that they are applicable on the protonmotive cytochrome bc_1 and NADH dehydrogenase complexes. It should be recalled (Section 3.2) that cytochrome oxidase is apparently *not only* a proton pump, but probably consists of a proton pump and a directly coupled proton well (or electron translocator) type of mechanism arranged in series (Fig. 1). Perhaps we should seriously consider whether all protonmotive redox complexes of the respiratory chain may primarily be arranged according to the redox loop principle, to which certain complexes (such as cytochrome oxidase) may have added a proton pump during evolution. The latter principle may have been adopted from the H$^+$-translocating ATPase (see Section 4; Wikström *et al.*, 1981b).

Apart from the fact that this speculation has some esthetic appeal and does not lack support from structural considerations, it might provide a point of reconciliation between proponents of completely direct or completely indirect types of coupling in oxidative phosphorylation.

ACKNOWLEDGMENTS. I would like to thank Dr. Angelo Azzi for providing the manuscript of his review prior to publication. This work has been supported by the Sigrid Jusélius Foundation and the Finnish Academy (Medical Research Council).

REFERENCES

Alexandre, A., and Lehninger, A. L. (1979). *J. Biol. Chem.* **254**, 11555–11560.
Artzatbanov, V. Yu., Konstantinov, A. A., and Skulachev, V. P. (1978). *FEBS Lett.* **87**, 180–185.
Azzi, A. (1980). *Biochim. Biophys. Acta*, **594**, 231–252.
Azzone, G. F., Pozzan, T., and Di Virgilio, F. (1979). *J. Biol. Chem.* **254**, 10206–10212.
Casey, R. P., Chappell, J. B., and Azzi, A. (1979). *Biochem. J.* **182**, 149–156.
Casey, R. P., Thelen, M., and Azzi A. (1980). *J. Biol. Chem.* **255**, 3994–4000.
Coin, J. T., and Hinkle, P. C. (1979). In *Membrane Bioenergetics* (C. P. Lee, G. Schatz, and L. Ernster, eds.), pp. 405–412, Addison–Wesley, Reading, Mass.
Erecińska, M., Wilson, D. F., Sato, N., and Nicholls, P. (1972). *Arch. Biochem. Biophys.* **151**, 188–193.
Krab, K., and Wikström, M. (1978). *Biochim. Biophys. Acta* **504**, 200–214.
Krab, K., and Wikström, M. (1979). *Biochim. Biophys. Acta* **548**, 1–15.
Lorusso, M., Capuano, F., Boffoli, D., Stefanelli, R., and Papa, S. (1979). *Biochem. J.* **182**, 133–147.
Malmström, B. G. (1979). *Biochim. Biophys. Acta* **549**, 281–303.
Mitchell, P. (1966). *Biol. Rev.* **41**, 445–502.
Mitchell, P. (1979). *Eur. J. Biochem.* **95**, 1–20.
Mitchell, P., and Moyle, J. (1979). *Biochem. Soc. Trans.* **7**, 887–894.
Papa, S., Guerrieri, F., Lorusso, M., Izzo, G., Boffoli, G., Capuano, F., Capitanio, N., and Altamura, N. (1980). *Biochem. J.* **192**, 203–218.
Penttilä, T., and Wikström, M. (1981). In *Vectorial Reactions in Electron and Ion Transport in Mitochondria and Bacteria* (F. Palmieri, E. Quagliariello, N. Siliprandi, and E. C. Slater, eds.), pp. 71–80, Elsevier/North-Holland, Amsterdam.
Prochaska, L. J., Steffens, G. C. M., Buse, G., Bisson, R., and Capaldi, R. A. (1981). *Biochim. Biophys. Acta* **637**, 360–373.
Saraste, M., Penttilä, T., and Wikström, M. (1981). *Eur. J. Biochem.*, **115**, 261–268.
Sebald, W., Machleidt, W., and Wachter, E. (1980). *Proc. Natl. Acad. Sci. USA* **77**, 785–789.
Sigel, E., and Carafoli, E. (1978). *Eur. J. Biochem.* **89**, 119–123.
Sigel, E., and Carafoli, E. (1979). *J. Biol. Chem.* **254**, 10572–10574.
Sigel, E., and Carafoli, E. (1980). *Eur. J. Biochem.* **111**, 299–306.
Sorgato, M. C., Furguson, S. J., Kell, D. B., and John, P. (1978). *Biochem. J.* **174**, 237–256.
Stryer, L. (1981). *Biochemistry*, 2nd ed., Freeman, San Francisco.
Van Verseveld, H. W., Krab, K., and Stouthamer, A. H. (1981). *Biochim. Biophys. Acta*, **635**, 525–534.
Wikström, M. (1977). *Nature (London)* **266**, 271–273.
Wikström, M. (1981a). In *Mitochondria and Microsomes* (C. P. Lee, G. Schatz, and G. Dallner, eds.), pp. 249–269. Addison–Wesley, Reading, Mass.
Wikström, M. (1981b). *Proc. Natl. Acad. Sci. USA*, **78**, 4051–4054.
Wikström, M., and Krab, K. (1979). *Biochim. Biophys. Acta* **549**, 177–222.
Wikström, M., and Krab, K. (1980). *Curr. Top. Bioenerg.* **10**, 51–101.
Wikström, M., and Saari, H. (1977). *Biochim. Biophys. Acta* **462**, 347–361.
Wikström, M., Krab, K., and Saraste, M. (1981a). *Annu. Rev. Biochem.* **50**, 623–655.
Wikström, M., Krab, K., and Saraste, M. (1981b). *Cytochrome Oxidase—A Synthesis*, Academic Press, London.

45

Mechanism of Active Proton Translocation by Cytochrome Systems

Sergio Papa

1. INTRODUCTION

The cytochrome systems of mitochondria and plasma membrane of prokaryotes catalyze the transfer of reducing equivalents from dehydrogenases to oxygen, or other oxidants, and convert the energy so made available into transmembrane thermodynamic potential difference of protons, $\Delta\tilde{\mu}_{H^+}$ (Mitchell, 1966, 1980a; Papa, 1976; Haddock and Jones, 1977; Wikström and Krab, 1979).

While it has become clear that transmembrane H^+ circulation is the major mode of energy transfer in membranes (Boyer *et al.*, 1977), the molecular mechanism by which $\Delta\tilde{\mu}_{H^+}$ is generated and utilized remains elusive, and conflicting data and ideas on this issue flourish in the literature.

The controversies on $\Delta\tilde{\mu}_{H^+}$ generation by the cytochrome system concern two problems. The first is whether cytochrome c oxidoreductase (EC 1.9.3.1), which is a well-established $\Delta\tilde{\mu}_{H^+}$ producer (Mitchell and Moyle, 1967; Hinkle, 1973; Papa, 1976), operates as an electronic $\Delta\tilde{\mu}_{H^+}$ generator, i.e., by catalyzing transmembrane electron flow (Mitchell, 1966, 1980a; Papa *et al.*, 1974; Lorusso *et al.*, 1979), or as a proton pump (Wikström and Krab, 1979). The second problem is whether the proton-pumping activity of the ubiquinone-cytochrome c segment results directly from vectorial H^+ translocation by ubiquinone (Mitchell, 1976) or if it involves, alternatively to or in conjunction with this, cooperative H^+ transfer by ionizable groups in hemeproteins and/or Fe–S proteins (Papa *et al.*, 1973, 1978, 1981a).

2. H+/e− STOICHIOMETRY

A basic postulate of direct chemiosmotic protonmotive models (Mitchell, 1966, 1980a) is that 2 H^+ are translocated across the membrane for each pair of electrons traversing an

Sergio Papa • Institute of Biological Chemistry, Faculty of Medicine, University of Bari, Bari, Italy.

energy-conserving site. Indirect mechanisms, on the other hand, do not necessarily require this rigid stoichiometry (Papa, 1976).

A series of determinations verified a general stoichiometry of $2H^+/2e^-$ *per site* in mitochondria and bacteria (Mitchell, 1980b; Papa, 1976; Haddock and Jones, 1977). Thus, an H^+/O ratio of 4 for electron flow in the cytochrome system (succinate oxidation) was observed (Mitchell, 1980b). These measurements were essentially obtained with the *oxygen pulse method* (Mitchell *et al.*, 1979). This was questioned by Lehninger and colleagues (Lehninger, 1978) who introduced, what they call, the *steady-state rate* method and found that the H^+/O ratio could reach, under conditions defined as optimal, a value of 8 with succinate and 12 with NAD-linked substrates. Others found ratios of 6 and 9, respectively (Brand, 1977; Wikström and Krab, 1979).

It was, however, soon pointed out that the steady-state rate method was probably introducing more uncertainties than it eliminated (Mitchell, 1980b; Papa *et al.*, 1980a,b). This method, which is based on activation of electron flow by addition of a reductant to the aerobic respiratory chain, depends on: (1) accurate determination of the initial respiratory rate elicited by reductants; (2) accurate measurement of the rate of proton translocation; (3) definition of the reductants and oxidants involved in the process; (4) absence of or correction for proton translocation unrelated to the redox process. Papa *et al.* (1980a,b) reported that it was probably the failure to meet one or another of these conditions that produced high H^+/O ratios. They showed that the H^+/O quotient for succinate oxidation is, under the conditions used by Lehninger and colleagues, exactly 4 at neutral pH, when computed on the basis of accurate spectrophotometric determination of respiration with hemoglobin. The higher H^+/O quotients previously found resulted, apparently, from underestimation of the initial respiratory rate. An H^+/O ratio of 4 was also obtained by replacing succinate with duroquinol, whose oxidation was measured spectrophotometrically.

It has been observed (Brand, 1977) that N-ethylmaleimide enhances the $H^+/2e^-$ (per site) quotient, measured with oxygen pulses, from 2 to 3. Hence, it was concluded that the quotient of 2 was underestimated because of proton back-flow through P_i/H^+ symport. Mitchell *et al.* (1979) rejected this claim and proposed that the effect of N-ethylmaleimide was due to activation of H^+-translocating oxidation of endogenous NAD(P)H, and/or to promotion of semiquinone dismutation by b cytochromes (Mitchell, 1980c).

Various measurements of the $H^+/2e^-$ quotient for electron flow from the substrate terminus of the cytochrome system to ferricytochrome c (Wikström and Krab, 1979; Mitchell, 1980a; Papa *et al.*, 1980b) show that 4 H^+ are released at the outer side of the mitochondrial membrane—2 H^+ equivalents being effectively contributed by succinate or quinol, the other 2, with their positive charges, being translocated from the matrix or negative side to the outer or positive side—as $2e^-$ flow from quinol to cytochrome c.

Thus, if the H^+/O quotient for electron flow along all the cytochrome system is 4, and, in fact, no objections have so far been raised against the measurements of Papa *et al.* (1980a,b), it follows that two positive charges (per two electrons), but no protons, are effectively translocated from the negative to the positive side by cytochrome c oxidase *"turning over"* in its natural environment in the membrane.

3. PROTON TRANSLOCATION IN CYTOCHROME c OXIDASE

There seems to be a consensus (Mitchell, 1980a; Hinkle, 1973; Papa *et al.*, 1974, 1978; Wikström and Saari, 1977) about the anisotropic reduction of O_2 to H_2O by cytochrome c oxidase, whereby electrons are delivered from cytochrome c located at the outer side of the membrane and protons and probably H_2O are exchanged from the matrix phase.

Investigations on proton translocation at the third site with artificial reductants have produced contrasting results. Mitchell and Moyle (1967) observed that aerobic oxidation of ferrocyanide or $N,N,N'N'$-tetramethyl-p-phenylenediamine (TMPD) plus ascorbate by antimycin-supplemented mitochondria produced no H^+ translocation other than that arising from consumption of H^+ from the matrix in the reduction of O_2 to H_2O. Wikström (1977) observed, however, that, under particular conditions, antimycin-insensitive H^+ ejection from mitochondria accompanied ferrocyanide oxidation. This and related observations (Wikström and Saari, 1977) were interpreted as evidence that the oxidase functions as a proton pump. Various investigators subsequently reproduced and extended Wikström's observations (Wikström and Krab, 1979; Sigel and Carafoli, 1978, 1980; Casey *et al.*, 1979).

The evidence in favor of a proton pump in cytochrome c oxidase comes from the following lines: (a) H^+ release from mitochondria associated with aerobic oxidation of reductants of cytochrome c or ferrocytochrome c itself; (b) measurement of transmembrane $\Delta\bar{\mu}_{H^+}$ in "inside-out" submitochondrial particles; (c) H^+ release associated with oxidation of ferrocytochrome c by purified cytochrome c oxidase incorporated into liposomes. The results produced along these lines are open to criticism (Mitchell, 1980a; Papa *et al.*, 1978; Lorusso *et al.*, 1979), and in the opinion of the reviewer, unless these uncertainties are resolved, they cannot be taken as proof for the operation of cytochrome c oxidase as a proton pump. This possibility is also made unlikely for the normally *turning over* oxidase by the H^+/e^- stoichiometries just discussed.

a. A number of observations (Moyle and Mitchell, 1978; Papa *et al.*, 1978, 1980b; Lorusso *et al.*, 1979) show that H^+ release from mitochondria, associated with the oxidation of reductants of cytochrome c, results from antimycin-insensitive rereduction of the oxidized products by endogenous hydrogenated reductants.

With ferrocyanide as reductant, it was found (Papa *et al.*, 1978; Moyle and Mitchell, 1978; Lorusso *et al.*, 1979) that: (1) When its oxidation resulted in H^+ release, the rate of net ferricyanide formation was lower than that of oxygen reduction. (2) The rate of scalar H^+ consumption for oxygen reduction, measured in the presence of FCCP, was lower that the rate of oxygen reduction. (3) 2-Heptyl-4-hydroxyquinoline-N-oxide inhibited, in the presence of antimycin, the H^+ release and the rereduction of ferricyanide by endogenous hydrogenated reductants (Lorusso *et al.*, 1979).

With TMPD plus ascorbate, the following observations were made (Papa *et al.*, 1980b). (1) When the respiratory rate was measured spectrophotometrically with hemoglobin, the H^+/e^- quotient for vectorial H^+ release was only 0.2, under the conditions where a quotient of 1 was found with polarographic measurements (Sigel and Carafoli, 1978). (2) The rate of scalar H^+ consumption was lower than that of oxygen reduction. (3) Aerobic preincubation with valinomycin suppressed almost completely vectorial H^+ release.

Oxidation of exogenous ferrocytochrome c was also found to result in H^+ release by mitochondria. It could, however, be shown that this acidification was scalar and not associated with electron flow in the oxidase (Moyle and Mitchell, 1978; Lorusso *et al.*, 1979).

b. Sorgato and Ferguson (1978) have shown that aerobic oxidation of TMPD plus ascorbate generates, in the presence of antimycin, membrane potential and ΔpH in inside-out submitochondrial particles and considered this as evidence in favor of a proton pump in the oxidase. Detection of membrane potential in such a system is equally consistent with vectorial electron flow or electrogenic H^+ translocation by cytochrome oxidase. On the other hand, the data of Sorgato and Ferguson (1978) do not exclude the possibility that the ΔpH generated during the long incubation period was due to H^+ production by oxidation of ascorbate and/or $TMPDH^+$ inside the vesicles, or to activation by TMPD of antimycin-insensitive electron flow through other segments of the respiratory chain.

c. A strong piece of evidence in favor of a proton pump in cytochrome oxidase is considered to be provided by the observed H^+ release caused by oxidation of ferrocytochrome c by cytochrome c oxidase in liposomes (Wikström and Krab, 1979; Sigel and Carofoli, 1980; Casey *et al.*, 1979). This observation presents, however, various drawbacks. The experiments that resulted in a reproducible H^+ release were those in which the oxidized enzyme was pulsed with ferrocytochrome c. Under these conditions, the oxidase is in an inactive or *resting state*, characterized by a low rate of internal electron flow and a different conformational state (Bonaventura *et al.*, 1978) compared with the pulsed reduced enzyme, which corresponds to the enzyme turning over at the *steady state*. It seems therefore essential that the observations on the resting enzyme be reproduced with the pulsed enzyme. Few experiments of this type have been attempted and have produced uncertain results. No H^+ release (Hinkle, 1973) or a small H^+ release could be detected (Wikström and Saari, 1977; Coin and Hinkle, 1979) that lasted for only one turnover and was equivalent to that expected for scalar proton release from redox Bohr effects of cytochrome c oxidase (Papa *et al.*, 1979, 1980c).

In addition, the available results still leave doubts whether the H^+ release caused by a ferrocytochrome c pulse of the reconstituted resting oxidase really represents a vectorial process and not a reversible or irreversible scalar acidification caused by interaction of cytochrome c with the lipid-oxidase system (Lorusso *et al.*, 1980).

Sigel and Carafoli (1980) have measured H^+/e^- and K^+/e^- ratios for H^+ release and valinomycin-mediated K^+ uptake in reconstituted oxidase vesicles oxidizing TMPD plus ascorbate, which would be consistent with a proton pump in the oxidase. These ratios appear, however, uncertain, for respiration, K^+ uptake, and H^+ release started with a lag after the addition of the reductant and rapidly declined to values consistent with simple vectorial electron flow.

4. PROTON TRANSLOCATION IN THE UBIQUINONE/CYTOCHROME c OXIDOREDUCTASE

The original protonmotive redox-loop model (Mitchell, 1966) required an $H^+/2e^-$ quotient of 2 and a $q^+/2e^-$ quotient of 0 for H^+ release from mitochondria associated with electron flow from quinol to cytochrome c. As already reported, these quotients were found to be 4 and 2, respectively. This and related observations were still explained in terms of direct ligand conduction by development of mechanisms like the protonmotive quinone cycle (Mitchell, 1976) or two quinone redox-loops in series (Crofts *et al.*, 1975). While important support is apparently lent to these mechanisms by identification of separate quinone systems and quinone-binding proteins (Trumpower, 1981), the models, in their present form, depend on certain conditions that remain highly speculative and do not take into account the protolytic characteristics of seminquinone and the cooperative linkage in b cytochromes and the Rieske Fe-S protein between the redox state of the metal and protolytic equilibria (Papa, 1976; Prince and Dutton, 1976; Papa *et al.*, 1979).

Papa *et al.* (1973, 1978, 1981a) have proposed that Bohr effects in metalloproteins of the bc_1 complex can result in redox-linked vectorial H^+ translocation across the membrane. The bc_1 complex was found to exhibit redox Bohr effects with an H^+/e^- coupling number (referenced to the Rieske Fe–S protein) of 1.5 at pH 7.2 (Papa *et al.*, 1980c, 1981a). This value is, however, higher than that expected from the pH dependence of the E_m of this carrier (Prince and Dutton, 1976). It is possible that additional H^+ release derives from aerobic oxidation of protein-bound QH_2 to $\overset{\cdot}{Q}^-$.

Investigations on the effect of pH on the H^+/e^- stoichiometry for H^+ release from mitochondria associated with electron flow from quinols to oxygen or to cytochrome c

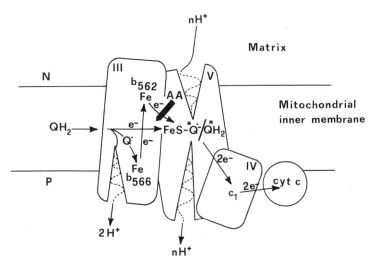

Figure 1. Schematic description of the molecular mechanism of "ubiquinone-gated proton pump" of the bc_1 complex of the respiratory chain. The Fe atoms of cytochromes b_{566} and b_{562} are shown to be located respectively on the outer and inner side of the insulating layer of the membrane (cf. Mitchell, 1976; Wikström, 1973; Papa *et al.*, 1981b). $*\dot{Q}^-$ and $*QH_2$ represent ubisemiquinone anion and ubiquinol species specifically bound to polypeptide(s) of the bc_1 complex (Yu and Yu, 1980; Trumpower, 1981). It should be noted that the pK_a of the ubisemiquinone radical is not higher than 6.5 (Trumpower, 1981). The polypeptide subunits of the bc_1 complex are numbered according to Bell *et al.* (1979).

showed that this first decreased below 2 upon raising the pH from 7.2 to 8, then increased above 2 at pH 9 (Papa *et al.*, 1977, 1981a). These observations cannot be explained by direct ligand conduction mechanisms like the quinone cycle or the two quinone redox-loops in series, which predict a fixed H^+/e^- quotient of 2 (however, see Mitchell, 1980c).

Papa *et al.* (1981a) have recently developed a mechanism for redox-linked proton translocation in the bc_1 complex (Fig. 1). They propose that the quinol is oxidized to semiquinone by the Rieske Fe–S protein in an antimycin-insensitive reaction, the semiquinone being oxidized by b cytochromes and Fe–S protein through an antimycin-sensitive reaction. Up to here, $2H^+/2e^-$ can be released at the positive side. The two electrons reunite at the level of the Fe–S protein, and both pass through ubisemiquinone associated with the Fe–S center and from this to cytochrome c_1. The cytochrome b shunt explains the antimycin-promoted oxidant-induced reduction of b cytochromes (Papa *et al.*, 1981b). Protonation of the $U\dot{Q}/UQH_2$ couple from the negative side of the membrane and deprotonation at the outer side will result in transmembrane translocation of 2–4 H^+ (depending on the ionization state of semiquinone) per $2e^-$ traversing the Fe–S–$U\dot{Q}$-cytochrome c_1 system. H^+ access to the Fe–S–$U\dot{Q}$ center from the negative and H^+ release to the positive domain would take place through a specific proton channel in the polypeptides of the complex, access of H^+ into the channel and their release on the opposite side being favoured by redoxlinked pK shifts of ionizable groups in the channel (vectorial Bohr mechanism). Net oxidation of the Fe–S–$U\dot{Q}$ center results in a decrease of the pK of ionizable groups, acting in series as proton transfer sites in the proton channel, with their deprotonation. This would explain higher H^+ release/e^- observed during the completion of the first turnover, with respect to the enzyme at steady-state (Papa *et al.*, 1981a), when, on the other hand, the stoichiometry is dictated by the $U\dot{Q}/UQH_2$ couple. This type of mechanism, involving specific protein-bound quinones and redox Bohr effects, could generally apply to other cytochrome systems and the NADH-ubiquinone oxidoreductase.

REFERENCES

Bell, R. L., Sweetland, J., Ludwig, B., and Capaldi, R. A. (1979). *Proc. Natl. Acad. Sci. USA,* **76,** 741–745.

Bonaventura, C., Bonaventura, J., Brunori, M., and Wilson, M. T. (1978). *FEBS Lett.* **85,** 30–34.

Boyer, P. D., Chance, B., Ernster, L., Mitchell, P., Racker, E., and Slater, E. C. (1977). *Annu. Rev. Biochem.* **46,** 955–1026.

Brand, M. D. (1977). *Biochem. Soc. Trans.* **5,** 1615–1620.

Casey, R. P., Chappell, J. B., and Azzi, A. (1979). *Biochem. J.* **183,** 149–156.

Coin, J. T., and Hinkle, P. C. (1979). In *Membrane Bioenergetics* (C. P. Lee, G. Schatz, and L. Ernster, eds.), pp. 405–412, Addison–Wesley, Reading, Mass.

Crofts, A. R., Crowther, D., and Tierney, G. V. (1975). In *Electron Transfer Chains and Oxidative Phosphorylation* (E. Quagliariello, S. Papa, F. Palmieri, E. C. Slater, and N. Siliprandi, eds.), pp. 233–241, North-Holland, Amsterdam.

Haddock, B. A., and Jones, C. W. (1977). *Bacteriol. Rev.* **41,** 47–99.

Hinkle, P. (1973). *Fed. Proc.* **32,** 1988–1992.

Lehninger, A. L. (1978). *Protons and Ions Involved in Fast Dynamic Phenomena,* pp. 435–452, Elsevier, Amsterdam.

Lorusso, M., Capuano, F., Boffoli, D., Stefanelli, R., and Papa, S. (1979). *Biochem. J.* **182,** 133–147.

Lorusso, M., Boffoli, D., Capuano, F., Capitanio, N., Pace, V., and Papa, S. (1980). *EBEC Short Reports* **1,** 99–100.

Mitchell, P. (1966). *Chemiosmotic Coupling in Oxidative and Photosynthetic Phosphorylation,* Glynn Research, Bodmin, England.

Mitchell, P. (1976). *J. Theor. Biol.* **62,** 327–367.

Mitchell, P. (1980b). In *Oxidases and Related Oxidation–Reduction Systems* (T. E. King, H. S. Mason, and M. Morrison, eds.), Vol. 3, Wiley, New York, in press.

Mitchell, P. (1980a). *Ann. N.Y. Acad. Sci.* **341,** 564–584.

Mitchell, P., and Moyle, J. (1967). In *Biochemistry of Mitochondria* (E. C. Slater, Z. Kaniuga, and L Wojtczak, eds.), pp. 55–74, Academic Press/PWN, London/Warsaw.

Mitchell, P., Moyle, J., and Mitchell, R. (1979). *Methods Enzymol.* **55**(F), 627–640.

Moyle, J., and Mitchell, P. (1978). *FEBS Lett.* **88,** 268–272.

Papa, S. (1976). *Biochim. Biophys. Acta* **456,** 39–84.

Papa, S., Guerrieri, F., Lorusso, M., and Simone, S. (1973). *Biochimie* **55,** 703–716.

Papa, S., Guerrieri, F., and Lorusso, M. (1974). *Biochim. Biophys. Acta* **357,** 181–192.

Papa, S., Guerrieri, F., Lorusso, M., Izzo, G., Boffoli, D., and Capuano, F. (1977). *Biochemistry of Membrane Transport* (G. Semenza and E. Carofoli, eds.), pp. 504–519, Springer-Verlag, Berlin.

Papa, S., Guerrieri, F., Lorusso, M., Izzo, G., Boffoli, D., and Stefanelli, R. (1978). In *Membrane Proteins* (P. Nicholls, J. V. Moller, P. L. Jorgensen, and A. J Moody, eds.), pp. 37–48. Pergamon Press, Elsmford, N.Y.

Papa, S., Capuano, F. Markert, M., and Altamura, N. (1980a) *FEBS Lett.* **111,** 243–248.

Papa, S., Guerrieri, F., Lorusso, M., Izzo, G., Boffoli, D., Capuano, F., Capitanio, N., and Altamura, N. (1980b). *Biochem. J.* **192,** 203–218.

Papa, S., Guerrieri, F., and Lorusso, M. (1980c). *EBEC Short Reports* **1,** 19–20.

Papa, S., Guerrieri, F., Lorusso, M., Izzo, G., and Capuano, F. (1981a). In *Function of Quinones in Energy Conserving Systems* (B. L. Trumpower, ed.), Academic Press, New York, in press.

Papa, S., Lorusso, M., Izzo, G., and Capuano, F. (1981b). *Biochem. J.* **194,** 395–406.

Prince, R. C., and Dutton, P. L. (1976). *FEBS Lett.* **65,** 117–119.

Sigel, E., and Carafoli, E. (1978). *Eur. J. Biochem.* **89,** 119–123.

Sigel, E., and Carafoli, E. (1980). *Eur. J. Biochem.* **111,** 299–306.

Sorgato, M. C., and Ferguson, S. J. (1978). *FEBS Lett.* **90,** 178–182.

Trumpower, B. L. (1981). *J. Bioenerg. Biomembr.,* **13,** 1–24.

Wikström, M. K. F. (1973). *Biochim. Biophys. Acta* **301,** 155–193.

Wikström, M. K. F. (1977). *Nature (London)* **266,** 271–273.

Wikström, M. K. F., and Krab, K. (1979). *Biochim. Biophys. Acta* **549,** 177–222.

Wikström, M. K. F., and Saari, H. T. (1977). *Biochim. Biophys. Acta* **462,** 347–361.

Yu, C. A., and Yu, L. (1980). *Biochem. Biophys. Res. Commun.* **96,** 286–292.

46

Structure of Cytochrome c Oxidase

Roderick A. Capaldi, Stephen D. Fuller, and Victor Darley-Usmar

1. INTRODUCTION

Cytochrome c oxidase (EC 1.9.3.1) is a multisubunit enzyme containing two heme a moieties and two copper atoms as prosthetic groups, functioning to catalyze the reduction of molecular oxygen to water in the reaction: 4 cytochrome $c^{2+} + 4H^+ + O_2 \rightleftharpoons 4$ cytochrome $c^{3+} + 2H_2O$. The energy released during this electron transfer reaction is conserved by the oxidase complex as a proton gradient (Wikström, 1977; Wikström and Saari, 1977; Casey et al., 1979) for subsequent use in ion transport or ATP synthesis. Our goals in studying the structure of cytochrome c oxidase are not only to understand the functioning of the enzyme but also to determine those features of the protein that allow it to integrate into the membrane, an environment where it is partly buried in an essentially hydrocarbon medium and partly exposed to water. In this short review we describe our recent progress in determining the structure of beef heart cytochrome c oxidase.

2. SUBUNIT STRUCTURE

Fractionation studies have shown that cytochrome c oxidases from mitochondria (as opposed to the enzyme from prokaryotic species) contain from 7 to 12 different polypeptides (Sebald et al., 1973; Poyton and Schatz, 1975; Downer et al., 1976; Steffens and Buse, 1976; Merle and Kadenbach, 1980; Darley-Usmar et al., 1981). There are three mitochondrially synthesized polypeptides with molecular weights of around 55,000, 28,000 and 35,000 (I–III in Fig. 1) and cytoplasmically synthesized polypeptides of 17,100, 12,500, and 8500 (IV–VI) (Schatz and Mason, 1974). These six polypeptides are seen in cytochrome c oxidases from all eukaryotic sources examined so far. The present uncertainty about the subunit structure lies in the fact that cytochrome c oxidases isolated from mam-

Roderick A. Capaldi, Stephen D. Fuller, and Victor Darley-Usmar • Institute of Molecular Biology, University of Oregon, Eugene, Oregon 97403.

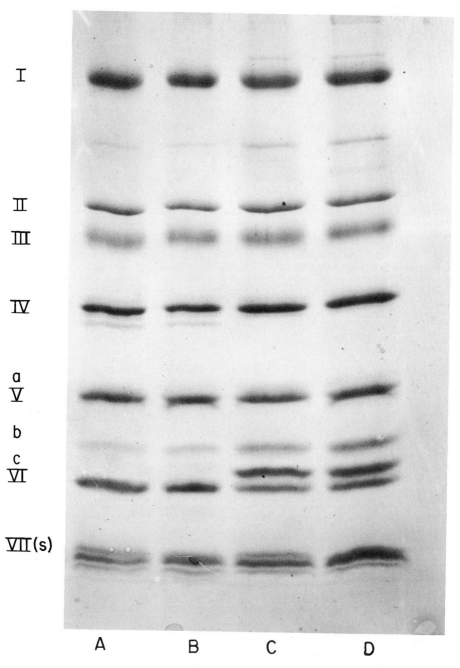

Figure 1. Subunit structure of beef heart cytochrome c oxidase as visualized by SDS-polyacrylamide gel electrophoresis. Samples of cytochrome c oxidase were incubated with trypsin (cytochrome aa_3 : trypsin = 20 : 1) for 2 hr at room temperature (channels A and B). Control samples were treated identically but without trypsin (channels C and D). After proteolytic digestion, approximately 100 μg of each sample was denatured in a solution containing 5% SDS, 8 M urea, 0.5 M Tris–HCl (pH 6.2) with (channels B and D) or without (channels A and C) 1% 2-mercaptoethanol. After denaturation for 1–2 hr at room temperature, samples were electrophoresed on a 16% discontinuous Laemmli-type gel containing 6 M urea (Laemmli, 1970). After electrophoresis, samples were stained and destained as described by Downer *et al.* (1976).

malian sources contain additional polypeptides, which could be true subunits or impurities adventitiously bound to the complex during purification. Three of these extra polypeptides are labeled a, b, and c in Fig. 1. In addition, there are three polypeptides all with molecular weights between 5000 and 6000 and labeled collectively as VII's. Some workers consider all of the above 12 polypeptides to be subunits of the enzyme complex. We have found that polypeptides a, b, and c are present in substoichiometric amount in preparations of high heme content (Downer *et al.*, 1976; Ludwig *et al.*, 1979). Components b and c can be removed by proteolysis (Fig. 1 gels C and D) without affecting cytochrome *c* oxidase structure as judged spectrally or by electron microscopy and image reconstruction studies (Fuller *et al.*, 1979) and without loss of electron transfer or proton pumping function. For these reasons we consider a, b, and c to be impurities. The situation with the three VII's is less clear. Each is a unique polypeptide with Ile, Ser, and Phe as NH_2-terminal amino acids (Buse *et al.*, 1981). One or more of these components are present in yeast, *Neurospora*, as well as the mammalian enzyme. One possibility is that they are the leader sequences of the cytoplasmically synthesized subunits (IV–VI) that are clipped but not further degraded in the membrane.

In the past 2 years the sequences of subunits I, II, and III from human placental beef and yeast cytochrome *c* oxidase have been obtained by sequencing the genes coding for these polypeptides (Barrell *et al.*, 1979; Bonitz *et al.*, 1980; Thalenfeld and Tzagoloff, 1980). Subunits IV, V, VII Ser, and VII Ile of the beef heart enzyme have been sequenced directly (Steffens and Buse, 1976; Sacher *et al.*, 1979; Buse *et al.*, 1981).

3. PROSTHETIC GROUPS

Cytochrome *c* oxidase contains four prosthetic groups per monomer. There is a "low-potential" heme *a* (+ 215 mV) and a low-potential copper (+ 215 mV), which together accept a pair of electrons from cytochrome *c*, probably in one-electron steps through heme *a*. The other two prosthetic groups are heme a_3 (+ 340 mV) and a second copper (+ 350 mV) (Mackey *et al.*, 1973). Heme a_3 and the high-potential copper bind oxygen. Heme a_3 also binds carbon monoxide and cyanide (Keilin and Hartree, 1938). The rate-limiting step appears to be the transfer of electrons from the heme *a*–Cu couple to the heme a_3–Cu pair for reaction with oxygen.

The locus of the prosthetic groups in the cytochrome *c* oxidase complex is presently under intensive study. The recent consensus is that subunits I and II contain all four prosthetic groups. This is based on the fact that these two subunits can be purified with the prosthetic groups attached, subunit I containing heme *a* and subunit II with heme *a* and copper bound (Winter *et al.*, 1980). Also, Buse *et al.* (1981) have claimed that subunit II contains copper, based on sequence comparisons between this polypeptide and copper-containing redox proteins. A further indication that subunits I and II contain the prosthetic groups comes from recent studies on cytochrome *c* oxidase from *Paracoccus denitrificans*. This prokaryotic cytochrome *c* oxidase is a two-heme, two-copper-containing complex made up of only two kinds of subunits (Ludwig and Schatz, 1980). These two subunits are similar in size and amino acid composition to subunits I and II of eukaryotic cytochrome *c* oxidases. Antibodies raised against the subunits of the *Paracoccus* enzyme cross-react with the two largest subunits of the yeast cytochrome *c* oxidase (B. Ludwig, personal communication).

4. SIZE AND SHAPE OF THE PROTEIN COMPLEX

Beef heart cytochrome c oxidase isolated in Triton X-100 or in deoxycholate can form two-dimensional arrays (Sun *et al.*, 1968; Seki *et al.*, 1970) that can be visualized by electron microscopy. We have obtained electron micrographs of these cytochrome c oxidase crystals under a variety of conditions such as in the absence of any stains, in weak negative stains such as metrazamide and gold glucose, and in uranyl acetate. Micrographs have been obtained at several different tilt angles, and these have been processed to give a low-resolution picture of the arrangement of cytochrome c oxidase molecules in the two crystal forms (Henderson *et al.*, 1977; Fuller *et al.*, 1979). The Triton X-100-derived crystal ($P22_12_1$) is a collapsed vesicle in which cytochrome c oxidase dimers are all oriented with the bulk of the protein within the interior of the vesicle (Henderson *et al.*, 1977; J. Deatherage, unpublished results). Antibody-binding experiments show that the outer (exposed) surface of these vesicles is equivalent to the matrix-facing side of the mitochondrial inner membrane (Frey *et al.*, 1978).

Both crystal forms give the same low-resolution structure for the cytochrome c oxidase molecule. The protein is seen as Y shaped, giving rise to three domains, two of which (M1 and M2 in Fig. 2) span the bilayer, the third domain (C) extending from the cytoplasmic side of the mitochondrial inner membrane. The distance between the two M domains has been measured at around 40 Å.

5. ARRANGEMENT OF POLYPEPTIDES

Our present studies are directed at understanding the folding of the subunits of cytochrome c oxidase within the three-dimensional structure shown in Fig. 2. Employing chemical labeling procedures, we have tagged those polypeptides outside the lipid bilayer or the cytoplasmic side of the mitochondrial inner membrane or on the matrix side of the membrane with water-soluble, lipid-insoluble reagents such as diazobenzene[^{35}S]sulfonate and N-(4-azido-2-nitrophenyl)-2-aminoethane[^{35}S]sulfonate (NAP-taurine) (Ludwig *et al.*, 1979). Polypeptides contributing to the bilayer-intercalated part of the complex have been labeled with radioactive arylazidophospholipids (Bisson *et al.*, 1979; Prochaska *et al.*, 1980) or [^{3}H]adamantane diazirine (G. Georgevitch, unpublished results). Subunits have then been purified and the sites of reaction of the various labeling reagents within the sequences of individual subunits are being determined. The results obtained so far show clearly that subunits I, II, III, and IV all span the bilayer (Bisson *et al.*, 1979; Prochaska *et al.*, 1980). The segments of these subunits within the bilayer are in all cases predominantly uncharged and mainly hydrophobic stretches of amino acids. There are two bilayer-spanning stretches in subunit II, only one in subunit IV.

6. CYTOCHROME c BINDING SITE

The site of interaction of substrate cytochrome c with cytochrome c oxidase includes subunit II as shown from cross-linking of the beef heart enzyme with arylazidocytochrome c derivatives (Bisson *et al.*, 1978, 1980) and from cross-linking of a preformed complex of cytochrome c and beef heart cytochrome c oxidase with dithiobissuccinimidylpropionate (Briggs and Capaldi, 1978). Binding of cytochrome c to subunit II has been shown to

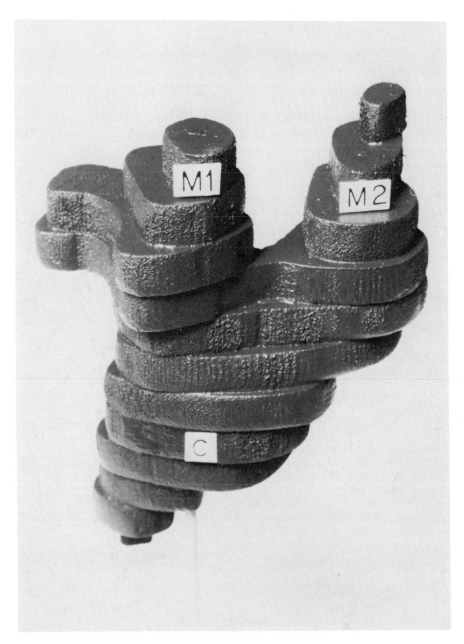

Figure 2. Balsa wood model of the monomeric unit of cytochrome *c* oxidase. (Reproduced with permission from Fuller *et al.*, 1979.)

involve the high-affinity binding site for this subunit ($K_d > 10^{-7}$ M) (Bisson *et al.*, 1980). Beef heart cytochrome *c* oxidase binds a second molecule of cytochrome *c* with lower affinity ($K_d \simeq 10^{-6}$ M) (Ferguson-Miller *et al.*, 1976). This low-affinity site has recently been shown to involve the tightly bound cardiolipins associated with the enzyme (Vik, 1980).

7. SUMMARY

The combination of electron microscopy and biochemical approaches has provided a low-resolution structure of cytochrome c oxidase. This shows the protein as spanning the membrane (mitochondrial inner membrane) and asymmetrically arranged as might be expected of an enzyme functioning for vectorial transfer of electrons and protons. The arrangement of individual subunits can in part be traced. The more detailed analysis in progress will hopefully help to localize prosthetic groups within the protein complex, and this in turn should lead to a better understanding of the functioning of the terminal oxidase in electron transfer and energy coupling.

ACKNOWLEDGMENTS. The authors' research reported herein was supported by National Institutes of Health Research Grant HL22050 and National Science Foundation Research Grant PCM 78 26258. R.A.C. is an Established Investigator of the American Heart Association.

REFERENCES

Barrell, B. G., Bankier, A. T., and Drouin, J. (1979). *Nature (London)* **282**, 189–194.

Bisson, R., Azzi, A., Gutweniger, H., Colonna, R., Montecucco, C., and Zanotti, A. (1978). *J. Biol. Chem.* **253**, 1874–1880.

Bisson, R., Montecucco, C., Gutweniger, H., and Azzi, A. (1979). *J. Biol. Chem.* **254**, 9962–9965.

Bisson, R., Jacobs, B., and Capaldi, R. A. (1980). *Biochemistry* **19**, 4173–4178.

Bonitz, S. G., Coruzzi, G., Thalenfeld, B. W., Tzagoloff, A., and Macino, G. (1980). *J. Biol. Chem.* **255**, 11927–11941.

Briggs, M. M., and Capaldi, R. A. (1978). *Biochem. Biophys. Res. Commun.* **80**, 553–559.

Buse, G., Steffens, G. J., Steffens, G. C. M., Sacher, R., and Erdeg, M. (1981). In *Interaction between Iron and Proteins in Electron Transport* (C. Ho, ed.), Elsevier, Amsterdam, in press.

Casey, R. P., Thelen, M., and Azzi, A. (1979). *Biochem. Biophys. Res. Commun.* **87**, 1044–1051.

Darley-Usmar, V. M., Alizai, N., Al-Ayash, A. I., Jones, G. D., Sharpe, A., and Wilson, M. T. (1981). *Comp. Biochem. Physiol.*, **68B**, 445–456.

Downer, N. W., Robinson, N. C., and Capaldi, R. A. (1976). *Biochemistry* **15**, 2930–2936.

Ferguson-Miller, S., Brautigan, D. L., and Margoliash, E. (1976). *J. Biol. Chem.* **251**, 1104–1115.

Frey, T. G., Chan, S. H. P., and Schatz, G. (1978). *J. Biol. Chem.* **253**, 4389–4395.

Fuller, S. F., Capaldi, R. A., and Henderson, R. (1979). *J. Mol. Biol.* **134**, 305–327.

Henderson, R., Capaldi, R. A., and Leigh, J. S. (1977). *J. Mol. Biol.* **112**, 631–648.

Keilin, D., and Hartree, E. F. (1938). *Nature (London)* **141**, 870–871.

Laemmli, V. K. (1970). *Nature* (London) **227**, 681–685.

Ludwig, B., and Schatz, G. (1980). *Proc. Natl. Acad. Sci. USA* **77**, 196–200.

Ludwig, B., Downer, N. W., and Capaldi, R. A. (1979). *Biochemistry* **18**, 1401–1407.

Mackey, L. N., Kuwana, T., and Hartzell, C. R. (1973). *FEBS Lett.* **36**, 326–329.

Merle, P., and Kadenbach, B. (1980). *Eur. J. Biochem.* **105**, 499–507.

Poyton, R. O., and Schatz, G. (1975). *J. Biol. Chem.* **250**, 752–761.

Prochaska, L., Bisson, R., and Capaldi, R. A. (1980). *Biochemistry* **19**, 3174–3179.

Sacher, R., Buse, G., and Steffens, C. M. (1979). *Hoppe-Seylers Z. Physiol. Chem.* **360**, 1377–1383.

Schatz, G., and Mason, T. L. (1974). *Annu. Rev. Biochem.* **43**, 51–87.

Sebald, W., Machleidt, W., and Otto, J. (1973). *Eur. J. Biochem.* **38**, 311–314.

Seki, S., Hayashi, H., and Oda, T. (1970). *Arch. Biochem. Biophys.* **138**, 110–121.

Steffens, G. J., and Buse, G. (1976). *Hoppe-Seylers Z. Physiol. Chem.* **357**, 1125–1137.

Sun, F. F., Prezbundowski, K. S., Crane, F. L., and Jacobs, E. E. (1968). *Biochim. Biophys. Acta* **153**, 804–818.

Thalenfeld, B. E., and Tzagoloff, A. (1980). *J. Biol. Chem.* **255**, 6173–6180.

Vik, S. B. (1980). Ph.D. thesis, University of Oregon.

Wikström, M. K. F. (1977). *Nature (London)* **266,** 271–273.

Wikström, M. K. F., and Saari, H. T. (1977). *Biochim. Biophys. Acta* **462,** 347–361.

Winter, D. B., Bruyninkx, W. J., Foulke, F. G., Grinich, N. P., and Mason, H. S. (1980). *J. Biol. Chem.* **255,** 11408–11414.

47

Structure of Cytochrome c Oxidase

Recent Progress

Angelo Azzi

1. INTRODUCTION

Cytochrome c oxidase (ferricytochrome c: oxygen oxidoreductase, EC 1.9.3.1) is a complex enzyme that can catalyze the transfer of four electrons from its substrate ferricytochrome c to molecular oxygen and at the same time pump protons across the inner mitochondrial membrane from the matrix to the cytosol space (Capaldi and Briggs, 1976; Wikström and Krab, 1979; Azzi and Casey, 1979; Azzi, 1981). In this process a proton gradient is generated in which the free energy of the redox process is conserved (Wikström, 1977; Casey *et al.*, 1979a; Sigel and Carafoli, 1979; Mitchell, 1966). The heme content of cytochrome c oxidase indicates a minimum molecular weight for the enzyme of approximately 140,000. The same molecular weight can be calculated for the monomer structure by summing the molecular weights of the seven subunits of the enzyme. It contains 4 one-electron redox centers, namely two a hemes and two coppers.

Although it is rather probable that the active form of cytochrome oxidase consists of a dimer, it cannot be excluded that a monomer also can function at a lower rate (Robinson and Capaldi, 1977).

The different purification procedures for cytochrome c oxidase, which have been applied to *Saccharomyces, Neurospora,* bacteria, and mammalian tissues, are all based on the solubilization by detergent of the membrane-bound enzyme and its subsequent precipitation by salts. It is worth mentioning a new purification procedure for cytochrome c oxidase based on chromatography on an affinity column (Bill *et al.*, 1980). *S. cerevisiae* cytochrome c is the affinity ligand attached to an activated Thiol-Sepharose 4B column via its cysteinyl-103 residue. If a Triton X-100 extract of beef heart mitochondrial membrane is loaded on such a column, specific binding of cytochrome c oxidase and bc_1 complex results.

Angelo Azzi • Medizinisch-chemisches Institut der Universität, CH-3012 Bern, Switzerland.

The bound oxidase is eluted at 50 mM salt concentrations in the presence of 0.1% Triton X-100. The novelty of this column is that the lysine residues of cytochrome c, which are important points of binding with the cytochrome c oxidase, are left free for interacting with the oxidase. Starting from mitochondrial membranes, such a technique yields a highly purified enzyme in a very short time.

2. SUBUNIT COMPOSITION

Different preparation techniques and different analytical methods have produced a lack of agreement concerning the number of subunits that comprise cytochrome c oxidase (Capaldi and Briggs, 1976; Merle and Kadenbach, 1980). The anomalous behavior of the migration of cytochrome oxidase polypeptides from different species or in different gel systems has created remarkable nomenclature difficulties. Another possible source of complications comes from the presence in certain preparations of cytochrome oxidase of co-purified polypeptides, which are considered sometimes to be impurities and at other times intrinsic enzyme components.

The enzymes prepared from *S. cerevisiae* and *N. crassa* contain 7 and 8 subunits, respectively, and represent a simple case compared to the enzyme from beef heart (Poyton and Schatz, 1975; Werner, 1977). This enzyme clearly contains more than 7 subunits, up to a maximum of 12 (Ludwig *et al.*, 1979; Steffens *et al.*, 1979; Merle and Kadenbach, 1980).

It is not clear, however, which of these subunits represent intrinsic components of the enzyme and which are instead copurified proteins. It is clear, however, that the three highest molecular weight subunits are present in all types of cytochrome c oxidase of eukaryotic cells, and that they have several features in common. They are coded for by the mitochondrial DNA, they are more hydrophobic than the others, and they possibly include the binding site of the redox centers. In fact, the sequence of subunit II indicates a strong analogy with plastocyanine, stellacyanine, and azurin (Steffens and Buse, 1979). The location of heme and copper groups at the level of the three highest molecular weight subunits is also supported by the studies of cytochrome c oxidases isolated from bacteria. These oxidases (e.g., *Paracoccus denitrificans,* Ludwig and Schatz, 1980) contain a smaller number of subunits with high molecular weights analogous to the larger molecular weight subunits of mammalian cytochrome c oxidases. As it is improbable that large structural differences exist between enzymes that are spectrally and functionally very similar, it is tempting to speculate that hemes and coppers are accommodated in the high molecular weight subunits. Subunit III is labeled specifically by dicyclohexylcarbodiimide (Casey *et al.*, 1980), and simultaneously with the labeling an inhibition of the proton pump results. Extraction of subunit III also results in inhibition of the proton pump (Saraste *et al.*, 1980) without evident changes in the enzyme absorbance spectrum. The data thus indicate in subunit III a component of the H^+-pumping mechanism and not a redox-center-carrying subunit.

3. SUBUNIT TOPOLOGY

The gross molecular structure of cytochrome c oxidase from beef heart as seen in Triton X-100- or deoxycholate-prepared enzymes resembles that of a Y where the arms of the Y are 55 Å in length and have a center-to-center separation of 40 Å the total monomeric molecule is roughly 110 Å long. The arms of the Y are located on the matrix side of the molecule, the remainder being embedded in the phospholipid bilayer membrane. A large domain, the stem of the Y, protrudes at the other side of the membrane (Henderson *et al.*, 1977; Fuller *et al.*, 1979). The fine topology of the enzyme, i.e., the location of the different subunits with respect to each other, is not fully established. The most widely employed techniques for exploring this type of relationship are the following: (1) use of hydrophilic impermeable reagents to label exposed subunits; (2) use of lipid-soluble molecules to label those parts of the subunits that interact with phospholipid; (3) use of antibodies directed to individual subunits and study of their binding or the inhibition of function.

Although the three approaches have provided conflicting results in some cases, they are in general in good agreement. It is in fact possible to conclude that subunits II and III are exposed largely to the cytoplasmic water phase (Ludwig *et al.*, 1979) but that they also interact (Bisson *et al.*, 1979a) with the membrane lipids. Subunits II and III can also be seen protruding at the matrix face of the mitochondrial membrane (Ludwig *et al.*, 1979). Subunit IV is seen only at the matrix face and interacts only to a small extent with membrane lipids (Bisson *et al.*, 1979b). Subunit I cannot be labeled by hydrophilic probes, and on this basis it was attributed to the interior of the membrane. Recent subunits also protrude to a certain extent toward the cytosolic membrane face (Prochaska *et al.*, 1980). The problems concerning the identification of the low molecular weight subunits of cytochrome oxidase are such that their precise localization with respect to each other remains problematic.

4. SUBUNIT INTERACTIONS

Neighboring subunits in multipeptide enzyme complexes can be cross-linked by reagents that have two reactive groups, namely bifunctional reagents. Using these reagents, it has been shown that subunit V cross-links with I, II, III, and VII, while subunit IV cross-links with VI and VII. It can be suggested that the mentioned subunits are organized in groups (Briggs and Capaldi, 1977). However, the group specificity of the reagents employed, their solvent specificity, and the perturbation they induce make this technique rather questionable for the identification of near-neighboring subunits of complex enzymes. Similarly, the use of fluorescence probes to measure intersubunit distances suffers from a number of assumptions. Uncertainties in the determination of the overlapping of donor and emission spectra, and of the relative orientation of the fluorochromes make the distance measurements of only qualitative interest (Dockter *et al.*, 1978).

5. INTERACTION WITH CYTOCHROME c

The interaction of cytochrome *c* with the oxidase occurs at the cytosolic site in the membrane-bound enzyme. Cross-linking studies using photolabeled cytochrome *c* have clarified which subunit of cytochrome oxidase is involved in the interaction with cytochrome *c* (Bisson *et al.*, 1977). Arylazidocytochrome *c* derivatives with the reactive group at lysine 13 and lysine 22 do not behave identically as far as the reaction with cytochrome *c* oxidase is concerned. The lysine-13 derivative cross-links after illumination with the oxidase, whereas the lysine-22 derivative does not. The cross-linked subunit is the second largest molecular weight polypeptide of cytochrome *c* oxidase (Bisson *et al.*, 1978). Subunit III may also be in contact with cytochrome *c*, as it has been shown to interact, at least in yeast cytochrome oxidase, with cytochrome c (Birchmeier *et al.*, 1976). The notion that there are structural analogies between subunit II of cytochrome oxidase and copper-binding proteins permits the speculation that copper is the first electron acceptor in cytochrome oxidase.

6. THE MOBILITY OF CYTOCHROME OXIDASE

The mobility of the membrane-bound enzyme is important because the catalytic events may ultimately be governed by the rate of lateral diffusion of the enzyme. Information on the macromolecular motion of cytochrome oxidase has been obtained essentially by two techniques. The first technique (Kawato *et al.*, 1980) is based on the observations of the time dependence of the dichroism changes following flash photolysis of a CO–a_3 complex in intact mitochondria. They are consistent with the rotational correlation time of cytochrome a_3 in the order of 300 to 400 nsec. This value can be considered the expression of fast rotational motion of the enzyme.

In experiments carried out using a spin-labeled enzyme and an ESR technique (saturation transfer), which allows determination of motion in the microsecond to millisecond time domain, a similar conclusion was reached (Ariano and Azzi, 1980; Swanson *et al.*, 1980). In this case a correlation time of 35 μsec for the enzyme reconstituted in phospholipid vesicles was obtained. It therefore can be concluded that cytochrome oxidase rotates fast in the membrane. This is consistent with the idea that the enzyme does not form large aggregates that are functionally important.

7. REQUIREMENT OF LIPIDS FOR CATALYTIC ACTIVITY

The concept of "boundary lipids" (Jost *et al.*, 1973) to indicate a layer of immobilized molecules bound to the enzyme, implied that the physical state of such a lipid "annulus" affects the functional characteristic of the membrane protein. This interpretation (Marsh *et al.*, 1978), largely based on spin-label studies of cytochrome oxidase, was contrasted by Seelig and Seelig (1978), who using deuterium and phosphorus NMR produced evidence that a more immobilized layer of lipid does not exist around cytochrome c oxidase. Rather, the lipids around the enzyme are in a more disordered state. Thus, the spin-label data indicating the existence of an "immobile component" should be considered for example in terms of regions of high microviscosity, of lipids trapped between protein aggregates or of specific interactions between the spin label and the enzyme (Chapman *et al.*, 1979). In any case it seems improbable that the lipids around the protein have an important meaning in terms of the catalytic function of the enzyme.

8. CONCLUSIONS

What has been reported until now is a personal rather than a comprehensive view on recent cytochrome oxidase studies. The structural aspects have been emphasized over the functional aspects of the enzyme as these have been studied extensively and specifically. To our view the most exciting findings are those that have provided a structural basis for cytochrome oxidase, namely the electron microscopic studies of Henderson *et al.* (1977). Toward an understanding of the structure of the enzyme, the amino acid sequence analysis, in particular that carried out by Buse *et al.* (1980), has considerable importance. The location of the cytochrome *c* binding site at the level of subunit II and the identification of subunit III (Casey *et al.*, 1979b, 1980) as a possible candidate in the function of proton pumping, are the first functional indications for subunits in cytochrome *c* oxidase. The crystallization of cytochrome oxidase in three-dimensional crystals (Ozawa *et al.*, 1980) is the beginning of a possibility of studying the structure of this enzyme with X ray diffraction. Finally, the comparison between bacterial oxidases and those from mammalian tissues has offered and will offer in the future a large and important number of results to clarify the problem of cytochrome oxidase and also in general the problems of membrane proteins and their structure–function relationships (Ludwig, 1981).

ACKNOWLEDGMENT. This work was supported in part by Grant 3.228.077 from the Schweizerischen Nationalfonds.

REFERENCES

Ariano, B. H., and Azzi, A. (1980). *Biochem. Biophys. Res. Commun.* **93**, 478–485.

Azzi, A. (1980). *Biochim. Biophys. Acta*, **594**, 231–252.

Azzi, A., and Casey, R. P. (1979). *Mol. Cell. Biochem.* **28**, 169–184.

Bill, K., Casey, R. P., Broger, C., and Azzi, A. (1980). *FEBS Lett.* **120**, 248–250.

Birchmeier, W., Kohler, C. E., and Schatz, G. (1976). *Proc. Natl. Acad. Sci. USA* **73**, 4334–4338.

Bisson, R., Gutweniger, H., and Montecucco, C. (1977). *FEBS Lett.* **81**, 147–150.

Bisson, R., Azzi, A., Gutweniger, H., Colonna, R., Montecucco, C., and Zanotti, A. (1978). *J. Biol. Chem.* **253**, 1874–1880.

Bisson, R., Montecucco, C., Gutweniger, H., and Azzi, A. (1979a). *J. Biol. Chem.* **254**, 9962–9999.

Bisson, R., Montecucco, C., Gutweniger, H., and Azzi, A. (1979b). *Biochem. Soc. Trans.* **7**, 156–160.

Briggs, M. M., and Capaldi, R. A. (1977). *Biochemistry* **16**, 73–77.

Buse, G., Steffens, G. J., Steffens, G. C. M., Sacher, R., and Erdweg, M. (1980). In *First European Bioenergetics Conference*, pp. 41–42, Patron, Bologna.

Capaldi, R. A., and Briggs, M. (1976). In *The Enzymes of Biological Membranes* (A. Martonosi, ed.), Vol. IV, pp. 87–102, Wiley, New York.

Casey, R. P., Chappell, J. B., and Azzi, A. (1979a). *Biochem. J.* **182**, 149–156.

Casey, R. P., Thelen, M., and Azzi, A. (1979b). *Biochem. Biophys. Res. Commun.* **87**, 1044–1051.

Casey, R. P., Thelen, M., and Azzi, A. (1980). *J. Biol. Chem.* **255**, 3994–4000.

Chapman, D., Gómez-Fernández, J. C., and Goñi, F. M. (1979). *FEBS Lett.* **98**, 211–223.

Dockter, M. E., Steinemann, A., and Schatz, G. (1978). *J. Biol. Chem.* **253**, 311–317.

Fuller, S. D., Capaldi, R. A., and Henderson, R. (1979). *J. Mol. Biol.* **134**, 305–327.

Henderson, R., Capaldi, R. A., and Leigh, J. S. (1977). *J. Mol. Biol.* **112**, 631–648.

Jost, P., Griffith, O. H., Capaldi, R. A., and Vanderkooi, G. (1973). *Biochim. Biophys. Acta* **311**, 141–152.

Kawato, S., Sigel, E., Carafoli, E., and Cherry, R. J. (1980). *J. Biol. Chem.* **255**, 5508–5510.

Ludwig, B. (1980). *Biochim. Biophys. Acta*, **594**, 117–189.

Ludwig, B., and Schatz, G. (1980). *Proc. Natl. Acad. Sci. USA* **77**, 196–200.

Ludwig, B., Downer, N. W., and Capaldi, R. A. (1979). *Biochemistry* **18**, 1401–1407.

Marsh, D., Watts, A., Maschke, W., and Knowles, P. F. (1978). *Biochem. Biophys. Res. Commun.* **81**, 397–402.

Merle, P., and Kadenbach, B. (1980). *Eur. J. Biochem.* **105**, 499–507.

Mitchell, P. (1966). *Biol. Rev.* **41**, 445–502.

Ozawa, T., Suzuki, H., and Tanaka, M. (1980). *Proc. Natl. Acad. Sci. USA* **77**, 928–930.

Poyton, R. O., and Schatz, G. (1975). *J. Biol. Chem.* **250**, 752–761.

Prochaska, C., Bisson, R., and Capaldi, R. A. (1980). *Biochemistry* **19**, 3174–3179.

Robinson, N. C., and Capaldi, R. A. (1977). *Biochemistry* **16**, 375–381.

Saraste, M., Penttilä, T., and Wikström, M. (1980). *FEBS Lett.* **114**, 35–38.

Seelig, A., and Seelig, J. (1978). *Hoppe-Seylers Z. Physiol. Chem.* **359**, 1747–1756.

Sigel, E., and Carafoli, E. (1979). *J. Biol. Chem.* **254**, 10572–10574.

Steffens, G., and Buse, G. (1979). *Hoppe-Seylers Z. Physiol. Chem.* **360,** 613–619.

Steffens, G. C. M., Steffens, G. J., and Buse, G. (1979). *Hoppe-Seylers Z. Physiol. Chem.* **360,** 1641–1650.

Swanson, M. S., Quintanilha, A. T., and Thomas, D. D. (1980). *J. Biol. Chem.* **255,** 7494–7502.

Werner, S. (1977). *Eur. J. Biochem.* **79,** 103–110.

Wikström, M. K. F. (1977). *Nature (London)* **266,** 271–273.

Wikström, M., and Krab, K. (1979). *Biochim. Biophys. Acta* **549,** 177–222.

48

On The Mechanism of Energy Coupling in Cytochrome Oxidase

D. E. Green, M. Fry, and H. Vande Zande

Highly purified cytochrome oxidase from beef heart mitochondria without any supplementation catalyzes the coupling of the oxidation of ferrocytochrome c by O_2 to the transport of monovalent or divalent cations (Green $et\ al.$, 1980; Fry and Green, 1980a). This finding established that cytochrome oxidase was a complete energy-coupling system and that coupling took only one form—the coupling of electron transfer to cation transport. Cytochrome oxidase was ideally suited to probe the mechanism of energy coupling. The components and structure of the electron transfer chain were well defined, the subunit structure of the complex was firmly established, the stoichiometry of the oxidative reaction (oxidation of ferrocytochrome c by O_2) was known, the assay of cytochrome oxidase activity was sufficiently selective that it was possible to measure this activity as readily in mitochondria as in dispersions of the purified oxidase, and finally and most importantly ion transport is the least complicated expression of a coupled process.

Using ion-specific electrodes for measuring changes in K^+ and H^+ concentrations and the oxygen pulse technique to initiate the energized coupling sequence, it was found that the profiles for ΔK^+ and ΔH^+ in such a sequence were U shaped. On energization, K^+ was transported into the particle at a linear rate until an equilibrium was abruptly established (a plateau region); when the oxygen tension diminished below a critical value, K^+ effluxed from the particle at a linear rate and continued until the gradient established during the energized phase was completely eliminated. It was clear from the U-shaped profile that cation transport (in absence of a permeant anion) was cyclical in nature. A balance was quickly established between the energized influx of K^+ and the nonenergized efflux of K^+. The total energized flux of K^+ was the sum of the observed influx rate (measured by the slope of the descending limb of the ΔK^+ profile) and the observed efflux rate (measured by the slope of the ascending limb). The ratio of the K^+ flux to the rate of O_2 consumption $\div 4$ provided a measure of the K^+/e^- ratio. This ratio was found consistently to be unity.

D. E. Green, M. Fry, and H. Vande Zande • Institute for Enzyme Research, University of Wisconsin, Madison, Wisconsin 53706.

The U-shaped profile for ΔH^+ during an oxygen pulse sequence was directionally opposite to the ΔK^+ profile. The $\Delta H^+/e^-$ ratio estimated by the two-slope method described above was also found to be unity within experimental error. Thus, for each electron traversing the electron transfer chain, one K^+ was transported into the particle and one proton was released into the external aqueous phase; for each K^+ that effluxed from the particle, one proton was taken up from the external phase. Two crucial points had to be clarified before these interrelationships of ionic changes could be rationalized. What was the source of the proton that was released when K^+ was transported inwardly during energization? What was the connection between K^+ efflux and H^+ uptake? Why were these two events causally related?

When ferrocytochrome c is oxidized by cytochrome oxidase, both an electron and a proton are released; the electron traverses the membrane via the electron transfer chain and the proton is released into the aqueous phase. In other words, bound ferrocytochrome c is both an electron and a proton donor (Green and Vande Zande, 1981). The proof of this postulated origin of the proton is as follows. When ferrocytochrome c is bound by delipidated cytochrome oxidase (in the presence of a high concentration of sodium azide to prevent enzymatic oxidation), one proton is taken up per molecule of cytochrome oxidase; when ferricytochrome c is bound by delipidated Complex III, one proton is released per molecule of Complex III. Apparently some ionizable group in the protein domain of cytochrome c provides the negative charge required for neutralization of the third positive charge in bound ferricytochrome c. In bound ferrocytochrome c this ionizable group is protonated. When bound ferrocytochrome c is oxidized, this ionizable group is deprotonated. In free cytochrome c, oxidoreduction is not accompanied by proton release or uptake but rather by anion uptake or release.

When H^+ is measured during an oxygen pulse in an electron transfer particle that has the opposite directionality of mitochondria or cytochrome oxidase, protons are taken up during energization and released when deenergization sets in by virtue of O_2 depletion. Thus, in energized transport of K^+, protons are released on the side where oxidation is initiated (the I side) and taken up on the side where oxidation is terminated (the M side). The electron transfer sequence begins with "dehydrogenation" on the I side of the membrane and terminates with hydrogenation of O_2 on the M side of the membrane. Given that ferrocytochrome c is the source of the proton released during oxidation, it necessarily follows that the H^+/e^- ratio cannot be greater than unity because of the 1 : 1 relation between proton and electron demonstrated by our binding studies.

It may be useful to consider the assumptions that underlie the directionality we have invoked for cytochrome oxidase, either in the inner membrane or in lysolecithin dispersions. On the basis of the direction of ion transport, we can infer that the I or outer side of cytochrome oxidase is the side where transport begins and the M or inner side is the side where transport ends. Because protons are released on the side of cytochrome oxidase where oxidation is initiated, we can conclude that proton release takes place on the I side. Similarly, because protons are taken up on the side where reduction takes place, we can conclude that proton uptake takes place on the M side. On the basis of these relationships we can infer that the directionality of cytochrome oxidase in lysolecithin dispersions (or in liposomes) is the same as that of cytochrome oxidase in the inner membrane and opposite to that of the electron transfer particle. When protonic changes in cytochrome oxidase are followed during an oxygen pulse, energization leads to uptake of a proton rather than release (Green *et al.*, 1980). This anomaly arises from the fact that free ferrocytochrome c takes up a proton when bound and the bound form releases a proton when oxidized. By virtue of this cancellation of protonic changes, the profile for ΔH^+ is out of phase with the profile for ΔK^+. In mitochondria this anomaly is not observed because cytochrome c is in the bound form throughout the oxygen pulse.

The inverse relation between nonenergized cation efflux and proton uptake is inextricably tied in to the mechanism by which cations are transported into and extruded from the particle. There were many hints that cardiolipin played an essential role in this mechanism. The first direct evidence for this involvement was the demonstration by Fry and Green (1980b) that cardiolipin was an essential and specific requirement of the activity of cytochrome oxidase. When the cardiolipin content of cytochrome oxidase was reduced below 2 wt%, activity was progressively lost; and activity could be recovered by adding back cardiolipin to the delipidated oxidase. This specific, catalytic effect of cardiolipin has to be distinguished from the general dispersive effect demonstrable with phospholipids generally and with some detergents (lysolecithin, Tween 20). Studies on the reversible resolution of cytochrome oxidase by Fry and Green (1980c) established that enzymatic activity was completely recoverable in a fragment that contained only four of the seven subunits—the four largest (I–IV). This fragment was found to contain the full complement not only of heme and copper but also of cardiolipin (Fry and Green, 1980c; Fry *et al.*, 1980). Resolution of this fragment with SDS into the individual subunits then showed that cardiolipin was associated predominantly with subunits I and II (I $>>>$ II). Winter *et al.* (1980) established concurrently that the entire complement of heme and copper present in native form was also linked to these same two subunits. Thus, the functional groups (heme, copper, and cardiolipin) were all attached to the same two subunits of which subunit I was transmembrane (Fry and Green, 1980d).

Cardiolipin has been shown by Tyson *et al.* (1976) to be a highly active ionophore that transports a wide variety of monovalent and divalent cations with equal efficiency. These studies were carried out in a two-phase system. Fry and Green (1980a) have examined the cation specificity of cytochrome oxidase and found it to be indistinguishable from that of cardiolipin. Cytochrome oxidase can transport amino acids as readily as cations; cardiolipin shows the same capability. Cytochrome oxidase can mediate K^+/H^+ exchange as we have noted; cardiolipin shows the same capability. Ruthenium red inhibits cation transport mediated either by cytochrome oxidase or by cardiolipin. Thus, by all criteria the transport patterns for cytochrome oxidase and cardiolipin are identical.

The tight association of cardiolipin with subunits of cytochrome oxidase and the presence of 2 molecules of cardiolipin per molecule of cytochrome oxidase are clear indications that cardiolipin cannot function as a mobile ionophore. The available evidence points to some type of cage structure comparable to that of a gramicidin pore (Urry *et al.*, 1975). A molecular model that we have constructed shows that 2 molecules of cardiolipin can be arranged to form a cage structure for the transmembrane passage of cations (Fry and Green, 1980e). Figure 1 is a diagramatic representation of the relationship of the two chains in cytochrome oxidase—the cardiolipin and electron transfer chain.

Now finally we can return to the second question posed earlier, namely why is there an inverse directional relation between nonenergized cation efflux and proton uptake? In energized influx the electron drives the cation inward via the cardiolipin cage, the electron being the negatively charged partner for the cation. In nonenergized efflux the cation successively displaces a proton from the four phosphate residues as it moves through the cardiolipin cage, and thus nonenergized efflux of the cation goes parallel with the inward flow of protons by successive displacement of a proton from the phosphate residues. In cyclical cation transport, protons are released on the I side, transferred by K^+/H^+ exchange to the M side, and then taken up on the M side. The net protonic change at the end of each oxygen pulse is thus zero.

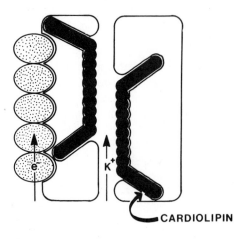

Figure 1. The ion transport and electron transfer chains of cytochrome oxidase.

The conclusion that we have reached is that the energized movement of the electron via its electron transfer chain is directly coupled to the movement of the cation via the cardiolipin chain. The two-chain direct-coupling thesis is formalized in Fig. 2. Coupling in Complexes I and III has been examined in the same fashion as coupling in cytochrome oxidase. As limitations of space will not permit a summary of these studies, suffice it to

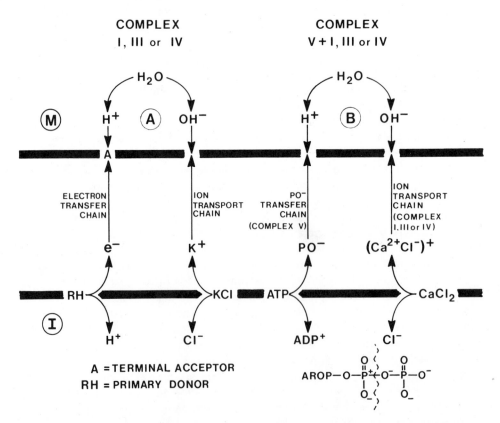

Figure 2. The two-chain direct-coupling thesis of energy coupling applied to an electron transfer complex (A) and to a combination of the oligomycin-sensitive ATPase and an electron transfer complex. The uptake of a hydroxyl ion by ADP^+ leads to the release of a proton.

say that precisely the same two-chain direct-coupling pattern has been shown to apply to the other two-electron transfer complexes, as well as to the combination of oligomycin-sensitive ATPase and an electron transfer complex that catalyzes the coupling of pyrophosphorolysis of ATP to the active transport of Ca^{2+}.*

REFERENCES

Fry, M., and Green, D. E. (1980a). *Biochem. Biophys. Res. Commun.* **95**, 1529–1535.

Fry, M., and Green, D. E. (1980b). *Biochem. Biophys. Res. Commun.* **93**, 1238–1246.

Fry, M., and Green, D. E. (1980c). *J. Bioenerg. Biomembr.* **13**, 61–87.

Fry, M., and Green, D. E. (1980d). *Proc. Natl. Acad. Sci. USA* **76**, 2664–2668.

Fry, M., and Green, D. E. (1980e). *Proc. Natol. Acad. Sci. USA* **77**, 6391–6395.

Fry, M., Blondin, G. A., and Green, D. E. (1980). *J. Biol. Chem.* **255**, 9967–9970.

Green, D. E., and Vande Zande, H. (1981). *Biochem Biophys. Res Commun.* **98**, 635–641.

Green, D. E., Vande Zande, H., Skopp, R., and Fry, M. (1980). *Biochem. Biophys. Res. Commun.* **95**, 1522–1528.

Tyson, C. A., Vande Zande, H., and Green, D. E. (1976). *J. Biol. Chem.* **251**, 1326–1332.

Urry, D. W., Long, M. M., Jacobs, M., and Harris, R. D. (1975). *Ann. N.Y. Acad. Sci.* **264**, 203–220.

Winter, D. E., Bruyninckx, W. J., Foulke, F. G., Grinich, N. P., and Mason, H. S. (1980). *J. Biol. Chem.*, **255**, 11408–11414.

---

*The studies of Jeng *et al.* [Proc. Natl. Acad. Sci. USA (1978) **75**, 2125–2129] strongly suggest that Ca^{2+} transport driven by ATP pyrophosphorolysis or electron transfer is mediated by an ionophore, calciphorin, the localization of which is still undetermined. The transported species would be a positively charged adduct of Ca^{2+} and calciphorin^{-1} [(calciphorin$^{-1}Ca^{2+}$)$^{1+}$]. The principle underlying coupling of ATP pyrophosphorolysis to Ca^{2+} transport is independent of whether the ionophoric entity is attributed to the cardiolipin channel or to calciphorin.

49

Insights Gained by EPR Spectroscopy into the Composition and Function of the Mitochondrial Respiratory Chain

Helmut Beinert

Ever since Keilin's fundamental discovery—or rediscovery—in the 1930s of the cytochromes (Keilin, 1966), which have been and still are considered by many to be the mainstays of the mitochondrial respiratory chain, optical spectroscopy has played a paramount role in this field of research. The past two or three decades have witnessed the introduction into biochemistry of additional and different forms of spectroscopy, using frequencies of the electromagnetic spectrum considerably higher or lower than traditional spectrophotometry, which of course often required radically different technologies. In its impact on our knowledge of what has been called the respiratory chain, one form of spectroscopy stands unequaled, viz. electron paramagnetic resonance (EPR) spectroscopy. This might have been expected, for EPR spectroscopy is *the* spectroscopic technique to directly look at unpaired electrons in organic molecules—e.g., cofactors such as flavin or ubiquinone (UQ)—and transition metal ions, which are electron carriers in oxidative enzymes. Fortunately, other virtues of the technique have joined this most important quality to make EPR spectroscopy the thus far most successful tool in advancing our knowledge far beyond the status as essentially set by Keilin's pioneering work and by logical extensions thereof in the 1940s and 1950s. The virtues are as follows. (1) The sensitivity of EPR spectroscopy, when carried out at low temperatures, is sufficient to let us clearly recognize even in whole tissues practically all the electron carriers that have been identified by EPR in more concentrated, purified fractions. (2) While the sensitivity of EPR spectroscopy, in most instances, lies one or two orders of magnitude below that of spectrophotometry, its power of discrimination is far superior, particularly because effects of the intensity of the impinging radiation (saturation), observation temperature, and coupling of the observed paramagnet to other magnetic species can be exploited. (3) The great obstacle in the application of spectrophotometry to integrated systems such as particulate and membranous material, viz.

Helmut Beinert • Institute for Enzyme Research, University of Wisconsin, Madison, Wisconsin 53706.

opacity, is nonexistent for EPR; only magnetic opacity most commonly originating from spurious copper, iron, or manganese ions or metal particles is of concern. (4) EPR spectroscopy does not need the "extinction coefficients" required in spectrophotometry for quantitative determination; one unpaired electron has the same extinction coefficient no matter in what substance it occurs as long as it is not subjected to moderate to strong interaction with neighboring (<1 nm) magnetic species. Thus, the concentration of the species observed can be determined, often from a single spectral line. (5) Because it is of relatively high sensitivity and applicable to material in a crude state and because—once instrumentation is set up—of the ease and speed of recording spectra on series of samples, EPR is well suited for exploratory experimentation, much more so than, e.g., Mössbauer, CD, or X-ray spectroscopy, although these techniques may far surpass EPR in detailed information content in the realm of their applicability.

Equipped with this brief outline of the advantages of EPR spectroscopy, we will select from the many successful applications a few prominent examples illustrating the capabilities of the technique as well as the kind of knowledge that has been gained with it.

It is fair to say that the rapid surge of the field of iron-sulfur proteins (Lovenberg, 1977) is in large measure due to EPR spectroscopy. We know now that the Fe–S components in the respiratory chain outnumber the cytochromes and are more diversified in the range of oxidation–reduction potentials they cover and in their function. EPR spectroscopy has also been instrumental in establishing copper as an active component in cytochrome c oxidase (Beinert *et al.*, 1976). With the advent of Fe–S proteins, copper, and ubiquinone, the idea of a respiratory chain, along which electrons move in sequence, is no longer realistic. Rather, we must visualize a complex three-dimensional network with different domains that probably have prescribed entrance and exit ports and in which are located rapidly responding, redox-poising and -buffering and electron storage components, perhaps also regulatory ones and components providing links to functions that are coupled to electron transfer.

A glance at EPR spectra of various frozen whole tissues and subcellular fractions probably gives the most convincing introduction to the multiple capabilities of EPR. In a single spectrum of excised frozen heart tissue (Fig. 1C) (Beinert, 1977, 1978; Orme-Johnson *et al.*, 1974) in which the electron acceptors will largely be in the reduced state, we can readily recognize several Fe–S components of NADH dehydrogenase (Complex I) at $g = 2.10, 2.05, 1.93, 1.92, 1.89, 1.86$ (Beinert, 1978)—that of ETF-UQ oxidoreductase (Ruzicka and Beinert, 1977) (at $g = 2.08$) and the so-called Rieske Fe–S protein (Rieske *et al.*, 1964) of the cytochrome bc_1 complex (Complex III, at $g = 1.89$ and 1.78)—whereas in more oxidized states, which are more readily produced with isolated mitochondria (Fig. 1A) (Beinert, 1977, 1978), the high-potential iron protein (HiPIP)-type cluster of succinate dehydrogenase ($g = 2.01$) (Beinert *et al.*, 1975; Ohnishi *et al.*, 1976), the UQ semiquinone pair (Ruzicka *et al.*, 1975; Ingledew *et al.*, 1976) linked to this enzyme (Fig. 1B at $g = 2.03$ and 1.98), and some of the copper of cytochrome c oxidase (Fig. 1A, insert at left) can be seen. On further oxidation and at higher sensitivity, the spectra of several oxidized cytochromes appear (Fig. 2 at $g \sim 6, 3.8, 3.4, 3.0, 2.2, 1.5$) (Beinert *et al.*, 1976; Orme-Johnson *et al.*, 1974), those of transferrin (Fig. 3 at $g = 4.2$) and occasionally of metmyoglobin and (or) methemoglobin (Fig. 3 at $g = 6$). Whole liver tissue (Fig. 3) will, in addition, show the unambiguous absorptions of cytochrome P-450 (Fig. 3 at $g = 2.43$ and 2.26), of catalase (Fig. 3 at $g = 6.5$ and 5.3), Mn^{2+} (Fig. 3, sharp lines between $g = 2$ and 2.26), and under some conditions of sulfite reductase. Needless to say that in purified fractions (particulate or soluble), both the sensitivity and the possibilities of fine discrimination and quantitative determination are only rising.

There have in fact been at least two instances where a mitochondrial protein has been purified simply following a single EPR resonance (Fig. 1A, $g = 2.01$; Fig. 1C, $g = 2.08$)—an Fe–S flavoprotein and the ''mitochondrial HiPIP,'' which eventually were identified as ETF-UQ oxidoreductase (Ruzicka and Beinert, 1977) and aconitase (Ruzicka and Beinert, 1978), respectively. As may be imagined, finding an activity for a protein is not a trivial task; and again in this undertaking observation of the EPR signal was the only ''assay'' available. The study of both these proteins had its surprises. ETF-UQ oxidoreductase seems to be the terminal protein in a cascade of three flavoproteins, which links the β-oxidation of fatty acyl derivatives of coenzyme A to the respiratory chain: acyl-CoA dehydrogenase, electron-transferring flavoprotein, and ETF-UQ oxidoreductase, with the last protein being an Fe–S protein. Thus, the pathways of electron flow from the three principal direct feeder substances of the respiratory chain, viz. NADH, succinate, and fatty acids, are linked into the chain at the UQ level via Fe–S clusters. Is this a general requirement? We do not know yet. An additional example, the membrane-bound flavoprotein α-glycerophosphate dehydrogenase, is under study. The second mitochondrial Fe–S protein purified to homogeneity by following its EPR signal (i.e., aconitase) has brought even more unexpected findings. However, as it is not a component of the respiratory chain, it may suffice to say here that the Fe–S cluster in this protein seems to be of a novel type, viz. a [3Fe–3S] cluster (Kent *et al.*, 1981). Its function is not in electron transfer but apparently in the regulation of the activity of aconitase; only after reduction of the cluster can the protein assume the active conformation.

Another unexpected finding was made concerning UQ. Although an EPR signal of mitochondria attributed to UQ semiquinone was reported in 1970 (Bäckström *et al.*, 1970), little significance had generally been allotted to UQ semiquinone as it had been shown that compounds of closely related structure underwent very rapid disproportionation of the semiquinone to the oxidized and reduced forms (Kröger, 1976). At low temperatures ($< 25°K$), mitochondria show an EPR signal centered at $g \sim 2$, which has the general behavior of signals involving transition metal ions, viz. strong temperature dependence and spin relaxation, but behaves in reductive titrations as if it were due to a two-electron acceptor, which rather pointed to a quinone or flavin-type compound. By extraction of UQ from submitochondrial particles and reconstitution with UQ, this signal was then shown to be due to two interacting UQ semiquinones (Ruzicka *et al.*, 1975), oriented such that they are aligned edge-on, not with planes parallel (Ruzicka *et al.*, 1975; Ingledew *et al.*, 1976). This was carried further (Salerno *et al.*, 1979a) in studies of the signal in stacked oriented membrane multilayers by EPR. It was concluded from this work that the quinone pair is oriented with the oxygen–oxygen axes perpendicular to the membrane plane, as if it might partly span the membrane. It is of particular interest here that the concentration of UQ semiquinone in this pair is approximately that of the succinate dehydrogenase components, still only a fraction of the UQ pool but locally rather a remarkable concentration. This UQ pair seems to function as the immediate acceptor of electrons donated by one of the Fe–S centers of succinate UQ reductase (Ackrell *et al.*, 1977).

The same technique as applied to the UQ pair, namely studies by EPR on stacked membranes, was also used to determine the orientation of cytochromes and Fe–S centers (Salerno *et al.*, 1979a; Erecińska *et al.*, 1978). These species were generally found to have a fixed position relative to the plane of the membrane so that most Fe–S clusters and heme groups can be said to be oriented with respect to each other and the membrane plane in at least one direction. When, in addition, the orientation of the **g** tensor of the paramagnet with respect to the molecular axes of the Fe–S cluster or heme compound is known—as it is, for instance, for [2Fe–2S] clusters (Gibson *et al.*, 1966)—the orientation of the whole structure respective to the membrane is defined.

In addition to magnetic interactions detected between UQ semiquinone molecules,

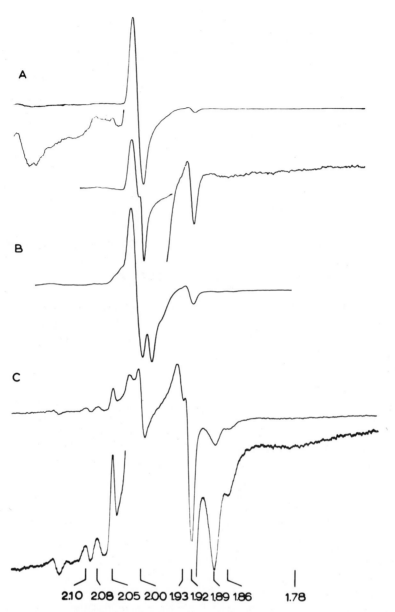

2.10 2.08 2.05 2.00 1.93 1.92 1.89 1.86   1.78

Figure 1. EPR spectra of mitochondria or whole tissue at different states of reduction. (A) Blowfly mitochondria, approximately 40 mg protein/ml, as prepared in 0.15 M KC1, 10 mM Tris (pH 7.3), 1 mM EDTA, and 0.5% bovine serum albumin. (B) Beef heart mitochondria, approximately 50 mg/ml, as prepared in 0.25 M sucrose, 0.02 M Tris (pH 7.4). (C) Whole pigeon heart frozen immediately after removal of the heart. All EPR spectra shown in this chapter represent the first derivative of the absorption (ordinate) with linearly increasing magnetic field. Unless otherwise mentioned, prominent peaks or shoulders are indicated on a g-factor scale, although this does not mean that the values given are true g values. In many instances they will be very close to actual g values.

Conditions of EPR spectroscopy: (A) Microwave power 2.7 mW; modulation amplitude 0.63 mT; scanning rate 40 mT/min; temperature 13.3°K; enlarged wings left and right, modulation amplitude 0.8 mT, 10× amplified; center inset 0.09 mW; modulation of 0.63 mT; 2.5× amplified. This inset recorded at low power shows that a radical signal is superimposed on the HiPIP signal. (B) Conditions as the main spectrum of (A). (C) Conditions as (A) and (B), except microwave power 0.27 mW and scanning rate 20 mT/min. The enlarged wings are 5× amplified.

Unless stated otherwise, the modulation frequency was 100 kHz and the microwave frequency 9.2 GHz for all figures. For more detail see Beinert (1978). [Reproduced with permission from Beinert, 1977.]

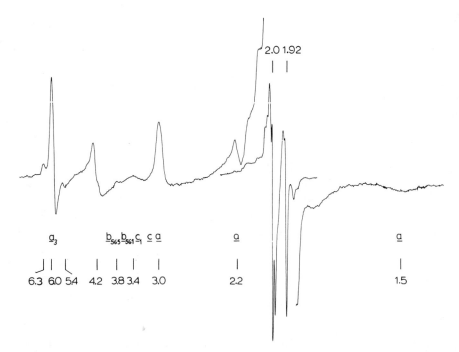

Figure 2. EPR spectrum of beef heart mitochondria. Approximately 60 mg protein/ml was washed twice with sucrose–Tris (pH 7.5), stirred under exposure to O_2 for 30 min to deplete endogenous substrates, and then frozen in isopentane at 120°K. Conditions of EPR spectroscopy: microwave power 2.7 mW; modulation amplitude 1 mT; scanning rate 100 mT/min; temperature 13°K. The inset in the center was recorded at 0.27 mW and at an amplification 0.32 times that of the main spectrum. Features of the center part are better resolved in Figs. 1A and B and will therefore not be analyzed for this figure. The cytochromes detectable (all in their oxidized state) are the following: a_3 at $g = 6.3$, 6.0, and 5.4; b_{565} at $g = 3.8$; b_{561} and c_1, overlapping at $g = 3.4$; a at $g = 3.0$, 2.2, and 1.5; c lies under the a peak at $g = 3$ and causes the low-field skewing.

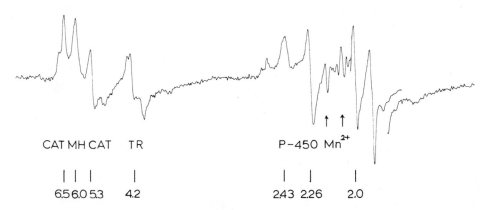

Figure 3. EPR spectrum of whole rat liver. Conditions of EPR spectroscopy: microwave power 2.7 mW; modulation amplitude 0.8 mT; scanning rate 125 mT/min; temperature 12.5°K. The inset in the center was recorded at 0.4 times the amplification of the main spectrum. The minor peaks at $g \sim 6$ indicate the presence of an alternative form of catalase (CAT). It cannot be determined from this spectrum what fraction of catalase and transferrin (TR) may be due to imbibed blood, as methemoglobin (MH) certainly is. Mn^{2+} has six sharp lines of roughly equal distance. Because of overlap with other resonances, only two lines are clearly seen in the spectrum shown.

more detailed studies on EPR signal shapes and saturation behavior (spin relaxation) have shown that multiple interactions between components of the respiratory chain do, in fact, occur, allowing conclusions as to the distances separating them. It has long been known that the spin relaxation of flavin semiquinones changes as neighboring Fe–S clusters become reduced (Beinert and Hemmerich, 1965). With succinate dehydrogenase, a drastic increase in spin relaxation of Fe–S cluster 1 is observed on extensive reduction of the enzyme, e.g., by dithionite (Beinert *et al.*, 1977; Salerno *et al.*, 1979b). In most purified preparations, additional signal intensity arises under these conditions, which has been attributed to an Fe–S cluster 2 (Beinert, 1977, 1978; Beinert *et al.*, 1977; Salerno *et al.*, 1979b). It is thought that this cluster is itself more effectively relaxed and thus increases the relaxation rate of cluster 1 by spin–spin interaction. In support of this is the fact that extra resonances, of the general nature as also found for the UQ semiquinone pair, were observed as well as so-called half-field signals (Salerno *et al.*, 1979b), which are diagnostic of spin–spin interaction. However, while all preparations show a more or less pronounced increase of relaxation of cluster 1 when an excess of dithionite or other strong reductant is added, not all preparations show extra signal intensity of a cluster 2 (Albracht, 1980). While the interpretation of these observations is still in dispute (Albracht, 1980; Coles *et al.*, 1979), there is no question that some kind of interaction between electron acceptors is the basis for the observed behavior.

Indications of spin–spin interactions have also been observed with the electron acceptors of NADH dehydrogenase (Ohnishi, 1979). Current structural studies indicate that this complex of the respiratory chain is one of its most complicated structures (Heron *et al.*, 1979). It possesses at least four, and possibly as many as six, different Fe–S clusters. While chemical analysis for iron and labile sulfide indicated that a number of Fe–S clusters was to be expected, it was EPR spectroscopy that established the identities and individual properties of these multiple clusters. A multifaceted picture emerged. The oxidation–reduction midpoint potentials of the clusters range over a span of -250 (possibly -380) to -20 mV; thus, there are clusters close in potential to the substrate (NADH), and there is at least one cluster with a potential close to the acceptor, viz. UQ. This supports the notion of a principle increasingly recognized with complex multicomponent electron transfer enzymes or systems, namely they are organized in such a fashion that the actual potential jump lies within the enzyme itself, e.g., between low- and high-potential Fe–S clusters of NADH dehydrogenase and not between substrate and enzyme or enzyme and acceptor. The same principle is seen with succinate UQ reductase, cytochrome bc_1 complex (Complex III), and cytochrome oxidase. The large potential gap between Fe–S clusters of NADH dehydrogenase and the pH dependence of the midpoint potential of at least one cluster have given rise to inquiry and speculation about the possibility that these very features of NADH dehydrogenase are related to proton translocation and energy conservation linked to NADH oxidation via the respiratory chain (Ohnishi, 1979; Gutman *et al.*, 1972). Effects of ATP, uncouplers, and pH on the potentials have been sought and in some cases found (Ohnishi, 1979; Gutman *et al.*, 1972; Ingledew and Ohnishi, 1980). The interpretation of the observations have, however, remained a difficult and partly controversial subject so that it does not lend itself to discussion in the framework of this review. Yet the challenge and the definite possibility remain that the observations made are indeed related to energy conservation in a manner not fully understood at this time.

Cytochrome c oxidase has attracted much attention in recent years because it offers challenges to the inorganic chemist and the student of electron transfer paths and mechanisms as well as to those interested in membrane proteins and their structure and properties, and also because the enzyme has a number of useful spectroscopic features and can be prepared in high purity and large quantity. EPR spectroscopy has contributed significantly to the progress of our knowledge and understanding of the four metal components of the oxidase and their behavior and relationships. EPR was instrumental in establishing copper not only as a constituent of the enzyme but also as a constituent active in electron transfer (Beinert *et al.*, 1976). Quantitative evaluation of EPR spectra has also shown unequivocally that in the resting oxidized form of the enzyme, the two heme as well as the two copper components have different properties and functions (Beinert *et al.*, 1976; Hartzell and Beinert, 1976; Aasa *et al.*, 1976). EPR gave the first indications that one of the hemes and one of the copper ions were closely coupled so as to make them undetectable by EPR (Van Gelder and Beinert, 1969). Verification that this is correct had to come from other techniques (Tweedle *et al.*, 1978; Moss *et al.*, 1978), for a negative result, i.e., lack of signals for these components, can only serve as a clue, not proof. EPR in combination with low-temperature reflectance spectroscopy was decisive in assigning features of the optical spectra to individual components, such as the 830-nm absorption to the EPR-detectable copper (Beinert *et al.*, 1980) and the 655-nm absorption to some feature of the cytochrome a_3–copper interaction (Hartzell *et al.*, 1973). Rapid reaction techniques in combination with EPR and low-temperature reflectance spectroscopy have furnished clues on reaction rates and a number of transient intermediate states that were unexpected (Shaw *et al.*, 1978a,b). These states point to changes in spin state and/or conformations in the enzyme following more slowly after sudden exposure to ligands, reductant, oxidant, H^+, or OH^-, or combinations thereof.

As repeatedly pointed out here, many of these observations did not provide definitive answers, as EPR is limited in its information content to the immediate surroundings of the paramagnetic centers observed; but it did in all instances provide a wealth of clues. And because of the sensitivity and ease of application, EPR has in general preceded the application of more demanding, heavy-handed techniques, which may have been more decisive, in some instances. EPR is extremely sensitive as to the electronic structure of the paramagnetic species that give rise to a signal. From the measurements of line shifts in the spectrum of the cytochrome a component of cytochrome oxidase, it was, for instance, possible to say whether the enzyme was frozen for observation in a vacuum, in the presence of a gas such as O_2, N_2, or CO, or in the presence of He, indicating that there was a gas pocket or channel in the enzyme that transmitted some message of its state to the heme of cytochrome a.

Clearly, EPR as well as some of the other spectroscopic techniques have a power of discrimination that far exceeds our (bio)chemical knowledge of these proteins and our ability to interpret such fine detail. Thus, in a way, low-temperature EPR is much too sensitive a technique for our crude understanding of the chemistry of the systems with which we are dealing; and it requires a fair amount of judgment and experience on the part of the experimenter not to get lost in uninterpretable detail, however real and interesting it may be, but at the same time not to miss subtle but essential and potentially interpretable clues.

ACKNOWLEDGMENTS. I am indebted to all my colleagues and collaborators who have supplied materials and information during the development of EPR as a tool in the study of oxidative enzymes, particularly to Dr. Richard H. Sands for his introduction to the technique and background and for his continued counsel and to Mr. R. E. Hansen for his collaboration and assistance with instrumentation. Continued support by a Research Career Award (5-K06-GM-18,442) from the National Institute of General Medical Sciences is gratefully acknowledged.

REFERENCES

Aasa, R., Albracht, S. P. J., Falk, K. E., Lanne, B., and Vänngård, T. (1976). *Biochim. Biophys. Acta* **422,** 260–272.

Ackrell, B. A. C., Kearney, E. B., Coles, C. J., Singer, T. P., Beinert, H., Wan, Y.-P., and Folkers, K. (1977). *Arch. Biochem. Biophys.* **182,** 108–117.

Albracht, S. P. J. (1980). *Biochim. Biophys. Acta* **612,** 11–28.

Bäckström, D., Norling, B., Ehrenberg, A., and Ernster, L. (1970). *Biochim. Biophys. Acta* **197,** 108–111.

Beinert, H. (1977). In *Iron–Sulfur Proteins* (W. Lovenberg, ed.), Vol. III, pp. 61–100, Academic Press, New York.

Beinert, H. (1978). *Methods Enzymol.* **54,** 133–150.

Beinert, H., and Hemmerich, P. (1965). *Biochem. Biophys. Res. Commun.* **18,** 212–220.

Beinert, H., Ackrell, B. A. C., Kearney, E. B., and Singer, T. P. (1975). *Eur. J. Biochem.* **54,** 185–194.

Beinert, H., Hansen, R. E., and Hartzell, C. R. (1976). *Biochim. Biophys. Acta* **423,** 339–355.

Beinert, H., Ackrell, B. A. C., Vinogradov, A. D., Kearney, E. B., and Singer, T. P. (1977). *Arch. Biochem. Biophys.* **182,** 95–106.

Beinert, H., Shaw, R. W., Hansen, R. E., and Hartzell, C. R. (1980). *Biochim. Biophys. Acta* **591,** 458–470.

Coles, C. J., Holm, R. H., Kurtz, D. M., Jr., Orme-Johnson, W. H., Rawlings, J., Singer, T. P., and Wong, G. B. (1979). *Proc. Natl. Acad. Sci. USA* **76,** 3805–3808.

Erecińska, M., Wilson, D. F., and Blasie, J. K. (1978). *Biochim. Biophys. Acta* **501,** 53–62, 63–71.

Gibson, J. F., Hall, D. O., Thornley, J. F., and Whatley, F. (1966). *Proc. Natl. Acad. Sci. USA* **56,** 987–990.

Gutman, M., Singer, T. P., and Beinert, H. (1972). *Biochemistry* **11,** 556–562.

Hartzell, C. R., and Beinert, H. (1976). *Biochim. Biophys. Acta* **423,** 323–338.

Hartzell, C. R., Hansen, R. E., and Beinert, H. (1973). *Proc. Natl. Acad. Sci. USA* **70,** 2477–2481.

Heron, C., Smith, S., and Ragan, C. I. (1979). *Biochem. J.* **181,** 435–443.

Ingledew, W. J., and Ohnishi, T. (1980). *Biochem. J.* **186,** 111–117.

Ingledew, W. J., Salerno, J. C., and Ohnishi, T. (1976). *Arch. Biochem. Biophys.* **177,** 176–184.

Keilin, D. (1966). In *The History of Cell Respiration and Cytochrome,* (J. Keilin, ed.), Cambridge Univ. Press, Cambridge.

Kent, T. A., Dreyer, J.-L., Emptage, M. H., Moura, I., Moura, J. J. G., Huynh, B. H., Xavier, A. V., LeGall, J., Beinert, H., Orme-Johnson, W. H., and Münck, E. (1981). In *Interaction between Iron and Proteins in Oxygen and Electron Transport* (C. Ho, ed.), Elsevier/North-Holland, Amsterdam, in press.

Kröger, A. (1976). *FEBS Lett.* **65,** 278–280.

Lovenberg, W. (ed.) (1977). In *Iron–Sulfur Proteins,* Vol. I–III, Academic Press, New York.

Moss, T. H., Shapiro, E., King, T. E., Beinert, H., and Hartzell, C. (1978). *J. Biol. Chem.* **253,** 8072–8073.

Ohnishi, T. (1979). In *Membrane Proteins in Energy Transduction* (R. A. Capaldi, ed.), pp. 1–87, Dekker, New York.

Ohnishi, T., Lim, J., Winter, D. B., and King, T. E. (1976). *J. Biol. Chem.* **251,** 2105–2109.

Orme-Johnson, N. R., Hansen, R. E., and Beinert, H. (1974). *J. Biol. Chem.* **249,** 1928–1939.

Rieske, J. S., Hansen, R. E., and Zaugg, W. S. (1964). *J. Biol. Chem.* **239,** 3017–3022.

Ruzicka, F. J., and Beinert, H. (1977). *J. Biol. Chem.* **252,** 8440–8445.

Ruzicka, F. J., and Beinert, H. (1978). *J. Biol. Chem.* **253,** 2514–2517.

Ruzicka, F. J., Beinert, H., Schepler, K. L., Dunham, W. R., and Sands, R. H. (1975). *Proc. Natl. Acad. Sci. USA* **72,** 2886–2890.

Salerno, J. C., Blum, H., and Ohnishi, T. (1979a). *Biochim. Biophys. Acta* **547,** 270–281.

Salerno, J. C., Lim, J., King, T. E., Blum, H., and Ohnishi, T. (1979b). *J. Biol. Chem.* **254,** 4828–4835.

Shaw, R. W., Hansen, R. E., and Beinert, H. (1978a). *J. Biol. Chem.* **253,** 6637–6640.

Shaw, R. W., Hansen, R. E., and Beinert, H. (1978b). *Biochim. Biophys. Acta* **504,** 187–199.

Tweedle, M. F., Wilson, L. J., García-Iñiguez, L., Babcock, G. T., and Palmer, G. (1978). *J. Biol. Chem.* **253,** 8065–8071.

Van Gelder, B. F., and Beinert, H. (1969). *Biochim. Biophys. Acta* **189,** 1–24.

50

Orientations of the Mitochondrial Redox Components

Maria Erecińska

1. INTRODUCTION

Transfer of electrons in mitochondrial and photosynthetic systems occurs with a high degree of specificity and with velocities that, in some cases, exceed the limits of diffusion-controlled reactions. It has been speculated, therefore, that functional integrity of electron transport may require precise orientation of the redox components both within the supporting membrane and with respect to one another. The importance of this structural relationship was recognized about a decade ago in photosynthesis, and orientations of chromophores in both chromatophores and chloroplasts have been investigated by a number of laboratories (see e.g. Geacintov *et al.*, 1972, 1974; Breton *et al.*, 1973; Junge and Eckhof, 1974; Junge *et al.*, 1977; Paillotin and Breton, 1977; Paillotin *et al.*, 1979; Vermeglio and Clayton, 1976).

In the early 1970s, Junge and co-workers (Junge, 1972; Junge and DeVault, 1975) made the first attempt to determine symmetry, orientation, and rotational mobility of mitochondrial cytochrome a_3 heme by analyzing photo-induced linear absorption changes resulting from the photolysis of the cytochrome a_3–carbon monoxide complex. They postulated on the basis of their findings that cytochrome c oxidase does not undergo Brownian diffusion around any axis in the mitochrondrial membrane [but see Kawato *et al.* (1980) for an opposite conclusion derived from essentially the same data] and suggested that the plane of cytochrome a_3 heme is coplanar to the membrane (Junge and DeVault, 1975; Kunze and Junge, 1977). During the past few years, techniques have been developed (Blasie *et al.*, 1978; Clark *et al.*, 1980) to evaluate the orientation of the mitochondrial enzymes and their chromophores in a more direct and accurate way. This short review summarizes briefly the results of these efforts and compares them with the data obtained on other systems.

Maria Erecińska • Department of Pharmacology and Department of Biochemistry and Biophysics, University of Pennsylvania, Philadelphia, Pennsylvania 19104.

2. TECHNIQUES

Techniques developed for the study of orientation of the mitochondrial redox carriers fall into two classes: those concerned with investigation of entire proteins and those confined to evaluation of the chromophore active sites such as hemes and nonheme irons. Both types of methods rely on the preparation of specimens that are either crystalline in nature or uniformly oriented with respect to the support. None of the tightly bound components of the mitochondrial respiratory chain has been isolated in a truly crystalline state. [The three-dimensional crystals of cytochrome *c* oxidase reported recently by Ozawa *et al.* (1980) appear as very thin needles and may be unsuitable for such studies as those described here.] However, there are two types of preparations that have been used successfully to date. The first is the two-dimensional, "crystalline" preparation of cytochrome *c* oxidase (Vanderkooi *et al.*, 1972), the second, the hydrated oriented multilayers that can be formed through slow partial dehydration at known humidities from membranous enzymes (Blasie *et al.*, 1978; Clark *et al.*, 1980; Erecińska *et al.*, 1977, 1978a), proteins incorporated into lipid vesicles (Erecińska and Wilson, 1979), vesicular membranes [mitochondria (Erecińska *et al.*, 1977, 1978b, 1979; Salerno *et al.*, 1977, 1979); microsomes (Rich *et al.*, 1979)], submitochondrial particles (Blum *et al.*, 1978a,b), and bacterial vesicles (Erecińska *et al.*, 1979; Tiede *et al.*, 1978; Prince *et al.*, 1980).

Techniques concerned with the study of isolated proteins include electron diffraction (Henderson *et al.*, 1977; Frey *et al.*, 1978) and X-ray diffraction (Blasie *et al.*, 1978); with regard to mitochondrial redox components, they have thus far been applied only to isolated cytochrome *c* oxidase. The former method involves analysis of electron micrographs of two-dimensional vesicle crystals of cytochrome oxidase (Vanderkooi *et al.*, 1972) stained with uranyl acetate taken at different tilt angles. The nominal resolution of the analysis was reported to be 30 Å (Henderson *et al.*, 1977). X-Ray diffraction (Blasie *et al.*, 1978) has been performed on hydrated oriented multilayers of membranous cytochrome *c* oxidase (Sun *et al.*, 1968) and gave a resolution of about 20 Å.

Techniques concerned with the study of chromophores include polarized absorption spectroscopy, ESR spectroscopy, and anomalous X-ray scattering. The first was applied to the cytochromes of the mitochondrial respiratory chain (Erecińska *et al.*, 1977, 1978a,b, 1979). The experiments involve measurements of absorption spectra of samples oriented on a glass support and directed at various angles between the multilayers and the incident light, which is polarized either vertically or horizontally with respect to the laboratory axes. As electronic transitions responsible for the spectral changes arise from absorption of photons whose electric vector has a component parallel to the direction of the molecular dipole moment, alignment of the chromophores with respect to the supporting membrane will induce linear dichroism in the absorption of the oriented multilayers. Spectra taken with differently polarized light (Fig. 1) therefore exhibit differences in absorbance intensities, which depend on the manner in which the chromophores are oriented with respect to their support. Interpretation of the results relies on the assumption that electronic transitions responsible for the visible and Soret absorption of the hemeproteins are polarized in the plane of the molecule (i.e., $x–y$ polarized) and approximately doubly degenerate (i.e., the x and y axes being approximately equivalent).

ESR spectroscopy applied to hemeproteins (Erecińska *et al.*, 1977, 1978a,b, 1979; Blum *et al.*, 1978a,b; Tiede *et al.*, 1978) and iron–sulfur proteins (Salerno *et al.*, 1979; Prince *et al.*, 1980) relies on the paramagnetic properties of the iron in its ligand field. Most mitochondrial hemeproteins in the fully oxidized state contain iron in the low–spin form ($S = \frac{1}{2}$) and exhibit an ESR absorbance characterized by three g values, corresponding to the three principal components of the **g** tensor. Determination of the heme orientation involves measurements of ESR absorbance in chromophore-containing oriented multilayers rotated at various angles in the magnetic field (Fig. 1). As in the case of optical spectra, electromagnetic radiation is absorbed only by transitions with dipole moments parallel to the direction of the magnetic field. Thus, the individual signals that arise from the three components of the **g** tensor increase and decrease in intensity depending on the orientation of the active group when the sample is rotated in the magnetic field. It should be kept in mind that when the oriented multilayers are rotated through 360° in the magnetic field, the magnitudes of the signals corresponding to the **g**-tensor axes that lie in the plane of the multilayers, or at 90° with respect to it, will exhibit two maxima 180° apart. On the other hand, for those axes of the **g** tensor that are directed at any other angle, there will be four maxima during 360° rotation because for each maximum at an angle x between 0 and 90°, there are corresponding maxima at $180° \pm x$ and $360° \pm x$ (or $0° \pm x$). Interpretation of the results relies on the assumption that the component of the **g** tensor that gives rise to the largest g values, g_z, is approximately normal to the plane of the heme, in analogy with single crystals of cytochrome c (Hori, 1971; Mailer and Taylor, 1972; Taylor, 1977) and low-spin derivatives of myoglobin (Bennett *et al.*, 1957; Helcke *et al.*, 1968). On the other hand, although the orientations of the g_x and g_y axes with respect to the plane of the multilayers can be determined, predictions of their relation to the planar heme structure (i.e., to the N–Fe–N direction) are more difficult. This is because the in-plane components of the **g** tensor are differently oriented with respect to the N–Fe–N axis even in the various crystalline low-spin ferric hemes or heme derivatives (Bennett *et al.*, 1957; Hori, 1971; Mailer and Taylor, 1972), and no general rules exist that would help to interpret the results.

In the case of iron–sulfur proteins with two iron atoms, it is generally assumed that the z axis passes through the two iron atoms (i.e., Fe–Fe axis) and gives rise to the $g \simeq 2.0$ resonance (Gibson *et al.*, 1966; Sands and Dunham, 1974).

Anomalous X-ray scattering studies, currently in progress (Tavormina *et al.*, 1980) on oriented membrane multilayers of cytochrome c oxidase, are based on the differences in the lamellar X-ray scattering of the specimen when the wavelength of the incident photons is near the K-absorption edge of either Cu or Fe or far away from it. Such studies should allow the determination of the positions of the four metal atoms of cytochrome c oxidase *across* the profile of the membrane. (The positions of the metal atoms in the plane of the membrane can be determined from differences in the scattering in the equatorial direction.)

3. STRUCTURAL STUDIES ON CYTOCHROME c OXIDASE

Electron diffraction studies (Henderson *et al.*, 1977; Frey *et al.*, 1978) have shown that all cytochrome c oxidase molecules in the crystalline membranes are asymmetrically oriented, with the orientation opposite to that found in intact mitochondria (Frey *et al.*, 1978). The vesicles appeared to be closed and impermeable to externally added α-globulin (Frey *et al.*, 1978). Similar results have been reported from electron density profiles obtained by X-ray diffraction of hydrated oriented multilayers of noncrystalline cytochrome c oxidase (Blasie *et al.*, 1978). The cytochrome c oxidase molecule extended through the entire thickness of the membrane and was asymmetrically oriented in the membrane profile

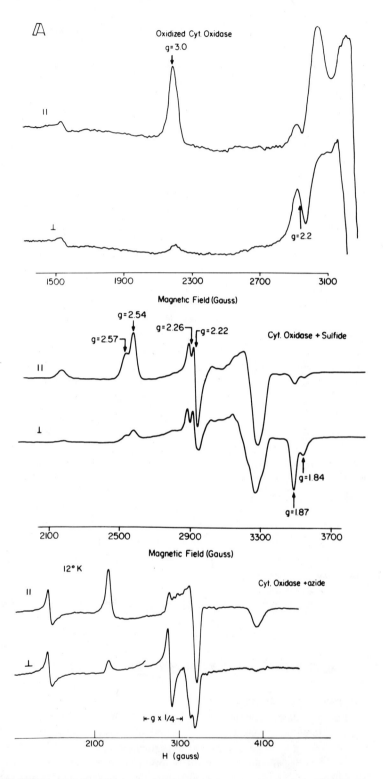

Figure 1. (A) EPR spectra of oxidized cytochrome c oxidase and its sulfide and azide derivatives in frozen, oriented multilayers of the membranous enzyme recorded at 0 and 90° with respect to the magnetic field direction. For detailed experimental conditions, see Erecińska *et al.* (1978a,b). (B) Polarized optical absorption spectra of oxidized and reduced cytochrome c oxidase and of its formate derivative in frozen, oriented multilayers of the

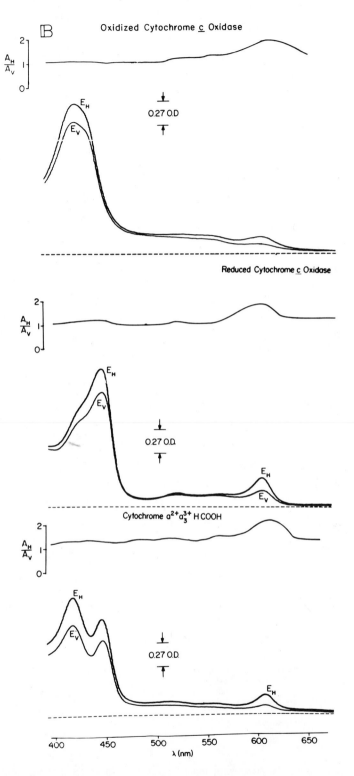

membranous enzyme. The spectra were recorded with the light beam polarized vertically and horizontally with respect to the laboratory axes at an angle of 45° between the incident bean and the normal to the plane of oriented multilayers. For detailed experimental conditions, see Erecińska *et al.* (1977, 1978a,b).

Table I. Orientations of the Electron Transport Proteins[a]

Component	Orientation of the heme plane or Fe–Fe axis with respect to the plane of the membrane	References
Mitochondrial respiratory chain		
Cytochromes		
a	⊥	Erecińska et al. (1977, 1978a,b), Blum et al. (1978a)
a_3	⊥	Erecińska et al. (1977, 1978a,b), Blum et al. (1978a)
c	~70°	Erecińska et al. (1977, 1978b), Vanderkooi et al. (1977)
c_1	⊥	Erecińska and Wilson (1979)
b_{561}	⊥	Erecińska and Wilson (1979)
b_{566}	⊥	Erecińska and Wilson (1979)
Ubiquinone		
Q–Q pair	⊥	Salerno et al. (1977)
Q ring	⊥	Salerno et al. (1977)
Iron–sulfur centers		
S–1	‖	Salerno et al. (1979)
N–1	‖	Salerno et al. (1979)
Rieske's	‖	Erecińska and Wilson (1979), Salerno et al. (1979)
Drug-metabolizing systems		
Cytochrome P-450 (adrenal glands, mitochondria)	‖	Blum et al. (1978b)
Cytochrome P-450 (microsomes)	‖	Rich et al. (1979)
Cytochrome b_5	Random	Rich et al. (1979)
Other systems		
Cytochrome d from Tetrahymena pyriformis	⊥	Kilpatrick and Erecińska (1979)
Cytochrome aa_3 in Paracoccus denitrificans	⊥	Erecińska et al. (1979)
Cytochrome c_{553} from Chromatium vinosum	‖	Tiede et al. (1978)
Cytochrome c_{555} from Chromatium vinosum	⊥	Tiede et al. (1978)
Chlorophyll in photosynthetic membranes	‖	Geacintov et al. (1972), Breton et al. (1973), Junge and Eckhof (1974), Vermeglio and Clayton (1976)
Rieske iron–sulfur cluster in spinach chloroplasts	‖	Prince et al. (1980)

[a]The hydrated multilayers show a certain degree of disorientation or mosaic spread. The Gaussian mosaic spread determined by optical polarization measurements in the model system of membranous cytochrome c oxidase was 8–17° (half-width at half-maximum) (Blasie et al., 1978). The disorder parameter in the spectra of iron–sulfur clusters was estimated from computer simulation to be about 30°, of which 20° was probably due to mosaic spread (Salerno et al., 1979). The hemes of the mitochondrial cytochromes are within 5° of the normal to the multilayer, whereas the vector connecting the two quinone radicals lies within 15° of the normal to the multilayer.

with a large portion of the scattering mass being present in the extravesicular surface of the membrane. Bundles of α-helical polypeptide chains were also detected with the average orientation normal to the plane of the membrane.

4. ORIENTATIONS OF THE ELECTRON TRANSPORT PROTEINS

Orientations of the chromophores of the various electron transport proteins are summarized in Table I and only a few comments are added here.

1. The chromophores of the mitochrondrial electron transport carriers were always found to be oriented similarly in phospholipid–protein model membranes (i.e., lipid vesicles containing the appropriate proteins) and in native membranes (mitochondria and submitochondrial particles).

2. The orientation was unaffected by the redox and liganded state of the various proteins (see e.g. Erecińska *et al.*, 1978a,b, 1979; Rich *et al.*, 1979; Salerno *et al.*, 1979).

3. The orientations of the g_x and g_y axes of the **g** tensor in cytochrome c oxidase with respect to the plane of the multilayers (investigated in some detail by Erecińska *et al.*, 1979) were shown to be different in the fully oxidized enzyme (interpreted as cytochrome a_3^+) and in various liganded derivatives (attributed to cytochrome a_3^+–ligand).

4. The in-plane components of the **g** tensor (g_x and g_y) in the liganded derivatives of cytochrome a_3 had the same orientation in the oxidase from various sources (pigeon breast, beef heart, *Paracoccus denitrificans*), whereas those in the fully oxidized enzyme showed some variability (Erecińska *et al.*, 1979; Kilpatrick and Erecińska, 1979).

5. From studies on the NO derivative of reduced cytochrome c oxidase, it was deduced that NO makes an angle of about 135° with the plane of oriented multilayers (Barlow and Erecińska, 1979).

6. The tetranuclear iron–sulfur clusters of mitochondrial succinic dehydrogenase and NADH dehydrogenase were also found to be oriented with respect to the plane of the multilayers (Salerno *et al.*, 1979). Unfortunately, no model exists thus far that relates the magnetic anisotropy to their structure.

The general conclusion that can be drawn from these studies is that mitochondrial electron transport carriers have precise orientation with respect to the plane of the membrane. However, our limited knowledge of the mechanics of electron transfer precludes any further speculations at this time.

ACKNOWLEDGMENT. The author's research reported herein was supported by USPHS Grants GM122-2 and HL18704. M.E. is the William Stroud Established Investigator of the American Heart Association.

REFERENCES

Barlow, D., and Erecińska, M. (1979). *FEBS Lett.* **98**, 9–12.

Bennett, J. E., Gibson, J. F., and Ingram, D. J. E. (1957). *Proc. R. Soc. London Ser. A* **240**, 67–82.

Blasie, J. K., Erecińska, M., Samuels, S., and Leigh, J. S. (1978). *Biochim. Biophys. Acta* **501**, 33–52.

Blum, H., Harmon, J. J., Leigh, J. S., Salerno, J. C., and Chance, B. (1978a). *Biochim. Biophys. Acta* **502**, 1–10.

Blum, H., Leigh, J. S., Salerno, J. C., and Ohnishi, T. (1978b). *Arch. Biochem. Biophys.* **187**, 153–157.

Breton, J., Michel-Villaz, M., and Paillotin, G. (1973). *Biochim. Biophys. Acta* **316**, 42–56.

Clark, N. A., Rothschild, K. J., Luippold, D. A., and Simon, B. A. (1980). *Biophys. J.* **31**, 65–96.

Erecińska, M., and Wilson, D. F. (1979). *Arch. Biochem. Biophys.* **192**, 80–85.

Erecińska, M., Wilson, D. F., and Blasie, J. K. (1977). *FEBS Lett.* **76**, 235–239.

Erecińska, M., Wilson, D. F., and Blasie, J. K. (1978a). *Biochim. Biophys. Acta* **501**, 53–62.

Erecińska, M., Wilson, D. F., and Blasie, J. K. (1978b). *Biochim. Biophys. Acta* **501**, 63–71.

Erecińska, M., Wilson, D. F., and Blasie, J. K. (1979). *Biochim. Biophys. Acta* **545**, 352–364.

Frey, T. G., Chan, S. H. P., and Schatz, G. (1978). *J. Biol. Chem.* **253**, 4389–4395.

Geacintov, N. E., Van Nostrand, F. Becker, J. F., and Tinkel, J. B. (1972). *Biochim. Biophys. Acta* **267,** 65–79.

Geacintov, N. E., Van Nostrand, F., and Becker, J. F. (1974). *Biochim. Biophys. Acta* **347,** 443–463.

Gibson, J. F., Hall D. O., Thornley, J. H. M., and Whatley, F. R. (1966). *Proc. Natl. Acad. Sci. USA* **56,** 987–990.

Helcke, G. A., Ingram, D. J. E., and Slade, E. F. (1968). *Proc. R. Soc. London Ser. B* **169,** 275–288.

Henderson, R., Capaldi, R. A., and Leigh. J. S. (1977). *J. Mol. Biol.* **112,** 631–648.

Hori, H. (1971). *Biochim. Biophys. Acta* **251,** 227–235.

Junge, W. (1972). *FEBS Lett.* **25,** 109–112.

Junge, W., and DeVault, D. (1975). *Biochim. Biophys. Acta* **408,** 200–214.

Junge, W., and Eckhof, A. (1974). *Biochim. Biophys. Acta* **357,** 103–117.

Junge, W., Schaffernicht, H., and Nelson, N. (1977). *Biochim. Biophys. Acta* **462,** 73–85.

Kawato, S., Sigel, E., Carafoli, E., and Cherry, R. J. (1980). *J. Biol. Chem.* **255,** 5508–5510.

Kilpatrick, L., and Erecińska, M. (1979). *Arch. Biochem. Biophys.* **197,** 1–9.

Kunze, U., and Junge, W. (1977). *FEBS Lett.* **80,** 429–434.

Mailer, C., and Taylor, C. P. S. (1972). *Can. J. Biochem.* **50,** 1048–1055.

Ozawa, T., Suzuki, H., and Tanaka, M. (1980). *Proc. Natl. Acad. Sci. USA* **77,** 928–930.

Paillotin, G., and Breton, J. (1977). *Biophys. J.* **18,** 63–80.

Paillotin, G., Vermeglio, A., and Breton, J. (1979). *Biochim. Biophys. Acta* **545,** 249–264.

Prince, R. C., Crowder, M. S., and Bearden, A. J. (1980). *Biochim. Biophys. Acta* **532,** 323–337.

Rich, P. R., Tiede, D. J., and Bonner, W. D. (1979). *Biochim. Biophys. Acta* **546,** 307–315.

Salerno, J. C., Harmon, H. J., Blum, H., Leigh, J. S., and Ohnishi, T. (1977). *FEBS Lett.* **82,** 179–182.

Salerno, J. C., Blum, H., and Ohnishi, T. (1979). *Biochim. Biophys. Acta* **547,** 270–281.

Sands, R. H., and Dunham, W. R. (1974). *Q. Rev. Biophys.* **7,** 443–504.

Sun, F. F., Prezbindowski, K. S., Crane, F. L., and Jacobs, E. E. (1968). *Biochim. Biophys. Acta* **153,** 804–818.

Tavormina, A., Pachence, J., Erecińska, M., Dutton, P. L., Blasie, J. K., Stamatoff, J., Eisenberger, P., and Brown, G. (1980). *Fed Proc.* **39,** 2147.

Taylor, C. P. S. (1977). *Biochim. Biophys. Acta* **431,** 137–149.

Tiede, D. M., Leigh, J. S., and Dutton, P. L. (1978). *Biochim. Biophys. Acta* **503,** 524–544.

Vanderkooi, G., Senior, A. E., Capaldi, R. A., and Hayashi, H. (1972). *Biochim. Biophys. Acta* **503,** 524–544.

Vanderkooi, J. M., Landesberg, R., Hayden, G. W., and Owen, C. S. (1977). *Eur. J. Biochem.* **81,** 339–347.

Vermeglio, A., and Clayton, R. K. (1976). *Biochim. Biophys. Acta* **449,** 500–515.

51

Functional and Topological Aspects of the Mitochondrial Adenine Nucleotide Carrier

P. V. Vignais, M. R. Block, F. Boulay, G. Brandolin, and G. J. M. Lauquin

1. INTRODUCTION

The vectorial exchange between the mitochondrial and the cytosolic pools of ADP and ATP is catalyzed by a specific carrier protein located in the inner mitochondrial membrane (Klingenberg, 1976; Vignais, 1976). This function appears to be a general attribute of eukaryotic cells. Even in respiration-deficient mutants of the yeast *Saccharomyces cerevisiae* that are incapable of oxidative phosphorylation, the ADP/ATP carrier is still functional, and its molecular properties appear to be preserved intact (Kolarov *et al.*, 1972; Šubík *et al.*, 1974). After a brief survey of the literature on the functioning of ADP/ATP transport in the cell, we shall focus on recent physical and chemical approaches to investigate the topography of the ADP/ATP carrier in an attempt to better comprehend its mechanism.

2. GENERAL PRINCIPLES OF THE FUNCTIONING OF ADP/ATP TRANSPORT

Under physiological conditions, when mitochondria are both respiring and phosphorylating, the ADP/ATP carrier catalyzes the exchange of cytosolic ADP for mitochondrial ATP with a stoichiometry of 1 : 1 (see Klingenberg, 1976; Vignais, 1976). The strict specificity of the carrier for ADP and ATP and its stoichiometric nature ensure that the oxidative phosphorylation system is supplied with cytosolic ADP in exchange for mitochondrial ATP. Only in exceptional cases (for example, respiration-deficient mutants of

P. V. Vignais, M. R. Block, F. Boulay, G. Brandolin, and G. J. M. Lauquin • Laboratoire de Biochimie (INSERM U. 191 et CNRS/ERA 903), Département de Recherche Fondamentale, Centre d'Etudes Nucléaires, 38041 Grenoble Cedex, France, et Faculté de Médecine de Grenoble, France.

yeast) does the carrier function in the reverse direction, i.e., cytosolic ATP made by gly-colysis is supplied to mitochondria.

The ADP/ATP carrier is thought to catalyze an electrogenic exchange (Laris, 1977; LaNoue *et al.*, 1978; Villiers *et al.*, 1979), for at neutral pH, ADP and ATP bear three and four negative charges, respectively. The driving force has been identified as the mito-chondrial membrane potential from two types of evidence. (1) Under steady-state condi-tions, the ratio of ATP/ADP outside to ATP/ADP inside mitochondria is related in a linear fashion to the membrane potential of mitochondria (Klingenberg and Rottenberg, 1977). (2) Reagents that collapse the membrane potential, such as valinomycin, alter the kinetics of ADP/ATP transport whereas reagents that collapse pH gradients, such as nigericin, are ineffective (Villiers *et al.*, 1979). Thus, membrane potential is required to drive the asym-metric exchange, and in turn, some energy is required to maintain the membrane potential. According to Duszynski *et al.*, (1979), one-third of the energy delivered by the mitochon-drial respiratory chain appears to be needed for exporting ATP against ADP, the remaining energy being used for ATP synthesis. The asymmetric nature of the ATP/ADP exchange complemented by the finding that respiring mitochondria have a higher affinity (about 10 times) for ADP than for ATP (Souverijn *et al.*, 1973) explains why the ATP/ADP ratio is higher outside than inside mitochondria (Heldt *et al.*, 1972).

Phosphorylation of $[^{14}C]$-ADP added to mitochondria is faster than its diffusion into and its mixing with the adenine nucleotide pool of the matrix compartment; and similarly, the synthesized $[^{14}C]$-ATP is transferred to the outside without complete mixing with the unlabeled ATP of the matrix compartment (P. V. Vignais *et al.*, 1975; Out *et al.*, 1976). It is probable that the extreme viscosity of the highly concentrated protein solution within the matrix compartment [more than 50% protein w/v (Srere, 1980)] slows down the rapid diffusion of even small molecules like ADP or ATP, and that consequently the external ADP has more direct access to the oxidative phosphorylation system than to the pool of internal adenine nucleotides. However, because the entire pool of intramitochondrial ad-enine nucleotides is exchangeable with externally added ADP or ATP (Duée and Vignais, 1969), it is unlikely that physical compartmentation of internal adenine nucleotides exists. Therefore, the views expressed above are quite compatible with the finding that the ADP/ATP exchange is an electrical process driven by the membrane potential, and the pool of adenine nucleotides in liver mitochondria is accessible to the mitochondrial AT-Pase, the ADP/ATP carrier, and the carbamyl phosphate synthetase (Letko *et al.*, 1979).

The activity of ADP/ATP transport has to cope with the high capacity of the oxidative phosphorylation system (Stubbs *et al.*, 1978). This may explain why the low turnover of ADP/ATP transport, 1000 to 2000 per min at 20°C (due possibly to the bulky substrates to transfer), is compensated by the high concentration of the carrier in mitochondria (5 to 10% of the protein of the inner mitochondrial membrane consists of the ADP/ATP carrier). Other considerations on the control of ADP/ATP transport in the cell economy are dealt with in greater detail in recent reviews (Stubbs, 1979; Vignais and Lauquin, 1979).

3. TOPOLOGICAL PROBES OF THE ADP/ATP CARRIER PROTEIN IN THE MITOCHONDRIAL MEMBRANE AND IN ARTIFICIAL PHOSPHOLIPID VESICLES

Two classes of inhibitory ligands of exquisite specificities and high affinities are able to bind to the ADP/ATP carrier protein (Stubbs, 1979). To the first class belong atracty-loside (ATR) and carboxyatractyloside (CATR), two heteroglucosides in which the aglycone is a diterpene and the sugar moiety a glucose disulfate. The other class consists of bongkrekic acid (BA) and its isomer (IsoBA); these molecules are long unsaturated fatty acids with three carboxylic groups whose pKs range between 4 and 6. CATR and ATR do not readily penetrate the matrix space of mitochondria; they bind to the outer surface of the inner mitochondrial membrane, most likely to the exposed region of the ADP/ATP carrier (Vignais *et al.*, 1973). On the contrary, BA has to penetrate the mitochondrial membrane to inhibit the ADP/ATP carrier. In fact, inhibition by BA requires that the pH of the medium be slightly acidic, which results in protonation of a fraction of the BA molecules; only the protonated BA molecules are thought to readily penetrate the mitochondrial membrane (Kemp *et al.*, 1971). The reverse situation holds with inside-out submitochondrial particles, i.e., BA inhibits ADP/ATP transport even at alkaline pH; on the other hand, externally added CATR and ATR are ineffective, but they inhibit transport when they are entrapped in the particles (Lauquin *et al.*, 1977a,b). These data show that the difference in binding and inhibitory properties of CATR (ATR) and BA is not due simply to differences in permeability, but actually reflects anisotropic properties of the carrier protein. The effects of CATR, ATR, and BA have also been tested on a reconstituted system made of the purified carrier protein inserted in phospholipid vesicles (Brandolin *et al.*, 1980). The behavior of CATR and BA in this system is similar to that observed in mitochondria, indicating that the purified carrier protein after insertion into liposomes has the same asymmetrical arrangement as in mitochondria.

There is another class of inhibitors of potential physiological importance, the long-chain acyl CoAs. They inhibit ADP/ATP transport when externally added to mitochondria or to inside-out submitochondrial particles (Lauquin *et al.*, 1977b). A number of spin-labeled long-chain acyl CoAs have been prepared with the nitroxide radical placed at different positions of the acyl chain (Devaux *et al.*, 1975). These molecules, which have the same inhibitory properties as the original ones, have been used to probe the shape and the lipid environment of the ADP/ATP carrier protein in the mitochondrial membrane. Similar experiments have been conducted with spin-labeled acyl atractyloside (Lauquin *et al.*, 1977a). A detailed analysis of the interaction of the spin label with the carrier protein led us to conclude that (1) the carrier protein (considered as a channel spanning the membrane) is bulkier near the aqueous surface than in the middle of the membrane, and (2) the ADP/ATP carrier protein is in direct contact with the lipid core of the membrane.

By freeze-fracture electron microscopy of ADP/ATP carrier proteoliposome preparations, it has been possible to investigate the size distribution of the protein particles (Brandolin *et al.*, 1980). The mean value of the diameters of the particles was found to be close to 75 Å, which by calibration with water-soluble proteins of known molecular weight corresponds to a molecular weight of the order of 60,000. As the minimum molecular weight of the ADP/ATP carrier protein is 30,000, the reconstituted carrier protein is probably in a dimeric state; the same conclusion may hold for the native carrier protein in the mitochondrial membrane.

The ADP/ATP carrier can be specifically labeled with photoactivable derivatives of radiolabeled ATR and ADP (Lauquin *et al.*, 1976, 1978). Data have been obtained, showing that covalent photolabeling is an appropriate tool for mapping the ADP/ATP carrier (Boulay *et al.*, 1979). The principle, illustrated for the ATR site, is the following. Intact mitochondria are photolabeled by a radiolabeled photoactivable derivative of ATR. This results in covalent radiolabeling of the ADP/ATP carrier at, or close to, the ATR site. Then the covalently labeled carrier is extracted and purified. Finally, peptide bonds in the isolated carrier protein are specifically cleaved by chemical or enzymatic means, and the resulting peptides are isolated. The peptide that is labeled presumably contains the ATR site. In a preliminary experiment, with cyanogen bromide as cleavage reagent, one single radiolabeled peptide of molecular weight 23,000 was isolated. Hopefully, a more precise localization of the ATR site may be obtained by using other cleavage procedures.

4. EVIDENCE FOR LIGAND-INDUCED CONFORMATIONAL CHANGES IN THE ADP/ATP CARRIER PROTEIN

When ADP (or ATP) is externally added to the mitochondrial suspension, the reactivity of the carrier protein is perturbed, probably reflecting structural modifications of the protein. The first indication of substrate-induced conformational changes in the ADP/ATP carrier came from studies of the effect of SH-group reagents on kinetic and binding properties. In mitochondria, a penetrant SH-reagent, like *N*-ethylmaleimide (NEM), is able to inhibit ADP/ATP transport and ATR binding, provided that mitochondria are pretreated with micromolar amounts of ADP or ATP (Leblanc and Clauser, 1972; Vignais and Vignais, 1972; P. M. Vignais *et al.*, 1975). The inhibition by NEM in the presence of ADP or ATP is markedly enhanced when the mitochondria are energized by preincubation with an oxidizable substrate (Vignais *et al.*, 1976). It is remarkable that energization, i.e., the development of a protonmotive force, is able to favor a conformational state of the ADP/ATP carrier protein characterized by SH groups accessible and reactive to NEM.

Another example of substrate-induced conformational change is the modification of BA binding upon addition of ADP. By preincubation of mitochondria with micromolar amounts of ADP, a new plateau of saturation for high-affinity BA sites is obtained (30 to 50% higher than the control in the absence of ADP), and the affinity for BA is increased by about 10 times (Lauquin and Vignais, 1976). Conversely, bound ADP is increased upon addition of BA (Klingenberg and Buchholz, 1973). Preincubation with ADP also leads to an increased sensitivity of ADP/ATP transport to inhibition by ATR or CATR (Silva Lima and Denslow, 1979). Presumably, the carrier protein responds to externally added ADP (or ATP) by conformational changes resulting in a larger number of exposed sites to BA or CATR and in a greater affinity for these ligands.

Conformational changes of the carrier protein caused by CATR and BA binding have also been shown by immunochemical experiments in the following way. The ADP/ATP carrier protein is prepared in the presence of either CATR or BA, and because of the very high affinity of these inhibitors, it is finally isolated as a complex made of the protein and CATR or BA (Riccio *et al.*, 1975). The antibody that can then be raised against the CATR–carrier protein complex reacts specifically against the same complex, but not against the BA–carrier protein complex (Buchanan *et al.*, 1976), indicating different conformational determinants in CATR and BA complexes.

Finally, recent studies on the intrinsic fluorescence of the ADP/ATP carrier protein demonstrate changes induced by substrates and inhibitors. The carrier fluorescence, mainly due to tryptophanyl residues, is increased by ADP or ATP. Addition of BA leads to a further increase of fluorescence whereas CATR has the opposite effect (Brandolin *et al.*, 1981).

5. ADP/ATP TRANSPORT MECHANISMS: THE SHUTTLING SUBSTRATE SITE

As discussed by Singer (1977), transport mechanisms fall into two broad categories: (1) The mobile-carrier mechanism in which the carrier is a small rotating and shuttling protein (reorienting carrier) exposing its substrate site alternatively to the outside and the inside; (2) the fixed-pore mechanism in which the transport protein spans the membrane, the substrate site being translocated forward and backward across the membrane within the protein-lined pore (reorienting site). Based on thermodynamic arguments, the second type of transport is more likely. Singer (1977) has further suggested that it may be a general feature of transport proteins to be arranged in aggregates of two or more identical or similar subunits spanning the membrane, each subunit having the same orientation so that the aggregate is asymmetrically positioned. On the basis of data of ultracentrifugation of the purified ADP/ATP carrier in detergent (Hackenberg and Klingenberg, 1980) and of electron micrographs of freeze-fractured proteoliposomes after reconstitution (Brandolin *et al.*, 1980), the ADP/ATP carrier appears to be arranged as a dimer made of two 30,000-molecular-weight subunits.

Assuming a dimeric organization of the ADP/ATP carrier, several models of ADP/ATP transport can be proposed (Fig. 1). In model A, the substrate (ADP or ATP) is supposed to move through a pore limited by the two subunits. However, it must not be overlooked that, if the two subunits of the ADP/ATP carrier are symmetrical, the space that they limit has to accommodate the asymmetrical molecule of ADP or ATP. One is therefore led to assume that the ADP/ATP carrier is either a pseudodimer made of two similar (same molecular weight) but not identical subunits (model B), or a true homodimer, each of the subunits containing a binding site for ADP or ATP (model C). The carrier in models A and B contains one substrate binding site and in model C two substrate sites. The ADP/ATP exchange in model C occurs as follows: ADP binds to the cytosolic face of one subunit of the carrier, and ATP to the matrix face of the other subunit; the binding sites are mobile crevices, and the substrates that they bind are transported in opposite ways by a transversal movement of the crevices. Models A and B satisfy a ping-pong mechanism in which the carrier binds alternately its substrates to form only binary complexes: carrier–external ADP,

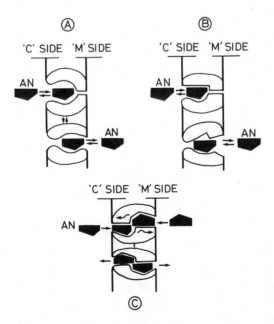

Figure 1. Models for adenine nucleotide (AN) transport. (A) Homodimer with one substrate site. (B) Heterodimer with one substrate site. (C) Homodimer with two substrate sites. The curved arrows indicate the transversal motion of the crevice with its bound substrate (for a detailed explanation see text).

carrier–internal ATP. Recent kinetic data (Duyckaerts *et al.*, 1980; Barbour and Chan, 1981) support a sequential mechanism of transport, involving a ternary complex consisting of the carrier and two nucleotides, one nucleotide coming from the outside, the other from the inside, as depicted in model C.

It is a consequence of the presumed anisotropy of fixed transport proteins that their outward- and inward-facing sites differ by conformation. The first clear evidence that the substrate site in a transport protein changes its conformation when shuttling between the two sides of the membrane was reported by Edwards (1973) in his study of choline transport in human red cells (see also Devès and Krupka, 1981). The different affinities of the ADP/ATP carrier for substrates, in the case of mitochondria (Souverijn *et al.*, 1973) and inside-out submitochondrial particles (Lauquin *et al.*, 1977b), also speak in favor of two different conformational states for the outward- and inward-facing substrate sites.

6. THE ASYMMETRICAL BINDING OF CATR (ATR) and BA: ONE SINGLE SITE FOR INHIBITORS (IDENTICAL WITH THE SUBSTRATE SITE) OR TWO DISTINCT SITES?

As there is a binding anisotropy for the inhibitory ligands CATR (ATR) and BA (cf. Section 3), it is central to the ADP/ATP transport mechanism to determine whether CATR (ATR) and BA share the same locus as the substrates ADP and ATP. Two cases can be envisaged. In the first one, the same site is recognized by substrates and inhibitors, and all these ligands compete for the same locus made by a well-defined set of amino acids (single site hypothesis) (Klingenberg, 1976). In the second case, it is proposed that CATR (ATR) and BA bind to different sites (either distinct or overlapping) and that the interaction between the two sites is indirect [double inhibitor site hypothesis (Block *et al.*, 1979)].

As the structures of CATR (ATR) and BA molecules are very different, the binding site in the single site hypothesis must assume two conformational states that are utilized for specific binding of either CATR (ATR) or BA, and that depend on the exposure of the site to the outside [cytosolic or "C" state for CATR (ATR) binding] and the inside [matrix or "M" state for BA binding]. The single site hypothesis originated from the finding that BA binding is able to increase the amount of ADP bound to mitochondria (Klingenberg and Buchholz, 1973); it was formulated as follows. In the absence of externally added ADP, all ADP/ATP sites of the mitochondrial membrane are exposed and immobilized in their empty form to the outside (C state). Upon addition of ADP or ATP, the sites are filled with nucleotides, and are rendered mobile. The subsequent addition of BA results in (1) trapping of the ADP (ATP)-filled sites inside in a new conformational state, the M state; (2) release by BA of the bound ADP to the matrix, which explains the excess ADP binding; and (3) immobilization of the sites in the M state by the bound BA. Two variants (A and B) of the single site hypothesis are shown in Fig. 2. Model A corresponds to a mobile reorienting carrier, model B to a mobile site in a fixed pore. That CATR (ATR) and BA bind to the same site (characterized by the same set of amino acids with two possible conformations) is quite plausible in model A, but hardly feasible in model B, because of the considerable constraints to which the reorienting site might be subjected in the transversal motion. On the other hand, as discussed in Section 5, the reorienting mobile carrier in model A is unlikely on the basis of thermodynamical arguments (Singer, 1977).

In the double inhibitor site hypothesis, it is proposed that the ADP/ATP transport protein is a fixed gated channel spanning the membrane with the CATR (ATR) binding site exposed to the outside and the BA site exposed to the inside (Fig. 2, model C). The

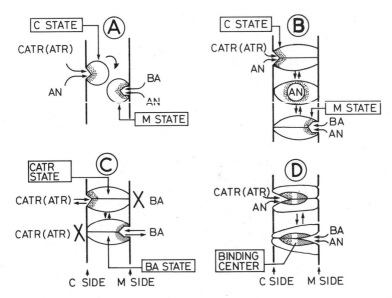

Figure 2. Binding anisotropy of CATR (ATR) and BA. Models A and B (single site hypothesis): Model A corresponds to a mobile reorienting carrier, model B to a fixed transport protein with a mobile reorienting site. The same site assumes two conformationally distinct states (C and M states). Both states accept the substrates ADP and ATP. The C state only accepts CATR and ATR, and the M state only accepts BA. Model C (double inhibitor site hypothesis): The same protein contains two distinct preexisting binding sites, one for CATR (ATR) on the C side, one for BA on the M side. The protein is in either the CATR or the BA conformational state. The two states are in equilibrium. The binding of CATR (ATR) or BA shifts the equilibrium toward the CATR or BA conformation, respectively. Model D: Common binding center hypothesis with two inhibitor sites distinct from (or overlapping) the ADP/ATP site.

double inhibitor site hypothesis does not preclude that the ADP/ATP site is not able to reorient (this basic transport mechanism is in fact postulated), nor that part of the binding sites of CATR (ATR) or BA is shared by bound ADP or ATP. It only means that CATR (ATR) and BA binding involve some distinct amino acids. The interaction between CATR (ATR) and BA in this model is indirect; at any time, the carrier is either in the CATR conformation [site available for CATR (ATR) outside, but not for BA inside] or in the BA conformation [site available for BA inside, but not for CATR (ATR) outside]. The two conformations are in equilibrium, and this equilibrium is shifted in one direction or the other by the binding of either of the inhibitors. The fact that the interaction between CATR (ATR) and BA is enhanced by ADP (ATP) suggests that the CATR (ATR) or BA binding requires an ADP/ATP activated state of the carrier protein.

A way to discriminate between the two hypotheses (single site and double inhibitor site) is to compare the effect of chemical modifiers on the binding of CATR (ATR) and BA. As mentioned earlier (Leblanc and Clauser, 1972; Vignais and Vignais, 1972), upon addition of minute amounts of ADP (ATP), one or two SH groups in the ADP/ATP carrier are unmasked. These SH groups are readily alkylated by NEM, resulting in inhibition of both ATR binding and ADP/ATP transport; BA binding remains unaltered. These data may be interpreted by the single site hypothesis. In the absence of added ADP, all carrier sites are empty and immobilized in the C state; upon addition of ADP, they are filled and move inside (M state), and concomitantly SH groups are unmasked. Covalent alkylation of these unmasked SH groups by NEM results in immobilization of the sites in the M state, allowing BA binding, but preventing CATR binding. The same type of explanation could be applied to the data on the differential inactivation by UV light of ATR and BA binding (Block *et al.*, 1979). However, the single site hypothesis is unable to explain the results of chemical modifications by phenylglyoxal and butanedione, two rather selective reagents for arginyl residues, and those obtained with hydroxynitrobenzyl bromide, a modifier of tryptophanyl residues (Block *et al.*, 1981). These reagents inhibit more readily CATR binding than BA binding, like NEM and UV light do, but their inactivation effect, in contrast to that of NEM or UV, is protected by ADP. This protection by ADP is just opposite to what is predicted in the single site hypothesis. Thus, the effect of added ADP is not necessarily linked to the substrate site reorientation, as postulated in the single site hypothesis. More probably, it is due to a broad change of conformation conferred to the carrier protein by bound ADP, so that different strategic functional groups become either accessible or masked to chemical modifiers. For example, ADP or ATP increases the number of SH groups that can be alkylated by NEM (P. M. Vignais *et al.*, 1975) and decreases the number of arginyl residues accessible to butanedione or phenylglyoxal (Block *et al.*, 1981). Another fact that supports the idea of distinct ATR and BA sites is that ATR and BA preincubated with mitochondria prior to the addition of butanedione or phenylglyoxal protect against the inhibition of the binding of ATR and BA, respectively (Block *et al.*, 1981).

As mentioned above, the double inhibitor site model does not preclude that the ADP/ATP site may share part of the CATR (ATR) site and (or) part of the BA site to form a common binding center. The idea that the ADP/ATP site may overlap the CATR (ATR) site is in fact suggested by the similar rates of UV light inactivation of ADP/ATP transport and ATR binding (Block *et al.*, 1979). A compromise between the single site hypothesis and the two inhibitor site hypothesis is given by model D, which supposes a common binding center for ADP, ATP, CATR (ATR), and BA, with specific sites for CATR (ATR) and BA. In this model, which is in agreement with all experimental data, CATR (ATR) and BA sites are distinct, but close to (and possibly overlapping) the ADP/ATP site.

REFERENCES

Barbour, R. L., and Chan, S. H. P. (1981). *J. Biol. Chem.* **256,** 1940–1948.

Block, M. R., Lauquin, G. J. M., and Vignais, P. V. (1979). *FEBS Lett.* **104,** 425–430.

Block, M. R., Lauquin, G. J. M., and Vignais, P. V. (1981). *Biochemistry* **20,** 2692–2699.

Boulay, F., Lauquin, G. J. M., and Vignais, P. V. (1979). *FEBS Lett.* **108,** 390–394.

Brandolin, G., Doussière, J., Gulik, A., Gulik-Krzywicki, T., Lauquin, G. J. M., and Vignais, P. V. (1980). *Biochim. Biophys. Acta* **592,** 592–614.

Brandolin, G., Dupont, Y., and Vignais, P. V. (1981). *Biochem. Biophys. Res. Commun.* **98,** 28–35.

Buchanan, B. B., Eiermann, W., Riccio, P., Aquila, H., and Klingenberg, M. (1976). *Proc. Natl. Acad. Sci. USA* **73,** 2280–2284.

Devaux, P. F., Bienvenüe, A., Lauquin, G. J. M., Brisson, A., Vignais, P. M., and Vignais, P. V. (1975). *Biochemistry* **14,** 1272–1280.

Devès, R., and Krupka, R. M. (1981). *J. Membrane Biol.* **61,** 21–30.

Duée, E. D., and Vignais, P. V. (1969). *J. Biol. Chem.* **244,** 3920–3931.

Duszynski, J., Bogucka, K., Letko, G. H., Küster, U., and Wojtczak, L. (1979). In *Function and Molecular Aspects of Biomembrane Transport* (E. Quagliariello, F. Palmieri, S. Papa, and M. Klingenberg, eds.), pp. 309–312, Elsevier/North-Holland, Amsterdam.

Duyckaerts, C., Sluse-Goffart, C. M., Fux, J. P., Sluse, F. E., and Liebecq, C. (1980). *Eur. J. Biochem.* **106,** 1–6.

Edwards, P. A. W. (1973). *Biochim. Biophys. Acta* **311,** 123–140.

Hackenberg, H., and Klingenberg, M. (1980). *Biochemistry* **19,** 548–555.

Heldt, H. W., Klingenberg, M., and Milovancev, M. (1972). *Eur. J. Biochem.* **30,** 434–440.

Kemp, A., Souverijn, J. H. M., and Out, T. A. (1971). In *Energy Transduction in Respiration and Photosynthesis* (E. Quagliariello, S. Papa, and C. S. Rossi, eds.), pp. 959–969, Adriatica Editrice, Bari.

Klingenberg, M. (1976). In *The Enzymes of Biological Membranes* (A. N. Martonosi, ed.), Vol. 3, pp. 383–438, Plenum Press, New York.

Klingenberg, M., and Buchholz, M. (1973). *Eur. J. Biochem.* **38,** 346–358.

Klingenberg, M., and Rottenberg, H. (1977). *Eur. J. Biochem.* **73,** 125–130.

Kolarov, J., Šubík, J., and Kováč, L. (1972). *Biochim. Biophys. Acta* **267,** 465–478.

LaNoue, K., Mizani, S. M., and Klingenberg, M. (1978). *J. Biol. Chem.* **253,** 191–198.

Laris, P. C. (1977). *Biochim. Biophys. Acta* **459,** 110–118.

Lauquin, G. J. M., and Vignais, P. V. (1976). *Biochemistry* **15,** 2316–2322.

Lauquin, G. J. M., Brandolin, G., and Vignais, P. V. (1976). *FEBS Lett.* **67,** 306–311.

Lauquin, G. J. M., Devaux, P. F., Bienvenüe, A., Villiers, C., and Vignais, P. V. (1977a). *Biochemistry* **16,** 1202–1208.

Lauquin, G. J. M., Villiers, C., Michejda, J. W., Hryniewiecka, L. V., and Vignais, P. V. (1977b). *Biochim. Biophys. Acta* **460,** 331–345.

Lauquin, G. J. M., Brandolin, G., Lunardi, J., and Vignais, P. V. (1978). *Biochim. Biophys. Acta* **501,** 10–19.

Leblanc, P., and Clauser, H. (1972). *FEBS Lett.* **23,** 107–113.

Letko, G., Markefski, M., and Bohnensack, R. (1979). *Acta Biol. Med. Ger.* **38,** 1081–1090.

Out, T. A., Valeton, E., and Kemp, A. (1976). *Biochim. Biophys. Acta* **440,** 697–710.

Riccio, P., Aquila, H., and Klingenberg, M. (1975). *FEBS Lett.* **56,** 133–138.

Silva Lima, M., and Denslow, N. D. (1979). *Arch. Biochem. Biophys.* **193,** 368–372.

Singer, S. J. (1977). *J. Supramol. Struct.* **6,** 313–323.

Souverijn, J. H. M., Huisman, L. A., Rosing, J., and Kemp, A. (1973). *Biochim. Biophys. Acta* **305,** 185–198.

Srere, P. A. (1980). *Trends Biochem. Sci.* **5,** 120–121.

Stubbs, M. (1979). *Pharmacol. Ther.* **7,** 323–349.

Stubbs, M., Vignais, P. V., and Krebs, H. A. (1978). *Biochem. J.* **172,** 333–342.

Šubík, J., Kolarov, J., and Kováč, L. (1974). *Biochim. Biophys. Acta* **357,** 453–456.

Vignais, P. M., Chabert, J., and Vignais, P. V. (1975). In *Biomembranes: Structure and Function* (G. Gàrdos and J. Szasz, eds.), pp. 307–313, North-Holland, Amsterdam.

Vignais, P. V. (1976). *Biochim. Biophys. Acta* **456,** 1–38.

Vignais, P. V., and Lauquin, G. J. M. (1979). *Trends Biochem. Sci.* **4,** 90–92.

Vignais, P. V., and Vignais, P. M. (1972). *FEBS Lett.* **26,** 27–31.

Vignais, P. V., Vignais, P. M., and Defaye, G. (1973). *Biochemistry* **12,** 1508–1519.

Vignais, P. V., Vignais, P. M., and Doussière, J. (1975). *Biochim. Biophys. Acta* **376,** 219–230.

Vignais, P. V., Lauquin, G. J. M., and Vignais, P. M. (1976). In *Mitochondria: Bioenergetics, Biogenesis and Membrane Structure* (L. Packer and A. Gomez-Puyou, eds.), pp. 109–125, Academic Press, New York.

Villiers, C., Michejda, J. W., Block, M., Lauquin, G. J. M., and Vignais, P. V. (1979). *Biochim. Biophys. Acta* **546,** 157–170.

52

Unconventional Iron–Sulfur Clusters in Mitochondrial Enzymes

Thomas P. Singer

1. INTRODUCTION

It is a common misconception that the more we learn about a scientific phenomenon, the simpler its explanation becomes. At least in the field of membrane-bound Fe–S enzymes, the opposite seems to have happened. When metal flavoproteins were first discovered, it was often concluded that nonheme Fe was an integral component of an enzyme, solely on the basis of finding it in purified preparations, and if the enzyme was inhibited by Fe clusters, this seemed to justify the conclusion that the metal functioned in catalysis. When it was recognized that such evidence is usually insufficient and frequently misleading (Singer and Massey, 1957), new and less ambiguous approaches were sought. The discovery that succinate and NADH dehydrogenases in membrane preparations exhibited an EPR signal at $g \sim 1.94$ on reduction by substrate and the recognition that an iron compound was probably responsible for this signal (Beinert and Sands, 1960) provided a powerful new tool for the study of nonheme iron function in enzymes. When combined with the freeze-quench method (Bray, 1961) for the study of rapid reactions, this permitted following the changes in the redox state of the Fe during catalysis. In a few cases it could even be shown that the rate of appearance of this reduced iron signal agreed well with the turnover rate measured in steady-state assays (Rajagopalan and Handler, 1968).

The conceptual framework for understanding the structure and chemistry of nonheme iron in proteins started with Massey's (1957) accidental observation that a labile form of sulfur accompanied the Fe in succinate dehydrogenase and culminated in the unraveling of the structure of what became known as "iron–sulfur clusters" by X-ray crystallography (reviewed by Beinert, 1973) and the chemical synthesis of model compounds (Holm and Ibers, 1977). It was widely accepted that mitochondria contain two types of Fe–S clusters:

Thomas P. Singer • Molecular Biology Division, Veterans Administration Medical Center, San Francisco, California 94121, and Department of Biochemistry and Biophysics, University of California, San Francisco, California 94143.

a binuclear ([2Fe–2S]) and a tetranuclear ([4Fe–4S]) type and that, except for instances where spin coupling could be demonstrated, each Fe–S cluster could be shown to be paramagnetic (i.e., give an EPR signal) in either the oxidized or the reduced state under appropriate experimental conditions.

Recent advances have shown that neither of these conclusions is always correct and have brought to light some fascinating new features of Fe–S clusters in some long-known and thoroughly studied mitochondrial enzymes.

2. NADH DEHYDROGENASE

Since the first isolation of mitochondrial NADH dehydrogenase in reasonably intact form (Ringler *et al.*, 1963), it has been known to contain 16 to 18 g-atoms of Fe per mole of flavin. The same ratio was found in Complex I (Hatefi *et al.*, 1962), the simplest membrane-bound form of the enzyme. Later, it was shown that in Complex I the ratio of labile S to Fe is 1 : 1 (Lusty *et al.*, 1965), while in the purified enzyme an excess of labile S over Fe has been found, regardless of the method of preparation. For many years the reason for this high S-to-Fe ratio was not understood, although it was always recognized to be some type of artifact. Recent, unpublished experiments by Dr. Paech in this laboratory have provided evidence that during isolation of the enzyme some Fe–S clusters are destroyed, resulting in mineralization of the Fe and recombination of the labile S with the remaining native NADH dehydrogenase to form a persulfide. The latter, of course, registers as labile S in the usual colorimetric reaction.

At 77°K, where EPR studies used to be customarily conducted, only one cluster of the enzyme is seen (Center 1) and its rate of reduction by substrates agrees gratifyingly with steady-state kinetic data (Beinert *et al.*, 1965). At 4 to 20°K, however, four EPR-detectable Fe–S components were later recognized both in Complex I (Orme-Johnson *et al.*, 1971) and in the purified, soluble enzyme (Gutman *et al.*, 1971), each at about the same concentration as the flavin. Taking 16 as the number of Fe atoms present, a simple conclusion would have been that the metal enzyme contains four [4Fe–4S] clusters. This conclusion would have contradicted, however, experience in the field that the EPR signal of [4Fe–4S] clusters is too broad at 77°K to be readily detected. Moreover, Ohnishi (1975) postulated the presence of six EPR-detectable Fe–S centers in the enzyme in pigeon heart mitochondria on the basis of potentiometric titrations in the presence of mediator dyes and also believed that the lowest potential center (Center 1) consists of two components, with different redox potentials. Later (Ingledew and Ohnishi, 1980), the number of centers detected was stated to be five. Albracht *et al.* (1977, 1979) could only find four centers by EPR, but they also found Center 1 to consist of two components, but—to complicate matters further—not the same two as Ohnishi's group. Moreover, each of the two components appeared to be present at well below the concentration of the flavin, for which no reasonable explanation has been advanced.

Because EPR techniques, in the hands of competent specialists, have led to contradictory and confusing data, we turned to a chemical method—the cluster extrusion technique (Holm, 1977)—in order to determine the number and type of Fe–S clusters in the enzyme. In this method Fe–S proteins are unfolded anaerobically in a water-miscible organic solvent [usually 80% (v/v) hexamethylphosphoramide], in the presence of a large excess of an aromatic thiol, which combines with the Fe–S clusters as they are extruded from the protein. The adducts formed with bi- and tetranuclear clusters being different, they may be

Table I. Cluster Extrusion of NADH Dehydrogenase [a]

Experiment	Initial concentration after dilution (μM) [b]		Concentration after extrusion (μM) [b]		Extrusion %	Ratio d^{2-}/t^{2-}
	Flavin	Fe	d^{2-}	t^{2-}		
1	8.6	154	37	18	95	2.1
2	11	190	49	25	100	2.0
3	8.6	150	35	17	92	2.0
4	15	250	—	—	—	2.0

[a] Experimental conditions: 80% (v/v) hexamethylphosphoramide/aqueous buffer during ligand exchange in 30 mM potassium phosphate, pH 7.8 (experiments 1 and 2), and in 33 mM potassium phosphate, 30 mM Tris–phosphate, pH 8.5 (experiments 3 and 4). The molar ratio of o-xylyl-α,α'-dithiol : Fe was ~ 125 : 1 and that of p-trifluoromethylbenzenethiol : o-xylyl-α,α'-dithiol $\simeq$ 4 : 1. The designations d^{2-} and t^{2-} refer to the [2Fe–2S] and [4Fe–4S] clusters, respectively.
[b] Samples were diluted fivefold in the unfolding step. From Paech *et al.* (1982).

distinguished and each quantitated by a variety of techniques. The data in Table I were obtained by two cycles of extrusion, the second using [^{19}F]-p-trifluoromethylbenzenethiol and a high-resolution Fourier-transform NMR apparatus.

It is seen that six Fe–S clusters were detected per mole of flavin (i.e., per mole of enzyme) and that both [2Fe–2S] and [4Fe–4S] clusters were present, in the ratio of 2 : 1. This means that four binuclear and two tetranuclear clusters are present per mole of flavin. (Trinuclear clusters, to be discussed below, need not be considered, for in all known cases they show an EPR signal at $g = 2.01$ in the oxidized state, which has never been seen in NADH dehydrogenase.)

As summation of the integrated EPR signals reported by Orme-Johnson *et al.* (1974) or Albracht *et al.* (1977, 1979) accounts for only four clusters per mole of FMN, it seems that two of the clusters revealed by the extrusion studies might be EPR silent.

3. SUCCINATE DEHYDROGENASE

For many years there was general agreement that succinate dehydrogenase contains 8 g-atoms each of Fe and labile S per mole of flavin in both soluble and particulate preparations (Lusty *et al.*, 1965; Baginsky and Hatefi, 1969; Davis and Hatefi, 1971; Ackrell *et al.*, 1977). As the enzyme can be relatively easily prepared in homogeneous form, unlike NADH dehydrogenase, there was no ambiguity as to whether these are components of the enzyme or represent impurities. Two variants of the cluster extrusion technique showed that these are made up of one [4Fe–4S] and two [2Fe–2S] clusters (Coles *et al.*, 1979).

The roots of the confusion that has arisen are conflicting reports on the number of EPR-detectable Fe–S clusters in the enzyme. The first one discovered was a ferredoxin-type cluster ($g = 1.93$), which is binuclear because it is detected at 77°K, and was named Center 1. A second cluster (Center 3) is seen only at very low temperature and in the oxidized form of the enzyme and gives a $g = 2.01$ EPR signal, which disappears on reduction with succinate (Beinert *et al.*, 1975; Ohnishi *et al.*, 1976). The conformation of the tetranuclear cluster is quite labile, so that its EPR signal is only seen in membrane preparations and in the most carefully prepared soluble ones. Quantitation of the EPR signals of Center 1 (in the reduced enzyme) and of Center 3 (in the oxidized form) yield a 1 : 1 ratio of each component to the flavin in membrane preparations, including Complex II, thus accounting for six of the eight Fe–S groups.

Evidence for another [2Fe–2S] cluster, needed to account for the chemically determined Fe and S content, was first reported by Ohnishi *et al*. (1973) and was also seen by Beinert *et al*. (1975, 1977) in Complex II and a variety of soluble preparations. This cluster is thought to have a very low redox potential (-400 mV), so that it is not reduced by succinate but becomes EPR detectable on reduction with dithionite and manifests itself as an intensification of the $g = 1.93$ signal, with a change in lineshape. At present, both groups agree that the signal intensity of this Center 2 is only 20 to 50% greater on reduction with dithionite than with succinate. This has been interpreted as being due to spin sharing between Centers 1 and 2 (Beinert *et al*., 1975).

The logic and coherence of these conclusions were gratifying, all the more so because the consensus it represented is not common in this field. The accord turned out to be short-lived, however, for in a recent paper Albracht (1980) denied the existence of Center 2 and suggested that the enzyme contains 6, not 8, g-atoms of Fe and labile S. His conclusions stem from the fact that in submitochondrial particles and membrane fragments he found the same intensity of the $g = 1.93$ signal on reduction with either succinate or dithionite.

Center 2, in fact, is also not detectable in the soluble preparation of Ackrell *et al*. (1977), the only homogeneous preparation that is 100% active in reconstitution tests. Significantly, this preparation and inner-membrane samples also have the highest catalytic center activity, a fact that may have prompted Albracht to conclude that Center 2, when seen, is a preparative artifact.

In order to understand how Albracht rationalizes the finding of eight Fe–S units per mole of flavin by chemical analysis in several laboratories, it is necessary to review the findings of Coles *et al*. (1972) on the ratio of the two subunits of the enzyme. These authors reported that while in the membrane-bound form (Complex II) the 30K and 70K subunits of the enzyme are equimolar by quantitative analysis of the bands corresponding to these subunits, separated on polyacrylamide gels, in all soluble preparations tested the ratio was higher (1.35–1.5 moles 30K/mole 70K subunit). An excess of 30K subunit over 70K subunit was also noted later in the 100% reconstitutively active enzyme (Ackrell *et al*., 1977). As no obvious explanation could be found for these curious findings, it was suggested (Coles *et al*., 1972) that during extraction of the enzyme some dissociation of the subunits may occur; the insoluble 70K subunit would then precipitate and be discarded with the unextracted residue, while the 30K subunit might recombine with the native holoenzyme, yielding some molecules of succinate dehydrogenase containing an extra 30K subunit. It must be emphasized that this was sheer speculation, for which no evidence exists.

Albracht (1980) adopted this hypothesis and added two further assumptions: that during the putative dissociation of the holoenzyme the Fe–S clusters remain attached to the individual subunits and that the [4Fe–4S] cluster (Center 3) is located in the 70K subunit where the flavin is located, while the [2Fe–2S] cluster (Center 1) is in the 30K subunit. If the molar ratio of 30K/70K subunits is 2, one would then arrive at a ratio of 8 Fe : 8 S : 1 flavin.

Unfortunately, there are internal contradictions in this ingenious explanation. One is that in Complex II the subunit ratio was reported to be 1 : 1 (Coles *et al*., 1972), although the Fe : S : flavin ratio is 8 : 8 : 1. Second, the subunit ratio in soluble samples in no case was found to be as high as 2, the ratio required to account for the analytical finding for Fe, S, and flavin, but more like 1.2 to 1.5, which, using Albracht's assumption on the location of the Fe-S clusters, would give the following ratios: at 30K/70K = 1.5,

Fe/flavin = 7, [2Fe–2S]/[4Fe–4S] = 1.5, [2Fe–2S]/flavin = 1.5, which contradict Albracht's (1980) and our own analytical data. Third, although there is no conclusive evidence that the tetranuclear cluster is in the 30K subunit, there are significant indications that this is the case (Beinert *et al.*, 1977). Fourth, preparations of the dehydrogenase extracted from acetone powders without succinate contain both subunits but the Fe : S : flavin ratio is 4 : 4 : 1 (Beinert *et al.*, 1975; Ohnishi *et al.*, 1976) and lack a HiPIP EPR signal, hence, Center 3. This would suggest that the enzyme prepared anaerobically had lost its [4Fe–4S] cluster but contains the two [2Fe–2S] clusters, Centers 1 and 2.

It is possible to explain all known facts about the stoichiometry of the prosthetic group of the enzyme without invoking the assumptions made by Albracht. We believe that both particulate and most soluble preparations contain 8 g-atoms of Fe and S per mole, which exist as one tetranuclear and two binuclear clusters. One of the [2Fe–2S] clusters, Center 2, is EPR silent in the inner membrane and in the most intact soluble samples but, because of the different conformation of the protein, is EPR detectable (on reduction with dithionite) in Complex II and in certain soluble preparations. When the latter are reinserted into the inner membrane, Center 2 once again becomes EPR silent, because the conformation characteristic of the inner membrane is again acquired. The reversible conformation changes of the enzyme proposed here are also reflected in corresponding changes in the turnover number. But what about the excess 30K subunit reported by Coles *et al.* (1972)? Recent studies in collaboration with Dr. Ackrell have provided a rational explanation for that finding. It is seen that in the Coomassie blue staining of the separated subunits on polyacrylamide gel electrophoresis gels, the larger subunit does not obey Beer's Law, except at very low concentrations, while the smaller subunit does obey the law. Consequently, except at very low protein concentrations, as had been used with Complex II, an artificially high ratio is obtained. Interestingly, the color yield of the 30K subunit in the biuret and Lowry reactions is also unusually high.

4. ACONITASE

The facts that aconitase contains 1 to 2 g-atoms of nonheme Fe per mole and that it can be activated by incubation with Fe^{2+} in the presence of a thiol have long been known (Villafranca and Mildvan, 1971; Kennedy *et al.*, 1972), but opinions were divided as to whether the Fe moiety functions in catalysis (Glusker, 1971) or not (Villafranca and Mildvan, 1971). That iron is actually incorporated during activation of aconitase in the presence of Fe^{2+} was assumed by several of these workers, although the finding of labile S in association with the Fe in yeast aconitase (Suzuki *et al.*, 1976) raised questions about how Fe–S clusters could be built up without adding a source of sulfide along with the iron. Although homogeneous preparations of the enzyme have been available for a decade or so, reports on the Fe and labile S content of the enzyme varied from 1 to 3 g-atoms per mole, in part because of widely divergent values used for the molecular weight of the enzyme (Kurtz *et al.*, 1979).

An indication of the complexity of aconitase, once considered to be a relatively simple hydrase, came to light when Ruzicka and Beinert (1978) reported that beef heart aconitase gives a typical HiPIP EPR signal, $g = 2.01$, in the oxidized (inactive) state, which disappears on reduction to the active state. The puzzling aspect of this finding was that a HiPIP signal was thought to be associated exclusively with [4Fe–4S] clusters, while the preparations of Ruzicka and Beinert (1978) contained only 2 to 2.5 g-atoms of Fe per mole. Applications of the cluster extrusion technique to aconitase (Kurtz *et al.*, 1979) showed that 1 mole of binuclear cluster was extruded per mole of enzyme. This implied that, despite the characteristic $g = 2.01$ signal in the oxidized state, the Fe–S components are present in a binuclear cluster, although the excess of Fe and labile S (~ 0.5 g-atom per mole) present in the enzyme over that required for a binuclear cluster, remained unexplained.

An answer to these paradoxes emerged shortly thereafter. Mössbauer, X-ray crystallographic, and EPR studies on ferredoxins from *Azotobacter vinelandii* (Emptage et al., 1980; Stout et al., 1980) and from *Desulfovibrio gigas* (Huynh et al., 1980) revealed the presence of a novel Fe–S cluster in these proteins, with the composition [3Fe–3S], which gives a $g = 2.01$ EPR signal in the oxidized state and decomposes to a [2Fe–2S] cluster under the conditions of extrusion. Mössbauer experiments (Kent et al., 1982) show that beef heart aconitase may also contain a trinuclear cluster. In accord with this, aconitase isolated in rigorous anaerobiosis contains very nearly 3 g-atoms of Fe and S per mole (based on a molecular weight of 83,000), while exposure to oxygen during isolation seems to yield samples containing more nearly 2 g-atoms per mole (Ramsay et al., 1981). Collectively, these observations appear to explain why a HiPIP EPR signal is seen in the oxidized enzyme, although the iron content is insufficient for a tetranuclear cluster, and why the extrusion technique gives a nearly quantitative recovery of a binuclear cluster.

The unexpected failure of the cluster extrusion technique in the case of aconitase in no way detracts from the proven reliability of the method for the identification and quantitation of bi- and tetranuclear clusters. Rather, this is one of the indications of the extreme lability of trinuclear clusters. The work of Kent et al. (1982) has provided additional evidence for the latter conclusion, for Mössbauer and EPR spectra indicate that on reduction of aconitase with dithionite a substantial part of the [3Fe–3S] cluster is transformed into an entity with the properties of a [4Fe–4S] cluster. On reconstituting aconitase from the apoenzyme, Fe^{2+}, and S^-, the cluster reinserted appears to be different from the original one and resembles a tetranuclear cluster, although catalytic activity is regained. It appears that trinuclear clusters can collapse to binuclear ones or perhaps even be built up to tetranuclear ones and that enzyme activity may not be uniquely dependent on the geometry of the cluster.

There remain other fascinating questions about aconitase, such as the role of the Fe–S cluster and the chemistry of the activation–deactivation process. Current work (Ramsay et al., 1981) has given some insight into these questions, but much remains for further research. Activation clearly involves reduction of the Fe–S cluster and not necessarily insertion of Fe atoms, as full activation can be reached coulometrically starting with the oxidized, inactive enzyme, and, conversely, the same technique can be used to oxidize and deactivate the enzyme. However, catalytic activity always lags considerably behind the absorbance and EPR changes, suggesting that the redox state of the cluster only initiates a conformation change and the latter determines the activity. The redox changes are not fully reversible, however. It also appears that both the activated and the deactivated (or unactivated) forms can bind the substrate, but, as judged from binding studies with a transition-state analog, only the activated form can undergo the ES$\rightarrow$ES* transition, i.e., reach the transition-state step in the catalytic cycle (Ramsay et al., 1981).

Thus, although the geometry and electronic states of the Fe–S cluster in the active and inactive forms have not been completely clarified, it seems established that activation and deactivation are consequences of conformational changes in aconitase, which are only triggered by oxidation or reduction of the cluster. Recent evidence (Ramsay, 1982) for this view is the finding that during activation tryptophan fluorescence increases at the same rate as does catalytic activity.

REFERENCES

Ackrell, B. A. C., Kearney, E. B., and Coles, C. J. (1977). *J. Biol. Chem.* **252,** 6963–6965.

Albracht, S. P. J. (1980). *Biochim. Biophys. Acta* **612,** 11–28.

Albracht, S. P. J., Dooijewaard, G., Leeuwerik, F. J., and Van Swol, B. (1977). *Biochim. Biophys. Acta* **459,** 300–317.

Albracht, S. P. J., Leeuwerik, F. J., and Van Swol, B. (1979). *FEBS Lett.* **104,** 197–200.

Baginsky, M. L., and Hatefi, Y. (1969). *J. Biol. Chem.* **244,** 5313–5319.

Beinert, H. (1960). *Biochem. Biophys. Res. Commun.* **3,** 41–46.

Beinert, H. (1973). In *Iron–Sulfur Proteins* (W. Lovenberg, ed.), Vol. III, pp. 1–36, Academic Press, New York.

Beinert, H., Palmer, G., Cremona, T., and Singer, T. P. (1965). *J. Biol. Chem.* **240,** 475–480.

Beinert, H., Ackrell, B. A. C., Kearney, E. B., and Singer, T. P. (1975). *Eur. J. Biochem.* **54,** 185–194.

Beinert, H., Ackrell, B. A. C., Vinagradov, A. D., Kearney, E. B., and Singer, T. P. (1977). *Arch. Biochem. Biophys.* **182,** 95–106.

Bray, R. C. (1961). *Biochem. J.* **81,** 189–193.

Coles, C. J., Tisdale, H. D., Kenney, W. C., and Singer, T. P. (1972). *Physiol. Chem. Phys.* **4,** 301–316.

Coles, C. J., Holm, R. H., Kurtz, D. M., Orme-Johnson, W. H., Rawlings, J., Singer, T. P., and Wong, G. B. (1979). *Proc. Natl. Acad. Sci. USA* **76,** 3805–3808.

Davis, K. H., and Hatefi, Y. (1971). *Biochemistry* **10,** 2507–2516.

Emptage, M. H., Kent, T. A., Huynh, B. H., Rawlings, J., Orme-Johnson, W. H., and Münck, E. (1980). *J. Biol. Chem.* **255,** 1793–1796.

Glusker, J. P. (1971). In *The Enzymes* (P. D. Boyer, ed.), Vol. V, pp. 413–439, Academic Press, New York.

Gutman, M. Singer, T. P., and Beinert, H. (1971). *Biochem. Biophys. Res. Commun.* **44,** 1572–1578.

Hatefi, Y., Haavik, A. G., and Griffiths, D. E. (1962). *J. Biol. Chem.* **237,** 1676–1685.

Holm, R. H. (1977). *Acc. Chem. Res.* **10,** 427–434.

Holm, R. H., and Ibers, J. A. (1977). In *Iron–Sulfur Proteins* (W. Lovenberg, ed.), Vol. III, pp. 206–281, Academic Press, New York.

Huynh, B. H., Moura, J. J. G., Moura, I., Kent, T. A., LeGall, J., Xavier, A. V., and Münck, E. (1980). *J. Biol. Chem.* **255,** 3242–3244.

Ingledew, W. J., and Ohnishi, T. (1980). *Biochem. J.* **186,** 111–117.

Kennedy, C., Sr., Rauner, R., and Gawron, O. (1972). *Biochem. Biophys. Res. Commun.* **47,** 740–745.

Kent, T. A., Dreyer, J.-L., Kennedy, M. C., Huynh, B. H., Emptage, M. H., Beinert, H., and Munck, E. (1982). *Proc. Nat. Acad Sci. USA* Submited for publication.

Kurtz, D. M., Holm, R. H., Ruzicka, F. J., Beinert, H., Coles, C. J., and Singer, T. P. (1979). *J. Biol. Chem.* **254,** 4967–4969.

Lusty, C. J., Machinist, J. M., and Singer, T. P. (1965). *J. Biol. Chem.* **240,** 1804–1810.

Massey, V. (1957). *J. Biol. Chem.* **229,** 763–770.

Ohnishi, T. (1975). *Biochim. Biophys. Acta* **387,** 475–490.

Ohnishi, T., Winter, D. B., Lim, J., and King, T. E. (1973). *Biochem. Biophys. Res. Commun.* **53,** 231–237.

Ohnishi, T., Lim, J., Winter, D. B., and King, T. E. (1976). *J. Biol. Chem.* **251,** 2105–2109.

Orme-Johnson, N. R., Orme-Johnson, W. H., Beinert, H., and Hatefi, Y. (1971). *Biochem. Biophys. Res. Commun.* **44,** 446–452.

Orme-Johnson, N. R., Hansen, R. E., and Beinert, H. (1974). *J. Biol. Chem.* **249,** 1922–1927.

Paech, C., Friend, A., and Singer, T. P. (1982). *Biochem. J.* Submitted for publication.

Rajagopalan, K. V., and Handler, P. (1968). In *Biological Oxidations* (T. P. Singer, ed.), pp. 301–337, Wiley, New York.

Ramsey, R. R. (1982). To be published.

Ramsay, R. R., Dreyer, J.-L., Schloss, J. V., Jackson, R. H., Coles, C. J., Beinert, H., Cleland, W. W., and Singer, T. P. (1981). *Biochemistry,* in press.

Ringler, R. L., Minakami, S., and Singer, T. P. (1963). *J. Biol. Chem.* **238,** 801–810.

Ruzicka, F. J., and Beinert, H. (1978). *J. Biol. Chem.* **253,** 2514–2517.

Singer, T. P., and Massey, V. (1957). *Rec. Chem. Prog.* **18,** 201–244.

Stout, C. D., Ghosh, D., Vasantha, P., and Robbins, A. H. (1980). *J. Biol. Chem.* **255,** 1797–1800.

Suzuki, T., Akiyama, S., Fujimoto, S., Ishikawa, M., Nakao, Y., and Fukuda, H. (1976). *J. Biochem. (Japan)* **80,** 799–804.

Villafranca, J. J., and Mildvan, A. S. (1971). *J. Biol. Chem.* **246,** 772–779.

53

Transport of Ions and the Membrane Potential of Mitochondria

Henry Tedeschi

1. THE MEMBRANE POTENTIAL OF MITOCHONDRIA

In contemporary thought, the electrochemical gradient of H^+ in mitochondria, the so-called *protonmotive force* (μ_{H^+}), is considered directly responsible for the transport of ions and the phosphorylation of ADP.

A variety of techniques including P NMR studies of intact cells (Cohen *et al.*, 1978; Ogawa *et al.*, 1980) indicate that in the absence of valinomycin, the ΔpH is generally less than 0.5. Because the synthesis of ATP requires a much larger gradient, it has been generally assumed that the major component of the protonmotive force is the electric potential across the mitochondrial membrane, the $\Delta\Psi$, thought to be in the range of -200 to -250 mV (negative inside).

Ions of weak acids and bases are thought to distribute at steady state as a function of the ΔpH. This dependence is directly predictable from a model requiring that the membrane be permeable primarily to the undissociated form of the ion. In contrast, the distribution of ions of strong acids and bases after an appropriate equilibration has been thought to reflect the $\Delta\Psi$, which is calculated using the Nernst equation.

Calculations of $\Delta\Psi$ from the distribution of the ions face a number of difficulties. An unequal distribution of ions is no proof of the presence of a membrane potential. All present indications are that the H^+ efflux is electroneutral, being matched stoichiometrically by the entry of cations or the exit of anions. This fact was well recognized in some of the earlier studies for the case of the $Ca^{2+}/2K^+$ exchange (Scarpa and Azzone, 1970) and K^+/H^+ exchange in the presence of valinomycin (Massari and Azzone, 1970). Furthermore, presently available evidence indicates a lack of correspondence between the phosphorylative capacity and the calculated protonmotive force in either mitochondria (see below) or submitochondrial particles (Azzone *et al.*, 1978b). Therefore, the basic premises of the chemiosmotic model are in question.

Henry Tedeschi • Department of Biological Sciences, State University of New York, Albany, New York 12222.

In the early experiments, the distribution of K^+ in the presence of valinomycin was used to calculate the $\Delta\Psi$. Values as high as -200 mV were calculated by Mitchell and Moyle (1969), whereas Rottenberg (1973), under different experimental conditions, calculated -130 mV. The experimental results, however, do not support a significant difference between the $[K^+]_i$* before and after energization, a shift that produces only minor changes in calculated $\Delta\Psi$. For example, in the experiments of Mitchell and Moyle (1969), after K^+ depletion, the internal K^+ content under anaerobic conditions and in the presence of valinomycin was reported as 10 to 15 μmoles/g protein (Mitchell and Moyle, 1969, p. 476). Therefore, the $\Delta\Psi$ in the absence of metabolism corresponds to -140 mV as calculated using the Nernst equation and the other assumptions of Mitchell and Moyle. After energization, Mitchell and Moyle (1969) calculated -199 mV, a difference of only -59 mV. In fact, even this small calculated difference is not likely to be correct. Mitchell and Moyle (1969) assumed the absence of osmotic volume change when K^+ is taken up. However, the uptake of K^+ is generally matched by a corresponding uptake of water with no significant change in $[K^+]_i$ (Rottenberg and Solomon, 1969, Figs. 1 and 4; Rottenberg, 1973, Table 5). In fact, experiments of Rottenberg (1973, Table 5) and Padan and Rottenberg (1973, Table 1) show that there is no significant change in $[K^+]_i/[K^+]_o$ with metabolic changes.

Other studies have used the Rb^+ distribution in the presence of valinomycin assuming that $[K^+]_i/[K^+]_o = [Rb^+]_i/[Rb^+]_o$ as required by the chemiosmotic hypothesis. This assumption is almost certainly incorrect, for in some experiments the calculated $[K^+]_i$ would have to be as high as 950 mM where the osmotic pressure of the external medium corresponds approximately to 300 mM (Klingenberg and Rottenberg, 1977, Fig. 5).

The strict electroneutrality of the exchanges is also clearly indicated by the exact correspondence between the increases in internal anionic groups paralleling the net cationic uptakes. This has been shown for the K^+ uptake in the presence of valinomycin calculated from published results (see Tedeschi, 1980, Fig. 1 and Table 2). A precise correspondence between the Ca^{2+} and the anions taken up by the mitochondria has also been demonstrated for the cases of β-hydroxybutyrate, lactate, and acetate (Fig. 5 of Harris, 1978). The inconsistencies in the treatment by the chemiosmotic model can be shown most clearly in the case of the uptake of the cations accompanied by weak acid. For example, succinate distributes following the ΔpH (Rottenberg, 1973). Postulating that the $\Delta\Psi$ determines the K^+ distribution cannot be correct because $[\text{succinate}] = 2[K^+]$. Similar arguments can be used for the case of other cations taken up in conjunction with anions of weak acids (e.g., see the considerations of Lehninger, 1974).

The distributions of several cations including lipophilic ions have been used to calculate the metabolically dependent membrane potential in mitochondria. Although some studies calculate membrane potentials in excess of -200 mV, much lower estimates have also been published ranging from approximately 0 to above -200 mV regardless of technique used (e.g., Azzone *e.g.*, 1976; Deutsch *et al.*, 1979; Walsh Kinnally *et al.*, 1978). More importantly, at least with some of the standard techniques, the protonmotive force is very small although the phosphorylation proceeds normally (Walsh Kinnally and Tedeschi, 1976; Walsh Kinnally *et al.*, 1978; Azzone *et al.*, 1978b). Commonly, lipophilic cations have been used to estimate the potential. All indications are that the steady-state distributions of the ions result from electroneutral ion exchanges with H^+ for these cases as well (see Tedeschi, 1980, pp. 185–186 for a discussion of the earlier literature). Higuti and colleagues (see 1980 for tetraphenylarsonium results and references) have found evidence for

---

*Subscripts i and o indicate respectively the concentration internal and external to the mitochondrial membrane.

a metabolically dependent binding of the cationic inhibitors of phosphorylation—ethidium, acriflavine, and most recently tetraphenylarsonium (TPA$^+$), an analog of tetraphenylphosphonium (TPP$^+$) commonly used to estimate $\Delta\Psi$. The TPA$^+$ uptake corresponds quantitatively to the H$^+$ efflux where the H$^+$/TPA$^+$ ranges from 0.4 to 0.7 with succinate and from 0.9 to 1.0 with ATP. As concluded by Higuti *et al.* (1980), the uptake of TPA$^+$ and the other cations used in these studies cannot be explained on the basis of $\Delta\Psi$ but it can be accounted for by binding. The magnitude of the binding could account for the uptake of all ions generally used to estimate the $\Delta\Psi$. Similar binding sites have been demonstrated for the dye safranine and cyanine dyes (Colonna *et al.*, 1973), also commonly used to estimate membrane potential in mitochondria. Indications of the unreliability of the TPP$^+$ distribution as an indicator of membrane potential is shown by the fact that the [TPP$^+$]$_i$/[TPP$^+$]$_o$ is much greater than the [Rb$^+$]$_i$/[Rb$^+$]$_o$ (Shen *et al.*, 1980). It is interesting to note in this respect that a cation uptake to maintain electric neutrality (which would result in a Gibbs–Donnan distribution) would formally resemble a binding by exhibiting saturation when the cation concentration is increased. Based on the asymmetry of the inhibition, Higuti *et al.* (1980) interpret the uptakes as the result of a binding at the outer surface of the inner mitochondrial membrane. It is possible, however, that a substantial portion of the uptake could be occurring inside the mitochondria. The energy-dependent surface charges (Kamo *et al.*, 1976; Quintinilha and Packer, 1977) seem to account for fewer charged sites than the values calculated with TPA$^+$.

Results obtained with microelectrodes in giant mitochondria have found no metabolically dependent membrane potential. These results have recently been reviewed (Tedeschi, 1980, pp. 186–194). The mitochondria apparently function normally after impalement in the production of ATP from ADP and P_i and the accumulation of calcium phosphate. The microelectrodes appear to be in the internal mitochondrial space as shown by a variety of validations including the reversible microinjections of a water soluble dye (Bowman and Tedeschi, 1980).

In conclusion,

1. Present evidence from microelectrode studies or from the distribution of ions does not support the presence of a significant metabolically dependent membrane potential.
2. The net uptake of cations of strong bases, whether synthetic or natural, appears to be largely explainable by an electroneutral H$^+$/cation counterexchange.
3. In the cases of several lipophilic cations, there is considerable evidence for an energy-dependent binding to surface groups.

2. THE TRANSPORT OF CATIONS: GENERAL CONSIDERATIONS

Generally, the initiation of mitochondrial metabolism in resting mitochondria or the addition of certain agents (such as valinomycin or P_i) to metabolizing mitochondria induces the uptake of cations, until eventually the internal concentration reaches a steady state. Although a net influx takes place under a variety of conditions, there is no reason to believe that the converse process, i.e., a metabolically dependent efflux, requires an entirely separate mechanism. In the experiments of Diwan and Tedeschi (1975) and Diwan and Lehrer (1978) with well-coupled rat liver mitochondria (respiratory control ratio = 5–6), the net K$^+$ efflux is outward even in the presence of metabolism. When metabolism is blocked, the unidirectional influx and efflux are both decreased (Diwan and Tedeschi, 1975). Similarly, in rat liver mitochondria (Diwan *et al*., 1977) and beef heart mitochondria (Jung *et al*., 1977); Chávez *et al*., 1977), mersalyl stimulates both the unidirectional fluxes approximately equally. P_i in the experiments of Jung *et al*. (1977) and Chávez *et al*. (1977) also increases both fluxes substantially. In this respect it is interesting to note that P_i was found to increase the net influx of K^+ or Na^+ (Anagnosti and Tedeschi, 1970; Izzard and Tedeschi, 1970). In contrast to these agents, *N*-ethylmaleimide also stimulates the unidirectional fluxes at pH 8 but has a differential effect (an inhibition of influx and not efflux) at pH 7 (Diwan and Lehrer, 1978) at low concentrations of K^+. Diwan and Lehrer (1978) consider the effect of *N*-ethylmaleimide as indirect, involving a block of the P_i transport mechanism. 2,4-Dinitrophenol (Diwan and Tedeschi, 1975) also blocks the influx markedly, with little effect on the efflux. This latter effect may perhaps be a function of how the process is coupled to the transducing reactions.

When a metabolically dependent net influx of ions take place, the mitochondria swell. The increase in mitochondrial volume has generally been shown to correspond closely to the ion uptake (for references see Tedeschi, 1980, p. 174). Metabolic blocks, e.g., cyanide (Izzard and Tedeschi, 1970) or cyanide in the presence of oligomycin (Rottenberg, 1973), and usually the initiation of phosphorylation or the addition of 2,4-dinitrophenol (Izzard and Tedeschi, 1970) release the ions taken up during the active period and concomitantly the mitochondria shrink. Under these conditions the metabolic blocks must decrease the unidirectional influx more than the efflux.

Under yet another set of conditions the mitochondria exhibit a metabolically dependent net efflux of ions accompanied by shrinkage. In these experiments the mitochondia are first swollen, generally by the passive uptake of ions (e.g., with metabolic inhibition with rotenone at pH 8 or in the presence of EDTA) in rat liver mitochondria (Massari *et al*., 1972), or alternatively at alkaline pH with KNO_3 or $NaNO_3$ in beef heart mitochondria (Brierley *et al*., 1977). When metabolism is initiated (e.g., by the addition of succinate), these mitochondria exhibit a net efflux. The unidirectional efflux must now be greater than the influx. The mitochondrial shrinkage and ion efflux are blocked by the addition of ADP and P_i or alternatively by the addition of metabolic inhibitors or uncouplers. Several investigators (Brierley *et al*., 1977; Chávez *et al*., 1977; Azzone *et al*., 1978a,c) have postulated that the net efflux obtained under these conditions corresponds to an electroneutral H^+/cation exchange, independent of the mitochondrial membrane potential. The metabolic shrinkage probably reflects an active transport of ions, i.e., against an electrochemical gradient, for it is reversed when the suspension becomes anaerobic or antimycin is added in the presence of valinomycin (Figs. 1 and 5 of Azzi and Azzone, 1967). Similar results are available for beef heart mitochondria (Fig. 5 of Brierley *et al*., 1977).

3. K^+ TRANSPORT

Rottenberg (1973) has presented data that have been considered supportive of the concept that the K^+ influx is driven by a metabolically dependent membrane potential, either in the presence or in the absence of valinomycin. The $\Delta\Psi$ was calculated from the K^+ distribution using the Nernst equation. As already discussed, this approach is questionable. In fact, there is no evidence from these data that the $[K^+]_i/[K^+]_o$ changes with the metabolism. Furthermore, as the cytoplasmic concentration of K^+ in liver is approximately 170 mM, the application of the considerations of Rottenberg (equilibrium and use of the Nernst equation) would imply that the $[K^+]_i$ of the liver mitochondria *in situ* would have to be approximately 27 M.

In the absence of valinomycin, the inhibition of both influx and efflux when metabolism is blocked (Diwan and Tedeschi, 1975) clearly conflicts with the notion that the presumed $\Delta\Psi$ plays a role in the ion fluxes.

Several studies of the transport of K^+ in mitochondria either in the presence or in the absence of valinomycin have been published. The results are generally in agreement with the following conclusions. (1) The influx of K^+ exhibits saturation kinetics either in the presence or in the absence of valinomycin. This is evident from experiments carried out with rat liver mitochondria (Harris *et al.*, 1967; Massari and Azzone, 1970; Rottenberg, 1973; Diwan and Lehrer, 1977, 1978; Diwan *et al.*, 1977; Gauthier and Diwan, 1979) or beef heart mitochondria (Jung *et al.*, 1977). Conclusions not in agreement with this statement (e.g., Rottenberg, 1973) result from the improper plotting of the data. (2) The apparent K_m of the influx is approximately 5 mM for rat liver mitochondria (above references) and 10 to 11 mM for beef heart mitochondria (Jung *et al.*, 1977). (3) Generally, the apparent K_m is not significantly altered by activators of transport such as valinomycin (Harris *et al.*, 1967; Massari and Azzone, 1970; Rottenberg, 1973), mersalyl (Diwan *et al.*, 1977; Jung *et al.*, 1977), N-ethylmaleimide (Diwan and Lehrer, 1978), or P_i (Jung *et al.*, 1977). There may be some small decreases of the apparent K_m with valinomycin in some of the experiments. (4) Changes in the apparent V_{max} are responsible for the increases in influx. These considerations suggest that a variety of activators of K^+ transport modulate the activity of the same carrier molecule without affecting its affinity for K^+. The assumption that valinomycin simply increases the permeability of mitochondria to K^+ (e.g., Mitchell and Moyle, 1969; Rottenberg, 1973) is unlikely from these considerations. However, it is entirely possible that the natural transport system is in series with the valinomycin-induced K^+ transport as originally proposed by Pressman (1968).

4. Mg^{2+} TRANSPORT

Mg^{2+} is transported into metabolizing rat liver mitochondria (Judah *et al.*, 1965; Johnson and Pressman, 1969) and beef heart mitochondria (Brierley *et al.*, 1962). The unidirectional influx of Mg^{2+} exhibits saturation kinetics with an apparent K_m of 0.7 mM (Diwan *et al.*, 1979). Mg^{2+} competes with the influx of K^+ both in the presence (Ligeti and Fonyó, 1977) and in the absence of valinomycin (Jung *et al.*, 1977). Furthermore, Tl^+, a competitive inhibitor of the K^+ influx, also competes in the influx of Mg^{2+} (Diwan *et al.*, 1979). Similarly to the K^+ influx, the apparent K_m of Mg^{2+} influx is not affected in a major way by agents that affect the V_{max}, e.g., mersalyl (Diwan *et al.*, 1980). As in the case of K^+, changes in metabolic state or other conditions affect both the unidirectional influx and the efflux of Mg^{2+} (Diwan *et al.*, 1979, 1980). It would seem that the Mg^{2+} transport system resembles the K^+ transport system and the two might perhaps involve the same molecular mechanism.

Mg^{2+} competes with Ca^{2+} transport in both liver and heart mitochondria (Parr and Harris, 1976). However, there are significant differences between the two transport systems; for example, La^{3+} inhibits Ca^{2+} transport but not Mg^{2+} transport.

5. Ca^{2+} TRANSPORT

It is generally agreed that the flux of Ca^{2+} into mitochondria is the result of metabolically dependent H^+ ejection, which takes place in a $2H^+/Ca^+$ stoichiometry. The Ca^{2+}

influx is considered to be driven by the $\Delta\Psi$ (for references see Carafoli, 1979; Fiskum and Lehninger, 1980). In the presence of a weak-acid anion such as phosphate or acetate, an anion is taken up secondarily reducing or eliminating the H^+ efflux (e.g., see Lehninger, 1974). The conclusion that the uptake is driven by the membrane potential is based on the fact that in the absence of the transport of anions, it is accompanied by the stoichiometric H^+ efflux. In addition, when a K^+ gradient is imposed in the presence of valinomycin and in the absence of metabolism, the K^+ gradient (possibly in the form of a diffusion potential) drives the Ca^{2+} influx, presumably electrophoretically (Rossi *et al.*, 1967; Pozzan and Azzone, 1976; Åkerman, 1978; Fiskum *et al.*, 1979). However, the conclusion is entirely arbitrary. The $2H^+/Ca^{2+}$ stoichiometry in the presence of metabolism or the $2K^+/Ca^{2+}$ stoichiometry in its absence corresponds to an electroneutral exchange. Furthermore, the influx of Ca^{2+} in response to a K^+ gradient (whether resulting in a diffusion potential or not) may simply be a reflection of a favorable electrochemical gradient, satisfying a thermodynamic requirement. It should be noted that in the earlier studies (Rossi *et al.*, 1967), the results were explained without invoking a membrane potential.

Crompton *et al.*, (1978) and Carafoli (1979) also conclude that Ca^{2+} is driven by the membrane potential. Their conclusion is based on their observation that the Ca^{2+} influx is not kinetically equivalent to the PO_4 influx, although the internal P_i does determine the steady-state Ca^{2+} level. However, the two uptakes need not correspond exactly. A $2H^+/Ca^{2+}$ exchange or an exchange with other ions could also guarantee electroneutrality. The study does not address this question. It is therefore difficult to justify their conclusion.

REFERENCES

Åkerman, K. E. O. (1978). *FEBS Lett.* **93**, 293–296.
Anagnosti, E., and Tedeschi, H. (1970). *J. Cell Biol.* **47**, 520–525.
Azzi, A., and Azzone, G. F. (1967). *Biochim. Biophys. Acta* **135**, 444–453.
Azzone, G. F., Bragadin, M., Pozzan, T., and Dell'Antone, P. (1976). *Biochim. Biophys. Acta* **459**, 96–109.
Azzone, G. F., Borlotto, F., and Zanotti, A. (1978a). *FEBS Lett.* **96**, 135–140.
Azzone, G. F., Pozzan, T., Viola, E., and Arslan, P. (1978b). *Biochim. Biophys. Acta* **501**, 317–329.
Azzone, G. F., Zanotti, A., and Colonna, R. (1978c). *FEBS Lett.* **96**, 141–147.
Bowman, C., and Tedeschi, H. (1980). *Science* **209**, 1251–1252.
Brierley, G. P., Bachman, E., and Green, D. E. (1962). *Proc. Natl. Acad. Sci. USA* **48**, 1928–1935.
Brierley, G. P., Jurkowitz, M., Chávez, E., and Jung, D. W. (1977). *J. Biol. Chem.* **252**, 7932–7939.
Carafoli, E. (1979). *FEBS Lett.* **104**, 1–5.
Chávez, E., Jung, D. W., and Brierley, G. P. (1977). *Arch. Biochem. Biophys.* **183**, 460–470.
Cohen, S. M., Ogawa, S., Rottenberg, H., Glynn, P., Yamane, T., Brown, T. R., and Shulman, R. G. (1978). *Nature (London)* **273**, 554–556.
Colonna, R., Massari, A., and Azzone, G. F. (1973). *Eur. J. Biochem.* **34**, 577–585.
Crompton, M., Hediger, M., and Carafoli, E. (1978). *Biochem. Biophys. Res. Commun.* **80**, 540–546.
Deutsch, C., Erecińska, M., Werrlein, R., and Silver, I. A. (1979). *Proc. Natl. Acad. Sci. USA* **76**, 2175–2179.
Diwan, J. J., and Lehrer, P. H. (1977). *Biochem. Soc. Trans.* **5**, 203–204.
Diwan, J. J., and Lehrer, P. H. (1978). *Membr. Biochem.* **1**, 43–60.
Diwan, J. J., and Tedeschi, H. (1975). *FEBS Lett.* **60**, 176–179.
Diwan, J. J., Markoff, M., and Lehrer, P. H. (1977). *Indian J. Biochem. Biophys.* **14**, 342–346.
Diwan, J. J., Dazé, M., Richardson, R., and Aronson, D. (1979). *Biochemistry* **18**, 2590–2595.
Diwan, J. J., Aronson, D., and Gonsalves, N.O. (1980). *J. Bioenerg. Biomembr.* **12**, 205–212.
Fiskum, G., and Lehninger, A. L. (1980). *Fed. Proc.* **39**, 2432–2436.
Fiskum, G., G., Reynafarje, B., and Lehninger, A. L. (1979). *J. Biol. Chem.* **254**, 6288–6295.
Gauthier, L. M., and Diwan, J. J. (1979). *Biochem. Biophys. Res. Commun.* **87**, 1072–1079.
Harris, E. J. (1978). *Biochem. J.* **176**, 983–991.
Harris, E. J., Catlin, G., and Pressman, B. C. (1967). *Biochemistry* **6**, 1360–1369.

Higuti, T., Arakaki, N., Niimi, S., Nakasima, S., Saito, R., Tani, I., and Ota, F. (1980). *J. Biol. Chem.* **255,** 7631–7636.

Izzard, S., and Tedeschi, H. (1970). *Proc. Natl. Acad. Sci. USA* **67,** 702–709.

Johnson, J. H., and Pressman, B. C. (1969). *Arch. Biochem. Biophys.* **132,** 139–145.

Judah, D. W., Ahmed, K., McLean, A. E. M., and Christie, G. M. (1965). *Biochim. Biophys. Acta* **94,** 452–460.

Jung, D. W., Chávez, E., and Brierley, G. P. (1977). *Arch. Biochem. Biophys.* **183.** 452–459.

Kamo, N., Muratsugu, M., Kurihara, K., and Kobatake, Y. (1976). *FEBS Lett.* **72,** 247–250.

Klingenberg, M., and Rottenberg, H. (1977). *Eur. J. Biochem.* **73,** 125–130.

Lehninger, A. L. (1974). *Proc. Natl. Acad. Sci. USA* **71,** 1520–1524.

Ligeti, E., and Fonyó, A. (1977). *FEBS Lett.* **79,** 33–36.

Massari, S., and Azzone, G. F. (1970). *Eur. J. Biochem.* **12,** 310–318.

Massari, S., Frigeri, L., and Azzone, G. F. (1972). *J. Membr. Biol.* **9,** 71–82.

Mitchell, P., and Moyle, J. (1969). *Eur. J. Biochem.* **7,** 471–484.

Ogawa, S., Shen, C., and Castillo, C. L. (1980). *Biochim. Biophys. Acta* **590,** 159–169.

Padan, E., and Rottenberg, H. (1973). *Eur. J. Biochem.* **40,** 431–437.

Parr, D. R., and Harris, E. J. (1976). *Biochem. J.* **158,** 289–294.

Pozzan, T., and Azzone, G. F. (1976). *FEBS Lett.*, **71,** 62–66.

Pressman, B. C. (1968). *Fed. Proc.* **27,** 1283–1288.

Quintinilha, A. T., and Packer, L. (1977). *FEBS Lett.* **78,** 161–165.

Rossi, C., Azzi, A., and Azzone, G. F. (1967). *J. Biol. Chem.* **242,** 951–957.

Rottenberg, H. (1973). *J. Membr. Biol.* **11,** 117–137.

Rottenberg, H., and Solomon, A. K. (1969). *Biochim. Biophys. Acta* **93,** 48–57.

Scarpa, A., and Azzone, G. F. (1970). *Eur. J. Biochem.* **12,** 328–335.

Shen, C., Boens, C. C., and Ogawa, S. (1980). *Biochem. Biophys. Res. Commun.* **93,** 243–249.

Tedeschi, H. (1980). *Biol. Rev.* **55,** 171–206.

Walsh Kinnally, K., and Tedeschi, H. (1976). *FEBS Lett.* **62,** 41–46.

Walsh Kinnally, K., Maloff, B. L., and Tedeschi, H. (1978). *Biochemistry* **17,** 3419–3428.

54

The Significance of Protein and Lipid Mobility for Catalytic Activity in the Mitochondrial Membrane

Heinz Schneider and Charles R. Hackenbrock

1. INTRODUCTION

The growing knowledge of the properties and interactions of membrane lipids and proteins permits a better understanding of the role of these components in membrane structure and function. Although a generally complete picture is still lacking, it is becoming increasingly clear that diffusional and rotational motion of membrane lipids and proteins are associated with some membrane functions. Clearly, such motion is a force favoring randomization of lipids and proteins in the plane of the membrane. Although motion and randomization may occur, the principle of structural order is satisfied with a polar lipid bilayer providing for precise orientation of integral and peripheral proteins relative to the membrane plane.

2. SIGNIFICANCE OF LIPID AND PROTEIN MOBILITY

The lateral mobility of lipids in the membrane bilayer can support diffusion of membrane proteins. The rate of diffusion of the proteins is largely dependent on the degree of membrane bilayer viscosity, which is determined by the extent of motion of the lipids (Scandella et al., 1971). The diffusion of some proteins, such as enzymes, carriers, and receptors, may affect membrane-mediated catalytic functions directly.

In several membrane systems it appears that specific catalytic activities are associated with lateral diffusion of specific membrane proteins, as for example in microsomal membrane electron transfer between NADH-cytochrome b_5 reductase and cytochrome b_5 (Strittmatter and Rogers, 1975; Taniguchi et al., 1979), in electron transfer in the mitochondrial

Heinz Schneider and Charles R. Hackenbrock • Department of Anatomy, Laboratories for Cell Biology, University of North Carolina School of Medicine, Chapel Hill, North Carolina 27514.

inner membrane (Hackenbrock, 1976; Heron *et al.*, 1978), and in hormone receptor–adenylate cyclase interaction in the plasma membrane (Hanski *et al.*, 1979).

3. THE MITOCHONDRIAL INNER MEMBRANE AS A MODEL MEMBRANE

The mitochondrial inner membrane has been utilized as a model system in studies of the motion of membrane lipids and proteins related to catalytic activities (Hackenbrock, 1976). The inner membrane is the site of a variety of metabolic functions, the most significant being the transfer of electrons from various respiratory substrates to oxygen along a sequence of oxidation–reduction components collectively known as the respiratory chain and the synthesis of ATP derived from the free energy produced by such electron transfer. Related to these significant and specialized functions, the inner membrane mediates transport and exchange activities representative of similar functions carried out by a variety of cell membranes. For studying the relationships between mobility of membrane components and membrane-mediated functions, the mitochondrial inner membrane represents a superior model system, for it contains rather high concentrations of various catalytically interacting oxidation–reduction proteins, and ubiquinone, the relative proportions of which are well known (Hatefi and Galante, 1978). The catalytic activities between these various components, i.e., the rates of electron transfer, can be monitored with excellent precision (Chance *et al.*, 1967).

4. MEMBRANE STRUCTURE AND COMPOSITION

Although recent studies indicate that mitochondrial electron transfer might be diffusion mediated (Hackenbrock, 1976; Heron *et al.*, 1978; Schneider *et al.*, 1980a), the rapid and sequential character of the electron transfer reactions tends to support the widely held notions that the structural organization of the mitochondrial inner membrane is an essentially solid-state system and that the specific oxidation–reduction proteins are immobilized in a rigid protein–protein lattice (Fleischer *et al.*, 1967; Klingenberg, 1968; Lehninger, 1970; Sjöstrand and Barajas, 1970; Capaldi and Green, 1972). These notions have been supported by the unusually high protein content (75%) in the mitochondrial inner membrane (Colbeau *et al.*, 1971; Levy *et al.*, 1969), as compared to other eukaryotic cell membranes. Although a high protein content could generally support the idea of a solid-state structural organization of the membrane, only about 50% of the protein is integral to the membrane (Capaldi and Tan, 1974; Harmon *et al.*, 1974). Recent ultrastructural observations reveal that the proteins of the inner membrane occupy only one-third to one-half of the total membrane area (Hackenbrock *et al.*, 1976; Sowers and Hackenbrock, 1980). Further, the lipid component of the inner membrane is highly fluid primarily because it consists largely of highly unsaturated phospholipids and because cholesterol is virtually absent (Colbeau *et al.*, 1971; Comte *et al.*, 1976). As reviewed by Hackenbrock (1976), these observations indicate that there is considerable lateral space in the plane of the mitochondrial inner membrane in which diffusion of proteins may occur and suggest that the inner membrane is considerably more plastic in its macromolecular organization than previously recognized. In the following sections we will focus on the structural and functional evidence demonstrating the fluid nature of the inner membrane and relate this structural fluidity to membrane-mediated catalytic functions.

5. FREEZE/FRACTURE OBSERVATIONS OF THE MOBILITY OF MEMBRANE COMPONENTS

5.1. Lipid Phase Transition and Lipid–Protein Separation

Freeze-fracture electron microscopy has revealed a random distribution of intramembrane particles (integral proteins) in mitochondrial inner membranes frozen rapidly from 30°C (Höchli and Hackenbrock, 1976). Equilibration of the membranes at temperatures below the phase transition temperature of the bilayer lipids demonstrated a striking lateral separation between smooth, intramembrane particle-free regions (rich in gel-state lipid) and particle-dense regions (rich in integral protein) (Hackenbrock *et al.*, 1976). Upon warming to a physiological temperature, this lateral phase separation is reversed, with the intramembrane particles rerandomizing rapidly. Such thermotropic reversible lipid–protein separations do not destroy the capacity of the membrane for electron transfer and oxidative phosphorylation (Höchli and Hackenbrock, 1976). These results demonstrate that the integral proteins of the inner membrane have a high capacity for lateral diffusion in the plane of the membrane.

5.2. Mobility of Cytochrome c Oxidase

Thermotropic, lateral separation between lipids and proteins has been used to examine the motional freedom of a specific oxidation–reduction component—cytochrome *c* oxidase, a transmembranous integral protein (Hackenbrock and Miller Hammon, 1975). Cytochrome *c* oxidase was specifically cross-linked below the lipid phase transition temperature with immunoglobulin monospecific for the oxidase. Upon warming the membrane above the lipid phase transition temperature, clusters of large oxidase-related intramembrane particles are retained while other smaller particles rerandomize (Höchli and Hackenbrock, 1978). Such clustering of cytochrome *c* oxidase was also apparent after incubation of inner membranes with monospecific immunoglobulin at room temperature (Höchli and Hackenbrock, 1978), clearly demonstrating that the oxidase can diffuse laterally and independently in the plane of the membrane. In addition, these studies reveal that many other as yet unidentified integral proteins can diffuse independently of the oxidase.

5.3. Liposome Fusion with the Inner Membrane

A recently developed low-pH method has been used to fuse liposomes with mitochondrial inner membranes, which results in the enrichment of the membrane lipid bilayer with exogenous lipid (Schneider *et al.*, 1980a). Freeze-fracture electron microscopy revealed that the surface area of the membrane bilayer increased upon lipid incorporation (Schneider *et al.*, 1980a,b). As the bilayer surface area increased, the particle density in the membrane decreased in proportion to the increase in bilayer lipid. This decrease in particle density due to distribution of the integral proteins into the newly expanded lipid bilayer resulted in an average increase in the distance between the integral proteins (Schneider *et al.*, 1980a). These observations suggest that the catalytically active integral proteins in the mitochondrial inner membrane are free to diffuse laterally and independently of one another, resulting in a random distribution into the lateral space available in the membrane.

6. MOBILITY OF COMPONENTS AND FUNCTION

The rates of specific catalytic events in the mitochondrial inner membrane have been determined after enriching the membrane lipid bilayer with exogenous phospholipid. Analysis of the electron transfer rates from NADH and succinate to oxygen revealed decreases in these rates proportional to the degree of phospholipid enrichment of the membrane lipid bilayer (Schneider *et al.*, 1980a,b). These decreases in the electron transfer rates from the membrane dehydrogenases to the hemeproteins of the membrane appear to be related to the increase in the average distance between the integral proteins as the membrane surface area increases upon incorporation of additional phospholipid. Indeed, rate analysis of single-electron-transferring enzymes and their sequential interactions along various segments of the eiectron transfer sequence identified a diffusion-mediated step between the membrane dehydrogenases and the bc_1 cytochromes (Schneider *et al.*, 1980b). Ubiquinone incorporation along with phospholipid resulted in a partial restoration of the electron transfer rates between the membrane dehydrogenases and the bc_1 cytochromes as compared with phospholipid incorporation alone (Schneider *et al.*, 1981). The degree of this partial restoration appears to be dependent on the chain length of the newly incorporated ubiquinone. These observations have led to the conclusion that ubiquinone diffusion mediates electron transfer between independently diffusing membrane dehydrogenase complexes and cytochrome bc_1 complexes. Other evidence suggests that electron transfer from cytochrome c_1 to cytochrome c oxidase might occur through diffusion of cytochrome c along the membrane surface (Lee *et al.*, 1965; Roberts and Hess, 1977). It was reported earlier that one cytochrome c molecule can interact with several cytochrome c oxidases (Wohlrab, 1970) and that only one common binding site exists on cytochrome c for the interaction with both cytochrome c_1 and cytochrome c oxidase (Rieder and Bosshard, 1978; Speck *et al.*, 1979). Further, it has been established that cytochrome c oxidase rotates on an axis normal to the plane of the mitochondrial inner membrane (Kawato *et al.*, 1980), indicating its motional freedom with respect to other oxidation–reduction protein complexes.

In conclusion, these new observations demonstrate the significance of lipid and protein mobility in the mitochondrial inner membrane for eliciting the specific catalytic reactions along the mitochondrial electron transfer sequence. Random diffusion of the various specific electron transfer components and energetically favorable collisions between these components along an electropositive gradient most likely account for the rapid and sequential character of the mitochondrial electron transfer reactions from the respiratory substrates to oxygen. Lateral mobility of proteins as well as lipids may play an important role, not only in the mitochondrial inner membrane, but in any membrane requiring collisional interaction between membrane components for specific catalytic activities.

ACKNOWLEDGMENTS. This work was supported by research grants from the National Science Foundation (PCM-8040778) and the National Institutes of Health (GM 28704).

REFERENCES

Capaldi, R. A., and Green, D. E. (1972). *FEBS Lett.* **25**, 205–209.
Capaldi, R. A., and Tan, R.-F. (1974). *Fed. Proc.* **33**, 1515.

Chance, G., DeVault, D., Legallais, V., Mela, L., and Yonetani, T. (1967). *Nobel Symp.* **5**, 437–468.

Colbeau, A., Nachbaur, J., and Vignais, P. M. (1971). *Biochim. Biophys. Acta* **249**, 462–492.

Comte, J., Maisterrena, B., and Gautheron, D.C. (1976). *Biochim. Biophys. Acta* **419**, 271–284.

Fleischer, S., Fleischer, B., and Stoeckenius, W. (1967). *J. Cell Biol.* **32**, 193–208.

Hackenbrock, C. R. (1976). *Nobel Symp.* **34**, 199–234.

Hackenbrock, C. R., and Miller Hammon, K. (1975). *J. Biol. Chem.* **250**, 9185–9197.

Hackenbrock, C. R., Höchli, M., and Chau, R. M. (1976). *Biochim. Biophys. Acta* **455**, 466–484.

Hanski, E., Rimon, G., and Levitzki, A. (1979). *Biochemistry* **18**, 846–853.

Harmon, H. J., Hall, J. D., and Crane, F. L. (1974). *Biochim. Biophys. Acta* **344**, 119–155.

Hatefi, Y., and Galante, Y. M. (1978). In *Energy Conservation in Biological Membranes* (G. Schäfer and M. Klingenberg, eds.), pp. 19–30, Springer-Verlag, Berlin.

Heron, C., Ragan, C. I., and Trumpower, B. L. (1978). *Biochim. J.* **174**, 791–800.

Höchli, M., and Hackenbrock, C. R. (1976). *Proc. Natl. Acad. Sci. USA* **73**, 1636–1640.

Höchli, M., and Hackenbrock, C. R. (1978). *Proc. Natl. Acad. Sci. USA* **76**, 1236–1240.

Kawato, S., Sigel, E., Carafoli, E., and Cherry, R. J. (1980). *J. Biol. Chem.* **255**, 5508–5510.

Klingenberg, M. (1968). In *Biological Oxidations* (T. P. Singer, ed.), pp. 3–54, Interscience, New York.

Lee, C. P., Estabrook, R. W., and Chance, B. (1965). *Biochim. Biophys. Acta* **99**, 32–45.

Lehninger, A. L. (1970). *Biochemistry*, Worth Publishers, New York.

Levy, M., Toury, R., Sannes M.-T., and Andre, J. (1969). In *Mitochondria: Structure and Function* (L. Ernster and Z. Drahota, eds.), pp. 33–42, Academic Press, New York.

Rieder, R., and Bosshard, H. R. (1978). *FEBS Lett.* **92**, 223–226.

Roberts, H., and Hess, B. (1977). *Biochim. Biophys. Acta* **462**, 215–234.

Scandella, C. J., Devaux, P., and McConnell, H. M. (1972). *Proc. Natl. Acad. Sci. USA* **69**, 2056–2060.

Schneider, H., Lemasters, J. J., Höchli, M., and Hackenbrock, C. R. (1980a). *Proc. Natl. Acad. Sci. USA* **77**, 442–446.

Schneider, H., Lemasters, J. J. Höchli, M., and Hackenbrock, C. R. (1980b). *J. Biol. Chem.* **255**, 3748–3756.

Schneider, H., Lemasters, J. J., and Hackenbrock, C. R. (1981). In *Functions of Quinones in Energy Conserving Systems* (B. L. Trumpower, ed.), Academic Press, New York, in press.

Sjöstrand, F. S., and Barajas, L. (1970). *J. Ultrastruct. Res.* **32**, 293–306.

Sowers, A. E., and Hackenbrock, C. R. (1980). In *38th Annual Proceedings of the Electron Microscopy Society of America* (G. W. Bailey, ed.), pp. 620–621.

Speck, S. H., Ferguson-Miller, S., Osheroff, N., and Masgoliash, E. (1979). *Proc. Natl. Acad. Sci. USA* **76**, 155–159.

Strittmatter, P., and Rogers, M. J. (1975). *Proc. Natl. Acad. Sci. USA* **72**, 2658–2661.

Taniguchi, H., Imai, Y., Iyanagi, T., and Sato, R. (1979). *Biochim. Biophys. Acta* **550**, 341–356.

Wohlrab, W. (1970). *Biochemistry* **9**, 474–479.

VI

Energy-Transducing ATPases and Electron Transport in Microorganisms

55

Properties of Isolated Subunits of H$^+$-ATPase

Yasuo Kagawa

1. INTRODUCTION

1.1. F$_0$ and F$_1$: General Properties

H$^+$-ATPase (F$_0$F$_1$) is an H$^+$ pump present in almost all cells (Mitchell, 1979; Racker, 1976; Fillingame, 1980; Kagawa *et al.*, 1979) and is composed of an ATPase (F$_1$) and an H$^+$ channel (F$_0$). F$_0$ and F$_1$ have been found in mitochondria (MF$_0$ and MF$_1$) (Penefsky, 1979), chloroplasts (CF$_0$ and CF$_1$) (Shavit, 1980), and plasma membranes of prokaryotes (Downie *et al.*, 1979) such as *Escherichia coli* (EF$_0$ and EF$_1$) and the thermophilic bacterium PS3 (TF$_0$ and TF$_1$) (Kagawa, 1978). As the reviews cited above describe details of F$_0$F$_1$, only recent reports on isolated subunits of F$_0$F$_1$ will be discussed here.

F$_1$ of all species studied has a molecular weight of 35,000–40,000 and consists of five different subunits (α, β, γ, δ and ϵ). The molecular weights of the respective subunits of MF$_1$, CF$_1$, EF$_1$, and TF$_1$ were shown to be very similar (Yoshida *et al.*, 1979). The stoichiometric relationship between the different subunits ($\alpha_3\beta_3\gamma\delta\epsilon$ or $\alpha_2\beta_2\gamma_2\delta_2\epsilon_2$) is still controversial, although it has been examined in many ways (see above reviews) such as by determination of subunit cross-linking (Bragg and Hou, 1980) and molecular weight estimation (Todd *et al.*, 1980; Yoshida *et at.*, 1979). Mammalian MF$_1$ contains a sixth subunit that is a natural ATPase inhibitor bound to the β subunit (Klein *et al.*, 1980), but this subunit is not needed for ATP synthesis. The β and γ subunits of EF$_1$ and TF$_1$ are functionally interchangeable in reconstitution studies (Futai *et al.*, 1980), whereas those of MF$_1$ and CF$_1$ are not (Kagawa, unpublished, 1980). There is a homology in amino acid sequence of the β subunits of TF$_1$ and MF$_1$ (Yoshida *et al.*, 1981). Immunological cross-reactions were detected between MF$_1$, CF$_1$, EF$_1$, and F$_1$ of chromaffin granules (GF$_1$) and between the corresponding subunits of MF$_1$ and GF$_1$ (Apps and Schatz, 1979). Thus, F$_1$'s have very similar subunits.

Yasuo Kagawa • Department of Biochemistry, Jichi Medical School, Minamikawachi, Tochigi-ken 329-04, Japan.

The numbers and properties of subunits of F_0 are still controversial. TF_0 contains only three subunits (Sone *et al.*, 1975), and a similar number has been found in other prokaryotic F_0's such as EF_0 (Fillingame, 1980; Foster and Fillingame, 1980; Negrin *et al.*, 1980). About five subunits have been found in MF_0 of yeast (Todd *et al.*, 1980). However, there is one common subunit of all F_0's, the dicyclohexylcarbodiimide (DCCD)-binding protein. The amino acid sequences of this subunit obtained from MF_0, EF_0, CF_0, and TF_0 were shown to be homologous (Sebald *et al.*, 1979).

The subunit organization of F_0F_1 is summarized as follows:

$$\text{H}^+\text{-ATPase} \atop (F_0F_1) \atop (\text{H}^+ \text{ pump}) \left\{ \begin{array}{l} F_1 \\ (\text{ATPase}) \\ \\ \\ \\ \\ F_0 \\ (\text{H}^+ \text{ channel}) \end{array} \right.$$

nucleotide-binding subunits	α (allosteric) β (catalytic)
connecting subunits	γ ($\alpha\beta$-connecting) δ (F_0-$\alpha\beta\gamma$-connecting) ϵ (F_1-inhibitory)
	DCCD-binding protein F_1-binding protein OSCP and other proteins

1.2. Reconstitution of F_0F_1 from Its Subunits

F_0F_1 was reconstituted from F_1 and crude F_0 (Kagawa and Racker, 1966). Because F_0 is hydrophobic and embedded in biomembranes, its H^+-channel activity was measured after incorporating it into liposomes (Okamoto *et al.*, 1977; Negrin *et al.*, 1980). The flow of H^+ through F_0 was blocked by energy transfer inhibitors, such as DCCD, which specifically binds to a carboxyl group of DCCD-binding protein (Sebald *et al.*, 1979). When F_1 was bound to F_0 in the liposomes, the H^+ flow was coupled to ATP synthesis or hydrolysis by F_1 (Kagawa, 1972, 1978; Racker, 1976). Reconstitution of F_0 from its subunits is still incomplete. However, OSCP (oligomycin sensitivity conferring protein) was purified from MF_0, and when MF_0 was depleted of OSCP its function could be restored by adding OSCP (MacLennan and Tzagoloff, 1968). Active F_1-binding protein was isolated from TF_0 (Sone *et al.*, 1978).

To date, complete reconstitution of MF_1 and CF_1 from their five subunits has not been achieved. Thus, the properties of isolated subunits of MF_1 and CF_1, such as their conformation and antigenicity, may not represent those of the native subunits. In contrast, TF_1 and TF_0 are stable, reconstitutable, and highly antigenic (Yoshida *et al.*, 1979; Kagawa *et al.*, 1976). Subunits of EF_1 are unstable, but reconstitutable in the presence of MgATP (Dunn and Futai, 1980), and genetically well defined (Downie *et al.*, 1979). Thus, the properties of subunits of F_0F_1 described in this review are mainly based on experiments on subunits of TF_1, EF_1, and TF_0. In the notation employed, $E\alpha$, $E\beta$, etc. refer to the α, β, and other subunits of EF_1; $T\alpha$ etc. for those of TF_1; $M\alpha$ etc. for those of MF_1; and $C\alpha$ etc. for those of CF_1.

2. SUBUNITS OF F_1

2.1. Nucleotide-Binding Subunits (α and β)

The molecular weights of $T\alpha$ and $T\beta$—54,000 and 51,000, respectively (Yoshida *et al.*, 1979)—are very similar to those reported for other α's and β's. The contents of α

helix of Tα and Tβ are 31 and 34%, respectively, and those of β sheet 19 and 23% (Yoshida *et al.*, 1979). Owing to denaturation or conformational change of the subunits, these parameters are difficult to determine in other species.

ATPase must interact with nucleotides, and isolated Tα and Tβ were each shown to bind 1 ATP or 1 ADP per mole (Ohta *et al.*, 1980b). Eα bound tightly to 1 ATP per mole, but Eβ did not (Dunn and Futai, 1980). $^{32}P_i$ bound to MF_1 (Kasahara and Penefsky, 1978), TF_1, and Tβ (Kagawa, unpublished).

None of the isolated subunits of F_1's alone showed ATPase activity, but the antibodies against Tα and Tβ both completely abolished ATP hydrolysis and ATP synthesis by TF_1 (Yoshida *et al.*, 1979). However, antibody against Cβ did not inhibit CF_1 (Nelson, 1976), and those against α and β of *Micrococcus lysodeikticus* F_1 only slightly inhibited F_1 of the same origin (Mollinedo *et al.*, 1980), perhaps owing to flexibility of these subunits after isolation.

There are many complicated hypotheses on the conformational change associated with nucleotide binding to F_1 during ATP synthesis. The conformational changes on addition of subunits were estimated with both a circular dichroic spectrometer (250- to 310-nm region) and a Fourier-transform infrared spectrometer (1550 cm^{-1}, for measurement of the ^1H–^2H exchange rate), because the α-helix and β-sheet contents were not affected by nucleotides (no spectral changes around 220 nm, or 1650 cm^{-1}). The relaxation spectrum of ^1H–^2H exchange in Tα and Tβ clearly showed that both subunits were stabilized by the addition of nucleotides (Ohta *et al.*, 1980a). Upon binding of ATP, the radius of gyration of Eα changed from 2.8 nm to 2.6 nm, and the volume from 1.24×10^2 nm^3 to 0.95×10^2 nm^3 (Kuhlmeyer and Paradies, 1980).

Experiments with predeuterated Tα and Tβ (Tα* and Tβ*) showed that in hybrid complexes (Tα*Tβ and TαTβ*), Tβ* stabilized Tα, and in the presence of AT(D)P, Tα* stabilized Tβ (Ohta *et al.*, 1980a). In fact, the binding of ATP to F_1 changed the molecular shape of TF_1 crystals as revealed by computerized image reconstruction (Wakabayashi, 1978), and the space group of CF_1 crystals from $C222$ to $P422$ (Kuhlmeyer and Paradies, 1980).

The catalytic site in the β subunit is composed of at least two parts: one is a $[^{14}C]$-DCCD-reactive glutamyl site and the other an ATP-binding tyrosine site. When DCCD is added to F_0F_1, it is bound specifically to the carboxyl residue of DCCD-binding protein in F_0, but at higher concentrations of DCCD, it is also bound to the β subunit of F_1 and inactivates ATPase activity of the isolated F_1. The DCCD-binding glutamyl site in Mβ was shown to have the following sequence (Yoshida *et al.*, unpublished):

Glu-Leu-Ile-Asn-Val-Ala-Lys-Ala-His-Gly-Gly-Tyr-Ser-Val-Phe-Ala-Gly-Val-Gly-Glu*-
Arg-Thr-Arg-Glu-Gly-Asn-Asp-Leu-Tyr-His-Glu**-Met

where Glu* is the DCCD-binding residue of Tβ and Glu** is the DCCD-binding residue of Mβ. The underlined sequence is identical to that of Tβ, and this homology suggests the importance of this segment in the catalytic site. Glu* and Glu** may cooperate when both are close, and DCCD may label only one of the two in a "charge relay system." However, it is still too early to discuss the molecular events during ATP synthesis, because there are about 40 polypeptides in the tryptic digest of Tβ, and the complete sequencing of the Eβ gene and crystallographic analysis of F_1 may take a few years.

The other key lies in the ATP-binding tyrosine site.

The circular dichroic spectrum of the Tα–nucleotide complex showed the fixation of nucleotide in an *anti* form, while that of the Tβ–nucleotide complex showed interaction of a protonated tyrosine residue with the base ring of the nucleotide (Ohta *et al.*, 1980b).

There have been many reports that the catalytic site of F_1's is blocked by tyrosine reagents (see reviews in Section 1.1), and the ATP-binding site in Mβ was shown to have the following sequence (Esch and Allison, 1978):

-Ile-Met-Asp-Pro-Asn-Ile-Val-Gly-Ser-Glu-His-Tyr*-Asp-Val-Ala-Arg-

where Tyr* is the O-[^{14}C]-sulfonylated derivative of the tyrosine residue. It is interesting that this tyrosine residue is surrounded by an imidazole, an arginyl, and two carboxyl groups that may transfer H$^+$ during the interaction of ADP, P_i, and Mg^{2+}. Chemical modification of TF$_1$ with arginyl and carboxyl reagents resulted in complete inactivation (Arana *et al.*, 1980). ADP or ATP prevented the modification of one arginine per mole with [^{14}C]phenylglyoxal. Substrates of F_1, including GDP and IDP, protected TF$_1$ against both reagents, but CTP, which is not a substrate, did not (Arana *et al.*, 1980). Tβ bound all the substrates of TF$_1$, while Tα bound CTP (Ohta *et al.*, 1980b). In contrast to modifications of Mβ, Cβ, Eβ, and Tβ with ATP analogs, modification of Mα with a dialdehyde derivative of ADP did not inhibit the ATPase activity of MF$_1$ (Kozlov and Milgrom, 1980). Aurovertin, a direct F_1 inhibitor, bound to Mβ (Verschoor *et al.*, 1977) and Eβ, but not to mutated Eβ resistant to aurovertin (Satre *et al.*, 1980) and Tβ (Kagawa and Nukiwa, 1981). A natural ATPase-inhibiting peptide was also shown to bind to Mβ (Klein *et al.*, 1980). These and other findings support the idea that the catalytic and allosteric sites of F_1 are localized on α and β, respectively.

2.2. Connecting Subunits (γ, δ, and ϵ)

The molecular weights of Tγ, Tδ, and Tϵ are 30,200, 21,000, and 16,000, respectively, and similar values have been obtained for other γ's and δ's, except Cγ (37,000) (Yoshida *et al.*, 1979). The molecular weights of ϵ's from other sources are about 10,000 (see reviews in Section 1.1). The contents of α helix of Tγ, Tδ, and Tϵ and 49, 65, and 33%, respectively, and those of β sheet 4, 15, and 24% (Yoshida *et al.*, 1979). Cδ is a rodlike molecule ($25 \times 28 \times 90$ Å^3) and Cϵ an ellipsoid ($25.4 \times 25.4 \times 50$ Å^3) (Schmidt and Paradies, 1977a,b).

Tγ, Tδ, and Tϵ were all required for H$^+$-translocation and ATP-synthesis activities in reconstitution studies, but none of the antibodies against these subunits inhibited either activity of TF$_0$F$_1$ liposomes (Yoshida *et al.*, 1979). This discrepancy may be explained by hypothesizing that an antibody is not accessible to its antigen when organized into a quaternary structure. Various results have been reported for antibodies against subunits of other F_1's. (1) Anti-Cγ, but not anti-Cβ, inhibited the ATPase activity of CF$_1$ (Nelson, 1976), but none of the antibodies against the subunits of F_1 of *M. lysodeikticus* alone inhibited its ATPase activity (Mollinedo *et al.*, 1980). (2) The $\alpha\beta\gamma$ complexes of TF$_1$ and EF$_1$, obtained by reconstitution, showed ATPase activities similar to those of the original TF$_1$ and EF$_1$, respectively (Yoshida *et al.*, 1977b; Dunn and Futai, 1980; Futai *et al.*, 1980). As the $\alpha\beta\gamma$ complex is much more stable than the $\alpha\beta$ complex, γ may be an organizing subunit between α and β. Interactions between $\alpha\gamma$ and $\beta\gamma$ have been demonstrated as follows: By partial dissociation of EF$_1$, the $\alpha\gamma\epsilon$ complex (molecular weight 100,000) and the $\alpha\gamma\delta\epsilon$ complex were obtained (Vogel and Steinhart, 1976). The $\beta\gamma$ complex of TF$_1$ has no activity (Kagawa and Nukiwa, 1981). Cross-linked products of EF$_1$ included the $\beta\gamma$ complex (Bragg and Hou, 1980). Binding of ATPase-inhibiting peptide to Mβ in MF$_1$ prevented formation of cross-linked $\beta\gamma$, suggesting that the inhibitor may interact with Mβ at a site close to Mγ (Klein *et al.*, 1980).

Both Tδ and Tϵ were shown to be essential for binding of the T$\alpha\beta\gamma$ complex to TF$_0$ (Yoshida *et al.*, 1977a), and this fact was confirmed in Eδ and Eϵ (Sternweis, 1978). Any F$_1$ blocks H$^+$ leakage through F$_0$, and this H$^+$-gate activity is essential for maintaining the electrochemical potential difference of H$^+$ across the membrane during ATP synthesis (Kagawa, 1978). The T$\gamma\delta\epsilon$ complex was shown to have this H$^+$-gate activity (Yoshida *et al.*, 1977a). Eδ is attached to the NH$_2$ terminus of Eα, and interaction between Eγ and Eϵ was demonstrated by direct isolation of the E$\gamma\epsilon$ complex (Dunn, 1980).

3. SUBUNITS OF F$_0$

3.1. DCCD-Binding Protein

The DCCD-binding protein of TF$_0$, the smallest of all DCCD-binding proteins reported, has a molecular weight of 7300 (Kagawa, 1980). There are 72 amino acid residues in the DCCD-binding protein of TF$_0$, 79 in that of EF$_0$, 81 in that of CF$_0$, 75 in that of MF$_0$ from bovine heart, and 76 and 81 in those of MF$_0$ from *Saccharomyces cerevisiae* and *Neurospora crassa*, respectively (Sebald *et al.*, 1979).

The DCCD-binding protein of CF$_0$ was shown to translocate H$^+$ when it was reconstituted into bacteriorhodopsin liposomes (Nelson *et al.*, 1977). However, the DCCD-binding proteins from EF$_0$ (Fillingame, 1980) and TF$_0$ did not translocate H$^+$ in the same conditions, but required the presence of F$_1$-binding protein, perhaps to organize the H$^+$ channel (Sone *et al.*, 1979).

The DCCD-binding proteins from all F$_0$'s tested are highly hydrophobic and soluble in chloroform–methanol. Complete sequencing of these proteins revealed that a few polar residues are clustered in the middle of the sequence of MF$_0$ of *Neurospora*. Between these clusters there are two long hydrophobic sequences of about 25 residues each. This clustering of residue suggests that two hydrophobic segments traverse the lipid bilayer of the membrane (Sebald *et al.*, 1979). Six residues (Gly 27, Gly 31, Gly 42, Arg 45, Pro 47, Ala 66) are identical in all sequences tested. There is also homology in the polar segment of this protein, when isofunctional substitutions (Glu/Asp, Arg/Lys, Gln/Asn, and hydrophobic residues) are considered: -Ala^{35}-Arg-Gln-Pro-Glu-Leu-Arg^{41}-hydrophobic residues-Glu^{65}-Ala-Leu-. DCCD is bound to Glu 65 in all cases except in that of EF$_0$, where it is bound to Asp 65 (Sebald *et al.*, 1979). The most invariant parts of the sequences of MF$_0$ occur in the two hydrophobic segments (16–40 and 53–78), and these may play a specific role besides having hydrophobic interactions with the membrane lipid bilayer.

There are only eight ionizable residues in this protein from TF$_0$ (four Arg, three Glu, and one Tyr). Chemical modification of Glu 65 with DCCD (the other two Glu were not modified at neutral pH, because they were dissociated), Arg with glyoxal, or Tyr with tetranitromethane completely abolished H$^+$ translocation through TF$_0$ without impairing TF$_1$-binding activity (Sone *et al.*, 1979). The pH profile of the rate of H$^+$ translocation showed that there is a monoprotic H$^+$-binding site with a pK_a of 6.8 (Okamoto *et al.*, 1977). As there is no histidine residue in any known DCCD-binding protein, DCCD-reactive Glu 65, which is surrounded by hydrophobic residues, is a strong candidate for this H$^+$-binding site (H$^+$ filter). In fact, water-soluble carbodiimide, 1-ethyl-3-(3-dimethylaminopropyl)carbodiimide, did not modify Glu 65 (Kagawa, unpublished, 1979). A similar shift of the pK_a of the carboxyl group from 4.3 (in a water phase) to 6.0 was observed in Glu 35 at the catalytic site in a hydrophobic region of the lysozyme molecule. The ^{1}H

NMR spectrum of DCCD-binding protein dissolved in organic solvents showed relatively free rotation of polar residues (Nagayama and Kagawa, unpublished, 1980).

3.2. F_1-Binding Protein and Other Factors

Many coupling factors other than F_0 and F_1 of oxidative phosphorylation have been reported (Racker, 1976; Joshi *et al.*, 1979). However, net synthesis of ATP by an electro-chemical potential of H^+ was demonstrated in proteoliposomes containing pure F_0F_1 of a thermophilic bacterium (Kagawa, 1978; Sone *et al.*, 1977). Thus, any factor essential for the coupling should be a component of F_0F_1 in this organism. The subunits of TF_0 other than DCCD-binding protein are the F_1-binding protein and an OSCP-like protein.

The F_1-binding protein was isolated in the presence of SDS as an active protein and has a molecular weight of 13,500 (Sone *et al.*, 1978). DCCD-binding protein from all sources tested did not bind F_1. In contrast to DCCD-binding protein, F_1-binding protein in liposomes is sensitive to treatment with trypsin and acetic anhydride. F_1-binding protein scientifically binds $T\delta$ and $T\epsilon$.

OSCP purified from MF_0 has a molecular weight of 18,000, and is a basic protein easily extracted with dilute alkali (MacLennan and Tzagoloff, 1968). OSCP confers oli-gomycin sensitivity on F_1 only when both F_1 and OSCP are bound to OSCP-depleted crude F_0. OSCP-like protein is also found in TF_0F_1 precipitated with anti-TF_1 antibody in the presence of detergents.

EF_0 contains three subunits with molecular weights of 24,000, 19,000, and 8400 (DCCD-binding protein), although no active subunits of EF_0 have been isolated (Foster and Fillingame, 1979). The genetics of EF_0 suggest that all three subunits are essential for phosphorylation (Fillingame, 1980) and a subunit with molecular weight of 14,000 was also found near the promotor of F_0 which may serve as an organizer of F_0 (Futai, Walker *et al.*, personal communication, 1981).

MF_0 may contain more subunits than the three found in TF_0 and EF_0. Two peptides have been purified and shown to be essential for restoration of oxidative phosphorylation of crude H^+-ATPase preparations. One is coupling factor 6 (F_6), which was extracted from submitochondrial particles with silicotungstate and has a molecular weight of 8000 (Racker, 1976). The other is factor B, or F_2. This protein is a monomeric SH protein, and has a molecular weight of 13,000 (Joshi *et al.*, 1979). It is not a component of MF_0, but it is essential for oxidative phosphorylation of a crude membrane preparation. Factor B, like dilute oligomycin solution, has been suggested to render the crude membrane less perme-able to H^+, and thus to improve phosphorylation activity (Racker, 1976).

4. CONCLUSIONS

The isolation of eight subunits of H^+-ATPase (F_0F_1) from a thermophilic bacterium and *E. coli* has enabled us to determine the physicochemical and biological properties of these subunits. Of the five subunits of F_1, α and β are the nucleotide-binding sites, δ and ϵ are the connecting bridge of F_1 and to F_0, and γ links the $\alpha\beta$ complex to the $\delta\epsilon$ complex. It is suggested that α is an allosteric site, β is a catalytic site, and $\gamma\delta\epsilon$ is an H^+ gate. The binding of nucleotides to α and β changes the conformation of the latter. The catalytic site of β contains a base-binding protonated tyrosyl residue surrounded by an arginyl and car-boxyl residues. Of the three subunits of F_0, the DCCD-binding protein forms a highly hydrophobic part of the H^+ channel, and the F_1-binding protein specifically binds δ and ϵ

of F_1. The role of the OSCP-like protein remains to be elucidated. The DCCD-binding site of the DCCD-binding protein is an undissociated carboxyl residue in the middle of a hydrophobic sequence, and this residue, assisted by arginyl and tyrosyl residues of the same protein, appears to be an H^+ filter.

Most results on H^+-ATPase obtained by indirect methods, such as studies with chemical modifiers, antisubunit antibodies, and bacterial mutants, support the results obtained by direct reconstitution of H^+-ATPase from its subunits. Details of the mechanism of this enzyme can be studied by crystallographic analysis, electrical rapid kinetics, and DNA sequencing of the *unc* operon.

REFERENCES

Apps, D. K., and Schatz, G. (1979). *Eur. J. Biochem.* **100**, 411–419.

Arana, L. J., Yoshida, M., Kagawa, Y., and Vallejos, R. H. (1980). *Biochim. Biophys. Acta* **593**, 11–16.

Bragg, P. D., and Hou, C. (1980). *Eur. J. Biochem.* **106**, 495–503.

Downie, J. A., Gibson, F., and Cox, G. R. (1979). *Annu. Rev. Biochem.* **48**, 103–131.

Dunn, S. D. (1980). *Fed. Proc.* **39**, 1979.

Dunn, S. D., and Futai, M. (1980). *J. Biol. Chem.* **255**, 113–118.

Esch, S. F., and Allison, W. S. (1978). *J. Biol. Chem.* **253**, 6100–6106.

Fillingame, R. H. (1980). *Annu. Rev. Biochem.* **49**, 1079–1113.

Foster, D. L., and Fillingame, R. H. (1979). *J. Biol. Chem.* **254**, 8230–8236.

Futai, M., Kanazawa, H., Takeda, K., and Kagawa, Y. (1980). *Biochim. Biophys. Res. Commun.* **96**, 227–234.

Joshi, S., Hughes, J. B., Shaikh, F., and Sanadi, R. (1979). *J. Biol. Chem.* **254**, 10145–10152.

Kagawa, Y. (1972). *Biochim. Biophys. Acta* **265**, 297–338.

Kagawa, Y. (1978). *Biochim. Biophys. Acta* **505**, 45–93.

Kagawa, Y. (1980). *J. Membr. Biol.* **55**, 1–8.

Kagawa, Y., and Nukiwa, N. (1981). *Biochem. Biophys. Res. Commun.* **100**, 1370–1376.

Kagawa, Y., and Racker, E. (1966). *J. Biol. Chem.* **241**, 2467–2474.

Kagawa, Y., Sone, N., Yoshida, M., Hirata, H., and Okamoto, H. (1976). *J. Biochem. (Tokyo)* **80**, 141–151.

Kagawa, Y., Sone, N., Hirata, H., and Yoshida, M. (1979). *J. Bioenerg. Biomembr.* **11**, 39–78.

Kasahara, M., and Penefsky, H. S. (1978). *J. Biol. Chem.* **253**, 4180–4187.

Klein, G., Stare, M., Dianoux, A.-C., and Vignais, P. V. (1980). *Biochemistry* **19**, 2919–2923.

Kozlov, I. A., and Milgrom, Y. M. (1980). *Eur. J. Biochem.* **106**, 451–462.

Kuhlmeyer, J., and Paradies, H. H. (1980). *Eur. J. Cell Biol.* **22**, 277.

MacLennan, D. H., and Tzagoloff, A. (1968). *Biochemistry* **7**, 1603–1610.

Mitchell, P. (1979). *Science* **206**, 1148–1159.

Mollinedo, F., Larraga, V., Coll, F. J., and Muñoz, E. (1980). *Biochem. J.* **186**, 713–723.

Negrin, R. S., Foster, D. L., and Fillingame, R. H. (1980). *J. Biol. Chem.* **255**, 5643–5648.

Nelson, N. (1976). *Biochim. Biophys. Acta* **456**, 314–338.

Nelson, N., Eytan, E., Notsani, B., Sigrist, H., Sigrist-Nelson, K., and Gutler, C. (1977). *Proc. Natl. Acad. Sci. USA* **74**, 2375–2378.

Ohta, S., Tsuboi, M., Yoshida, M., and Kagawa, Y. (1980a). *Biochemistry* **19**, 2160–2165.

Ohta, S., Tsuboi, M., Oshima, T., Yoshida, M., and Kagawa, Y. (1980b). *J. Biochem. (Tokyo)* **87**, 1609–1617.

Okamoto, H., Sone, N., Hirata, H., Yoshida, M., and Kagawa, Y. (1977). *J. Biol. Chem.* **252**, 6125–6131.

Penefsky, H. S. (1979). *Adv. Enzymol.* **49**, 223–280.

Racker, E. (1976). *A New Look at Mechanisms in Bioenergetics,* Academic Press, New York.

Satre, M., Bof, M., and Vignais, P. V. (1980). *J. Bacteriol.* **142**, 768–776.

Schmidt, U. D., and Paradies, H. H. (1977a). *Biochem. Biophys. Res. Commun.* **78**, 383–390.

Schmidt, U. D., and Paradies, H. H. (1977b). *Biochem. Biophys. Res. Commun.* **78**, 1043–1052.

Sebald, W., Hoppe, J., and Wachter, E. (1979). In *Function and Molecular Aspects of Biomembrane Transport* (E. Quagliariello, F. Palmieri, S. Papa, and M. Klingenberg, eds.), pp. 63–74, Elsevier, Amsterdam.

Shavit, N. (1980). *Annu. Rev. Biochem.* **49**, 111–138.

Sone, N., Yoshida, M., Hirata, H., and Kagawa, Y. (1975). *J. Biol. Chem.* **250**, 7919–7923.

Sone, N., Yoshida, M., Hirata, H., and Kagawa, Y. (1977). *J. Biol. Chem.* **252**, 2956–2960.

Sone, N., Yoshida, M., Hirata, H., and Kagawa, Y. (1978). *Proc. Natl. Acad. Sci. USA* **75**, 4219–4223.

Sone, N., Ikeba, K., and Kagawa, Y. (1979). *FEBS Lett.* **97**, 61–64.

Sone, N., Hamamoto, T., and Kagawa, Y. (1981). *J. Biol. Chem.* **256**, 2873–2877.

Sternweis, P. C. (1978). *J. Biol. Chem.* **253**, 3123–3128.

Todd, R. D., Greisenbeck, T. A., and Douglas, M. G. (1980). *J. Biol. Chem.* **255**, 5461–5467.

Verschoor, G. J., Van der Sluis, P. R., and Slater, E. C. (1977). *Biochim. Biophys. Acta* **462**, 438–449.

Vogel, G., and Steinhart, R. (1976). *Biochemistry* **15**, 208–216.

Wakabayashi, T. (1978). In *Diffraction Studies of Biomembranes and Muscles, and Synchrotron Radiation* (T. Mitsui, ed.), pp. 315–339, Taniguchi Foundation, Tokyo.

Yoshida, M., Okamoto, H., Sone, N., Hirata, H., and Kagawa, Y. (1977a). *Proc. Natl. Acad. Sci. USA* **74**, 936–940.

Yoshida, M., Sone, N., Hirata, H., and Kagawa, Y. (1977b). *J. Biol. Chem.* **252**, 3480–3485.

Yoshida, M., Sone, N., Hirata, H., Kagawa, Y., and Ui, N. (1979). *J. Biol. Chem.* **254**, 9525–9533.

Yoshida, M., Poser, J. W., Allison, W. S., and Esch, F. S. (1981). *J. Biol. Chem.* **256**, 148–1.

56

Proton-Translocating ATPase (F_1F_0) of Escherichia coli

Masamitsu Futai and Hiroshi Kanazawa

1. INTRODUCTION

Studies on proton-translocating ATPase (F_1F_0) from *Escherichia coli* started much later than those on F_1F_0 from eukaryotic organelles such as mitochondria and chloroplasts. The F_1-ATPase of *E. coli* was purified after solubilization from cytoplasmic membranes and was shown to have essentially the same function and five-subunit structure (α, β, γ, δ, ϵ) as other F_1's. The entire F_1F_0 complex and F_0 portion were also purified (for review see Futai and Kanazawa, 1980). In this review we discuss studies on F_1F_0 from *E. coli*, particularly the isolation of its subunits, the reconstitution of the complex, the identification of defective subunits in mutants, and the physical location of the genes in the bacterial chromosome. These recent findings show that techniques of molecular biology have led to detailed understanding of the complex.

2. SUBUNITS OF F_1

F_1 preparations of five subunits and four subunits (without δ) but with equal ATPase activity have been obtained. Futai *et al.* (1974) suggested that the δ subunit is essential for binding of the catalytic portion of F_1 to F_0, because four-subunit F_1 could not bind to membranes depleted of F_1. Smith and Sternweis (1977) purified the δ and ϵ subunits from five-subunit F_1, and showed that four-subunit F_1 plus purified δ could bind to F_0 in membranes previously depleted of F_1. Both δ and ϵ were required for the binding of the major subunit complex ($\alpha\beta\gamma$), which can be prepared by passing δ-deficient F_1 through a column containing immobilized antibody against ϵ (Sternweis, 1978) or by reconstitution from isolated subunits (Futai, 1977; Dunn and Futai, 1980). A sequence of 15 amino acids from

Masamitsu Futai and Hiroshi Kanazawa • Department of Microbiology, Faculty of Pharmaceutical Sciences, Okayama University, Okayama 700, Japan.

the NH_2 terminus of the α subunit was shown to be necessary for the binding of the δ subunit to the major subunit assembly, suggesting that this portion of α interacts with the δ subunit (Dunn *et al.*, 1980). Direct interaction of ϵ with the γ subunit was suggested from measurement of intrinsic fluorescence and results of gel filtration (Dunn, 1980a).

The three major subunits (α, β, γ) of F_1 from thermophilic bacterium PS3 (abbreviated TF_1) and from *E. coli* (EF_1) have been isolated in reconstitutively active forms (Yoshida *et al.*, 1977a,b; Futai, 1977; Dunn and Futai, 1980). The procedure used for dissociation of TF_1 was not suitable for EF_1, mainly because EF_1 dissociated irreversibly in the reagents used. As an alternative procedure, F_1 was dialyzed overnight in the cold against buffer of high salt concentration and then the dialyzate was rapidly frozen (Futai, 1977). The thawed fraction seemed to be completely dissociated, and practically pure preparations of α, β, and γ subunits could be obtained from it by two different procedures (Futai, 1977; Dunn and Futai, 1980). The isolated α subunit had a high-affinity binding site for ATP or ADP (1 mole/mole α) with K_d values of 0.1 and 0.9 μM, respectively (Dunn and Futai, 1980). Detailed studies recently conducted (Dunn, 1980b) on the interaction of the α subunit with ATP revealed the following. (1) The $s_{20,w}^0$ value of α increased 14% in the presence of ATP, indicating a large conformational change of the polypeptide on binding of ATP. (2) The dissociation rate constant of ATP from α is larger than those of most known protein–ligand complexes, indicating a slow rate of dissociation of the nucleotide from α. These properties of the interaction between α and ATP make it unlikely that the binding site in α is the catalytic site. This site may be the site of the tightly bound nucleotide found in purified EF_1 (2 moles of ATP and 1 mole of ADP) (Maeda *et al.*, 1976). These nucleotides were also found in inactive F_1's from mutants (ATPase⁻) including *uncA401* (Maeda *et al.*, 1977a). *In vivo* experiments showed that ATP in the F_1 molecule turns over very slowly, if at all, while ADP turns over fairly rapidly (Maeda *et al.*, 1977b). The β subunit of EF_1 had a site for aurovertin (Dunn and Futai, 1980) as found in the β subunits of mitochondrial F_1 (for review see Futai and Kanazawa, 1980). The functions and properties of the subunits of F_1's have been discussed in detail (Futai and Kanazawa, 1980).

3. RECONSTITUTION OF F₁

Single subunits of EF_1 or mixtures of any two subunits had no ATPase activity, but ATPase could be reconstituted by dialyzing a mixture of α, β, and γ subunits against buffer containing ATP and Mg^{2+} (Futai, 1977). A functional F_1 was reconstituted from three-subunit ATPase ($\alpha\beta\gamma$) and isolated δ and ϵ (Dunn and Futai, 1980). Mixtures containing the subunits in a molar ratio 3 : 3 : 1 ($\alpha : \beta : \gamma$) resulted in reconstitution of the highest specific activity. This result is consistent with the stoichiometry of $\alpha_3\beta_3\gamma$ proposed for EF_1 (Bragg and Hou, 1975).

Activity for hydrolysis of ATP was reconstituted from the α and β subunits from EF_1 and γ from TF_1, α and β from TF_1 and γ from EF_1, α and γ from EF_1 and β from TF_1 (Futai *et al.*, 1980). It is of interest that subunits of the energy-transducing apparatus from entirely different bacteria could form a hybrid molecule. In fact, β and γ from EF_1 and TF_1 seem to have similar functions and structures, because these two subunits from EF_1 could be replaced by those from TF_1. As discussed above, ATP is required for reconstitution of EF_1 (Futai, 1977), whereas ATP has no effect on reconstitution of TF_1 (Yoshida *et al.*, 1977a). ATP was also essential for reconstitution of ATPase from combinations including the α subunit from EF_1, while significant activity was reconstituted from a combination of α and β from TF_1 and γ from EF_1 without ATP. These results suggest that interaction of α of EF$_1$ with nucleotide may be essential for its assembly with other subunits.

4. SUBUNITS OF THE F$_0$ PORTION

The F$_0$ portion has been purified from *E. coli* in reconstitutively active form by Negrin *et al.* (1980) and Schneider and Altendorf (1980). The two preparations both contained polypeptides with molecular weights of about 19,000 and 8000. However, they also had different additional polypeptides with molecular weights of 14,000 (Schneider and Altendorf, 1980) and 24,000 (Negrin *et al.*, 1980). The polypeptide with the lowest molecular weight is similar to the dicyclohexylcarbodiimide (DCCD)-binding protein found in other organelles, which has been shown to be a constituent of the proton channel. The primary sequence of this protein was found to be similar to those of the protein from other sources (Sebald and Wachter, 1978). An oligomer, probably a hexamer, of DCCD-binding proteins constitutes one functional proton pathway (Altendorf, 1977; Sebald and Wachter, 1978). The other polypeptides found in the F$_0$ preparations have not yet been established to be functional components, and the reason for the discrepancy between the two preparations is unknown. However, experiments with a transducing phage (λasn-5) carrying ATPase genes suggested that polypeptides with molecular weights of 19,000 and 24,000 are subunits of F$_0$ (Foster *et al.*, 1980), because these two polypeptides were overproduced with increase of the gene dosage upon induction of phage lysogen. However, it must be noted that the protein of molecular weight of 24,000 was present in the preparation of Schneider and Altendorf (1980) only in minute amount. Unequivocal proof of the polypeptide composition of F$_0$ requires more refined experiments.

5. MUTANTS AND GENES FOR F$_1$F$_0$

Mutants of F$_1$F$_0$ have also been isolated in various laboratories, and mutations in F$_1$ subunits were assigned to a specific polypeptide by two independent procedures. (See Chapter 57 for details of genetic studies.) As discussed above, ATPase activity of EF$_1$ can be reconstituted by mixing the α, β, and γ subunits, and the entire F$_1$ molecule can be reconstituted by adding δ and ϵ to the $\alpha\beta\gamma$ complex. Thus, *in vitro* complementation assay for the mutant F$_1$ can be established by mixing dissociated mutant F$_1$ and one of the isolated individual subunits from the wild type. Applying this approach, AN120 (*uncA401*) was shown to be defective in the α subunit (Dunn, 1978; Kanazawa *et al.*, 1978). A mutation in the β subunit (*uncD11*) was also identified by the same procedure (Kanazawa *et al.*, 1980a). Tryptic peptide analysis of the subunit of *uncD11* showed a difference in a single peptide from that of the wild-type strain, confirming the mutation. It is noteworthy that mutation in the α subunit gave F$_1$ without ATPase activity, although many experiments suggest that the β subunit carries the catalytic site (Futai and Kanazawa, 1980). The second approach is to analyze mutant membrane proteins by two-dimensional gel electrophoresis. By this approach, AN463 (*unc*D409) was shown to have an altered β subunit (Fayle *et al.*, 1978), and results on the *uncA* mutant were confirmed (Senior *et al.*, 1979).

Mutation in a single subunit seems to change the assembly properties of the entire F$_1$ molecule, probably because alteration of the one polypeptide affects the interaction of the subunit with other subunits. Several examples of this type of alteration have been reported recently. In the *uncD* mutant, the altered β subunit is tightly bound to the membrane (Fayle *et al.*, 1978). Only the α subunit was released from the membranes of one kind of mutant by washing them with buffer under conditions that release the entire F_1 complex from wild-type membranes; the other polypeptides (β, γ, and ϵ) were demonstrated immunochemically in a detergent extract of the membranes (Futai and Kanazawa, 1979; Kanazawa and Futai, 1980).

Mutations in DCCD-binding protein have been reported (Fillingame, 1975; Hoppe *et al.*, 1980). DCCD binds to an aspartyl residue at position 61 of the polypeptide chain. Mutant (DG7/1) protein in which this residue was replaced by a glycine residue did not show functions as a proton channel (Wachter *et al.*, 1980; Hoppe *et al.*, 1980). The mutant isolated by Fillingame (1975) was resistant to DCCD because the reactivity of the protein was decreased, although the ATPase activity coupled to energy was normal. It is of interest to study amino acid replacement in the mutant protein. The DCCD-resistant strain DC1 also has functional F_1F_0, although its ATPase activity is resistant to DCCD. Analysis of the mutant protein showed the presence of a valine residue at position 28 instead of isoleucine in the wild type (Wachter *et al.*, 1980). This residue may be very close to the aspartyl residue (position 61) in the native conformation, thus having a profound effect on DCCD sensitivity. Recently, Friedl *et al.* (1980) constructed diploid strains carrying the wild-type allele of DCCD-binding protein on the chromosome and the mutant (DG7/1) allele on an F' plasmid. They found that the mutant allele was partially dominant over the wild-type allele, suggesting that DCCD-binding proteins are functionally dependent on each other. We have essentially confirmed their results using transducing phage lysogens (F. Tamura *et al.*, 1981). These findings are consistent with results showing that an oligomer of the proteins constitutes one functional proton pathway, as mentioned above. The F_0 portion of the complex changes when the growth conditions are varied. Differences have been found in the proton permeability of F_0 from cells of the F_1 mutant (NR70, DL54) grown aerobically and anaerobically (Boonstra *et al.*, 1975; Hasan and Rosen, 1977) or grown in a synthetic medium and a rich medium (Kanazawa and Futai, 1979).

6. PHYSICAL ORGANIZATION OF GENES FOR F_1F_0

Recent improvements in techniques of molecular biology have made it possible to determine the sequence of nucleotides in DNA much more easily than that of amino acids in proteins. If the primary amino acid sequence can be deduced from DNA of the entire complex, more refined studies can be made on the function and assembly of the complex. For this purpose the exact physical locations of the genes on the chromosome were determined in a series of experiments. First we showed that all the genes for F_1F_0 are located between *glmS* and *oriC* in the *E. coli* chromosome and are carried by a defective transducing phage λasn-5 ($cI857S7$). When lysogenic cells carry λasn-5 were induced at high temperature, synthesis of F_1F_0 increased to several times that of noninduced cells (Kanazawa *et al.*, 1979; Foster *et al.*, 1980). Analysis of λasn-5 and two other transducing phages suggested that the structural genes for F_1F_0 are located on DNA of molecular weight about 10,400,000. This segment is much longer than the DNA required for the F_1F_0 complex. We obtained a number of new transducing phages and constructed plasmids carrying parts of the DNA of this region. Their analysis by genetic complementation and with restriction endonucleases suggested that all the structural genes for F_1F_0 are present in a DNA segment of molecular weight approximately 4,500,000 (Kanazawa *et al.*, 1980b). This value for the length of DNA is close to the value required for coding all the polypeptides of F_1F_0, as calculated from the molecular weights of the subunits determined by polyacrylamide gel electrophoresis. We also located 41 mutant alleles on this DNA and obtained a detailed physical map. Enough information has been obtained to allow us to start studies on the sequencing of DNA coding for the α and β subunits of F_1. Further studies on F_1F_0 of this organism by techniques of molecular biology with biochemical methods should contribute greatly to our understanding of the function and assembly of the complex.

7. PRIMARY STRUCTURES OF THE GENES FOR F_1F_0

The DNA sequence of the structural genes for DCCD-binding protein, δ subunit, and amino terminal portion of α subunit have been obtained (Kanazawa *et al.*, 1981a; Mabuchi *et al.*, 1981). Mabuchi *et al.* (1981) also found a frame coding for protein of 18,000 dalton molecular weight in the region between genes for DCCD-binding protein and δ subunit. Further, the complete DNA sequence of the rest of the subunits was determined very recently by two groups independently (Kanazawa *et al.*, 1981b; Gay and Walker, 1981a,b; Saraste *et al.*, 1981).

ACKNOWLEDGMENTS. Research from this laboratory cited herein was supported by grants-in-aid from the Ministry of Education, Science and Culture of Japan. We thank Drs. S. D. Dunn and R. H. Fillingame for information on their unpublished results.

REFERENCES

Altendorf, K. (1977). *FEBS Lett.* **73**, 271–275.
Boonstra, H., Gutnick, D. L., and Kaback, H. R. (1975). *J. Bacteriol.* **124**, 1248–1255.
Bragg, P. D., and Hou, C. (1975). *Arch. Biochem. Biophys.* **167**, 311–321.
Dunn, S. D. (1978). *Biochem. Biophys. Res. Commun.* **82**, 596–602.
Dunn, S. D. (1980a). *Fed. Proc.* **39**, 1979.
Dunn, S. D. (1980b). *J. Biol. Chem.*, **255**, 11857–11860.
Dunn, S. D., and Futai, M. (1980). *J. Biol. Chem.* **255**, 113–118.
Dunn, S. D., Heppel, L. A., and Fullmer, C. S. (1980). *J. Biol. Chem* **255**, 6891–6896.
Fayle, D. R. H., Downie, J. A., Cox, G. B., Gibson, F., and Radik, J. (1978). *Biochem. J.* **172**, 523–531.
Fillingame, R. H. (1975). *J. Bacteriol.* **124**, 870–883.
Foster, D. L., Mosher, M. E., Futai, M., and Fillingame, R. H. (1980). *J. Biol. Chem.*, **255**, 12037–12041.
Friedl, P., Friedl, C., and Schairer, H. U. (1980). *FEBS Lett.* **119**, 254–256.
Futai, M. (1977). *Biochem. Biophys. Res. Commun.* **79**, 1231–1237.
Futai, M., and Kanazawa, H. (1979). In *Cation Flux across Biomembranes* (Y. Mukohata and L. Packer, eds.), pp. 291–298, Academic Press, New York.
Futai, M., and Kanazawa, H. (1980). *Curr. Top. Bioenerg.* **10**, 181–215.
Futai, M., Sternweis, P. C., and Heppel, L. A. (1974). *Proc. Natl. Acad. Sci. USA* **71**, 2725–2729.
Futai, M., Kanazawa, H., Takeda, K., and Kagawa, Y. (1980). *Biochem. Biophys. Res. Commun.* **96**, 227–234.
Gay, N. J., and Walker, J. E. (1981a). *Nucleic Acid Res.* **9**, 2187–2194.
Gay, N. J., and Walker, J. E. (1981b). *Nucleic Acid Res.* **9**, 3919–3926.
Hasan, S. M., and Rosen, B. P. (1977). *Biochim. Biophys. Acta* **459**, 225–240.
Hoppe, J., Schairer, H. U., and Sebald, W. (1980). *FEBS Lett.* **109**, 107–111.
Kanazawa, H., and Futai, M. (1979). *FEBS Lett.* **105**, 275–277.
Kanazawa, H., and Futai, M. (1980). *FEBS Lett.* **109**, 104–106.
Kanazawa, H., Saito, S., and Futai, M. (1978). *J. Biochem.* **84**, 1513–1517.

Kanazawa, H., Miki, T., Tamura, F., Yura, T., and Futai, M. (1979). *Proc. Natl. Acad. Sci. USA* **76,** 1126–1130.

Kanazawa, H., Horiuchi, Y., Takagi, M., Ishino, Y., and Futai, M. (1980a). *J. Biochem.* **88,** 695–703.

Kanazawa, H., Tamura, F., Mabuchi, K., Miki, T., and Futai, M. (1980b). *Proc. Natl. Acad. Sci. USA* **77,** 7005–7009.

Kanazawa, H., Mabuchi, K., Kayano, T., Tamura, F., and Futai, M. (1981a). *Biochem. Biophys. Res. Comm.* **100,** 219–225.

Kanazawa, H., Kayano, T., Mabuchi, K., and Futai, M. (1981b). *Biochem. Biophys. Res. Commun.*, in press.

Kanazawa, H., Mabuchi, K., Kayano, T., Nouoni, T., Sekiya, T., and Futai, M. (1981c). *Biochem. Biophys. Res. Commun.*, in press.

Mabuchi, K., Kanazawa, H., Kayano, T., and Futai, M. (1981). *Biochem. Biophys. Res. Commun.* **102,** 172–179.

Maeda, M., Kobayashi, H., Futai, M., and Anraku, Y. (1976). *Biochem. Biophys. Res. Commun.* **70,** 228–234.

Maeda, M., Futai, M., and Anraku, Y. (1977a). *Biochem. Biophys. Res. Commun.* **76,** 331–338.

Maeda, M., Kobayashi, H., Futai, M., and Anraku, Y. (1977b). *J. Biochem.* **82,** 311–314.

Negrin, R. S., Foster, D. L., and Fillingame, R. H. (1980). *J. Biol. Chem.* **255,** 5643–5648.

Saraste, M., Gay, N. J., Eberle, A., Runswick, M. J., and Walker, J. E. (1981). *Nucleic Acid Res.* **9,** 5287–5296.

Schneider, E., and Altendorf, K. (1980). *FEBS Lett.* **116,** 173–176.

Sebald, W., and Wachter, E. (1978). In *Energy Conservation in Biological Membranes* (G. Schäfer and M. Klingenberg, eds.), pp. 228–236, Springer-Verlag, Berlin.

Senior, A. E., Downie, J. A., Cox, G. B., Gibson, F., Langman, L., and Fayle, D. R. H. (1979). *Biochem. J.* **180,** 103–109.

Smith, J. B., and Sternweis, P. C. (1977). *Biochemistry* **16,** 306–311.

Sternweis, P. C. (1978). *J. Biol. Chem.* **253,** 3123–3128.

Tamura, F., Kanazawa, H., Tsuchiya, T., and Futai, M. (1981). *FEBS Lett.* **127,** 48–52.

Wachter, E., Schmid, R., Deckers, G., and Altendorf, K. (1980). *FEBS Lett.* **113,** 265–270.

Yoshida, M., Okamoto, H., Sone, N., Hirata, H., and Kagawa, Y. (1977a). *Proc. Natl. Acad. Sci. USA* **74,** 936–940.

Yoshida, M., Sone, N., Hirata, H., and Kagawa, Y. (1977b). *J. Bioll. Chem.* **252,** 3480–3485.

57

Genetics of the Adenosine Triphosphatase Complex of Escherichia coli

J. Allan Downie, Frank Gibson, and Graeme B. Cox

1. INTRODUCTION

The ATPase complex from *Escherichia coli*, like that from many sources, can be dissociated into two portions known as the F_1-ATPase, which contains the catalytic site for ATP hydrolysis, and the F_0 portion, which is located within the membrane and appears to act as a proton channel (Fillingame, 1980).

There is now general agreement that the F_1-ATPase consists of five dissimilar subunits (α, β, γ, δ, and ϵ in order of decreasing molecular weights), forming a complex of aggregate molecular weight about 360,000 (see Downie *et al.*, 1979). The stoichiometry of the subunits appears to be $\alpha_3\beta_3\gamma_1\delta_1\epsilon_1$ (Bragg and Hou, 1975) although there is some disagreement (Vogel and Steinhart, 1976).

The subunit composition of the F_0 portion of the ATPase from *E. coli* has not yet been as clearly defined as that of the F_1. Foster and Fillingame (1979) purified an F_1F_0-ATPase complex and concluded that, in addition to the F_1 components, three membrane components of molecular weights 24,000, 19,000, and 8400* were present. Subsequently (Negrin *et al.*, 1980), a preparation of the F_0 portion of the ATPase was derived from this F_1F_0 complex, and the polypeptides of molecular weights 24,000, 19,000, and 8400 were present along with a low level of F_1 subunits. Friedl *et al.* (1979) reported the purification of an F_1F_0 complex that contained, in addition to the F_1 subunits, polypeptides of molecular weights 28,000, 19,000, and 8500. Their preparation also contained a low amount of polypeptides of about 24,000 and 14,000 molecular weight. Schneider and Altendorf (1980) purified an F_1F_0-ATPase complex containing the same polypeptides as found by Friedl *et al.* (1979), but when the F_0 portion of the complex was solubilized using urea treatment,

---

*The F_0 subunit of molecular weight 8400 is also known as the dicyclohexylcarbodiimide-binding protein.

J. Allan Downie, Frank Gibson, and Graeme B. Cox • Biochemistry Department, John Curtin School of Medical Research, Australian National University, Canberra, Australian Capitol Territory 2601, Australia.

the only major polypeptides found were of molecular weights 19,000, 14,000, and 8500. They concluded that the protein of molecular weight 28,000 in the F_1F_0 complex was probably a dimer that was converted to the monomer by urea treatment. Rosen and Hasan (1979) concluded that the F_0 portion of their F_1F_0 complex contained only two polypeptides, of molecular weights 10,000 and 8300. It is difficult to ascertain which polypeptides comprise the F_0, and it is possible, for example, that proteolytic cleavage of the 24,000-molecular-weight component may occur, giving rise to a product of 14,000 molecular weight.

2. ISOLATION AND CHARACTERIZATION OF MUTANTS AFFECTED IN THE E. COLI ATPase COMPLEX

Mutants of *E. coli* lacking a functional F_1F_0-ATPase complex (*unc* mutants) have been isolated in many laboratories, and membranes prepared from such strains do not carry out oxidative phosphorylation and cannot couple ATP hydrolysis to membrane energization (see Downie *et al.*, 1979).

Initially two classes of *unc* mutants were described. The first, designated *uncA*, lacked ATPase activity (Butlin *et al.*, 1971), whereas the second, designated *uncB*, retained this activity (Gutnick *et al.*, 1972; Butlin *et al.*, 1973). Using the F_1-ATPase purified from the *uncB* mutant strain, it was possible to reconstitute oxidative phosphorylation activity in membranes from the *uncA* mutant strain from which an inactive F_1 had been removed by washing with low-ionic-strength buffer (Cox *et al.*, 1973). Thus, it was established that the *uncA* mutation affected a subunit of the F_1-ATPase and the *uncB* mutation a subunit of the F_0 portion of the ATPase. Further classification required a genetic complementation system, and a method was devised to transfer mutant *unc* alleles onto an F plasmid (Gibson *et al.*, 1977). Complementation between *unc* mutant alleles on the F plasmids and *unc* alleles on the chromosomes was assessed in partially diploid strains by measuring their ability to grow on succinate, their growth yields on limiting glucose, and ATP-dependent energization in membrane preparations.

Seven different genetic complementation groups designated *uncABCDEFG* have thus far been identified, and these all occur at the same locus on the *E. coli* chromosome (Downie *et al.*, 1979, 1980, 1981). Using polarity mutants resulting from the insertion of the bacteriophage Mu into *unc* genes (Gibson *et al.*, 1978; Downie *et al.*, 1979, 1980, 1981), it has been shown that the *unc* genes form an operon and that the order of the genes is *uncBFEAGDC** with the *uncB* gene at the promoter end of the operon and proximal to the *ilv* gene cluster (see Fig. 1).

Mutant strains affected in the *uncB*, *uncF*, or *uncE* genes all retain ATPase activity and have membranes that cannot be made proton-permeable by extensive washing, and cannot be reconstituted using purified normal F_1-ATPase to give ATP-dependent membrane energization. Therefore, it was concluded that all three genes code for subunits that form the functional proton pore of the F_0 (Downie *et al.*, 1979, 1981).

Mutants affected in the *uncA*, *uncG*, or *uncD* genes lack ATPase activity and are thus affected in the F_1 subunits. In *uncC* mutants, the F_0 appears to be normal and a cytoplasmic

---

Note added in Proof. It has recently been shown unequivocally, using DNA sequencing, that the order of the genes coding for the proteins of the F_0 ATPase is *uncBEF*, not *uncBFE* as in Fig. 1. In addition, there is no 14,000 molecular weight protein as indicated in Fig. 1 and that the 14,000 molecular weight protein as seen in *in vitro* protein synthesis is probably an artifact as discussed in the Introduction. For details and references see Gay and Walker (1981), and Kanazawa *et al.* (1981).

ATPase is observed, indicating that a component of the F_1-ATPase not essential for ATP hydrolytic activity has been affected (J. A. Downie, unpublished observations).

3. GENE–POLYPEPTIDE RELATIONSHIPS

3.1. uncB

In vitro transcription and translation of a cloned fragment of DNA carrying the *uncBFEA* genes identified three polypeptides that were translated from *unc* genes and became incorporated into membranes after translation (Downie *et al.*, 1981). These polypeptides, of molecular weights 24,000, 18,000,* and 8400, were concluded to be subunits of the F_0 portion of the ATPase (cf. above). A plasmid carrying a deletion within the *uncB* gene did not form the 24,000-molecular-weight subunit but did form the 18,000- and 8400-molecular-weight subunits as well as a peptide that presumably corresponded to the shortened *uncB* gene product. In addition, only the 24,000-molecular-weight subunit was formed from a plasmid carrying the *uncB* gene but not the *uncF*, *uncE*, or *uncA* genes. Thus, it was concluded that the 24,000-molecular-weight F_0 polypeptide was coded for by the *uncB* gene.

3.2. uncE and uncF

Because the *uncB* gene codes for the F_0 subunit of molecular weight 24,000, the other two F_0 subunits must be coded for by the *uncF* and *uncE* genes. It was shown (Downie *et al.*, 1981) that membrane preparations from two *uncF* mutant strains lacked the 18,000-molecular-weight subunit but retained the 8400-molecular-weight subunit. Conversely, the 8400-molecular-weight subunit was found to be absent from membranes of an *uncE* mutant strain and the 18,000-molecular-weight subunit was present. Thus, it was concluded that the *uncF* and *uncE* genes code for the 18,000- and 8400-molecular-weight subunits, respectively.

3.3. uncA

The *uncA* gene codes for the α subunit of the F_1-ATPase. This was shown by dissociation of the inactive F_1-ATPase from an *uncA* mutant and subsequent reassociation in the presence of a purified normal α subunit, which alone among the purified F_1 subunits restored ATPase activity (Dunn, 1978; Kanazawa *et al.*, 1978). Further, analysis of a number of *uncA* mutants revealed two strains that had electrophoretically altered α subunits (Senior *et al.*, 1979a).

3.4. uncG

The genes of the *unc* operon were cloned as two separate *HindIII* restriction endonuclease fragments, one carrying the *uncBFEA* genes and the other carrying the *uncDC* genes (Downie *et al.*, 1980). When the two fragments were ligated, the resultant plasmid complemented an *uncG* mutant, indicating that there is a *HindIII* restriction endonuclease site within the *uncG* gene. Analysis of the *in vitro* translation products of the plasmids carrying the *uncBFEA* or *uncDC* genes indicated that no γ subunit was formed. In similar experiments with the recombinant plasmid carrying the *uncBFEAGDC* genes, a normal γ subunit

---

*The F_0 subunits of reported molecular weights 18,000 and 19,000 (see also Introduction) are presumably the same.

was formed (Downie *et al.*, 1980). It was concluded that the γ subunit is coded for by the *uncG* gene.

3.5. *uncD*

Membranes from mutant strains carrying the *uncD409* allele were found to contain a β subunit of abnormal net charge (Fayle *et al.*, 1978). A hybrid F_1-ATPase containing both normal and abnormal β subunits was formed in a partially diploid strain (*unc⁺/uncD409*), purified, and found to contain the abnormal β subunit. A range of *uncD* mutants with altered β subunits has also been described (Senior *et al.*, 1979b), including one that formed an F_1 complex, which upon purification was found to have low ATPase activity (see also Kanazawa *et al.*, 1980).

3.6. *uncC*

The plasmid carrying the *uncDC* genes was found to code for only two F_1 subunits, the β and ε subunits, implying the ε subunit is coded for by the *uncC* gene (Downie *et al.*, 1980).

3.7. δ *Subunit*

No mutations affecting the δ subunit have yet been described, but this subunit was found to be coded for by a gene on the plasmid carrying the *uncBFEA* genes (Downie *et al.*, 1980). The exact location of this gene in the operon has not been ascertained (see Fig. 1).

4. CONCLUSIONS

The order of the *unc* genes and the subunits for which they code are summarized in Fig. 1. It is of interest that the genes coding for the three F_0 subunits are transcribed first. Further work on the molecular biology of the *unc* operon in *E. coli* should yield valuable information in a number of areas. In the case of membrane assembly, *uncD* mutants have altered β subunits firmly attached to the membrane (Fayle *et al.*, 1978); an *uncA* mutant

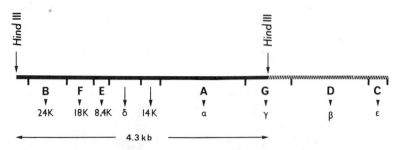

Figure 1. Gene–polypeptide relationships in the *unc* operon. The length of DNA for each gene is estimated from the known molecular weight of the corresponding subunit. The function, if any, of the polypeptide of 14,000 molecular weight is not known. Such a polypeptide was formed during experiments on *in vitro* protein synthesis, but was not seen in gels from membrane preparations. The gene order in that region of DNA coding for the δ subunit is not known with certainty. See note added in proof, p. 454.

has only normal β and altered α subunits of the F_1 attached to the membrane (Senior *et al.*, 1979a). Extension of this work should shed light on the problems of the assembly of the F_1F_0-ATPase. It has already been shown that in an *uncA* mutant a nucleotide, binding site is altered (Bragg and Hou, 1980; Verheijen *et al.*, 1980), and further examination of the biochemical properties of mutants should provide additional insight into the molecular mechanisms of ATP synthesis.

Finally, if the stoichiometry of the F_1-ATPase is in fact $\alpha_3\beta_3\delta_1\gamma_1\epsilon_1$, the regulation of the amounts of the individual subunits presents an intriguing problem, given the order of the genes on the operon.

REFERENCES

Bragg, P. D., and Hou, C. (1975). *Arch. Biochem. Biophys.* **167**, 311–321.

Bragg, P. D., and Hou, C. (1980). *Biochem. Biophys. Res. Commun.* **95**, 952–957.

Butlin, J. D., Cox, G. B., and Gibson, F. (1971). *Biochem. J.* **124**, 75–81.

Butlin, J. D., Cox, G. B., and Gibson, F. (1973). *Biochim. Biophys. Acta* **292**, 366–375.

Cox, G. B., Gibson, F., and McCann, L. (1973). *Biochem. J.* **134**, 1015–1021.

Downie, J. A., Gibson, F., and Cox, G. B. (1979). *Annu. Rev. Biochem.* **48**, 103–131.

Downie, J. A., Langman, L., Cox, G. B., Yanofsky, C., and Gibson, F. (1980). *J. Bacteriol.* **143**, 8–17.

Downie, J. A., Cox, G. B., Langman, L., Ash, G., Becker, M., and Gibson, F. (1981). *J. Bacteriol.* **145**, 200–210.

Dunn, S. D. (1978). *Biochem. Biophys. Res. Commun.* **82**, 596–602.

Fayle, D. R. H., Downie, J. A., Cox, G. B., Gibson, F., and Radik, J. (1978). *Biochem. J.* **172**, 523–531.

Fillingame, R. H. (1980). *Annu. Rev. Biochem.* **49**, 1079–1113.

Foster, D. L., and Fillingame, R. H. (1979). *J. Biol. Chem.* **254**, 8230–8236.

Friedl, P., Friedl, C., and Schairer, H. U. (1979). *Eur. J. Biochem.* **100**, 175–180.

Gay, N. J., and Walker, J. E. (1981). *Nucleic Acids Research* **9**, 3919–3926.

Gibson, F., Cox, G. B., Downie, J. A., and Radik, J. (1977). *Biochem. J.* **164**, 193–198.

Gibson, F., Downie, J. A., Cox, G. B., and Radik, J. (1978). *J. Bacteriol.* **134**, 728–736.

Gutnick, D. L., Kanner, B. I., and Postma, P. W. (1972). *Biochim. Biophys. Acta* **283**, 217–222.

Kanazawa, H., Saito, S., and Futai, M. (1978). *J. Biochem.* **84**, 1513–1517.

Kanazawa, H., Horiuchi, Y., Takagi, M., Ishino, Y., and Futai, M. (1980). *J. Biochem.* **88**, 695–703.

Kanazawa, H., Mabuchi, K., Kayano, T., Noumi, T., Sekiya, T., and Futai, M. (1981). *Biochem. Biophys. Res. Commun.* **103**, 613–620.

Negrin, R. S., Foster, D. L., and Fillingame, R. H. (1980). *J. Biol. Chem.* **255**, 5643–5648.

Rosen, B. P., and Hasan, S. M. (1979). *FEBS Lett.* **104**, 339–342.

Schneider, E., and Altendorf, K. (1980). *FEBS Lett.* **116**, 173–176.

Senior, A. E., Downie, J. A., Cox, G. B., Gibson, F., Langman, L., and Fayle, D. R. H. (1979a). *Biochem. J.* **180**, 103–109.

Senior, A. E., Fayle, D. R. H., Downie, J. A., Gibson, F., and Cox, G. B. (1979b). *Biochem. J.* **180**, 110–118.

Verheijen, J. H., Postma, P. W., and Van Dam, K. (1980). *FEBS Lett.* **116**, 307–309.

Vogel, G., and Steinhart, R. (1976). *Biochemistry* **15**, 208–216.

58

The Genetics of Electron Transport in Escherichia coli

Bruce A. Haddock

1. INTRODUCTION

The gram-negative bacterium *Escherichia coli* is a facultative anaerobe that can derive energy for growth both fermentatively, via glycolysis, and oxidatively using either oxygen or, under anaerobic conditions, fumarate and nitrate as terminal electron acceptors. The ease with which mutants can be generated, analyzed, and manipulated genetically makes it hardly surprising that this organism has received considerable experimental attention over the last decade (for reviews see Gibson and Cox, 1973; Cox and Gibson, 1974; Haddock and Jones, 1977; Haddock, 1977).

It is important to appreciate that not all the membrane-bound redox enzymes synthesized by *E. coli* are necessarily involved in energy conservation. Many serve simply for the reoxidation of reduced coenzymes, the removal of potentially toxic metabolic products, or the reduction of intermediates required for biosynthetic reactions. For example, under appropriate conditions, *E. coli* can synthesize an energy-dependent membrane-bound pyridine nucleotide transhydrogenase. Under physiological conditions, this enzyme is required for NADPH synthesis and, for thermodynamic reasons, cannot be energy conserving (Jones, 1977). The purification of a lipid-depleted transhydrogenase enzyme has been reported (Houghton *et al.*, 1976), but the physiological structure of the protein remains in some doubt (Liang and Houghton, 1980). Mutants lacking transhydrogenase activity (*pnt*) have been isolated and characterized genetically and biochemically (Zahl *et al.*, 1978; Hanson and Rose, 1979, 1980).

The energy-conserving redox reactions can be divided operationally into low- and high-potential segments or enzyme complexes that serve separately or together to generate a proton gradient across the cytoplasmic membrane. This can be used subsequently for ATP synthesis or other energy-requiring processes such as solute transport, motility, and NADPH production.

Bruce A. Haddock • Biogen S.A., 1227 Carouge/Geneva, Switzerland.

2. LOW-POTENTIAL SEGMENTS

Three reductants, NADH, hydrogen, and formate, reduce either ubiquinone (under aerobic and anaerobic conditions) or fumarate (under anaerobic conditions) and, in so doing, translocate protons across the cytoplasmic membrane.

2.1. NADH Dehydrogenase

This enzyme has been purified from the cytoplasmic membrane as a single polypeptide containing FMN and an iron–sulfur center (Dancey *et al.*, 1976). Mutants with defects in the structural gene (*ndh*) have been described (Young and Wallace, 1976), and the gene has been cloned (Young *et al.*, 1978) and sequenced (I. G. Young, personal communication). Under sulfate-limited growth conditions aerobically or under anaerobic growth conditions (Poole and Haddock, 1975), the NADH dehydrogenase becomes non-proton translocating and, therefore, non-energy conserving. It is not clear whether there are two distinct enzymes produced under different growth conditions or whether the proton-translocating NADH dehydrogenase is converted, by modification, into a non-proton-translocating form. There is immunological evidence for two forms of the enzyme in membrane preparations of *E. coli* (Owen and Kaback, 1979a,b).

2.2. Hydrogenase

The membrane-bound hydrogenase of *E. coli* is involved in the energy-conserving oxidation of hydrogen (Bernard and Gottschalk, 1978; Yamamoto and Ishimoto, 1978; Jones, 1980a), and also in the formate hydrogenlyase pathway that additionally requires formate dehydrogenase (see below) and some periplasmically located polypeptides (Gray and Guest, 1965). The enzyme has been purified and shown to be a dimer of identical subunits with a molecular weight of 113,000, containing 12 iron and 12 acid-labile sulfide atoms per molecule (Adams and Hall, 1979). Mutants lacking hydrogenase activity (*hyd*) have been isolated and characterized (Glick *et al.*, 1980; Graham *et al.*, 1980a).

2.3. Formate Dehydrogenase

The partial purification of this protein has been documented by several workers, but there is only one report of purification to homogeneity (Enoch and Lester, 1975). As isolated, the enzyme had a molecular weight of about 600,000 and contained heme, molybdenum, selenium, nonheme iron, and acid-labile sulfide. The enzyme is composed of three polypeptides designated α, β, and γ in approximately a 2 : 2 : 1 molar ratio and of molecular weights 110,000, 32,000, and 20,000, respectively. Only the α subunit contains significant amounts of selenium, and it has been suggested that the heme is associated with the γ subunit. The distribution of the other redox centers in the polypeptides was not reported. The topography of the polypeptides in the membrane has been studied functionally (Jones, 1980b) and structurally, by covalent modification of spheroplasts and membrane vesicles using non-membrane-permeant reagents, and the data indicate that both the α and the β polypeptides span the cytoplasmic membrane (Graham, 1980). Genetically, the assembly of a functional formate dehydrogenase is complex. Certain mutants designated *chlA, chlB,* and *chlD* are pleiotropic in that they result in the inability of the cell to synthesize a functional nitrate reductase (see below) as well as a functional formate dehydrogenase (see Begg *et al.*, 1977). The *chlA, chlB,* and *chlD* gene products are required for the insertion of molybdenum into the respective apoproteins. Other mutants (*fdhA*, *fdhB*) have been described that specifically lack formate dehydrogenase activity and immunologically detectable polypeptides (Mandrand-Berthelot *et al.*, 1978; Graham *et al.*, 1980b). When *E. coli* is grown in the absence of added molybdate to the growth medium or, more dramatically, in the presence of tungstate, it has been shown that inactive forms of formate dehydrogenase are produced (Sperl and DeMoss, 1975; Scott and DeMoss, 1976; Scott *et al.*, 1979). These inactive forms of the enzyme have an unusual subunit composition in that there is considerably less α subunit than β subunit present as compared with equivalent preparations from molybdenum-sufficient cells (Giordano *et al.*, 1980). In addition, no immunologically detectable formate dehydrogenase polypeptides are present in membranes prepared from *chlD* (Giordano *et al.*, 1980), *chlA*, or *chlB* mutants (Graham *et al.*, 1980b). These results lead to the conclusion that demolybdo-forms of formate dehydrogenase are unstable.

These three proton-translocating dehydrogenases reduce the quinone pool in the membrane, ubiquinone under aerobic growth conditions, and either ubiquinone or menaquinone under anaerobic conditions. In turn, the quinol can be reoxidized by fumarate via fumarate reductase under anaerobic conditions, with the concomitant synthesis of 1 mole of ATP per mole of donor oxidized. Fumarate reductase has been extensively studied by Guest and his colleagues. Mutants in the structural gene *frdA* have been identified (Spencer and Guest, 1973), strains are available with multiple chromosomal copies of the gene (Cole and Guest, 1979a,b), and two types of $\lambda frdA$ transducing phage have been obtained (Cole and Guest, 1980). In addition, *lac–frdA* gene fusions have been reported, which should prove of use in studies on the genetic regulation of fumarate reductase synthesis (Ruch *et al.*, 1979).

Alternatively, the quinol produced by these low-potential segments can be reoxidized by three high-potential proton-translocating segments, with the concomitant synthesis of 2 moles of ATP per mole of initial donor oxidized.

3. HIGH-POTENTIAL SEGMENTS

Two of the three high-potential segments use oxygen as terminal electron acceptor, the third utilizes nitrate under anaerobic conditions. The quinone pool may be reduced by other donors such as succinate and D-lactate via their respective dehydrogenases; in this case, however, only 1 mole of ATP is produced per mole of donor oxidized.

3.1. Cytochrome Oxidases

E. coli synthesizes two kinetically competent terminal oxidases, cytochrome *o* and cytochrome *d*, under appropriate growth conditions (Haddock *et al.*, 1976; Poole *et al.*, 1979), both receiving reducing equivalents from quinol via, probably different, *b*-type cytochromes. The factors regulating the synthesis of the two routes are not known, but coordinate synthesis of cytochromes b_{558} and *d* occurs under a variety of growth conditions (see Haddock and Jones, 1977; Reid and Ingledew, 1979). The experimental evidence suggests that the stoichiometry of respiration-driven proton translocation is similar via each of the two routes. So far no mutants have been described with specific lesions in their ability to synthesize specific apocytochromes, but mutants unable to synthesize heme (*hem*), ubiquinone (*ubi*), or menaquinone (*men*) have been used extensively in biochemical studies (see Haddock and Jones, 1977; Downie and Cox, 1979; Jones *et al.*, 1980). Attempts to isolate the different cytochromes from the membrane in pure form have been rather unsatisfactory, but recent reports on the isolation of cytochrome b_{556} (Kita *et al.*, 1978) and a respiratory oxidase complex (Reid and Ingledew, 1980) encourage optimism for the future.

3.2. Nitrate Reductase

Under anaerobic conditions and in the presence of nitrate, *E. coli* synthesizes a respiratory nitrate reductase. As isolated from the cytoplasmic membrane, nitrate reductase contains molybdenum, nonheme iron, and acid-labile sulfide, and is composed of two nonidentical subunits of molecular weight 155,000 (α) and 63,000 (β). A third polypeptide, the γ subunit (molecular weight 19,000) or cytochrome $b_{556}^{NO_3-}$ is required for functional ubiquinol-dependent nitrate reductase activity. The *chlC* gene is the structural gene for the α subunit as indicated from an analysis of membrane-bound proteins of a *chlC* mutant using specific antisera (MacGregor, 1975) and from the isolation of temperature-sensitive *chlC* mutants (DeMoss, 1978). The genetic regulation of the *chlC* gene has been studied in *lac–chlC* gene fusions (Fimmel and Haddock, 1979). The structural gene for the β subunit has not been identified, but a recent report suggests that the *chlI* gene is the structural gene for apocytochrome $b_{556}^{NO_3-}$ (Orth *et al.*, 1980). Gene-polypeptide relationships have not yet been established unambiguously for the *chlE* (Graham *et al.*, 1980b), *chlF* (if it exists), and *chlG* (Jenkins *et al.*, 1979; Jenkins and Haddock, 1980) genes whose expression is also required for functional nitrate reductase activity. In addition, and as mentioned above, expression of the *chlA*, *chlB*, and *chlD* genes is required for the insertion of molybdenum into aponitrate reductase. Interestingly, and unlike the situation for formate dehydrogenase, demolybdo-forms of nitrate reductase appear to be stable in *chlA*, *chlB*, and *chlD* mutants as well as in molybdenum-limited or tungstate-grown wild-type cells (Graham *et al.*, 1980b; Giordano *et al.*, 1980).

The topography of the nitrate reductase polypeptides has been studied both functionally and structurally. The use of artificial permeant and nonpermeant reductants (Jones and Garland, 1977; Jones *et al.*, 1980) has established that cytochrome $b_{556}^{NO_3-}$ can accept electrons at the periplasmic surface and at least part of nitrate reductase is able to donate electrons to oxidized dyes at the internal surface of the cytoplasmic membrane. Using nonmembrane-permeant reagents for covalent labeling of polypeptides, it has shown that cytochrome $b_{556}^{NO_3-}$ is located at the periplasmic face (Boxer and Clegg, 1975), whereas the α (Boxer and Clegg, 1975; MacGregor and Christopher, 1978; Graham and Boxer, 1978) and β (Graham and Boxer, 1980) subunits are at the cytoplasmic face of the membrane.

4. CONCLUDING COMMENTS

The respiration systems of bacteria exhibit great diversity, not only in their redox composition but also in the number of potential energy conservation sites that they contain. In the last decade considerable information has been obtained on the factors regulating the synthesis, assembly, and functional expression of specific energy-conserving multienzyme complexes that will in turn be valuable for describing the mechanism of respiratory-driven proton translocation at the molecular level.

REFERENCES

Adams, M. W. W., and Hall, D. O. (1979). *Biochem. J.* **183**, 11–22.
Begg, Y. A., Whyte, J. N., and Haddock, B. A. (1977). *FEMS Microbiol. Lett.* **2**, 47–50.

Bernard, T., and Gottschalk, G. (1978). *Arch. Microbiol.* **116**, 235–238.

Boxer, D. H., and Clegg, R. A. (1975). *FEBS Lett.* **60**, 54–57.

Cole, S. T., and Guest, J. R. (1979a). *FEMS Microbiol. Lett.* **5**, 65–67.

Cole, S. T., and Guest, J. R. (1979b). *Eur. J. Biochem.* **102**, 65–71.

Cole, S. T., and Guest, J. R. (1980). *Mol. Gen. Genet.* **178**, 409–418.

Cox, G. B., and Gibson, F. (1974). *Biochim. Biophys. Acta* **346**, 1–25.

Dancey, G. F., Levine, A. E., and Shapiro, B. M. (1976). *J. Biol. Chem.* **251**, 5911–5920.

DeMoss, J. A. (1978). *J. Bacteriol.* **133**, 626–630.

Downie, J. A., and Cox, G. B. (1979). *J. Bacteriol.* **133**, 477–484.

Enoch, H. G., and Lester, R. L. (1975). *J. Biol. Chem.* **250**, 6693–6705.

Fimmel, A. L., and Haddock, B. A. (1979). *J. Bacteriol.* **138**, 726–730.

Gibson, F., and Cox, G. B. (1973). *Essays Biochem.* **9**, 1–29.

Giordano, G., Haddock, B. A., and Boxer, D. H. (1980). *FEMS Microbiol. Lett.* **8**, 229–235.

Glick, B. R., Wang, P. Y., Schneider, M., and Martin, W. G. (1980). *Can. J. Biochem.* **58**, 361–367.

Graham, A., (1980). Ph.D. thesis, University of Dundee.

Graham, A., and·Boxer, D. H. (1978). *Biochem. Soc. Trans.* **6**, 1210–1211.

Graham, A., and Boxer, D. H. (1980). *FEBS Lett.* **113**, 15–20.

Graham, A., Boxer, D. H., Haddock, B. A., Mandrand-Berthelot, M. A., and Jones, R. W. (1980a). *FEBS Lett.* **113**, 167–172.

Graham, A., Jenkins, H. E., Smith, N. H., Mandrand-Berthelot, M. A., Haddock, B. A., and Boxer, D. H. (1980b). *FEMS Microbiol. Lett.* **7**, 145–151.

Gray, C. T., and Guest, J. R. (1965). *Science* **148**, 186–192.

Haddock, B. A. (1977). *Symp. Soc. Gen. Microbiol.* **27**, 95–120.

Haddock, B. A., and Jones, C. W. (1977). *Bacteriol. Rev.* **41**, 47–99.

Haddock, B. A., Downie, J. A., and Garland, P. B. (1976). *Biochem. J.* **154**, 285–294.

Hanson, R. L., and Rose, C. (1979). *J. Bacteriol.* **138**, 783–787.

Hanson, R. L., and Rose, C. (1980). *J. Bacteriol.* **141**, 401–404.

Houghton, R. L., Fisher, R. J., and Sanadi, D. R. (1976). *Biochem. Biophys. Res. Commun.* **73**, 751–757.

Jenkins, H. E., and Haddock, B. A. (1980). *FEMS Microbiol. Lett.*, **9**, 293–296.

Jenkins, H. E., Graham, A., and Haddock, B. A. (1979). *FEMS Microbiol. Lett.* **6**, 169–173.

Jones, C. W. (1977). *Symp. Soc. Gen. Microbiol.* **27**, 23–59.

Jones, R. W. (1980a). *Biochem. J.* **188**, 345–350.

Jones, R. W. (1980b). *FEMS Microbiol. Lett.* **8**, 167–171.

Jones, R. W., and Garland, P. B. (1977). *Biochem. J.* **164**, 199–211.

Jones, R. W., Lamont, A., and Garland, P. B. (1980). *Biochem. J.* **190**, 79–94.

Kita, K., Yamato, I., and Anraku, Y. (1978). *J. Biol. Chem.* **253**, 8910–8915.

Liang, A., and Houghton, R. L. (1980). *FEBS Lett.* **109**, 185–188.

MacGregor, C. H. (1975). *J. Bacteriol.* **121**, 1117–1121.

MacGregor, C. H., and Christopher, A. R. (1978). *Arch. Biochem. Biophys.* **185**, 204–213.

Mandrand-Berthelot, M. A., Wee, M. Y. K., and Haddock, B. A. (1978). *FEMS Microbiol. Lett.* **4**, 37–40.

Orth, V., Chippaux, M., and Pascal, M. C. (1980). *J. Gen. Microbiol.* **177**, 257–262.

Owen, P., and Kaback, H. R. (1979a). *Biochemistry* **18**, 1413–1421.

Owen, P., and Kaback, H. R. (1979b). *Biochemistry* **18**, 1422–1426.

Poole, R. K., and Haddock, B. A. (1975). *Biochem. J.* **152**, 537–546.

Poole, R. K., Waring, A. J., and Chance, B. (1979). *Biochem. J.* **184**, 379–389.

Reid, G. A., and Ingledew, W. J. (1979). *Biochem. J.* **182**, 465–472.

Reid, G. A., and Ingledew, W. J. (1980). *FEBS Lett.* **109**, 1–4.

Ruch, F. E., Kuritzkes, D. R., and Lin, E. C. C. (1979). *Biochem. Biophys. Res. Commun.* **91**, 1365–1370.

Scott, R. H., and DeMoss, J. A. (1976). *J. Bacteriol.* **126**, 478–486.

Scott, R. H., Sperl, G. T., and DeMoss, J. A. (1979). *J. Bacteriol.* **137**, 719–726.

Spencer, M. E., and Guest, J. R. (1973). *J. Bacteriol.* **114**, 563–570.

Sperl, G. T., and DeMoss, J. A. (1975). *J. Bacteriol.* **122**, 1230–1238.

Yamamoto, I., and Ishimoto, M. (1978). *J. Biochem. (Tokyo)* **84**, 673–679.

Young, I. G., and Wallace, B. J. (1976). *Biochim. Biophys. Acta* **449**, 376–385.

Young, I. G., Jaworowski, A., and Poulis, M. (1978). *Gene* **4**, 25–36.

Zahl, K. J., Rose, C., and Hanson, R. L. (1978). *Arch. Biochem. Biophys.* **190**, 598–602.

59

The N,N′-Dicyclohexylcarbodiimide-Sensitive ATPase in Streptococcus faecalis Membranes

Adolph Abrams and Richard M. Leimgruber

1. INTRODUCTION

Much has been learned in recent years about the structure of the proton-translocating AT-Pases (H^+-ATPase) in mitochondria, chloroplasts, and bacteria, although the molecular basis for its action remains obscure. This membrane-associated enzyme may contain up to 20 polypeptide chains of 7 to 10 kinds and is one of the most complicated multisubunit complexes in nature. Hopefully some useful insights may emerge by comparing various structural features of the enzyme obtained from widely different sources. Although H^+-ATPases are reversible, there could be some fundamental differences, particularly at the level of the subunit structure, between an H^+-ATPase that functions physiologically mainly as an ATP synthetase, as in mitochondria, chloroplasts, and aerobic bacteria, and the H^+-ATPase that operates only in the hydrolytic direction, as in *Streptococcus faecalis*. With comparative aspects in mind, we will survey in this review the current status of the subunit structure of the N,N′-dicyclohexylcarbodiimide (DCCD)-sensitive ATPase in *S. faecalis*. As far as we are aware, *S. faecalis* is the only strictly fermentative organism whose H^+-ATPase has been studied in considerable detail. This organism is a homolactic fermenter and lacks a respiratory chain. Consequently, ATP generated glycolytically is the sole energy source and the H^+-ATPase is used solely for the purpose of coupling ATP hydrolysis to solute transport (Harold *et al.*, 1969; Abrams *et al.*, 1972; Harold and Spitz, 1975). The ATPase is firmly associated with the plasma membrane ghosts that are formed when protoplasts are subjected to osmotic or metabolic lysis (Abrams, 1965). However, washing the membranes with low-ionic-strength buffers causes release of the ATPase in a water-soluble form provided that multivalent cations are absent (Abrams, 1965). Complete reat-

Adolph Abrams and Richard M. Leimgruber • Department of Biochemistry/Biophysics/Genetics, University of Colorado School of Medicine, Denver, Colorado 80262.

tachment of the soluble ATPase to depleted membranes can occur when Mg^{2+} ions are added (Abrams and Baron, 1968). These effects of ionic strength and Mg^{2+} suggest that both electrostatic interactions and hydrophobic forces are involved in maintaining the stability of the ATPase–membrane complex (Abrams and Smith, 1974). Notably, a similar low-salt wash procedure is effective for solubilizing the ATPase in a variety of bacterial membranes (Abrams and Smith, 1974; Downie *et al.*, 1979), but for some unknown reason it seems to be ineffective for the eukaryotic H^+-ATPase.

The native ATPase–membrane complex in *S. faecalis,* and the reconstituted complex as well, is inhibited by DCCD, an energy-transfer inhibitor discovered by Beechey *et al.* (1967). The soluble ATPase, however, is insensitive to this reagent (Harold *et al.*, 1969; Abrams *et al.*, 1972). By the use of a DCCD-resistant mutant of *S. faecalis* (SFdcc-8), it was possible to prove unequivocally that DCCD reacts selectively with a membrane component rather than by a direct action on the attached ATPase. The evidence was based on *in vitro* genetic hybridization experiments using components obtained from the wild-type and the DCCD-resistant strain (Abrams *et al.*, 1972). These findings demonstrated that the DCCD-sensitive ATPase complex in *S. faecalis* consists of two sectors, a water-soluble catalytic sector (termed SF_1 or SF_1-ATPase) and an insoluble membrane sector (termed SF_0), the latter containing the DCCD-reactive site. The association between the two sectors to form the SF_1F_0 complex confers sensitivity of the ATPase to DCCD in some indirect way. The features of DCCD inhibition observed in *S. faecalis* are typical of all H^+-ATPases (Simoni and Postma, 1975). Interestingly, the antibiotic oligomycin, like DCCD, inhibits the ATPase in the F_1F_0 complex in eukaryotic organelles by acting on the F_0 sector. It is important to note oligomycin does not affect the F_1F_0 in *S. faecalis* or other prokaryotes. The reason for this intriguing difference between eukaryotic and bacterial F_1F_0 is not known. DCCD is known to block the proton channel in the F_0, and as a consequence it inhibits energy-coupled ATP hydrolysis or synthesis (Simoni and Postma, 1975). Hence, a DCCD-sensitive ATPase is considered to be a proton-translocating ATPase. The molecular basis of DCCD inhibition will be discussed below.

2. PURIFICATION AND CHARACTERIZATION OF THE DCCD-SENSITIVE ATPase IN S. faecalis

The F_1F_0 complex from wild-type *S. faecalis* (ATCC 9790) and a DCCD-resistant strain has been isolated by extraction of a reconstituted ATPase–membrane preparation with deoxycholate (Leimgruber *et al.*, 1981). The SF_1F_0 was purified by $(NH_4)_2SO_4$ fractionation and polyacrylamide gel electrophoresis (Fig. 1). We used electrophoretic homogeneity as the criterion of purity of the F_1F_0 complex, while others have employed immunoprecipitation (Todd *et al.*, 1980), ion-exchange chromatography (Sone *et al.*, 1975), and zonal sedimentation (Foster and Fillingame, 1979).

To purify as well as characterize the SF_1F_0 complex, we employed a two-dimensional polyacrylamide gel electrophoretic procedure utilizing 2.5–27% acrylamide gradient gels in both dimensions. Electrophoresis in the first dimension under nondenaturing conditions permitted the isolation of a catalytically active form of SF_1F_0 as a single band that was well resolved from various impurities including several proteins, phospholipid, and a carbohydrate-containing substance we believe to be lipoteichoic acid (a major component of gram-positive bacterial membranes). Also observed was a catalytically active δ-less form of SF_1, which detached from the SF_1F_0 as a consequence of the loss of the δ subunit. Analysis of the purified SF_1F_0 by SDS-PAGE in the second dimension showed the presence

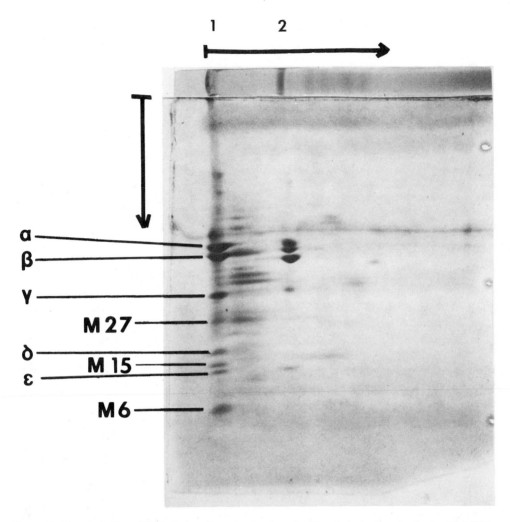

Figure 1. Characterization of SF₁F₀ by two-dimensional polyacrylamide gel electrophoresis. Semipurified SF₁F₀ was electrophoresed in the first-dimension gel under nondissociating conditions in 3 mM Tris, 70 mM glycine, 2 mM MgSO₄, pH 7.5. After incubation of the first-dimension gel in 2% (w/v) SDS, 2% (v/v) 2-mercaptoethanol, 10 mM sodium phosphate (pH 7.0), SDS-PAGE was performed in the second dimension. Electrophoresis in each dimension was performed at 25 V at 18°C for about 20 hr. Note that in the first dimension the electrophoretically purified SF₁F₀, lane 1, resolves from a δ-less form of SF₁, lane 2. The SF₁-sector subunits are labeled αβγδε, and the SF₀-sector subunits are labeled M27, M15, and M6. As the Coomassie blue staining intensities of M6 (molecular weight 6000) and the γ subunit (molecular weight 35,000) are nearly equivalent, there may be as many as six copies of M6 in the SF₁F₀ complex.

of the five subunits of the SF₁ sector, αβγδε (Abrams *et al.*, 1976a) and three additional proteins derived from the membrane sector, SF₀ (Fig. 1). The SF₀ subunits have molecular weights of 27,000, 15,000, and 6000, and have been designated M27, M15, and M6, respectively, to denote their membrane sector origin and their approximate size (Fig. 1). Two of the SF₀ proteins, M15 and M6, are tightly associated with SF₁F₀ as seen by the fact that they are confined to a single electrophoretic track. The M27 protein, however, appears to be a loosely bound component of the SF₁F₀ as it is not confined to a single track. The subunit compositions of SF₁F₀ isolated and purified from the wild-type and the

Table I. Subunit Composition of the Bacterial Membrane Sector, F_o, of F_1F_o

Source	Solubilizing detergent	F_0 subunit composition $(M_r \times 10^{-3})$	Reference
E. coli	Deoxycholate	29, 9[a]	Hare (1975)
E. coli[b]	Deoxycholate	26, 15, 8.4[a]	Foster and Fillingame (1979)
E. coli	Aminoxid WS-35	28, 19, 8.5[a]	Friedl et al. (1979)
PS3	Triton X-100	19, 13.5, 5.4[a]	Sone et al. (1975)
M. phlei	Triton X-100	24, 18, 8	Cohen et al. (1978)
S. faecalis	Triton X-100	16, 9.5	Babakov and Vasilov (1979)
S. faecalis[b,c]	Deoxycholate	27, 15, 6[a]	Leimgruber et al. (1981)

[a] Identified as the DCCD-reactive protein.
[b] F_1F_o complex of a DCCD-resistant strain had the same components. The smallest F_o component did not react with [^{14}C]-DCCD.
[c] The 27,000-molecular-weight protein may be a loosely bound subunit of SF_1F_o.

DCCD-resistant strain are identical. The proteins of the streptococcal F_0 sector are very similar in size to those found in other bacterial systems (Table I).

The function of M27 and M15 is unknown, but M6 has been identified as the DCCD-reactive protein. After treatment of the SF_1F_0 complexes from the wild-type and the DCCD-resistant mutant strain with [^{14}C]-DCCD, only the wild-type SF_1F_0 complex was inhibited (80–90%) and only the wild-type M6 protein was labeled (Fig. 2). From the relative amount of M6 present in the complex, there may be as many as six copies of the M6 protein in SF_1F_0 (Fig. 1). Sebald et al. (1979) have found that there are multiple copies of the DCCD-reactive protein in mitochondria also. Babakov and Vasilov (1979) have characterized an F_1F_0 complex from another strain of S. faecalis and found only two components in the F_0 sector (Table I). The DCCD-reactive protein was not identified. As shown in Table I, the F_0's from certain bacterial sources have two types of subunits while others have three. Eukaryotic F_0's apparently consist of four or more subunits. At present, it seems that the F_1 sector from all sources is composed of five different proteins while the number of different F_0 components varies. However, it is possible that the method of preparation may affect the number of proteins found in the F_0 sector. In this connection it is known that the δ subunit is easily removed from the F_1-ATPase (Abrams et al., 1976a) and the M27 component in SF_0 tends to dissociate from the complex (Fig. 1).

The amino acid sequence of the DCCD-reactive protein has been reported for yeast and Neurospora mitochondria (Sebald et al., 1980) and in E. coli (Hoppe et al., 1980). In the mitochondria, DCCD reacts covalently with a single glutamyl carboxyl group, whereas in E. coli the target is an aspartyl residue. Presumably, the DCCD-reactive COOH group mediates energy-linked H^+ translocation, for the Asp is replaced by Gly in an uncoupled E. coli mutant and the mutant is DCCD resistant. By contrast, the spontaneous DCCD-resistant mutant of S. faecalis, SFdcc-8, is not defective in energy coupling (Abrams et al., 1972). This behavior implies that a proton-translocating carboxyl group is present in SFdcc-8 and suggests that DCCD resistance is due to a genetic modification that lowers the accessibility or reactivity of the DCCD-reactive carboxyl group (Fig. 2).

3. THE F_1 SECTOR IN S. faecalis

The molecular weight of the S. faecalis F_1 sector, determined by analytical ultracentrifugation, is 385,000 (Schnebli et al., 1970). A subunit stoichiometry of $\alpha_3\beta_3\gamma\delta\epsilon$ is

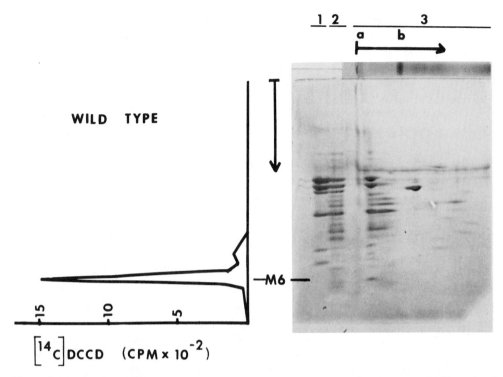

Figure 2. Identification of M6 as the DCCD-reactive protein in the SF_1F_0 complex. Semipurified SF_1F_0 isolated from the wild-type strain of *S. faecalis* was reacted with [^{14}C]-DCCD and was then analyzed by two-dimensional electrophoresis as described in Fig. 1. Only the M6 component was significantly labeled. Under identical conditions, the SF_1F_0 complex of the DCCD-resistant mutant exhibited an identical subunit composition but there was no detectable ^{14}C label in M6 or in any of the other proteins (data not shown). Reference marker proteins are shown on the left side of the slab gel: lane 1, SF_1; lane 2, semipurified SF_1F_0. Lane 3a of the two-dimensional analysis is electrophoretically purified SF_1F_0 and lane 3b is trace amounts of SF_1 and a contaminating protein.

consistent with the molecular weight of the native molecule and the values of the subunits as determined by SDS-PAGE, namely $\alpha = 55,000$, $\beta = 50,000$, $\gamma = 35,000$, $\delta = 20,000$, and $\epsilon = 12,000$ (Abrams *et al.*, 1976a,b). A value of about 385,000 has also been found for the F_1-ATPase from such widely diverse sources as liver mitochondria (Catterall *et al.* 1973), yeast mitochondria (Todd *et al.*, 1980), chloroplasts, and thermophilic bacterium PS3 (Yoshida *et al.*, 1979). Moreover, in all these cases and in *E. coli* as well (Bragg and Hou, 1980), the subunit stoichiometry seems to be $\alpha_3\beta_3\gamma\delta\epsilon$.

Electron microscopy of the F_1-ATPase from *S. faecalis* and other sources revealed a hexagonal array of six globules with an overall diameter of approximately 100 Å (Abrams and Smith, 1974). This suggested the possibility of an alternating array of the major subunits, α and β, but cross-linking experiments with the *E. coli* F_1-ATPase seem to exclude such an arrangement (Bragg and Hou, 1980).

All F_1-ATPases isolated have been found to contain one or a few molecules of tightly bound ADP and ATP. These so-called endogenous tightly bound nucleotides in *S. faecalis* (Abrams, 1976; Abrams *et al.*, 1973) and in *E. coli* (Maeda *et al.*, 1977) do not turn over either *in vitro* or in metabolizing intact cells. Therefore, it seems unlikely that they are involved in energy-coupled phosphorylation or hydrolysis. Their physiological significance remains unknown. Speculatively, they may play a role in regulation, structural stabilization, or bioassembly of the ATPase complex.

4. BINDING OF F_1 TO F_0: ROLE OF δ, α, AND Mg^{2+}

Considerable attention has been given to the factors involved in maintaining the association between the F_1-ATPase and the membrane sector in *S. faecalis* (see introduction). In *S. faecalis* the need for Mg^{2+} in reattachment of F_1 to depleted membranes is related to the function of the δ subunit in the F_1-ATPase. That there is a protein attachment factor in the *S. faecalis* F_1-ATPase was first realized when it was found that gel filtration of the enzyme in the absence of Mg^{2+} yielded a form of the enzyme that would not attach to depleted membranes (Baron and Abrams, 1971). A fraction eluted from the gel filtration column contained a factor that restored the capacity of the deficient F_1-ATPase to reattach. The factor was called nectin (Baron and Abrams, 1971; Abrams and Smith, 1974; Abrams *et al.*, 1974). In subsequent studies (Abrams *et al.*, 1976a), it became apparent that nectin corresponds to the δ subunit, which was identified as an attachment factor in other F_1-ATPases (Senior, 1979a). The removal of the δ subunit from SF_1-ATPase is readily accomplished by preparative polyacrylamide gel electrophoresis of the native SF_1-ATPase in the absence of Mg^{2+}. The resultant δ-less F_1-ATPase fails to reattach to depleted membranes even in the presence of Mg^{2+} (Abrams *et al.*, 1976a,b). However, if Mg^{2+} is present in the gel during preparative electrophoresis, a five-subunit F_1 is obtained that reattaches normally to depleted membranes. As both δ and Mg^{2+} are required for membrane attachment, the subunit structure of SF_1-ATPase was formulated as $\alpha_3\beta_3\gamma\epsilon(Mg^{2+})_n\delta$ (Abrams *et al.*, 1976a). This speculative formulation implies that Mg^{2+} ligands form an ionic bridge between δ and the remainder of the enzyme. In this connection it is interesting that Senior (1979b) has reported that Mg^{2+} is an integral part of the purified mitochondrial F_1-ATPase, although the role of the bound Mg^{2+} was not established. Except for the *S. faecalis* system, there have been no reports that Mg^{2+} serves to stabilize the association between δ and the rest of the F_1-ATPase. Therefore, the *S. faecalis* F_1-ATPase may be unique in this respect.

There is some evidence that the α chains as well as the δ subunit are involved in attachment of the SF_1-ATPase to F_0 (Abrams *et al.*, 1976b; Leimgruber *et al.*, 1978). Limited proteolysis of SF_1 by chymotrypsin selectively cleaves a short segment of about 2000 molecular weight from each α chain (molecular weight 55,000) with no apparent effect on the other subunits. The altered enzyme remains fully active but fails to bind to depleted membranes even in the presence of Mg^{2+}, suggesting that the three α-chain "tails" are involved in membrane attachment. Support for this notion was obtained by showing that the α tails are unaffected by chymotrypsin treatment of reconstituted F_1-ATPase–membrane complex (Leimgruber *et al.*, 1978). A reasonable explanation for this finding is that the α-chain tails are "buried" in the F_0 sector. Using diazobenzene[^{35}S]sulfonate, Ludwig *et al.* (1980) have also found that the binding of F_1 to F_0 shields the α subunits in mitochondrial $F_1 F_0$. The remarkable susceptibility of the terminal region of the α chains in SF_1 to protease attack has also been observed in the F_1-ATPase in *Mycobacterium phlei* (Ritz-Gold *et al.*, 1979) and *E. coli* (Dunn *et al.*, 1980) and thus may be a universal feature of prokaryotic ATPases. Notably, in each case the enzyme remains active but fails to rebind to the membrane sector. The studies of the effect of partial proteolysis represent the first attempts to relate primary structure to function in H^+-ATPase. Dunn *et al.* (1980) determined that the chymotryptic cleavage product of the *E. coli* α chain consists of 19 amino acids and originates from the NH_2-terminal end. Its amino acid sequence was also determined. As yet it is not known whether the α chains in eukaryotic F_1-ATPase also have a protease-sensitive tail.

ACKNOWLEDGMENT. This work was supported by Grant 05810 from the National Institutes of Health.

REFERENCES

Abrams, A. (1965). *J. Biol. Chem.* **240**, 3675–3681.

Abrams, A. (1976). In *The Enzymes of Biological Membranes* (A. Martonosi, ed.), Vol. 3, pp. 57–73, Plenum Press, New York.

Abrams, A., and Baron, C. (1968). *Biochemistry* **7**, 501–506.

Abrams, A., and Smith, J. B. (1974). In *The Enzymes* (P. D. Boyer, ed.), Vol. X, 3rd ed., pp. 395–429, Academic Press, New York.

Abrams, A., Smith, J. B., and Baron, C. (1972). *J. Biol. Chem.* **247**, 1484–1488.

Abrams, A., Nolan, E. A., Jensen, C., and Smith, J. B. (1973). *Biochem. Biophys. Res. Commun.* **55**, 22–29.

Abrams, A., Baron, C., and Schnebli, H. (1974). In *Methods in Enzymology* (S. Fleischer and L. Packer, eds.), Vol. 32, Part B, pp. 428–439, Academic Press, New York.

Abrams, A., Jensen, C., and Morris, D. (1976a). *Biochem. Biophys. Res. Commun.* **69**, 804–811.

Abrams, A., Morris, D., and Jensen, C. (1976b). *Biochemistry* **15**, 5560–5566.

Babakov, A. V., and Vasilov, R. G. (1979). *Bio-Organic Chem.* **5**, 119–125.

Baron, D., and Abrams, A. (1971). *J. Biol. Chem.* **246**, 1542–1544.

Beechey, R. B., Roberton, A. M., Holloway, C. T., and Knight, I. G. (1967). *Biochemistry* **6**, 3867–3879.

Bragg, P. D., and Hou, C. (1980). *Eur. J. Biochem.* **106**, 495–503.

Catterall, W. A., Coty, A., and Pedersen, P. D. (1973). *J. Biol. Chem.* **248**, 7427–7431.

Cohen, N. S., Lee, S. H., and Brodie, A. F. (1978). *J. Supramol. Struct.* **8**, 111–117.

Downie, J. A., Gibson, F., and Cox, G. B. (1979). *Annu. Rev. Biochem.* **48**, 103–131.

Dunn, S. D., Heppel, L. A., and Fullmer, C. S. (1980). *J. Biol. Chem.* **255**, 6891–6896.

Foster, D. L., and Fillingame, R. H. (1979). *J. Biol. Chem.* **254**, 8230–8236.

Friedl, P., Friedl, C., and Schairer, H. U. (1979). *Eur. J. Biochem.* **100**, 175–180.

Hare, J. F. (1975). *Biochem. Biophys. Res. Commun.* **66**, 1329–1337.

Harold, F. M., and Spitz, E. (1975). *J. Bacteriol.* **122**, 266–277.

Harold, F. M., Baarda, J. R., Baron, C., and Abrams, A. (1969). *J. Biol. Chem.* **244**, 2261–2268.

Hoppe, J. Schairer, H. U., and Sebald W. (1980). *FEBS Lett.* **109**, 107–111.

Leimgruber, R. M., Jensen, C., and Abrams, A. (1981). *J. Bacteriol.* **147**, 363–372.

Leimgruber, R. M., Jensen, C., and Abrams, A. (1978). *Biochem. Biophys. Res. Commun.* **81**, 439–447.

Ludwig, B., Prochaska, L., and Capaldi, R. A. (1980). *Biochemistry* **19**, 1516–1523.

Maeda, M., Kobayashi, H., Futai, M., and Anraku, Y. (1977). *J. Biochem.* **82**, 311–314.

Ritz-Gold, C. J., Gold, C. M., and Brodie, A. F. (1979). *Biochim. Biophys. Acta* **547**, 1–17.

Schnebli, H., Vatter, A. E., and Abrams, A. (1970). *J. Biol. Chem.* **245**, 1122–1127.

Sebald, W., Graf, T., and Luckins, H. B. (1979). *Eur. J. Biochem.* **93**, 587–599.

Sebald, W., Machleidt, W., and Wachter, E. (1980). *Proc. Natl. Acad. Sci. USA* **77**, 785–789.

Senior, A. E. (1979a). In *Membrane Proteins in Energy Transduction* (R. A. Capaldi, ed.), pp. 233–278, Dekker, New York.

Senior, A. E. (1979b). *J. Biol. Chem.* **254**, 11319–11323.

Simoni, P. D., and Postma, P. W. (1975). *Annu. Rev. Biochem.* **44**, 523–554.

Sone, N., Yoshida, M., Hirata, H., and Kagawa, Y. (1975). *J. Biol. Chem.* **250**, 7917–7923.

Todd, R. D., Griesenbeck, T. A., and Douglas, M. G. (1980). *J. Biol. Chem.* **255**, 5461–5467.

Yoshida, M., Sone, N., Hirata, H., Kagawa, Y., and Ui, Y. (1979). *J. Biol. Chem.* **254**, 9525–9533.

60

Energy-Transducing Coupling Factor F_0F_1 from Mycobacterium phlei

Arnold F. Brodie and Vijay K. Kalra

1. INTRODUCTION

An understanding of the molecular mechanism(s) of the energy transduction for the synthesis of ATP and active transport of solute requires a basic knowledge of the component parts necessary to carry out these functions and is, thus, automatically concerned with the little-understood area where chemical function is intimately involved with physical structure at the molecular level of membrane organization. Methods used to study biochemical processes depend on the resolution of the process into component parts. Solubilized coupling factors were first resolved from *Mycobacterium phlei* and described as protein component(s) necessary for the restoration of phosphorylation (Brodie and Gray, 1956; Brodie, 1959); however, the latent ATPase associated with the coupling factor was not recognized.

A number of different types of membrane structures and membrane-associated components have been resolved from whole cells of *M. phlei* (Brodie *et al.*, 1979). These membrane structures have been characterized with regard to the nature of the respiratory components (Asano and Brodie, 1964), the sequence of electron transport carriers, sites of phosphorylation, membrane orientation, and active transport of metabolites (Brodie and Adelson, 1965; Brodie *et al.*, 1972). Sonic treatment of whole cells results in the formation of membrane vesicles that are mostly oriented inside-out. These membrane structures are capable of carrying out oxidative phosphorylation. However, centrifugation of these membranes in the absence of ions yields membranes that are capable of oxidation with succinate and NAD^+-linked substrates but devoid of coupled phosphorylation (Higashi *et al.*, 1969).

Restoration of phosphorylation occurs on addition of the resolved coupling-factor-latent ATPase (BCF_1) to the depleted membranes. Similar procedures, involving washing with low-ionic-strength buffer in the presence of EDTA, have been used to solubilize the

Arnold F. Brodie and Vijay K. Kalra • Department of Biochemistry, University of Southern California School of Medicine, Los Angeles, California 90033.

membrane-bound ATPase from a number of bacterial species (recent reviews: Downie *et al.*, 1979; Futai and Kanazawa, 1980).

The membrane-bound coupling factor ATPase from *M. phlei* has been shown to elicit ATPase activity upon treatment with trypsin (Higashi *et al.*, 1975).

2. SUBUNIT COMPOSITION AND STOICHIOMETRY OF BCF$_1$

The solubilized BCF$_1$ was purified to homogeneity in a single step by affinity chromatography on Sepharose coupled to ADP (Kalra *et al.*, 1975). Purified latent ATPase exhibited coupling factor activity and latent ATPase activity when unmasked by trypsin (Higashi *et al.*, 1975). The molecular weight of purified ATPase was determined to be $404,000 \pm 6\%$. It has five subunits: α, 64,000; β, 53,000; γ, 33,000; δ, 14,000; and ϵ, 8000 (Ritz and Brodie, 1977). Subunit stoichiometry revealed a ratio of 3 : 3 : 1 for α, β, and γ subunits and a probable ratio of 1 : 1 for δ and ϵ subunits (Ritz and Brodie, 1977). The exact number of δ and ϵ subunits are difficult to determine because these subunits dissociate from the enzyme rather easily.

3. PROTEOLYTIC DIGESTION OF BCF$_1$

Trypsin treatment of solubilized BCF$_1$ stimulates ATPase activity and inactivates its capacity for coupled phosphorylation. In contrast, trypsin treatment of membrane vesicles unmasks ATPase activity without loss of coupled phosphorylation (Bogin *et al.*, 1970). Trypsin treatment of BCF$_1$ that results in maximal stimulation of ATPase activity occurs following a molecular weight decrease of about 20,000, concomitant with a complete loss of its ability to rebind to BCF$_1$ depleted membranes (Ritz and Brodie, 1977; Ritz-Gold and Brodie, 1979). Analysis of the subunit composition revealed that the native α subunit (molecular weight 64,000) is simultaneously converted to an α' intermediate (molecular weight 61,000) upon trypsin treatment. Limited proteolysis of BCF$_1$ (Ritz and Brodie, 1977) was carried out in an attempt to understand its affect on ATPase activity, rebinding ability, and coupling factor activity. During initial stages of activation of ATPase (24%) by trypsin treatment, the loss of coupling factor activity (16% decrease) did not parallel the loss of rebinding activity (5%). The major structural change obtained at this stage was the conversion of the α subunit to α'. At a later stage of ATPase activation (87%) by trypsin, there was a large decrease in both coupling factor activity (93%) and rebinding. The concomitant major structural change was conversion of α' to the stable α'' species (molecular weight 58,000).

As coupling factor activity is dependent upon the ability of the molecule to rebind, one can suggest from these data that the observed structural change in the α subunit induces a conformational change with the result that the enzyme cannot rebind to the membrane. It is also possible that a subtle conformational change responsible for loss of binding might have occurred in other subunits that is not apparent. Theoretical modeling studies (Ritz-Gold *et al.*, 1979) were carried out in an effort to obtain information concerning the intermediate structural states leading to the fully activated ATPase ($\alpha_3^{||}\beta_3\gamma\delta\epsilon$) state under different conditions of trypsin treatment. The theoretical models of structure–function relationships of BCF$_1$ that best represented the experimental data predicted that the native BCF$_1$ molecule contains three copies of the α (64,000) form of the alpha subunit, that the α'' (58,000) species contributes maximally, and the α' (61,000) form about half-maximally to ATPase activity. Also, the membrane rebinding ability is proportional to the number of native alpha subunits in BCF_1, and at least one native α subunit molecule is required for full expression of coupled phosphorylation. Hanson and Kennedy (1973) have isolated from *E. coli* an F_1 that contains four subunits, the δ subunit being absent. This four-subunit enzyme was unable to reconstitute ATP-driven transhydrogenase activity in membranes depleted of F_1, whereas the F_1 enzyme isolated by Bragg *et al.* (1973), which contained the δ subunit, was able to bind to membrane and restore the activity. These studies as well as others suggest that the δ subunit is responsible for attachment to the membrane F_0 complex.

4. BINDING AND AFFINITY LABELING OF NUCLEOTIDES

Tightly bound nucleotides have been observed in highly purified BCF_1 preparations. The amount of ADP and ATP bound per mole of the enzyme was 24 and 26 mmoles, respectively (Lee *et al.*, 1977), which is comparatively small compared to the ATPase from *S. faecalis* (Abrams *et al.*, 1973) and beef heart mitochondria (Garrett and Penefsky, 1975). Equilibrium binding studies show that there are 3 moles of ADP bound per mole of BCF_1 enzyme with an apparent equilibrium constant of 68 μM. AMP-P(NH)P, adenyl-5'-yl-imidodiphosphate, a nonhydrolyzable analog of ATP, exhibited one binding site per molecule of the enzyme. This analog inhibited ATPase activity but did not affect oxidative phosphorylation in membrane vesicles of *M. phlei*. The binding of ADP decreased about one-third, while AMP-P(NH)P binding remained unchanged in trypsin-treated BCF_1. The binding of AMP-P(NH)P to BCF_1 was not displaced by ADP, suggesting that ATP and ADP do not share a common binding site.

Studies have been carried out to delineate the site(s) of nucleotide binding in the ATPase molecule (McCarty, 1978) in order to identify the site(s) involved in ATP synthesis and ATP hydrolysis. A photoaffinity analog of ATP—8-azidoadenosine-5'-triphosphate (8-azido-ATP)—has been shown to bind to the β subunit of ATPase from *E. coli, Micrococcus* sp., and beef heart mitochondrial F_1. However, a photoaffinity probe of the ADP-binding site, *N*-4-azido-2-nitrophenyl-aminobutyryl-2'-ADP, labeled both the α and the β subunit of mitochondrial F_1 (Lunardi *et al.*, 1977). 8-Azido-ATP has been found to be a competitive inhibitor for ATP in latent ATPase of *M. phlei* but failed to bind to ATPase after photoactivation (Kumar *et al.*, 1979). The 2',3'-dialdehyde derivative of ATP (dial-ATP) has been shown to be an affinity label for the ATP-binding site of BCF_1 from *M. phlei* (Kumar *et al.*, 1979). This analog of ATP caused inactivation of both latent and unmasked ATPase. Stoichiometric studies of the inactivation process indicated one ATP binding site per molecule of purified ATPase and the binding occurred on the α subunit. Although BCF_1 has three α subunits, the binding of only 1 mole of dial-ATP per mole of enzyme is sufficient to block ATPase activity. One explanation is that the interactions between the ATP-binding site(s) in the ATPase show strong negative cooperativity such that only one of the three subunits can bind ATP at a time; this leads to the concept that three α subunits are functionally asymmetric in nature. The other possibility is that the binding of ATP to one α subunit reduces the affinity of other subunits for ATP. Support for the functional asymmetry of the α subunits in BCF_1 from *M. phlei* has been documented using theoretical computer modeling studies (Ritz-Gold *et al.*, 1979).

Affinity labeling of the ADP-binding sites with dial-ADP revealed that the ADP-binding site was present on both α and β subunits of BCF_1 (Kumar *et al.*, 1979). The binding of dial-ADP has been shown to be reduced in both α and β subunits after trypsin treatment of the enzyme. However, it is not clear at the present time whether the α or the β subunit, or both, is involved in the phosphorylation of ADP to ATP.

5. PURIFICATION AND PROPERTIES OF F_0F_1 COMPLEX

N,N'-dicyclohexylcarbodiimide (DCCD) has been shown to inhibit ATPase activity when BCF_1 was bound to the membrane. It did not inhibit the solubilized BCF_1-ATPase activity (Kalra and Brodie, 1971) as has been observed in other bacterial membranes, chloroplasts, or mitochondria. The BCF_0-BCF_1 complex has been solubilized from membranes of *M. phlei* by cholate treatment followed by solubilization with 2% Triton X-100 (Lee *et al.*, 1976). Further purification by affinity chromatography on Sepharose-ADP yielded a homogeneous BCF_0-BCF_1 complex. This complex exhibited ATPase activity that was sensitive to DCCD. It also restored oxidative phosphorylation in Triton X-100-treated membranes, devoid of BCF_0-BCF_1 complex. Coupled phosphorylation was inhibited by DCCD in reconstituted membranes. Vesicles reconstituted from DCCD-sensitive ATPase complex (TF_0F_1) and phospholipids were able to synthesize ATP from ADP and P_i with energy from an electrochemical gradient formed by a pH gradient and membrane potential across the membrane (Sone *et al.*, 1977). The F_0 portion from BCF_1-BCF_0 complex has been resolved by chromatography on Sepharose-ADP. Whereas F_1F_0 remained bound when the elution buffer contained 0.15 M KCl, elution with the same buffer lacking KCl resulted in elution of F_0 (Cohen *et al.*, 1978). Because of the hydrophobic nature of F_0, it exhibited anomalous behavior on gel electrophoresis as well as on gel filtration. BCF_0 is a lipoprotein complex with a molecular weight of 60,000 and appears to contain three polypeptides with approximate molecular weights of 24,000, 18,000, and 8000. The polypeptide molecular weights are somewhat higher than those reported by Sone *et al.* (1975) for thermophilic bacterium PS3, TF_0 (19,000, 13,500, and 5400). Further purification of TF_0 revealed that TF_1-binding protein (13,500) and DCCD-binding protein (5400) could mediate proton translocation in liposomes, suggesting that these two subunits are functionally sufficient as F_0 subunits (Okamoto *et al.*, 1977).

In conclusion, BCF_0-BCF_1 complex of *M. phlei* exhibits ATPase activity, coupling factor activity, and DCCD-sensitive proton-forming channel. Subunit composition of resolved BCF_1 from BCF_1F_0 revealed that it has five subunits. The α subunit is involved in the binding of ATP and plays a role in the hydrolysis of ATP. There are three α subunits and 1 molecule of ATP binds per molecule of the enzyme, suggesting that α subunits are functionally asymmetric. Both α and β subunits bind ADP. It is not clear at the present time which is specifically involved in coupled phosphorylation. Studies of purified F_0 revealed that it has three subunits. This portion is involved in the binding of DCCD as has been observed in ATPase from other systems. Studies have shown that the F_0 portion of BCF_1F_0-ATPase is involved in the passage of protons. However, we do not know the molecular mechanism of how proton-translocating ATPases utilize the electrochemical proton gradient to synthesize ATP. Further studies on the function and assembly of each subunit of the BCF_1F_0 complex should provide an insight into the molecular aspects of ATP synthesis and oxidative phosphorylation.

REFERENCES

Abrams, A., Nolan, E. A., Jensen, C., and Smith, J. F. (1973). *Biochem. Biophys. Res. Commun.* **55**, 22–29.
Asano, A., and Brodie, A. F. (1964). *J. Biol. Chem.* **240**, 4002–4010.

Bogin, E., Higashi, T., and Brodie, A. F. (1970). *Arch. Biochem. Biophys.* **136,** 337–351.

Bragg, P. D., Davies, P. L., and Hou, C. (1973). *Arch. Biochem. Biophys.* **167,** 311–321.

Brodie, A. F. (1959). *J. Biol. Chem.* **234,** 398–404.

Brodie, A. F., and Adelson, J. (1965). *Science* **149,** 265–269.

Brodie, A. F., and Gray, C. T. (1956). *Biochim. Biophys. Acta* **19,** 384–386.

Brodie, A. F., Hirata, H., Asano, A., Cohen, N. S., Hinds, T. R., Aithal, H. N., and Kalra, V. K. (1972). In *Membrane Research* (C. F. Fox, ed.), pp. 445–472, Academic Press, New York.

Brodie, A. F., Kalra, V. K., Lee, S. H., and Cohen, N. S. (1979). *Methods Enzymol.* **LV,** 175–200.

Cohen, N. S., Lee, S. H., and Brodie, A. F. (1978). *J. Supramol. Struct.* **8,** 111–117.

Downie, J. A., Gibson, F., and Cox, G. B. (1979). *Annu. Rev. Biochem.* **48,** 103–131.

Futai, M., and Kanazawa, H. (1980). *Curr. Top. Bioenerg.* **10,** 181–215.

Garrett, N. E., and Penefsky, H. S. (1975). *J. Biol. Chem.* **250,** 6640–6647.

Hanson, R. L., and Kennedy, E. P. (1973). *J. Bacteriol.* **114,** 772–781.

Higashi, T., Bogin, E., and Brodie, A. F. (1969). *J. Biol. Chem.* **244,** 500–502.

Higashi, T., Kalra, V. K., Lee, S. H., Bogin, E., and Brodie, A. F. (1975). *J. Biol. Chem.* **250,** 6541–6548.

Kalra, V. K., and Brodie, A. F. (1971). *Arch. Biochem. Biophys.* **147,** 653–659.

Kalra, V. K., Lee, S. H., Ritz, C. J., and Brodie, A. F. (1975). *J. Supramol. Struc.* **3,** 231–241.

Kumar, G., Kalra, V. K., and Brodie, A. F. (1979). *J. Biol. Chem* **254,** 1964–1971.

Lee, S. H., Cohen, N. S., and Brodie, A. F. (1976). *Proc. Natl. Acad. Sci. USA* **73,** 3050–3053.

Lee, S. H., Kalra, V. K., Ritz, C. J., and Brodie, A. F. (1977). *J. Biol. Chem.* **252,** 1084–1091.

Lunardi, J., Lauquin, G. J. M., and Vignais, P. V. (1977). *FEBS Lett.* **80,** 317–323.

McCarty, R. E. (1978). *Curr. Top. Bioenerg.* **7,** 245–278.

Okamoto, H., Sone, N., Hirata, H., Yoshida, M., and Kagawa, Y. (1977). *J. Biol. Chem.* **252,** 6125–6131.

Ritz, C. J., and Brodie, A. F. (1977). *Biochem. Biophys. Res. Commun.* **75,** 933–939.

Ritz-Gold, C. J., and Brodie, A. F. (1979). *Biochim. Biophys. Acta* **547,** 18–26.

Ritz-Gold, C. J., Gold, C. M., and Brodie, A. F. (1979). *Biochim. Biophys. Acta* **547,** 1–17.

Sone, N., Yoshida, M., Okamato, H., Hirata, H., and Kagawa, Y. (1975). *J. Biol. Chem* **250,** 7917–7923.

Sone, N., Yoshida, M., Hirata, H., and Kagawa, Y. (1977). *J. Biol. Chem.* **252,** 2956–2960.

61

Biochemical Characterization of the Fungal Plasma Membrane ATPase, An Electrogenic Proton Pump

Carolyn W. Slayman

1. INTRODUCTION

From a variety of flux and electrophysiological studies, it has become clear that there is a rapid and physiologically important circulation of protons across the fungal plasma membrane (Eddy, 1978; Goffeau and Slayman, 1981). Protons are pumped outwards by a primary ATPase, generating both a pH gradient (up to 3 pH units) and a membrane potential (150 to 250 mV, inside negative), and return inwards by way of H^+-dependent cotransport systems for sugars, amino acids, and inorganic ions (Fig. 1). A major goal of research during the past few years has been to establish the molecular structure of the fungal plasma membrane ATPase. In particular, there has been reason to ask whether it belongs to the F_0F_1 class of mitochondrial, chloroplast, and bacterial ATPases (which are capable of ATP-driven proton translocation, even though they usually function in the opposite direction to synthesize ATP), or whether it is more closely related to the Na^+/K^+-ATPase, Ca^{2+}-ATPase and H^+/K^+-ATPase of animal cells (which act physiologically in the "transport" direction to pump cations across their respective membranes).

2. PROPERTIES OF THE MEMBRANE-BOUND ENZYME

Before the ATPase could be characterized biochemically, a method was needed to prepare plasma membranes in good yield and free of contamination by other cellular membranes. This proved to be a particularly difficult problem in the fungi, which are surrounded by a rigid polysaccharide cell wall and contain large numbers of membranous

Carolyn W. Slayman • Departments of Human Genetics and Physiology, Yale University School of Medicine, New Haven, Connecticut 06510.

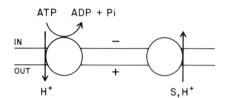

Figure 1. Model of the fungal plasma membrane, containing the H^+-translocating ATPase and a series of H^+-dependent co-transport systems.

organelles (especially mitochondria). In the past few years, however, several successful methods have been developed. (1) Starting with a cell-wall-less strain of *Neurospora*, Scarborough showed that the plasma membrane can be stabilized with concanavalin A (Scarborough, 1975; Stroobant and Scarborough, 1979). Hypotonic lysis of the coated cells yields large sheets of membrane that sediment at low speed; later, if desired, the concanavalin A can be removed by treatment with α-methylmannoside, and the membrane sheets form vesicles. (2) Alternatively, if untreated protoplasts are lysed gently in a medium designed to keep organelles intact, the plasma membrane breaks up into small fragments and vesiculates. After removal of mitochondria and other large organelles by centrifugation, plasma membrane vesicles can be collected at higher speed, identified by suitable markers, and if necessary, further purified by density-gradient centrifugation (Bowman *et al.*, 1981b). (3) Finally, intact cells can be disrupted mechanically, yielding a heterogeneous array of membrane and organelle fragments from which plasma membranes can be purified by several cycles of differential and density-gradient centrifugation (Fuhrman *et al.*, 1976; Delhez *et al.*, 1977; Dufour and Goffeau, 1978; Serrano, 1978; Peeters and Borst-Pauwels, 1979).

Membranes prepared in these ways from four species of fungi have been found to contain substantial ATPase activity, amounting to 0.4–1.8 μmoles/min · mg protein in *Saccharomyces cerevisiae* (Serrano, 1978; Willsky, 1979; Peeters and Borst-Pauwels, 1979), 4.0 in *Candida tropicalis* (F. Blasco *et al.*, 1981), 0.4–2.8 in the cell-wall-less strain of *Neurospora* (Scarborough, 1977; Bowman and Slayman, 1977), 3.5–8.0 in wild-type *Neurospora* (Bowman *et al.*, 1981b), and 5.6–25.7 in *Schizosaccharomyces pombe* (Delhez *et al.*, 1977; Dufour and Goffeau, 1978). What is especially striking is the qualitative similarity of the fungal ATPases to one another. In all cases, the pH optimum (5.0–6.7) is significantly below that of mitochondrial ATPase (8.0–8.6). Furthermore, the plasma membrane enzymes share a high degree of specificity for ATP as substrate, with GTP and ITP hydrolyzed less than 5% as well (in membrane preparations free from mitochrondrial contamination; Delhez *et al.*, 1977; Bowman *et al.*, 1981a,b). Among the divalent cations that have been tested, Mg^{2+}, Mn^{2+}, and Co^{2+} give appreciable activity at equimolar concentrations of cation and ATP. When the ratio of divalent cation to ATP is varied, complex patterns are obtained that suggest the existence of a second cation-binding site (Dufour and Goffeau, 1980; Borst-Pauwels and Peeters, 1981). Monovalent cations including K^+, Na^+, and NH_4^+ have only minor effects on ATPase activity, inconsistent with any direct role in the transport cycle (Bowman and Slayman, 1977, 1979). Finally, the plasma membrane ATPase is resistant to most classical inhibitors of mitochondrial ATPase, including oligomycin, venturidicin, and azide, but it is sensitive to vanadate, diethylstilbestrol, and N,N'-dicyclohexylcarbodiimide (DCCD). Thus, from the properties of the membrane-bound enzyme, even before purification, the picture has emerged of a distinctive ATPase that differs conspicuously from its mitochondrial counterpart.

3. PURIFICATION AND SUBUNIT COMPOSITION

Early attempts at purification made it clear that the fungal plasma membrane ATPase is an integral membrane protein, requiring detergents for solubilization and phospholipids for maximal activity. The ATPase has now been purified nearly to homogeneity from *Schizosaccharomyces pombe* with the use of lysolecithin (Dufour and Goffeau, 1978) and from *Neurospora* with the use of deoxycholate in the presence of glycerol (Bowman *et al.*, 1978a, 1981b), in both cases followed by centrifugation through a density gradient of sucrose or glycerol. The purified enzymes have specific activities of 35 and 90 μmoles/min $\cdot$ mg protein, respectively (assayed in the presence of added phospholipids); and they retain the kinetic properties of the membrane-bound state, with only minor changes in pH optimum, nucleotide and divalent cation specificity, and inhibitor sensitivity (Dufour and Goffeau, 1980; Bowman *et al.*, 1981b).

Upon SDS-polyacrylamide gel electrophoresis, the purified ATPases from *Schizosaccharomyces pombe* and *Neurospora* display a single polypeptide band of molecular weight approximately 100,000 (Dufour and Goffeau, 1978; Bowman *et al.*, 1978b, 1981b); a similar band was found to be enriched in a partially purified enzyme preparation from *Saccharomyces cerevisiae* (Malpartida and Serrano, 1980). The $M_r = 100,000$ polypeptide is nearly identical in electrophoretic mobility to the major subunits of Na^+/K^+-ATPase, Ca^{2+}-ATPase from sarcoplasmic reticulum, and H^+/K^+-ATPase from gastric mucosa. By contrast, it bears no resemblance to mitochondrial, chloroplast, and bacterial ATPases, which contain about 10 polypeptides ranging in molecular weight from 7000 to 60,000.

The quaternary structure of the purified ATPase has recently been investigated in the case of *Schizosaccharomyces pombe*. From measurements of phospholipid content (only 74 μmoles/mole polypeptide), sedimentation coefficient ($s_{20,w}^{0.784} = 24$), and Stokes radius (7.0–7.7 nm), Dufour and Goffeau (1980) concluded that the enzyme had been purified as an oligomeric complex of eight to ten $M_r = 100,000$ monomers. This form displays little if any enzymatic activity. Upon addition of suitable phospholipid, however, spontaneous reactivation takes place, with the oligomer disassociating and smaller molecular forms inserting into the phospholipid bilayer (Dufour and Goffeau, 1980; Dufour and Tsong, 1981). In at least one case (when microvesicles of dimyristoylphosphatidylcholine were used), maximal ATPase activity was observed at a ratio of one $M_r = 100,000$ monomer per microvesicle, indicating that a single polypeptide chain is sufficient for enzymatic function.

4. REACTION MECHANISM

Kinetic studies on both the membrane-bound and the purified forms of the ATPase have provided useful clues to the reaction mechanism, identifying MgATP as substrate (Bowman and Slayman, 1977; Scarborough, 1977; Delhez *et al.*, 1977; Dufour and Goffeau, 1980; Ahlers *et al.*, 1978; Borst-Pauwels and Peeters, 1981) and pointing to the possibility of additional binding sites for ATP (Bowman and Slayman, 1979), vanadate (Bowman and Slayman, 1979; Borst-Pauwels and Peeters, 1981), and divalent cations (Dufour and Goffeau, 1980; Borst-Pauwels and Peeters, 1981). The most significant finding has been that the ATPase reacts by way of a covalent phosphorylated intermediate. Following several earlier reports that the exposure of plasma membranes to [^{32}P]-ATP leads to the appearance of one or more phosphorylated bands in the high-molecular-weight region of SDS gels (Willsky, 1979; Malpartida and Serrano, 1980), a careful analysis of phosphorylation has been carried out by Dame and Scarborough (1980), using plasma membranes from the cell-wall-less strain of *Neurospora*, and by Amory *et al.* (1980), using plasma membranes and purified ATPase from *Schizosaccharomyces pombe*. In both cases, at millimolar concentrations of [^{32}P]-ATP, the $M_r = 100,000$ polypeptide was rapidly labeled, with a turnover time sufficient to account for the ATPase activity of the preparation and with a K_m very close to that of the ATPase. Once formed, the phosphorylated intermediate was acid stable, alkali labile, and hydroxylamine sensitive, suggesting that—as in the case of Na$^+$/K$^+$-, Ca^{2+}-, and H$^+$/K$^+$-ATPases—it is likely to be an acyl phosphate. The amino acid residue that is phosphorylated remains to be identified, as do the quantitative characteristics of the phosphorylation and dephosphorylation steps.

5. PROTON TRANSPORT

In flux studies on suspensions of fungal cells, there have been numerous reports of net proton ejection that is inhibited by compounds such as DCCD, Dio-9, and diethylstilbestrol and is presumably mediated by the plasma membrane ATPase (reviewed by Goffeau and Slayman, 1981). More persuasive evidence has come recently from studies on isolated plasma membrane vesicles, where the functioning of the ATPase can be observed without the complication of possible metabolic changes in pH or ATP concentration. The most extensive series of experiments has been carried out by Scarborough on inverted plasma membrane vesicles from the cell-wall-less strain of *Neurospora*. Depending upon the experimental conditions, ATP hydrolysis by these vesicles produces either a membrane potential, positive inside (monitored by [^{14}C]-SCN uptake or ANS fluorescence; Scarborough, 1976), or a pH gradient, acid inside (measured by the accumulation of the weak base [^{14}C] imidazole or by the quenching of intravesicular fluorescein-labeled dextran; Scarborough, 1980); both processes are sensitive to vanadate and to CCCP. Taken together, these results provide strong support for the notion of electrogenic proton translocation into the inverted vesicles, mediated by the plasma membrane ATPase. Similar results have since been obtained for vesicles from wild-type *Neurospora* (D. S. Perlin, and C. W. Slayman, 1981) and from the yeast *Candida tropicalis* (F. Blasco *et al.*, 1981).

Several difficulties exist in using plasma-membrane vesicles for a full exploration of the mechanism of proton translocation, however. For one thing, the vesicles from both *Neurospora* and *C. tropicalis* are quite leaky to protons, and undoubtedly contain multiple pathways (either naturally occurring or produced during vesicle isolation) by which protons can cross the membrane. In addition, it is intrinsically impossible to measure unidirectional proton fluxes by means of radioisotopes because of the exceedingly rapid exchange with water. For these reasons, efforts are shifting to the reconstitution of the ATPase into artificial lipid vesicles or bilayers, which should be free of other proton-translocating pathways, and to current–voltage analysis of intact cells (and eventually reconstituted preparations), from which unidirectional rate constants for the pump can be calculated.

6. SUMMARY

At the present stage of knowledge, by far the most intriguing property of the fungal plasma-membrane ATPase is its similarity to the transport ATPases of animal cells. Like Na$^+$/K$^+$-ATPase, Ca^{2+}-ATPase, and H$^+$/K$^+$-ATPase, it consists of a $M_r = 100,000$ polypeptide that is phosphorylated by ATP during the reaction cycle and inhibited by vanadate.

On the other hand, the fungal ATPase appears to operate as a simple electrogenic proton pump, providing energy for H^+-dependent cotransport of substrates. The molecular details of proton transport are not yet known, but as information begins to emerge from studies on the reconstituted ATPase, it will be interesting to compare with data from other primary proton pumps such as bacteriorhodopsin and the F_0F_1 ATPases. The outcome of such a comparison may shed light on the way in which energy-conserving systems have developed during evolution.

REFERENCES

Ahlers, J., Ahr, E., and Seyforth, A. (1978). *Mol. Cell. Biochem.* **22**, 39–49.
Amory, A., Foury, F., and Goffeau, A. (1980). *J. Biol. Chem.* **255**, 9353–9357.
Blasco, F., Chapuis, J. P., and Giordani, R. (1981). *Biochimie* **63**, 507–514.
Borst-Pauwels, G. W. F. H., and Peeters, P. H. J. (1981). *Biochim. Biophys. Acta,* **642**, 173–186.
Bowman, B. J., and Slayman, C. W. (1977). *J. Biol. Chem.* **252**, 3357–3363.
Bowman, B. J., and Slayman, C. W. (1979). *J. Biol. Chem.* **254**, 2928–2934.
Bowman, B. J., Blasco, F., and Slayman, C. W. (1978a). In *Frontiers of Biological Energetics* (P. L. Dutton, J. S. Leigh, and A. Scarpa, eds.), Vol. 1, pp. 525–533, Academic Press, New York.
Bowman, B. J., Mainzer, S. E., Allen, K. E., and Slayman, C. W. (1978b). *Biochim. Biophys. Acta* **512**, 13–28.
Bowman, B. J., Blasco, F., and Slayman, C. W. (1981a), *J. Biol. Chem.*, **256**, 12343–12349.
Bowman, E. J., Bowman, B. J., and Slayman, C. W. (1981b). *J. Biol. Chem.*, **256**, 12336–12342.
Dame, J. B., and Scarborough, G. A. (1980). *Biochemistry* **19**, 2931–2937.
Delhez, J., Dufour, J.-P., Thines, D., and Goffeau, A. (1977). *Eur. J. Biochem.* **79**, 319–328.
Dufour, J.-P., and Goffeau, A. (1978). *J. Biol. Chem.* **253**, 7026–7032.
Dufour, J.-P, and Goffeau, A. L. (1980). *Eur. J. Biochem.* **105**, 145–154.
Dufour, J.-P, and Goffeau, A. L. (1981). *J. Biol. Chem.*, **255**, 10591–10598.
Dufour, J.-P., and Tsong, T. Y. (1981). *J. Biol. Chem.*, **256**, 1801–1808.
Eddy, A. A. (1978). *Curr. Top. Membr. Transp.* **10**, 279–360.
Fuhrman, G. F., Boehm, C., and Theuvenet, A. P. R. (1976). *Biochim. Biophys. Acta* **433**, 583–596.
Goffeau, A. L., and Slayman, C. W. (1981). *Biochim. Biophys. Acta*, in press.
Malpartida, F., and Serrano, R. (1980). *FEBS Lett.* **111**, 69–72.
Peeters, P. H. J., and Borst-Pauwels, G. W. F. H. (1979). *Physiol. Plant.* **46**, 330–337.
Perlin, D. S., and Slayman, C. W. (1981). *Fed. Proc.* **40**, 1784.
Scarborough, G. A. (1975). *J. Biol. Chem.* **250**, 1106–1111.
Scarborough, G. A. (1976). *Proc. Natl. Acad. Sci. USA* **73**, 1485–1488.
Scarborough, G. A. (1977). *Arch. Biochem. Biophys.* **180**, 384–393.
Scarborough, G. A. (1980). *Biochemistry* **19**, 2925–2931.
Serrano, R. (1978). *Mol. Cell. Biochem.* **22**, 51–62.
Stroobant, P., and Scarborough, G. A. (1979). *Anal. Biochem.* **95**, 554–558.
Willsky, G. R. (1979). *J. Biol. Chem.* **254**, 3326–3332.

62

Charge-Transport Characteristics of a Plasma Membrane Proton Pump

Clifford L. Slayman

1. INTRODUCTION

In the mycelial fungus, *Neurospora*, the plasma membrane—like that of most other non-animal cells—is organized in a "chemiosmotic" fashion (Slayman, 1974; Eddy, 1978). This means that the major expenditure of metabolic energy at the membrane occurs via a proton efflux pump, which creates a standing difference of eletrochemical potential for protons ($\Delta\tilde{\mu}_{H^+}$). And it is this $\Delta\tilde{\mu}_{H^+}$ that supplies energy for most other transport processes. A similar arrangement is found in animal cells, but there the major transport energy is used to create a $\Delta\tilde{\mu}_{Na^+}$, which in turn drives other processes. (The structure and chemical properties of the membrane ATPase, which underlies H^+ pumping in *Neurospora*, are described in Chapter 61.)

A major respect in which transport architectures based on H^+ differ from those based on Na^+ is that the electrical term, $F\Delta\psi$, in the total electrochemical potential for protons is often overwhelmingly dominant. In *Neurospora*, for example, the normal H^+-concentration ratio is about 25, inwardly directed; but the membrane potential ($\Delta\psi$) is about -200 mV (cell interior negative), equivalent to a concentration ratio of nearly 2500. Because up to 90% of the membrane potential can be promptly abolished, either by respiratory inhibitors such as cyanide (Slayman *et al.*, 1973) or by specific pump inhibitors such as vanadate (Kuroda *et al.*, 1980), the proton pump appears strongly electrogenic under steady-state conditions. Indeed, as with most other proton pumps, that in the *Neurospora* membrane appears to be *obligatorily* electrogenic.

2. KINETIC PROPERTIES OF THE PROTON PUMP

To the extent that the *Neurospora* proton pump *is* obligatorily electrogenic, it is an enzyme in which one reaction product is charge, or electric current. As charges necessarily

Clifford L. Slayman • Department of Physiology, Yale University School of Medicine, New Haven, Connecticut 06510.

interact with the electric field of the membrane, a third dimension—the membrane potential—must be added to the conventional velocity–concentration plot in order to obtain a complete kinetic description of the enzyme. And when the enzyme velocity (current) is plotted against membrane potential alone, the resulting curve is termed the current–voltage (I–V) relationship.

By an amalgamation of the arguments of Britton (1966) with those of Läuger and Stark (1970), it is possible to show (Gradmann *et al.*, 1982; Hansen *et al.*, 1981) that most pump I–V relationships can be completely described by a very simple reaction diagram (Fig. 1) having only two forms of the "carrier" (N_i^+, N_o^+) and four empirical reaction constants: two k_{io}^o, k_{oi}^o) associated with the charge-transfer process, and two (κ_{oi}, κ_{io}) associated with the ensemble of voltage-independent steps (including binding and unbinding of the ion-ligand, plus reaction with ATP or another energy source). The physical meaning of these reaction constants is distorted by the presence of inaccessible carrier forms in the voltage-independent pathway, but under most conditions that distortion should be small and it can be neglected for present purposes.

Application of this model to I–V data obtained from *Neurospora*, in the presence and absence of 1 mM potassium cyanide, is shown in Fig. 2. Membrane I–V curves (●, ○) were calculated from microelectrode data obtained *via* a computer-controlled current-pulse sequence (Gradmann *et al.*, 1978). In order to obtain an internally consistent analysis, it has been necessary to assume that the membrane contains—in addition to the proton pump—an ohmic leak, which presumably represents all the true leaks, channels, and cotransport systems normally draining $\Delta\bar{\mu}_{H^+}$. The resulting I–V fits are given by the solid curves in Fig. 2A, and the separate pump and leak components are drawn in Fig. 2B. Least-squares

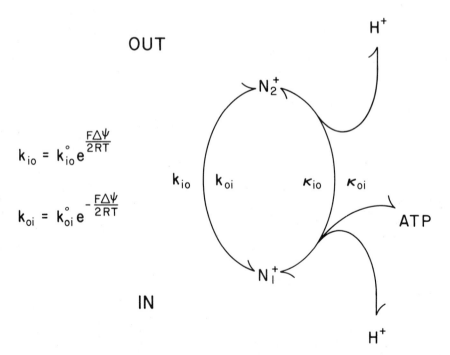

Figure 1. Diagram of pseudo-two-state kinetic model representing an electrogenic proton pump. Reaction constants for charge-transit of the membrane (k_{io}, k_{oi}) incorporate voltage dependence via a symmetric Eyring barrier. Reaction constants for all other steps are assumed to be voltage independent, and are lumped together (κ_{io}, κ_{oi}).

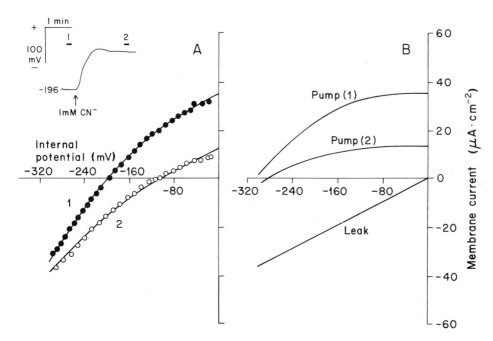

Figure 2. Fit of the kinetic model in Fig. 1 to current–voltage (I–V) curves of the *Neurospora* membrane ± potassium cyanide. (Inset) Time course of membrane potential (measured with conventional microelectrodes) upon addition of 1 mM KCN. Depolarization upward. Bars (1, 2) indicate intervals in which I–V scans were run. (A) Plotted points (●, ○) show I–V curves calculated by the method of Gradmann *et al.* (1978) from pulse data. Solid curves (———) give the least-squares fit of the pseudo-two-state kinetic model, plus a parallel and linear leak, to the plotted points. Fit carried out with all model parameters, except κ_{oi}, held in common for both curves. (B) Component I–V curves for the H⁺-pump and the leak, extracted from (κA). Replotted after Gradmann *et al.* (1982).

Parameters of the fit are given in Table I.

estimates for the empirical reaction constants are listed in Table I, along with several other calculated parameters.

The most important features of these results are the following. (1) The saturating pump current is $i_{sat+} = 35$ μA/cm² (the range for all experiments is 10-50 μA/cm²), which could consume 40% of total metabolic energy. Pump current for resting conditions

Table I. Properties of an Electrogenic Proton Pump, From I–V Analysis [a]

	k_{io}°	k_{oi}°	κ_{oi}	κ_{io}	i_{sat+}	g_{leak}
Control	2.8×10^{6}	1.8×10^{-1}	3.5×10^{2}	2.4×10^{4}	35	120
Cyanide	2.8×10^{6}	1.8×10^{-1}	1.3×10^{2}	2.4×10^{4}	13	120
	$E_c = \dfrac{RT}{F} \ln \dfrac{k_{oi}^{\circ}}{k_{io}^{\circ}}$		$E_n = \dfrac{RT}{F} \ln \dfrac{\kappa_{io}}{\kappa_{oi}}$		$E_r = E_c + E_n$	E_{leak}
Control	−415		107		−308	≡0
Cyanide	−415		132		−283	0

[a] Units as follows: k, $\kappa = $ sec⁻¹; $i = \mu$A/cm²; $g = \mu$S/cm²; $E = $ mV. The minus sign designates the cell interior as negative. $i_{sat+} = $ saturating current through the pump, with strong depolarization. See Fig. 1 and legend to Fig. 2 for details.

($\Delta\psi = -200$mV) would be one-half to two-thirds of that value. (2) The effect of ATP withdrawal by cyanide can be completely described by a change only of the single empirical reaction constant, κ_{oi}, which characterizes the "reloading" segment of the pump cycle. (3) The decrease of κ_{oi} is 2.7-fold, corresponding to the observed decrease in i_{sat+}. That represents a change of only 25 mV (out of more than 300mV), or 8%, in the pump reversal potential (E_r), which must mean that pumps can be *kinetically* controlled, by individual reaction steps, so their transport velocity is not directly related to the total driving force. In such circumstances linear thermodynamic descriptions are clearly invalid. (4) The major energetic asymmetry in the pump occurs at the charge-transit step, and is represented by the ratio $k^{\circ}_{io}/k^{\circ}_{oi} = 1.56 \times 10^7$, corresponding to a membrane potential of -415mV. As the total energy available from hydrolysis of ATP is not much larger (ca. 500 mV; Warncke and Slayman, 1980), charge-transit must be the step at which energy from phosphate anhydride bonds is actually converted to an electrochemical gradient.

3. INTEGRATION OF THE PROTON PUMP INTO CELLULAR FUNCTION

A second respect in which transport architectures based on H^+ may differ from those based on Na^+ is that processes that drain $\Delta\tilde{\mu}_{H^+}$ are very sensitive to metabolic status (Slayman, 1980). This may also be viewed, at least in part, as an evolutionary correlate of the dominance of $F\Delta\psi$ in $\Delta\tilde{\mu}_{H^+}$; membrane potential is a global parameter subject to local disasters. In *Neurospora*, almost any sustained, nonlethal, metabolic insult is followed shortly by restoration of ATP levels and membrane potential. This applies to temperature downshifts, partial respiratory blockade, carbon or nitrogen starvation, and osmotic shock. And in each of these cases a predominant element in the compensatory process is modulation of the ionic leakiness of the membrane.

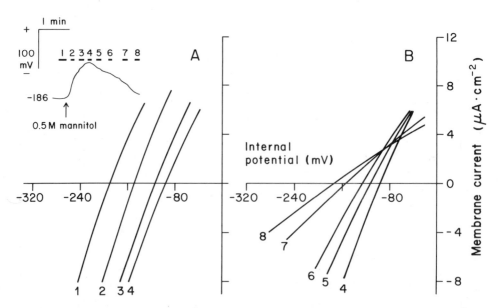

Figure 3. Effect of osmotic shock upon the membrane I–V curve in *Neurospora*. (Inset) Time course of membrane potential. Depolarization upward. Numbered bars indicate intervals of I–V scan. (A) I–V curves during depolarization. (B) I–V curves during repolarization. Experiment in collaboration with Dr. Martin Pall.

Osmotic shock treatment, analyzed by means of current–voltage relationships, provides an excellent example. As shown in Fig. 3 (inset), 0.5 M mannitol, added to the normal recording medium (ca. 90 mOsM), produces transient depolarization that lasts 40–50 sec (the osmotic shift was complete in 10 sec), then gives way to sustained repolarization. Membrane I–V curves, determined at various intervals (numbered bars in Fig. 3 inset), show an almost parallel shift during depolarization (curves $1 \rightarrow 4$, Fig. 3A), which is most simply interpreted as a scaling down of the pump current (see curve 1, Fig. 2B, for the span -180 to -90 mV). During repolarization, however, the conspicuous change is a diminishing slope (curves $4 \rightarrow 8$, Fig. 3B; compare the leak curve in Fig. 2B). More detailed analysis reveals that the pump and leak both shut down with time constants of 60–90 sec, but the response of the leak is delayed by 30–40 sec. The physiological consequence of this complex behavior is to restore the normal steady-state value of $\Delta\psi$ at a fourfold reduced level of transport-linked energy turnover.

The exact combination of changes in membrane potential, pump current, and leak conductance that occurs, during any metabolic challenge, depends upon the exact nature of that challenge. Sanders *et al.* (1981) have observed, for example, that mild cytoplasmic acidification—as produced by addition of 5 mM sodium butyrate to the normal recording medium—results, paradoxically, in sustained depolarization. I–V analysis demonstrates that, although pump current does increase—as it should because of increased $[H^+]_i$, the effect is short-circuited by increased membrane leak conductance. The result makes excellent physiologic sense, again because $F\Delta\psi$ dominates $\Delta\bar{\mu}_{H^+}$. Enhanced electrogenic extrusion of protons alone cannot combat cytoplasmic acid loading, because the expected increase of membrane potential would proportionately enhance H^+ reentry. But increased leak conductance to *ions other than protons* (K^+?, butyrate?) can combat the acid load by providing a non-H^+ reentry current.

4. CONCLUDING REMARKS

One awkward legacy of the early formulation of active transport mechanisms in terms of carrier diffusion has been a tendency in recent years to regard charge separation as epiphenomenological. Specific ramifications of this tendency, such as the notion that energy coupling in the mitochondrial membrane could be driven by simple diffusion of protons to the active site of the ATP synthetase, have been countered on thermodynamic grounds (Morowitz, 1978). The results given in Table I and Fig. 2 above now supply a kinetic argument: because the major energetic transition appears in the charge-transit step, not in the voltage-independent path, the carrier's energy charge must decay during charge-transit (or in a step that follows it extremely rapidly). Therefore, charge separation should be regarded as integral to the enzymatic mechanism of active proton transport. It is no surprise, then, that the consequent capability for storage and distribution of electrical energy should have been both utilized and physiologically regulated in the course of evolution.

ACKNOWLEDGMENTS. This work was supported by Research Grant GM-15858 from the National Institute of General Medical Sciences. The author is indebted to several collaborators who contributed to various portions of the worked cited: Dr. U.-P. Hansen, Dr. Dale Sanders, Dr. Martin Pall, and Dr. D. Gradmann.

REFERENCES

Britton, H. G. (1966). *Arch. Biochem. Biophys.* **117,** 167–183.

Eddy, A. A. (1978). *Curr. Top. Membr. Transp.* **10,** 279–360.

Gradmann, D., Hansen, U.-P., Long, W. S., Slayman, C. L., and Warncke, J. (1978). *J. Membr. Biol.* **39,** 333–367.

Gradmann, D., Hansen, U.-P, and Slayman, C. L. (1982). In *Electrogenic Ion Pumps* (C. L. Slayman, ed.), Academic Press, New York.

Hansen, U.-P., Gradmann, D., Sanders, D., and Slayman, C. L. (1981). *J. Membr. Biol.,* **63,** 165–190.

Kuroda, H., Warncke, J., Sanders, D., Hansen, U.-P., Allen, K. E., and Bowman, B. J. (1980). In *Plant Membrane Transport: Current Conceptual Issues* (R. M. Spanswick, W. J. Lucas, and J. Dainty, eds.), pp. 507–508, Elsevier, Amsterdam.

Läuger, P., and Stark, G. (1970). *Biochim. Biophys. Acta* **211,** 458–466.

Morowitz, H. J. (1978). *Am. J. Physiol.* **235,** R99–R114.

Sanders, D., Hansen, U.-P., and Slayman, C. L. (1981). *Proc. Natl. Acad. Sci. USA* **78,** 5903–5907.

Slayman, C. L. (1974). In *Membrane Transport in Plants* (U. Zimmermann and J. Dainty, eds.), pp. 107–119, Springer-Verlag, Berlin.

Slayman, C. L. (1980). In *Plant Membrane Transport: Current Conceptual Issues* (R. M. Spanswick, W. J. Lucas, and J. Dainty, eds.), pp. 179–190, Elsevier, Amsterdam.

Slayman, C. L., Long, W. S., and Lu, C. Y.-H. (1973). *J. Membr. Biol.* **14,** 305–338.

Warncke, J., and Slayman, C. L. (1980). *Biochim. Biophys. Acta* **591,** 224–233.

63

Biosynthesis of the Yeast Mitochondrial ATPase Complex

Contribution of the Mitochondrial Protein-Synthesizing System

Sangkot Marzuki, Henry Roberts, and Anthony W. Linnane

1. INTRODUCTION

Elucidation of the mechanism of assembly of multimeric enzyme complexes of the mitochondrial inner membrane is a major challenge in biochemistry. One of the most intensively studied of these enzyme complexes is the mitochondrial H^+-translocating ATPase. The ATPase complex is a mosaic consisting of mitochondrially made protein subunits and subunits imported from the extra mitochondrial cytoplasm. The formation of this enzyme complex, therefore, is a complicated process involving the assembly of the mitochondrially synthesized subunits as well as the synthesis of the other subunits on the cytoplasmic ribosomes, transport of these subunits to and across the mitochondrial membranes, and their subsequent assembly into a functional H^+-ATPase.

Recent studies in several laboratories have greatly advanced our understanding of the role played by the mitochondrial genetic system in the formation of the mitochondrial ATPase in the yeast *Saccharomyces cerevisiae*. In this review we discuss briefly some of the more recent data to illustrate the approaches used in these studies.

Sangkot Marzuki, Henry Roberts, and Anthony W. Linnane • Department of Biochemistry, Monash University, Clayton, Victoria 3168, Australia.

2. MITOCHONDRIALLY SYNTHESIZED SUBUNITS OF THE YEAST MITOCHONDRIAL H^+-ATPase

An essential first step in elucidating the mechanism of the assembly of the H^+-ATPase is to determine the structure of the enzyme complex. As with H^+-ATPase from a wide variety of other organisms, the yeast mitochondrial ATPase can be divided into two functional sectors: the F_1 sector, which contains the catalytic site for the synthesis and hydrolysis of ATP, and the F_0 sector, which is an integral part of the mitochondrial inner membrane and is thought to act as a proton channel, linking a transmembrane proton gradient generated by respiratory enzymes to the synthesis of ATP on the F_1 sector. The F_1 sector is now fairly well defined, and is known to consist of five different protein subunits, all imported from the extramitochondrial cytoplasm (see Criddle *et al.*, 1979, and Tzagoloff *et al.*, 1979, for the most recent reviews). On the other hand, the subunit composition of F_0 is uncertain. Purified preparations of the yeast mitochondrial ATPase complex appear to contain, in addition to the five F_1 subunits, up to five more polypeptides (Ryrie and Gallagher, 1979; Orian *et al.*, 1981). Whether all of these polypeptides are subunits of the enzyme complex, however, has not been established.

The problem associated with establishing the subunit composition of an integral membrane enzyme complex, such as the F_0 sector, is illustrated by the fact that although it has been known for many years that several of the membrane subunits of the mitochondrial ATPase are synthesized by the mitochondria, the exact number of these subunits is still subject to some disagreement. It was initially suggested that in yeast up to four of the F_0-sector subunits are synthesized by the mitochondria (Tzagoloff and Meagher, 1971). These mitochondrially synthesized polypeptides, which have apparent molecular weights of 29,000, 22,000, 12,000, and 7500 (designated subunits 5, 6, 8, and 9, respectively, in Tzagoloff's nomenclature), are associated with the ATPase when the enzyme complex is immunoprecipitated from Triton-solubilized mitochondria, using a specific rabbit antiserum raised against the purified H^+-ATPase. Recent studies in our laboratory, however, indicated that two of the above polypeptides (subunits 5 and 8, apparent molecular weights 32,000 and 10,000 in our gel system) are not part of the mitochondrial ATPase (Orian *et al.*, 1981; Orian and Marzuki, 1981). The 10,000-molecular-weight mitochondrial translation product is occasionally observed when the enzyme complex is isolated by immunoprecipitation using an ATPase-specific antiserum. However, because it is always absent in preparations of oligomycin-sensitive ATPase obtained by glycerol gradient centrifugation, this polypeptide appears to be a contaminant that has been adsorbed to the immunoprecipitate.

The 32,000-molecular-weight polypeptide (subunit 5) is present when the enzyme complex is purified on glycerol gradients. However, this polypeptide is only loosely associated with the enzyme complex. When the complex is isolated by immunoprecipitation, and the immunoprecipitates are washed extensively, the 32,000-molecular-weight polypeptide is absent.

A comparison of polyacrylamide gel profiles of mitochondrial translation products from different *mit⁻* mutant strains (see Section 3 for definition of the *mit⁻* mutation) lacking either the cytochrome *b* apoprotein or the cytochrome oxidase subunit II, shows that these two polypeptides are the only mitochondrial translation products that have molecular weights of around 30,000–32,000. However, mtATPase purified on glycerol gradients from each of these *mit⁻* strains contains a 32,000-molecular-weight polypeptide, presumably apocytochrome *b* in the mitochondrial ATPase from the *mit⁻* mutant lacking cytochrome oxidase subunit II and cytochrome oxidase subunit II in the apocytochrome *b*-less mutant. It appears, therefore, that the 32,000-molecular-weight mitochondrially synthesized polypeptide of the wild-type ATPase is a mixture of the cytochrome *b* apoprotein and subunit II of cytochrome oxidase contaminating the ATPase preparation. This suggestion is confirmed by the finding that the 32,000-molecular-weight component of the wild-type mitochondrial ATPase cross-reacts with antisera monospecific either for apocytochrome *b* or for cytochrome oxidase subunit II (Orian *et al.*, 1981). A combination of these two antisera was needed to completely precipitate the 32,000-molecular-weight component.

Our results suggest that only two subunits of the ATPase complex, with estimated molecular weights of about 20,000 and 7600, are synthesized in the mitochondrion. Genetic and biochemical studies of yeast mitochondrial gene mutations affecting the H^+-ATPase support this conclusion, and have enabled the structural genes for the two subunits to be located on the mitochondrial DNA.

3. MITOCHONDRIAL MUTANTS IN THE STUDY OF THE BIOGENESIS OF THE MITOCHONDRIAL ATPase

The key to our present knowledge of the contribution of the mitochondrial genetic system to the biogenesis of the mitochondrial ATPase is the isolation of yeast mutants, with specific lesions in their mtDNA affecting the function, synthesis, or assembly of the enzyme complex.

The first types of ATPase mutants isolated were the oligomycin-resistant strains of yeast. Genetically, three different classes of oligomycin-resistance mutations can be distinguished. These classes of mutations were designated A, B, and C in our laboratory (Trembath *et al.*, 1976), corresponding to the O_{III}, O_I, and O_{II} loci (subsequently renamed *oli3*, *oli1*, and *oli2*) described by Griffiths (Lancashire and Griffiths, 1975). The class C mutations are widely separated from mutations of the other two classes by several other genetic loci including those concerned with the specification of the cytochrome *b* apoprotein. Class A and B mutations, on the other hand, are closely linked, separated by recombination frequencies of only about 1–2%. Phenotypically, however, these two classes of mutation are distinguishable; class A mutations also confer cross-resistance to venturicidin, in contrast to mutations of class B, which do not affect the sensitivity of the mutant strains to this inhibitor. Although mutations of both A and B classes are located in the structural gene of the same mtATPase subunit (the proteolipid subunit 9, see below), the mutations are apparently in different parts of the structural gene and therefore might be useful in the study of the function of different regions of the proteolipid molecule.

Two classes of *mit⁻* mutants have been isolated that are defective in the mtATPase. These mutants are not capable of growth by oxidative metabolism as the result of a defect in the assembly or the function of the enzyme complex. The first class of mutants carries lesions linked to the *oli2* mutations that confer oligomycin resistance to the mtATPase, as indicated by various genetic and physical mapping procedures. These mutants include the *pho1* mutants of Tzagoloff (Foury and Tzagoloff, 1976) and the *oli2 mit⁻* mutants described by our laboratory (Roberts *et al.*, 1979). The other class of *mit⁻* mutants is linked to the *oli1* mutations and include the *pho2* mutations (Foury and Tzagoloff, 1976).

More recently, the range of mitochondrial mutants available for the study of the mtATPase has been extended with the isolation of mutants with a respiratory-deficient phenotype conditional upon growth at low or elevated temperatures. As with the *mit⁻* mutations, the lesions in the temperature-conditional mutants can be grouped into mutations linked to the *oli1* loci and those linked to the *oli2* mutations.

In addition to the type of mutants described above, petite mutants of yeast, which are

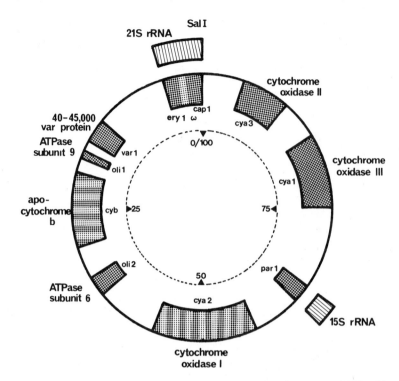

Figure 1. Location of the *oli1* and *oli2* regions on the physical map of the yeast mtDNA. The map of the mtDNA was constructed according to the principles discussed by Nagley *et al.* (1977) and Linnane and Nagley (1978). The 0/100 point on the map corresponds to the unique *Sal*I restriction enzyme cleavage site in the mtDNA. One map unit is about 800 base pairs. Genetic loci and the gene products associated with specific regions of the mtDNA are indicated inside and outside the circle, respectively. [From Linnane *et al.*, 1980b.]

the result of large deletions of the mtDNA, have also been very useful especially in the genetic and physical mapping of the mitochondrial ATPase mutations. The extent of these deletions, which occur randomly in different regions of the mtDNA, range from less than 60% to more than 99% of the mitochondrial genome. Thus, the location of a mitochondrial mutation can be physically mapped by genetically determining the genetic loci retained or lost in a series of petites, and by characterizing the segment of mtDNA retained in each petite using hybridization techniques or the restriction enzyme technology (see Nagley *et al.*, 1977; Linnane and Nagley, 1978; Locker *et al.*, 1979). Our recent version of the physical map of the yeast mitochondrial genome is shown in Fig. 1.

4. BIOSYNTHESIS OF THE MITOCHONDRIAL ATPase SUBUNIT 9

Unlike the *Neurospora* and the mammalian proteolipids, which are imported from the extramitochondrial cytoplasm, the yeast proteolipid subunit 9 is synthesized in the mitochondria. Despite the difference in their coding origin, however, a high degree of homology exists between the yeast subunit 9 and the proteolipid isolated from *Neurospora* or beef heart mitochondria (see Sebald *et al.*, 1979).

The structural gene of the ATPase subunit 9 has been shown to be located in the *oli1* region of the mtDNA (see Fig. 1). The amino acid sequences of proteolipids from several *oli1* mutants and a revertant of a *pho2* mutant have been determined and found to contain amino acid substitutions (Wachter *et al.*, 1977; Sebald *et al.*, 1979). Furthermore, the nucleotide sequence of mtDNA of petites retaining the *oli1* gene (Hensgens *et al.*, 1979; Macino and Tzagoloff, 1979) have been determined and found to agree with the known amino acid sequence. It is of interest that in one residue the DNA sequence predicts a leucine instead of a threonine, but this discrepancy has recently been resolved. It was found that in yeast mitochondria, the codon CUA is recognized by a threonyl tRNA (Li and Tzagoloff, 1979).

The *oli3* locus (class A oligomycin-resistance mutation), which is separable by recombination from *oli1,* has been shown to also lie in the structural gene of the proteolipid. Several class A oligomycin-resistance mutations have been shown to affect the mobility of the proteolipid in SDS-polyacrylamide gels (Murphy *et al.*, 1978), or to affect the p*I* of the proteolipid in isoelectric focusing gels (Partis *et al.*, 1979). One temperature-sensitive class A (*oli3*) mutation was reported to affect the synthesis or integration of a 20,000-molecular-weight mitochondrially synthesized ATPase subunit (Groot Obbink *et al.*, 1976). However, it was subsequently shown in our laboratory that the mutation in this strain is also in the proteolipid structural gene and that the effect observed on subunit 6 is an indirect result of the mutation (Murphy *et al.*, 1978). This observation suggests that a close interaction exists between subunit 6 and subunit 9 in the membrane sector of the mitochondrial H$^+$-ATPase. Consistent with this suggestion, a mutation in the *oli1* region of the mtDNA has recently been reported to restore oxidative metabolism in a *mit$^-$* strain with a defective subunit 6 (Linnane *et al.*, 1980a).

So far a proteolipid from only one *oli3* mutant has been sequenced. An amino acid substitution was found, but was located in a position between those residues affected by two *oli1* mutations (Sebald *et al.*, 1979).

5. BIOSYNTHESIS OF THE MITOCHONDRIAL ATPase SUBUNIT 6

A considerable advance in our understanding of the synthesis of subunit 6 was made possible by the isolation of yeast mutants in which this protein is altered. Several *mit$^-$* mutations in the *oli2* region of the yeast mtDNA resulted in the loss of subunit 6, and in some strains its replacement by new mitochondrial translation products with apparent molecular weights of less than 20,000 (Roberts *et al.*, 1979). These new polypeptides appear to be mutationally altered forms of the subunit, for, like the wild-type subunit 6, they were immunoprecipitated with the ATPase complex by antiserum against purified oligomycin-sensitive ATPase.

One of the *oli2 mit$^-$* mutations has been precisely located on the mitochondrial genome (Linnane *et al.*, 1980a). Restriction endonuclease digestion of mtDNA of a yeast strain carrying the *oli2* mutation (strain Mb12) showed that it lacked the *Eco*RI restriction site between the *Eco*RI fragments 7 and 8 found in the wild-type parent (cf. Morimoto and Rabinowitz, 1979). This enabled the mutation to be placed within the hexanucleotide recognition site for this enzyme, and is consistent with the position of the *oli2* locus established by earlier mapping studies (Choo *et al.*, 1977; Morimoto *et al.*, 1978).

Some other restriction sites have also been identified in the region of the *Eco*RI site missing from strain Mb12 (Linnane *et al.*, 1980a) (Fig. 2). These restriction sites are consistent with the restriction sites of the mtDNA from a petite strain (DS14) that retains a very short segment of the mitochondrial genome (4.1 kb), containing the *oli2* loci (Macino and Tzagoloff, 1980). The nucleotide sequence of this petite has been determined

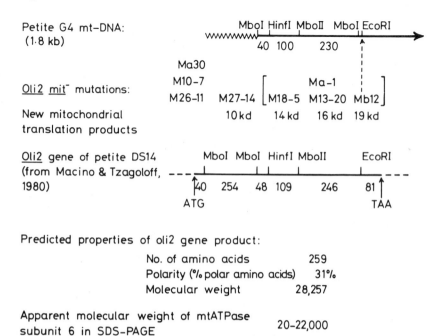

Figure 2. Comparison of the structural gene of mtATPase subunit 6 with the *oli2* gene nucleotide sequence of Macino and Tzagoloff (1980). The map of the restriction site in petite G4 mtDNA is adapted from Linnane *et al.* (1980a). The zigzag line indicates the uncertainty in mapping the left-hand end of the petite mtDNA. The broken line and arrow indicates the position of the *Eco*RI restriction site, which is missing in the *mit⁻* strain Mb12. The order of the four *mit⁻* mutations in the parentheses is based only on the sizes of the new polypeptides produced in each mutant, assuming that the new polypeptides are the result of an early termination during translation of the subunit 6 protein. The restriction sites in the petite DS14 mtDNA are deduced from the *oli2* gene sequence reported by Macino and Tzagoloff (1980).

(Macino and Tzagoloff, 1980), and found to contain a potential coding sequence, in which the restriction sites in Fig. 2 were identified. The fact that *mit⁻* mutations in this segment of DNA affect the synthesis of subunit 6 suggests that the coding sequence is the structural gene for subunit 6. The sequence can be translated in one reading frame starting with an AUG initiator, in the same direction as that suggested by the positions of the *oli2 mit⁻* mutations in Fig. 2.

The amino acid sequence deduced from the nucleotide sequence suggests that the gene codes for a protein containing 259 amino acid residues, equivalent to a molecular weight of approximately 28,000. The protein is predicted to be highly hydrophobic, with only 31% polar amino acids, which occur in several clusters. The gene product deduced cannot yet be compared directly to subunit 6 of the ATPase, as the amino acid composition and sequence of this subunit have not yet been determined. The predicted molecular weight of 28,000 is considerably larger than the apparent molecular weight of 20,000 for subunit 6 observed in polyacrylamide gels. A similar discrepancy has been found between the apparent molecular weight of cytochrome oxidase subunit III (20,000, like that of ATPase subunit 6) and the size of the protein predicted from the nucleotide sequence of its structural gene (30,000) (Thalenfeld and Tzagoloff, 1980).

There are several possible explanations of the discrepancy between the predicted and the observed subunit 6 molecular weight. First, the true molecular weight of subunit 6 may be larger than 20,000, for it is a hydrophobic protein and thus may behave abnormally during electrophoresis. Second, it is possible that not all of the possible codons are translated into protein, if either the first AUG codon is not the true initiator or the gene contains intervening sequences. Third, the protein may be synthesized as a 28,000-molecular-weight precursor, which is cleaved to produce a smaller mature protein. Determination of at least the NH_2- and COOH-terminal amino acid sequences of subunit 6 would help to answer some of these questions.

The determination of the amino acid sequence of subunit 6 might also explain the observation in our laboratory that subunit 6 can be separated by isoelectric focusing into two components of different p*I* but similar apparent molecular weight (Stephenson *et al.*, 1980). Both components are altered by single *mit⁻* mutations in the *oli2* gene. Although at present one could not rule out the possibility that the two subunit 6 components are due to technical artifact, one of the alternative explanations for this observation is that subunit 6 is composed of two slightly different polypeptides that are genetically related; one, for example, could arise from posttranslational modification of the other (e.g., by limited proteolysis or phosphorylation). Consistent with our observation, it has recently been reported that subunit 6 can occasionally be resolved into two bands when a purified ATPase preparation is analyzed on a 4–17% SDS-polyacrylamide gel (Todd *et al.*, 1980).

The transcripts of the *oli2* gene have recently been analyzed in our laboratory (M. W. Beilharz and G. S. Cobon, personal communication). The most abundant *oli2* transcript was found to be of the order of either 4500 or 3500 nucleotides in different wild-type strains. This difference can be correlated with deletions of mtDNA sequences 1–2 kb downstream of the gene, which in some cases delete both the *Pst*I site and a downstream *Eco*RI site. Less abundant transcripts were also detected that were larger or smaller than the putative mRNA. In a number of *oli2 mit⁻* mutants, alterations in the relative abundance of transcripts of this region of the genome have been observed.

6. MITOCHONDRIALLY SYNTHESIZED SUBUNITS OF THE MITOCHONDRIAL H⁺-ATPase ARE FUNCTIONAL COMPONENTS OF THE ENZYME COMPLEX

The involvement of the proteolipid subunit 9 in the proton translocation steps of oxidative phosphorylation is well established, and the evidence for this is well documented in two recent reviews (Criddle *et al.*, 1979; Fillingame, 1980). However, the mechanism of proton translocation through F_0 remains the subject of speculation. It is clear that the transport step must be coupled to some secondary events that can trap the potential energy of the protons. These events probably involve at least one other F_0 subunit that interacts with the proteolipid itself. It is also by no means certain that the proteolipid is the only component of the proton channel of F_0. A detailed characterization of F_0 and the mode of attachment of F_1 to F_0 may provide some of the insight needed to solve the problem of coupling proton translation to ATP synthesis.

The function of the 20,000-molecular-weight subunit 6 has not been fully established. Criddle *et al.* (1979) have suggested that thioacyl intermediates are formed during ATP synthesis, and have reported that purified yeast ATPase catalyzes the hydrolysis of acylthioesters (Criddle *et al.*, 1979). This activity was inhibited by oligomycin and venturicidin. It was postulated that the thioacyl intermediate could involve pantothenate, and it was found that [³H]pantothenate could be covalently attached to subunit 6 of the ATPase complex (Criddle *et al.*, 1977). It was thus proposed that subunit 6 may be the active site for thioesterase activity, and cited in support of the proposal was the observation that *oli3* mutant strains resistant to venturicidin yielded ATPase preparations in which the thioesterase activity was resistant to venturicidin (Criddle *et al.*, 1979). However, it is now known that the *oli3* venturicidin-resistance mutations are not in the structural gene of subunit 6, which is in the *oli2* region of the genome (Roberts *et al.*, 1979).

More recently, the function of the mtATPase subunit 6 has been studied using yeast mutants that carry lesions in the *oli2* region of the mtDNA. Most of these mutants have gross alterations of the subunit 6 structure, and as a result contain only very low ATPase activity that is insensitive to oligomycin, presumably because F_1 is not correctly attached to the membrane. However, two *oli2* mutants that show a leaky *mit⁻* phenotype were found to have a defect in the coupling of oxidative phosphorylation as a result of a small modification of subunit 6 (Murphy *et al.*, 1980). Unlike the other *oli2 mit⁻* strains, these *oli2* mutants contained an assembled F_0F_1-ATPase complex, as their ATPase activity could be fully inhibited by oligomycin.

When the mutant strains were forced to grow with a generation time of 7 hr in glucose-limited chemostat cultures, the mitochondria had a very low ATP-$^{32}P_i$ exchange activity and P : O ratio, indicating a major defect in oxidative phosphorylation. Mitochondrial respiration in these mutants was inhibited by oligomycin as effectively as in the wild-type parent (W. M. Choo and S. Marzuki, unpublished observation). This inhibition was reversed by uncouplers of oxidative phosphorylation. Thus, it appears that the alteration of subunit 6 in these strains has allowed the proton gradient to be dissipated through the proton channel of F_0 without being coupled to ATP synthesis.

REFERENCES

Choo, K. B., Nagley, P., Lukins, H. B., and Linnane, A. W. (1977). *Mol. Gen. Genet.* **153,** 279–288.

Criddle, R. S., Edwards, T., Partis, M., and Griffiths, D. E. (1977). *FEBS Lett.* **84,** 278–282.

Criddle, R. S., Johnston, R. F., and Stack, R. J. (1979). *Curr. Top. Bioenerg.* **9,** 89–145.

Fillingame, R. H. (1980). *Annu. Rev. Biochem.* **49,** 1079–1113.

Foury, F., and Tzagoloff, A. (1976). *Eur. J. Biochem.* **68,** 113–119.

Groot Obbink, D. J., Hall, R. M., Linnane, A. W., Lukins, H. B., Monk, B. C., Spithill, T. W., and Trembath, M. K. (1976). In *The Genetic Function of Mitochondrial DNA* (C. Saccone and A. M. Kroon, eds.), pp. 163–173, North-Holland, Amsterdam.

Hensgens, L. A. M., Grivell, L. A., Borst, P., and Bos, J. L. (1979). *Proc. Natl. Acad. Sci. USA* **76,** 1663–1667.

Lancashire, W. E., and Griffiths, D. E. (1975). *Eur. J. Biochem.* **51,** 403–413.

Li, M., and Tzagoloff, A. (1979). *Cell* **18,** 47–53.

Linnane, A. W., and Nagley, P. (1978). *Arch. Biochem. Biophys.* **187,** 277–289.

Linnane, A. W., Astin, A. M., Beilharz, M. W., Bingham, C. G., Choo, W. M., Cobon, G. S., Marzuki, S., Nagley, P., and Roberts, H. (1980a). In *The Organization and Expression of the Mitochondrial Genome* (A. M. Kroon and C. Saccone, eds.), pp. 253–263, Elsevier/North-Holland, Amsterdam.

Linnane, A. W., Marzuki, S., Nagley, P., Roberts, H., Beilharz, M. W., Choo, W. M., Cobon, G. S., Murphy, M., and Orian, J. M. (1980b). In *The Plant Genome* (D. R. Davies and D. A. Hopwood, eds.), pp. 99–110, The John Innes Charity, Norwich.

Locker, J., Lewin, A., and Rabinowitz, M. (1979). *Plasmid* **2,** 155–181.

Macino, G., and Tzagoloff, A. (1979). *J. Biol. Chem.* **254,** 4617–4623.

Macino, G., and Tzagoloff, A. (1980). *Cell* **20,** 507–517.

Morimoto, R., and Rabinowitz, M. (1979). *Mol. Gen. Genet.* **170,** 11–23.

Morimoto, R., Merten, S., Lewin, A., Martin, N. C., and Rabinowitz, M. (1978). *Mol. Gen. Genet.* **163,** 241–255.

Murphy, M., Gutowski, S. J., Marzuki, S., Lukins, H. B., and Linnane, A. W. (1978). *Biochem. Biophys. Res. Commun.* **85,** 1283–1290.

Murphy, M., Roberts, H., Choo, W. M., Macreadie, I., Marzuki, S., Lukins, H. B., and Linnane, A. W. (1980). *Biochim. Biophys. Acta* **592,** 431–444.

Nagley, P., Sriprakash, K. S., and Linnane, A. W. (1977). *Adv. Microbiol. Physiol.* **16,** 157–277.

Orian, J. M., and Marzuki, S. (1981). *J. Bacteriol.*, **146,** 813–815.

Orian, J. M., Murphy, M., and Marzuki, S. (1981). *Biochim. Biophys. Acta* **652,** 234–239.

Partis, M. D., Bertoli, E., Zanders, E. D., and Griffiths, D. E. (1979). *FEBS Lett.* **105,** 167–170.

Roberts, H., Choo, W. M., Murphy, M., Marzuki, S., Lukins, H. B., and Linnane, A. W. (1979). *FEBS Lett.* **108,** 501–504.

Ryrie, I. J., and Gallagher, A. (1979). *Biochim. Biophys. Acta* **545,** 1–14.

Sebald, W., Hoppe, J., and Wachter, E. (1979). In *Function and Molecular Aspects of Biomembrane Transport* (E. Quagliariello, E. Palmieri, S. Papa, and M. Klingenberg, eds.), pp. 63–74, Elsevier/North-Holland, Amsterdam.

Stephenson, G., Marzuki, S., and Linnane, A. W. (1980). *Biochim. Biophys. Acta* **609,** 329–341.

Thalenfeld, B. E., and Tzagoloff, A. (1980). *J. Biol. Chem.* **255,** 6137–6180.

Todd, R. D., Griesenbeck, T. A., and Douglas, M. G. (1980). *J. Biol. Chem.* **255,** 5461–5467.

Trembath, M. K., Molloy, P. L., Sriprakash, K. S., Cutting, G. J., Linnane, A. W., and Lukins, H. B. (1976). *Mol. Gen. Genet.* **145,** 43–52.

Tzagoloff, A., and Meagher, P. (1972). *J. Biol. Chem.* **247,** 594–603.

Tzagoloff, A., Macino, G., and Sebald, W. (1979). *Annu. Rev. Biochem.* **48,** 419–441.

Wachter, E., Sebald, W., and Tzagoloff, A. (1977). In *Mitochondria 1977: Genetics and Biogenesis of Mitochondria* (W. Bandlow, R. J. Schweyen, K. Wolf, and F. Kaudewitz, eds.), pp. 441–449, de Gruyter, Berlin.

64

Specification and Expression of Mitochondrial Cytochrome b

Henry R. Mahler

Recent studies on the structure, organization, and expression of the mitochondrial gene for (apo)cytochrome b in *Saccharomyces cerevisiae* and related species have provided answers to some old questions and revealed some hitherto unexpected complexities of regulatory and, possibly, evolutionary significance.* To the first set belong, most importantly, the primary structure of the protein, a problem that had defied solution by protein sequencing techniques, but was solved by Nobrega and Tzagoloff (1980) by DNA sequencing: the polypeptide chain consists of 385 amino acids, corresponding to a molecular weight of 44,000. Another question that has been resolved without ambiguity concerns the type and number of polypeptides responsible for the multiplicity of cytochrome b species implicated in the bc_1 segment (Complex III, coenzyme QH_2 : cytochrome b oxidoreductase) of the mitochondrial respiratory (electron transfer) chain (reviewed by von Jagow and Sebald, 1980). Because single, defined, revertible mutational lesions in the coding segments of a unique and discrete gene, known as *cob*, are now known to result in the complete elimination or alteration of all forms of the cytochrome in question (Claisse *et al.*, 1978; Mahler *et al.*, 1978; Alexander *et al.*, 1979; Haid *et al.*, 1979; Kreike *et al.*, 1979), all such cytochromes must have been synthesized with the same primary sequence. The different species observed may therefore be the consequence of placing the same polypeptide in different environments, of posttranslational modification, or a combination of these two effects. Related to this problem, and equally accessible by a combination of genetic and biochemical techniques, is the localization of sites of interaction between the polypeptide on the one hand, and heme or various characteristic inhibitors of Complex III (such as antimycin A, diuron, funiculosin and mucidin) on the other. For instance, separate and discrete sites of interaction with antimycin A (Roberts *et al.*, 1980) have been shown to exist and are localized to the exon segments (see below and Fig. 1) E_{1l} and E_{3l}, respec-

*For two recent symposia summarizing most of the current work, see Kroon and Saccone (1980) and Slonimski *et al.* (1982).

Henry R. Mahler • Department of Chemistry, and the Molecular, Cellular, and Developmental Biology Program, Indiana University, Bloomington, Indiana 47405.

tively (Colson and Slonimski, 1978; Colson *et al.*, 1979). Now that the primary sequence of the DNA is known, it should be relatively easy and straightforward to determine which clusters of amino acids in the protein are responsible for the binding of these and other inhibitors, for attachment of heme, and for interaction with the other polypeptides of Complex III.

Finally, it should be mentioned that—at least in yeast and *Neurospora*—the mitochondrial contribution to the NADH : and succinate : cytochrome *c* reductase segments of the respiratory chain (Complexes I, II, and III) is restricted entirely to this single polypeptide. In other words, all other polypeptides in these segments, including the cytochrome *b* in succinate : coenzyme Q reductase (Complex II) (Weiss *et al.*, 1979), are specified by nuclear genes, synthesized on cytoplasmic ribosomes and imported into the mitochondria (reviewed by Schatz and Mason, 1974; Tzagoloff *et al.*, 1979; von Jagow and Sebald, 1980, Neupert and Schatz, 1981). This discovery explains why interference with the synthesis of cytochrome *b* either by genetic or by biochemical means, i.e., exposure to inhibitors of mitochondrial transcription or translation, such as ethidium bromide or chloramphenicol, does not result in a complete block of the accumulation of the cytochrome or an inhibition of formation of Complex II (succinate : coenzyme Q reductase) activity (Mahler and Perlman, 1971).

The second set consists, principally, of five items and their corollaries and consequences (reviewed by Dujon, 1979; Jacq *et al.*, 1980; Grivell *et al.*, 1980; Perlman *et al.*, 1980a; Lewin, 1980; Jacq *et al.*, 1982; Mahler *et al.*, 1982). First, the gene exhibits the mosaic organization previously established for, and now recognized as one of the most salient features of, eukaryotic nuclear and viral genes (for recent reviews see Darnell, 1978; Crick, 1979; Gilbert, 1979; Abelson, 1979). The cytochrome *b* genes in different wild-type strains of *S. cerevisiae* fall into two limit classes depending on the number of coding (expressed or *exon*) and intervening (*intron*) sequences they contain (Nobrega and Tzagoloff, 1980; Jacq *et al.*, 1980; Lazowska *et al.*, 1980; Perlman *et al.*, 1980b). The "long form" extends over a total of ~ 6000 base pairs and is composed of six exons (E_{1l}, E_{2l}, E_{3l}, E_{4l}, E_{5l}, E_{6l}) and five introns (I_{1l}, I_{2l}, I_{3l}, I_{4l}, I_{5l}), while in the "short form" (3300 base pairs long), the first (promoter or NH_2 terminus proximal) four exons are fused. This arrangement reduces their number to three (E_{1s}, E_{2s} and E_{3s}) and the number of introns to two (I_{1s} and I_{2s}), 1414 and 733 base pairs in length, respectively (Fig. 1).

Second, several of the introns, and in particular the second and fourth in long-form (I_{2l} and I_{4l} in Fig. 1) and the first (I_{1s}) in short-form strains (which as shown is homologous to I_{4l}), contain extensive open reading frames in phase with those of the preceding exons (Nobrega and Tzagoloff, 1980; Jacq *et al.*, 1980; Lazowska *et al.*, 1980). That this arrangement of base sequences is actually utilized in producing interlaced, hybrid polypeptides becomes apparent from studies on intron mutants. These result in faulty, or arrested,

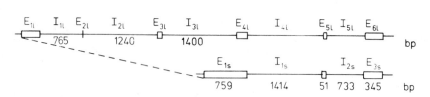

Figure 1. Structure and organization of *cob*, the mitochondrial gene for apocytochrome *b* in *Saccharomyces* strains. Two forms are shown: "long form" (*cob^l*) on top, "short form" (*cob^s*) on the bottom. The long form contains six exons designated by E and the appropriate subscript, and five introns designated I; the short form consists of three exons and two introns. Lengths are approximately to scale and are given in base pairs (bp): total length is 6721 bp for the long form and 3302 bp for the short form of the gene.

processing (splicing) of the primary transcript—which retains intervening sequences—to the mature mRNA, which does not (van Ommen *et al.*, 1980; Grivell *et al.*, 1980; Halbreich *et al.*, 1980; Church and Gilbert, 1980; Schmelzer *et al.*, 1981). In consequence, such mutants, in contrast to the wild type, accumulate incompletely spliced transcripts that are translated into mutationally altered proteins, containing combinations of amino acid sequences specified by exon as well as intron species (Solioz and Schatz, 1979; Hanson *et al.*, 1979; Kreike *et al.*, 1979; Alexander *et al.*, 1980; Claisse *et al.*, 1980; Jacq *et al.*, 1980; Perlman *et al.*, 1980b).

Third, at least some of the intervening sequences define separate complementation groups and are implicated in the specification of discrete *trans*-acting elements (Lamouroux *et al.*, 1980; Dhawale *et al.*, 1981). This set of observations attests to the functional competence of these elements, presumably at some step(s) in the splicing sequence referred to in the previous paragraph. Whether this function involves the wild-type version of some of the hybrid proteins described (as hypothetical maturases) (Jacq *et al.*, 1980; Church and Gilbert, 1980) or segments of the RNA transcripts (in providing complementary guides) (Church *et al.*, 1979; Dujon, 1979) remains to be established.

Fourth, mutational alterations in the gene for cytochrome b can lead to pleiotropic defects in the expression of a second, unlinked, mosaic mitochondrial gene, the gene *oxi3* responsible for the specification of the largest subunit (I) of cytochrome oxidase (cytochrome c : O_2 oxidoreductase, cytochrome aa_3, Complex IV) (Pajot *et al.*, 1977; Claisse *et al.*, 1978; Mahler *et al.*, 1978; Alexander *et al.*, 1979, 1980; Grivell *et al.*, 1980). Analysis of the effect (Church *et al.*, 1979; Grivell *et al.*, 1980; Dhawale *et al.*, 1981) suggests positive regulation of the processing of the primary transcript of that gene by a product specified, at least in part, by base sequences in I_{1s} of short-form strains, and the structurally homologous I_{4l}, (see Fig. 1) acting on the structurally similar fourth intron of *oxi3*.

Fifth, it has recently become evident (Mahler *et al.*, 1982; Jacq *et al.*, 1982) that the "indispensable" penultimate intron (I_{4l} or I_{1s}) exhibits pecularities in structure, function and expression not shared by I_{2s}. Mutants in it have been subjected to fine structure genetic and physical mapping, and the nature of the sequence alterations responsible for the mutation established in many of them. They appear to fall into at least three clusters of mutational sites. The first (*box9*) appears to be localized within a region a dozen or so base pairs in length, located some 330 bases downstream (3') from the exon (E_{4l} or E_{1s})-intron boundary; the second (*box7*) consists of the remaining segment of the open reading frame, a region some 800 base pairs in length, while the third (*box2*) is located within the closed reading frame, close to the boundary with the next, downstream exon (E_{5l} or E_{2s}). All of these mutants are deficient in their ability to remove (splice out) the intron and accumulate mRNA precursors containing it. But only mutants in *box7* are complemented by exon mutations and fail to accumulate P27, a protein identical with, or related to, the maturase required for the splicing reaction. The conclusion is therefore inescapable that successful splicing cannot occur unless provision is made for both an intron-encoded protein, as well as successful recognition by it of RNA sequences as signals for the set of reactions required for splicing; these are not restricted to the canonical splice sites proper located in the immediate vicinity of the intron-exon boundaries.

These results raise the intriguing question of the prevalence and significance of intervening sequences in these and other mitochondrial genes. As concerns the gene for cytochrome b, all available information suggests that the long- or short-form strains (see Fig. 1), containing either five or two intervening sequences, exhibit no significant differences in such physiological parameters as growth rate or yield, or biochemical parameters such as the amount of cytochrome b, of cytochrome b-linked enzyme activities, or in the regulation of expression of the *oxi3* gene. The same holds true also of other strains constructed by us (Dhawale *et al.*, 1981; Perlman *et al.*, 1980b) that contain either three or four intervening sequences. Of course, the possibility exists and needs to be explored that other, more subtle forms of regulatory interplay may be lost in this graduated reduction in the number of intervening sequences. That this is a real possibility is suggested by the observation that a conditional form of regulation of expression of the *oxi3* gene is absent in strains that lack intervening sequences I_{1l} through I_{3l} (Dhawale *et al.*, 1981). This regulation becomes manifest when certain mutants in E_{1l} but not in the corresponding segment of E_{1s} (Fig. 1), all specifying the NH_2-terminal portion of the protein, are grown on glucose rather than galactose.

Perhaps more interesting, and profound, is the question whether intervening sequences are required at all, and its corollary, the nature of the consequences of their removal. Future studies will show whether such surgery can be performed successfully on the cytochrome *b* gene in *Saccharomyces*. However, some inferences are already possible on the basis of existing information. Even in *Saccharomyces* the majority of mitochondrial genes are *not* split (reviewed in Borst and Grivell, 1978; Tzagoloff *et al.*, 1979; Perlman *et al.*, 1980a). It might of course be argued that apocytochrome *b*, because of its unique structure and function, might be the one protein that absolutely requires (or at least required, sometime during its evolution) this particular organization. For instance, the segments of its primary structural domains flanking the introns might exhibit some unusual properties (Crick, 1979; Gilbert, 1979; Slonimski, 1980; Craik *et al.*, 1980). But there is already a counterexample for this contention: there are no intervening sequences anywhere in the mitochondrial DNA of animals, including man, and the putative segments responsible for encoding cytochrome *b* are no exception to the rule (Attardi *et al.*, 1980). Vertebrate animals are far removed in an evolutionary sense from yeast, and the mitochondrial genome is known to be subject to very rapid and profound evolutionary changes (reviewed in Mahler and Raff, 1976; Mahler and Perlman, 1979; Mahler, 1980). So, a critical test will come from determinations whether different yeast species, in particular ones with a mitochondrial DNA of considerably smaller size, such as *Schizosaccharomyces pombe* (mtDNA molecular weight 12,000,000; P. S. Perlman, private communication), retain *any* intervening sequences. The question has been answered in the case of another fungus, *Asperigullus nidulans*, by studies of Waring *et al.* (1981). In this case the gene for cytochrome *b* is split; it contains a single intron, structurally homologous to intron I_{3l} of yeast (Lazowska *et al.*, 1981), which we will recall is indispensable in that organism.

ACKNOWLEDGMENTS. Studies in the author's laboratory have been supported by Research Grant GM 12228 from the National Institute of General Medical Sciences; he holds a Research Career Award (KO6 05060) from the same Institute. It is a pleasure to acknowledge continued productive collaboration and fruitful discussions with Professor P. S. Perlman, Department of Genetics, Ohio State University.

REFERENCES

Abelson, J. (1979). *Annu. Rev. Biochem.* **48**, 1035–1069.

Alexander, N. J., Vincent, R. D., Perlman, P. S., Miller, D. H., Hanson, D. K., and Mahler, H. R. (1979). *J. Biol. Chem.* **254**, 2471–2479.

Alexander, N. J., Perlman, P. S., Hanson, D. K., and Mahler, H. R. (1980). *Cell* **20**, 199–206.

Attardi, G., Cantatore, P., Ching, E., Crews, S., Gelfand, R., Merkel, C., Montoya, J., and Ojala, D. (1980). In *The Organization and Expression of the Mitochondrial Genome* (A. M. Kroon and C. Saccone, eds.), pp. 103–120, North-Holland, Amsterdam.

Borst, P., and Grivell, L. A. (1978). *Cell* **15**, 705–723.

Church, G. M., and Gilbert, W. (1980). In *Mobilization and Reassembly of Genetic Information* (D. R. Joseph, J. Schultz, W. A. Scott, and R. Werners, eds.), pp. 379–395, Academic Press, New York.

Church, G. M., Slonimski, P. P., and Gilbert, W. (1979). *Cell* **18**, 1209–1215.

Claisse, M. L., Spyridakis, A., Wambier-Kluppel, M. L., Pajot, P., and Slonimski, P. P. (1978). In *Biochemistry and Genetics of Yeast* (M. Bacila, B. L. Horecker, and A. O. M. Stoppani, eds.), pp. 369–390, Academic Press, New York.

Claisse, M. L., Slonimski, P. P., Johnson, J., and Mahler, H. R. (1980). *Mol. Gen. Genet.* **177**, 375–387.

Colson, A. M., and Slonimski, P. P. (1978). *Mol. Gen. Genet.* **167**, 287–298.

Colson, A. M., Michaelis, G., Pratje, E., and Slonimski, P. P. (1979). *Mol. Gen. Genet.* **167**, 299–300.

Craik, C. S., Buchman, S. B., and Beychok, S. (1980). *Proc. Natl. Acad. Sci. USA* **77**, 1384–1388.

Crick, F. H. C. (1979). *Science* **204**, 264–271.

Darnell, J. E., Jr. (1978). *Science* **202**, 1257–1260.

Dhawale, S., Hanson, D. K., Alexander, N. J., Perlman, P. S., and Mahler, H. R. (1981). *Proc. Natl. Acad. Sci. USA*, **78**, 1778–1782.

Dujon, B. (1979). *Nature (London)* **282**, 777–778.

Gilbert, W. (1979). In *Eucaryotic Gene Regulation* (R. Axel, T. Maniatis, and C. F. Fox, eds.), pp. 1–12, Academic Press, New York.

Grivell, L. A., Arnberg, A. C., Hensgens, L. A. M., Roosendall, E., van Ommen, G.-J. B., and van Bruggen, E. F. J. (1980). In *The Organization and Expression of the Mitochondrial Genome* (A. M. Kroon and C. Saccone, eds.), pp. 37–50, North-Holland, Amsterdam.

Haid, A., Schweyen, R. J., Bechmann, H., Kaudewitz, F., Solioz, M., and Schatz, G. (1979). *Eur. J. Biochem.* **94**, 451–464.

Haid, A., Grosch, G., Schmelzer, C., Schweyen, R. S., and Kaudewitz, F. (1980). *Curr. Genet.* **1**, 155–161.

Halbreich, A., Pajot, P., Foucher, M., Grandchamp, C., and Slonimski, P. P. (1980). *Cell* **19**, 321–329.

Hanson, D. K., Miller, D. H., Mahler, H. R., Alexander, N. J., and Perlman, P. S. (1979). *J. Biol. Chem.* **254**, 2480–2490.

Jacq, C., Lazowska, J., and Slonimski, P. P. (1980). In *The Organization and Expression of the Mitochondrial Genome* (A. M. Kroon and C. Saccone, eds.), pp. 139–152, North-Holland, Amsterdam.

Jacq, C., Pajot, P., Lazowska, J., Dujardin, G., Claisse, M., Groudinsky, O., de la Salle, H., Grandchamp, C., Labouesse, M., Gargouri, A., Guiard, B., Spyridakis, A., Dreyfus, M., and Slonimski, P. P. (1982). In *Mitochondrial Genes* (P. Slonimski, P. Borst, and G. Attardi, eds.), in press.

Kreike, J., Bechmann, H., Van Hemert, F. J., Schweyen, R. J., Boer, P. H., Kaudewitz, F., and Groot, G. S. P. (1979). *Eur. J. Biochem.* **101**, 607–617.

Kroon, A. M., and Saccone, C. (eds.) (1980). *The Organization and Expression of the Mitochondrial Genome*, North-Holland, Amsterdam.

Lamouroux, A., Pajot, P., Kochko, A., Halbreich, A., and Slonimski, P. P. (1980). In *The Organization and Expression of the Mitochondrial Genome* (A. M. Kroon and C. Saccone, eds.), pp. 139–152, North-Holland, Amsterdam.

Lazowska, J., Jacq, C., and Slonimski, P. P. (1980). *Cell* **22**, 333–348.

Lazowska, J., Jacq, C., and Slonimski, P. P. (1981). *Cell* **27**, 12–14.

Lewin, B. (1980). *Cell* **22**, 324–326.

Mahler, H. R. (1980). *Ann. N.Y. Acad. Sci.*, **361**, 53–75.

Mahler, H. R., and Perlman, P. S. (1971). *Biochemistry* **10**, 2979–2990.

Mahler, H. R., and Perlman, P. S. (1979). In *Extrachromosomal DNA* (D. Cummings, P. Borst, I. Dawid, S. Weissman, and C. F. Fox, eds.), Vol. XV, pp. 11–33, Academic Press, New York.

Mahler, H. R., and Raff, R. A. (1976). *Int. Rev. Cytol.* **43**, 1–124.

Mahler, H. R., Hanson, D., Miller, D., Lin, C. C., Alexander, N. J., Vincent, R. D., and Perlman, P. S. (1978). In *Biochemistry and Genetics of Yeasts* (M. Bacila, B. Horecker, and A. O. M. Stoppani, eds.), pp. 513–547, Academic Press, New York.

Mahler, H. R., Perlman, P. S., Hanson, D. K., Lamb, M. R., Anziano, P. G., Glaus, K. R., and Haldi, M. L. (1982). In *Mitochondrial Genes* (P. Slonimski, P. Borst, and G. Attardi, eds.), in press, Cold Spring Harbor Laboratory.

Neupert, W., and Schatz, G. (1981). *Trends in Biochemical Sciences* **6**, 1–4.

Nobrega, F. C., and Tzagoloff, A. (1980). *J. Biol. Chem.* **255**, 9828–9837.

Pajot, P., Wambier-Kluppel, M. L., and Slonimski, P. P. (1977). In *Mitochondria 1977* (W. Bandlow, R. J. Schweyen, K. Wolf, and F. Kaudewitz, eds.), pp. 173–184, de Gruyter, Berlin.

Perlman, P. S., Alexander, N. J., Hanson, D. K., and Mahler, H. R. (1980a). In *Gene Structure and Expression* (D. H. Dean, L. F. Johnson, P. C. Kimbal, and P. S. Perlman, eds.), pp. 210–253, Ohio State University, Columbus.

Perlman, P. S., Mahler, H. R., Dhawale, S., Hanson, D., and Alexander, N. J. (1980b). in *The Organization and Expression of the Mitochondrial Genome* (A. M. Kroon and C. Saccone, eds.), pp. 161–172, North-Holland, Amsterdam.

Roberts, H., Smith, C. C., Marzuki, S., and Linnane, W. A. (1980). *Arch. Biochem. Biophys.* **200,** 387–395.

Schatz, G., and Mason, T. (1974). *Annu. Rev. Biochem.* **43,** 51–87.

Schmelzer, C., Haid, A., Grosch, G., Schweyen, R. J., and Kaudewitz, F. (1981). *J. Biol. Chem.* **256,** 7610–7619.

Slonimski, P. P. (1980). *C. R. Acad. Sci.* **290,** 331–334.

Slonimski, P. P., Borst, P., and Attardi, G., eds. (1982). *Mitochondrial Genes,* in press.

Solioz, M., and Schatz, G. (1979). *J. Biol. Chem.* **254,** 9331–9334.

Tzagoloff, A., Macino, G., and Sebald, W. (1979). *Annu. Rev. Biochem.* **49,** 419–441.

van Ommen, G.-J. B., Boer, P. H., Groot, G. S. P., de Haan, M., Roosendall, E., Grivell, L. A., Haid, A., and Schweyen, R. J. (1980). *Cell* **20,** 172–183.

von Jagow, G., and Sebald, W. (1980). *Annu. Rev. Biochem.* **49,** 281–314.

Waring, R. B., Davies, R. W., Lee, S., Grisi, E., McPhail Berks, M., and Scazzocchio, C. (1981). *Cell* **27,** 4–11.

Weiss, H., Wingfield, P., and Leonard, K. (1979). In *Membrane Bioenergetics* (C. P. Lee, G. Schatz, and L. Ernster, eds.), pp. 119–132, Addison–Wesley, Reading, Mass.

65

The Structure of Mitochondrial Ubiquinol : Cytochrome c Reductase

Kevin Leonard and Hanns Weiss

Cytochrome reductase (ubiquinol : cytochrome c reductase, EC 1.10.2.2) is a major enzyme of the mitochondrial oxidative phosphorylation system (Rieske, 1976; Crane, 1977; Hatefi and Galante, 1978; von Jagow and Sebald, 1980). The enzyme catalyzes electron transfer from ubiquinone to cytochrome c and utilizes the energy of the redox reaction to translocate protons across the mitochondrial inner membrane (Mitchell, 1975; Papa, 1976). The transmembranous proton gradient can be used by the ATP synthetase to drive the synthesis of ATP from ADP and phosphate (Boyer et $al.$, 1977).

Cytochrome reductase has a molecular weight of approximately 550,000. The enzyme is a dimer, the monomeric unit consisting of at least eight different subunits. Three of the subunits carry redox centers; these are the cytochromes b (molecular weight $\sim 30,000$) and c_1 ($\sim 31,000$), and the iron–sulfur subunit ($\sim 25,000$). Five subunits (50,000 45,000 14,000 12,000 and 8000) probably do not have prosthetic groups (Weiss and Kolb, 1979; Engel et $al.$, 1980). The enzyme contributes about 10% of the inner mitochondrial membrane protein.

Cytochrome reductase is also of interest with respect to the genetics and biogenesis of mitochondria because the cytochrome b subunit belong to the small group of proteins that are translated on mitochondrial ribosomes (Weiss and Ziganke, 1974; Katan $et.$, 1976) and coded by mitochondrial DNA (Tzagoloff et $al.$, 1975). The DNA of the cytochrome b gene from mammalian and yeast mitochondria has recently been sequenced (Barrell et $al.$, 1980; Tzagoloff and Nobrega, 1980).

We have isolated cytochrome reductase from $Neurospora$ $crassa$. The enzyme was first released from the cytochrome c-depleted mitochondrial membranes by Triton X-100. A micelle (Molecular weight $\sim 100,000$) of this nonionic detergent replaces (most of) the phospholipids. The enzyme–detergent complex was then purified by affinity chromatography on immobilized cytochrome c and gel chromatography. It has a Stokes radius of

Kevin Leonard and Hanns Weiss • European Molecular Biology Laboratory, 6900 Heidelberg, West Germany.

86 Å and a sedimentation coefficient $s_{20,w}$ of 15.6 (Weiss and Kolb, 1979). The enzymatic activity is retained when (most of) the phospholipid in the vicinity of the hydrophobic part of the enzyme is replaced by a micelle of nonionic [alkyl (phenyl) polyoxyethylene] detergent. However, the following three precautions must be taken into account in order to measure accurately the Michaelis constants and maximal activity (Weiss and Wingfield, 1979). (1) In nonionic detergents, the enzymes and the hydrophobic substrate ubiquinone-10 are inserted into discrete micelles. The substrate will partition freely among all micelles, namely enzyme-bound micelles and free micelles, and so will mostly be in a discontinuous phase with respect to the enzyme. For an enzymatic reaction to occur, the substrate must transfer from free micelles to enzyme-bound micelles. (2) The substrate transfer reaction may be slower than the enzymatic reactions. The kinetics of the transfer reaction are, however, negligible when the enzymatic reaction is coupled to an auxiliary nonenzymatic reaction that rapidly converts the ubiquinone back into ubiquinol within the enzyme-bound micelle. If duroquinol, for example, is used for the reduction of ubiquinone, the substrate transfer reaction, which determines the concentration of substrate within the enzyme-bound micelle, will be under thermodynamic rather than kinetic control. (3) Using nonionic detergents with varying polyoxyethylene and alkyl chain lengths, the kinetics of the substrate transfer and the intrinsic enzymatic activity may both be changed in different ways.

After purification, the cytochrome reductase in detergent solution was sequentially dissociated into subunit complexes and subunits. By mild salt treatment in Triton X-100 solution, three parts were obtained: (1) A subunit complex that contains the cytochromes b and c_1 and the three small subunits without redox centers; (2) a subunit complex without redox centers that contains the 45,000- and 50,000-molecular-weight subunits; (3) an iron–sulfur subunit (Hovmöller et al., 1981). By salt treatment in deoxycholate solution, the cytochrome bc_1 subunit complex was dissociated and the cytochromes b and c_1 were obtained in monodisperse form (unpublished result).

According to their solubility in aqueous buffer or detergent solution, the subunits can be subdivided into three groups. (1) Hydrophobic subunits: These bind detergent micelles after isolation and are therefore assumed to span the membrane. The cytochrome b and iron–sulfur subunits belong to this group. (2) Amphiphilic subunits: In this case only a small part of the subunit is bound to a detergent micelle and after this has been cleaved off by proteolysis the subunit is water soluble. The subunit is therefore assumed to extend into the water and to be anchored to the membrane by a small hydrophobic part. The cytochrome c_1 subunit is of this type (Li et al., 1981). (3) Hydrophilic subunits: These are water soluble without detergent and are assumed to extend into the aqueous phase. The 50,000- and 45,000-molecular-weight subunits without redox centers belong to this group.

The amphiphilic cytochrome c_1 subunit (Molecular weight 31,000) and the water-soluble cytochrome c_1 (molecular weight 24,000) preparations bind ferricytochrome c with high affinity. Cytochrome c_1 is therefore assumed to be located at the outer membrane surface (Li et al., 1981).

The complete amino acid sequence of bovine heart cytochrome c_1 was recently published (Wakabayashi et al., 1981). The results indicate that the heme is located near the NH_2-terminal region and that the only continuous hydrophobic fragment is located near the COOH terminus. As this hydrophobic fragment consists of only 15 residues, it is too small to span the bilayer. It is probable that it penetrates the bilayer in such a way that it bends back upon itself to form a hairpinlike bend. Thus, both the NH_2 and the COOH termini of cytochrome c_1 would be located on the outer surface of the mitochondrial inner membrane.

Membrane crystals have been prepared from the cytochrome reductase and the cytochrome bc_1 subunit complex both of *Neurospora crassa*. This was carried out by mixing the enzyme–Triton X-100 complex with mixed phospholipid–Triton X-100 micelles and subsequently removing the Triton (Wingfield *et al.*, 1979; Hovmöller *et al.*, 1981) either by adsorption to polystyrene beads (Holloway, 1973) or by dialysis. When polystyrene beads were stirred in the solution, 98% of the Triton was removed within 2 hr; when the solution was dialyzed, it took 2–3 days to remove 90% of the Triton. The formation of the crystals did not require a specific phospholipid composition. The ratio of phosphatidyl-serine to phosphatidylcholine was varied from 0 to 60% without affecting the size or quality of the crystals. Membrane crystals were also obtained using synthetic dimyristoylphosphatidylcholine (Sigma) as the sole phospholipid (Leonard *et al.*, 1981).

The electron micrographs of the negatively stained membrane crystals diffract to 2.5 nm. The diffraction pattern of the cytochrome reductase crystals showed systematic absences along the h and k axes where h $(k) = 2n + 1$, which is consistent with pgg symmetry. The only possible symmetry of biological molecules that gives rise to a pgg plane group in projection is the two-sided plane group $p22_12_1$. In this symmetry the alternate dimeric enzymes are packed up-and-down across the membrane bilayer, i.e., the glide planes seen in projection in the image correspond to twofold screw axes in the plane of the membrane. The diffraction pattern of the cytochrome bc_1 membrane crystals showed only the two-fold symmetry imposed by Friedel's law. The computer-calculated phases for the diffraction pattern show an average deviation from 0° to 180° of 9° for the 13 strongest reflections, which is consistent only with $p21$ symmetry. This means that the molecules are related by a two-fold axis perpendicular to the plane of the membrane and are facing the same way. This is in contrast to the membrane crystals of cytochrome reductase where alternate dimeric molecules face up and down. The unit-cell dimensions for the cytochrome reductase are 13.7 nm $\times$ 17 nm = 240 nm² and for the cytochrome bc_1 subunit complex 8.1 nm $\times$ 14.2 nm = 115 nm². Crystals of the whole enzyme have two dimers per unit cell, crystals of the cytochrome bc_1 complex only one per unit cell.*

The dimensions of the cytochrome reductase in projection are roughly the same as those of the cytochrome bc_1 complex, namely 10 nm $\times$ 7 nm, although the cytochrome bc_1 complex (molecular weight $\sim$250,000) is only about one-half as big as the cytochrome reductase (molecular weight 550,000). The dimension of the cytochrome bc_1 complex in the direction perpendicular to the membrane, therefore, can only be about one-half that of the whole enzyme. Therefore, we assume that, in contrast to the cytochrome reductase, most of the protein of the cytochrome bc_1 complex is embedded in the bilayer (see below).

The three-dimensional structure of the cytochrome reductase (Fig. 1) was calculated by combining tilted views of the membrane crystals. The structure shows that the monomeric units of the enzyme are related by a twofold axis perpendicular to the membrane. They are elongated, extending approximately 15 nm across the membrane. The protein is unequally distributed with about 30% of the total mass located in the bilayer, 50% in a section that extends 7 nm from one side of the bilayer, and 20% in a section that extends 3 nm from the opposite side of the bilayer. The two monomeric units are in contact only in the membranous section.

By comparing this cytochrome reductase structure with the biochemical properties of the isolated subunits, we propose the following model. The cytochrome c_1 subunit contributes to that section of the enzyme that extends 3 nm from the membrane. Because cytochrome c_1 interacts with cytochrome c on the cytoplasmic side of the mitochondrial inner membrane (König *et al.*, 1980; Bosshard *et al.*, 1979), this section extends into the intermembrane space of mitochondria. The cytochrome b and iron–sulfur subunits are located

---

*This apparent symmetry is a result of unequal negative staining of the two sides of the crystal. A three-dimensional reconstruction of cytochrome bc₁ complex (Karlsson *et al.*, in preparation) has now shown that the true symmetry of these crystals is p22₁2₁ with two dimeric molecules per unit cell packed in the same way as cytochrome reductase.

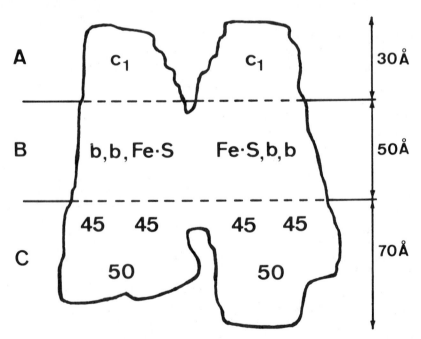

Figure 1. (Top) A model for one cytochrome reductase dimer, showing the two aqueous domains (dark) and the membrane domain (light). The section thickness is 1 nm and the overall height of the structure about 15 nm. (Bottom) A schematic drawing of the way in which the biochemical data for cytochrome reductase can be fitted to the three-dimensional model. The upper aqueous region (A) would correspond to the cytochrome c_1 on the cytoplasmic side of the inner mitochondrial membrane. The central region (B) corresponds to the membrane subunits, mainly cytochrome b and the iron–sulfur protein. The lower region (C) corresponds to the large water-soluble subunits on the matrix side of the membrane.

in the membrane. A complex containing the two large subunits (molecular weights 50,000 and 45,000) without redox centers accounts for that section of cytochrome reductase that extends 7 nm from the membrane into the mitochondrial matrix phase. We do not have the corresponding information about the 8000-, 12,000- and 14,000-molecular-weight subunits.

We are at present analyzing the three-dimensional structure of the cytochrome bc_1 subunit complex. One important difference between the structures of cytochrome reductase and cytochrome bc_1 complex might be the absence of the iron–sulfur subunit in the cytochrome bc_1 complex. This iron–sulfur subunit was suggested to be the electron donor for cytochrome c_1 (Trumpower and Edwards, 1979). A comparison between the two structures might give us new information about the mutual arrangement of cytochrome c_1 and the iron–sulfur subunit.

If the subunit topography that we propose is correct, all subunits of cytochrome reductase that take part in redox reactions are either located in the membrane or extend into the intermembrane space of mitochondria. A large subunit complex without redox centers extends into the matrix space. Whether it is involved in the proton-translocating activity of the enzyme, or has a regulatory function, is a challenging question.

REFERENCES

Barrell, B. G., Anderson, S., Bankier, A. T., de Bruijn, M. H. L., Chen, E., Coulson, A. R., Drouin, J., Epheron, J. C., Nierlich, D. P., Roe, B., Sanger, F., Schreier, P. H., Smith, A. J. H., Staden, R., and Young, J. G. (1980). In *Biological Chemistry of Organelle Formation* (T. Bücher, W. Sebald, and H. Weiss, eds.), pp. 11–26, Springer-Verlag, Berlin.

Bosshard, H. R., Zurrer, M., Schägger, H., and von Jagow, G. (1979). *Biochem. Biophys. Res. Commun.* **89,** 250–258.

Boyer, P. D., Chance, B., Ernster, L., Mitchell, P., Racker, E., and Slater, E. C. (1977). *Annu. Rev. Biochem.* **46,** 955–1026.

Crane, F. L. (1977). *Annu. Rev. Biochem.* **46,** 439–469.

Engel, W. D., Schägger, H., and von Jagow, G. (1980). *Biochim. Biophys. Acta* **592,** 211–222.

Hatefi, Y., and Galante, Y. M. (1978). In *Energy Conservation in Biological Membranes* (G. Schäfer and M. Klingenberg, eds.), pp. 19–30, Springer-Verlag, Berlin.

Holloway, P. S. (1973). *Anal. Biochem.* **53,** 304–308.

Hovmöller, S., Leonard, K., and Weiss, H. (1981). *FEBS Lett.* **123,** 118–122.

Katan, M. B., Pool, L., and Groot, G. S. P. (1976). *Eur. J. Biochem.* **65,** 95–105.

König, B. W., Schilder, L. T. M., Tervoort, M. J., and Van Gelder, B. F. (1980). *Biochim. Biophys. Acta,* **621,** 283–295.

Leonard, K., Wingfield, P., Arad, T., and Weiss, H. (1981). *J. Mol. Biol.* **149,** 259–274.

Li, Y., Leonard, K., and Weiss, H. (1981). *Eur. J. Biochem.* **116,** 199–205.

Mitchell, P. (1975). *FEBS Lett.* **56,** 1–6.

Papa, S. (1976). *Biochim. Biophys. Acta* **456,** 39–84.

Rieske, J. S. (1976). *Biochim. Biophys. Acta* **456,** 195–247.

Trumpower, B. L., and Edwards, C. A. (1979). *J. Biol. Chem.* **254,** 8697–8706.

Tzagoloff, A., and Nobrega, G. F. (1980). In *Biological Chemistry of Organelle Formation* (T. Bücher, W. Sebald, and H. Weiss, eds.), pp. 1–10, Springer-Verlag, Berlin.

Tzagoloff, A., Akai, A., Needleman, R. B., and Zulch, G. (1975). *J. Biol. Chem.* **250,** 8236–8242.

von Jagow, G., and Sebald, W. (1980). *Annu. Rev. Biochem.* **49,** 281–314.

Wakabayashi, S., Matsubara, H., Kim, C. H., Kawai, K., and King, T. E. (1980). *Biochem. Biophys. Res. Commun.* **97,** 1548–1554.

Weiss, H., and Kolb, H. J. (1979). *Eur. J. Biochem.* **99,** 139–149.

Weiss, H., and Wingfield, P. (1979). *Eur. J. Biochem.* **99,** 139–149.

Weiss, H., and Ziganke, B. (1974). *Eur. J. Biochem.* **41,** 63–71.

Wingfield, P., Arad, T., Leonard, K., and Weiss, H. (1979). *Nature (London)* **280,** 696–697.

VII

Ion Transport Systems in Animal Cells

66

Rules and the Economics of Energy Balance in Coupled Vectorial Processes

William P. Jencks

1. INTRODUCTION

We would like to understand what a coupled vectorial process (CVP) is and how it works. Progress toward this goal demands that the right questions be asked and answers to them found; we should know what questions need to be answered in order that we can say that we understand how a CVP works. It is frequently suggested that coupling in active transport and other CVPs results from changing affinities for ligands (e.g., Ca^{2+}, Na^+, H^+), "energized states," and particular steps with "energy coupling" (e.g., a "power stroke"). Such concepts may be useful if they are clearly defined, but can slow progress if they give a sense that a mechanism is understood when it is not.

Two separate and distinct questions are outlined here; more complete statements are available (Hill, 1977; Jencks, 1980a).

1. What are the *rules* that define the coupling process? The coupling commonly results from a stepwise, stoichiometric cyclic mechanism that is defined by a set of specificity rules, not from changing affinities for ligands.
2. What is the *energy balance* of the system that (a) keeps intermediates at concentrations that permit rapid turnover in the steady state, (b) conserves energy by avoiding premature irreversible steps, such as ATP hydrolysis, and (c) drives the overall process, which may be reversible or irreversible?

An understanding of these rules and of the utilization of binding energies in different states of the system might be said to constitute an understanding of the mechanism of a CVP.

William P. Jencks • Graduate Department of Biochemistry, Brandeis University, Waltham, Massachusetts 02254.

2. COUPLING

A cyclic CVP can be coupled by having two sets of states, each of which has a particular specificity [equation (1)]. For example, E_1 and E_2 can differ

$$E_1 \rightleftharpoons E_1^*$$
$$\Updownarrow \quad \Updownarrow$$
$$E_2 \rightleftharpoons E_2^* \tag{1}$$

in a geometric or vectorial property, such as the side of a membrane on which a ligand-binding site is exposed, and in their catalytic specificity; E and E* can differ in their chemical nature and specificity for ligand binding. These specificities are implicit in most models for such CVPs; they need to be explicitly identified and characterized. There is no particular step that can be identified with the coupling; the entire cyclic process gives stoichiometric transport or motion with each turnover.

The cycle for active transport by the Ca^{2+} (or Na^+)-ATPase provides an example for coupling in which the vectorial position and catalytic specificity are defined by E_1 (ion-binding site outside) and E_2 (binding site inside) and the chemical structure and specificity for ligand binding by E and E*, with E* a phosphoenzyme (Hasselbach, 1974). The basic rules are:

1. E_1 catalyzes the reversible transfer of phosphate from E–P to ADP (to give ATP), but not to water (to give P_i); E_2 has the converse specificity. This rule prevents the hydrolysis of ATP without ion transport.
2. E* undergoes reversible in–out exposure of its binding site *only* when the site is occupied by Ca^{2+} (or Na^+); E undergoes this transition only without Ca^{2+} (or with K^+ bound). This rule prevents ion transport without hydrolysis or synthesis of ATP.

In the coupling of mechanical work and ATP hydrolysis in muscle contraction, upon movement of a myosin head from one actin monomer, A_1, to another, A_2, the vectorial specificity is defined by A_1 and A_2 and the catalytic specificity by $A \cdot M$ and $M^* \cdot$ nucleotide [equation (2); Taylor, 1979]. The basic rules for coupling are:

$$A_1 \cdot M \xrightleftharpoons{\pm ATP} A_1 + M^* \cdot ATP$$
$$\Updownarrow$$
$$A_2 \cdot M \xrightleftharpoons{\pm ADP, P_i} A_2 + M^* \cdot ADP \cdot P_i \tag{2}$$

1. M* catalyzes the (reversible) hydrolysis of ATP and can move between sites A_1 and A_2, but does not dissociate nucleotides; AM does not catalyze hydrolysis and does not move between sites A_1 and A_2, but does reversibly bind ATP, ADP, and P_i. The catalytic specificities can be enforced by the kinetics of the cycle rather than by enzyme specificity per se; for example, A_1M can give M*ATP faster than it hydrolyzes ATP.
2. M*ATP combines only with A_1 (as it must, from microscopic reversibility), whereas M*ADP $\cdot P_i$ combines only with A_2. The difference can result from a conformation change upon ATP hydrolysis and, possibly, from movement of the thick filament to which M* is attached (the latter would be another example of kinetic specificity).

3. ENERGETICS

The primary function of the energy balance in a CVP is kinetic, as indicated in a, b, and c (Section 1). In order to maintain a reasonable turnover rate, it is necessary that the concentrations of all intermediate states be roughly comparable, so that the product of the concentrations and forward rate constants for each state are adequate for the overall steady-state rate. A very-low-energy intermediate will accumulate and tie up most of E, and a very-high-energy intermediate is difficult to form. ATP hydrolysis is used to drive the cycle and, sometimes, to make it irreversible (as in proofreading; Hopfield, 1974). Hydrolysis can occur at any point in the cycle if the kinetic requirements are satisfied.

These requirements are met by the operation of interaction energies to maintain the energy balance (Wyman, 1964; Weber, 1975). This is basically simple but is not widely understood, perhaps because it is basically a matter of economics, an unpleasant subject. It is necessary to pay for what you get, and in each case the energy balance is maintained by a payment: the intrinsic binding energy (i.b.e.) is *utilized* or not expressed in one state through some sort of destabilization mechanism, so that the observed binding energy in this state is correspondingly decreased (Jencks, 1975).

A simple example is the ATP-mediated dissociation of actomyosin, which is shown in the Gibbs energy diagram of Fig. 1A (these diagrams are drawn for standard states of 1 M). Strong intrinsic binding energies of myosin for actin and for ATP are expressed in $A \cdot M$ and $M \cdot ATP$, but in the ternary complex $A \cdot M \cdot ATP$ there is a mutual destabilization of the binding of ATP by A and of A by ATP by the amount ΔG_I (Highsmith, 1976). If there were no such destabilization, there would be equally tight binding in the ternary complex (dashed lines, which are parallel to the solid lines), $A \cdot M \cdot ATP$ would accumulate, and dissociation would not occur.

More complex and interesting examples are found in CVPs that bring about the reversible synthesis of energy-rich compounds through the utilization of binding energy, with no net input of energy. For example, the Ca^{2+}-ATPase brings about the spontaneous synthesis of an acyl phosphate from inorganic phosphate ($\Delta G° \cong -3$ kcal mole^{-1}) although the corresponding reaction in solution is unfavorable by $+10$ kcal mole^{-1} (Fig. 1B; Punzengruber *et al.*, 1978; Gerstein and Jencks, 1964). The enzymatic reaction may be divided into two steps. In the first step the high-energy bond of the acyl phosphate (ECOO~P) is formed without allowing contact of the phosphate group with its binding site, so that $\Delta G°$ for this reaction is the same as for the nonenzymatic reaction. In the second step the phosphate group is allowed to interact with its binding site (indicated by the dot in Fig. 1B) so that its intrinsic binding energy, ΔG_P^j, is expressed. This ΔG_P^j of ~ -13 kcal mole^{-1} provides the driving force that makes the reaction favorable. The system works because this ΔG_P^j is not expressed in the noncovalent complex with inorganic phosphate, $ECOO^- \cdot P_i$. Instead, it is *utilized* to overcome a destabilization energy, ΔG_D, and to freeze P_i with a loss of entropy, $-T\Delta S$, so that it is in the correct position to react. The bound P_i is in a relatively high (Gibbs) energy state, from which it can readily form the acyl phosphate. The difference in the expression of the i.b.e. of phosphate in E–P and $E \cdot P_i$ is the interaction energy, $\Delta G_I = \Delta G_D - T\Delta S$ (Jencks, 1975, 1980a).

In order to maintain comparable concentrations of intermediate states and reversibility under physiological conditions, it is necessary that (1) the covalent E–P intermediate of the Ca^{2+}-ATPase must exist in two states, one of which is at equilibrium with the energy-rich phosphate of ATP and the other with low-energy P_i, and (2) the binding of calcium to E must be strong outside and weak inside the vesicle. These requirements are nicely met by the balance of binding energies shown in Fig. 1C. The i.b.e. of Ca^{2+} is expressed

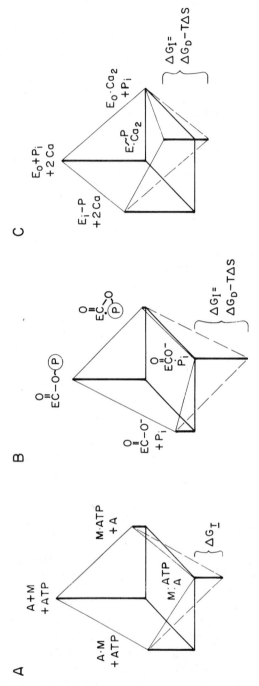

Figure 1. Gibbs free-energy diagrams for: (A) the binding of actin and ATP to myosin; (B) the reaction of inorganic phosphate with the Ca^{2+}-ATPase to form a noncovalent complex (bottom), an acyl phosphate with no noncovalent binding to the enzyme (top), and an acyl phosphate with expression of the intrinsic binding energy of the phosphate group (right); (C) binding to the Ca^{2+}-ATPase of calcium and of phosphate, to give E–P, separately and together. The diagrams are drawn for standard states of 1 M; the Gibbs energies are different under physiological concentrations.

outside to give tight binding to E, and the i.b.e. of phosphate is expressed to give a stable E–P in the absence of Ca^{2+}, as described above. However, there is a mutual destabilization, ΔG_I, in the ternary complex, E$\sim$$P$ · Ca^{2+}_2, so that phosphorylation of E · Ca^{2+}_2 gives an unstable, energy-rich phosphate. The system has the important properties that (1) the species E$\sim$$P$ ·$2Ca^{2+}$ does not undergo hydrolysis and (2) Ca^{2+} cannot dissociate from this species to the outside; it can dissociate to the inside and the destabilization energy makes the dissociation constant large so that this occurs readily.

The facile synthesis of bound ATP from bound ADP and P_i at the active site of myosin ATPase reflects the same kind of expression of binding energy in one state but not the other state of the system. The i.b.e. of the adenine, ribose, and phosphate groups is expressed in M* · ATP and stabilizes this species; in M* · ADP ·P much of the i.b.e. is *utilized* to bring together ADP and P_i in the correct position ($-T\Delta S$) and overcome destabilization energy (ΔG_D) that is relieved upon ATP formation. The sum of these two terms is the interaction energy that makes the system work, $\Delta G_I = \Delta G_D - T\Delta S$. In this case, ΔG_I results from approximately equal contributions of (1) "freezing" and destabilization when ADP and P_i are bound individually and (2) an unfavorable mutual interaction when they are bound together (Jencks, 1981).

These examples illustrate how enzymes catalyze CVPs by acting as simple machines that express i.b.e. in one state but not in another state; the difference is an interaction Gibbs energy composed of ΔG_D and $-T\Delta S$. The same mechanism accounts for catalysis of reactions of specific substrates by enzymes through utilization of the i.b.e. of nonreacting groups to stabilize the transition state but not the ES complex; the interaction energy in the ES complex, $\Delta G_I = \Delta G_D - T\Delta S$, makes it easier to reach the transition state (Pauling, 1948; Jencks, 1975, 1980b). The same mechanism also provides the driving force for conversion of a receptor from an inactive to an active form (or vice versa) when a hormone, neurotransmitter, drug, or other effector binds. The i.b.e. of the effector is expressed in the active, high-energy state of the receptor and is not expressed in the inactive state; if the i.b.e. is expressed in both states, the receptor remains in the low-energy, inactive state and the effector is an antagonist.

However, the most important problem in characterizing CVPs is to identify the rules that are responsible for coupling and to determine how they are enforced. These rules are basically expressions of enzyme specificity, but CVPs differ from most enzymatic reactions in that different states of the enzyme have at least two *different* specificities that are coupled to a vectorial component.

Acknowledgment. All of the above was made possible by numerous helpful discussions with investigators of CVPs.

REFERENCES

Gerstein, J., and Jencks, W. P. (1964). *J. Am. Chem. Soc.* **86,** 4655–4663.

Hasselbach, W. (1974). In *The Enzymes* (P. D. Boyer, ed.), 3rd ed., Vol. 10, pp. 431–467, Academic Press, New York.

Highsmith, S. (1976). *J. Biol. Chem.* **251,** 6170–6172.

Hill, T. L. (1977). *Free Energy Transduction in Biology,* Academic Press, New York.

Hopfield, J. J. (1974). *Proc. Natl. Acad. Sci. USA* **71,** 4135–4139.

Jencks, W. P. (1975). *Adv. Enzymol.* **43,** 219–410.

Jencks, W. P. (1980a). *Adv. Enzymol.* **51,** 75–106.

Jencks, W. P. (1980b). In *Molecular Biology, Biochemistry and Biophysics* (F. Chapeville and A.-L. Haenni, eds.), Vol. 32, pp. 3–25, Springer-Verlag, Berlin.

Jencks, W. P. (1981). *Proc. Natl. Acad. Sci. USA* **78,** 4046–4050.

Pauling, L. C. (1948). *Am. Sci.* **36,** 51–88.

Punzengruber, C., Prager, R., Kolassa, N., Winkler, F., and Suko, J. (1978). *Eur. J. Biochem.* **92,** 349–359.

Taylor, E. W. (1979). *Crit. Rev. Biochem.* **6,** 103–164.

Weber, G. (1975). *Adv. Protein Chem.* **29,** 1–83.

Wyman, J., Jr. (1964). *Adv. Protein Chem.* **19,** 223–286.

67

Na^+/K^+-ATPase

Structure of the Enzyme and Mechanism of Action of Digitalis

Arnold Schwartz and John H. Collins

1. INTRODUCTION

There seems to be no doubt that the Na^+/K^+-ATPase is the enzyme that represents the machinery of active transport of sodium and potassium across cell membranes. Very early in evolution, critical ion gradients must have been established; movement involving the expenditure of energy across membrane barriers must also have represented an early development, for without this mechanism, important biological processes could not exist. These include nerve excitability, motility, maintenance and possible regulation of cell volume, excretion, and reabsorption, a variety of sodium-dependent or -coupled transport processes, and electrical and mechanical activity of muscle. While an ATP-dependent process for transport was recognized for many years, it was not until the work of J. C. Skou (1957, 1960) that a specific enzyme system was recognized. Among the criteria for recognition of this enzyme is the very potent inhibition of ATP hydrolysis by cardiac glycosides. Thousands of papers have been published using ouabain, a water-soluble glycoside, as a marker for this important enzyme. When we examine the structure of a typical cardiac glycoside, we are struck with its apparent similarity to hormonal steroids. The geometrical configuration, however, is quite different. Like the steroids, the glycosides are used in very low concentrations to produce their effects. The primary effect on the intact organism is an increased force of contraction of heart muscle, a discovery made by William Withering more than 200 years ago. This class of drugs is still the most widely used in the treatment of heart failure. The fascination with this drug also arises from its very potent inhibitory effect on Na^+/K^+-ATPase, so it is not illogical that it is employed as a "chemical tool"

Arnold Schwartz and John H. Collins • Department of Pharmacology and Cell Biophysics, University of Cincinnati College of Medicine, Cincinnati, Ohio 45267.

to, on the one hand, dissect the mechanism of action of the Na^+/K^+-ATPase and on the other hand to search for the mechanism of its action on the heart. Since the early studies of Solomon *et al.* (1956) and Repke (1963), a concept has developed that the inhibition of the Na^+/K^+-ATPase by ouabain is directly associated with and indeed causes the positive inotropic effect. The evidence that ouabain in low concentrations inhibits the enzyme activity by a very specific binding inducing a conformation that is resistant to the action of potassium, is solid (Schwartz *et al.*, 1975; Wallick *et al.*, 1979; Akera and Brody, 1978; Schwartz and Adams, 1980). The number of experiments that have been carried out in attempts to prove or disprove that the Na^+/K^+-ATPase is the pharmacological receptor for digitalis is overwhelming, which is certainly an indication that the problem is far from solved. Whether or not the receptor turns out to be Na^+/K^+-ATPase, elucidation of the exact molecular interaction of ouabain with the enzyme would certainly provide information regarding the structure, function, and mechanism of enzyme action. In this article, we summarize recent developments in elucidation of the structure of this enzyme as it bears on the site(s) of ouabain binding. We will also summarize the current status on the mechanism of pharmacological action.

2. SUBUNIT STRUCTURE AND STOICHIOMETRY

Na^+/K^+-ATPase has been purified from mammalian kidney, dogfish rectal gland, and eel electric organ (Wallick *et al.*, 1979). In all cases the highly purified enzyme has been found to contain two major protein subunits: α, with a molecular weight of 85,000–140,000, and β, with a molecular weight of 35,000–67,000 (for references see Craig and Kyte, 1980). In the past 3 years, evidence has been accumulating that there may be a third, γ subunit of molecular weight about 12,000 (Forbush *et al.*, 1978; Rogers and Lazdunski, 1979; Reeves *et al.*, 1980). The question of the existence and the function of a possible γ subunit in the enzyme is considered below. The α subunit, also known as the catalytic subunit, spans the cell membrane and is oriented asymmetrically, with at least part of the receptor site for cardiac glycosides on the external surface and the site for ATP hydrolysis on the internal surface. The β subunit is a glycoprotein of unknown function, although there is some evidence (Hall and Ruoho, 1980) that β may be involved in binding of cardiac glycosides. On the basis of chemical cross-linking and ligand-binding studies, the functional unit of the enzyme appears to be $(\alpha\beta)_2$, although there is still some question on this matter (for review and discussion see Craig and Kyte, 1980).

The α and β subunits have been isolated from dog kidney (Kyte, 1972), shark rectal gland and eel electric organ (Perrone *et al.*, 1975), duck nasal gland (Hopkins *et al.*, 1976), lamb kidney (Lane *et al.*, 1979), and brine shrimp (Peterson and Hokin, 1980). The β subunits from the various sources are similar in size and amino acid composition, and in each case the NH_2-terminal amino acid residue is alanine. The α subunits from these sources are also very similar in size and amino acid composition, although they do not all contain the same NH_2-terminal amino acid (see below). In brine shrimp and eel electric organ, α has been found to contain carbohydrate (Churchill *et al.*, 1979; Peterson and Hokin, 1980); this has not been reported to be the case in the kidney and duck salt gland enzymes. The β subunit is always a glycoprotein, and there is considerable species variation in its carbohydrate composition.

3. PRIMARY STRUCTURE OF THE α SUBUNIT

Information on the amino acid sequences of the protein subunits of Na$^+$/K$^+$-ATPase is at present very limited. We are currently engaged in a determination of the complete sequence of the lamb kidney α subunit; to the best of our knowledge, other laboratories are not working in this area. The NH$_2$-terminal amino acid of α is glycine in the case of lamb kidney (Collins, unpublished data), dog kidney (Kyte, 1972), and duck nasal salt gland (Hopkins *et al.*, 1976), alanine in shark rectal gland, and serine in eel electric organ (Perrone *et al.*, 1975). The sequence of the 11 NH$_2$-terminal residues of duck salt gland α is Gly-Arg-Asn-Lys-Tyr-Glu-Thr-Ala-X-Gly, where X is an unidentified residue (Hopkins *et al.*, 1976). A high degree of conservation is indicated by our finding (unpublished) that the NH$_2$-terminal sequence of lamb kindey α is Gly-Arg-Asn-Lys-Tyr-Glu, and by the finding of Castro and Farley (1979) that the NH$_2$-terminal sequence of dog kidney α is Gly-Arg-Asx.

Recent studies (Castro and Farley, 1979; Farley *et al.*, 1980) on limited tryptic and chymotryptic digests of purified dog kidney Na$^+$/K$^+$-ATPase have yielded information on the location of functionally important amino acid residues along the α polypeptide chain. Depending on whether Na$^+$ or K$^+$ was present in the medium during digestion, the α subunit was degraded to produce six well-defined fragments of molecular weight 35,000–77,000. These overlapping fragments could be aligned within the sequence of α, and as a result the approximate locations (within a span of about 160 amino acid residues) of the phosphorylation site and a reactive SH group could be determined. It was also determined that those parts of α that are embedded within the lipid bilayer are probably all located in the COOH-terminal two-thirds of the polypeptide chain. These results were interpreted as suggesting that there may be two structural domains on α: an NH$_2$-terminal, cytoplasmic domain that contains the phosphorylation site and a COOH-terminal domain that contains the membrane-embedded segment(s) of the molecule.

Our own work on the sequencing of lamb kidney α (Collins and Lane, 1978; Collins, unpublished data) has to date mostly been confined to the tryptic peptides of the carboxy-methylated and succinylated protein. We prepared these peptides by procedures very similar to those used by Allen (1980) in his studies on the primary structure of the Ca^{2+}-ATPase of sarcoplasmic reticulum. At this writing, we have purified 17 (out of an expected 41) small peptides and are in the process of sequencing them. The completion of this project will require several more years.

4. PROTEOLIPID (γ SUBUNIT)

Proteolipids are small, very hydrophobic proteins that are soluble in organic solvents, such as chloroform–methanol mixtures. In 1972, De Robertis and colleagues (Rivas *et al.*, 1972) provided data that ouabain may bind to a proteolipid of eel electroplax Na$^+$/K$^+$-ATPase. Several years later, Racker (1976) suggested that Na$^+$/K$^+$-ATPase contains a proteolipid of molecular weight about 10,000 that may form an ion channel through the membrane, as has been found with other membrane ATPases. More recently, Forbush *et al.* (1978) found that photoaffinity labeling of the cardiac glycoside-binding site of pig kidney Na$^+$/K$^+$-ATPase resulted in about half of the label being bound to α and half to a proteolipid of molecular weight about 12,000, suggesting that both proteins may be involved in the formation of the glycoside receptor site. In a similar study, Rogers and Lazdunski

(1979) also observed a labeled small component in the Na^+/K^+-ATPase of eel electric organ and of cardiac plasma membranes. More recent photoaffinity labeling studies (Forbush and Hoffman, 1979; Hall and Ruoho, 1980; Rossi *et al.*, 1980) suggest that the primary region which recognizes the lactone ring and steroid portions of the molecule of cardiac glycoside binding is on α, while α, β, and the proteolipid all form a secondary region that binds to the carbohydrate portion.

As part of our own study on the structure of lamb kidney Na^+/K^+-ATPase, we devised a procedure for the large-scale isolation of the proteolipid, which we have designated the γ component of Na^+/K^+-ATPase (Reeves *et al.*, 1980). We separated γ into two similar components ($\gamma1$ and $\gamma2$); our recent data (Collins *et al.*, unpublished) indicate that $\gamma1$ is an aggregated form of $\gamma2$. The γ is obtained in a total yield of 0.9–2.1 moles per mole of α recovered, suggesting that γ is present in stoichiometric amounts in the enzyme. In an earlier report from our laboratory, Dowd *et al.* (1976) found a small (molecular weight 11,700) protein component of unknown function in beef cardiac Na^+/K^+-ATPase preparations that is phosphorylated by protein kinase. The proteolipid nature of this component was not investigated, but its amino acid composition is similar to our lamb kidney γ.

5. *DIGITALIS–Na^+/K^+-ATPase INTERACTION*

The early studies of Schatzmann and others using intact systems (Schwartz *et al.*, 1975) established that cardiac glycosides specifically inhibit sodium and potassium transport in intact systems. In 1957 and 1960, Skou correlated the inhibition of ion transport with an inhibition of Na^+/K^+-ATPase activity associated with microsomal membranes of crab nerve. Since that time, many experiments have corroborated these studies and extended them to the purified enzyme. We first showed that the inhibition of enzyme activity was related to the concentration of sodium and potassium with the former increasing the apparent affinity and the latter decreasing the affinity of the enzyme for the drug (Matsui and Schwartz, 1966a,b, 1968). This is certainly consistent with the well-known "antagonistic" interaction between digitalis and potassium at the outer surface of the cell membrane. A plausible mechanism of action based upon these studies is described as follows:

$$E_1 + ATP \overset{Na^+}{\rightleftharpoons} E_1P + ADP$$
$$E_1P \rightleftharpoons E_2P \overset{Dig}{\longrightarrow} Dig \cdot E_2P \text{ (stable conformer)}$$
$$\downarrow K^+$$
$$E_2 + P_i$$

Ouabain, interacting with a phosphorylated conformation of the Na^+/K^+-ATPase, converts it to some type of digitalis-stable Na^+/K^+-ATPase conformer whose affinity for potassium may be much less than the non-digitalis-treated enzyme. By this means the *rate* of dephosphorylation of the digitalis–enzyme complex would be very slow. In addition, ouabain prevents phosphorylation. In this way an inhibition of ATPase as well as of ion transport in intact systems occurs. To test this hypothesis, we prepared a crude ATPase enzyme from brain and from heart and carried out a radioligand experiment that revealed saturable binding. The rate was markedly stimulated by sodium and was inhibited by potassium. An analysis revealed that the apparent affinity of the enzyme for digitalis in the presence of magnesium, ATP, and sodium was on the order of $10^8 M^{-1}$ and that this affinity could be lowered by including potassium in the binding medium (Matsui and Schwartz, 1968;

Schwartz *et al.*, 1968). It is of interest with respect to pharmacological action that this study represents the first radioligand drug binding to an isolated putative receptor. Albers *et al.* (1968) reproduced these results using Na$^+$/K$^+$-ATPase isolated from the electroplax of the eel, and extended these studies to suggest that digitalis interacted with the highest affinity to a phosphorylated "intermediate" of the enzyme. Further work showed that ouabain can bind to the enzyme in the presence of a variety of ligands even in the absence of a phosphorylated enzyme. Moreover, it appears that under certain conditions the *regulation* of ouabain binding may occur at a monovalent cation site. At this site sodium, which stimulates the rate of binding, has a K_a of 14 mM while potassium, which prevents binding, has a K_a of 0.13 mM (Lindenmayer and Schwartz, 1973). These results were obtained on isolated membrane preparations and purified Na$^+$/K$^+$-ATPase. It was postulated that this site in intact systems would be an external activation site (K$^+$) for ion transport. Bodeman and Hoffman (1976) have indicated, however, that in systems with "sidedness," such as the red blood cell, external Na$^+$ does not stimulate the rate of binding of ouabain, unless external K$^+$ is also present. The opposite would be true at intracellular sites. Moreover, the data suggest that intracellular Na$^+$ and K$^+$ sites may "regulate" extracellular K$^+$ site(s) in antagonizing binding of ouabain. Robinson (1980), using a purified canine kidney Na$^+$/K$^+$-ATPase, recently presented data consistent with this notion. It is possible that K$^+$ at intracellular sites ("α sites") can somehow antagonize the binding of ouabain to external sites. The external K$^+$ site ("β site"), which has a higher affinity than the internal α site, activates Na$^+$/K$^+$-ATPase activity and perhaps may indirectly regulate digitalis action, by changing the "states" of the enzyme. The α-site for K$^+$, presumably on the inside of the membrane, appears to be the site that activates the K$^+$-phosphatase reaction (Blostein *et al.*, 1979). Some of the evidence Robinson used in his argument includes a protection by strophanthidin against enzyme modification by acetic anhydride and trinitrobenzenesulfonic acid, in a manner similar to protection by K$^+$ presumably at an intracellular (α) site, and a 20% activation of phosphatase activity by 10^{-4} M ouabain, in the absence of added K$^+$. While the observations and suggestions are interesting though complex, it is difficult to imagine how *intracellular* K$^+$, which is 150 mM unless severely compartmented, could modulate digitalis binding to an outside site.

6. PHARMACOLOGICAL ACTION OF CARDIAC GLYCOSIDES

The positive inotropic action of cardiac glycosides has been an intriguing topic for over a century. Evidence to date suggests an involvement of Na$^+$/K$^+$-ATPase. If there is an active sodium–calcium "carrier" in cardiac cell membranes, then a binding of the glycoside to the enzyme system could lead to a conformational change and inhibition as described above. This could produce either a steady-state increase in internal sodium or perhaps a transient increase in sodium (Akera, 1977) such that calcium from an internal site is displaced with each beat of the heart, resulting in an increased force. There are many problems with this concept, not the least of which is that a number of investigators have evidence dissociating the inhibition of the enzyme from a positive inotropic effect (for review see Noble, 1980). We suggested that one solution to this problem may involve the following: The drug would specifically bind to the membranes that contain Na$^+$/K$^+$-ATPase and this leads to a change in the affinity of phospholipids for calcium, which would provide an increased availability of intracellular calcium per beat (Schwartz and Adams, 1980). Lüllman and his co-workers have developed a very interesting kinetic model that could explain the suggestion that calcium for contraction is mobilized by cardiac glycosides from lipids and that the inhibition of the pump is not directly connected with inotropic activity (Lüllman and Peters, 1977).

Another problem is that a number of investigators have found that very low concentrations of cardiac glycosides appear to *stimulate* the Na^+/K^+-ATPase pump (Noble, 1980), and this interesting action may be associated with a *negative* inotropic action, no change at all in contraction or increased contraction. This highly complicated situation may be explained by the existence of two forms of the enzyme or two sites for glycoside–membrane interaction, as has been suggested by a number of investigators (for review see Wellsmith and Lindenmayer, 1980).

Because cardiac glycosides are so specific for Na^+/K^+-ATPase, it appears unlikely that the enzyme is not somehow connected with the positive inotropic action and hence is at least part of the pharmacological receptor. If this is not true, one wonders how to resolve the plethora of data that invoke a pharmacological role for this enzyme. If it is so, the parallel between the action of morphine and its congeners, and an endogenous receptor is striking. We have two plant principles, opioids and the cardiac glycosides, both of which produce effects at very low concentrations, and yet are not specifically found in the organism. It seems logical to expect that certain chemical principles are elaborated that may regulate the sodium pump. Such a substance might also have a positive inotropic action on the heart. This is not a new suggestion (Thorp and Cobbin, 1967). Hence, a search for such endogenous digitalis-like principles, which we call ENDODIGIN, that would interact with an external site on the membrane, is the subject of intense activity (Schwartz and Adams, 1980).

ACKNOWLEDGMENTS. The original studies cited were supported in part by Grants PO1 HL 22619 and AM 20875. We are very grateful to Dr. Lois K. Lane and Dr. Earl T. Wallick for their critical review of the manuscript and to Ms. Pam Smith for her expert typing.

REFERENCES

Akera, T. (1977). *Science* **198**, 569–574.
Akera, T., and Brody, T. M. (1978). *Pharmacol. Rev.* **29**, 187–220.
Albers, R. W., Koval, G. J., and Seigel, G. J. (1968). *Mol. Pharmacol.* **4**, 324–336.
Allen, G. (1980). *Biochem. J.* **187**, 545–563.
Blostein, R., Pershadsingh, M. A., Drapeau, P., and Chu, L. (1979). In *Na,K-ATPase: Structure and Kinetics* (J. C. Skou and J. B. Norby, eds.), Academic Press, New York.
Bodeman, H. H., and Hoffman, J. F. (1976). *J. Gen. Physiol.* **67**, 497–525.
Castro, J., and Farley, R. A. (1979). *J. Biol. Chem.* **254**, 2221–2228.
Churchill, L., Peterson, G. L., and Hokin, L. E. (1979). *Biochem. Biophys. Res. Commun.* **90**, 488–490.
Collins, J. H., and Lane, L. K. (1978). *Fed. Proc.* **37**, 1301.
Craig, W. S., and Kyte, J. (1980). *J. Biol. Chem.* **255**, 6262–6269.
Dowd, F., Pitts, B., and Schwartz, A. (1976). *Arch. Biochem. Bophys.* **175**, 321–331.
Farley, R. A., Goldman, D. W., and Bayley, H. (1980). *J. Biol. Chem.* **255**, 860–864.
Forbush, B., and Hoffman, J. F. (1979). *Biochim. Biophys. Acta* **555**, 299–306.
Forbush, B., Kaplan, J. H., and Hoffman, J. F. (1978). *Biochemistry* **17**, 3667–3676.
Hall, C., and Ruoho, A. (1980). *Proc. Natl. Acad. Sci. USA* **77**, 4529–4533.
Hopkins, B. L., Wagner, H. W., Jr., and Smith, T. W. (1976). *J. Biol. Chem.* **251**, 4365–4371.
Kyte, J. (1972). *J. Biol. Chem.* **247**, 7642–7649.
Lane, L. K., Potter, J. D., and Collins, J. H. (1979). *Prep. Biochem.* **9**, 157–170.
Lindenmayer, G. E., and Schwartz, A. (1973). *J. Biol. Chem.* **248**, 1291–1300.
Lüllman, H., and Peters, T. (1977). *Clin. Exp. Pharmacol. Physiol.* **4**, 49–57.

Matsui, H., and Schwartz, A. (1966a). *Biochem. Biophys. Res. Commun.* **25,** 147–150.

Matsui, H., and Schwartz, A. (1966b). *Biochim. Biophys. Acta* **128,** 380–390.

Matsui, H., and Schwartz, A. (1968). *Biochim. Biophys. Acta* **151,** 655–663.

Noble, D. (1980). *Cardiovasc. Res.* **14,** 495–514.

Perrone, J. R., Hackney, J. F., Dixon, J. F., and Hokin, L. C. (1975). *J. Biol. Chem.* **250,** 4178–4184.

Peterson, G. L., and Hokin, L. E. (1980). *Biochem. J.* **192,** 107–118.

Racker, E. (1976). *Trends Biochem. Sci.* **1,** 244.

Reeves, A. S., Collins, J. H., and Schwartz, A. (1980). *Biochem. Bophys. Res. Commun.* **95,** 1591–1598.

Repke, K. R. H. (1963). In *New Aspects of Cardiac Glycosides* (W. Wilbrandt, ed.), Vol. III, pp. 47–73, Pergamon Press, Elmsford, N.Y.

Rivas, E., Lew, V., and De Robertis, E. (1972). *Biochim. Biophys. Acta* **290,** 419–423.

Robinson, J. D. (1980). *Biochem. Pharmacol.* **29,** 1995–2000.

Rogers, T. B., and Lazdunski, M. (1979). *FEBS Lett.* **98,** 373–376.

Rossi, B., Viulleumier, P., Gache, C., Balerna, M., and Lazdunski, M. (1980). *J. Biol. Chem.* **255,** 9936–9941.

Schwartz, A., and Adams, R. (1980). *Circ. Res.* **46**(Suppl. I)**,** 154–160.

Schwartz, A., Matsui, H., and Laughter, A. H. (1968). *Science* **159,** 323–325.

Schwartz, A., Lindenmayer, G. E., and Allen, J. C. (1975). *Pharmacol. Rev.* **27,** 3–134.

Skou, J. C. (1957). *Biochim. Biophys. Acta* **23,** 394–401.

Skou, J. C. (1960). *Biochim. Biophys. Acta* **42,** 6–23.

Solomon, A. K., Gill, T. J., 3rd, and Lennard, G. (1956). *J. Gen. Physiol.* **40,** 327–350.

Thorp, R. H., and Cobbin, L. B. (1967). In *Cardiac Stimulant Substances*, pp. 263–268, Academic Press, New York.

Wallick, E. T., Lane, L. K., and Schwartz, A. (1979). *Annu. Rev. Physiol.* **41,** 397–411.

Wellsmith, N. V., and Lindenmayer, G. E. (1980). *Circ. Res.* **47,** 710–720.

68

Conformational Changes Associated with K^+ Transport by the Na^+/K^+-ATPase

I. M. Glynn and S. J. D. Karlish

*I am reminded of the remark attributed to Dean Inge
. . . that the word 'bloody' had become simply a sort
of notice that a noun may be expected to follow:
'conformational' has become simply a notice that the
word 'change' may be expected to follow.*
A. F. Huxley (1974), speaking of conformational
changes in myosin.

1. INTRODUCTION

In this article, we discuss two changes of the Na^+/K^+-ATPase protein: in the first, the phosphorylated enzyme is hydrolyzed and K^+ ions bound to the extracellular surface appear to become trapped within; in the second, the trapped K^+ ions gain access to the intracellular surface of the enzyme whence they can be released to the cell interior. In each case, the alteration in the accessibility of the K^+-binding sites justifies the term conformational change. The interest of these reactions, however, is not that they provide a proper use for an overworked phrase but that there are reasons for believing that the two changes in sequence play a central role in the transport of K^+ ions into the cell, which is one of the two primary functions of Na^+/K^+-ATPase.

In 1972, Post and colleagues described experiments in which they looked at the rate of phosphorylation of Na^+/K^+-ATPase that had just been dephosphorylated. They found that the rate of phosphorylation by ATP depended on whether Li^+ or Rb^+ ions (both congeners of K^+ ions) had been used to catalyze the hydrolysis of the phosphoenzyme. This was true even if the experiment was done in such a way that the composition of the medium during rephosphorylation was identical; in other words, the enzyme appeared to

I. M. Glynn • Physiological Laboratory, University of Cambridge, Cambridge CB2 3EG, England.
S. J. D. Karlish • Department of Biochemistry, Weizmann Institute, Rehovoth, Israel.

remember which K^+ congener had catalyzed the hydrolysis. To explain this "memory," Post and colleagues suggested that, during hydrolysis, the catalyzing ions became occluded within the dephosphoenzyme, and were released only later after a slow conformational change. Furthermore, as ATP in high concentrations accelerated the rephosphorylation, they reasoned that the binding of ATP to a low-affinity site accelerated the change in conformation.

At the time these experiments were done, it was known that the unphosphorylated Na^+/K^+-ATPase could exist in states (E_1 and E_2) with higher or lower affinity for ATP, depending on the relative amounts of sodium and potassium in the medium. At that time, however, there was no reason to think that the form of Na^+/K^+-ATPase (E_2) that was stable in predominantly potassium media was identical with the hypothetical occluded-K^+ form.

Unequivocal evidence that enzyme in a potassium medium differs in its conformation from enzyme in a sodium medium was provided by Jørgensen's (1975) experiments showing two distinct patterns of attack by trypsin depending on the nature of the medium. High concentrations of ATP tended to push the enzyme into the E_1 (sodium) pattern even in a potassium medium; on the other hand, when enzyme in a sodium medium was phosphorylated by ATP and Mg^{2+}, the E_2 (potassium) pattern was obtained. These interactions made it attractive to suppose that the conformational changes revealed by the patterns of tryptic digestion were relevant to sodium and potassium transport.

2. EVIDENCE THAT THE FORM OF ENZYME STABLE IN POTASSIUM MEDIA IS IDENTICAL WITH THE HYPOTHETICAL OCCLUDED-K^+ FORM

The earliest evidence suggesting this identity was the finding of Post *et al.* (1975) that enzyme in a high-sodium medium, but newly dephosphorylated in the presence of potassium, resembled enzyme in a potassium medium in being susceptible to phosphorylation by inorganic phosphate.

Strong evidence for the identity came first from studies using formycin nucleotides (Karlish *et al.*, 1978a,b), and subsequently from measurements of changes in the intrinsic fluorescence of the enzyme protein (Karlish and Yates, 1978). Formycin triphosphate and formycin diphosphate are fluorescent analogs of ATP and ADP. They are treated by the Na^+/K^+-ATPase very like their adenine counterparts, but because their fluorescence increases when they are bound to the enzyme, the net rate at which they are bound or released can be followed with a stopped-flow fluorimeter. By choosing concentrations of FTP or FDP such that the enzyme in the E_1 form was mostly bound to nucleotide, whereas the enzyme in the E_2 form was mostly free, we were able to use the fluorescence as an indicator of the enzyme conformation. A more direct, though less sensitive indication of the conformation is given by the intrinsic fluorescence of the protein, which, at appropriate wavelengths, increases when the medium is changed in such a way that the E_1 form is converted to the E_2 form. Using both of these techniques, we found that the conversion of the E_2 to the E_1 form of the enzyme was remarkably slow ($k \cong 0.2$ sec^{-1} at 20°C) at very low nucleotide concentrations, but increased linearly with nucleotide concentration at least up to 100 μM ATP. The very slow conformational change, and the acceleration of that change by ATP at high concentrations, are of course precisely the features postulated by Post and his colleagues (1972) for the occluded-K^+ form.

The rate of conversion of the E_2 form of the enzyme to the E_1 form has also been followed by observing the increase in fluorescence of fluorescein-labeled enzyme (Karlish, 1979; 1980). This method, too, gave a rate constant of about 0.2 sec^{-1} at 20°C in the absence of nucleotides. Because nucleotides are not bound by the fluorescein-labeled enzyme, an accelerating effect on the conformational change was not to be expected, and none was seen.

Further evidence for the identity of the occluded-K^+ form and the form of the unphosphorylated enzyme in potassium media came from experiments in which the enzyme was allowed to hydrolyze FTP or ATP. When these substrates at very low concentrations are hydrolyzed by the Na+/K+-ATPase in the presence of potassium, the rate-limiting step should be the breakdown of the occluded-K^+ form, and the bulk of the enzyme should therefore be in that form. If the occluded-K^+ form is identical with the stable form of the enzyme in potassium media, it follows that, during turnover at low substrate concentrations, FTP fluorescence should be low and the intrinsic fluorescence of the enzyme should be high. Both predictions have been verified (Karlish *et al.*, 1978a; Karlish and Yates, 1978).

2.1. Direct Evidence for Occlusion

Because the rate of conversion of the E_2 form to the E_1 form is so slow in the absence of nucleotides, it is possible to demonstrate the occluded state of the cation when the enzyme is in the E_2 form by forcing enzyme suspended in a low-Rb^+ Na-free medium down an ion-exchange column, and looking at the amount of Rb^+ emerging with the enzyme at the bottom of the column (Beaugé and Glynn, 1979; Glynn and Richards, 1980). Experiments of this kind have shown that, if the initial conditions are such that the enzyme starts in the E_2 form, Rb^+ is carried through the column by the enzyme. By varying the flow rate, and therefore the time elapsing before the enzyme emerged, Glynn and Richards (1980) showed that the rate constant for the release of the occluded Rb^+ was about 0.2 sec^{-1} at 21°C, in excellent agreement with the rate constant for the conformational change, $E_2 \rightarrow E_1$, estimated by the fluorescence methods. The rate of release of the occluded Rb^+ was not affected by the presence of Na^+ or K^+ ions, but was greatly increased by ATP (or ADP) in high concentrations. Vanadate, at concentrations sufficient to prevent the change from E_2 to E_1 in the fluorescein-labeled enzyme (Karlish *et al.*, 1979), prevented the release of occluded Rb^+ (Glynn and Richards, 1980). These findings, together with those described above, make it virtually certain that the form of the enzyme that is stable in potassium media contains occluded K^+ ions and is identical with the occluded-K^+ form postulated by Post and colleagues in 1972. Interestingly, no occluded cation could be detected in the presence of ouabain (Glynn and Richards, 1980). We do not know whether this is because ouabain allows the occluded ion to escape or stabilizes a form of the enzyme in which the K^+-binding sites are occluded but empty.

2.2 Conversion of E_1 to E_2

When potassium is added to a suspension of the enzyme in a medium containing little or no sodium, the enzyme is converted from the E_1 to the E_2 form, as can be shown by any of the three fluorescence techniques described above. An interesting feature of this conversion is that the apparent affinity for K^+ ions is high, judged by their effect on the equilibrium, but low, judged by their effect on the rate of the reaction. To explain this paradox, we postulated that K^+ ions bind to low-affinity sites on E_1, and that the E_1–K^+ complex then changes its conformation, the change being slow compared with the binding reaction and poised heavily in favor of the E_2 form:

$$K^+ + E_1 \overset{K_1}{\rightleftharpoons} E_1K \overset{K_2}{\longrightarrow} E_2K \tag{1}$$

(where E_1K represents the complex of enzyme in the E_1 form with K^+; E_2K represents enzyme in the E_2 form containing occluded K^+). Because the second reaction is rate limiting, the rate of the overall reaction will be determined by the fraction of the E_1 form that binds K^+, and that will reflect the low affinity of the binding sites on E_1. Because the second reaction is poised heavily to the right, low concentrations of K^+ will suffice to convert most of the enzyme into the E_2K form.

The ratio of the rate constant for the conversion of E_1 to E_2 at saturating K^+ concentrations (ca. 290 sec^{-1}, estimated by extrapolation from observations on the fluorescein-labeled enzyme) to the rate constant for the conversion of E_2 to E_1 (ca. 0.2 sec^{-1}, measured by any of the three fluorescence techniques) gives the value of K_2, which works out as a little over 10^3.

Independent evidence for the hypothesis summarized by equation (1) comes from experiments in which the equilibrium between the E_1 and the E_2 forms (estimated by their intrinsic fluorescence) was studied as a function of ATP concentration at different potassium concentrations (Beaugé and Glynn, 1980; Jørgensen and Karlish, 1980). For Na^+/K^+-ATPase from pig kidney outer medulla, the results could be fitted by assuming that, in the absence of ATP, K_2 (defined as $[E_2K]/[E_1K]$) was between 10^2 and 10^3, and probably nearer the latter, that K_2 at high ATP concentrations was not far from unity, and that K_1—the dissociation constant of the $E_1–K^+$ complex—was between 10 and 10^2. Interestingly, enzyme pretreated with trypsin in a sodium medium until the end of the fast phase of inactivation—the so-called "invalid" enzyme of Jørgensen (1975)—behaved as though the equilibrium were poised less in favor of the E_2 form. The equilibrium between the two forms of the phosphoenzyme (E_1P, the ADP-sensitive form, and E_2P, the K^+-sensitive form) was also shifted (toward the E_1P form) in the "invalid" enzyme (Jørgensen and Karlish, 1980).

3. A HYPOTHESIS FOR K^+ TRANSPORT

Because E_2K, assuming it to be identical with the occluded-K^+ form, can also be generated by hydrolysis of the K^+-sensitive phosphoenzyme (E_2P), we may expand equation (1) to read:

$$K^+ + E_1 \rightleftharpoons E_1K \rightleftharpoons E_2K \underset{\Big\updownarrow}{\overset{}{\rightleftharpoons}} E_2P \cdot K \rightleftharpoons E_2P + K^+ \qquad (2)$$
$$P_i$$

where the two routes to the occluded-K^+ form have been coupled back to back. (In this scheme, the conversion of E_2K to E_1K is supposed to be accelerated by the binding of ATP, but ATP and P_i do not necessarily compete.)

We know, from a great deal of indirect evidence and from recent direct evidence (Blostein and Chu, 1977), that the K^+ ions that catalyze the hydrolysis of the phosphoenzyme act at the extracellular surface, and at sites with a high affinity. If the low-affinity K^+-binding sites on E_1 are accessible from the intracellular surface, the sequence of reactions summarized by equation (2) could therefore transport K^+ ions across the membrane in either direction. Furthermore, the movements of potassium would show many of the features characteristic of potassium fluxes associated with the normal or reversed running of the pump or with $K^+–K^+$ exchange, namely: (1) a high affinity for K^+ at the extracellular surface, (2) a low affinity for K^+ at the intracellular surface, (3) a requirement for inorganic phosphate when K^+ was being moved outwards (because the exchange of K^+ ions between E_2K and the extracellular medium is assumed to be very slow), and (4) a requirement for ATP or a nonphosphorylating analog when K^+ was being moved inwards—see Glynn and Karlish, 1975, and Glynn *et al.*, 1979, for references.

3.1. The Sidedness of the Low-Affinity K⁺-Binding Sites on E₁

Indirect evidence suggesting that, in the absence of turnover, K^+ ions can enter the occluded-K^+ form of the enzyme only after combining with intracellular sites comes from observations on the ATPase and p-nitrophenyl phosphatase activities of inside-out red cell vesicles and on sodium fluxes in dialyzed squid axons (Blostein and Chu, 1977; Blostein *et al.*, 1979; Beaugé and DiPolo, 1979). Stronger and more direct evidence comes from recent experiments in which Na^+/K^+-ATPase has been incorporated into tight artificial lipid vesicles and subjected to the actions of trypsin or vanadate (Karlish and Pick, 1981). The enzyme in artificial lipid vesicles is oriented in either direction, but if ATP is present only in the suspending medium the rate at which labeled Na^+ ions enter the vesicles reflects the activity solely of those pumps that are oriented with their ATP-binding sites facing outwards. By exposing the vesicles containing Na^+/K^+-ATPase to trypsin under different conditions, and then measuring ATP-dependent Na^+ uptake under standard conditions (which, where appropriate, included the presence of ionophores to allow K^+ ions to enter or leave the vesicles freely), Karlish and Pick (1981) showed that only when K^+ was present *outside* the vesicles did the action of trypsin follow the pattern characteristic of potassium media. As that pattern is thought to indicate the presence of the occluded-K^+ form of the enzyme, it follows that K^+ ions can generate that form (in the absence of turnover) only from the intracellular (extravesicular) surface.

The vanadate experiments took advantage of the fact that vanadate inhibits Na^+/K^+-ATPase by stabilizing the occluded-K^+ form of the enzyme (Karlish *et al.*, 1979; Glynn and Richards, 1980), an effect that is only slowly reversible (Cantley *et al.*, 1978). By exposing vesicles containing Na^+/K^+-ATPase to vanadate under different conditions, removing the vanadate, and assaying the degree of inhibition under standard conditions, Karlish and Pick showed that only K^+ ions *in the medium* were effective in promoting inhibition by vanadate. The inference is that only K^+ ions in the medium, i.e., at the original intracellular face of the enzyme, led to the formation of the occluded-K^+ form.

4. STOICHIOMETRY

Equation (2) does not define the number of K^+ ions involved in each reaction, but, if the reaction sequence is to account for the fluxes of K^+ through the pump, it must allow more than one K^+ ion to be transported in each cycle. The relation between the K^+ concentration and the fraction of enzyme that is in the occluded-K^+ form should reflect the number of ions that must bind to E_1 to cause the conformational change. Surprisingly, different methods for looking at that relation have given different results. The fraction of enzyme in the E_2 or occluded-K^+ form has been estimated from the observed pattern of tryptic digestion (Jørgensen, 1975), from the measured ability of the enzyme to be phosphorylated by inorganic phosphate (Post, 1977), from direct measurements of the amount of rubidium occluded (D. E. Richards and I. M. Glynn, unpublished experiments), and from measurements of intrinsic protein fluorescence or of the fluorescence of fluorescein-labeled protein (Karlish and Yates, 1978; Karlish, 1980). The trypsin method gave a distinctly sigmoid curve at low potassium concentrations; the phosphorylation method, and the direct measurement of occluded rubidium, gave curves that were slightly sigmoid at low potassium concentrations (Hill coefficients about 1.27); the fluorescence methods gave results that could be fitted by rectangular hyperbolae. Measurements of the rates of conversion of E_1 to E_2 at different K^+ concentrations, using formycin nucleotides or fluorescein-labeled enzyme (Karlish *et al.*, 1978b; Karlish and Yates, 1978), showed an apparently linear dependence of rate on concentration, but the results were not accurate enough to exclude slight curvature at low K^+ concentrations. Although the direct measurements of Rb^+ occlusion as a function of Rb^+ concentration gave curves that were only slightly sigmoid, at saturating Rb^+ concentrations about three Rb^+ ions were occluded per phosphorylation site (or ouabain-binding site). Unless the kidney Na^+/K^+-ATPase preparation used in these experiments contained a large amount of enzyme damaged in such a way that it was able to occlude Rb^+ but unable to accept a phospho group from ATP (or to bind ouabain), it follows that the Na^+/K^+-ATPase molecule can occlude more than one K^+ ion. It is possible, however, that not all of the available sites have to be occupied for the conformational change to occur. Furthermore, if the sites have different affinities, the filling of a particular site may be critical in a given range of concentrations.

5. RATES OF THE CONFORMATIONAL CHANGES

If the reactions of equation (2) are to account for the fluxes of K^+ through the pump, they must be fast enough. At 20°C, the maximal turnover of Na^+/K^+-ATPase from pig kidney outer medulla running in the forwards direction, is about 30 sec^{-1} (Karlish and Yates, 1978). When the enzyme is incorporated into artificial lipid vesicles, the maximal rate of K^+–K^+ exchange, under optimal conditions at 20°C, is only about 20% of the maximal rate of Na^+–K^+ exchange (S. J. D. Karlish and W. D. Stein, unpublished experiments); we therefore have to account for a turnover, in the reverse direction [i.e., left to right in equation (2)], of about 6 sec^{-1}.

5.1. K^+ Influx

Mårdh (1975) showed that the rate constant for the hydrolysis of E_2P in brain Na^+/K^+-ATPase, at 21°C in the presence of 10 mM K^+, was at least 230 sec^{-1}. Even allowing for possible species differences, the hydrolysis of E_2P is therefore unlikely to be rate limiting, and the crucial question is whether the conversion of $E_2(K)$ to E_1 at high ATP concentrations is fast enough. There are no data at physiological levels of ATP, but measurements of intrinsic fluorescence of the kidney Na^+/K^+-ATPase gave a rate constant of about 12 sec^{-1} (22°C) at 100 μM ATP, where the relation between rate and ATP concentration was still linear (Karlish and Yates, 1978). The K_d for ATP binding to the $E_2(K)$ form of the enzyme is uncertain, but it cannot be less than the observed K_m for Na^+/K^+-ATPase activity (200 μM), and has been estimated to be about 450 μM (Karlish and Yates, 1978). If 450 μM is correct, the rate constant for the conversion of $E_2(K)$ to E_1 at saturating concentrations of ATP works out as about 60 sec^{-1}. That figure requires that during maximal turnover about 50% of the enzyme be in the $E_2(K)$ form, a fraction that agrees well with experimental estimates based on measurements of intrinsic fluorescence (Karlish, 1980). Still higher figures for the rate constant for the conversion of $E_2(K)$ to E_1 in the presence of ATP are suggested by the computer simulations of Lowe and Smart (1977) and of

Mårdh and Lindahl (1977) based on studies of rates of phosphorylation and dephosphorylation.

5.2. K⁺ Efflux

Extrapolation of observations on fluorescein-labeled enzyme suggest that the rate constant for the conversion of E_1 to $E_2(K)$ at saturating K^+ concentrations and at 22°C is about 290 sec^{-1} (Karlish, 1980). At K^+ concentrations low enough for measurements to be possible, the rate determined using normal enzyme and formycin nucleotides (Karlish *et al.*, 1978b) is only a little greater than the rate determined using the fluorescein-labeled enzyme. The rate constant for the conversion of E_1 to $E_2(K)$ under physiological conditions is therefore, almost certainly, far greater than 6 sec^{-1}.

The rate of phosphorylation of $E_2(K)$ by inorganic phosphate is not known, but when kidney cortex Na$^+$/K$^+$-ATPase was exposed to 1 mM orthophosphate at 0°C in suitable K^+-containing media, about 25% of the enzyme became phosphorylated in the steady state (Post *et al.*, 1975). This implies that the pseudo-first-order rate constant for phosphorylation was about one-third of the rate constant for hydrolysis. If we assume (1) that the two rate constants are in the same ratio at room temperature, and (2) that Mårdh's minimum figure of 230 sec^{-1} for the hydrolysis of E_2P from brain enzyme applies to the preparation used by Post and colleagues, we can deduce that the rate constant for the phosphorylation of E_2K by orthophosphate at a concentration of 1 mM is at least 77 sec^{-1}. The assumptions may well be slightly wrong, but it seems likely that the rate constant for phosphorylation is great enough to account for the observed K^+ efflux during K^+–K^+ exchange.

We conclude that, so far as they can be estimated, the rate constants for both conformational changes in both directions are compatible with the hypothesis that equation (2) summarizes the sequences of events that carry K^+ ions inwards and outwards across the membrane. The way in which the changes responsible for transporting K^+ are related to the rest of the Na$^+$/K$^+$-ATPase cycle is beyond the scope of this article but is discussed by Karlish *et al.* (1978b) and by Glynn *et al.* (1979).

REFERENCES

Beaugé, L. A., and DiPolo, R. (1979). *Biochim. Biophys. Acta* **553**, 495–500.
Beaugé, L. A., and Glynn, I. M. (1979). *Nature (London)* **280**, 510–512.
Beaugé, L. A., and Glynn, I. M. (1980). *J. Physiol. (London)* **299**, 367–383.
Blostein, R., and Chu, L. (1977). *J. Biol. Chem.* **252**, 3035–3043.
Blostein, R., Pershadsingh, H. A., Drapeau, P., and Chu, L. (1979). In *Na,K-ATPase: Structure and Kinetics* (J. C. Skou and J. G. Nørby, eds.), pp. 233–245, Academic Press, New York.
Cantley, L. C., Cantley, L. G., and Josephson, L. (1978). *J. Biol. Chem.* **253**, 7361–7368.
Glynn, I. M., and Karlish, S. J. D. (1975). *Annu. Rev. Physiol.* **37**, 13–55.
Glynn, I. M., and Richards, D. E. (1980). *J. Physiol. (London)* **308**, 58P.
Glynn, I. M., Karlish, S. J. D., and Yates, D. W. (1979). In *Na,K-ATPase: Structure and Kinetics* (J. C. Skou and J. G. Nørby, eds.), pp. 101–113, Academic Press, New York.
Huxley, A. F. (1974). *J. Physiol. (London)* **234**, 1–43.
Jørgensen, P. L. (1975). *Biochim. Biophys. Acta* **401**, 399–415.
Jørgensen, P. L., and Karlish, S. J. D. (1980). *Biochim. Biophys. Acta* **597**, 305–317.
Karlish, S. J. D. (1979). In *Na,K-ATPase: Structure and Kinetics* (J. C. Skou and J. G. Nørby, eds.), pp. 115–128, Academic Press, New York.
Karlish, S. J. D. (1981). *J. Bioenergetics and Biomembranes* **12**, 111–136.
Karlish, S. J. D., and Pick, U. (1981). *J. Physiol. (London)*, **312**, 505–529.

Karlish, S. J. D., and Yates, D. W. (1978). *Biochim. Biophys. Acta* **527**, 115–130.
Karlish, S. J. D., Yates, D. W., and Glynn, I. M. (1978a). *Biochim. Biophys. Acta* **525**, 230–251.
Karlish, S. J. D., Yates, D. W., and Glynn, I. M. (1978b). *Biochim. Biophys. Acta* **525**, 252–264.
Karlish, S. J. D., Beaugé, L. A., and Glynn, I. M. (1979). *Nature (London)* **282**, 333–335.
Lowe, A. G., and Smart, J. W. (1977). *Biochim. Biophys. Acta* **481**, 695–705.
Mårdh, S. (1975). *Biochim. Biophys. Acta* **391**, 448–463.
Mårdh, S., and Lindahl, S. (1977). *J. Biol. Chem.* **252**, 8058–8061.
Post, R. L. (1977). In *Biochemistry of Membrane Transport* (G. Semenza and E. Carafoli, eds.), Springer-Verlag, Berlin.
Post, R. L., Hegyvary, C., and Kume, S. (1972). *J. Biol. Chem.* **247**, 6530–6540.
Post, R. L., Toda, G., and Rogers, F. N. (1975). *J. Biol. Chem.* **250**, 691–701.

69

Topology in the Membrane and Principal Conformations of the α Subunit of Na$^+$/K$^+$-ATPase

Peter Leth Jørgensen

> *Proteins cannot be described by one three-dimensional structure. They have as a fourth dimension their conformational adaptability. This is going to offer interesting problems to scientists in all the future we can imagine.*
> Hugo Theorell, Harvey Lectures, 1965.

1. INTRODUCTION

A significant gap in our understanding of the reaction mechanism of the Na$^+$/K$^+$ pump is the lack of information about the molecular structure of the protein. Pure preparations of the proteins of Na$^+$/K$^+$-ATPase have been available for a decade, but analysis of the primary structure is complicated by aggregation of hydrophobic segments and the three-dimensional structure is unknown. Some provisional information is available about the organization of the protein in the membrane (Jørgensen, 1977; Karlish *et al.*, 1977; Castro and Farley, 1979). In addition, controlled proteolysis of the α subunit (Jørgensen, 1975, 1977; Giotta, 1975) and fluorescence analysis (Karlish and Yates, 1978; Jørgensen and Karlish, 1980; Karlish, 1980; Hegyvary and Jørgensen, 1981) show that the protein can assume two principal conformations.

This topical review is concerned mainly with two aspects of the Na$^+$/K$^+$-pump structure: First, the information about the topology of the α subunit that can be derived from the straightforward combination of controlled proteolysis with selective chemical labeling; and second, the evidence for conformational transitions is examined to see if polypeptide chain rearrangements in the α subunit are of sufficient magnitude to support a mechano-chemical hypothesis for the ion translocation process.

Peter Leth Jørgensen • Institute of Physiology, University of Aarhus, 8000 Aarhus C, Denmark.

The purified membrane-bound Na$^+$/K$^+$-ATPase from mammalian kidney provides an ideal system for examination of these problems, as it can be isolated in its native environment, within the membrane structure, without perturbing lipoprotein associations of the native membrane (Jørgensen, 1974). The proteins of this preparation—the α subunit (molecular weight 104,000) and β subunit (a glycoprotein of molecular weight close to 40,000)—are seen as tightly packed, asymmetrically arranged particles in membrane discs with both surfaces exposed to the medium (Deguchi *et al.*, 1977).

2. TOPOLOGY OF THE PROTEINS IN THE MEMBRANE

Electron microscopy after negative staining shows that the proteins of renal Na$^+$/K$^+$-ATPase are organized in particles of maximum diameter 50 Å, each containing one α subunit and one β subunit. The particles are free to move in the plane of the membrane and the protein protrudes above the plane of the bilayer on both surfaces (Deguchi *et al.*, 1977; Maunsbach *et al.*, 1979). Both the α subunit and the β subunit possess extracellular portions, whereas the cytoplasmic protrusion appears to be formed exclusively by the α subunit. The mass of protein protruding above the plane of the membrane has not been determined, but 20–40% of the protein is estimated to be embedded in the lipid bilayer by infrared spectroscopy (Chetverin *et al.*, 1980) or detergent binding (Clarke, 1975). It is uncertain whether the ion-conducting pathway is formed in a cleft between subunits in an oligomer structure ($\alpha_2\beta_2$) or whether a channel passes through the center of a protomer $\alpha\beta$ unit.

In the following sections it is shown how combination of controlled proteolysis with specific chemical labeling can provide information about the topology of the protein in the membrane. The available information is limited, but in principle this technique can be pursued to high resolution by combining other techniques for site-specific hydrolysis of proteins with selective chemical labeling.

2.1. Tryptic Cleavage of the α Subunit

The α subunit comprised of approximately 1000 residues, contains 44 lysines and 27 arginines. Accordingly, after denaturation in SDS the protein is split into numerous tryptic fragments. In the membrane-bound Na$^+$/K$^+$-ATPase, the protection by lipid and a tight structure of the protein can explain that only one or two bonds are exposed at one time to primary tryptic cleavage. Figure 1 depicts a model for the arrangement of the major tryptic fragments. Primary cleavage depends on the ligands present in the medium. In media containing KCl, the primary cleavage of bond 1 near the middle of the chain is followed by cleavage of bond 2 near the NH$_2$ terminus. In NaCl media, the rate of cleavage of bond 2 becomes very high, whereas bond 3 is cleaved slowly and bond 1 is protected to cleavage.

Tryptic digestion of Na$^+$/K$^+$-ATPase in the red cell membrane (Giotta, 1975) and in reconstituted vesicles (Karlish and Pick, 1981) show that the three trypsin-sensitive bonds of the α subunit are exposed on the cytoplasmic surface. The splits have well-defined catalytic consequences. Splits 1 and 3 are accompanied by loss of ATP binding and split 2 induces a well-defined modification of the Na$^+$/K$^+$-ATPase (Jørgensen, 1977; Jørgensen and Karlish, 1980). The major fragments are membrane bound, but cleavage of bond 2 releases a peptide with molecular weight close to 2000 from the NH$_2$-terminal end of the

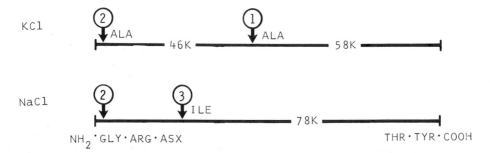

Figure 1. Model of tryptic cleavage in KCl or NaCl of α subunit (molecular weight 104,000) in membrane-bound Na⁺/K⁺-ATPase. Arrows mark sites of primary cleavage. In KCl (E₂K), cleavage of bond 1 produces fragments of 46K and 58K and precedes cleavage of bond 2. In NaCl (E₁Na), rapid cleavage of bond 2 causes loss of 85% of K-phosphatase and 50% of Na⁺/K⁺-ATPase activity, while bond 3 is cleaved at a low rate (Jørgensen, 1975, 1977). NH₂ terminals (left) and COOH terminals (right) as determined by Castro and Farley (1979).

α subunit (Castro and Farley, 1979). As all primary splits are cytoplasmic, the NH₂ terminus may therefore be located on the cytoplasmic surface of the membrane. The location of the COOH terminus is unknown.

2.2. Labeling of Fragments from Extracellular and Cytoplasmic Membrane Surfaces

Residues from different parts of the α subunit seem to contribute to formation of the ATP-binding area at the cytoplasmic membrane surface. The aspartyl-^{32}P formed from [γ-^{32}P]-ATP is located on the 46K fragment (Giotta, 1975; Castro and Farley, 1979), whereas the 58K fragment contains an amino group in the ATP-binding area that reacts with fluorescein isothiocyanate, FITC (Karlish, 1980).

Cardiac glycosides bind to the protein on the extracellular surface. Probing of the membrane-bound Na⁺/K⁺-ATPase with various photolabeled cardiac glycosides has shown that the α subunit forms the primary part of the binding area that receives the aglycone part of the glycoside (Forbush and Hoffman, 1979; Hall and Rouho, 1980). ³H-labeled 2-nitro-5-azidobenzoylouabain (NAB-ouabain) covalently labels the α subunit but not the β subunit. Labeling of both α and β subunits is observed when the reactive groups are placed in the region of the third sugar moiety of digitoxin, i.e., 5 Å further away from the aglycone than the azide group in NAB-ouabain. These experiments show that the binding area for the sugar moiety of the glycosides is formed by a part of the α subunit that borders the β subunit at the extracellular surface (Hall and Rouho, 1980).

After tryptic cleavage, the major 46K, 58K, and 78K fragments are covalently labeled with NAB-ouabain (Forbush *et al.*, 1978). In the 46K fragment, ouabain is bound to the region between split 1 and 3. Forbush *et al.* (1979) showed that NAB-ouabain and ^{32}P attach to the same α subunit in the Na⁺/K⁺-ATPase molecule. The observations therefore indicate that the part of fragment 46K between bond 1 and 3 of the α subunit traverses the bilayer. Because the tryptic splits are located on the cytoplasmic surface, it follows that the 46K fragment must traverse the bilayer twice. The 58K segment must traverse at least once, because it has its NH₂ terminus exposed on the cytoplasmic surface and interacts with ouabain from the extracellular surface.

2.3. Labeling from within the Lipid Bilayer

For identification of lipid-associated segments, hydrophobic compounds are equipped with photolabels. [^{125}I]iodonaphthylazide (INA) (Bercovici and Gitler, 1978) is one example. In the dark, the unreactive INA partitions into the membranes due to the hydrophobicity of the naphthyl moiety. Subsequent irradiation converts the azide into a reactive nitrene that attaches covalently into lipids and proteins in the membrane bilayer. At low concentrations, INA labels preferentially the 46K fragment, but at higher concentrations both the 46K and the 58K fragments are labeled (Karlish et al., 1977). Generally, nitrenes are less reactive than carbenes. The carbene precursor adamantane diazirine (AD) was therefore expected to be more effective than INA (Bayley and Knowles, 1978). However, AD inserts exclusively into the 58K fragment (Farley et al., 1980) and does not label the 46K fragment. INA appears to be the more successful of the two labels, for it attaches to the transmembrane segments of both the 46K and the 58K fragments, whereas AD seems to be sterically or chemically prevented from reaction with the lipid-associated segments of the 46K fragment. The experience with the hydrophobic labels is therefore that the positive labeling of a peptide may be regarded as evidence for lipid association, but the lack of reaction of AD with the 46K fragment should not be used as evidence for a model in which the entire 46K fragment forms a cytoplasmic NH$_2$-terminal domain of the α subunit.

Together the combination of controlled proteolysis and chemical labeling suggests a model in which the α subunit traverses the bilayer at least three times with the NH$_2$ terminus and the three trypsin-sensitive bonds located at the cytoplasmic surface. Considering the available information about the structure of other intrinsic membrane proteins, like the Ca^{2+}-ATPase from sarcoplasmic reticulum (Allen et al., 1980) and bacteriorhodopsin (Ovchinnikov et al., 1979), it is likely that additional transmembrane segments of the α subunit will be identified.

3. PRINCIPAL CONFORMATIONS OF THE α SUBUNIT

Two cation-induced conformations of the α subunit, the Na-bound form (E$_1$Na) and the K-bound form (E$_2$K), were detected by the two patterns of tryptic cleavage of the α subunit and inactivation of Na$^+$/K$^+$-ATPase and K-phosphatase activities. As sensed with trypsin, the two conformations are interconvertible through the action of ATP. Specifically, in the absence of K$^+$, phosphorylation of the α subunit in the presence of Mg^{2+} and Na$^+$ (MgE$_1$P or MgE$_2$P) produces a conformation in which the α subunit is digested in the same way as after binding of K$^+$ (E$_2$K). Alternatively, ATP in the presence of K$^+$ (E$_1$K ATP) produces a conformation similar to that found with Na$^+$ alone (E$_1$Na) (Jørgensen, 1975). Differences in fluorescence intensity of intrinsic and extrinsic probes and in reactivity of sulfhydryl groups show that residues in addition to the trypsin-sensitive bonds are involved in changes in conformation of the α subunit.

This treatise is limited to a discussion of the extent and mechanism of the structural changes within the α subunit. For detailed considerations of kinetic parameters and the relationship to ion transport, the reader is referred to other articles (Karlish and Yates, 1978; Jørgensen and Petersen, 1979; Karlish, 1980; Jørgensen and Karlish, 1980; Beaugé and Glynn, 1980; Skou and Esmann, 1980; Hegyvary and Jørgensen, 1981). The techniques for monitoring differences in structure of the α subunit offer the advantage that transitions can be followed after addition of each of the specific ligands, one by one. To evaluate the extent of the polypeptide chain rearrangements in the α subunit, the relatively simple relationships after addition of the cations Na^+ or K^+ (Rb^+) are considered first. Next the structure of the α subunit in the cation-induced conformations, E_1Na and E_2K, is compared with the structure in the ATP-bound form (E_1ATP), the phosphoforms $(MgE_1P$ and $MgE_2P)$, and the inhibitor-stabilized forms $(MgE_2K$-vanadate or MgE_2P-ouabain) to see if there is evidence for changes in structure of the α subunit in addition to those described for the cation-induced conformations E_1Na and E_2K.

3.1. Cation-Stabilized Conformations of the α Subunit, E₁Na and E₂K

3.1.1. Tryptic Digestion

The structure of the α subunit in E_1Na differs from that in E_2K in at least three positions along the path of the peptide. The Na-bound form (E_1Na) is defined by its biphasic pattern of digestion. Bond 2 is cleaved at a much higher rate than bond 3 (Fig. 1). The time course of inactivation of the Na^+/K^+-ATPase is described by two exponentials: $EA = A \cdot e^{-\alpha t} + B \cdot e^{-\beta t}$, where EA is the activity remaining at time t, A and B are the fractions of the activity in phase A and B, and α and β are the rate constants for cleavage of bond 2 and 3, respectively (Jørgensen, 1975, 1977). The ratio of the rate constants α/β is 15–25 at 37°C, but it is higher at 20°C (Karlish and Pick, 1981). The reason for this is that the cleavage of bond 3 (β) is more temperature sensitive than cleavage of bond 2 (α) (Jørgensen, 1982).

Addition of excess K^+, Rb^+, or Cs^+ to media containing Na^+, Tris, or choline changes the biphasic pattern to a monoexponential pattern of tryptic inactivation of the Na^+/K^+-ATPase activity in which bond 1 is cleaved at a rate that is intermediate between the rates of cleavage in phase A and B with NaCl in the medium. In E_2K, bond 2 is still exposed, but it is cleaved at a slower rate than bond 1 and cleavage of bond 1 must precede cleavage of bond 2 within the same α subunit (Jørgensen, 1975, 1977).

For the evaluation of the mechanism of the change in position of the three bonds, it is important to know the relationship to cation binding and to determine the sidedness of the cation-binding sites. The affinity of the Na^+/K^+-ATPase for Na^+ is too low for equilibrium binding experiments, but it is possible to determine the binding of $^{86}Rb^+$ (Matsui *et al.*, 1977). Recently, we analyzed tryptic cleavage of the α subunit and binding of $^{86}Rb^+$ under identical conditions. The maximum capacity for the native Na^+-K^+-ATPase is 12.3 ± 0.5 nmoles $^{86}Rb^+$/mg protein with $K_D = 7.7$ μM corresponding to about two Rb^+-binding sites per α subunit. Addition of Rb^+ in concentrations between 0 and 100 μM to the medium gradually alters the biphasic pattern of tryptic cleavage in Tris chloride to a monoexponential pattern. The changes in the rate constants α and β for cleavage of bond 2 and 3 are closely related to the saturation of cation-binding sites as determined by equilibrium binding of $^{86}Rb^+$ to the native Na^+/K^+-ATPase. It is therefore the binding of Rb^+ or K^+ ions to the Na^+/K^+-ATPase that stabilizes the E_2K or E_2Rb forms of the α subunit (Jørgensen, 1982).

Recently Karlish and Pick (1981) examined the sidedness of the cation effects in reconstituted vesicles using a centrifugation technique to control the cation composition on both sides of the vesicle membrane. Tryptic digestion of Na^+/K^+-ATPase in the reconstituted vesicles results in losses of active coupled Na^+/K^+ transport that are parallel to those seen for the loss of Na^+/K^+-ATPase activity. With adequate controls the results show that Na^+ and K^+ stabilize the alternative conformations of the α subunit by binding to cytoplasmic sites.

Although both the cation-binding sites and the trypsin-sensitive sites are cytoplasmic, it is unlikely that the change in α subunit structure is due to simple shielding by cations of sites from trypsin, for similar changes in structure of the α subunit can be induced by dissimilar ligands. In other proteins, primary tryptic splits usually occur between domains of more compact substructures (Zurawski *et al.*, 1975). It is therefore proposed that the change in position of the trypsin-sensitive bonds is due to movements of unstructured links between more firmly structured domains of the α subunit.

3.1.2. Intrinsic Protein Fluorescence of E_1Na and E_2K

Intensity of intrinsic tryptophan fluorescence of the E_2K form is 2–3% higher than that of the E_1Na form of the renal Na^+/K^+-ATPase (Karlish and Yates, 1978). In spite of the small size of the fluorescence change, it is possible to show that the fluorescence intensity levels are altered in parallel to saturation of the two sites for binding of $^{86}Rb^+$ per α subunit in the range from 0 to 100 μM RbCl (Jørgensen, 1982). The intrinsic fluorescence intensity level is therefore a reliable index of the binding of Rb^+ (or K^+) in the absence of other ligands.

The changes in fluorescence intensity may reflect real differences in the quantum yield of fluorescence, for the minute changes in K^+ or Rb^+ concentrations are unlikely to alter sample turbidity. As the magnitude of the fluorescence change is only 2–3% of the total fluorescence, it can be explained by a change in the state of no more than one or two tryptophanyls. For understanding the mechanism, it is important that the change in intensity is accompanied by a shift to shorter wavelengths of the fluorescence of E_2K in comparison with E_1Na (Karlish and Yates, 1978). This shift suggests that the dielectric constant of the environment is reduced and may reflect movement of indole chromophores from a polar to a nonpolar environment. There is no structural evidence for localization of the tryptophan residues involved in this change to the α subunit, although this can be expected because the changes in fluorescence are elicited by the same ligands and take the same direction as the changes in tryptic digestion patterns.

3.1.3. FITC Fluorescence of E_1Na and E_2K

Fluorescein isothiocyanate (FITC) reacts with an amino group in the ATP-binding area and is covalently attached to the 58K fragment of the α subunit. ATP binding is abolished, but other catalytic functions of the protein are retained (Karlish, 1980). With one fluorescein molecule bound per α subunit, the transition from E_1Na to E_2K is accompanied by 20–25% quenching of fluorescence. The large size of this signal allows very detailed titrations of Na^+ and K^+ interactions with Na^+/K^+-ATPase (Karlish, 1980; Hegyvary and Jørgensen, 1981). Also, for this extrinsic probe the change in fluorescence intensity levels is directly related to saturation of $^{86}Rb^+$-binding sites, but attachment of the probe alters the cation affinity. The maximum capacity for binding of $^{86}Rb^+$ is two Rb^+ ions per α subunit as for native Na^+/K^+-ATPase, but the apparent affinity is lower ($K_D = 29.2$ μM) than for native Na^+/K^+-ATPase ($K_D = 7.7$ μM) (Jørgensen, 1982). This reduction in apparent affinity for Rb^+ can be due to an ATP-like effect of the fluorescein bound to an amino group in the ATP-binding area.

Although the change in fluorescence intensity level is considerable, it can be explained by minor changes in position of the fluorescein molecule relative to another chromophore or by a change in dielectric constant of the environment of the fluorescein.

3.1.4. The Role of Sulfhydryl Groups

Titration with N-ethylmaleimide (NEM) reveals only minor changes in reactivity of sulfhydryl groups when K^+ is exchanged for Na^+ (Hart and Titus, 1973). The fluorescence of eosin bound to sulfhydryl groups in Na^+/K^+-ATPase from shark rectal glands was quenched about 20% by exchanging Na^+ for K^+ (Skou and Esmann, 1980). In renal Na^+/K^+-ATPase, the cation-induced change in eosin fluorescence intensity is limited to 2–3% whereas phosphorylation and ouabain binding cause a 20–30% quenching (Jørgensen, 1982). The observation that thimerosal (ethylmercurithiosalicylate) reversibly blocks the K^+-inducted change in conformation of the α subunit does, however, confirm that sulfhydryl groups also are involved in cation-induced structural changes in renal Na^+/K^+-ATPase (Hegyvary and Jørgensen, 1981), but the localization of these groups is unknown.

3.1.5. Infrared Spectroscopy

The Na^+- or K^+-induced conformational changes are accompanied by very limited changes in infrared spectra, suggesting that the proportion of α helices, β-pleated sheets, and random coils in the protein remains unaltered (Chetverin *et al.*, 1980). This does not exclude that even large movements or changes in shape and relative positions of these secondary structure elements can occur.

3.2. ATP-Bound Form of the α Subunit

High-affinity ATP binding to the E_1Na form causes only minor changes in structure of the α subunit. The tryptic digestion patterns and the tryptophan fluorescence levels of E_1Na and E_1ATP are identical (Jørgensen, 1975; Karlish and Yates, 1978). The affinity of the E_2K form for ATP is low, and the binding of ATP and K^+ or Rb^+ are antagonistic, for ATP stabilizes the alternative conformation E_1ATP. Titration of the reactivity of sulfhydryl groups shows that the binding of ATP protects two groups in the α subunit to reaction with NEM (Hart and Titus, 1973) or fluorescein mercuriacetate (Jesaitis and Fortes, 1980). Apart from this change in reactivity of sulfhydryl groups, there is no evidence for differences between the E_1Na and the E_1ATP forms with respect to the structure of the α subunit.

3.3. Phosphoforms of the α Subunit

3.3.1. Mg-Bound Forms

Phosphorylation from ATP or P_i requires Mg^{2+} and so do the transitions between the phosphoforms and the dephosphorylation reaction. Only recently has information about structural effects of Mg^{2+} binding been obtained. In the absence of other ligands, Mg^{2+} reacts with low affinity to expose bond 3 to trypsin, but the Mg-bound form is classified as an E_1 form because bonds 2 and 3 are exposed to cleavage (Jørgensen and Petersen, 1979). Mg^{2+} delays the rate of transition from E_2K or E_2Rb to E_1Na in a way suggesting that complex formation with Mg^{2+} may be a part of the explanation for the "occlusion" of K^+ or Rb^+ ions (Hegyvary and Jørgensen, 1981).

3.3.2. E_1P and E_2P

The major change in α-subunit structure occurs when E_1P is transformed into E_2P and not when phosphate is transferred from ATP to the protein. Effects of the individual ligands can be resolved in some detail. Binding of ATP in the presence of Mg^{2+} increases the rate constant β, but the digestion pattern remains biphasic with exposure of bond 2. The rate constant α for cleavage of bond 2 is drastically reduced by Na^+ (Jørgensen and Petersen, 1979). This protection of bond 2 is related to the transition from E_1P to E_2P rather than to the Na-induced transfer of phosphate from ATP to the protein, for it can be shown that cleavage of bond 2 alters the poise of the equilibrium between E_1 and E_2 forms in the direction of the E_1 forms for both dephospho- and phosphoforms of the protein (Jørgensen and Karlish, 1980).

As sensed with trypsin (Jørgensen, 1975) and tryptophan fluorescence (Jørgensen and Karlish, 1980), E_2K and E_2P appear to be identical. Formation of E_2P is also accompanied by a 20% quenching of fluorescein fluorescence (Karlish, 1980). The cytoplasmic aspects of the structure of the α subunit therefore seem to be homogeneous in E_2K and E_2P. However, other experiments suggest that notable differences in structure between these conformations should be detectable. In the fluorescein enzyme, formation of E_2P is unaffected while the transition from E_1Na to E_2K is reversibly blocked with thimerosal. Formation of E_2P therefore involves chemical reactions that are different from those involved in formation of E_2K (Hegyvary and Jørgensen, 1981). Phosphorylation also exposes sulfhydryl groups to reaction with NEM (Hart and Titus, 1973) and greatly increases the affinity for ouabain at the extracellular surface (Hansen, 1978). The cation-induced conformational changes do not alter ouabain affinity. It is therefore probable that the formation of E_2P involves structural changes of intramembranous and extracellular portions of the α subunit in addition to the structural changes that can be demonstrated in the cytoplasmic portions.

3.4. Inhibitor-Stabilized Conformations of the α Subunit

Analysis by fluorescence of intrinsic tryptophan and fluorescein bound to the α subunit reveals that the protein must complete the transition to the structure in the Mg^{2+}-bound form of E_2K before vanadate can form the stable complex MgE_2K : vanadate and prevent further motion within the protein (Karlish, 1980; Jørgensen and Karlish, 1980).

High-affinity ouabain binding to E_2P exposes a peptide bond within the α subunit to cleavage by chymotrypsin. This split divides the 78K fragment into 35K and 40K fragments (Castro and Farley, 1979). This ouabain-bound form is characterized by a very low level of fluorescence, 66% of E_1Na, from fluorescein bound to the α subunit (Karlish, 1980; Hegyvary and Jørgensen, 1981). These observations show that ouabain stabilizes a unique conformation of the α subunit that can be classified as a subconformation of E_2P.

4. CONCLUSIONS; EXTENT OF THE STRUCTURAL CHANGES IN THE α SUBUNIT

The data discussed in this article show unequivocally that significant structural changes of the α subunit are coupled to the biochemical reactions of the Na^+/K^+-ATPase. Drastic structural changes like rotation or translation of major protein portions do not occur. The responses are subtle in the sense that a limited fraction of about 1000 residues in the α

subunit are involved. As a minimum, the binding of Na⁺ or K⁺ (Rb⁺) to cytoplasmic sites alters the positions of residues forming lysyl or arginyl peptide bonds at three different positions along the path of the α subunit. The fluorescence signals show that one to three tryptophanyls and an amino group in the ATP-binding area change their positions. Sulfhydryl groups of unknown number and location are also involved in the structural change. The two principal conformations are interconvertible through the action of ATP as sensed with trypsin and the fluorescence techniques, but phosphorylation and transition between the phosphoforms involve additional sulfhydryl groups and the change in α-subunit structure appears to be transmitted through the membrane. Ouabain binding to extracellular aspects of the phosphoenzyme induces a further change in structure to produce a unique conformation of the α subunit.

A rigorous description of the extent of these structural changes must await determination of the exact number and coordinates of the groups involved in the conformational responses of this protein. It is clear, however, that molecular explanations of the principal conformations that the α subunit can assume will be essential for description of the mechanism for the transformation of the energy in ATP to translocation of Na⁺ and K⁺.

ACKNOWLEDGMENT. Work in the author's laboratory was supported by the Danish Medical Research Council.

REFERENCES

Allen, G., Trinnaman, B. J., and Green, N. M. (1980). *Biochem. J.* **187**, 591–616.

Bayley, H., and Knowles, J. R. (1978). *Biochemistry* **17**, 2420–2423.

Beaugé, L. A., and Glynn, J. M. (1980). *J. Physiol.* **299**, 367–383.

Bercovici, T., and Gitler, C. (1978). *Biochemistry* **17**, 1484–1489.

Castro, J., and Farley, R. A. (1979). *J. Biol. Chem.* **254**, 2221–2228.

Chetverin, A. B., Venyaminov, S. Y., Emelyanenko, V. I., and Burstein, E. A. (1980), *Eur. J. Biochem.* **108**, 149–156.

Clarke, S. (1975). *J. Biol. Chem.* **250**, 5459–5469.

Deguchi, N., Jørgensen, P. L., and Maunsbach, A. B. (1977). *J. Cell Biol.* **75**, 619–634.

Farley, R. A., Goldman, D. W., and Bayley, H. C. (1980). *J. Biol. Chem.* **255**, 860–864.

Forbush, B., and Hoffman, J. F. (1979). *Biochemistry* **18**, 2308–2315.

Forbush, B., Kaplan, I., and Hoffman, J. F. (1978). *Fed. Proc.* **37**, 239.

Giotta, G. (1975). *J. Biol. Chem.* **250**, 5159–5164.

Hall, C. C., and Rouho, A. E. (1980). *Proc. Natl. Acad. Sci. USA* **77**, 4529–4533.

Hansen, O. (1978). *Biochim. Biophys. Acta* **511**, 10–22.

Hart, W. M., and Titus, E. O. (1973). *J. Biol. Chem.* **248**, 4674–4681.

Hegyvary, C. and Jørgensen, P. L. (1981). *J. Biol. Chem.*, **256**, 6296–6303.

Jesaitis, A. I. and Fortes, D. A. G. (1980). *J. Biol. Chem.* **255**, 459–467.

Jørgensen, P. L. (1974). *Biochim. Biophys. Acta* **356**, 36–52.

Jørgensen, P. L. (1975). *Biochim. Biophys. Acta* **401**, 399–415.

Jørgensen, P. L. (1977). *Biochim. Biophys. Acta* **466**, 97–108.

Jørgensen, P. L. (1982). Submitted for publication.

Jørgensen, P. L., and Karlish, S. J. D. (1980). *Biochim. Biophys. Acta* **597**, 305–317.

Jørgensen, P. L., and Petersen, J. (1979). In *Na,K-ATPase* (J. C. Skou and J. G. Nørby, eds.), pp. 143–155. Academic Press, New York.

Karlish, S. J. D. (1980). *J. Bioenerg. Biomembr.* **12**, 111–136.

Karlish, S. J. D., and Pick, U. (1981). *J. Physiol.*, **312**. 505–529.

Karlish, S. J. D., and Yates, D. W. (1978). *Biochim. Biophys. Acta* **527**, 115–130.

Karlish, S. J. D., Jørgensen, P. L., and Gitler, C. (1977). *Nature (London)* **269**, 715–717.

Matsui, H., Hayaski, Y., Humoredo, H., and Kimimura, M. (1977). *Biochem. Biophys. Res. Commun.* **75**, 373–380.

Maunsbach, A. B., Skriver, E., and Jørgensen, P. L. (1979). In *Na,K-ATPase* (J. C. Skou and J. G. Nørby, eds.), pp. 3–13, Academic Press, New York.

Ovchinnikov, Y. A., Abdulaev, N. G., Feigina, M. Y., Kiselev, A. V., and Lobanov, N. A. (1979). *FEBS Lett.* **100,** 219–224.

Skou, I.C., and Esmann, M. (1980). *Biochim. Biophys. Acta* **601,** 386–402.

Zurawski, V. R., Koth, W. I., and Foster, J. F. (1975). *Biochemistry* **14,** 5579–5586.

70

Reconstitution of the Sodium/Potassium Pump

Lowell E. Hokin

Several aspects of the Na^+/K^+-activated adenosine triphosphatase (Na^+/K^+-ATPase) are reviewed in this volume. In order to prove that the Na^+/K^+-ATPase is the Na^+/K^+ pump, it was necessary to insert the purified enzyme in a bilayer (liposomes or black lipid membranes) and reconstitute Na^+ and K^+ transport.

Reconstitution of Na^+ transport was accomplished in 1974 by Goldin and Tong (1974) using the pure Na^+/K^+-ATPase of Kyte (1971) and by Hilden et al. (1974) using the pure enzyme from *Squalus acanthias* of Hokin et al. (1973). Reconstitution of coupled Na^+ and K^+ transport from the same enzyme was accomplished by Hilden and Hokin (1975). Earlier that year, in a brief note, reconstitution of Na^+ and K^+ transport was demonstrated with a highly impure enzyme fraction (microsomes) from brain (Sweadner and Goldin, 1975). In a later paper, Goldin (1977) reconstituted coupled Na^+ and K^+ transport from the purified enzyme of Kyte (1971). Reconstitution of sodium but not potassium transport from a microsomal fraction from the electroplax of *Electrophorus electricus* was reported in a brief note by Racker and Fisher (1975). Anner et al. (1977) reported reconstitution of Na^+ and K^+ transport from Na^+/K^+-ATPase purified from the outer medulla of lamb kidney. In all of these earlier studies, with the exception of the study by Racker and Fisher (1975) in which lipids and enzyme were sonicated together, reconstitution was done essentially by the cholate–dialysis method. Cholate was rather harsh on the enzyme, and enzyme specific activities of the proteoliposomes containing the purified Na^+/K^+-ATPase were only of the order of 100 μmoles/P_i per mg protein per hr. More recently, Hokin and Dixon (1979) reconstituted Na^+ and K^+ transport from the Na^+/K^+-ATPase of *E. electricus* (Dixon and Hokin, 1978) by the freeze–thaw–sonication technique of Kasahara and Hinkle (1977). This enzyme could not be reconstituted by the cholate–dialysis technique because of denaturation by cholate. Proteoliposomes with equal transport rates could also be made from the Na^+/K^+-ATPase of *S. acanthias*. With this technique, transport rates at least one order

Lowell E. Hokin • Department of Pharmacology, University of Wisconsin Medical School, Madison, Wisconsin 53706.

of magnitude higher than the highest rates previously reported with the cholate–dialysis technique were obtained. The liposomes formed by the above methods appear to have single bilamellar membranes (Goldin, 1977; Kasahara and Hinkle, 1977), and the Na^+/K^+-ATPase appears to be oriented randomly so that approximately 50% of the substrate sites face outward and 50% face inward (Goldin, 1977). The enzyme molecules with substrate sites facing outward would have their ouabain sites facing inward, and the enzymes with their substrate sites facing inward would have their ouabain sites facing outward. The coupled transports of Na^+ and K^+ have only been successfully demonstrated by adding MgATP to the outside, which activates Na^+ transport into the liposome and K^+ transport out of the liposome. This is, of course, opposite to the situation in cells. This means that ouabain will only inhibit if it is introduced inside the liposome, for it inhibits only from the outer surface of cells. Vanadate (Josephson and Cantley, 1977; Cantley *et al.*, 1973, 1977; Beaugé and Glynn, 1978; Karlish *et al.*, 1979; Quist and Hokin, 1978), on the other hand, will only inhibit on the outside of the liposome, for it inhibits at or near the ATP site on the inner surface of the membrane. Table I shows some of the characteristics of the reconstituted Na^+/K^+ pump established from earlier work. The fact that the Na^+/K^+ pump reconstituted from purified Na^+/K^+-ATPase shows the same characteristics as the Na^+/K^+ pumps in the erythrocyte and the squid axon is convincing evidence that the isolated Na^+/K^+-ATPase is the Na^+/K^+ pump.

The function of the approximately 45,000- to 55,000-molecular-weight glycoprotein subunit of the Na^+/K^+-ATPase is unclear, for the binding sites for most of the ligands involved in Na^+/K^+-ATPase reside on the large chain (MgATP site, ouabain site, vanadate site, and possibly the ion-binding sites). It is known that antibody against the glycoprotein inhibits enzyme activity (Rhee and Hokin, 1975) and ouabain binding (Rhee and Hokin, 1979). Wheat germ agglutinin (WGA) also agglutinates and inhibits catalytic activity of the Na^+/K^+-ATPase from *S. acanthias* (Perrone and Hokin, 1978). At low molar ratios of WGA to enzyme, reconstituted Na^+ and K^+ transport is partially inhibited by WGA but "transport" Na^+/K^+-ATPase (that component not inhibited by external ouabain) is not (Pennington and Hokin, 1979). This indicates that the transports can be partially uncoupled from catalytic activity and raises interesting questions concerning reaction mechanisms of the Na^+/K^+-ATPase in which ion translocations are tightly coupled to partial reactions of the Na^+/K^+-ATPase. Interestingly, Na^+ transport was inhibited more than K^+ transport as

Table I. Characteristics of the Na^+/K^+ Pump Reconstituted from Purified Na^+/K^+-ATPase[a]

1. Ouabain-inhibitable Na^+ and K^+ transport dependent on external MgATP.
2. Transports of Na^+ and K^+ in opposite directions to those in cells.
3. Transports of Na^+ and K^+ coupled (omission of external Na^+ blocks K^+ exit).
4. Stoichiometry of Na^+ transport to K^+ transport 3 : 2.
5. Ouabain inhibits Na^+ and K^+ transport only from the inside (K^+ side).
6. Gradients approaching the physiological obtained for Na^+ and K^+.
7. Effectiveness of various nucleoside triphosphates for Na^+/K^+-ATPase activity and for reconstituted Na^+/K^+ transport parallel each other.
8. Ouabain-inhibitable exchange diffusion of Na^+ demonstrable in the presence of ATP and in the absence of K^+.
9. Ouabain-inhibitable exchange diffusion of K^+ demonstrable in the presence of ATP and P_i in the absence of Na^+.
10. Na^+ and K^+ transport inhibited by external vanadate.

[a]From Hokin (1979) by permission of Addison–Wesley.

the concentration of WGA was increased. This indicates that Na^+ transport and K^+ transport can be partially uncoupled by WGA. Uncoupling of Na^+ and K^+ transport has been observed in reconstituted systems under other conditions, e.g., by trypsin treatment (Anner and Jørgensen, 1979). This suggests that there may be separate "channels" or "gates" for Na^+ and K^+. Lectins bind to oligosaccharides on the exterior-facing site of the cell membrane. One would therefore expect WGA to inhibit transport in the reconstituted Na^+/K^+-ATPase liposomes only when presented to the inside-facing surface of the membrane, as occurs for ouabain. This was found to be the case (Pennington and Hokin, 1979). It could be shown by SDS-PAGE that WGA bound specifically to the glycoprotein subunit and not to the catalytic subunit (Pennington and Hokin, 1979). Oligosaccharides rich in N-acetyl-glucosamine residues, such as β-1,4-di-N-acetylglucosamine and ovomucoid, reversed inhibition of rectal gland Na^+/K^+-ATPase by WGA (Perrone and Hokin, 1978) and inhibition of reconstituted Na^+ and K^+ transport (Pennington and Hokin, 1979). The mechanism of inhibition of reconstituted Na^+ and K^+ transport by WGA is not clear. It is possible that binding of a relatively large molecule such as WGA [molecular weight of dimeric WGA = 35,000 (Nagata and Burger, 1974; Rice and Etzler, 1974)] covers or distorts certain sites on the pump molecule. This may account for the "uncouplings" that have been observed with WGA.

One would predict from the stoichiometry of $3Na^+ : 2K^+$ that the reconstituted Na^+/K^+ pump would be electrogenic with a positive potential from inside to outside. The Na^+/K^+ pump in erythrocytes has in fact been shown to be electrogenic (Hoffman *et al.*, 1979). One technique for measuring potentials is the use of the lipid-permeant anion $[^{14}C]$-SCN^-, which rapidly equilibrates across biomembranes and distributes itself according to the Gibbs–Donnan potential. If one assumes that the concentration of $[^{14}C]$-SCN^- is the same inside as outside in the absence of ATP, due to diffusional equilibration, one would expect to see a higher concentration of $[^{14}C]$-SCN^- inside than outside when MgATP is added because of the positive potential set up by the Na^+/K^+ pump. This was in fact found (Hokin, 1979; Dixon and Hokin, 1980). The $[^{14}C]$-SCN^- reached an equilibrated value in about 7 min with the "plus" ATP value considerably higher than the "minus" ATP value. The calculated potential in the presence of ATP from six experiments was 14 mV. Part of this potential was due to a diffusion potential due to the higher K^+ gradient than Na^+ gradient, because when the ionophore nigericin was used to collapse the gradients, the ratio of SCN^- dropped from 1.73 to 1.42. Calculation of the potential from the Nernst equation with this latter ratio gave an electrogenic potential of 9 mV, which is the same found by Hoffman *et al.* (1979) for the erythrocytes from *Amphiuma* and close to that found for sheep and human erythrocytes. This potential was abolished both by external vanadate and by internal ouabain, which inhibit Na^+/K^+ transport on the external and internal surface of the proteoliposome, respectively.

If three sodium ions move in for every two potassium ions moving out, either one anion must accompany the three sodium ions or one cation must accompany the two potassium ions to maintain charge equilibrium. This in fact was suggested by the finding that when all Cl^- was substituted with SO_4^{2+}, the stoichiometry of $Na^+ : K^+$ became 1 : 1 (using $^{35}SO_4^{2+}$, the permeability of sulfate was found to be only 1/200 that of chloride). Thus, in the presence of sulfate, the Na^+/K^+ pump can be converted from an electrogenic pump to an electrically neutral Na^+/K^+ pump. This strongly suggested that the $3 Na^+ : 2K^+$ stoichiometry was made possible by the cotransport of Cl^- with Na^-. This was shown directly with $^{36}Cl^-$. There was a net movement of $^{36}Cl^-$ into the liposome. This was equal to the difference between the higher Na^+ influx and the lower K^+ outflux so that the net movement of charge in the two directions was the same. These observations also make it highly unlikely that endogenous ions, possibly present in small amounts in the system (H^+, HCO_3^-, etc.), could take the place of Cl^-.

Considerable efforts have been made to detect ionophore activity in liposomes containing the holoenzyme, the glycoprotein [obtained by trypsin treatment in the absence of SDS (Churchill and Hokin, 1976; White and Hokin, unpublished observations)], and the 12,000-molecular-weight protein seen on SDS gels of most purified Na^+/K^+-ATPase preparations (Pennington and Hokin, unpublished observations). The latter could be somewhat selectively extracted into protein-free liposomes by incubating the holoenzyme with the liposomes. Although these proteins markedly enhanced the permeability of Na^+ and K^+ into the liposomes, they were not selective. It appeared that the permeability of all monovalent cations (Cl^-, Br^-, cations) was enhanced nonspecifically by both the glycoprotein and the 12,000-molecular-weight protein.

Recently, Jørgensen and Anner (1979), using the cholate–dialysis technique for reconstitution from purified kidney medulla Na^+/K^+-ATPase that had been modified by a single tryptic split, found that after reconstitution, the active transport of Na^+ in vesicles containing the trypsinized Na^+/K^+-ATPase was reduced to 30–40% of the nontrypsinized preparation. K^+ transport was not affected. Passive Na^+ and K^+ fluxes were unaffected. The residual Na^+ transport was resistant to vanadate, apparently as a result of reduction of vanadate affinity in the trypsinized preparation.

Reports of the phospholipid requirements for the Na^+/K^+-ATPase and the reconstituted Na^+/K^+ pump are conflicting (see above reviews and Racker and Fisher, 1975). Part of the confusion may be due to the fact that the ''boundary lipids'' in contrast to the bilayer lipids have, in most instances, not been removed in studies involving lipid requirements for the enzyme or the pump. Using the double-substitution technique of Warren et al. (1974), Hilden and Hokin (1976) exchanged all boundary lipid in the enzyme with egg phosphatidylcholine, and they reconstituted Na^+/K^+ transport by the cholate–dialysis technique using only egg phosphatidylcholine. Transport was as good or better than transport reconstituted with the native enzyme containing the endogenous lipids. No lipids other than phosphatidylcholine were seen by thin-layer chromatography of the enzyme doubly substituted with phosphatidylcholine.

REFERENCES

Anner, B. M., and Jørgensen, P. L. (1979). In *Na,K-ATPase: Structure and Kinetics* (J. C. Skou and J. G. Nørby, eds.), p. 87, Academic Press, New York.

Anner, B. M., Lane, L. K., Schwartz, A., and Pitts, B. J. R. (1977). *Biochim. Biophys. Acta* **467**, 340.

Beaugé, L. A., and Glynn, I. M. (1978). *Nature (London)* **272**, 551.

Cantley, L. C., Jr., Cantley, L. G., and Josephson, L. (1973). *J. Biol. Chem.* **253**, 7361.

Cantley, L. C., Jr., Josephson, L., Warner, R., Yanagisawa, M., Lechene, C., and Guidotti, G. (1977). *J. Biol. Chem.* **252**, 7421.

Churchill, L., and Hokin, L. E. (1976). *Biochim. Biophys. Acta* **434**, 258.

Dixon, J. F., and Hokin, L. E. (1978). *Anal. Biochem.* **86**, 378.

Dixon, J. F., and Hokin, L. E. (1980). *J. Biol. Chem.* **255**, 10681.

Goldin, S. M. (1977). *J. Biol. Chem.* **252**, 5630.

Goldin, S. M., and Tong, S. W. (1974). *J. Biol. Chem.* **249**, 5907.

Hilden, S., and Hokin, L. E. (1975). *J. Biol. Chem.* **250**, 6296.

Hilden, S., and Hokin, L. E. (1976). *Biochem. Biophys. Res. Commun.* **69**, 521.

Hilden, S., Rhee, H. M., and Hokin, L. E. (1974). *J. Biol. Chem.* **249**, 7432.

Hoffman, J. F., Kaplan, J. H., and Callahan, T. J. (1979). *Fed. Proc.* **38**, 2440.

Hokin, L. E. (1979). In *Membrane Bioenergetics* (C. P. Lee, G. Schatz, and L. Ernster, eds.), p. 281, Addison–Wesley, Reading, Mass.

Hokin, L. E., and Dixon, J. F. (1979). In *Na,K-ATPase: Structure and Kinetics* (J. C. Skou and J. G. Nørby, eds.), p. 47, Academic Press, New York.

Hokin, L. E., Dahl, J. D., Deupree, J. D., Dixon, J. F., Hackney, J. F., and Perdue, J. R. (1973). *J. Biol. Chem.* **248,** 2593.

Jørgensen, P. L., and Anner, B. M. (1979). *Biochim. Biophys. Acta* **555,** 485.

Josephson, L., and Cantley, L. C., Jr. (1977). *Biochemistry* **16,** 4572.

Karlish, S. J. D., Beaugé, L. A., and Glynn, I. M. (1979). *Nature (London)* **282,** 335.

Kasahara, M., and Hinkle, P. C. (1977). *J. Biol. Chem.* **252,** 7384.

Kyte, J. (1971). *J. Biol. Chem.* **246,** 4156.

Nagata, Y., and Burger, M. M. (1974). *J. Biol. Chem.* **249,** 3116.

Pennington, J., and Hokin, L. E. (1979). *J. Biol. Chem.* **254,** 9754.

Perrone, J. R., and Hokin, L. E. (1978). *Biochim. Biophys. Acta* **525,** 446.

Quist, E. E., and Hokin, L. E. (1978). *Biochim. Biophys. Acta* **511,** 202.

Racker, E., and Fisher, L. W. (1975). *Biochem. Biophys. Res. Commun.* **67,** 1144.

Rhee, H. M., and Hokin, L. E. (1975). *Biochem. Biophys. Res. Commun.* **63,** 1139.

Rhee, H. M., and Hokin, L. E. (1979). *Biochim. Biophys. Acta* **558,** 108.

Rice, R. H., and Etzler, M. E. (1974). *Biochem. Biophys. Res. Commun.* **59,** 414.

Sweadner, K. J., and Goldin, S. M. (1975). *J. Biol. Chem.* **250,** 4022.

Warren, G. B., Toon, P. A., Birdsall, N. J. M., Lee, A. G., and Metcalfe, J. C. (1974). *Proc. Natl. Acad. Sci. USA* **71,** 622.

71

Active and Passive Cation Transport and Its Association with Membrane Antigens in Sheep Erythrocytes

Developments and Trends

Peter K. Lauf

Erythrocyte membrane transport physiology, in part, owes its recent advance to research on cation polymorphic erythrocytes of ruminants. Investigations on the genetics, physiology, biochemistry, and immunology of these cells, as reported up to 1978, have been extensively reviewed by Tucker (1971), Lauf (1975, 1978a,b), and Ellory (1977). The purpose of this article is to briefly update and summarize the present state of the art, drawing particular attention to recent investigations on the nature of the M antigen, on a Cl^--activated passive K^+ permeability, and on transport changes accompanying the development of the cation polymorphism in the adult erythrocytes of sheep.

Since Tosteson and Hoffman (1960) based their model of red cell volume control on an analysis of cation transport parameters of high-potassium (HK) and low-potassium (LK) sheep erythrocytes, much of the subsequent work has attempted to shed light on the mechanisms by which the two types of red cells maintain their cation steady states. Considerable advance in this work occurred when a fortuitous genetic and functional association of membrane surface antigens with active (pump) and passive (leak) Na^+/K^+ transport in these cells was exploited.

The differences in the Na^+/K^+-pump activities originally noted by Tosteson and Hoffman (1960) to exist between HK and LK sheep red cells are of kinetic and quantitative origin. Hoffman and Tosteson (1971) showed that cellular K^+ ions normally competing with Na^+ ions for the three cytoplasmic Na^+-loading sites of the Na^+/K^+ pump were much more competitive in LK sheep red cells with the effect that even at very low cellular K^+

Peter K. Lauf • Department of Physiology, Duke University Medical Center, Durham, North Carolina 27710.

concentrations the K^+-pump activity, measured by ouabain-sensitive K^+ influx, was only a fraction of that seen in HK cells. Joiner and Lauf (1975) measured bound [³H]ouabain and found that LK sheep red cells had only some 40–60 Na^+/K^+ pumps as opposed to 120 in HK cells. On the basis of a direct correlation of unity between [³H]ouabain binding and K^+-pump inhibition, Na^+/K^+ turnover was found to be about 50% lower in LK cells than in HK cells (Joiner and Lauf, 1978a). This difference may be even larger if one takes into account that in HK cells most of the Na^+/K^+-pump activity is carried by about 85 "dominant" pumps (Joiner and Lauf, 1978a). The combined quantitative–kinetic differences between the two types of pumps were underscored by the observations that the rate of [³H]ouabain binding, which ordinarily is highest in high-Na^+ cells and hence related to the Na^+-loading step of the Na^+/K^+ pump, was much lower in fresh LK than in HK cells (Joiner and Lauf, 1978b). As expected, when in LK cells internal K^+ was exchanged for Na^+ by the nystatin method, the rate of ouabain binding increased, further consistent with the view of a functional and perhaps structural difference between HK and LK pumps.

A structural expression of the altered functional state of the Na^+/K^+ pump in LK cells may be the presence of the L antigen (Rasmusen, 1969; Ellory and Tucker, 1969), which is associated with the genetically dominant LK red cell property (Evans and King, 1955), whereas HK cells have only the M antigen (Rasmusen and Hall, 1966) detected by hemolytic antibody techniques (Lauf and Tosteson, 1969). When anti-L, an IgG_1-type antibody (Snyder *et al.*, 1971; Ellory *et al.*, 1973) with low hemolytic efficiency (Lauf and Dessent, 1973) was added to LK cells, their K^+-pump influx dramatically increased (Ellory and Tucker, 1969), which was explained in terms of a reduction of the above-described kinetic inhibition of the Na^+/K^+ pump (Lauf *et al.*, 1970a; Glynn and Ellory, 1972). Because anti-L increased the apparent affinity of the Na^+-loading site of the Na^+/K^+ pump, the rate of [³H]ouabain binding was elevated but not the total equilibrium number of ouabain molecules bound per cell (Joiner and Lauf, 1978b). This is consistent with the hypothesis of important functional differences between LK and HK pumps, which was also supported by the observation of significantly different temperature dependencies of the LK and HK pump activities (Joiner and Lauf, 1979). In part, these differences were alleviated by the action of anti-L pointing to differences in the lipid microenvironments of the HK and LK pumps or to differences in their protein structures interacting with the same lipids. As a wider application of this work, the finding of a Na^+-supported ouabain-binding rate corroborates earlier concepts on cycle-dependent conformational changes of the Na^+/K^+ pump supported by cellular Na^+ (Joiner and Lauf, 1978b).

Naturally, the M and L antigens became the targets of a search for a clue to the structure–function differences between HK and LK pumps. While Kropp and Sachs (1974) reported a unity relationship between the number of Na^+/K^+ pumps and L antigens in LK goat red cells, work in the sheep system showed that there was a substantial excess of M antigens (up to 6000/cell) and L antigens (up to 2000/cell) with regard to the number of Na^+/K^+ pumps in either cell (Lauf and Sun, 1976; Tucker *et al.*, 1976). Although the quantities of these two antigens are prohibitively small to facilitate an easy characterization, two different approaches have provided some insight into the nature of these peculiar surface antigens.

The first approach involved the application of the proteolytic enzyme trypsin, which had been successfully used to remove sialoglycopeptides from human red cell membranes (Winzler *et al.*, 1967). Trypsin, which did not touch the M antigens (Lauf *et al.*, 1970b), abolished the response of the LK pump to the stimulatory action of anti-L without inactivating the complement-mediated lytic action of the antiserum (Lauf *et al.*, 1971). The conclusion that there must be at least two L antigens on the surface of LK cells—the L_p

or pump-associated L antigen, and the L_{LY} antigen reacting with hemolytic L_{LY} antibody—was supported by Dunham (1976a) who correlated the effect of the L antiserum on ouabain-insensitive K^+-leak fluxes with the second L antibody. The two L-antigen specificities related to pumps and leaks were separated by showing that anti-L_p did not bind to trypsinized LK red cells, while anti-L_{LY}, apparently identical with anti-L_L against the K^+-leak-associated L_L-antigen, still bound to and affected K^+-leak flux in trypsinized LK sheep red cells (Lauf *et al.*, 1977). These findings were not entirely consistent with Dunham's hypothesis that anti-L somehow interconverts K^+-leak sites into active Na^+/K^+ pumps. A subsequent report (Dunham *et al.*, 1980a) clarified this controversy by showing that in heterozygous LK red cells, for a reason yet to be explained, trypsin inactivated the L_p antigens only partially. The effect of trypsin on LK cells has evaded any useful explanation thus far. In homozygous LK cells, the L_p antigen apparently was destroyed, and hence anti-L_p could not bring about the conformational change in the LK pump resulting in its elevated turnover number. In heterozygous LK cells, the presence of the M antigen may modulate the attack mode of trypsin (Dunham *et al.*, 1980a), although both M and L antigens seem to be sterically far enough apart as evident by lack of binding interference of the two antibodies (Lauf and Sun, 1976). An interpretation of the trypsin action in the sheep system must consider a recent report by Dunham and Ellory (1980) that trypsin had an opposite effect on the Na^+/K^+-pump activity in LK goat red cells. Obviously, more work is required to untangle these apparent incongruencies between the two red cell systems.

The second approach to the nature of the two membrane antigens is solubilization and biochemical characterization. It was early apparent that both M and L substances must be lipid-dependent proteins (Lauf, 1974). As the hemolytic M antibody assay is economically preferable over the L antibody technique using transmembranous fluxes with ion tracers of short half-life, initial work was done on the M substance. Based on earlier work (Shrager *et al.*, 1972) that clearly established the protein nature of the M antigen and the requirement of phospholipids and cholesterol for its full antigenic activity, Wiedmer and Lauf (1981) employed solubilization techniques involving Triton X-100. This approach was chosen in part because Triton X-100 can be used to separate band 3 protein from other red cell membrane proteins, such as glycoproteins or the sheep-specific band 2.2 protein. That band 3 protein was not the structure carrying the M/L antigenic determinants became evident earlier in the studies on the effect of proteolytic enzymes on the two antigens. It was found that, in contrast to the lack of action on human red cells, trypsin hydrolyzed almost completely the band 3 protein of intact HK or LK sheep red cells into several smaller membrane fragments that remained anchored within the membrane (Lauf, 1977). Neither anti-L nor anti-M prevented this enzyme effect. Other enzymes such as chymotrypsin or pronase had similar effects on band 3 protein (Wiedmer and Lauf, 1981) as shown by ion-exchange chromatography on Affi-Gel in Triton X-100 (England *et al.*, 1980). However, the M substance copurified with band 2.2 (a sheep-specific protein) and with band 6 protein and glycoproteins (Wiedmer and Lauf, 1981). The small quantities of M substance as well as its apparent lability in Triton X-100 when not bound to anti-M (Wiedmer and Lauf, 1981) preclude at present further identification with any of the specific proteins which apparently are unrelated to band 3 proteins. Studies with covalently bound isotopic labels should clarify whether or not the M and L substances are part of the membrane pump complex. Clarification of this relationship is paramount to constructing a useful model explaining the molecular mechanisms by which anti-L stimulates the Na^+/K^+ pump in LK cells and anti-M leaves the HK pump unaffected.

Curiously, much less work has been done to characterize the passive, ouabain-insensitive Na^+/K^+ transport in these cells. At the time when Tosteson and Hoffman (1960) wrote their paper on a model red cell, ouabain-insensitive Na^+/K^+ transport comprised the Na^+- and K^+-leak transport, presumably occurring across the lipid bilayer, and, in sheep red cells, the Na^+–Na^+ exchange system accounted for 90% of the passive Na^+ fluxes (Tosteson and Hoffman, 1960; Motais, 1973). Evolutionarily, the Na^+–Na^+ exchanger may be unrelated to the Na^+/K^+ pump as its activity is about equal in HK and LK sheep red cells and anti-L is without effect in LK cells (Duhm *et al.*, 1980). As there is no ouabain-insensitive K^+–K^+ exchange in sheep red cells (Tosteson and Hoffman, 1960), one would expect passive K^+ transport to be of purely electrodiffusional nature occurring through the lipid bilayer. However, already in 1976, Dunham showed that passive K^+ flux saturated at higher K^+ concentrations and was selectively reduced in LK red cells by the action of anti-L_L (Dunham, 1976b). That anti-L_L reduced K^+-leak flux also raised the question whether the antibody affected the lipid bilayer "ground permeability" for K^+ ions or a different, rather K^+-specific transport system that still had to be characterized in these cells.

An interesting advance distinguishing between the two possibilities was recently made by Lauf and Theg (1980a) who described that in homozygous or heterozygous LK red cells, N-ethylmaleimide activated a halide-anion-dependent, ouabain-insensitive K^+ flux that, apparently with a low activity, is present also in untreated control cells. There is growing evidence that this K^+-transport component is identical with the halide-ion-dependent, volume-sensitive K^+ flux reported by Ellory and Dunham (1980) in sheep red cells. The characteristics of this K^+ transporter, such as low Cl^- or Br^- and low K^+ affinity, are reminiscent of K^+–Cl^- symporters observed in human red cells (Dunham *et al.*, 1980b; Chipperfield, 1980), in duck red cells (Kregenow and Caryk, 1979), in toadfish red cells (Lauf and Theg, 1980b), and in Ehrlich ascites tumor cells (Geck *et al.*, 1980). Although much needs to be done to characterize kinetically and biochemically the Cl^--activated K^+ transport in sheep red cells, the discovery of its presence raises important questions about its role and the molecular mechanism by which it is switched on and off. Furthermore, in a not too distant future, a revision of the original volume control model of Tosteson and Hoffman (1960) is required, taking into account the K^+–Cl^- transporter as well as other ion-exchange systems recently defined in human red cells.

During the past 5 years, renewed interest has been devoted to the problem of the HK–LK transition, i.e., of the development of the LK steady-state cell. This work went far beyond the only, older studies in this field (Lee *et al.*, 1966) using novel cell separation techniques and volume measuring devices. Valet and Lauf (Valet *et al.*, 1978; Lauf and Valet, 1980) showed that cellular replacement determines the gradual HK–LK transition in newborn lambs, which, irrespective of their genotype, are born with fetal HK red cells. In the LK genotype, the fetal cell population is replaced by two consecutive cell populations, one containing rather small, fully developed LK cells and the other, initially of reticulocyte-rich HK-type red cells, rapidly maturing into the final LK red cells. The fact that the kinetics of cell polymorphism dominates the HK–LK transition in the growing lamb diminishes the chance to investigate in the young animal the maturation process actually occurring in red cells before or when they enter the peripheral circulation.

One would think that the massively bled LK sheep with its reticulocyte burst on day 6 after hemorrhage is a better experimental model to study the HK–LK transition. Recently, it was shown that both K^+-pump and -leak fluxes are more than 10-fold higher in the newly released, large reticulocyte-rich cell fraction and that both activities fall over a few days, closely accompanying the trend of cellular cations toward the LK steady state (Kim *et al.*, 1980). An exact evaluation of the above changes in terms of their temporal sequence of contribution to the final attainment of the LK steady state is complicated by the difficulty in obtaining the same cell population for transport studies. The newly formed cells changed their density and hence migrated out of the top 10% fraction as analyzed by density centrifugation (Kim *et al.*, 1980). To overcome this problem of "loosing" the maturing red cells, Lauf and Valet (1981) applied the elutriator technique separating at various times after bleeding the large new cells from the older smaller ones. Interestingly, the macrocytic red cell population, although its cells lost some volume, never joined the adult population as evidenced by monitoring hemoglobin C (Lauf and Valet, 1981). In spite of the uncertainty stemming from the possibility that entrance and destruction rates of these cells may be high and hence again will influence a kinetic interpretation of the transport changes, the data on hand indicate that quantitative (reduction of pump sites) and kinetic changes occur that lead to the development of the LK status (Lauf and Valet, 1981). Furthermore, the high apparent K^+ permeability reported earlier (Kim *et al.*, 1980) appears to be due to the presence of Cl^--activated K^+ transport, as NO_3^- replacement of Cl^- ions leads to a 75% reduction of ouabain-insensitive K^+ net flux (Lauf, 1981). With the caveat that the erythrocytes, released 6 days posthemorrhage, are emergency cells, there is sufficient evidence to conclude that red cell maturation and attainment of the LK steady state involves major modulations of active and passive cation transport, changes that hopefully will be unraveled by *in vitro* techniques (Tucker and Young, 1980; Kim *et al.*, 1980).

Measured by the wealth of information gathered, the cation polymorphic sheep red cell offers to be a major experimental and theoretical model system for further studies on physiology, biochemistry, and immunology of membrane ion transport.

ACKNOWLEDGMENT. This work was supported by USPHS Grant HL 2PO1-12,157.

REFERENCES

Chipperfield, A. R. (1980). *Nature (London)* **286**, 281–282.
Duhm, J., Becker, B. F., and Lauf, P. K. (1980). *Life Sci.* **26**, 1217–1222.
Dunham, P. B. (1976a). *Biochim. Biophys. Acta* **443**, 219–226.
Dunham, P. B. (1976b). *J. Gen. Physiol.* **68**, 567–581.
Dunham, P. B., and Ellory, J. C. (1980). *J. Physiol. (London)* **301**, 25–37.
Dunham, P. B., Tucker, E. M., Simonsen, E., and Ellory, J. C. (1980a). *J. Gen. Physiol.* **75**, 345–350.
Dunham, P. B., Stewart, G. W., and Ellory, J. C. (1980b). *Proc. Natl. Acad. Sci. USA* **77**, 1711–1715.
Ellory, J. C. (1977). In *Membrane Transport in Red Cells* (J. C. Ellory and V. L. Lew, eds.), pp. 363–381, Academic Press, New York.
Ellory, J. C., and Dunham, P. B. (1980). In *Alfred Benzon Symposium 14: Membrane Transport in Erythrocytes,* Munksgaard, Copenhagen.
Ellory, J. C., and Tucker, E. M. (1969). *Nature (London)* **222**, 477–478.
Ellory, J. C., Feinstein, A., and Herbert, J. (1973). *Immunochemistry* **10**, 785.
England, B. J., Gunn, R. B., and Steck, T. L. (1980). *Biochim. Biophys. Acta* **623**, 171–182.
Evans, J. V., and King, J. W. B. (1955). *Nature (London)* **176**, 171.
Geck, C., Pietrzyk, C., Burckhardt, B. C., Pfeiffer, B., and Heinz, E. (1980). *Biochim. Biophys. Acta* **600**, 432–447.
Glynn, L. M., and Ellory, J. C. (1972). In *Role of Membranes in Secretory Processes* (L. Bolis, R. D. Keynes, and W. Wilbrandt, eds.), pp. 224–237, North-Holland/American Elsevier, New York.
Hoffman, P. G., and Tosteson, D. C. (1971). *J. Gen. Physiol.* **58**, 438–466.
Joiner, C. H., and Lauf, P. K. (1975). *J. Membr. Biol.* **21**, 99–112.
Joiner, C. H., and Lauf, P. K. (1978a). *J. Physiol. (London)* **283**, 155–176.
Joiner, C. H., and Lauf, P. K. (1978b). *J. Physiol. (London)* **283**, 177–196.
Joiner, C. H., and Lauf, P. K. (1979). *Biochim. Biophys. Acta* **552**, 340–454.
Kim, H. D., Theg, B. E., and Lauf, P. K. (1980). *J. Gen. Physiol.* **76**, 101–121.

Kregenow, F. M., and Caryk, T. (1979). *Physiologist* **22,** 73.

Kropp, D. L., and Sachs, J. R. (1974). *Nature (London)* **252,** 244–246.

Lauf, P. K. (1974). *Ann. N.Y. Acad. Sci.* **242,** 324–342.

Lauf, P. K. (1975). *Biochim. Biophys. Acta* **415,** 173–229.

Lauf, P. K. (1977). *J. Gen. Physiol.* **70,** 11a.

Lauf, P. K. (1978a). In *Membrane Transport in Biology* (G. Giebisch, D. C. Tosteson, and H. H. Ussing, eds.), Vol. I, pp. 291–348, Springer-Verlag, Berlin.

Lauf, P. K. (1978b). In *Physiology of Membrane Disorders* (T. E. Andreoli, J. F. Hoffman, and D. D. Fanestil, eds.), pp. 369–398, Plenum Press, New York.

Lauf, P. K. (1981). *Fed. Proc.,* **40,** 484.

Lauf, P. K., and Dessent, M. P. (1973). *Immunol. Commun.* **2,** 193–212.

Lauf, P. K., and Sun, W. W. (1976). *J. Membr. Biol.* **28,** 351–372.

Lauf, P. K., and Theg, B. E. (1980a). *Biochem. Biophys. Res. Commun.* **92,** 1422–1428.

Lauf, P. K., and Theg, B. E. (1980b). *Physiologist* **23,** 616.

Lauf, P. K., and Tosteson, D. C. (1969). *J. Membr. Biol.* **1,** 177–193.

Lauf, P. K., and Valet, G. (1980). *J. Cell. Physiol.* **104,** 283–293.

Lauf, P. K., and Valet, G. (1981). *Biophys. J.,* **33,** 3a.

Lauf, P. K., Rasmusen, B. A., Hoffman, P. G., Dunham, P. B., Cook, P., Parmelee, M. L., and Tosteson, D. C. (1970a) *J. Membr. Biol.* **3,** 1–13.

Lauf, P. K., Rasmusen, B. A., and Tosteson, D. C. (1970b). In *Blood and Tissue Antigens,* p. 341.

Lauf, P. K., Parmelee, M. L., Snyder, J. J., and Tosteson, D. C. (1971). *J. Membr. Biol.* **4,** 52–67.

Lauf, P. K., Stiehl, B. J., and Joiner, C. H. (1977). *J. Gen. Physiol.* **70,** 221–242.

Lee, P., Woo, A., and Tosteson, D. C. (1966). *J. Gen. Physiol.* **50,** 379–390.

Motais, R. (1973). *J. Physiol. (London)* **233,** 395–422.

Rasmusen, B. A. (1969). *Genetics* **61,** 49s.

Rasmusen, B. A., and Hall, J. G. (1966). *Science* **151,** 1551–1552.

Shrager, P., Tosteson, D.C., and Lauf, P. K. (1972). *Biochim. Biophys. Acta* **290,** 186–199.

Snyder, J. J., Rasmusen, B. A., and Lauf, P. K. (1971). *J. Immunol.* **107,** 772–781.

Tosteson, D. C., and Hoffman, J. F. (1960). *J. Gen. Physiol.* **44,** 169–194.

Tucker, E. M. (1971). *Biol. Rev.* **46,** 341–386.

Tucker, E. M., and Young, J. D. (1980). *Biochem. J.* **192,** 33–39.

Tucker, E. M., Ellory, J. C., Wooding, F. B. P., Morgan, G., and Herbert, J. (1976). *Proc. R. Soc. London Ser. B* **194,** 271–277.

Valet, G., Franz, G., and Lauf, P. K. (1978). *J. Cell. Physiol.* **94,** 215–228.

Wiedmer, T., and Lauf, P. K. (1981). *Membr. Biochem.,* **4,** 31–47.

Winzler, R. J., Harris, E. D., Pekas, D. J., Johnson, C. A., and Weber, P. (1967). *Biochemistry* **6,** 2195.

72

How Do ATP-Driven Ion Membrane Transport Systems Realize High Efficiency in Both Energy Conversion and Flow Rate?

Kurt R. H. Repke and Richard Grosse

What is needed are specific models and axioms which convey ideas of the nature of the components of living systems. These are not necessarily the full truth, but by providing a guide to experimentation they may lead to the truth.

H. A. Krebs (1966)

1. STATEMENT OF PROBLEM

One method universally available to organisms for improving their survival value in evolution was to improve the efficient use of the high-grade energy of ATP. Roughly, the faster the process of energy release, the more high-grade energy is likely to be wasted in useless entropy production. In sudden irreversible processes, high-grade energy is degraded to heat. In smooth reversible processes, however, the energy release takes an infinite length of time. Hence, both energetic and kinetic efficiency are required to yield improved survival value (cf. Lumry and Biltonen, 1969; Welch, 1977).

In an isothermal, two-process system, the fraction of the Gibbs energy expenditure of the input process that undergoes conversion into the output process can be defined according to Kedem and Caplan (1965) by the efficiency function:

Kurt R. H. Repke and Richard Grosse • Biomembrane Section, Central Institute of Molecular Biology, Academy of Sciences of the German Democratic Republic, 1115 Berlin–Buch, G.D.R. This chapter is dedicated to the discoverer of the Lohmann reaction.

$$\eta = -\frac{\text{output power}}{\text{input power}} = -\frac{X_1 J_1}{X_2 J_2} = 1 - \frac{d_i S/dt}{X_2 J_2}$$

In ion membrane transport systems, X_1 is the developed electrochemical potential of the ion gradient over the membrane, X_2 the applied phosphate potential $= \Delta G°_{ATP} + RT$ $\ln[ADP][P_i]/[ATP]$, J_1 the net rate of ion transport, J_2 the net rate of ATP hydrolysis, and $d_i S/dt$ the entropy production per unit time. In principle, there are two borderline cases to meet the general evolutionary requirement of both minimal entropy production and high flow rate, i.e., either an increase of energetic efficiency without loss of kinetic efficiency or an increase of kinetic efficiency without loss of energetic efficiency. How the latter evolutionary achievement is realized will be shown in the present account for the first time. It is mainly based on ideas and observations accumulated by the members of our group— Dittrich, Eckert, Grosse, Repke, Schön, Schönfeld, Spitzer, Streckenbach—and published since 1973 in scattered articles as listed in the references. The ATP-driven Na^+/K^+ antiport system of the plasma membrane will be primarily treated here because for this system the most insight into basic principles is available.

2. PRINCIPLES OF HIGH EFFICIENCY IN ENERGY CONVERSION

The Na^+/K^+ antiport system of erythrocytes is estimated to work close to the thermodynamic equilibrium using the total concentrations of ATP and ADP for the calculation of the total Gibbs energy (Repke, 1977). The near-equilibrium state seems to be a necessary and sufficient condition (cf. Kedem and Caplan, 1965) to account for the 80% energetic efficiency of this transport system (Caldwell, 1969). Such a high energetic efficiency requires a correspondingly high degree of Gibbs energy conservation through tight coupling in time, but not necessarily in space, of all exergonic and endergonic events in transport work; the dimeric Na^+/K^+-ATPase thus acts as a Gibbs energy coupling device (cf. Kemeny, 1974). These relations were the starting point for the design of the ''flip-flop'' model of the Na^+/K^+-ATPase mechanism for converting the scalar Gibbs energy of ATP into the vectorial Gibbs energy of electrochemical gradients over the membrane (Repke and Schön, 1973). As any real understanding of a basic scientific problem includes the capacity to predict by generalization, Repke and Schön (1974) felt challenged to deduce from the scarce data then available the flip-flop model of energy interconversion by ATP synthetase, which predicted that exergonic ADP binding to one catalytic center drives the endergonic ATP removal from another catalytic center, whereas the electrochemical potential over the membrane is utilized for the synthesis of ATP from ADP and P_i. This prediction is experimentally supported by observations of Adolfsen and Moudrianakis (1976) and of Kayalar et al. (1977), to cite only the first supportive publications.

The proper thermodynamic analysis of the Na^+/K^+-ATPase pump requires the application of nonequilibrium thermodynamics which treats the complete cycle as basic unit in Gibbs energy transduction (cf. Hill, 1977). To probe the energetic relations of the flip-flop model, the Gibbs energy conversion was decomposed in a step-by-step fashion. Thus, the pump was treated as being in momentary microscopic equilibrium states at the various intermediary steps. Hence, the apparent standard Gibbs energy changes were calculated from the equilibrium states of the single events along the reaction pathway. This practice is justified by the findings that the pump works close to thermodynamic equilibrium and that the K_m or K_d values for ATP, Na^+, and K^+ found in the steady-state and equilibrium systems, respectively, are rather similar (Grosse et al., 1978, 1979). The application of the principle of Gibbs energy conservation, realized through coincidence of complementary exergonic and endergonic events along the reaction pathway of Na^+/K^+-ATPase (cf. Lumry and Biltonen, 1969), led to the conclusion that there is coupling between the subunits of the enzyme as to ATP binding and ADP release, phosphorylation and dephosphorylation, translocation of Na^+ and K^+ from the intracellular to the extracellular space and vice versa, combined with the change of cation affinities from high Na^+ to high K^+ and from high K^+ to high Na^+ affinity and followed by the cation exchanges (Schön *et al.*, 1974; Repke *et al.*, 1974, 1975; Repke, 1977).

In the absence of K^+, or at low [ATP], there is no coupling between the subunits. The V_{max} of ATP cleavage is low because the release of tightly bound ADP limits the rate. In the presence of K^+ and high [ATP], the anticooperative reactivity of the enzyme to ATP and ADP, produced by intersubunit communication, guarantees the faster release of ADP from the enzyme. The resulting higher V_{max} value is "paid for" by the utilization and the corresponding weakening of the binding energy of ATP that explains the concurrence of the high K_m and the high V_{max} value in Na^+/K^+-ATPase turnover. These relations were deduced by Grosse *et al.* (1978, 1979) to be the very nature of the "activating" effect of higher [ATP], which is well documented not only for ATP cleavage, but also for the Na^+/K^+ pump (cf. e.g. Brinley and Mullins, 1968).

Clearly, the binding energy of the adenine moiety of ATP provides the driving force for product removal. Thus, in erythrocyte ghosts, the rate of Na^+/K^+ antiport with ATP as substrate is 10, 40, or 100 times higher than with CTP, ITP, or UTP, respectively (Karlish and Glynn, 1974). In internally dialyzed squid axons, GTP, CTP, UTP, and acetyl phosphate are essentially ineffective (Brinley and Mullins, 1968). The indicated graduation of flow rate efficiency in the Na^+/K^+ antiport parallels roughly the graduation of the affinities to and of the cleavage rate by the transport enzyme as determined by Hegyvary and Post (1971). As generalized by Gutfreund and Trentham (1975), the reactions that appear to differentiate in the energy of binding of substrates and products are the isomerizations of the initial complex. In line with this reasoning, we assume that the nature of the purine moiety of the nucleoside triphosphates determines the occurrence and frequency of the isomerization which is produced by the binding of the nucleoside triphosphate to one catalytic center and removal of the nucleoside diphosphate from the other catalytic center. This "flip-over" on–off step may be related to a sliding of the subunit interfaces relative to each other (Dittrich and Repke, 1979; Repke and Dittrich, 1979) which effects the exclusion of ADP, and the exchange of the bound cations (Repke, 1977; Grosse *et al.*, 1978). In line with this explanation is the finding that the ability of the various nucleoside triphosphates to drive Na^+ uniport is less different than their ability to drive Na^+/K^+ antiport (Karlish and Glynn, 1974), for the former operation mode does not require subunit coupling (Grosse *et al.*, 1978).

The steady-state rates of catalysis by a number of enzymes are known to be limited by the rate of product removal from the enzyme. It is not uncommon, therefore, that substrate binding steps (i.e., on–off steps involving protein isomerizations) are rate limiting, and this requires that the actual catalytic steps for these enzymes are relatively fast (cf. Albery and Knowles, 1976). This appears to apply to the Na^+/K^+-ATPase pump system, too, as may be deduced from two major findings. First, the acceleration of the on–off step by increasing the [ATP] : [ADP] ratio increases the rate of Na^+/K^+ antiport (cf. Section 3). Second, near 0°C, as amply documented, the phosphorylation and dephosphorylation of the enzyme continue to proceed rapidly, whereas the Na^+/K^+-ATPase turnover and the Na^+/K^+ antiport are nearly completely stopped, apparently due to the high temperature dependence of the protein isomerization.

Similar relations and interpretations appear to fit the various Ca^{2+}-ATPase pumps as illustrated by a few examples.—The sarcoplasmic reticulum Ca^{2+}-ATPase shows the co-existence of two catalytic centers with high and low affinity for ATP and ADP. In generalizing the flip-flop model, Eckert *et al.* (1977) concluded from this finding that through anticooperative subunit interaction the energy of ATP binding to one catalytic center is utilized for the removal of ADP from the other catalytic center, thus accounting for the "activating effect" of higher ATP concentrations. In line with this reasoning, Verjovski-Almeida and Inesi (1979) suggested that the activation of Ca^{2+} transport at high ATP concentration is related to dimeric subunit interaction with occupancy of a second site accelerating the turnover of the first. The activating ATP effect on Ca^{2+}-ATPase is realized through anticooperative subunit interaction in the oligomeric enzyme and therefore absent in the monomeric enzyme (Møller *et al.*, 1980). The ease with which detergents transform the oligomeric enzyme into its monomers indicates the weakness of the intersubunit attraction forces. Hence, the sliding of the subunit interfaces in the isomerization may require much lower Gibbs energies from the binding of the substrates to Ca^{2+}-ATPase than to Na^+/K^+-ATPase. Actually, the sarcoplasmic reticulum Ca^{2+} pump can be fueled by almost any phosphoryl derivatives, although the rates of Ca^{2+} transport are different decreasing in the order $ATP > ITP = GTP > CTP > UTP >$ acetyl phosphate (Hasselbach, 1979).—The rate of the Ca^{2+} transport across the plasma membrane of erythrocytes (Mualem and Karlish, 1979) and of squid axon (DiPolo, 1979) is likewise accelerated at high ATP concentration. Contrary to the sarcoplasmic reticulum Ca^{2+} pump, their specificity for ATP as fueling substrate is very high (Schatzmann and Bürgin, 1978; DiPolo, 1979).

The overall Gibbs energy expenditure for driving Na^+/K^+ antiport must come from the difference between the Gibbs energies of ATP and its hydrolysis products released to the medium. The amount of Gibbs energy that is entropically lost in the phosphorylation and dephosphorylation of the enzyme must be almost negligible judged from the fact that even during maximal Na^+/K^+-ATPase activity relatively large back reactions occur (cf. Dahms and Boyer, 1973). In the closing remarks to a symposium on energy transformation in biological systems, Huxley (1975) stated that the answer to the question of how splitting a bond in ATP produces active transport still eludes us. Mitchell (1979) suggested that the circulation of the anionic adenine nucleotides and inorganic phosphate may be coupled to cation translocation in the Na^+/K^+-ATPase by what may be described as a semidirect electrovalently coupled ligand-conduction type of mechanism; the transformation of the chemical energy of ATP hydrolysis to the osmotic work of cation translocation could depend on electrovalent coupling between the cations and the anionic ATP and phosphate groups traveling through specific ligand-conducting pathways leading into and out of the active site region of the enzyme. Such a mechanism, however, appears to us to be unlikely from what is known on the structure and topology of the Na^+/K^+-ATPase (Repke, 1980).

The requirement of a highly efficient and reversible release and transduction of Gibbs energy from the high-energy arrangement of the oxygen-ligands of phosphorus atom in the terminal phosphoryl group of ATP and in the phosphoryl residue of phosphoenzyme intermediate can be met by the associative pathway of the phosphoryl transfer reactions in Na^+/K^+-ATPase turnover (Dittrich *et al.*, 1974; Repke *et al.*, 1975; Repke, 1977). In this mechanism, the pseudorotations of the oxygen-ligands in the indicated phosphoryl residues are proposed to be the basic processes of the conversion of scalar chemical Gibbs energy into the vectorial mechanical Gibbs energy of polypeptide segments produced by the pseudorotation of the interacting oxygen-ligands. This mechanical Gibbs energy must be transported along the polypeptide chain from the catalytic centers to the ionophoric centers (cf. Repke, 1977).—The proposed associative pathway of phosphoryl and energy transfer is supported by five lines of argument. First, this mechanism can account for the results of O-exchange studies (cf. Dahms and Boyer, 1973; Dahms *et al.*, 1973; Shaffer *et al.*, 1978), although other interpretations seem to be possible (Knowles, 1980). Second, the highly different facility of the pseudorotations of the oxygen-ligands of the pentacoordinated phosphorus and arsenic atom accounts for the finding that acid denaturation of the protein stabilizes enzyme carboxyl phosphate but labilizes enzyme carboxyl arsenate (Schönfeld and Streckenbach, 1977). Third, spin-labeled ATP (SL-ATP) arrests the catalytic action of Na$^+$/K$^+$-ATPase after the formation of the covalent enzyme–SL-ATP complex because the oxygen-ligands of the terminal phosphorus atom, owing to steric hindrance by the bulky reporter group at the ribose moiety, cannot undergo the pseudorotation-requiring rearrangements of both SL-ADPO-residue and enzyme polypeptide chain (Streckenbach *et al.*, 1980). Fourth, the associative mechanism meets the requirement that in bond splitting not much energy is wasted because through the pseudorotations strong P–O bonds become transformed into easily dissociable bonds. Fifth, the pseudorotation of the oxygen-ligands of phosphorus atom may be accompanied by large changes of bond energies as calculated by Gillespie *et al.* (1971).—With regard to the concerted interplay between Na$^+$/K$^+$-ATPase and creatine kinase to be described in Section 3, it is interesting to note that there is evidence for an associative mechanism also in the phosphoryl transfer step catalyzed by creatine kinase (Lowe and Sproat, 1980).

3. PRINCIPLES OF HIGH EFFICIENCY IN FLOW RATE

For the Na$^+$/K$^+$ antiport in erythrocytes, the total Gibbs energy of the forces calculated from the physiological concentrations of its five effectors in the intra- and extracellular space is close to thermodynamic equilibrium (Repke, 1977). Hence, besides net Na$^+$/K$^+$ antiport, ATP synthesis must also occur, and the observed flow rate efficiency must be well below the theoretic maximum. The latter evaluation, not considering the possible existence of a functional linkage between Na$^+$/K$^+$-ATPase and phosphoglycerate kinase at the cytoplasmic membrane surface (cf. Fossel and Solomon, 1979), was canceled as the outcome of studies on Na$^+$/K$^+$ antiport in vesicles formed from plasma membranes of cardiac muscle cells (Grosse *et al.*, 1980).

As shown by Saks *et al.* (1977), a kinetic coupling of Na$^+$/K$^+$-ATPase and creatine kinase, as far as the latter resides in the plasma membrane, permits the effective utilization of phosphocreatine (CrP) for the phosphorylation of ADP produced in the Na$^+$/K$^+$-ATPase reaction. When [ATP] is raised at saturating [CrP], the quantity of ADP released from Na$^+$/K$^+$-ATPase into the medium remains low and constant with time, whereas the quantity and rate of Cr production by creatine kinase become greater and increase linearly with time. Ouabain inhibits the activities of both Na$^+$/K$^+$-ATPase and creatine kinase in this coupled system, although only the former enzyme exhibits an inhibitory ouabain site. The coordinate activation and inhibition of the two enzymatic activities shows that they constitute an integrated unit of catalytic action ("enzyme cluster"). Under physiological conditions, the regeneration of ATP from ADP is favored due to the favorable value of the overall equilibrium constant $K = ([\text{MgADP}][\text{CrP}])/([\text{MgATP}][\text{Cr}])$ near 0.1 (Cohn, 1979) and the high [CrP] (near 15 mM).

Using the plasma membrane vesicles characterized by Saks *et al.* (1977), Grosse *et al.* (1980) showed that the rate of Na$^+$/K$^+$ antiport can be maximized either with the ATP-regenerating system consisting of the creatine kinase in the enzyme cluster and saturating [CrP], or with an ATP-regenerating system consisting of pyruvate kinase added to the bulk phase and saturating [phosphoenolpyruvate]; however, 0.3 mM MgATP or as much as 20 mM MgATP, respectively, is required in the two cases. This disparity is rationalized in terms of different flow rate efficiencies of sequentially working enzymes when they cooperate either physically associated in an organized enzyme cluster as an integrated unit of catalytic activity (for a review see Welch, 1977) or separated as a bulk reaction diffusion system. As evidenced by the similar V_{max} values of Na^+/K^+ antiport, the intrinsic kinetic constants of the component enzymes are not improved upon their association. Hence, the higher flow rate efficiency of the integrated catalytic unit is bound to arise from an increase of the thermodynamic driving force. In nonequilibrium thermodynamics, the chemical affinity A is defined by the equation $A = -\Sigma_i \nu_i d\mu_i$ (De Donder and van Rysselberghe, 1936). In the case of the Na^+/K^+ antiport system, ν_i represents the stoichiometric coefficients and μ_i the chemical potential of the substrates (ATP, intracellular Na^+, extracellular K^+) and of the products (ADP, P_i, extracellular Na^+, intracellular K^+).

The deduced increase of chemical affinity A in the coordinate interplay between Na^+/K^+-ATPase and creatine kinase may result from the convergence of four elements. First, the channeling of the ATP supply to one and the ADP removal from the other catalytic center of Na^+/K^+-ATPase increase the [ATP] : [ADP] ratio near the catalytic centers. Second, the diffusion transit times of ATP and ADP to or from the catalytic centers are reduced. Third, the directed diffusion flow reduces the translational and rotational entropy of ATP and ADP, resulting in negentropy gain. Fourth, these channeling effects decrease or eliminate the partial reversal reactions (including enzyme phosphorylation from medium P_i and excluding ATP dissociation from the enzyme) that exist even under conditions of maximal Na^+/K^+-ATPase activity as evidenced by ^{18}O-exchange studies (Dahms and Boyer, 1973).

As shown by Degani and Degani (1980), creatine kinase is a dimeric enzyme composed of two primarily identical polypeptide chains with two catalytic centers. In the enzyme-catalyzed reaction, there are anticooperative interactions between the subunits. The intersubunit communication in the heterologous homodimer appears to be rather similar to that of Na^+/K^+-ATPase, as described by Repke and Dittrich (1979). In the formation of the cluster between the two enzymes, optimally fitting conformations of their respective subunits may be induced by protein–protein interaction. Their noncovalent association may allow a transconformational synchronization of the reciprocal interaction of the opposed catalytic centers. They may form two rather close microcompartments, in which a cyclic channeling of ADP and ATP between the centers of the coordinate enzyme cluster occurs. The low [ATP] for maximization of Na^+/K^+ antiport may then only serve to replace those nucleotide molecules that escaped from that microcompartment. This model accounts for the observation (Grosse *et al.*, 1980) that the MgATP concentration required to maximize the flow rate efficiency of Na^+/K^+ antiport (near 0.3 mM) lies well below the MgATP concentration necessary for maximal activity of Na^+/K^+-ATPase of nonvesicular membrane fragments (near 3 mM).

The ATP-driven Ca^{2+} pump of cardiac plasma membrane vesicles was also shown to be coupled with membrane-associated creatine kinase and to act as an integrated unit of Ca^{2+} transport with a high rate at low [MgATP] and saturating [CrP] (Spitzer and Grosse, 1980). In this enzyme cluster, the [MgATP] required for half-maximal Ca^{2+} transport was found to be 0.007 and 0.17 mM with the endogenous or added ATP-regenerating systems, respectively. These findings can be accounted for in the same terms as outlined above.— In dialyzed squid axons, the rate of Na^+/K^+ antiport across the plasma membrane is accelerated with the increase of ATP concentration from 1 μM to 4000 μM by factors of 50 or 2 in the absence or presence of phosphoarginine, respectively (Brinley and Mullins, 1968;

Mullins and Brinley, 1969). These relations suggest again that Na^+/K^+-ATPase and arginine kinase form an integrated catalytic unit that in the presence of phosphoarginine produces nearly maximal rates of Na^+/K^+ antiport at very low [ATP].

The cyclic channeling of ATP and ADP between the catalytic centers of Na^+/K^+-ATPase and creatine kinase makes Na^+/K^+-ATPase inaccessible to metabolic oscillations in ATP concentration or phosphate potential, i.e., creatine kinase acts like a "thermodynamic buffer enzyme" (cf. Stucki, 1980). Therefore, the rate of Na^+/K^+ antiport appears to be solely dependent on the work load determined by the intracellular $[Na^+] : [K^+]$ ratio. Actually, in frog skeletal muscle, elevated ADP levels have little effect on the rate of Na^+/K^+ antiport, but cause Na^+/Na^+ exchange to appear through the plasma membrane (Kennedy and de Weer, 1977).—Creatine kinase may alternately interact either with the Na^+/K^+ pump or with the Ca^{2+} pump, thus controlling their flow rate efficiency. A similar coupling may also exist between the Na^+/K^+ antiport and membrane-bound phosphoglycerate kinase in red blood cells (Parker and Hoffman, 1967).

REFERENCES

Adolfsen, R., and Moudrianakis, E. N. (1976). *Arch. Biochem. Biophys.* **172**, 425–433.

Albery, W. J., and Knowles, J. R. (1976). *Biochemistry* **15**, 5632–5640.

Brinley, F. J., and Mullins, L. J. (1968). *J. Gen. Physiol.* **52**, 181–211.

Caldwell, P. C. (1969). *Curr. Top. Bioenerg.* **3**, 251–268.

Cohn, M. (1979). In *NMR and Biochemistry* (S. J. Opella and P. Lu, eds.), pp. 7–27, Dekker, New York.

Dahms, A. S., and Boyer, P. D. (1973). *J. Biol. Chem.* **248**, 3155–3162.

Dahms, A. S., Kanazawa, T., and Boyer, P. D. (1973). *J. Biol. Chem.* **248**, 6592–6595.

De Donder, T., and van Rysselberghe, P. (1936). *Thermodynamic Theory of Affinity.* Stanford University Press, Stanford, Calif.

Degani, C., and Degani, Y. (1980). *J. Biol. Chem.* **255**, 8221–8228.

DiPolo, R. (1979). *J. Gen. Physiol.* **66**, 795–813.

Dittrich, F., and Repke, K. R. H. (1979). *Acta Biol. Med. Ger.* **38**, K5–K11.

Dittrich, F., Schön, R., and Repke, K. R. H. (1974). *Acta Biol. Med. Ger.* **33**, K17–K25.

Eckert, K., Grosse, R., Levitsky, D. O., Kuzmin, A. V., Smirnov, V. N., and Repke, K. R. H. (1977). *Acta Biol. Med. Ger.* **36**, K1–K10.

Fossel, E. T., and Solomon, A. K. (1979). *Biochim. Biophys. Acta* **553**, 142–153.

Gillespie, P., Hoffmann, P., Klusacek, H., Marquarding, D., Pfohl, S., Ramirez, F., Tsolis, E. A., and Ugi, J. (1971). *Angew. Chem.* **83**, 691–721.

Grosse, R., Eckert, K., Malur, J., and Repke, K. R. H. (1978). *Acta Biol. Med. Ger.* **37**, 83–96.

Grosse, R., Rapoport, T., Malur, J., Fischer, J., and Repke, K. R. H. (1979). *Biochim. Biophys. Acta* **550**, 500–514.

Grosse, R., Spitzer, E., Kupriyanov, V. V., Saks, V. A., and Repke, K. R. H. (1980). *Biochim. Biophys. Acta* **603**, 142–156.

Gutfreund, H., and Trentham, D. R. (1975). *Ciba Found. Symp.* **31**, 69–86.

Hasselbach, W. (1979). *Top. Curr. Chem.* **78**, 1–56.

Hegyvary, C., and Post, R. L. (1971). *J. Biol. Chem.* **246**, 5234–5240.

Hill, T. L. (1977). *Trends Biochem. Sci.* **1977**, 204–207.

Huxley, A. F. (1975). *Ciba Found. Symp.* **31**, 401.

Karlish, S. J. D., and Glynn, I. M. (1974). *Ann. N.Y. Acad. Sci.* **242**, 461–470.

Kayalar, C., Rosing, J., and Boyer, P. D. (1977). *J. Biol. Chem.* **252**, 2486–2491.

Kedem, O., and Caplan, S. R. (1965). *Trans. Faraday Soc.* **21**, 1897–1911.

Kemeny, G. (1974). *Proc. Natl. Acad. Sci. USA* **71**, 3669–3671.

Kennedy, B. G., and de Weer, P. (1977). *Nature (London)* **268**, 165–167.

Knowles, J. R. (1980). *Annu. Rev. Biochem.* **49**, 877–919.

Krebs, H. A. (1966). In *Current Aspects of Biochemical Energetics* (N. O. Kaplan and E. P. Kennedy, eds.), pp. 83–95, Academic Press, New York.

Lowe, G., and Sproat, B. S. (1980). *J. Biol. Chem.* **255**, 3944–3951.

Lumry, R., and Biltonen, R. (1969). In *Structures and Stability of Biological Macromolecules* (S. N. Timasheff and G. D. Fassman, eds.), pp. 65–212, Dekker, New York.

Mitchell, P. (1979). *Eur. J. Biochem.* **95**, 1–20.

Møller, J. V., Lind, K. E., and Andersen, J. P. (1980). *J. Biol. Chem.* **255**, 1912–1920.

Mualem, S., and Karlish, S. J. D. (1979). *Nature (London)* **277**, 238–240.

Mullins, L. J., and Brinley, F. J. (1969). *J. Gen. Physiol.* **53**, 704–740.

Parker, J. C., and Hoffman, J. F. (1967). *J. Gen. Physiol.* **50**, 893–916.

Repke, K. R. H. (1977). In *Biochemistry of Membrane Transport* (G. Semenza and E. Carafoli, eds.), pp. 363–373, Springer-Verlag, Berlin.

Repke, K. R. H. (1980). In *Cell Compartmentation and Metabolic Channeling* (L. Nover, F. Lynen, and K. Mothes, eds.), pp. 33–46, VEB Gustav Fischer Verlag, Jena, and Elsevier/North-Holland, Amsterdam.

Repke, K. R. H., and Dittrich, F. (1979). In *Na,K-ATPase: Structure and Kinetics* (J. C. Skou and J. G. Nørby, eds.), pp. 487–500, Academic Press, New York.

Repke, K. R. H., and Schön, R. (1973). *Acta Biol. Med. Ger.* **31**, K19–K30.

Repke, K. R. H., and Schön, R. (1974). *Acta Biol. Med. Ger.* **33**, K27–K38.

Repke, K. R. H., Schön, R., Henke, W., Schönfeld, W., Streckenbach, B., and Dittrich, F. (1974). *Ann. N.Y. Acad. Sci.* **242**, 203–219.

Repke, K. R. H., Schön, R., and Dittrich, F. (1975). In *Proceedings of IXth FEBS Meeting* (G. Gárdos and J. Szasz, eds.), Vol. 35, pp. 241–253, Akadémiai Kiadó, Budapest.

Saks, V. A., Lipina, N. V., Sharov, V. G., Smirnov, V. N., Chazov, E., and Grosse, R. (1977). *Biochim. Biophys. Acta* **465**, 550–558.

Schatzmann, H. J., and Bürgin, H. (1978). *Ann. N.Y. Acad. Sci.* **307**, 125–147.

Schön, R., Dittrich, F., and Repke, K. R. H. (1974). *Acta Biol. Med. Germ.* **33**, K9–K16.

Schönfeld, W., and Streckenbach, B. (1977). *Ergeb. Exp. Med.* **24**, 201–205.

Shaffer, E., Azari, J., and Dahms, A. S. (1978). *J. Biol. Chem.* **253**, 5696–5706.

Spitzer, E., and Grosse, R. (1980). *J. Mol. Cell. Cardiol.* **12**, 158.

Streckenbach, B., Schwarz, D., and Repke, K. R. H. (1980). *Biochim. Biophys. Acta* **601**, 34–46.

Stucki, J. W. (1980). *Eur. J. Biochem.* **109**, 257–267.

Verjovski-Almeida, S., and Inesi, G. (1979). *J. Biol. Chem.* **254**, 18–21.

Welch, G. R. (1977). *Prog. Biophys. Mol. Biol.* **32**, 193–191.

73

The Structure of the Ca^{2+}/Mg^{2+}-ATPase of Sarcoplasmic Reticulum

David H. MacLennan and Reinhart A. F. Reithmeier

Insight into the structure of the Ca^{2+}/Mg^{2+}-ATPase of sarcoplasmic reticulum has come from a number of studies including analysis of primary structure, morphology, and biosynthesis and from deductions from kinetic analysis.

Analysis of primary structure began with the observation that the ATPase (molecular weight 110,000) in intact sarcoplasmic reticulum contains two sites readily sensitive to tryptic digestion, yielding three fragments of about 25,000, 30,000–33,000, and 45,000–55,000 molecular weight that remain tightly associated with the membrane (Migala et al., 1973; Thorley-Lawson and Green, 1973; Stewart and MacLennan, 1974). Figure 1 illustrates the positions of these tryptic sites (T) and of other sites within the sequence of the ATPase. The tryptic fragments have been isolated and partially characterized, and each of the fragments has been shown to contain membrane-associated sectors (Rizzolo and Tanford, 1978). Different functional sites have been associated with each fragment. The isolated 25,000 fragment acts as a Ca^{2+}-dependent and selective ionophore in black lipid membranes (Shamoo et al., 1976). Dicyclohexylcarbodiimide (DCCD), an inhibitor of ATP hydrolysis and Ca^{2+} uptake, binds competitively with Ca^{2+} to a site that can be located in the 25,000 fragment (Pick and Racker, 1979). These observations suggest that a Ca^{2+}-binding site (DCCD) and a Ca^{2+} ionophore (i) exist in this portion of the molecule. The 30,000 fragment contains the site of phosphorylation by ATP (p) (Thorley-Lawson and Green, 1973). The 45,000 fragment contains a binding site for the anion transport inhibitor 4,4'-diisothiocyano-2,2'-stilbene disulfonic acid (DIDS), suggesting that this segment of the molecule is involved in formation of an anion channel (Campbell et al., 1982).

The NH_2-terminal methionine of the ATPase (Tong, 1977) is acetylated. Studies on the NH_2- and COOH-terminal sequences in each of the tryptic fragments (Klip et al., 1980; Tong, 1980) showed that the NH_2 terminus of the 25,000 fragment was also blocked and

David H. MacLennan • Banting and Best Department of Medical Research, Charles H. Best Institute, University of Toronto, Toronto, Ontario M5G 1L6, Canada.   ***Reinhart A. F. Reithmeier*** • Department of Biochemistry, University of Alberta, Edmonton, Alberta T6G 2H7, Canada.

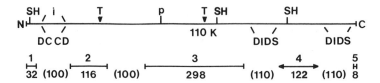

Figure 1. Location of various sites and known hydrophilic sequences (1–5) within the Ca^{2+}/Mg^{2+}-ATPase (molecular weight 110,000). The top line indicates the positions of the ionophoric site (i), tryptic cleavage sites (T), phosphorylation site (p), cytoplasmic cysteine residues (SH), and probable locations of the dicyclohexylcarbodiimide (DCCD) and 4,4'-diisothiocyano-2,2'-stilbene disulfonate (DIDS) binding sites, relative to the amino (N) and carboxyl (C) terminal amino acid residues. The lower line indicates the positions of five known hydrophilic sequences within the ATPase molecule. The exact position of sequence 4 is uncertain. The numbers below the lower lines indicate the number of amino acids present in each sequenced region and the number of amino acids (in parentheses) assigned to each unsequenced hydrophobic region. We have arbitrarily assigned 110 amino acids to unsequenced regions on either side of sequence 4.

proved that the alignment of the fragments is NH_2–25,000–30,000–45,000–COOH. These studies also showed that the fragments were created by two Arg-Ala cleavages and that no peptides were lost in the process.

In a series of elegant papers, Allen and Green (Allen, 1978, 1980a,b; Allen *et al.*, 1980a,b; Green *et al.*, 1980) described extensive analyses of the primary sequence of the ATPase. Five sequences, containing 575 residues, could be located in known regions of the molecule (Fig. 1). A 32-residue NH_2-terminal sequence and an 8-residue COOH-terminal sequence were readily located. A 116-residue sequence could be localized from the known sequence around the 25,000–30,000 cleavage point, and a 298-residue sequence was localized from known sequences around the 30,000–45,000 cleavage point. A 112-residue sequence could then be located only within the 45,000 fragment because of size constraints. However, its position within this fragment is unknown (Fig. 1).

Cleavage of the ATPase with proteolytic enzymes or cyanogen bromide produced two populations of peptides, soluble and insoluble. The five sequenced regions, composed of water-soluble peptides, contained about 52% nonpolar amino acids; the water-insoluble material contained 67% nonpolar amino acids. Thus, the five known sequences appeared to be composed of sections of the molecule residing in the aqueous phase on the membrane surface, whereas the bulk of the remaining peptides could arise from membrane-associated material. Support for this view came from an experiment in which the ATPase was digested while still in membranous form where hydrophobic, membrane-associated regions would be inaccessible to digestion. Peptides released under these conditions were all located in the five known sequences, confirming their location on the aqueous surfaces (Green *et al.*, 1980).

Once the hydrophilic sequences were located in the ATPase molecule, it was apparent that four regions, of about 100 amino acids each, intervened among the known sequences. If the assumption is made that the regions between the five known sequences represent membrane-associated sequences, then insight is gained into the way in which the polypeptide chain folds in the membrane.

Recent evidence suggests that the NH_2 terminus of the Ca^{2+}-ATPase is located on the cytoplasmic side of the sarcoplasmic reticulum membrane. The NH_2-terminal sequence of the ATPase is quite hydrophilic and contains a cysteine residue at position 12 (Allen, 1977). When membranes were reacted with ^{3}H-labeled N-ethylmaleimide (NEM) and the NH_2-terminal peptide (residues 1–31) was isolated, Cys 12 was shown to be labeled (Reithmeier and MacLennan, 1981). The labeling of all NEM-reactive sulfhydryls was blocked when the membranes were pretreated with membrane-impermeant sulfhydryl reagents such as glutathione maleimide. This demonstrates that all of the reactive sulfhydryls, including Cys 12, are located on the cytoplasmic surface of the membrane. Studies on the biosynthesis of the ATPase revealed that the ATPase is synthesized without a signal sequence and that its NH$_2$-terminal methionine is acetylated during translation (Reithmeier *et al.*, 1980). Therefore, one can conclude that the hydrophilic, NH$_2$-terminal portion does not interact with the membrane in the first stages of synthesis. Nevertheless, more than 92% of the bound mRNA for the ATPase is associated with membrane-bound polyribosomes (Greenway and MacLennan, 1978; Chyn *et al.*, 1979) so that a membrane-associating region must be synthesized soon after synthesis of the NH$_2$-terminal sequence. Thus, both biosynthetic data and direct labeling data suggest that the NH$_2$-terminal sequence lies on the cytoplasmic surface of the membrane.

Sequence 2 is clearly cytoplasmic as it contains a very sensitive trypsin cleavage site (T). Sequence 3 is cytoplasmic as it contains the other exposed trypsin site (T), the site of phosphorylation from ATP (p), and an NEM-reactive Cys in peptide E-13 (SH). Sequence 4 contains an NEM-reactive Cys in peptide E-9 (SH), suggesting that its location is cytoplasmic. The location of the COOH-terminal sequence has not been directly determined, but biosynthetic considerations suggest that it would be on the cytoplasmic surface at the time of chain termination. As it is hydrophilic, there would be no driving force for it to enter or cross the membrane. Therefore, we suppose that it is also cytoplasmic.

From these observations we can deduce a probable folding pattern of the peptide chain through the sarcoplasmic reticulum membrane. The NH$_2$-terminal 32 amino acids lie in the cytoplasm. The last few of these amino acids are basic (Allen, 1977), providing a site of interaction with negatively charged phospholipid head groups. The chain then enters the hydrophobic region of the membrane and passes through. Kennedy (1978) has provided compelling arguments for regularity in protein structures that pass through hydrophobic membrane sectors. The inability of the peptide backbone to react with H$_2$O forces intrachain hydrogen bonding, and this is optimal in regular structures such as an α helix or a β helix. There is, in fact, evidence that the membrane-associated sectors of sarcoplasmic reticulum are rich in α helices (Dunker *et al.*, 1980). Kennedy (1978) has also pointed out that it is unlikely that a peptide chain could turn within a hydrophobic environment. Therefore, we must assume that the chain does pass directly through the phospholipid bilayer and that it turns within the aqueous phase in the lumen of the membrane.

Having turned in the luminal space, the chain must immediately reenter the hydrophobic sector, passing through in a regular structure to create the cytoplasmic portion comprised of sequence 2. Between sequences 2 and 3, 3 and 4, and 4 and 5, this process is repeated so that the chain probably stitches back and forth through the membrane eight times in four transmembrane loops. If we assume that the four loops comprise some 400 amino acids and that at least 30 amino acids are required for each transmembrane passage, then very little material is left over for protein located on the luminal side of the membrane. There is no clear evidence that the ATPase chain does, in fact, extend to the luminal surface; however, because Ca^{2+} and anions do cross the membrane (Hasselbach, 1964), we can infer that a pathway must be formed all of the way across the lipid bilayer by the ATPase molecule. This pathway is probably formed by single ATPase molecules, for solubilized, monomeric preparations of the ATPase are capable of hydrolyzing ATP (Dean and Tanford, 1978; Møller *et al.*, 1980) and undergo the same Ca^{2+}-induced fluorescence changes as the membrane-associated form of the enzyme (Dean and Gray, 1980).

Morphological studies of the ATPase lend support to this proposed model of the enzyme. Hydrophobic globules, 9 nm in diameter, were observed by Deamer and Baskin

(1969) to be associated with the cytoplasmic leaflet of freeze-fractured sarcoplasmic reticulum membranes. These globules were later identified as consisting of buried sectors of the ATPase (MacLennan *et al.*, 1971). Were there any significant association of the ATPase protein with the luminal surface of the membrane, one would expect a more symmetrical cleavage pattern. In fact, when the ATPase was scrambled by reconstitution, the fracture patterns were symmetrical (MacLennan *et al.*, 1971; Packer *et al.*, 1974).

Ikemoto *et al.* (1968) and Inesi and Asai (1968) discovered surface particles about 4×6 nm in size in sarcoplasmic reticulum membranes. They were also shown to be comprised of the ATPase (Migala *et al.*, 1973; Thorley-Lawson and Green, 1973; Stewart and MacLennan, 1974). There are about three to four times as many surface particles as globules (Jilka *et al.*, 1975). This may reflect the location of more than one ATPase molecule in each intramembrane globule, or it may reflect the fact that at least three visible cytoplasmic extensions of the ATPase are expected for each transmembrane cluster of four loops. It is not possible to detect asymmetry of the membrane by this method, but positive staining with tannic acid (Saito *et al.*, 1978) showed that there is, indeed, asymmetry, with surface structures located on the cytoplasmic surface of the membrane only.

Studies with X-ray diffraction have also provided clear evidence for asymmetry of the sarcoplasmic reticulum membrane (Worthington and Liu, 1973; Dupont *et al.*, 1973; Herbette and Blasie, 1980). Herbette and Blasie (1980) have used a combination of X-ray and neutron diffraction to obtain details of the shape of the molecule in the membrane. It appears as a tapered rectangle, wider at the cytoplasmic surface, in which some 38% of the mass lies inside the bilayer with virtually no extension to the luminal surface but with 62% of the mass extending into the cytoplasm. Thus, all of these morphological studies are fully consistent with the model of polypeptide folding.

The question may be asked whether all of the functions of Ca^{2+} translocation occur within the ATPase molecule. In the original reconstitution of Ca^{2+} uptake with the ATPase (Racker, 1972), the enzyme also contained proteolipid and mixed phospholipids. Subsequent studies by Warren *et al.* (1974) showed that phosphatidylcholine could provide both enzyme activation and vesicle formation necessary to demonstrate Ca^{2+} uptake. Therefore, translocation did not appear to require any ionophoric component that might reside in the lipid fraction. Evidence was obtained, however, that the proteolipid could act as a Ca^{2+} ionophore (Racker and Eytan, 1975; Knowles *et al.*, 1980) and that it might constitute a Ca^{2+} uptake channel (Racker, 1977). This is unlikely, however, as the ATPase can be stripped of proteolipid and still retain Ca^{2+} transport function (MacLennan *et al.*, 1980; Knowles *et al.*, 1980). It is more likely that the proteolipid is involved in Ca^{2+} release (Shoshan *et al.*, 1981).

The sarcoplasmic reticulum like the Na^+/K^+-ATPase contains a 53,000-molecular-weight glycoprotein (Campbell and MacLennan, 1981a). In the Na^+/K^+-ATPase the glycoprotein appears to be an essential subunit of the ATPase (Kyte, 1972). The glycoprotein from sarcoplasmic reticulum is not essential to activity of the Ca^{2+}/Mg^{2+}-ATPase, for it is readily removed during fractionation and the glycoprotein-free ATPase can carry out its Ca^{2+} transport function after reconstitution (Racker, 1972). The glycoprotein may, however, play a regulatory role in Ca^{2+} transport, as it appears to be a protein kinase that might regulate transport through phosphorylation mechanisms (Campbell and MacLennan, 1981b).

REFERENCES

Allen, G. (1978). *FEBS Symp.*, **45,** 159–168.

Allen, G. (1980a). *Biochem. J.* **187,** 545–563.

Allen, G. (1980b). *Biochem. J.* **187,** 565–575.

Allen, G., Bottomley, R. C., and Trinnaman, B. J. (1980a). *Biochem. J.* **187,** 577–589.

Allen, G., Trinnaman, B. J., and Green, N. M. (1980b). *Biochem. J.* **187,** 591–616.

Campbell, K. P., and MacLennan, D. H. (1981a). *J. Biol. Chem.* **256,** 4626–4632.

Campbell, K. P., and MacLennan, D. H. (1981b). *Fed. Proc.* **40,** 1623.

Campbell, K. P., Reithmeier, R. A. F., Khanna, V. K., and MacLennan, D. H. (1982). *J. Biol. Chem.*, submitted.

Chyn, T. L., Martonosi, A. N., Morimoto, T., and Sabatini, D. D. (1979). *Proc. Natl. Acad. Sci. USA* **76,** 1241–1245.

Deamer, D. W., and Baskin, R. J. (1969). *J. Cell Biol.* **42,** 296–307.

Dean, W. L., and Gray, R. D. (1980). *J. Biol. Chem.* **255,** 7514–7516.

Dean, W. L., and Tanford, C. (1978). *Biochemistry* **17,** 1683–1690.

Dunker, A. K., Toogood, K. C., Stuart, G. W., and Williams, R. W. (1980). *Fed. Proc.* **39,** 1663.

Dupont, Y., Harrison, S. C., and Hasselbach, W. (1973). *Nature (London)* **244,** 555–558.

Green, N. M., Allen, G., and Hebdon, G. M. (1980). *Ann. N.Y. Acad. Sci.* **358,** 149–158.

Greenway, D. C., and MacLennan, D. H. (1978). *Can. J. Biochem.* **56,** 452–456.

Hasselbach, W. (1964). *Prog. Biophys. Mol. Biol.* **14,** 167–222.

Herbette, L., and Blasie, J. K. (1980). In *Calcium Binding Proteins: Structure and Function* (F. L. Siegel, E. Carafoli, R. H. Kretsinger, D. H. MacLennan, and R. H. Wasserman, eds.), pp. 115–120, Elsevier/North-Holland, Amsterdam.

Ikemoto, N., Streter, F. A., Nakamura, A., and Gergely, J. (1968). *J. Ultrastruct. Res.* **23,** 216–232.

Inesi, G., and Asai, H. (1968). *Arch. Biochem.* **126,** 469–477.

Jilka, R. L., Martonosi, A. N., and Tillack, T. W. (1975). *J. Biol. Chem.* **250,** 7511–7524.

Kennedy, S. J. (1978). *J. Membr. Biol.* **42,** 265–279.

Klip, A., Reithmeier, R. A. F., and MacLennan, D. H. (1980). *J. Biol. Chem.* **255,** 6562–6568.

Knowles, A., Zimniak, P., Alfonzo, M., Zinniak, A., and Racker, E. (1980). *J. Membr. Biol.* **55,** 233–239.

Kyte, J. (1972). *J. Biol. Chem.* **247,** 7642–7649.

MacLennan, D. H., Seeman, P., Iles, G. H., and Yip, C. C. (1971). *J. Biol. Chem.* **246,** 2702–2710.

MacLennan, D. H., Reithmeier, R. A. F., Shoshan, V., Campbell, K. P., and LeBel, D. (1980). *Ann. N.Y. Acad. Sci.* **358,** 138–148.

Migala, A., Agostini, B., and Hasselbach, W. (1973). *Naturforscher* **28,** 178–182.

Møller, J. U., Lind, K. E., and Andersen, J. P. (1980). *J. Biol. Chem.* **255,** 1912–1920.

Packer, L., Mehard, C. W., Meissner, G., Zahler, W. L., and Fleischer, S. (1974). *Biochim. Biophys. Acta* **363,** 159–181.

Pick, U., and Racker, E. (1979). *Biochemistry* **18,** 109–113.

Racker, E. (1972). *J. Biol. Chem.* **247,** 8198–8200.

Racker, E. (1977). In *Calcium Binding Proteins and Calcium Function* (R. S. Wasserman, R. A. Corradino, E. Carafoli, R. H. Kretsinger, D. H. MacLennan, and F. L. Seigel, eds.), pp. 115–163, North-Holland, Amsterdam.

Racker, E., and Eytan, E. (1975). *J. Biol. Chem.* **250,** 7333–7334.

Reithmeier, R. A. F., and MacLennan, D. H. (1981). *J. Biol. Chem.*, **256,** 5957–5960.

Reithmeier, R. A. F., de Leon, S., and MacLennan, D. H. (1980). *J. Biol. Chem.* **255,** 11839–11846.

Rizzolo, L. J., and Tanford, C. (1978). *Biochemistry* **17,** 4044–4048.

Saito, A., Wang, C. T., and Fleischer, S. (1978). *J. Cell Biol.* **79,** 601–616.

Shamoo, A. E., Ryan, T. E., Stewart, P. S., and MacLennan, D. H. (1976). *J. Biol. Chem.* **251,** 4147–4154.

Shoshan, V., MacLennan, D. H., and Wood, D. S. (1981). *Proc. Natl. Acad. Sci. USA,* **78,** 4828–4832.

Stewart, P. S., and MacLennan, D. H. (1974). *J. Biol. Chem.* **249,** 985–993.

Thorley-Lawson, D. A., and Green, N. M. (1973). *Eur. J. Biochem.* **40,** 403–413.

Tong, S. W. (1977). *Biochem. Biophys. Res. Commun.* **74,** 1242–1248.

Tong, S. W. (1980). *Arch. Biochem. Biophys.* **203,** 780–791.

Warren, G. B., Toon, P. A., Birdsall, N. J. M., Lee, A. G., and Metcalfe, J. C. (1974). *Proc. Natl. Acad. Sci. USA* **71,** 622–626.

Worthington, C. R., and Liu, S. C. (1973). *Arch. Biochem. Biophys.* **157,** 573–579.

74

Ca²⁺/Mg²⁺-Dependent ATPase in Sarcoplasmic Reticulum

Kinetic Properties in Its Monomeric and Oligomeric Forms

Taibo Yamamoto and Yuji Tonomura

1. INTRODUCTION

Since Hasselbach and Makinose (1961) showed the existence of the Ca^{2+}/Mg^{2+}-dependent ATPase in the membrane of isolated sarcoplasmic reticulum (SR), considerable progress has been made concerning the mechanism of the active transport of Ca^{2+}. The outline of the transport mechanism, which has been obtained mainly from kinetic studies of the Ca^{2+}/Mg^{2+}-dependent ATPase, is described briefly as follows: 1 mole of ATP and 2 moles of Ca^{2+} bind to 1 mole of the membrane-bound ATPase on the outside of the SR vesicle. The terminal phosphate of ATP is transferred to an aspartyl residue of the enzyme to form a phosphoenzyme. At the same time, Ca^{2+} is translocated from outside to inside the membrane. The phosphoenzyme is then hydrolyzed in the presence of Mg^{2+}. The entire process of Ca^{2+} transport can be reversed. When SR vesicles loaded with Ca^{2+} are reacted with P_i in the presence of Mg^{2+} and EGTA, 1 mole of phosphoenzyme is formed. When ADP is added to the phosphoenzymes, 2 moles of Ca^{2+} are released accompanying the formation of 1 mole of ATP. These studies have been reviewed in detail by Hasselbach (1979), Inesi (1979), Martonosi (1975), Tada et al. (1978), and Yamamoto et al. (1979).

However, to understand the exact mechanism of Ca^{2+} transport, it is important to know how the enzyme molecules are organized into the SR membrane to perform their functions. The main purpose of this review is to discuss how kinetic properties of the Ca^{2+}/Mg^{2+}-dependent ATPase are modified by the change of the structural features of the enzyme in the SR membrane.

Taibo Yamamoto and Yuji Tonomura • Department of Biology, Faculty of Science, Osaka University, Toyonaka, Osaka 560, Japan.

The ATPase protein accounts for more than two-thirds of the total protein of SR. Its purification was first achieved by MacLennan (1970). Later, similar attempts to purify the ATPase were made by several workers (Ikemoto *et al.*, 1971; LeMaire *et al.*, 1976; Meissner and Fleischer, 1971). Racker (1972) succeeded in reconstitution of the ATP-dependent Ca^{2+} transport by incorporating the purified ATPase into liposome. Warren *et al.* (1974a) reported that the activity of Ca^{2+} uptake can be reconstituted even when the intrinsic lipid of SR was completely substituted by a synthetic lipid. These results suggest that the active transport of Ca^{2+} requires only two components, the ATPase protein and one kind of phospholipid, and that the ATPase acts not only as an energy transducer for Ca^{2+} transport but also as a cation translocator. The ATPase molecule is a single polypeptide chain of 100,000 molecular weight. It can be cleaved in its native state by limited digestion with trypsin into two fragments of 50,000 and 45,000 molecular weight (Migala *et al.*, 1973; Thorley-Lawson and Green, 1973). The larger fragment can be cleaved by prolonged digestion into two fragments of about 30,000 and 20,000 molecular weight (Stewart *et al.*, 1976; Thorley-Lawson and Green, 1973). The 30,000-molecular-weight fragment contains a phosphorylation site, while the 20,000-molecular-weight fragment has an ionophoric activity. More recently, the structural studies of the ATPase protein has been advanced extensively by Allen *et al.* (1980), Klip *et al.* (1980), and Tong (1980).

2. ORGANIZATION OF THE ATPase MOLECULES WITHIN THE SR MEMBRANE

The problem of the structure–function relationships in the SR membrane can be separated into two subjects: the interaction between ATPase molecules and the interaction between the ATPase molecule and phospholipid.

Many observations on the SR membrane have been taken as evidence that interaction between ATPase molecules constitutes an important aspect of Ca^{2+} transport. The assembly of ATPase in intact and reconstituted membranes of SR has been investigated by electron microscopy (Jilka *et al.*, 1975; Scales and Inesi, 1976). The density of the 75- to 90-Å intramembrane particles revealed by freeze-etch electron microscopy was three to four times lower than that of the 35- to 45-Å surface particles seen after negative staining. These observations suggest that the hydrophobic ends of the ATPase chains aggregate to form oligomers within the membrane. Further analysis of the negative-stained electron micrographs revealed a pattern of four subunits on the surface membrane (Vanderkooi *et al.*, 1977). The interaction between ATPase molecules was analyzed in reconstituted SR vesicles by measuring the efficiency of energy transfer between two populations of the ATPase molecules modified with different fluorescent reagents (Vanderkooi *et al.*, 1977). A significant energy transfer occurred even after dilution of the lipid phase of the vesicles with phospholipid up to 10-fold. This result suggests that the ATPase molecule exists as an oligomer in the intact SR membrane, as has also been suggested by techniques such as analytical centrifugation of detergent-solubilized ATPase (LeMaire *et al.*, 1976) and laser-flash-induced photodichroism (Hoffman *et al.*, 1979).

SR ATPase is characterized by tight association with membrane lipids. Recently, the correlation between the structure of the enzyme and a possible role of lipid in maintaining the activity of SR ATPase has been more closely investigated. Warren *et al.* (1974b) found a simple procedure for the replacement of endogenous lipid with externally added lipid. Employing this technique, they showed that the enzyme retained full activity even after more than 99% of the original lipid was replaced with synthetic lipid. The possibility arises from the lipid-exchange experiment that a hydrophobic environment is essential for holding the enzyme protein in a reconstitutable state. This possibility was proven by Dean and Tanford (1978) who showed that most of the phospholipid can be replaced with any of a variety of nonionic detergents without loss of ATPase activity. In addition, they suggested on the basis of sedimentation equilibrium measurement that the ATPase exists mainly as a monomer in the detergent. By raising protein and salt concentrations and adding sucrose, Jorgensen *et al.* (1978) obtained SR ATPase stable even in the presence of deoxycholate, the ATPase being shown to exist in the monomeric form. More recently, Møller *et al.* (1980) have also reported that full enzymatic activity can be preserved for a long time in monomeric form under appropriate conditions for solubilization of SR ATPase with the nonionic detergent dodecyloctaethyleneglycolmonoether ($C_{12}E_8$). These observations indicate that the interaction between individual ATPase protein molecules in the oligomer may be easily disrupted by such detergents as deoxycholate and $C_{12}E_8$. LeMaire *et al.* (1978) have shown that when $C_{12}E_8$ was removed from ATPase by column chromatography, the ATPase protein emerged in variously sized aggregates. The aggregate size was found to be dependent on the ratio of detergent to SR ATPase. When $C_{12}E_8$ was added to SR membranes at a weight ratio of about 5 : 1, followed by removal of excess detergent and membrane lipid by column chromatography, a soluble mixture of trimeric and tetrameric oligomers of ATPase was obtained. Decrease in the ratio of $C_{12}E_8$ to SR protein resulted in the formation of larger oligomers. From these results, they suggested that phospholipid functions in oligomerization of ATPase molecules. Because the tendency for self-assembly of the ATPase molecules was enhanced by adding back phospholipids, they concluded that the enzyme exists as an oligomer in the intact SR membrane.

3. KINETIC PROPERTIES OF MEMBRANE-BOUND AND SOLUBILIZED ATPase

Kinetic properties of the Ca^{2+}/Mg^{2+}-dependent ATPase are altered when the membrane structure is destroyed or when the enzyme form is changed from oligomer to monomer.

Because ATPase molecules locate unidirectionally on the intact SR membrane, the ATPase reaction must be analyzed always taking into consideration the vectorial properties of the reaction. In addition, the ionic environments change continually during Ca^{2+} uptake, and this may also affect severely the kinetic behavior of the enzyme. To avoid these factors and to obtain the basic aspect of the reaction mechanism of the Ca^{2+}/Mg^{2+}-dependent ATPase, the presteady state and elementary steps of the reaction were analyzed using leaky SR membranes (Yamada *et al.*, 1971) or $C_{12}E_8$-solubilized oligomeric ATPase (Takisawa and Tonomura, 1978, 1979).

In the initial phase of the ATPase reaction, the time course of phosphoenzyme formation did not show a lag phase; the time course of P_i liberation from ATP clearly showed a lag phase corresponding to the exact period during which the amount of phosphoenzyme increases to a steady-state level. After the lag phase, P_i was liberated linearly with time. The ATP concentration dependence of both ATPase activity and phosphoenzyme formation could be fitted by straight lines on double reciprocal plots. Therefore, the ATPase reactions of solubilized or leaky SR membranes can be explained by the simple mechanism

$$E + ATP \rightleftharpoons E \cdot ATP \rightleftharpoons EP + ADP \rightarrow E + P_i$$

Thus, the reaction proceeds in a sequential route consisting of at least one enzyme–substrate complex and one phosphoenzyme (EP) without ADP. Recent works show that the ATPase reaction involves the sequential formation of at least two forms of EP, in the order ADP sensitive (E_1P) and insensitive (E_2P) (Shigekawa *et al.*, 1978; Takisawa and Tonomura, 1979; Yamada and Ikemoto, 1980). The conversion of E_1P into E_2P is accompanied by a great reduction in the affinity of the enzyme for Ca^{2+} (Shigekawa *et al.*, 1978; Takisawa and Tonomura, 1979).

In the case of the intact SR vesicle, the time courses of EP formation and P_i liberation are much more complicated than those of the solubilized ATPase. In the earlier studies of the transient kinetics on intact SR, Kanazawa *et al.* (1971) showed that the time course of P_i liberation consists of a lag phase, a burst phase, and a steady phase. EP formed without a lag phase, and its amount reached a steady-state level within a few seconds. The P_i burst is caused by a transient decrease in the rate constant of EP decomposition. Although the mechanism of this phenomenon is unknown, it appears to correlate with the structural change of the SR membrane, for this phenomenon was no longer observed when the SR membrane was treated with a detergent. With the use of a rapid quenching method, Froehlich and Taylor (1975, 1976) detected an overshoot in the phosphorylation reaction and the initial burst of P_i associated with the transient decay of EP to the steady-state level. To explain these phenomena, they proposed a flip-flop mechanism in which the enzyme acts as an oligomer for coupled transport of Ca^{2+} and Mg^{2+}. This possibility seems unlikely because the soluble oligomeric ATPase exhibits neither the overshoot of EP nor the P_i burst (Takisawa and Tonomura, 1978). The findings that the EP overshoot is dependent on whether the reaction is started by the addition of Ca^{2+} or ATP (Takisawa and Tonomura, 1978) and that the enzyme exists in two distinct forms, E_1 and E_2, which differ in affinity for Ca^{2+} (De Meis and Vianna, 1979; Dupont, 1978), indicate the following mechanism for the catalytic cycle by the ATPase: $E_2 \rightarrow E_1 \rightarrow E_1ATP \rightarrow E_1P \rightarrow E_2P \rightarrow E_2$. Consequently, the disappearance of the EP overshoot in the solubilized ATPase reaction can be easily explained by assuming that the transformation of E_2 into E_1 should be enhanced by destruction of the membrane.

As described in Section 2, the ATPase can be prepared as a monomer with retention of high enzymatic activity under appropriate conditions. It has been suggested by several investigators that the basic aspects of ATP hydrolysis are similar for the monomeric and the membranous ATPase (Dean and Tanford, 1978; Jorgensen *et al.*, 1978; Møller *et al.*, 1980). These results raise the possibility that most of the catalytic functions can be performed by the ATPase in a monomeric form. To test this possibility, we have investigated the reaction mechanism of the SR ATPase in the presence of sufficiently high concentration of $C_{12}E_8$. Under this condition, ATPase is assumed to exist dominantly as a monomer. We found a pronounced difference between monomeric ATPase and intact SR ATPase reactions in the effect of Mg^{2+} on EP decomposition. When the monomeric ATPase was phosphorylated by ^{32}P-labeled ATP in the presence of various concentrations of Mg^{2+}, then chased by nonradioactive ATP, the radioactivity in the EP decreased with a rate constant of about 0.3 sec^{-1}. The rate of EP decomposition was quite independent of Mg^{2+} in the range from 0 to 10 mM. On the other hand, the apparent rate constant of EP decomposition, calculated from the ratio of ATPase activity to the amount of EP at steady state, increased from about 0.3 to 1.5 sec^{-1} by raising the Mg^{2+} concentration from 0 to 10 mM. No activity was observed in the absence of Ca^{2+}. These results are summarized as follows:

$$\begin{array}{c} \text{stimulated by } Ca^{2+} \\ \overbrace{E + ATP \rightarrow E \cdot ATP \rightarrow {}^*E \cdot ATP \longrightarrow EP + ADP} \\ \uparrow \text{enhanced by } Mg^{2+} \text{---} \rightarrow \searrow ADP + P_i \searrow P_i \end{array}$$

Most of the EP formed under these conditions exists in the ADP-sensitive form, and its decomposition is the rate-limiting step of the reaction in the absence of Mg^{2+}. In the presence of Mg^{2+}, ATP is hydrolyzed directly into ADP and P_i via the enzyme–substrate complex. The existence of the Mg^{2+}-dependent uncoupled reaction was also supported by the fact that high concentrations of Mg^{2+} severely inhibit ATP formation from EP and ADP without preventing the ADP-stimulated decay of EP (Yamamoto and Tonomura, unpublished). In view of these facts, it is concluded that the existence of the ATPase molecules as an oligomer is required for the energy-coupled transport of Ca^{2+} in the intact SR membrane.

REFERENCES

Allen, G., Trinnaman, B. J., and Green, N. M. (1980). *Biochem. J.* **187,** 591–616.

Dean, W. L., and Tanford, C. (1978). *Biochemistry* **17,** 1683–1690.

De Meis, L., and Vianna, A. L. (1979). *Annu. Rev. Biochem.* **48,** 275–292.

Dupont, Y. (1978). *Biochem. Biophys. Res. Commun.* **82,** 893–900.

Froehlich, J. P., and Taylor, E. W. (1975). *J. Biol. Chem.* **250,** 2013–2021.

Froehlich, J. P., and Taylor, E. W. (1976). *J. Biol. Chem.* **251,** 2307–2315.

Hasselbach, W. (1979). *Top Curr. Chem.* **78,** 1–56.

Hasselbach, W., and Makinose, M. (1961). *Biochem. Z.* **333,** 518–528.

Hoffman, W., Sarzala, M. G., and Chapman, D. (1979). *Proc. Natl. Acad. Sci. USA* **76,** 3860–3864.

Ikemoto, N., Bhatnagar, G. M., and Gergely, J. (1971). *Biochem. Biophys. Res. Commun.* **44,** 1510–1517.

Inesi, G. (1979). In *Membrane Transport in Biology* (G. Giebisch, D. C. Tosteson, and H. H. Ussing, eds.), Vol. II, pp. 357–393, Springer-Verlag, Berlin.

Jilka, R. L., Martonosi, A. N., and Tillack, T. W. (1975). *J. Biol. Chem.* **250,** 7511–7524.

Jorgensen, K. E., Lind, K. E., Roigaard-Petersen, H., and Møller, J. V. (1978). *Biochem. J.* **169,** 489–498.

Kanazawa, T., Yamada, S., Yamamoto, T., and Tonomura, Y. (1971). *J. Biochem. (Tokyo)* **70,** 95–123.

Klip, A., Reithmeier, R. A. F., and MacLennan, D. H. (1980). *J. Biol. Chem.* **255,** 6562–6568.

LeMaire, M., Møller, J. V., and Tanford, C. (1976). *Biochemistry* **15,** 2336–2342.

LeMaire, M., Lind, K. E., Jorgensen, K. E., Roigaard, H., and Møller, J. V. (1978). *J. Biol. Chem.* **253,** 7051–7060.

MacLennan, D. H. (1970). *J. Biol. Chem.* **245,** 4508–4518.

Martonosi, A. N. (1975). In *Calcium Transport in Contraction and Secretion* (E. Carafoli, F. Clementi, W. Drabikowski, and A. Margreth, eds.), pp. 313–327, North-Holland, Amsterdam.

Meissner, G., and Fleischer, S. (1971). *Biochim. Biophys. Acta* **241,** 356–378.

Migala, A., Agostini, B., and Hasselbach, W. (1973). *Z. Naturforsch.* **28,** 178–182.

Møller, J. V., Lind, K. E., and Andersen, J. P. (1980). *J. Biol. Chem.* **255,** 1912–1920.

Racker, E. (1972). *J. Biol. Chem.* **247,** 8198–8200.

Scales, D., and Inesi, G. (1976). *Biophys. J.* **16,** 735–751.

Shigekawa, M., Doughery, J. P., and Katz, A. M. (1978). *J. Biol. Chem.* **253,** 1442–1450.

Stewart, P. S., MacLennan, D. H., and Shamoo, A. E. (1976). *J. Biol. Chem.* **251,** 712–719.

Tada, M., Yamamoto, T., and Tonomura, Y. (1978). *Physiol. Rev.* **58,** 1–79.

Takisawa, H., and Tonomura, Y. (1978). *J. Biochem. (Tokyo)* **83,** 1275–1284.

Takisawa, H., and Tonomura, Y. (1979). *J. Biochem. (Tokyo)* **86,** 425–441.

Thorley-Lawson, D. A., and Green, N. M. (1973). *Eur. J. Biochem.* **40,** 403–413.

Tong, S. W. (1980). *Arch. Biochem. Biophys.* **203,** 780–791.

Vanderkooi, J. M., Ierokomas, A., Nakamura, H., and Martonosi, A. N. (1977). *Biochemistry* **16,** 1262–1267.

Warren, G. B., Toon, P. A., Birdsall, N. J. M., Lee, A. G., and Metcalfe, J. C. (1974a). *Proc. Natl. Acad. Sci. USA* **71,** 622–626.

Warren, G. B., Toon, P. A., Birdsall, N. J. M., Lee, A. G., and Metcalfe, J. C. (1974b). *Biochemistry* **13,** 5501–5507.

Yamada, S., and Ikemoto, N. (1980). *J. Biol. Chem.* **255,** 3108–3119.

Yamada, S., Yamamoto, T., Kanazawa, T., and Tonomura, Y. (1971). *J. Biochem. (Tokyo)* **70,** 279–291.

Yamamoto, T., Takisawa, H., and Tonomura, Y. (1979). *Curr. Top. Bioenerg.* **9,** 179–236.

75

Isolation and Use of Ionophores for Studying Ion Transport across Natural Membranes

Tom R. Herrmann and Adil E. Shamoo

1. INTRODUCTION

One of the most important and interesting membrane-based processes is the transport of ions in and out of cells and cell organelles. In the study of ion transport across membranes, the concept of ionophoric activity has emerged. The term ionophoric activity has been used to mean various properties of a substance, depending on the method used to assay for the ionophorous material. Using the characteristic common to all methods, we may generally define an ionophore as a substance that enhances the movement or incorporation of an ion from an aqueous phase into a hydrophobic phase (Shamoo and Goldstein, 1977). An ionophore so defined is not necessarily involved in ion transport across biological membranes.

One object of this chapter is to describe how the concept of ionophoric activity can be applied in studying transport mechanisms, using the example of the Ca^{2+}/Mg^{2+}-ATPase from sarcoplasmic reticulum (SR). Before proceeding it would be well to compare and contrast two distinct approaches by which artificial systems are used in transport studies. In the first and most obvious approach, a protein or protein complex responsible for some ion transport function is isolated from a natural membrane and inserted as an intact unit into an artificial membrane. This eliminates parallel transport systems that may exist in the native membrane, and allows study of lipid specificity, the effect of removing specific components of a transport complex, and other such studies.

In the second approach, a protein or protein complex is physically broken down and the ion transport or ionophoric properties of the pieces are investigated. One then correlates ion transport characteristics of the pieces with properties of the intact system. For example, one measures the selectivity for transported ions of each piece and compares this with the

Tom R. Herrmann and Adil E. Shamoo • Department of Biological Chemistry, University of Maryland School of Medicine, Baltimore, Maryland 21201.

ion transport function of the intact system. Similarly, one observes whether any known inhibitors of the intact system cause inhibition of ionophoric activity of the pieces. The two approaches may be complementary in the information they provide.

As a tool in interpreting ionophoric activity studies, a simple conceptual framework has been proposed by Shamoo and Goldstein (1977). In this scheme, ion transport systems are made up of three basic parts: a nonselective channel, a selective gate, and an energy transducer. A given system may contain one, two, or all three of these. The ionophoric properties of these pieces can be inferred from their function. Thus, the nonselective channel should act as such in an artificial membrane. The gate should have a selective affinity for the transported ion and may or may not be lipid soluble. If the gate is lipid soluble, it will probably behave as an ion-selective ionophore. The energy transducer may have no ionophoric activity but should contain the binding site for the energy source substrate (such as ATP). Thus, the ionophoric activities of the pieces can give clues to their function in the intact system. The speculative nature of this framework allows only tentative functional assignments of the pieces, but even this opens up new experimental approaches to the study of energy transduction. It provides a starting place for designing experiments to check the functional assignments of the pieces, leading to more refined models.

The assays for ionophoric activity used in this work are mainly two: artificial phospholipid vesicles and planar bilayer lipid membranes (BLM). Of the two, the latter is more sensitive and is the only system in which determination of selectivity for ions is practical. A detailed description of the methodology has been given by Shamoo and Goldstein (1977).

2. IONOPHORICITY OF THE Ca^{2+}/Mg^{2+}-ATPase

The Ca^{2+}/Mg^{2+}-ATPase with which we work is an integral protein in the SR membrane of fast skeletal muscle. This membrane forms "sacs" surrounding muscle fibers, and its function is to control the Ca^{2+} level around the fibers. In the relaxed state, the SR sac has sequestered most of the Ca^{2+} away from the muscle fibers, while release of Ca^{2+} from SR results in muscle contraction. The Ca^{2+}/Mg^{2+}-ATPase has been reconstituted in vesicles and shown to actively pump Ca^{2+} with hydrolysis of ATP, indicating that the primary pump mechanism is contained in the protein molecule. This led Shamoo and MacLennan (1974) to test the intact system enzyme for ionophoric activity in a BLM assay, where it showed a Ca^{2+}-dependent and -selective ionophoric activity. Incorporation into a BLM was accomplished by either (1) treating the enzyme with succinic acid to make it soluble, (2) sonicating the insoluble protein to disperse it in the BLM bathing fluid, or (3) digesting the unsuccinylated enzyme with trypsin, yielding soluble fragments. In all of these cases the protein was denatured: its native *active* transport function was lost.

Incorporation of protein renders a BLM permeable to various ions. The selectivity for cations is $Ba^{2+} > Ca^{2+} > Sr^{2+} > Mg^{2+}$, Zn^{2+}, and alkali metals. The BLM conductance increase following Ca^{2+}/Mg^{2+}-ATPase protein addition to the bathing fluid was dependent on the presence of Ca^{2+} in the bathing fluid.

The above selectivity sequence and Ca^{2+} dependence exhibited by the protein suggest that whatever is giving rise to ionophoric activity is part of the pump mechanism. To further link Ca^{2+} ionophoric activity with the Ca^{2+} transport system, several common inhibitors were sought. It was found that Ca^{2+} conductance was inhibited by Zn^{2+}, Mn^{2+}, Hg^{2+}, and La^{3+}, agents that also inhibit Ca^{2+} pumping in SR (Shamoo and MacLennan, 1974). Further, at least two Na^+ ions are required to inhibit the Ca^{2+} conductance (Shamoo and Ryan, 1975). Na^+ is the only monovalent ion found to have an inhibitory effect; this is consistent with Na^+ inhibition of Ca^{2+} uptake into SR (Masuda and de Meis, 1974). It is noteworthy that Hg^{2+} inhibits the Ca^{2+} ionophore while CH_3Hg^{2+} does not. This correlates well with the observation that in SR vesicles, CH_3Hg^{2+} inhibits Ca^{2+} transport by acting at the ATP-binding site, while Hg^{2+} at low doses inhibits Ca^{2+} transport by competition with Ca^{2+} for the Ca^{2+} site (Shamoo and MacLennan, 1975).

3. IONOPHORIC FRAGMENTS OF THE Ca^{2+}/Mg^{2+}-ATPase

To localize the portion of the protein causing Ca^{2+} ionophoric activity, the protein may be broken up by digestion with trypsin. Thus, exposure of SR to trypsin in the presence of 1 M sucrose results in the cleavage of the 115,000-molecular-weight Ca^{2+}/Mg^{2+}-ATPase into two fragments of approximately 45,000 and 55,000 molecular weight. Further exposure to trypsin degrades the 55,000 piece into fragments of 30,000 and 25,000 (Thorley-Lawson and Green, 1975; Stewart *et al.*, 1976).

When one tests each of the purified fragments for ionophoric activity using the BLM assay, the ionophoric activity-conferring site can be narrowed down according to the scheme in Fig. 1. The 55,000 and 25,000 fragments show ionophoric activity, with the same divalent ion selectivity as the intact enzyme (Shamoo *et al.*, 1976). The 30,000 fragment shows no ionophoric activity. The 45,000 fragment causes a conductance increase in BLMs, but in contrast to the other fragments, it shows little selectivity between ions (Abramson and Shamoo, 1978).

Further digestion of the 25,000 fragment with cyanogen bromide produces four smaller fragments (Klip and MacLennan, 1978). Experiments with the BLM assay show that only the 13,000 fragment has Ca^{2+}-selective ionophoric activity (MacLennan *et al.*, 1980). This

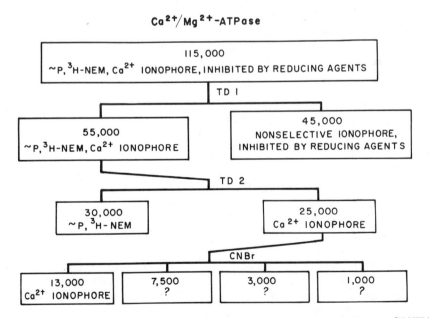

Figure 1. Diagram showing the relation of the Ca^{2+}/Mg^{2+}-ATPase to its digestion fragments. ^{3}H-NEM is *N*-[^{3}H]ethylmaleimide, a reagent reacting with sulfhydryl groups; $\sim$P is the site of acid-stable phosphorylation; CNBr is cyanogen bromide, a reagent that cleaves at methionine residues. The numbers represent molecular weight.

activity is sensitive to the inhibitors Hg^{2+}, Cd^{2+}, and Zn^{2+}. The cation selectivity sequence of the 13,000 fragment is changed from that of its parent fragments; it becomes $Mn^{2+} > Ca^{2+} > Ba^{2+} > Sr^{2+} > Mg^{2+}$. (Note that for the previous fragments, Mn^{2+} is inhibitory.) This change in selectivity suggests that the ionophoric site has been altered by CNBr cleavage.

As stated above, the 45,000 fragment acts as a nonselective ionophore for divalent cations in the BLM assay. Ionophoric activity is inhibited by low levels of $LaCl_3$ and $HgCl_2$ (Abramson and Shamoo, 1978). Furthermore, there are three disulfide bonds in the fragment (Thorley-Lawson and Green, 1977), which seem to be important to transport properties of this fragment: splitting the disulfides by preincubation with reducing agents completely inhibits the ionophoric properties of the fragment. This is in contrast to the 55,000 fragment, which is unaffected by these reducing agents. The ionophoric activity of the intact succinylated enzyme (115,000 molecular weight) is inhibited significantly by reducing agents. This suggests that in the intact enzyme, the 45,000 fragment acts in *series* with the 55,000 fragment.

During Ca^{2+} pumping in the functioning Ca^{2+}/Mg^{2+}-ATPase, ATP is hydrolyzed: the terminal phosphate of ATP is cleaved off and binds to the protein during part of the cycle. Using $[\gamma\text{-}^{32}P]$-ATP, the labeled phosphate was shown to bind to the 55,000 and 30,000 fragments (Stewart *et al.*, 1976). The site of ATP hydrolysis therefore appears to be in the 30,000 fragment. This fragment, as noted, exhibits no ionophoric activity.

4. RELATIONSHIP BETWEEN THE PIECES IN THE WORKING ENZYME

The interaction between the pieces obtained by tryptic digestion can be better understood by monitoring ATP hydrolysis activity and Ca^{2+}-pumping activity during digestion. This was done by Stewart and MacLennan (1974) and by Scott and Shamoo (1980; Scott, 1979) using SR vesicles. The first cleavage of the enzyme into 55,000 and 45,000 fragments has no effect on ATP hydrolysis or Ca^{2+} uptake activity. Apparently the enzyme is held together by strong forces other than the peptide chemical bond. However, the cleavage of the 55,000 piece into 30,000 and 25,000 fragments stops Ca^{2+} uptake into SR vesicles, while not affecting ATP hydrolysis. That is, Ca^{2+} pumping and ATP hydrolysis are uncoupled, so that the enzyme is no longer able to use energy from ATP hydrolysis to drive Ca^{2+} transport. Concomitantly, one of the two high-affinity Ca^{2+}-binding sites is lost (Scott, 1979). Coupling between ATP hydrolysis and Ca^{2+} transport can be disrupted by several other means, as well (Berman *et al.*, 1977; Hidalgo, 1980).

The tight coupling between ATP hydrolysis and Ca^{2+} transport in the SR membrane can be shown by running the pump backward. Normally the enzyme in the presence of ATP pumps Ca^{2+} into SR vesicles. However, high internal Ca^{2+} with ADP and P_i present externally causes the pump to run in reverse: Ca^{2+} ions flow out through the enzyme pathway, and $ADP + P_i$ combine to form ATP. (See Hasselbach, 1978, for a review of this.) Furthermore, Ratkje and Shamoo (1980) have recently demonstrated that ATP synthesis by the purified ATPase may be induced by sudden changes in the pH of the medium. They showed that the second tryptic cleavage stops such ATP synthesis, which is further evidence that the second tryptic cleavage between the hydrolytic site (30,000 fragment) and the Ca^{2+}-transport site (25,000 fragment) causes uncoupling of hydrolysis from transport.

The spatial relationship of the fragments has been studied by Stewart *et al.* (1976) who raised antibodies against the various fragments. These workers concluded that the

55,000 fragment is exposed on the surface of SR vesicles, while the 25,000 and 45,000 fragments are not exposed to the membrane surface.

Klip *et al.* (1980) have determined the alignment of the tryptic and CNBr fragments of the ATPase. The order of the tryptic fragments is Ac-NH$_2$–25,000–30,000–45,000–COOH. Within the 25,000 fragment, the order is Ac-NH$_2$–Met–13,000–1000–3000–7500–COOH. Thus, the smallest Ca^{2+}-ionophoric fragment is at the NH$_2$ terminus of the polypeptide. Klip and co-workers were able to align the sequenced regions determined by Allen (1977) with the protein, as well. This showed four regions of approximately 100 residues each, unsequenced because of their hydrophobicity. MacLennan *et al.* (1980) propose that these hydrophobic regions each span the membrane twice, totaling eight passes through the membrane, possibly as α-helices.

5. SUMMARY OF THE DATA: A MODEL

A schematic model consistent with the experimental data is shown in Fig. 2. This is a refinement of the models proposed by Shamoo and Ryan (1975) and Shamoo (1979). It shows the 45,000 fragment buried in the lipid matrix and the 55,000 fragment exposed on the surface, consistent with antibody experiments. Consistent with BLM work, the 45,000 fragment acts as a nonselective "pore" in series with a gating mechanism in the 55,000

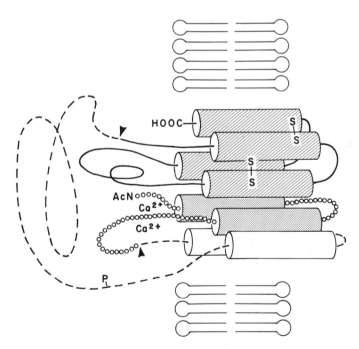

Figure 2. A hypothetical model for the Ca^{2+}/Mg^{2+}-ATPase, consistent with what is known about the polypeptide chain. The cylinders represent α-helical segments spanning the membrane, consistent with the model of MacLennan *et al.* (1980). The cross-hatched, dotted, and unmarked helices are in the 45,000, 25,000, and 30,000 tryptic fragments, respectively. The tryptic cleavage sites are represented by arrowheads. The 30,000 fragment is on the "outside," consistent with antibody studies, and contains the site of ATP binding and subsequent phosphorylation (P). At least one Ca^{2+}-binding site is near the Ac-NH$_2$ terminus of the protein, which is part of a selective gating mechanism. In series with the gate is a channel formed by the 45,000 fragment. This channel is disrupted when disulfide bridges (S–S) are reduced.

fragment. At least one site of Ca^{2+} binding is associated with the 13,000 fragment. The 30,000 fragment binds MgATP.

The spatial arrangement shown is only tentative at this time. Note that the model presents no details of the coupling mechanism between ATP hydrolysis and Ca^{2+} transport, as nothing is known about this mechanism. Nonetheless, this model is conceptually useful in summarizing experimental data and providing a starting place for speculation and design of experiments.

6. CONCLUSION

By using the example of the Ca^{2+}/Mg^{2+}-ATPase, we have described how the concept of ionophores can be used in studying ion transport. We have tried to assign the functions of ion selectivity, ion conduction, and energy transduction to specific pieces of the protein. However, some caution is clearly in order here. Even if each of the pieces exhibits a given function in artificial systems, we still have the problem of ensuring that each piece performs its assigned function in the native system. Thus, we look for characteristics (such as inhibitors and selectivity) common to both the pieces and the intact system. In the case of the Ca^{2+}/Mg^{2+}-ATPase, data of this type have provided good evidence for correlation of function in isolated pieces and in the intact enzyme. The application of these concepts to other ion transport systems leaves much work to be done.

REFERENCES

Abramson, J. J., and Shamoo, A. E. (1978). *J. Membr. Biol.* **44,** 233–257.

Allen, G. (1977). *Proceedings FEBS 11th Meeting*, Copenhagen, Symposium A4, Vol. 45, pp. 159–168.

Berman, M. D., McIntosh, D. B., and Kench, J. E. (1977). *J. Biol. Chem.* **252,** 994–1001.

Hasselbach, W. (1978). *Biochim. Biophys. Acta* **515,** 23–53.

Hidalgo, C. (1980). *Biochem. Biophys. Res. Commun.* **92,** 757–765.

Klip, A., and MacLennan, D. H. (1978). In *Frontiers of Biological Energetics: Electrons to Tissues* (A. Scarpa, F. Dutton, and J. Leigh, eds.), Academic Press, New York.

Klip, A., Reithmeier, R. A. F., and MacLennan, D. H. (1980). *J. Biol. Chem.* **255,** 6562–6568.

MacLennan, D. H., Reithmeier, R. A. F., Shoshan, V., Campbell, K. P., LeBel, D., Herrmann, T. R., and Shamoo, A. E. (1980). *Ann. N.Y. Acad. Sci.*, **358,** 138–148.

Masuda, H., and de Meis, L. (1974). *Biochim. Biophys. Acta* **332,** 313–315.

Ratkje, S. K., and Shamoo, A. E. (1980). *Biophys. J.* **30,** 523–530.

Scott, T. L. (1979). Ph.D. thesis, University of Rochester.

Scott, T. L., and Shamoo, A. E. (1980). in press.

Shamoo, A. E. (1979). In *Physical Chemical Aspects of Cell Surface Events in Cellular Regulation* (C. DeLisi and R. Blumenthal, eds.), pp. 59–68, Elsevier/North-Holland, Amsterdam.

Shamoo, A. E., and Goldstein, D. A. (1977). *Biochim. Biophys. Acta* **472,** 13–53.

Shamoo, A. E., and MacLennan, D. H. (1974). *Proc. Natl. Acad. Sci. USA* **71,** 3522–3526.

Shamoo, A. E., and MacLennan, D. H. (1975). *J. Membr. Biol.* **25,** 65–74.

Shamoo, A. E., and Ryan, T. E. (1975). *Ann. N.Y. Acad. Sci* **264,** 83–97.

Shamoo, A. E., Ryan, T. E., Stewart, P. E., and MacLennan, D. H. (1976). *J. Biol. Chem.* **251,** 4147–4154.

Stewart, P. S., and MacLennan, D. H. (1974). *J. Biol. Chem.* **249,** 985–993.

Stewart, P. S., MacLennan, D. H., and Shamoo, A. E. (1976). *J. Biol. Chem.* **251,** 712–719.

Thorley-Lawson, D. A., and Green, N. M. (1975). *Eur. J. Biochem.* **59,** 193–200.

Thorley-Lawson, D. A., and Green, N. M. (1977). *Biochem. J.* **167,** 739–748.

76

Ion-Transporting ATPases

Characterizing Structure and Function with Paramagnetic Probes

Charles M. Grisham

1. INTRODUCTION

ATP-hydrolyzing enzymes, which couple the free energy of hydrolysis of ATP to the transport of monovalent or divalent cations, are ubiquitous constituents of biological membranes. In our laboratory, we have been involved in a characterization of three of these transport systems, including the Na^+/K^+-ATPase and Mg^{2+}-ATPase from kidney plasma membrane and the Ca^{2+}-ATPase from sarcoplasmic reticulum. A specific transport function has not yet been assigned to the Mg^{2+}-ATPase, but the other two enzymes have become model systems for the study of structure–function relationships and energy coupling in membranes.

In recent years, important new information has been gained from studies on the reconstitution of the ATPases, from calorimetric studies of the binding of substrates and activators, and from characterization of proteolytic and chemical cleavage products of these enzymes. These areas of research are discussed by some of the principal contributors to the field in other chapters of this volume. On the other hand, spectroscopic methods, which can focus on discrete substrate- or activator-binding sites, have also provided important new structural and mechanistic information on these transport enzymes. This review will summarize recent results in the latter area for the three ATPases mentioned.

2. PLASMA MEMBRANE Na^+/K^+-ATPase

The Na^+/K^+-ATPase was once thought to consist of a large polypeptide of about 90,000 molecular weight, a glycoprotein of about 35,000 molecular weight, and a proteolipid of

Charles M. Grisham • Department of Chemistry, University of Virginia, Charlottesville, Virginia 22901.

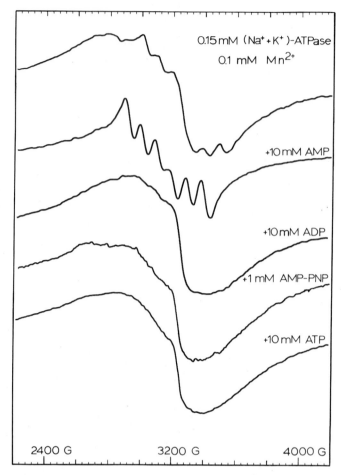

Figure 1. X-Band (9.0 GHz) EPR spectra of Mn^{2+} complexes with Na^+/K^+-ATPase. All solutions contained 20 mM TES, pH 7.5, 0.15 mM ATPase, 0.1 mM $MnCl_2$, and the concentrations of substrates shown. The top spectrum of the binary Na^+/K^+-ATPase–Mn^{2+} complex reflects a distorted coordination geometry for the enzyme-bound Mn^{2+}. AMP causes an increase in the symmetry of the Mn^{2+} ion, while ADP, AMP-PNP, and ATP all cause a greater distortion of the bound Mn^{2+}. Similar conclusions can be drawn from spectra at 35 GHz (O'Connor and Grisham, 1980a).

about 12,000 molecular weight (Hokin *et al.*, 1973; Lane *et al.*, 1973; Forbush *et al.*, 1978). The functional unit of the enzyme appeared to be a dimer of 90K chains and to contain glycoprotein and proteolipid as well. More recent studies (Craig and Kyte, 1980; Peters *et al.*, 1981) have shown the true mass of the "90K chain" to be 121,000 daltons, and that of the glycoprotein to be 56,000 daltons. In addition, the Lowry protein assay has been shown to give falsely high results for this enzyme. In view of these points, the stoichiometry of ligand binding to the purified Na^+/K^+-ATPase is such that for each 121K chain chains one ouabain (inhibitor) molecule or one ATP may be bound (Lane *et al.*, 1973; Jorgensen, 1974), or one ^{32}P may be covalently incorporated from $[^{32}P]$-ATP (Hokin *et al.*, 1973; Jorgensen, 1977). Our Mn^{2+} EPR studies have also shown that one divalent metal can be bound per 121K chain in the absence of ATP (O'Connor and Grisham, 1979), and kinetic studies have shown that the site detected by EPR is the divalent metal activating site of the ATPase reaction (Grisham and Mildvan, 1974).

In early studies, using Mn^{2+} at this site as a paramagnetic probe, a Na^+ site on the

ATPase was characterized in a series of $^{205}Tl^+$ NMR studies (Grisham *et al.*, 1974). For a variety of reasons, it is not chemically possible to use $^{205}Tl^+$ as an NMR probe of K^+ sites on the Na^+/K^+-ATPase. However, work in a number of laboratories has shown recently that $^7Li^+$ can be a superior nucleus for this type of study (Hutton *et al.*, 1977; Raushel and Villafranca, 1980), and we have used $^7Li^+$ NMR to identify and characterize a K^+ site on the Na^+/K^+-ATPase (Grisham and Hutton, 1978). The two most significant conclusions to be drawn from these studies are (1) the Na^+ and K^+ transport sites are close to the catalytic site of ATP hydrolysis (i.e., close to the Mn^{2+} site involved in ATP hydrolysis), and (2) the Na^+ and K^+ sites are close to each other (the $^{205}Tl^+$ and $^7Li^+$ NMR results set an upper limit on the Na^+; to K^+ distance of 12.6 Å).

It has also been shown that the EPR spectrum of Mn^{2+} bound to the single divalent cation site on the Na^+/K^+-ATPase can provide important structural information. Such spectra can yield data on the coordination geometry of the bound metal ion and about changes in same arising from binding of substrates and activators to the enzyme. Thus, it has been shown (O'Connor and Grisham, 1979, 1980a) that ATP, ADP, and the nonhydrolyzing substrate analog AMP-PNP cause a large rhombic distortion of the bound Mn^{2+}, whereas AMP causes an increase in the symmetry of the Mn^{2+} coordination geometry (Fig. 1). The similar effects of ATP, ADP, and AMP-PNP suggest that the terminal phosphate of ATP does not interact directly with the bound Mn^{2+} ion at the active site. Despite this lack of evidence for direct coordination, it has been shown that the Mn^{2+} and ATP sites are quite close on the Na^+/K^+-ATPase (O'Connor and Grisham, 1980b). These latter studies are discussed in more detail in Section 5.

3. SARCOPLASMIC RETICULUM Ca²⁺-ATPase

The Ca^{2+}-ATPase is a membrane-bound polypeptide of about 100,000 molecular weight and catalyzes the reversible, coupled hydrolysis of ATP and transport of Ca^{2+} into sarcoplasmic reticulum against a concentration gradient. The numerous chemical and physical studies of this complex transport system have all suffered from the indirect nature of the measurements. For example, Ca^{2+} uptake can be measured across a membrane, but the individual molecular events of Ca^{2+} transport have not been elucidated.

Hoping to provide a more direct means of characterizing the mechanism of Ca^{2+} transport and the active site structure of Ca^{2+}-ATPase from sarcoplasmic reticulum, we have employed gadolinium (Gd^{3+}), a trivalent lanthanide ion, as a paramagnetic probe of this system in a series of NMR and EPR investigations. It had previously been shown (Krasnow, 1972, 1977) that Gd^{3+} is transported across the sarcoplasmic reticulum membrane by Ca^{2+}-ATPase, albeit much more slowly than Ca^{2+} itself. Our kinetic studies of the Ca^{2+}- or Gd^{3+}-sensitive formation of phosphoenzyme from ATP have shown (Stephens and Grisham, 1979) that Gd^{3+} binds at the Ca^{2+} sites, with an affinity that is 10 times that of Ca^{2+} itself. A similar, very high affinity of Ca^{2+}-ATPase for Gd^{3+} has been observed by us in $^7Li^+$ NMR studies of the enzyme. These NMR experiments demonstrated that (1) the ATPase binds two Gd^{3+} ions per enzyme monomer in a sequential fashion at two nonidentical sites, and (2) the affinity of Gd^{3+} for these sites is 10 times that of Ca^{2+}, in excellent agreement with the kinetic studies. Completely independent water proton relaxation studies of the interaction of Gd^{3+} with this ATPase corroborated both these findings and provided compelling evidence that the Ca^{2+}-binding sites of this system are located in a hydrophobic pocket, perhaps deep in the sarcoplasmic reticulum membrane. This point has since been confirmed independently by Racker and his colleagues (Pick and Racker, 1979). In addition, these water proton relaxation studies have permitted us to characterize the time course of the binding of ATP and several ATP analogs to the Gd^{3+}–ATPase complex (Stephens and Grisham, 1979, 1980).

While these results with the Ca^{2+}-ATPase have been useful in their own right, they also provide a basis for the use of Gd^{3+} EPR measurements to characterize the structure of the Ca^{2+} transport sites and the mechanism of the transport process itself. The possibility of using the EPR properties of Gd^{3+} to probe its environment in and interactions with biological molecules has previously received little attention in the literature. However, we have now demonstrated that Gd^{3+} EPR can be a sensitive structural probe of the Ca^{2+}-ATPase (Stephens and Grisham, 1979, 1980; Grisham, 1980a,b). Several different enzyme states, which are sensitive to temperature, ligand additions, etc., can be identified from Gd^{3+} EPR spectra, and these unusual spectra can be interpreted on the basis of changes in ligand coordination geometry (Stephens and Grisham, 1981). These interpretations derive from simulations of Gd^{3+} EPR spectra based on the "independent systems crystal field" model of Faulkner and Richardson (1980). The time-dependent, temperature-dependent, and ATP-dependent changes we have observed with the Ca^{2+}-ATPase–Gd^{3+} complex are consistent with earlier observations by others of a temperature-dependent equilibrium of two functional forms of the ATPase (Inesi *et al.*, 1976) with different Ca^{2+}-binding sites and affinities (Ikemoto, 1975), different phosphorylated intermediates (Chiesi and Inesi, 1979) and different sensitivities to inhibition by lanthanides.

4. PLASMA MEMBRANE Mg^{2+}-ATPase

Early efforts to purify the Na^+/K^+-ATPase from plasma membrane were hampered by the presence of an ouabain-insensitive Mg^{2+}-ATPase. While this enzyme has previously received little attention in the literature and its transport function is unknown, it has now been purified using a digitonin extraction procedure (Gantzer and Grisham, 1979a). This enzyme differs from the Na^+/K^+-ATPase in several respects (ouabain sensitivity, subunit molecular weights, divalent cation selectivity, and sensitivity to monovalent cations), but also shows several similarities to the latter. Thus, the Mg^{2+}-ATPase displays both high- and low-affinity kinetic requirements for ATP, forms a covalent E–P intermediate from ATP, and exhibits a discontinuous Arrhenius plot with a break at 20°C, the same as that for the Na^+/K^+-ATPase (Grisham and Barnett, 1973). A thorough kinetic characterization of the divalent metal and ATP requirements of this Mg^{2+}-ATPase has been published (Gantzer and Grisham, 1979b). One of the most interesting points arising from this work is that this enzyme is not only activated by a single enzyme-bound divalent metal (characterized by Mn^{2+} EPR binding studies), but is also activated (six- to sevenfold) by binding of divalent metals to the lipid membrane surrounding the enzyme (Gantzer and Grisham, 1979b). This is the first known case of this type of activation of a membrane enzyme.

5. PROBING THE ATP SITES WITH METAL–NUCLEOTIDE ANALOGS

The past several years has witnessed the development of a number of ATP analogs that are beginning to provide important information on the ATP sites of the ion transport ATPases. For example, ATP analogs that incorporate SH-reactive moieties have been employed by several groups with the Na^+/K^+-ATPase (Patzelt-Wenczler *et al.*, 1975; Tobin and Akera, 1975). Two of the most useful types of analogs, however, have been the βS derivatives of ATP first described by Eckstein and Goody (1976) and later by Jaffe and

Table I. Summary of Kinetically Determined Binding Constants[a]

	Co(NH$_3$)$_4$ATP kinetics					
	High affinity (μM)			Low affinity (mM)		
	K_m^{MnATP}	K_i^{CoATP}	K_i/K_m	K_m^{MnATP}	K_i^{CoATP}	K_i/K_m
Na$^+$/K$^+$-ATPase	2.88	10	3.5	0.902	1.6	1.8
Mg^{2+}-ATPase	2.28	24	10.5	0.902	3.5	3.9
Ca^{2+}-ATPase	7.76	350	45.1	0.424	8.7	20.5

	Cr(H$_2$O)$_4$ATP kinetics					
	High affinity (μM)			Low affinity (mM)		
	K_m^{MnATP}	K_i^{CrATP}	K_i/K_m	K_m^{MnATP}	K_i^{CrATP}	K_i/K_m
Na$^+$/K$^+$-ATPase	10.7	256	23.9	0.368	0.46	1.3
Mg^{2+}-ATPase	7.4	16	2.2	0.19	0.17	0.89
Ca^{2+}-ATPase	10.4	545	52.4			

[a] These kinetic studies used [^{32}P]-ATP as a substrate, and thus when Co(NH$_3$)$_4$ATP displaces [^{32}P]-ATP, the analog appears to be an inhibitor of the ATPase. Nonetheless, Co(NH$_3$)$_4$ATP may still function as a legitimate substrate for the various ATPases. The same argument applies to Cr(H$_2$O)$_4$ATP.

Cohn (1978a,b), and the complexes of Co^{3+} and Cr^{3+} with ATP developed by Cleland and co-workers (Cleland and Mildvan, 1979). We have examined the ATPases described above using both ATPβS and the Co(NH$_3$)$_4$ATP and Cr(H$_2$O)$_4$ATP analogs (Gantzer and Grisham, 1981; Grisham, 1981). As can be seen from Table I, Co(NH$_3$)$_4$ATP and Cr(H$_2$O)$_4$ATP display competitive inhibition with respect to MnATP at both high- and low-affinity ATP sites on Na$^+$/K$^+$-ATPase and Mg^{2+}-ATPase. With the exception of a low-affinity requirement with Cr(H$_2$O)$_4$ATP, the same is true for the Ca^{2+}-ATPase.

Furthermore we have employed a combination of NMR, CD, and kinetic studies to demonstrate that the various analogs are in fact substrates for these enzymes. As shown in Table II, Cr(H$_2$O)$_4$ATP is a substrate for all three ATPases. Interestingly, while the Na$^+$/K$^+$-ATPase shows no substrate activity with Co(NH$_3$)$_4$ATP, both the Ca^{2+}-ATPase and the Mg^{2+}-ATPase can use the cobalt analog as a substrate. Moreover, CD studies show that both the Na$^+$/K$^+$-ATPase and the Ca^{2+}-ATPase use the Δ isomer of Cr(H$_2$O)$_4$ATP as a

TABLE II. ATP Analog Activity with Membrane ATPases

Enzymes	Analog	Substrate activity	Stereospecificity
Na$^+$/K$^+$-ATPase	ATPβS	Yes	Mg(A)
	Co(NH$_3$)$_4$ATP	No	—
	Cr(H$_2$O)$_4$ATP	Yes	Δ
Ca^{2+}-ATPase	ATPβS	Yes	No preference
	Co(NH$_3$)$_4$ATP	Yes	ND[a]
	Cr(H$_2$O)$_4$ATP	Yes	Δ
Mg^{2+}-ATPase	ATPβS	Yes	ND
	Co(NH$_3$)$_4$ATP	Yes	ND
	Cr(H$_2$O)$_4$ATP	Yes	Λ

[a] ND, not determined.

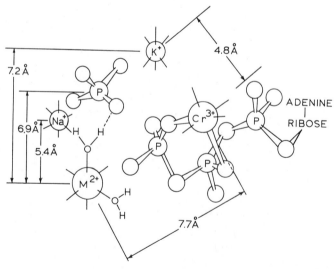

MgATPβS(A)

Δ-Cr(H$_2$O)$_4$ATP

MgATPβS(B)

Λ-Cr(H$_2$O)$_4$ATP

Figure 2. Absolute stereochemistry of the isomers of ATPβS and Cr(H$_2$O)$_4$ATP. The MgATPβS(A) complex has the same stereochemistry at the β phosphorus as the Δ isomer of Cr(H$_2$O)$_4$ATP. Thus, an enzyme such as Na$^+$/K$^+$-ATPase, which prefers Δ-Cr(H$_2$O)$_4$ATP as a substrate, should also prefer MgATPβS(A) over MgATPβS(B). The Co(NH$_3$)$_4$ATP isomers are analogous to those of Cr(H$_2$O)$_4$ATP.

substrate, while Mg^{2+}-ATPase uses the Λ isomer (Gantzer and Grisham, 1981). Figure 2 compares the structures of the Cr^{3+} nucleotides with those of MgATPβS(A) and MgATPβS(B). As would be predicted from this comparison, the Na$^+$/K$^+$-ATPase shows a strong preference for MgATPβS(A). However, the Ca^{2+}-ATPase shows no selectivity for Mg^{2+} complexes of the isomers of ATPβS. This intriguing anomaly could be explained if the rate-determining step for turnover of the Cr(H$_2$O)$_4$ATP by Ca^{2+}-ATPase was different

Figure 3. Active site structure of Na$^+$/K$^+$-ATPase as determined by ^{1}H, ^{205}Tl$^+$, ^{31}P, and ^{7}Li$^+$ NMR, Mn^{2+} EPR, and kinetic studies. The Mn^{2+}–Na$^+$ distance was determined from ^{205}Tl$^+$ NMR studies (Grisham *et al.*, 1974), while the Mn^{2+}–K$^+$ and Cr^{3+}–K$^+$ distances were obtained from ^{7}Li$^+$ NMR studies (Grisham and Hutton, 1978). The Mn^{2+}–Cr^{3+} distance of 7.7 Å is an average of results from Mn^{2+} EPR (O'Connor and Grisham, 1980b) and water proton NMR studies (O'Neal and Grisham, 1981).

from that for ATPβS. Of interest in this regard may be the studies of Pauls *et al.* (1980), who have shown that the Na$^+$/K$^+$-ATPase is very slowly inactivated by CrATP.

It is clear that Cr(H$_2$O)$_4$ATP will be an effective probe of structure and function with these ATPases. We have used Mn^{2+} EPR and water nuclear relaxation experiments to measure the Mn^{2+}–Cr^{3+} distance in ATPase–Mn^{2+}–Cr(H$_2$O)$_4$ATP complexes (O'Connor and Grisham, 1980b; O'Neal and Grisham, 1981). The ability to measure distances in such complexes is based upon the dipolar interaction between the unpaired electrons of Mn^{2+} and Cr^{3+}, which results in (1) a decrease in intensity with no apparent broadening of the Mn^{2+} EPR spectrum as first suggested by Leigh (1970) and (2) a cross-relaxation of Mn^{2+}-bound water protons as first described by Gupta (1977). As shown in Fig. 3, the Mn^{2+}–Cr^{3+} separation on the Na$^+$/K$^+$-ATPase in the absence of added cations is 7.7 $\pm$ 0.5 Å. This distance is decreased to 6.5 $\pm$ 0.3 Å by addition of Na$^+$ and K$^+$.

REFERENCES

Chiesi, M., and Inesi, G. (1979). *J. Biol. Chem.* **254**, 10370.
Cleland, W., and Mildvan, A. (1979). In *Advances in Inorganic Biochemistry* (G. Eichhorn and L. Marzilli, eds.), p. 163, Elsevier/North-Holland, Amsterdam.
Eckstein, F., and Goody, R. (1976). *Biochemistry* **15**, 1685.
Faulkner, R., and Richardson, F. (1980). *Mol. Phys.* **39**, 75.
Forbush, B., Kaplan, J., and Hoffman, J. (1978). *Biochemistry* **17**, 3667.
Gantzer, M., and Grisham, C. (1979a). *Arch. Biochem. Biophys.* **198**, 263.
Gantzer, M., and Grisham, C. (1979b). *Arch. Biochem. Biophys.* **198**, 268.
Gantzer, M., and Grisham, C. (1981). Submitted for publication.
Grisham, C. (1980a). *J. Biochem. Biophys. Methods* **3**, 39.
Grisham, C. (1980b). In *Polymer Characterization by ESR and NMR* (A. Woodward and F. Bovey, eds.), *ACS Symp. Ser.* **142**, 49.
Grisham, C. (1981). *J. Inorg. Biochem.* **14**, 45.
Grisham, C., and Barnett, R. (1973). *Biochemistry* **12**, 2635.
Grisham, C., and Hutton, W. (1978). *Biochem. Biophys. Res. Commun.* **81**, 1406.
Grisham, C., and Mildvan, A. (1974). *J. Biol. Chem.* **249**, 3187.
Grisham, C., Gupta, R., Barnett, R., and Mildvan, A. (1974). *J. Biol. Chem.* **249**, 6738.
Gupta, R. (1977). *J. Biol. Chem.* **252**, 5183.
Hokin, L., Dahl, J., Deupree, J., Dixon, J., Hackney, J., and Perdue, J. (1973). *J. Biol. Chem.* **248**, 2593.
Hutton, W., Stephens, E., and Grisham, C. (1977). *Arch. Biochem. Biophys.* **184**, 166.
Ikemoto, N. (1975). *J. Biol. Chem.* **250**, 7219.
Inesi, G., Cohen, J., and Coan, C. (1976). *Biochemistry* **15**, 5293.
Jaffe, E., and Cohn, M. (1978a). *Biochemistry* **17**, 652.
Jaffe, E., and Cohn, M. (1978b). *J. Biol. Chem.* **253**, 4823.
Jorgensen, P. (1974). *Biochim. Biophys. Acta* **356**, 53.
Jorgensen, P. (1977). *Biochim. Biophys. Acta* **466**, 97.
Krasnow, N. (1972). *Biochim. Biophys. Acta* **282**, 187.
Krasnow, N. (1977). *Arch. Biochem. Biophys.* **181**, 322.
Lane, L., Copenhaver, J., Lindenmayer, G., and Schwartz, A. (1973). *J. Biol. Chem.* **25**, 7197.
Leigh, J. (1970). *J. Chem. Phys.* **52**, 2608.
O'Connor, S., and Grisham, C. (1979). *Biochemistry* **18**, 2315.
O'Connor, S., and Grisham, C. (1980a). *Biochem. Biophys. Res. Commun.* **93**, 1146.
O'Connor, S., and Grisham, C. (1980b). *FEBS Lett.* **118**, 303.
O'Neal, A., and Grisham, C. (1981). Manuscript in preparation.
Patzelt-Wenczler, R., Pauls, H., Erdmann, E., and Schoner, W. (1975). *Eur. J. Biochem.* **53**, 301.
Pauls, H., Bredenbrocker, B., and Schoner, W. (1980). *Eur. J. Biochem.* **109**, 523.
Peters, W., Swarts, H., dePont, J., Schuurmans Stekhoven, F., and Bonting, S. (1981). *Nature* **290**, 338.
Pick, U., and Racker, E. (1979). *Biochemistry* **18**, 108.
Raushel, F., and Villafranca, J. J. (1980). *Biochemistry* **19**, 5481.

Stephens, E., and Grisham, C. (1979). *Biochemistry* **18,** 4876.
Stephens, E., and Grisham, C. (1980). *Fed. Proc.* **39,** 603.
Stephens, R., and Grisham, C. (1981). Manuscript in preparation.
Tobin, T., and Akera, T. (1975). *Biochim. Biophys. Acta* **389,** 126.

77

A Possible Role for Cytoplasmic Ca^{2+} in the Regulation of the Synthesis of Sarcoplasmic Reticulum Proteins

Anthony N. Martonosi

1. INTRODUCTION

The maintenance of intracellular free Ca^{2+} concentration in muscle cells depends upon a delicate balance between Ca^{2+} transport systems located in the surface membrane, mitochondria, sarcoplasmic reticulum, and several cytoplasmic Ca^{2+}-binding proteins, which control muscle contraction and key metabolic processes (Martonosi, 1980). The concentration and activity of the various systems must be accurately matched, which requires coordinate regulation of their rate of synthesis and degradation. Some features of this regulation are inherent properties of muscle cells, others are under neural control.

It is plausible to assume that the rate of synthesis of the protein components of Ca^{2+} transport systems and Ca^{2+} receptors would be influenced by the intracellular Ca^{2+} concentration, and that changes in the rate of synthesis of these proteins during embryonic development and in response to physiological stimuli in adult animals would be mediated by one common calcium-sensitive mechanism. We proposed that changes in cytoplasmic free Ca^{2+} concentrations during development or muscle activity may influence the rate of synthesis of the Ca^{2+}-ATPase of sarcoplasmic reticulum, and other Ca^{2+}-modulated proteins by Ca^{2+}-dependent regulation of the concentration of relevant classes of translatable mRNAs (Martonosi, 1975; Martonosi et al., 1978). The effect of cytoplasmic Ca^{2+} concentration may be mediated through nuclear Ca^{2+}-binding proteins serving as selective inducers or repressors of gene transcription (Schibeci and Martonosi, 1980a).

The purpose of this discussion is to evaluate the experimental evidence relating to the

Anthony N. Martonosi • Department of Biochemistry, SUNY, Upstate Medical Center, Syracuse, New York 13210.

role of Ca^{2+} concentration in the regulation of gene expression in developing chicken skeletal muscle, with emphasis upon the synthesis of membrane proteins.

2. *THE DEVELOPMENT OF SARCOPLASMIC RETICULUM IN CHICKEN MUSCLE*

The increase in the concentration of Ca^{2+}-transport ATPase in chicken pectoralis muscle cells during development *in vivo* or in tissue culture is initiated about the time of the fusion of myoblasts into multinucleated myotubes and roughly follows the time course of accumulation of myofibrillar proteins and ATP : creatine phosphotransferase (Martonosi *et al.*, 1972, 1977, 1980; Boland *et al.*, 1974; Tillack *et al.*, 1974). Lowering the Ca^{2+} concentration of the culture medium below 200 μM inhibits fusion (Shainberg *et al.*, 1969) and prevents the accumulation of Ca^{2+}-ATPase and other muscle-specific proteins (Martonosi *et al.*, 1977; Ha *et al.*, 1979).

We investigated the effect of extracellular Ca^{2+} concentration and Ca^{2+} ionophores, which are expected to influence cytoplasmic Ca^{2+} concentration, upon the synthesis of proteins in tissue culture (Roufa and Martonosi, 1980; Roufa *et al.*, 1981; Wu *et al.*, 1981).

Brief exposure of cultured chicken pectoralis muscle cells, grown in normal or low-Ca^{2+} media, to the calcium ionophores ionomycin (4×10^{-6} M for 3 hr) or A23187 selectively increases severalfold the rate of [^{35}S]methionine incorporation into an 80,000- and a 100,000-molecular-weight membrane protein, which sediment largely in the mitochondrial and microsomal fractions (Fig. 1). Increased synthesis of the 80K protein was also observed upon cell-free translation of poly(A)-enriched RNA isolated from ionomycin-treated as compared with control muscle cultures (Fig. 2). These observations suggest that ionomycin selectively increases the cellular concentration of mRNA, which codes for the 80K protein. The effect is probably mediated through an increase in cytoplasmic Ca^{2+} concentration caused by the ionophore. A similar effect of ionomycin was observed in cultured fibroblasts, HeLa cells, mouse LSP cells, and monkey kidney CVI cells (Fig. 1). The 80K proteins from the different cell types were similar when analyzed by two-dimensional gel electrophoresis.

The 100K protein is presumed to be the Ca^{2+}/Mg^{2+}-activated ATPase, which is a component of the endoplasmic reticulum in all types of cells, including muscle. Continuous growth of muscle cells in culture media containing about 10^{-8} M A23187 increases the steady-state concentration of the Ca^{2+}-transport ATPase determined by active site labeling with [^{32}P]-ATP (Martonosi *et al.*, 1977), and immunoprecipitation.

The 80K component is a phosphoprotein, as shown by its labeling in cultures grown with $^{32}P_i$. It may be related to an 80K phosphoprotein observed in microsomes isolated from embryonic chicken muscle after exposure to [^{32}P]-ATP (Martonosi, 1975). Its isolation and characterization are in progress.

It seemed important to test whether changes in cytoplasmic Ca^{2+} concentration in the absence of ionophore can be shown to influence the rate of protein and mRNA synthesis. The rate of cell-free translation of the 80K protein was increased using poly(A) RNA isolated 3 hr after transfer of muscle cultures grown in low-Ca^{2+} medium into media of normal Ca^{2+} concentration (Fig. 2). After continued incubation in normal medium for 22 hr, the rate of synthesis of the 80K protein decreased to the levels observed in control cultures. Because at low medium Ca^{2+} concentrations the Ca^{2+} permeability of the cells increases (Winegrad, 1971), such transfer is likely to produce a transient rise in cyto-

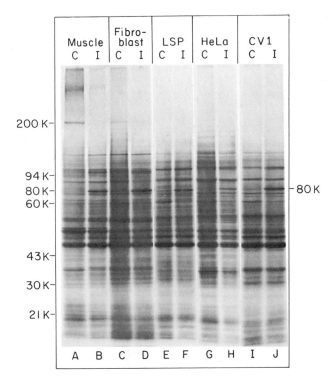

Figure 1. The effect of ionomycin on the incorporation of [³⁵S]methionine into proteins of cultured cells. The cells were grown in normal Ca^{2+} (1.8 mM) medium for 65 hr. One group of plates served as control (C), the other was treated with 4×10^{-6} M ionomycin (I). After 2 hr, both groups of cells were labeled with [³⁵S]methionine for 2 hr in a medium made from methionine-free minimum essential medium. The cells were washed with calcium-free phosphate-buffered saline (pH 7.2), and applied for electrophoresis. Electrophoresis was performed in 6–12% gradient polyacrylamide gel. Note the intense labeling of the 80K and 100K bands in the ionomycin-treated (I) samples. A–B, chicken muscle cells; C–D, chicken fibroblasts; E–F, mouse LSP cells; G–H, HeLa cells; I–J, monkey kidney CV1 cells (Wu and Martonosi, unpublished observations).

plasmic Ca^{2+}. Variation of the Na^+ and K^+ concentration of the medium or exposure of the cells to caffeine (2–8 mM), nigericine (10^{-7}–10^{-5} M), valinomycin (10^{-7}–10^{-5} M), or ouabain (10^{-6}–10^{-4} M) had no effect on the synthesis of the 80K and 100K proteins.

The effect of ionomycin and A23187 on protein synthesis is accompanied by an increase in cytoplasmic Ca^{2+} concentration, as evidenced by the contraction of the myotubes upon addition of the ionophore. Further evidence is required to prove that a change in cytoplasmic Ca^{2+} concentration induced by the ionophores is the direct cause of the observed changes in protein synthesis.

The activation of the synthesis of 80K and 100K proteins by ionomycin is accompanied by an inhibition of myosin synthesis (Roufa and Martonosi, 1980). This observation suggests that Ca^{2+} may act as a positive or negative regulator of protein synthesis, and optimum Ca^{2+} concentrations for the modulation of different systems may vary over a wide range.

Experiments are now in progress to test whether the selective gene expression observed in heat shock (Ashburner and Bonner, 1979; Kelley *et al.*, 1980), in glucose deprivation (Zala *et al.*, 1980), and upon transfer from anaerobic to aerobic incubation conditions (Lewis *et al.*, 1975) is also related to a transient increase in cytoplasmic Ca^{2+} concentration.

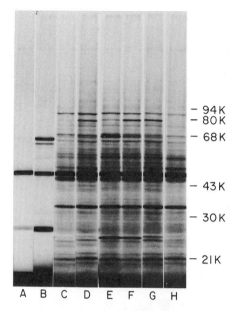

Figure 2. Autoradiogram of *in vitro* translation products of poly(A)-enriched RNA isolated from muscle cell cultures. Myoblasts isolated from 12-day-old chicken embryos were plated in normal Ca^{2+} medium for 24 hr followed by growth in either normal calcium (lanes C, D) or low-calcium medium (lanes E–H). After 64 hr of incubation, cultures D and F were exposed to 4×10^{-6} M ionomycin for 3 hr, while cultures C and E served as control. Cultures G and H grown in low-calcium medium were transferred into normal calcium medium after 64 hr incubation and incubated for 3 hr (lane G) or 22 hr (lane H) in normal medium prior to RNA extraction. At the end of the incubation, RNA was isolated from all cultures and fractionated into poly(A)-enriched RNA and non-poly(A) RNA on an oligo dT-cellulose column. Poly(A)-enriched RNA (0.87 μg) from each sample was translated in a rabbit reticulocyte lysate system *in vitro*. The polypeptide products were analyzed by electrophoresis on 6–12% polyacrylamide gradient gel slabs, dried, and autoradiographed. Note the intense labeling of the 80K band in lanes D, F, G. Poly(A) RNA isolated from low-Ca^{2+} cultures (E, F) is active in the synthesis of a 26–28K protein, irrespective of ionomycin treatment, but forms less myosin compared with cultures grown in normal medium (C, D). Less pronounced differences were observed in other protein bands. Lane A: Reticulocyte lysate without further addition. Lane B: Reticulocyte lysate with globin mRNA added. From Wu *et al.* (1981).

3. POSSIBLE MECHANISMS OF THE EFFECT OF Ca^{2+} UPON GENE EXPRESSION

The increase in the concentration of translatable mRNA coding for the 80K protein in cells treated with ionomycin is consistent with Ca^{2+} regulation either at the level of gene transcription or at some subsequent processing step.

The nuclear envelope in most cells is permeable to small molecules and ions, and the Ca^{2+} content of nuclei, analyzed *in situ* by electron microprobe (Somlyo *et al.*, 1978), is similar to that of the cytoplasm. It follows, therefore, that the free ionized Ca^{2+} concentration of the nucleoplasm reflects that of the cytoplasm. We propose that the effect of nucleoplasmic Ca^{2+} concentration upon gene transcription is mediated by nuclear Ca^{2+}-binding proteins, which bind Ca^{2+} with an affinity constant of 10^5–10^8, in the presence of 0.1 M KC1 and 5 mM $MgCl_2$. Such Ca^{2+}-binding proteins were recently observed among the nonhistone chromosomal proteins and in the insoluble fraction of the nuclei isolated from skeletal muscles of embryonic or adult chickens or rabbits and from rat liver (Schibeci and Martonosi, 1980a). The nuclear Ca^{2+}-binding proteins were separated into several distinct fractions by electrophoresis on polyacrylamide gels in the absence of detergents (Schibeci and Martonosi, 1980a,b); on the basis of electrophoretic mobility, these fractions are distinct from parvalbumin, prothrombin, and troponin C, but one of the fractions may contain calmodulin.

Among possible modes of involvement of Ca^{2+}-binding proteins in the regulation of mRNA synthesis and processing are:

1. *Ca^{2+}-dependent phosphorylation or dephosphorylation of histones or other chromatin components.* Marked activation of the phosphorylation of histone H3 by 0.1 μM Ca^{2+} was observed in isolated nuclei of butyrate-treated HeLa cells by Whitlock *et al.* (1980). Heat shock and deciliation of *Tetrahymena pyriformis* induces the phosphorylation of histone Hl possibly by a Ca^{2+}-dependent mechanism (Glover *et al.*, 1981). These changes in histone phosphorylation suggest the involvement of calmodulin-dependent proteinkinases. Interestingly, trifluoperazine, a known inhibitor of calmodulin, did not inhibit the effect of ionomycin on the synthesis of 80K protein; in fact, at higher trifluoperazine concentration, the relative rate of 80K synthesis increased (Wu and Martonosi, unpublished observations).

2. *Ca^{2+}-dependent attachment or dissociation of the Ca^{2+}-binding proteins to chromatin, with induction or repression of mRNA synthesis.* Although Ca^{2+} influenced the extraction of the Ca^{2+}-binding proteins from nuclei (Schibeci and Martonosi, 1980a), this effect was observed only at Ca^{2+} concentrations of 10^{-4}–10^{-3} M, and therefore it is unlikely to have physiological significance.

3. *A Ca^{2+}-dependent gating mechanism.* This may consist of a conformationally Ca^{2+}-sensitive chromosomal protein, which would regulate the transcription rate in a manner similar to the troponin–tropomyosin regulation of muscle contraction (Squire, 1981).

Other mechanisms of transcriptional or posttranscriptional control are also possible. In cell-free translation systems isolated from rat liver (Chyn *et al.*, 1979), the translation of Ca^{2+}-ATPase mRNA was not activated by Ca^{2+}.

The postulated mechanisms must account for the observation that the induction of the synthesis of 80K and 100K proteins by Ca^{2+} ionophores was accompanied by an inhibition of myosin synthesis (Roufa and Martonosi, 1980), while the synthesis of most other proteins was not affected. This implies that different structural genes are under the influence of distinct Ca^{2+}-dependent regulatory mechanisms, with presumably different Ca^{2+} affinities. Such differences in Ca^{2+} dependence are consistent with the heterogeneity of Ca^{2+}-binding sites in the nonhistone chromosomal protein fraction (Schibeci and Martonosi, 1980a). The existence of several nuclear Ca^{2+}-binding proteins with distinct Ca^{2+} affinities provides the opportunity for selective regulation of the synthesis of various proteins in response to distinct levels of nucleoplasmic Ca^{2+} concentration. At high concentrations of ionomycin and trifluoperazine, the synthesis of most proteins is inhibited, while the synthesis of 80K and 100K proteins continues.

4. THE POSSIBLE ROLE OF CALCIUM IN THE REGULATION OF PROTEIN SYNTHESIS DURING EMBRYONIC DEVELOPMENT

Although there are indications that Ca^{2+} may play a role in the control of gene expression during development of fertilized eggs (Gilkey *et al.*, 1978), in the transformation of lymphocytes (Freedman *et al.*, 1975; Hesketh *et al.*, 1977; Greene *et al.*, 1976), in the replication of hemopoietic stem cells (Gallien-Lartigue, 1976), in embryonic induction of *Rana pipiens* (Barth and Barth, 1974), and in the fusion and differentiation of muscle cells

(Bischoff, 1978), the evidence is usually indirect and the mechanism obscure. Much of the uncertainty arises from our lack of knowledge about cytoplasmic (and nucleoplasmic) free Ca^{2+} concentration during embryonic development and its relationship to extracellular Ca^{2+} concentration. Recent advances in the detection of resting levels of cytoplasmic Ca^{2+} using aequorin (Blinks *et al.*, 1976), Ca^{2+}-sensitive microelectrodes (Marban *et al.*, 1980), arsenazo III, antipyrilazo III, and other Ca^{2+} indicator dyes (Caswell, 1979) may eventually enable us to continually monitor cytoplasmic Ca^{2+} concentration in the vicinity of 10^{-7} M, for periods extending over several days, with a high level of accuracy. In the meantime, the following observations may support Ca^{2+} involvement in the regulation of muscle differentiation.

1. The fusion of muscle cells and the subsequent increase in the rate of synthesis of contractile proteins and Ca^{2+}-transport ATPase require an extracellular Ca^{2+} concentration in the physiological range (about 1.8 mM). Both fusion and muscle-specific protein synthesis are inhibited at a medium Ca^{2+} concentration of 0.2 mM or less. The structural rearrangement and eventual disruption of the surface membranes between muscle cells undergoing fusion may permit brief influx of extracellular Ca^{2+} that could initiate the synthesis of stable mRNAs and the accumulation of Ca^{2+}-specific proteins. This increased Ca^{2+} influx would not occur in cells grown in low-Ca^{2+} media with the result that their differentiation remains blocked at the prefusion stage. Direct analysis of cytoplasmic free Ca^{2+} during development of muscle cells is required to fully substantiate this hypothesis.

2. The total Ca^{2+} content of muscle cells is high just before the increased synthesis of Ca^{2+}-ATPase begins and decreases with age both *in vivo* and in tissue culture as development proceeds. In view of the small amount of sarcoplasmic reticulum (Boland *et al.*, 1974; Martonosi *et al.*, 1977) and parvalbumin (LePeuch *et al.*, 1979) in embryonic muscle and the rapid Ca^{2+} fluxes across the surface membrane of undifferentiated muscle cells (Roufa *et al.*, 1981), the free Ca^{2+} concentration in the cytoplasm of embryonic muscle is likely to be elevated, contributing to the induction of certain proteins and repression of others.

3. In chicken embryo muscle between 13 and 19 days of development, the action potential is largely due to Ca^{2+} current, indicating the presence of Ca^{2+} channels in the surface membrane (Spitzer, 1979). The influx of Ca^{2+} causes spontaneous contractions. This is the time when the accumulation of Ca^{2+}-ATPase begins. During subsequent development, the contribution of Ca^{2+} channels to the action potential sharply decreases, and the Ca^{2+}-ATPase concentration reaches a steady level. Ca^{2+} influx connected with muscle activity may also contribute to hypertrophy even in mature muscles.

4. Much of the increase in the Ca^{2+}-ATPase content of chicken muscle occurs around and after the time of hatching and coincides with the onset of muscle activity (Martonosi *et al.*, 1977; 1980). It may not be farfetched to assume that the Ca^{2+} pulses in the cytoplasm connected with nerve activity promote the synthesis and accumulation of Ca^{2+}-ATPase in sarcoplasmic reticulum. In line with this suggestion, the maximum steady-state concentration of Ca^{2+}-transport ATPase in sarcoplasmic reticulum *in vivo* is about 3–10 times greater than in cultured muscle fibers of similar age (Martonosi *et al.*, 1977). Chronic denervation of muscle decreases the Ca^{2+} transport activity of sarcoplasmic reticulum and produces muscle atrophy (for review see Martonosi, 1972). Cross-innervation of muscle changes both the amount and the composition of sarcoplasmic reticulum (Heilmann and Pette, 1979), and electrical stimulation of denervated muscle produces similar effects.

Repeated injections of curare into the chorioallantoic sacs of chicken embryos between days 8 and 14 of development markedly reduce the accumulation of Ca^{2+}-ATPase and other muscle proteins (Roufa and Martonosi, 1981).

While innervation and muscle activity apparently promote the synthesis and accumulation of several muscle-specific proteins, the extrajunctional acetylcholine receptor shows the opposite behavior; its density is low in the surface membranes of innervated muscle, and increases 20- to 30-fold after functional or surgical denervation. Electrical stimulation of muscle following denervation reduces or eliminates this increase in extrajunctional acetylcholine sensitivity (Fambrough, 1979; Edwards, 1979). Experiments are in progress in muscle cell cultures to test the involvement of Ca^{2+} in the regulation of acetylcholine receptor synthesis.

5. THE PROBLEM OF SPECIFICITY

There is a correlation between the speed of contraction and relaxation and the Ca^{2+} transport activity of sarcoplasmic reticulum in fast and slow skeletal, cardiac, and smooth muscles (Martonosi, 1972; Heilmann and Pette, 1979; Wang *et al.*, 1979). Some of these differences are only quantitative as the relative amount of sarcoplasmic reticulum and Ca^{2+}-transport ATPase is less in slow than in fast muscles. There are indications, however, that muscles of different phenotypes contain distinct isoenzymes of the Ca^{2+}-transport ATPase and myosin. Cross-innervation experiments indicate that neural influence selectively promotes the synthesis of certain phenotypically specific isoenzymes. The important regulatory influence appears to be the stimulus frequency, for in chronically stimulated denervated muscles the synthesis of "slow" or "fast" isoenzymes can be selectively promoted by imitating the firing pattern of the slow and fast nerves, respectively (Heilmann *et al.*, 1981). These observations suggest that the cells translate specific stimulus frequencies into selective metabolic signals for the synthesis of the two isoenzymes.

Each electric stimulus causes influx of Ca^{2+} into muscle cells and release of Ca^{2+} from intracellular sources, followed by a return to resting Ca^{2+} levels, which is complete at low stimulus frequencies. As the stimulus frequency increases, the cytoplasmic Ca^{2+} pulses begin to fuse and the cytoplasmic Ca^{2+} concentration rises to higher average steady-state levels.

Discrimination is possible if the Ca^{2+} receptors regulating the expression of the slow and fast isoenzymes have different Ca^{2+} affinities. In addition, there may be a stimulus-frequency-dependent discrimination based on the response time of the Ca^{2+} receptors (for example, the rate of phosphorylation and dephosphorylation of histones, or the rates of synthesis, processing, and degradation of mRNA, etc.) in comparison with the duration and frequency of Ca^{2+} pulses.

Essentially no information is available on the possible involvement of Ca^{2+} in signal processing of this type.

REFERENCES

Ashburner, M., and Bonner, J. J. (1979). *Cell* **17,** 241–254.
Barth, L. G., and Barth, L. J. (1974). *Dev. Biol.* **39,** 1–22.
Bischoff, R. (1978). In *Cell Surface Reviews* (G. Poste and G. L. Nicolson, eds.), Vol. 5, pp. 127–179, North-Holland, Amsterdam.
Blinks, J. R., Prendergast, F. G., and Allen, D. G. (1976). *Pharmacol. Rev.* **28,** 1–93.
Boland, R., Martonosi, A., and Tillack, T. W. (1974). *J. Biol. Chem.* **249,** 612–623.
Caswell, A. H. (1979). *Int. Rev. Cytol.* **56,** 145–181.
Chyn, T., Martonosi, A., Morimoto, T., and Sabatini, D. (1979). *Proc. Natl. Acad. Sci. USA* **76,** 1241–1245.

Edwards, C. (1979). *Neuroscience* **4**, 565–584.

Fambrough, D. M. (1979). *Physiol. Rev.* **59**, 165–227.

Freedman, M. H., Raff, M. C., and Gomperts, B. (1975). *Nature (London)* **255**, 378–382.

Gallien-Lartigue, O. (1976). *Cell Tissue Kinet.* **9**, 533–540.

Gilkey, J. C., Jaffe, L. F., Ridgway, E. B., and Reynolds, G. T. (1978). *J. Cell Biol.* **76**, 448–466.

Glover, C. V. C., Vavra, K. J., Guttman, D. S., and Gorovsky, M. A. (1981). *Cell* **23**, 73–77.

Greene, W. C., Parker, C. M., and Parker, C. W. (1976). *Cell. Immunol.* **25**, 74–89.

Ha, D. B., Boland, R., and Martonosi, A. (1979). *Biochim. Biophys. Acta* **585**, 165–187.

Heilmann, C., and Pette, D. (1979). *Eur. J. Biochem.* **93**, 437–446.

Heilmann, C., Muller, W., and Pette, D. (1981). *J. Membr. Biol.* **59**, 143–149.

Hesketh, T. R., Smith, G. A., Houslay, M. D., Warren, G. B., and Metcalfe, J. C. (1977). *Nature (London)* **267**, 490–494.

Kelley, P. M., Aliperti, G., and Schlesinger, M. J. (1980). *J. Biol. Chem.* **255**, 3230–3233.

LePeuch, C. J., Ferraz, C., Walsh, M. P., Demaille, J. G., and Fisher, E. H. (1979). *Biochemistry* **18**, 5267–5273.

Lewis, M., Helmsing, P. J., and Ashburner, M. (1975). *Proc. Natl. Acad. Sci. USA* **72**, 3604–3608.

Marban, E., Rink, T. J., Tsien, R. W., and Tsien, R. Y. (1980). *Nature (London)* **286**, 845–850.

Martonosi, A. (1972). *Curr. Top. Membr. Transp.* **3**, 83–197.

Martonosi, A. (1975). *Biochim. Biophys. Acta* **415**, 311–333.

Martonosi, A. (1980). *Fed. Proc.* **39**, 2401–2402.

Martonosi, A., Boland, R., and Halpin, R. A. (1972). *Cold Spring Harbor Symp. Quant. Biol.* **37**, 455–468.

Martonosi, A., Roufa, D., Boland, R., Reyes, E., and Tillack, T. W. (1977). *J. Biol. Chem.* **252**, 318–332.

Martonosi, A., Chyn, T. L., and Schibeci, A. (1978). *Ann. N.Y. Acad. Sci.* **307**, 148–159.

Martonosi, A., Roufa, D., Ha, D. B., and Boland, R. (1980). *Fed. Proc.* **39**, 2415–2421.

Roufa, D., and Martonosi, A. (1980). *Fed. Proc.* **39**, 954.

Roufa, D., and Martonosi, A. (1981). *Biochem. Pharmacol.* **30**, 1501–1505.

Roufa, D., Wu, F. S., and Martonosi, A. (1981). *Biochim. Biophys. Acta,* **674**, 225–237.

Schibeci, A., and Martonosi, A. (1980a). *Eur. J. Biochem.* **113**, 5–14.

Schibeci, A., and Martonosi, A. (1980b). *Anal. Biochem.* **104**, 335–342.

Shainberg, A., Yagil, G., and Jaffe, D. (1969). *Exp. Cell Res.* **58**, 163–167.

Somlyo, A. P., Shuman, H., and Somlyo, A. V. (1978). In *Frontiers of Biological Energetics* (P. L. Dutton, J. S. Leigh, and A. Scarpa, eds.), Vol. 1, pp. 742–751, Academic Press, New York.

Spitzer, N. C. (1979). *Annu. Rev. Neurosci.* **2**, 363–397.

Squire, J. F. (1981). *The Structural Basis of Muscular Contraction,* Plenum Press, New York.

Tillack, T. W., Boland, R., and Martonosi, A. (1974). *J. Biol. Chem.* **249**, 624–633.

Wang, T., Grassi de Gende, A. O., and Schwartz, A. (1979). *J. Biol. Chem.* **254**, 10675–10678.

Whitlock, J. P., Jr., Augustine, R., and Schulman, H. (1980). *Nature (London)* **287**, 74–76.

Winegrad, S. (1971). *J. Gen. Physiol.* **58**, 71–93.

Wu, F. S., Park, Y.-C., Roufa, D., and Martonosi, A. (1981). *J. Biol. Chem.* **256**, 5309–5312.

Zala, C. A., Salas-Prato, M., Yan, W.-T., Banjo, B., and Perdue, J. F. (1980). *Can. J. Biochem.* **58**, 1179–1188.

78

Active Calcium Transport in Human Red Blood Cells

H. J. Schatzmann

Human red blood cells are able to extrude Ca^{2+} across the plasma membrane against a gradient of 10^3 to 10^4 by a pump mechanism that directly depends on ATP hydrolysis and requires Mg^{2+} inside the cell. The Ca pump has also been found in red cells of dogs, pigs, cattle, and birds. The interest in this entity is justified by the discovery of a similar (or perhaps even identical) system in many, more highly organized cells such as nerve (Di Polo, 1978; Di Polo and Beaugé, 1979; Beaugé et al., 1981), cardiac muscle (Trumble et al., 1980; Caroni and Carafoli, 1980), kidney tubule (Gmaj et al., 1979), liver (van Rossum, 1970), and L cells (Lamb and Lindsay, 1971).

The absence of internal structures and of the Na–Ca exchange system (Porzig, 1970; Ferreira and Lew, 1977) makes the red cell a very convenient object to study the plasma membrane Ca^{2+} pump. There is no dearth in recent and extensive reviews on the subject (Schatzmann, 1975; Vincenzi and Hinds, 1976; Sarkadi and Tosteson, 1979; Roufogalis, 1979; Schatzmann and Bürgin, 1978; Sarkadi, 1980; Vincenzi and Hinds, 1981; Schatzmann, 1982).

The system transporting Ca^{2+} uphill is a protein of 130,000–150,000 molecular weight (see below) of which there are probably not more than 1000 copies per cell, amounting to something like 0.1% of the membrane proteins (Rega and Garrahan, 1975; Schatzmann, 1981). The affinity for Ca^{2+} (buffered with EGTA) at the inner surface is high ($K_{Ca} \sim 10^{-6} M$) (Schatzmann, 1973) if the protein is associated with the activator calmodulin,* which is present in the cytosol in excess over the pump sites in the membrane (Bond and Clough, 1973; Gopinath and Vincenzi, 1977; Vincenzi and Larsen, 1980). Removal of calmodulin

---

*Calmodulin is an ubiquitous Ca^{2+}-binding protein of molecular weight 16,700 with an isoelectric point of 3.5–4.5 (Jarrett and Penniston, 1978). It has four Ca^{2+}/Mg^{2+}-binding sites ($K_{Ca} \sim 10^{-6} M$) at least two of which must be occupied by Ca^{2+} for binding to the pump protein to occur (Wolff et al., 1977; Scharff, 1980; Vincenzi and Hinds, 1981). At saturating Ca^{2+} concentration, the K_{diss} for the pump–calmodulin complex is of the order of 10 nM (Jarrett and Kyte, 1979).

H. J. Schatzmann • Department of Veterinary Pharmacology, University of Bern, 3012 Bern, Switzerland.

shifts the K_{Ca} upwards by a factor of 10 or more (Sarkadi *et al.*, 1979; Scharff and Foder, 1978).

Sarkadi *et al.* (1981) removed by tryptic digestion a calmodulin-accepting peptide from the pump with the consequence that calmodulin neither bound nor was required for full transport activity. This is one reason among others to believe that it is not Ca^{2+} attached to calmodulin which is transported.

In *intact red cells* (without artificial Ca^{2+}-buffering), the apparent K_{Ca} at the inner surface was found to be 10^{-4} to 10^{-3} M (Ferreira and Lew, 1976). This may mean that there are calmodulin inhibitors present in the cell sap (Au, 1978; Sarkadi *et al.*, 1980a) or that the pump site for Ca^{2+} is not fully accessible (Schatzmann, 1982).

In order to translocate a Ca^{2+} ion through the hydrophobic region, against a Ca^{2+} gradient and the membrane potential, the protein must undergo a conformational change moving Ca^{2+} in space and simultaneously or afterwards its affinity for Ca^{2+} must drop. The K_{Ca} in fact is more than 10^{-3} M at the outer surface (Schatzmann, 1973). If the Ca^{2+} gradient from outside to inside is made large and the ATP/(ADP$\cdot P_i$) ratio sufficiently small, the system can be run backwards, forming ATP due to the running down of the Ca^{2+} gradient (Rossi *et al.*, 1978; Wüthrich *et al.*, 1979).

At saturating Ca^{2+} concentration the transport rate is of the order of 10 mmoles/(liter cells)·hr. The passive Ca^{2+} leak under the physiological gradient is less than 10 μmoles/(liter cells)·hr. Therefore, in intact cells the $[Ca^{2+}]_i$ is less than 10^{-6} M, a very small fraction of the pump capacity is used, the contribution of the Ca^{2+} transport to energy consumption by the cell is negligible, and calmodulin may partly uncouple from the pump, which possibly provides a mechanism for keeping internal Ca^{2+} within a narrow range (Scharff, 1980; Scharff and Foder, 1978).

The only other cation transported is Sr^{2+} (Schatzmann, 1975; Sarkadi *et al.*, 1978), but Ba, Mn, Zn, Ni, Co, Cu may somewhat stimulate the ATPase in the presence of Mg^{2+} (Pfleger and Wolf, 1975). Substrate specificity is high; transport can be run on ATP and less well on ITP or UTP (Sarkadi *et al.*, 1980a) but not on *p*-nitrophenylphosphate (which is slowly hydrolyzed in the presence of ATP and inhibits the ATPase) or acetylphosphate.

Ca^{2+} transport requires Mg^{2+} at the inner surface (Schatzmann, 1969), but Mg^{2+} is not translocated (Schatzmann, 1975). Nor is its role in providing MgATP as substrate, for free ATP can be used (see below). High internal Ca^{2+} ($>10^{-4}$ M) is inhibitory, possibly by competition with Mg^{2+}. The rate of Ca^{2+} transport observed in choline or Tris media is increased by K^+ or Na^+ (Sarkadi *et al.*, 1978).

The properties of the membrane ATPase requiring the simultaneous presence of Ca^{2+} and Mg^{2+} tally with those of the Ca^{2+} transport. With calmodulin, K_{Ca} is about 1 μM and K_{Mg} is 14 μM (for free Mg^{2+}) (Schatzmann, unpublished). The Ca^{2+}/Mg^{2+}-ATPase shows additional stimulation by K^+ or Na^+ ($K_K \sim 6$ mM, $K_{Na} \sim 33$ mM) (Schatzmann and Rossi, 1971; Bond and Green, 1971). Depletion of calmodulin (see above) reduces its Ca^{2+} affinity by a factor of more than 10 (Scharff and Foder, 1978). The ATP-accepting site is clearly on the internal side and P_i is liberated within the cell (Schatzmann and Roelofsen, 1977).

There is evidence that two Ca^{2+} ions are involved in the reaction cycle (Ferreira and Lew, 1976; Schatzmann and Roelofsen, 1977). Sarkadi *et al.* (1979) made the interesting suggestion that only one of the Ca^{2+} sites is influenced by calmodulin and that this site also accepts the Ca–EGTA complex [the latter assumption is meant to explain why in unfrozen membrane preparations the apparent Ca^{2+} affinity is 100 times higher in Ca–EGTA buffers than in $CaCl_2$ (Schatzmann, 1973; Sarkadi *et al.*, 1979), but this might also be explained by a hindrance of Ca^{2+} access by positive charges in the unfrozen membrane

(Schatzmann, 1982)]. The existence of two Ca^{2+} sites does not necessarily imply that the stoichiometric ratio (Ca^{2+} transported per ATP cleaved) is 2. Direct measurement gives the figure 1 (Schatzmann, 1973; Schatzmann and Roelofsen, 1977). However, the fraction inhibitable by externally applied La^{3+} has a ratio of 2 (Quist and Roufogalis, 1975; Sarkadi *et al.*, 1977), and the issue is further complicated by the finding that the ratio may depend on the Ca^{2+} concentration, increasing from below 1 to 2 with rising Ca^{2+} concentration (Sarkadi, 1980). Thus, the problem of stoichiometry is unsettled.

The protein spans the membrane and must be anchored in the membrane by hydrophobic interaction in its middle portion. Accordingly, there is a definite requirement for phospholipids (see Chapter 79). ATPase activity is abolished by enzymatic degradation of glycerophospholipids in the inner membrane leaflet, whereas removal of outer-leaflet phospholipids, including sphingomyelin, is without consequence (Roelofsen and Schatzmann, 1977). There is reason to believe that phosphatidylserine (PS) dispenses with the need for calmodulin in the isolated protein (Stieger and Luterbacher, 1981; Niggli *et al.*, 1981a,b; see below). If this reflects the *in situ* behavior, the conclusion is that PS is not normally surrounding the protein in the membrane.

The system undergoes rapid cyclic phosphorylation–dephosphorylation as demonstrated by the use of [γ-^{32}P]—ATP (Knauf *et al.*, 1974; Rega and Garrahan, 1975; Schatzmann and Bürgin, 1978; Niggli *et al.*, 1979; Szasz *et al.*, 1978). It is thought that this two-step reaction is the pathway of ATP hydrolysis by the system. It is an intrinsic property of the pump protein and no kinase or hydrolase is involved. Phosphorylation is only possible if Ca^{2+} is present. It proceeds in the absence of Mg^{2+} (Knauf *et al.*, 1974; Rega and Garrahan, 1975; Szasz *et al.*, 1978; Schatzmann and Bürgin, 1978), but Mg^{2+} accelerates the reaction elicited by Ca^{2+} (Garrahan and Rega, 1978). However, the important action of Mg^{2+} is to bring about a conformational change between a first ($E_1 \sim P$) and a second ($E_2 \sim P$) form of the phosphoprotein (Rega and Garrahan, 1975; Garrahan and Rega, 1978). Only $E_2 \sim P$ can react with water, and its hydrolysis is accelerated by ATP at concentrations exceeding those required at the enzymatic site (Garrahan and Rega, 1978). The double role of ATP reveals itself in biphasic curves relating ATPase to ATP concentration ($K_{m1} = 1$–4 μM, $K_{m2} = 120$–180 μM) (Richards *et al.*, 1978; Mualem and Karlish, 1979; Stieger and Luterbacher, 1981). Phosphorylation occurs at an acyl group [the bond is acid stable and can be decomposed by hydroxylamine (Rega and Garrahan, 1975; Wolf *et al.*, 1977; Lichtner and Wolf, 1980b)]. The point of attack of K (and Na) seems to be on the side of dephosphorylation (Enyedi *et al.*, 1980). Calmodulin accelerates the turnover (Rega and Garrahan, 1980) possibly by speeding up the $E_2 \rightarrow E_1$ transition (Mualem and Karlish, 1980), which is compatible with its increasing the apparent Ca^{2+} affinity.

Attempts at isolation of the protein have been successful in overcoming the low abundance and the extreme lability after disintegration of the membrane. Wolf and his associates (Wolf *et al.*, 1977; Lichtner and Wolf, 1980a,b), using gel filtration in the presence of lecithin of detergent-solubilized membranes, obtained a mixture of only three proteins associated with lecithin–detergent micelles. One of them of molecular weight 130,000–150,000 could be phosphorylated in a strictly Ca^{2+}-dependent way. The material was a Ca^{2+}/Mg^{2+}-activated ATPase with all the characteristics expected from previous knowledge about the Ca^{2+} transport protein. A much higher purification is achieved by calmodulin-affinity chromatography. Niggli *et al.* (1979) and Gietzen *et al.* (1980b) passed detergent-solubilized membranes over a column with fixed calmodulin. In the presence of Ca^{2+}, all proteins except the Ca pump were eluted. Next, the pump protein was recovered from the column in phospholipid–detergent micelles with EGTA (or EDTA) solution. In SDS gel electrophoresis, this material appears as a single peak at approximately 130,000 molecular weight or occasionally as two peaks. The heavier protein might be a dimer. The single protein thus obtained is phosphorylatable in the presence of Ca^{2+} and is a Ca^{2+}/Mg^{2+}-ATPase with a K_{Ca} of 3 μM (when a saturating amount of calmodulin is added), a K_{Mg} of 35 μM, and two K_m values for (total) ATP of 3.5 μM and 120 μM (Stieger and Luterbacher, 1981). It is activated by calmodulin if assembled with lecithin but fully active and refractory to calmodulin in phosphatidylserine (Luterbacher and Stieger, 1981; Carafoli and Niggli, 1981; Niggli *et al.*, 1981a,b).

The crowning of these efforts was the incorporation of the purified material into artificial lipid vesicles, which allows Ca^{2+} transport to be observed when ATP is added to the medium. Haaker and Racker (1979) achieved this with material from pig red cells purified essentially according to Wolf. The vesicles took up Ca^{2+} against a gradient. The ionophore A23187 released the accumulated Ca^{2+} and stimulated ATPase activity as expected. Carafoli and Niggli (1981) were successful with Niggli's isolated protein. Here again, addition of ATP to the medium led to a considerable deviation from equilibrium distribution of Ca^{2+}, which was reversed by A23187. These experiments prove beyond reasonable doubt that the 130,000-molecular-weight protein with its Ca^{2+}/Mg^{2+}-ATPase function, its phosphorylated intermediate, and its calmodulin sensitivity is the Ca pump.

Many inhibitors have been described, none of which seem to be very specific. They are sulfhydryl reagents, ruthenium red, quercetin, phloretin, suramin, and vanadate (see Schatzmann, 1975, 1981). Phenothiazine neuroleptics (Schatzmann, 1969; Raess and Vincenzi, 1981; Gietzen *et al.*, 1980a) and even more so vinblastine (Gietzen and Bader, 1980) are interesting in affecting preferentially the calmodulin stimulation (possibly by reacting with calmodulin), and La^{3+} in being active also from outside but with lower affinity than when applied inside (Szasz *et al.*, 1978).

REFERENCES

Au, K. S. (1978). *Int. J. Biochem.* **9**, 477–480.
Beaugé, L., Di Polo, R., Osses, L., Barnola, F., and Campos, M. (1981). *Biochim. Biophys. Acta,* **644**, 147–152.
Bond, G. H., and Clough, D. L. (1973). *Biochim. Biophys. Acta* **323**, 592–599.
Bond, G. H., and Green, J. W. (1971). *Biochim. Biophys. Acta* **241**, 393–398.
Carafoli, E., Niggli, V., and Penniston, J. T. (1981). *Ann. N.Y. Acad. Sci.,* **358**, 159–168.
Caroni, P., and Carafoli, E. (1980). *Nature (London)* **283**, 765–767.
Di Polo, R. (1978). *Nature (London)* **274**, 390–392.
Di Polo. R., and Beaugé, L. (1979). *Nature (London)* **278**, 271–273.
Enyedi, A., Szasz, I., Sarkadi, B., Bot, B., and Gardos, G. (1980). In *Proceedings IUPS, 28th International Congress,* Vol. 14, Abstract 1333.
Ferreira, H. G., and Lew, V. L. (1976). *Nature (London)* **259**, 47–49.
Ferreira, H. G., and Lew, V. L. (1977). In *Membrane Transport in Red Cells* (J. C. Ellory and V. L. Lew, eds.) pp. 53–91, Academic Press, New York.
Garrahan, P. J., and Rega, A. F. (1978). *Biochim. Biophys Acta* **513**, 59–65.
Gietzen, K., and Bader, H. (1980). *IRCS Med. Sci. Biochem.* **8**, 396–397.
Gietzen, K., Mansard, A., and Bader, H. (1980a). *Biochem. Biophys. Res. Commun.* **94**, 674–681.
Gietzen, K., Tejcka, M., and Wolf, H. U. (1980b). *Biochem. J.* **189**, 81–88.
Gmaj, P., Murer, H., and Kinne, R. (1979). *Biochem. J.* **178**, 549–557.
Gopinath, R. M., and Vincenzi, F. F. (1977). *Biochem. Biophys. Res. Commun.* **77**, 1203–1209.
Haaker, H., and Racker, E. (1979). *J. Biol. Chem.* **254**, 6598–6602.
Jarrett, H. W., and Kyte, J. (1979). *J. Biol. Chem.* **254**, 8237–8244.
Jarrett, H. W., and Penniston, J. T. (1978). *J. Biol. Chem.* **253**, 4676–4682.
Knauf, P. A., Proverbio, F., and Hoffman, J. F. (1974). *J. Gen. Physiol.* **63**, 324–336.
Lamb, J. F., and Lindsay, R. (1971). *J. Physiol.* **218**, 691–708.

Lichtner, R., and Wolf, H U. (1980a). *Biochim. Biophys. Acta* **598,** 472–485.

Lichtner, R., and Wolf, H. U. (1980b). *Biochim. Biophys. Acta* **598,** 486–493.

Mualem, S., and Karlish, S. J. D. (1979). *Nature (London)* **277,** 238–240.

Mualem, S., and Karlish, S. J. D. (1980). *Biochim. Biophys. Acta* **597,** 631–636.

Niggli, V., Penniston, J. T., and Carafoli, E. (1979). *J. Biol. Chem.* **254,** 9955–9958.

Niggli, V., Adunyali, E. S., Penniston, J. T., and Carafoli, E. (1981a). *J. Biol. Chem.* **256,** 395–401.

Niggli, V., Adunyali, E. S., and Carafoli, E. (1981b). *J. Biol. Chem.* **256,** 8588–8592.

Pfleger, H., and Wolf, H. U. (1975). *Biochem. J.* **147,** 359–361.

Porzig, H. (1970). *J. Membr. Biol.* **2,** 324–339.

Quist, E. E., and Roufogalis, B. D. (1975). *FEBS Lett.* **50,** 135–139.

Raess, B. U., and Vincenzi, F. F. (1981). *J. Mol. Pharmacol.*, **18,** 253–258.

Rega, A. F., and Garrahan, P. J. (1975). *J. Membr. Biol.* **22,** 313–327.

Rega, A. F., and Garrahan, P. J. (1980). *Biochim. Biophys. Acta* **596,** 487–489.

Richards, D. E., Rega, A. F., and Garrahan, P. J. (1978). *Biochim. Biophys. Acta* **511,** 194–201.

Roelofsen, B., and Schatzmann, H. J. (1977). *Biochim. Biophys. Acta* **464,** 17–36.

Rossi, J. P. F. C., Garrahan, P. J., and Rega, A. F. (1978). *J. Membr. Biol.* **44,** 37–46.

Roufogalis, B. D. (1979). *Can. J. Physiol. Pharmacol.* **57,** 1332–1349.

Sarkadi, B. (1980). *Biochim. Biophys. Acta* **604,** 159–190.

Sarkadi, B., and Tosteson, D. C. (1979). In *Transport across Single Biological Membranes* (G. Giebisch, D. C. Tosteson, and H. H. Ussing, eds.), pp. 117–160, Springer-Verlag, Berlin.

Sarkadi, B., Szasz, I., Gerloczy, A., and Gardos, G. (1977). *Biochim. Biophys. Acta* **464,** 93–107.

Sarkadi, B., MacIntyre, J. D., and Gardos, G. (1978). *FEBS Lett.* **89,** 78–82.

Sarkadi, B., Schubert, A., and Gardos, G. (1979). *Experientia* **35,** 1045–1047.

Sarkadi, B., Szasz, I., and Gardos, G. (1980a). *Biochim. Biophys. Acta* **598,** 326–338.

Sarkadi, B., Enyedi, A., and Gardos, G. (1980b). *Cell Calcium* **1,** 287–297.

Scharff, O. (1980). In *A. Benzon Symposium* (U. V. Lassen, H. H. Ussing, and J. O. Wieth, eds.), Vol. 14, pp. 236–254, Munksgaard, Copenhagen.

Scharff, O., and Foder, B. (1978). *Biochim. Biophys. Acta* **509,** 67–77.

Schatzmann, H. J. (1969). In *Symposium, Calcium and Cellular Function* (A. W. Cuthbert, ed.), pp. 85–95, Macmillan, London.

Schatzmann, H. J. (1973). *J. Physiol.* **235,** 551–569.

Schatzmann, H. J. (1975). *Curr. Top. Membr. Transp.* **6,** 125–168.

Schatzmann, H. J. (1981). In *Calcium Membrane Transport* (E. Carafoli, ed.), pp. 41–108, Academic Press, New York.

Schatzmann, H. J., and Bürgin, H. (1978). *Ann. N.Y. Acad. Sci.* **307,** 125–147.

Schatzmann, H. J., and Roelofsen, B. (1977). In *Biochemistry of Membrane Transport* (G. Semenza and E. Carofoli, eds.), pp. 389–400, Springer-Verlag, Berlin.

Schatzmann, H. J., and Rossi, G. L. (1971). *Biochim. Biophys. Acta* **241,** 379–392.

Stieger, J., and Luterbacher, S. (1981). *Biochim. Biophys. Acta,* **641,** 270–275.

Szasz, I., Hasitz, M., Sarkadi, B., and Gardos, G. (1978). *Mol. Cell. Biochem.* **22,** 147–152.

Trumble, W. R., Sutko, J. L., and Reeves, J. P. (1980). *Life Sci.* **27,** 207–214.

van Rossum, G. D. V. (1970). *J. Gen. Physiol.* **55,** 18–32.

Vincenzi, F. F., and Hinds, Th. R. (1976). In *The Enzymes of Biological Membranes* (A. Martonosi, ed.), Vol. 3, pp. 261–281, Plenum Press, New York.

Vincenzi, F. F., and Hinds, Th. R. (1981). In *Calcium and Cell Function* (W. Y. Cheung, ed.), Vol, 1, pp. 127–165. Academic Press, New York.

Vincenzi, F. F., and Larsen, F. L. (1980). *Fed. Proc.* **39,** 2427–2431.

Wolf, H. U., Dieckvoss, G., and Lichtner, R. (1977). *Acta Biol. Med. German.* **36,** 847–858.

Wolff, D. J., Poirier, P. G., Brostrom, C. D., and Brostrom, M. A. (1977). *J. Biol. Chem.* **252,** 4108–4117.

Wüthrich, A., Schatzmann, H. J., and Romero, P. (1979). *Experientia* **35,** 1589–1590.

79

The Lipid Requirement of the Ca^{2+}/Mg^{2+}-ATPase in the Erythrocyte Membrane

Ben Roelofsen

The Ca^{2+}/Mg^{2+}-ATPase found in erythrocyte membranes consists of a 140,000-molecular-weight protein that spans the lipid bilayer. This system has been shown to be identical to the "Ca pump," which maintains an outward-directed 10,000-fold Ca^{2+} gradient over the membrane (Schatzmann and Bürgin, 1978; Roufogalis, 1979).

Unlike other ATPase systems, studies on the lipid requirement of the Ca^{2+}/Mg^{2+}-ATPase in the erythrocyte membrane have been undertaken only recently. Coleman and Bramley (1975) observed that treatment of erythrocyte ghosts with a partially purified *C. welchii* phospholipase C caused a complete loss of activity when approximately 60% of the total phospholipids had been degraded. Reactivation by a lipid mixture derived from ox liver could only be achieved when a limited concentration of the phospholipase C had been used, thus resulting in a merely partial inactivation. Activity could also be restored by the addition of lysolecithin. Ronner *et al.* (1977) observed a slight activation of the enzyme when ghosts were treated with *N. naja* phospholipase A_2 in the absence of bovine serum albumin. However, the incubation had been done in the presence of only 0.1 mM $CaC1_2$, which is two orders of magnitude lower than the optimal concentration to activate this particular phospholipase A_2. Therefore, their results may be in agreement with a recent observation of Taverna and Hanahan (1980), showing that such an activation may occur when a very low concentration of free fatty acids (comprising about 1% of the total membrane fatty acids) is generated in the membrane. A complete loss of Ca^{2+}/Mg^{2+}-ATPase activity is observed when ghosts are extensively treated either with phospholipase A_2 in the presence of bovine serum albumin (Roelofsen and Schatzmann, 1977; Ronner *et al.*,

---

*The earlier observation of Coleman and Bramley (1975) that Ca^{2+}/Mg^{2+}-ATPase activity was completely abolished by treatment of ghosts with *C. welchii* phospholipase C has to be ascribed to the presence of theta toxin in the partially purified preparation they used (Roelofsen, 1977).

Ben Roelofsen • Laboratory of Biochemistry, State University of Utrecht, Transitorium 3, NL-3584 CH Utrecht, The Netherlands.

1977; Richards *et al.*, 1977a) or with *B. cereus* phospholipase C (Roelofsen and Schatz-mann, 1977; Richards *et al.*, 1977a). Both these treatments leave the sphingomyelin—comprising 25% of the total phospholipid—unaffected, indicating that this particular lipid does not play a role in maintaining the ATPase activity. This conclusion is supported by the observation that a complete and selective degradation of this phospholipid by treatment of ghosts with *S. aureus* sphingomyelinase C has no effect on the activity, whereas sphin-gomyelin is absolutely unable to reactivate the system upon complete delipidation (Roelof-sen and Schatzmann, 1977).

Richards *et al.* (1977a) studied the effect of phospholipases on the individual steps comprising the overall Ca^{2+}/Mg^{2+}-ATPase reaction in human erythrocyte ghosts. Pan-creatic phospholipase A_2 caused an inhibition of both phosphorylation and dephosphoryla-tion of the system. Quite interesting, however, treatment with *B. cereus* phospholipase C, although effectively blocking the overall ATPase reaction, was shown to cause an enhance-ment of the Ca^{2+}-dependent phosphatase activity. This led the authors to conclude that the removal of the polar head groups of the glycerophospholipids specifically inhibited the phosphorylation step. It is also of interest to note that comparable observations were made regarding the effect of this particular phospholipase C on the Na^+/K^+-ATPase system in the erythrocyte membrane (Richards *et al.*, 1977b). Furthermore, it is worthwhile mention-ing that this effect of the phospholipase C treatment seems to be unique for the system in the erythrocyte membrane, as the phosphatase reaction of the Ca^{2+}/Mg^{2+}-ATPase in sar-coplasmic reticulum is completely inhibited by the action of this particular phospholipase (Richards *et al.*, 1977a).

Detailed analyses of the results obtained by treatment of open ghosts with different concentrations of highly purified *B. cereus* phospholipase C in conjunction with informa-tion concerning the transbilayer distribution of each phospholipid class, have shown that a direct proportionality exists between the percentage residual *glycero*phospholipid in the *inner* monolayer of the membrane and the percentage residual Ca^{2+}/Mg^{2+}-ATPase activity (Roelofsen and Schatzmann, 1977). The observation that phosphatidylcholine and phos-phatidylethanolamine are degraded much faster than phosphatidylserine by this particular phospholipase C implies that the Ca^{2+}/Mg^{2+}-ATPase activity is equally well maintained by phosphatidylserine as it is by phosphatidylcholine and phosphatidylethanolamine. This is supported by the fact that extensive treatment of the ghosts with highly purified *C. welchii* phospholipase C, leaving phosphatidylserine as the only residual phospholipid, results in 36% residual activity (Roelofsen and Schatzmann, 1977).* Phosphatidylserine is known to comprise 35% of the total *glycero*phospholipid complement of the *inner* monolayer. The hypothesis that this particular ATPase system is only dependent upon the (gly-cero)phospholipids that form part of the *inner* half of the membrane is confirmed by the observation that a complete and selective degradation of the phospholipids in the *outer* monolayer does not have any effect on the Ca^{2+}/Mg^{2+}-ATPase activity (Roelofsen and Schatzmann, 1977). The surprising linear relationship between the amount of intact gly-cerophospholipids in the inner-membrane leaflet and the level of Ca^{2+}/Mg^{2+}-ATPase activ-ity suggests a "lateral long-range effect" of the lipid environment on this ATPase, rather than any kind of boundary or annular lipid. Consistent with the involvement of all three of the glycerophospholipids in this activity is the fact that after complete delipidation (residual activity $<5\%$), the system can be reactivated by either phosphatidylcholine, phosphatidyl-ethanolamine, or phosphatidylserine. That Ronner *et al.* (1977) could not reactivate the phospholipase A_2 plus bovine serum albumin delipidated sysem with phosphatidylcholine or phosphatidylethanolamine may be due to the fact that in their experiments sonicated lipid dispersions and delipidated membranes were simply added together, whereas a subsequent freeze-thawing step has appeared to be essential for proper reactivation (Roelofsen and Schatzmann, 1977). Reactivation could also be achieved by lysolecithin, but not by free fatty acids. On the contrary, free fatty acids partly inhibit the reactivation by lysolecithin, which explains the considerable (but not complete!) loss of activity by treatment of ghosts with phospholipase A_2 in the absence of bovine serum albumin. Contrastingly, Ronner *et al.* (1977) observed a very effective reactivation by oleic acid, whereas lysolecithin was almost ineffective. Although a particular reason for this discrepancy is not immediately obvious, it may be mentioned that those authors did not inhibit the phospholipase A_2 (used for the delipidation) prior to the extraction of the lipids, which makes the real extent of delipidation uncertain. Furthermore, they used an incubation medium that, in particular when the phospholipid degradation is only limited, is highly favorable for resealing of the ghosts. Exposure of latent activity may also have been the reason for the activation by Triton X-100 that they observed. Moreover, the conclusion that Triton X-100 is able to achieve a genuine reactivation is in conflict with the statement made by the same authors that the Ca^{2+}/Mg^{2+}-ATPase of the erythrocyte membrane has a specific requirement for negatively charged phospholipids, in particular phosphatidylserine (Ronner *et al.*, 1977). In further studies by this group, the ATPase was solubilized by treatment of ghosts with 0.3% Triton X-100 (Peterson *et al.*, 1978). Such treatment, however, not only solubilizes the intrinsic membrane proteins, including the ATPase, but also a great deal of the membrane (phospho)lipids. Therefore, the observed decrease in activity when the Triton is subsequently removed from the supernatant by adsorption to Bio-Beads might have been due to the formation of resealed vesicles, rather than to the removal of the "activator." In the same paper, Peterson *et al.* (1978) interpreted the results of their reconstitution/resolubilization experiments with the purified ATPase and various phospholipids to support the specific requirement of the enzyme for phosphatidylserine, although those results (at least to some extent) can equally well indicate that the existence of such a specificity is doubtful. Haaker and Racker (1979) reported a functionally successful incorporation of the Ca^{2+}/Mg^{2+}-ATPase, purified from pig erythrocytes, into phospholipid vesicles. However, because crude soybean phospholipids were used in these reconstitution studies, they could not give any information of the specificity of its lipid requirement. Gietzen *et al.* (1980a) recently solubilized the Ca^{2+}/Mg^{2+}-ATPase from human erythrocyte ghosts using deoxycholate in the presence of Tween 20, the latter added to stabilize the system. The solubilized enzyme was subsequently reconstituted into phosphatidylcholine vesicles, essentially according to the method of Meissner and Fleischer (1974). The vesicles obtained not only showed Ca^{2+}/Mg^{2+}-dependent ATPase activity, but were also able to accumulate Ca^{2+} ions when ATP was present in the medium. This still does not necessarily imply, however, that phosphatidylcholine can effectively act as the lipidic activator of the system. As a consequence of the procedures used in those experiments, the reconstituted system might have contained (essential) endogenous phospholipids. However, other recent studies from the same group (Geitzen *et al.*, 1980b) have shown that lecithin-containing buffers are quite effective in stabilizing the activity during purification of the enzyme by calmodulin affinity chromatography. This clearly indicates that even for such purposes there is no absolute need to use phosphatidylserine as has been claimed by Niggli *et al.* (1980). A most remarkable difference between the phosphatidylcholine/Triton X-100 (Gietzen *et al.*, 1980b) and phosphatidylserine/Triton X-100 (Niggli *et al.*, 1980) purified systems is, however, that the first one is highly sensitive to activiation by calmodulin, whereas the latter is already fully active without it. This interesting result may help to resolve the different phospholipid requirements reported in these studies. Of much greater importance, however, is that it also may provide a clue whereby further insight can be gained into both the nature of activation of the enzyme by calmodulin, as well as the function of phospholipids in this system. Further studies will be necessary to achieve these goals.

In summary, it can be concluded that the lipid requirement of the Ca^{2+}/Mg^{2+}-ATPase in the erythrocyte membrane can be satisfied by *mono-* and *di*acyl*glycero*phospholipids, neutral as well as negatively charged ones. In the native situation, however, it is only that fraction of the total *glycero*phospholipid complement of the membrane that forms part of the *inner* half of the bilayer that is actually involved.

Finally, it is of interest to emphasize the difference that appears to exist between the lipid requirement of this system and that of the Na^+/K^+-ATPase in the erythrocyte membrane. The latter has been shown to have an absolute requirement for *negatively* charged *di*acyl*glycero*phospholipids, of which phosphatidylserine has appeared to be the endogenous activator (Roelofsen and van Deenen, 1973; Roelofsen, 1977). Furthermore, there also appear to exist striking differences in its lipid requirement when compared to that of the Ca^{2+}/Mg^{2+}-ATPase in sarcoplasmic reticulum, which can be activated by almost every amphiphile, including nonionic detergents (compare: Meissner and Fleischer, 1972; The and Hasselbach, 1973; Knowles *et al.*, 1976; Bennett *et al.*, 1978; Dean and Tanford, 1978; Swoboda *et al.*, 1979).

ACKNOWLEDGMENT. The author wishes to thank Dr. Theodore Taraschi for critically reading the manuscript.

REFERENCES

Bennet, J. P., Smith, G. A., Houslay, M. D., Hesketh, T. R., Metcalfe, J. C., and Warren, G. B. (1978). *Biochim. Biophys. Acta* **513**, 310–320.

Coleman, R., and Bramley, T. A. (1975). *Biochim. Biophys. Acta* **382**, 565–575.

Dean, W. L., and Tanford, C. (1978). *Biochemistry* **17**, 1683–1690.

Gietzen, K., Seiler, S., Fleischer, S., and Wolf, H. U. (1980a). *Biochem. J.* **188**, 47–54.

Gietzen, K., Tejčka, M., and Wolf, H. U. (1980b). *Biochem. J.,* **189**, 81–88.

Haaker, H., and Racker, E. (1979). *J. Biol. Chem.* **254**, 6598–6602.

Knowles, A. F., Eytan, E., and Racker, E. (1976). *J. Biol. Chem.* **251**, 5161–5165.

Meissner, G., and Fleischer, S. (1972). *Biochim. Biophys. Acta* **255**, 19–33.

Meissner, G., and Fleischer, S. (1974). *J. Biol. Chem.* **249**, 304–309.

Niggli, V., Penniston, V. T., and Carafoli, E. (1980). *J. Biol. Chem.* **254**, 9955–9958.

Peterson, S. W., Ronner, P., and Carafoli, E. (1978). *Arch. Biochem. Biophys.* **186**, 202–210.

Richards, D. E., Vidal, J. C., Garrahan, P. J., and Rega, A. F. (1977a). *J. Membr. Biol.* **35**, 125–136.

Richards, D. E., Garrahan, P. J., and Rega, A. F. (1977b). *J. Membr. Biol.* **35**, 137–147.

Roelofsen, B. (1977). In *Membrane Proteins* (P. Nicholls, J. V. Møller, P. L. Jørgensen and A. J. Moody, eds.), pp. 183–190, Pergamon Press, Elmsford, N.Y.

Roelofsen, B., and Schatzmann, H. J. (1977). *Biochim. Biophys. Acta* **464**, 17–36.

Roelofsen, B., and van Deenen, L. L. M. (1973). *Eur. J. Biochem.* **40**, 245–257.

Ronner, P., Gazzotti, P., and Carafoli, E. (1977). *Arch. Biochem. Biophys.* **179**, 578–583.

Roufogalis, B. D. (1979). *Can J. Physiol. Pharmacol.* **57**, 1331–1349.

Schatzmann, H. J., and Bürgin, H. (1978). *Ann. N.Y. Acad. Sci.* **307**, 125–147.

Swoboda, G., Fritzsche, J., and Hasselbach, W. (1979). *Eur. J. Biochem.* **95**, 77–88.

Taverna, R. D., and Hanahan, D. J. (1980). *Biochem. Biophys. Res. Commun.* **94**, 652–659.

The, R., and Hasselbach, W. (1973). *Eur. J. Biochem.* **39**, 63–68.

80

The Transport of Ca^{2+} by Mitochondria

Ernesto Carafoli

1. INTRODUCTION

Most of the present interest in the field of mitochondrial Ca^{2+} transport is focused on the concept that the influx and efflux of Ca^{2+} occur via independent routes (Carafoli and Crompton, 1978; Pushkin *et al.*, 1976; see Carafoli, 1979, for a review). The idea that Ca^{2+} is continuously released from mitochondria was implicit in the early experiments of Drahota *et al.* (1965) on the steady-state maintenance of the accumulated Ca^{2+} in mitochondria, and in the more recent finding of Stücki and Ineichen (1974) that a portion of the normal State 4 respiration is due to the energy-dissipating reuptake of the lost Ca^{2+}. However, that the exit of Ca^{2+} was not due to the reversal of the uptake route, but proceeded through an independent pathway, was first indicated by experiments carried out by Rossi and colleagues some years ago (1973), in which it was shown that Ca^{2+} could be released from liver mitochondria in the presence of ruthenium red, i.e., under conditions in which the uptake pathway is blocked. This finding provided the experimental tool for most of the later experiments on the independence of the uptake and release pathway, which eventually led to the proposal of the "mitochondrial Ca^{2+} cycle" (Carafoli, 1979). It is obvious, indeed, that there would be no "cycle" if Ca^{2+} were taken up and released through the same route. In retrospect, it is now clear that the proposal that the uptake route operates essentially as a one-way process, and does not mediate the release of Ca^{2+}, had no alternative. Ca^{2+} uptake is driven, without charge compensation, by the transmembrane potential maintained by respiration (Rottenberg and Scarpa, 1974; Heaton and Nicholls, 1976; Crompton and Heid, 1978). Reversal of the uptake pathway would require extensive oscillations in the transmembrane potential, a highly improbable event in view of the central role of the potential in mitochondrial function.

The problem, however, is that mitochondria may have more than one Ca^{2+} release route, depending on the tissue and, possibly, on the experimental conditions. In this brief summary, the best identified among these release routes will be succinctly described. Prior

Ernesto Carafoli • Laboratory of Biochemistry, Swiss Federal Institute of Technology (ETH), 8092 Zürich, Switzerland.

to this, a summary will be given of the present state of knowledge on the mechanism of the uptake of Ca^{2+}.

2. THE UPTAKE OF Ca^{2+}

It is today generally accepted that the energy-linked uptake of Ca^{2+} is driven electrophoretically by the membrane potential component of the total protonmotive force (Rottenberg and Scarpa, 1974; Heaton and Nicholls, 1976; Crompton and Heid, 1978). Suggestions of partial charge compensation by H^+ antiport (Reed and Bygrave, 1975) or phosphate symport (Moyle and Mitchell, 1977a) have not been supported by recent experimental evidence (Crompton *et al.*, 1978a). The possibility of partial charge compensation by simultaneous uptake of β-hydroxybutyric acid (Moyle and Mitchell, 1977b), however, has not been explored in detail.

Recent developments in the field have seen a revival of interest in the Ca^{2+}-binding glycoprotein (Sottocasa *et al.*, 1972; Carafoli and Sottocasa, 1974), which is involved in the Ca^{2+} uptake process (Panfili *et al.*, 1976), but which has been suggested to function only as a superficial Ca^{2+}-recognition site (Carafoli, 1975). Very recently, Sottocasa and co-workers (Sottocasa *et al.*, 1980) have obtained evidence that the glycoprotein may mediate both the uptake *and* the release of Ca^{2+}. Another candidate for a role in the Ca^{2+} uptake process, possibly as the electrophoretic carrier, is a polypeptide (molecular weight ~ 3000) recently isolated by Jeng and co-workers (1978). It binds the Ca^{2+} analog Mn^{2+} with high affinity, and can reconstitute ruthenium red-sensitive Ca^{2+} permeability in liposomes and planar lipid bilayers.

3. THE RELEASE OF Ca^{2+}

It follows from the above discussion that the uptake of Ca^{2+} can be regarded as a passive process, in which Ca^{2+} penetrates in response to a preformed negative potential. In the presence of a negative-inside membrane potential, then, the exit of Ca^{2+} must be electrically silent, or even electrophoretic in the direction of the *net* import of positive charges. To this date, several mechanisms for promoting Ca^{2+} release have been described, but only one (Crompton *et al.*, 1976) has been characterized mechanistically in sufficient detail. It must be emphasized again that the insensitivity to ruthenium red has been instrumental in most studies aimed at identifying independent release routes.

3.1. The Na$^+$-Stimulated Release Pathway

That Na^+ could specifically release Ca^{2+} from mitochondria was first shown in 1974 by Carafoli and co-workers (Carafoli *et al.*, 1974). The process was characterized in considerable detail in a series of studies by Crompton and co-workers (Crompton *et al.*, 1976, 1977, 1978b, 1979; Crompton and Heid, 1978), and shown to be mediated by a specific Na^+–Ca^{2+} exchange, insensitive to ruthenium red, but sensitive to lanthanides, which inhibit also the electrophoretic Ca^{2+} uniporter. Interestingly, however, whereas the uptake process is maximally sensitive to Tm^{3+} (Crompton *et al.*, 1979), the Na^+-driven release is maximally sensitive to La^{3+}. The activity of the Na^+-promoted route varies in different tissues, heart, brain, and adrenal cortex having the highest activity, and kidney and liver

Table I. The Electrophoretic Ca²⁺ Uptake and the Na⁺-Induced Ca²⁺ Release

	Electrophoretic uptake	Na⁺-induced release
Energy requirements	Respiration or ATP	Faster in energized mitochondria
Inhibitors	Ruthenium red, lanthanides (max. Tm^{3+})	Lanthanides (max. La^{3+})
V_{max} (25°C)	Up to 10 nmoles/mg protein/sec	Up to 0.25 nmole/mg protein/sec
Rate at 1 μM Ca²⁺, and 1 mM external Mg^{2+} (25°C)	0.06 nmole/mg protein/sec	Up to 0.25 nmole/mg protein/sec
K_m (Ca²⁺)	2–13 μM in different tissues	13 μM
K_m (Na⁺)	—	6 mM
Tissue specificity	Practically all tissues (exceptions: blowfly flight muscle, yeast)	Highest in heart, brain, adrenal cortex, negligible in liver, kidney, lung
Mechanism	Electrophoretic, charge uncompensated	Electroneutral

the lowest. The route is activated half-maximally by about 6 mM external Na⁺, and its V_{max} is considerably lower than that of the electrophoretic uniporter. This would mean that the Na⁺-promoted route could provide efficient competition against the uptake route only if the latter were slowed down by some inhibitor: a good candidate here would be Mg^{2+}, which has been repeatedly shown to inhibit Ca²⁺ uptake (Jacobus *et al.*, 1975; Sordahl, 1974; Åkerman *et al.*, 1977). The essential parameters of the electrophoretic route, and of the Na⁺-promoted route, are compared in Table I. From their studies of the Na⁺-mediated process, Crompton *et al.* (1976) have extracted the concept of an energy-dissipating Na⁺–Ca²⁺ cycle in heart mitochondria. In their proposal, the cycle plays an important role in the maintenance of the appropriate Ca²⁺ homeostasis in heart cells.

3.2. The Fatty Acid-Induced Release of Ca²⁺

It is not surprising that fatty acids, which are well-known uncouplers, should induce Ca²⁺ release. When added in uncoupling concentrations, they evidently induce release by a reversal of the uptake uniporter. What is interesting, however, is that fatty acids, particularly polyunsaturated fatty acids, may induce Ca²⁺ release even when added in subuncoupling concentrations. This was documented some years ago for the case of prostaglandins (Carafoli *et al.*, 1973; Malmström and Carafoli, 1975) for which a Ca²⁺-ionophoric function has been suggested. More recently, a comprehensive study of the Ca²⁺-releasing property of fatty acids has been carried out by Roman *et al.* (1979) in kidney and liver mitochondria. They have found that a variety of polyunsaturated fatty acids induce an extensive and rapid Ca²⁺ release in both mitochondrial types, under conditions in which the coupling efficiency (i.e., the membrane potential) of mitochondria is not affected. Most interestingly, also in view of the release-promoting action of Na⁺ in heart mitochondria, they found that the releasing effect of fatty acids was counteracted by Na⁺. Roman *et al.* (1979) have concluded that fatty acids may act as Ca²⁺ ionophores, although an indirect effect on the fluidity state of the membrane has also been suggested as a possibility.

3.3. The Phosphate-Promoted Release Route

That phosphate can release Ca^{2+} from mitochondria has been known since Rossi and Lehninger (1964) observed that the respiration of rat liver mitochondria returned to the original State 4 rate after the accumulation of a pulse of Ca^{2+}, but failed to do so if phosphate was also accumulated. Rossi and Lehninger (1964) concluded that phosphate had a deleterious effect on the mitochondrial structure, leading to the release of the accumulated Ca^{2+}. This concept became generally accepted, and the possibility that phosphate could induce Ca^{2+} release without irreversibly damaging mitochondria was not seriously explored. There is little doubt, indeed, that irreversible damage to mitochondria can follow the accumulation of massive amounts of Ca^{2+} and phosphate. It is clear, however, that even under these conditions liver mitochondria maintain a reasonable ability to generate a membrane potential, as shown by the fact that their respiration becomes permanently activated after the accumulation of the massive Ca^{2+}–phosphate pulse. Evidently, they can take up continuously the Ca^{2+} that has been lost to the medium. More recently, the role of phosphate as a Ca^{2+}-releasing agent has been investigated under conditions where the functional (and structural) integrity of liver mitochondria has been continuously monitored, and found to be unaffected. These studies (Roos *et al.*, 1980; Siliprandi *et al.*, 1979, 1980) have employed low Ca^{2+} concentrations, which is apparently a prerequisite for avoiding irreversible damage. Roos *et al.* (1980) have used phosphate-depleted liver mitochondria, and have shown a rapid, phosphate-dependent Ca^{2+} efflux, which is sensitive to ruthenium red, and thus indicates the operation of an independent release pathway. The release requires the transfer of inorganic phosphate across the inner membrane, and is *not* accompanied by significant variations in the membrane potential with respect to controls not treated with phosphate. Roos *et al.* (1980) could also establish that phosphate, after penetration into liver mitochondria, leaves them again as Ca^{2+} is released. However, the releases of Ca^{2+} and phosphate are temporally out of phase, the latter preceding the former. This last finding rules out earlier tentative suggestions (Lötscher *et. al.*, 1979b) of a Ca–P_i symporter operating in the direction of Ca^{2+} release, and leaves open the problem of the mechanism of the release of Ca^{2+} under these conditions (a Ca^{2+}–H^+ exchanger activated by inorganic phosphate?, see below).

In the other recent study on the release of Ca^{2+} by inorganic phosphate (Siliprandi *et al.*, 1979, 1980), it has also been found that the coupling efficiency of liver mitochondria does not decrease during the induced release, as revealed by the safranine test. Interestingly, under these conditions mitochondrial Mg^{2+} is slowly released, leading to the progressive damage of the mitochondrial membrane, and to the stimulation of the ruthenium red-insensitive loss of Ca^{2+}. The latter evidently occurs on the reversed electrophoretic uptake uniporter. Also of interest is the finding that the oxidation of some membrane thiols is required for the slow loss of Mg^{2+}, and for the membrane damage consequent upon it.

4. THE RELEASE OF Ca^{2+} INDUCED BY VARIATIONS IN THE REDOX RATIO OF MITOCHONDRIAL PYRIDINE NUCLEOTIDES

In 1978, Lehninger and colleagues described experiments in which Ca^{2+} release from liver mitochondria could be promoted by inducing a transition of the redox ratio of endogenous pyridine nucleotides toward oxidation. Conditions that increase the reduction of mitochondrial $NAD(P)^+$, on the other hand, favor the retention of Ca^{2+}.

The observation of Lehninger *et al.* (1978) has been extended by Lötscher *et al.*

(1979a, 1980), who found that the redox level of mitochondrial pyridine nucleotides, and thus the balance of Ca^{2+} between mitochondria and medium, may be modulated by hydroperoxides and the combined action of glutathione peroxidase and glutathione reductase. They have also found that in the presence of accumulated Ca^{2+}, oxidized pyridine nucleotides are hydrolyzed inside mitochondria, and that nicotinamide is released to the outside medium. At variance with the conclusions of others (Nicholls and Brand, 1980; Wolkowicz and McMillin-Wood, 1980), Lötscher *et al.* (1980) found that during the process of Ca^{2+} release induced by the oxidation of pyridine nucleotides, liver mitochondria retain a considerable degree of functional integrity. Under appropriate conditions, even the release of nicotinamide is reversible, suggesting that the process may be physiologically meaningful. Also of interest is the finding (Lötscher *et al.*, 1980) that the release of Ca^{2+} from liver mitochondria induced by the oxidation of pyridine nucleotides is electroneutral. Possibly, the oxidation of pyridine nucleotides activates a Ca^{2+}–H^{+} exchange, which has been suggested to function in liver mitochondria in the direction of Ca^{2+} release (Fiskum and Lehninger, 1979).

5. CONCLUDING REMARKS

In these condensed notes, an effort has been made to review the aspects of the process of mitochondrial Ca^{2+} transport that have been of particular interest to specialists during the last 3 or 4 years. Due to space limitations, other aspects of interest have not been considered, but it is quite possible that they will become the focus of attention in the future. Among them, one can mention the molecular components of the transport process(es), and the role of mitochondria as metabolic regulators in the integrated cytosolic environment.

On the main point considered in this review, the release of Ca^{2+} by mitochondria, some concluding remarks are in order. One is the fact that mitochondria in different tissues may have (actually, they most likely have) different mechanisms for releasing Ca^{2+}. Mitochondria from heart, brain, and other tissues of the ''heart'' group (Crompton *et al.*, 1978b) release Ca^{2+} by the Na^{+}-activated pathway, those from liver *do not*, or they do it at a marginal rate (see Haworth *et al.*, 1980). Thus, the problem is to define the pathways for Ca^{2+} release from mitochondria (e.g., liver) that are Na^{+} insensitive. It seems important to stress this because in many studies of Ca^{2+} release from liver mitochondria, and of pathways different from the Na^{+}-activated one (see e.g. Wolkowicz and McMillin-Wood, 1980), the reader is left with the impression that the findings may be relevant to mitochondria in general, and not *only* to mitochondria where the activity of the Na^{+} pathways is negligible. To put it differently, the Ca^{2+}-releasing requirements of most mitochondria (the ''heart'' group) are satisfied by the Na^{+}-promoted pathway. For example, preliminary tests in this laboratory have shown that the balance of Ca^{2+} in heart mitochondria is little affected by changes in the redox state of pyridine nucleotides.

One other remark concerns the relationships among the different release routes so far described in mitochondria of the ''liver'' type. It seems important to establish whether these routes are independent, or whether they ''converge'' into a final common mechanism. If the second alternative is correct, then it will be important to characterize the common link among the different routes. At the present state of knowledge, the Ca^{2+}–H^{+} antiporter (Fiskum and Lehninger, 1979) may be a plausible candidate as the mediator of the Ca^{2+}-releasing signal activated by the different mechanisms described.

REFERENCES

Åkerman, K. E. O., Wikström, M. K. F., and Saris, N. O. (1977). *Biochim. Biophys. Acta* **464**, 287–294.

Carafoli, E. (1975). *Mol. Cell. Biochem.* **8**, 133–140.

Carafoli, E. (1979). *FEBS Lett.* **104**, 1–5.

Carafoli, E., and Crompton, M. (1978). *Curr. Top. Membr. Transp.* **10**, 152–216.

Carafoli, E., and Sottocasa, G. L. (1974). In *Dynamics of Energy Transducing Membranes* (L. Ernster, R. W. Estabrook, and E. C. Slater, eds.), pp. 455–469, Elsevier, Amsterdam.

Carafoli, E., Crovetti, F., and Ceccarelli, D. (1973). *Arch. Biochem. Biophys.* **154**, 40–46.

Carafoli, E., Tiozoo, R., Lugli, G., Crovetti, F., and Kratzing, C. (1974). *J. Mol. Cell. Cardiol.* **6**, 361–371.

Crompton, M., and Heid, I. (1978). *Eur. J. Biochem.* **91**, 599–608.

Crompton, M., Capano, M., and Carafoli, E. (1976). *Eur. J. Biochem.* **69**, 453–462.

Crompton, M., Künzi, M., and Carafoli, E. (1977). *Eur. J. Biochem.* **79**, 549–588.

Crompton, M., Hediger, M., and Carafoli, E. (1978a). *Biochem. Biophys. Res. Commun.* **80**, 540–546.

Crompton, M., Moser, R., Lüdi, H., and Carafoli, E. (1978b). *Eur. J. Biochem.* **82**, 25–31.

Crompton, M., Heid, I., Baschera, C., and Carafoli, E. (1979). *FEBS Lett.* **104**, 352–354.

Drahota, Z., Carafoli, E., Rossi, C. S., Gamble, R. L., and Lehninger, A. L. (1965). *J. Biol. Chem.* **240**, 2712–2720.

Fiskum, G., and Lehninger, A. L. (1979). *J. Biol. Chem.* **254**, 6236–6239.

Haworth, R. A., Hunter, D. R., and Berkoff, H. A. (1980). *FEBS Lett.* **110**, 216.

Heaton, G. M., and Nicholls, D. G. (1976). *Biochem. J.* **156**, 635–646.

Jacobus, W. E., Tiozzo, R., Lugli, G., Lehninger, A. L., and Carafoli, E. (1975). *J. Biol. Chem.* **250**, 7863–7870.

Jeng, A. Y., Ryan, T. E., and Shamoo, A. (1978). *Proc. Natl. Acad. Sci. USA* **75**, 2125–2130.

Lehninger, A. L., Vercesi, A., and Bababunmi, E. A. (1978). *Proc. Natl. Acad. Sci. USA* **79**, 1690–1694.

Lötscher, H. R., Winterhalter, K. H., Carafoli, E., and Richter, C. (1979a). *Proc. Natl. Acad. Sci. USA* **76**, 4340–4344.

Lötscher, H. R., Schwerzmann, K., and Carafoli, E. (1979b). *FEBS Lett.* **99**, 194–198.

Lötscher, H. R., Winterhalter, K. H., Carafoli, E., and Richter, C. (1980). *J. Biol. Chem.* **255**, 9325–9330.

Malström, K., and Carafoli, E. (1975). *Arch. Biochem. Biophys.* **171**, 418–423.

Moyle, J., and Mitchell, P. (1977a). *FEBS Lett.* **77**, 136–145.

Moyle, J., and Mitchell, P. (1977b). *FEBS Lett.* **73**, 131–136.

Nicholls, D. G., and Brand, M. D. (1980). *Biochem. J.* **188**, 113–118.

Panfili, E., Sandri, G., Sottocasa, G. L., Lunazzi, G., Liut, G., and Graziosi, G. (1976). *Nature (London)* **264**, 185–186.

Pushkin, J. S., Gunter, T. E., Gunter, K. K., and Russell, P. R. (1976). *Biochemistry* **15**, 3834–3842.

Reed, K. C., and Bygrave, F. L. (1975). *Eur. J. Biochem.* **55**, 497–504.

Roman, I., Gmaj, P., Nowicka, C., and Angielski, S. (1979). *Eur. J. Biochem.* **102**, 615–623.

Roos, I., Crompton, M., and Carafoli, E. (1980). *Eur. J. Biochem.* **110**, 319–325.

Rossi, C. S., and Lehninger, A. L. (1964). *J. Biol. Chem.* **239**, 3971–3980.

Rossi, C. S., Vasington, F. D., and Carafoli, E. (1973). *Biochem. Biophys. Res. Commun.* **50**, 846–852.

Rottenberg, H., and Scarpa, A. (1974). *Biochemistry* **13**, 4811–4819.

Siliprandi, N., Rugolo, M., Siliprandi, D., Toninello, A., and Zoccarato, F. (1979). In *Function and Molecular Aspects of Biomembrane Transport* (E. Quagliariello, F. Palmieri, S. Papa, and M. Klingenberg, eds.), pp. 147–155, Elsevier/North-Holland, Amsterdam.

Siliprandi, D., Rugolo, M., Toninello, A., and Zoccarato, F. (1980). *First European Bioenergetics Conference, Short Reports,* pp. 299–300.

Sordahl, L. A. (1974). *Arch. Biochem. Biophys.* **167**, 104–115.

Sottocasa, G. L., Sandri, G., Panfili, E., De Bernard, B., Gazzotti, P., Vasington, F. D., and Carafoli, E. (1972). *Biochem. Biophys. Res. Commun.* **47**, 808–813.

Sottocasa, G. L., Panfili, E., Sandri, G., and Liut, G. (1980). *First European Bioenergetics Conference, Short Reports,* pp. 267–268.

Stücki, J. W., and Ineichen, E. A. (1974). *Eur. J. Biochem.* **48**, 365–375.

Wolkowicz, P. E., and McMillin-Wood, J. (1980). *J. Biol. Chem.* **255**, 10348–10353.

81

Hormonal Control of Calcium Fluxes in Rat Liver

Fyfe L. Bygrave, Peter H. Reinhart, and Wayne M. Taylor

1. INTRODUCTION

The physiological responses of a number of hormones in a range of tissues are closely associated with the redistribution of intracellular Ca^{2+}. This redistribution appears to involve changes in the rate of Ca^{2+} transport in subcellular organelles known to contain appreciable ''pools'' of Ca^{2+} (Claret-Berthon *et al.*, 1977) and may result in an alteration in either the cytoplasmic or the intraorganellar Ca^{2+} concentration. As most mammalian cell types contain many Ca^{2+}-dependent reactions (Carafoli and Crompton, 1978), the hormonal regulation of Ca^{2+} transport activity in subcellular organelles may have an important role in mediating the responses to the hormones. Following a brief account of current information about mitochondrial and microsomal Ca^{2+} transport activities, this review proceeds to consider our understanding of how α-adrenergic agonists and glucagon may regulate these activities and examines the possible role this regulation may play in mediating cellular hormone responses. The discussion is purposely confined to a consideration of liver tissue only.

Both mitochondria and microsomes have been shown to actively sequester Ca^{2+} *in vitro*, and when incubated together are able to maintain the ambient Ca^{2+} concentration close to 0.2 μM (Becker *et al.*, 1980). As this concentration of Ca^{2+} was also maintained by suspensions of digitonin-treated hepatocytes (Becker *et al.*, 1980; Murphy *et al.*, 1980), this value may approximate the actual cytoplasmic free Ca^{2+} concentration *in situ*, and strongly implicates mitochondria and the endoplasmic reticulum in the maintenance of cytoplasmic Ca^{2+}.

One feature that most readily distinguishes mitochondrial from microsomal Ca^{2+} transport activities is the difference in sensitivity to ruthenium red; the former activity is almost totally abolished by low concentrations of the compound ($K_i \sim 30$ pmoles/mg protein) while

Fyfe L. Bygrave, Peter H. Reinhart, and Wayne M. Taylor • Department of Biochemistry, Faculty of Science, The Australian National University, Canberra, Australian Capital Territory 2600, Australia.

the latter is unaffected by concentrations several orders of magnitude greater. This fact has proved to be most useful in the experimental analysis of mitochondrial (i.e., ruthenium red-sensitive) and nonmitochondrial (i.e., ruthenium red-insensitive) Ca^{2+} transport in liver. In digitonin-treated hepatocytes for example, greater than 90% of the Ca^{2+} uptake activity is ruthenium red-sensitive (Babcock et al., 1979; Murphy et al., 1980) as is the case in a rat liver homogenate (Ash and Bygrave, 1977). This supports the idea that mitochondria play a major role in Ca^{2+} homeostasis. Further support derives from the observation that the initial rates of Ca^{2+} transport are significantly greater in mitochondria than microsomes.

The driving force for electrophoretic Ca^{2+} uptake by mitochondria occurs in response to the membrane potential negative inside, generated either by the hydrolysis of ATP or by electron transport activity (Nicholls and Crompton, 1980). In contrast, microsomal Ca^{2+} transport is coupled to the hydrolysis of MgATP only (Moore et al., 1975; Bygrave, 1978a).

Several features on the other hand appear to be common to both transport systems. These include the existence of specific carriers in the respective membranes, the cyclic nature of Ca^{2+} transport, the high affinity for free Ca^{2+}, and stimulation by permeant weak acids (Bygrave, 1978b; Carafoli and Crompton, 1978). Of special interest in the present context is the observation that both transport systems respond to physiological stimuli such as tissue development and changes in the circulating hormone concentration (Bygrave, 1978b; Reinhart and Bygrave, 1981).

2. ACTION OF α-ADRENERGIC AGONISTS AND GLUCAGON ON MITOCHONDRIAL Ca²⁺TRANSPORT

A number of laboratories have demonstrated that stable changes in Ca^{2+} transport occur in mitochondria following their isolation from liver administered in situ with glucagon or α-adrenergic agonists. Besides adding strength to a role of mitochondrial Ca^{2+} in the α-agonist and glucagon-induced responses, this phenomenon also provides a useful means of examining details of the mechanism involved in Ca^{2+} mobilization and in the action of the hormones.

Early experiments involved the intraperitoneal or intravenous injection of the hormone into an intact animal, followed by isolation of the liver mitochondria and assay of Ca^{2+} transport. More recent studies have adopted the perfused liver system or hepatocytes. Experiments with mitochondria isolated from liver 30 to 60 min after the intraperitoneal injection of glucagon (Hughes and Barritt, 1978; Prpic et al., 1978) and those with mitochondria isolated from liver 7 to 10 min after perfusion with hormone (Taylor et al., 1980a) have yielded similar information with respect to stable changes in Ca^{2+} transport. The most pronounced changes so far detected are an enhanced ability of the mitochondria to retain high concentrations of Ca^{2+} and an increase in the initial rate of Ca^{2+} influx, although these latter effects are not as reproducible as the former. This hormone-sensitive Ca^{2+} transport appears to be enriched in mitochondria sedimented at relatively low centrifugation forces (Prpic et al., 1978).

Other stable changes seen that are of potential relevance to the mechanism of mitochondrial Ca^{2+} fluxes include an increase in the transmembrane pH gradient, an increase in the adenine nucleotide concentration (Taylor et al., 1980a; Halestrap, 1978), and a reduced ability for oxaloacetate-induced NADPH oxidation (Prpic and Bygrave, 1980). These changes are seen also in mitochondria isolated from liver perfused with α-agonists for 7 to 10 min and suggest that the actions of the two hormones on several energy-linked mitochondrial reactions including Ca^{2+} fluxes, have many features in common.

Because the longer-term effects of the two hormones may not directly bear on the early action of the hormones, it has been important to assess the earliest times at which such effects are detectable. Recently we have investigated the action of cAMP and phenylephrine at very early times after their administration to the perfused liver (i.e., in the time span of 10 to 120 sec). These short-term perfusion studies, in contrast to the relatively longer-term ones discussed above, reveal important differences in the early action of the two hormones on energy-linked reactions in liver mitochondria (Taylor *et al.*, 1981). Phenylephrine is much more rapid in its action than glucagon or cAMP. By 30 sec after phenylephrine administration, levels of mitochondrial adenine nucleotides are elevated approximately 35%, while cAMP has little effect on this parameter even after 4 min. Whereas Ca^{2+} retention is increased by 50% 20 sec after administration of phenylephrine, no significant change is seen until 120 sec following perfusion with cAMP. Initial rates of Ca^{2+} transport are stimulated at about 120 sec after the administration of either hormone.

While these studies were in progress, Yamazaki *et al.* (1980) reported a rapid action of glucagon on hepatic mitochondrial metabolism. Within 1 min of the intravenous injection of glucagon into anesthetized rats, Ca^{2+} retention by isolated mitochondria was increased to near-maximal extents, an effect shown to be independent of the Ca^{2+} content of the mitochondria. The first-order rate constant for Ca^{2+} influx was also increased significantly.

Because of the strong possibility that changes in mitochondrial Ca^{2+} fluxes are closely associated with the cellular metabolic network (see below), it is of additional interest that a number of other rapid changes in mitochondrial energy-linked reactions occur following incubation of hepatocytes with glucagon or α-agonists. These include a stimulation of uncoupler-dependent ATPase (Titheradge *et al.*, 1979), increases in ADP-stimulated respiration and succinate dehydrogenase activity, as well as a shift to the reduced in the oxidation–reduction state (Siess and Wieland, 1980). The effects of the α-agonists were more rapid than those of glucagon (Titheradge *et al.*, 1979).

Thus, while the longer-term actions of the two hormones on mitochondrial energy-linked reactions, including Ca^{2+} fluxes, resemble each other in many respects, their early actions differ, particularly with regard to the rapidity with which they induce their responses.

3. ACTION OF HORMONES ON MICROSOMAL Ca^{2+} TRANSPORT

The experimental approach used to study the action of hormones on microsomal Ca^{2+} transport in liver, like that described in the previous section, has essentially involved administration of the hormone to the intact animal or perfused liver followed by isolation of the microsomal fraction and the measurement of Ca^{2+} transport. Isolated liver cells have been equally useful in this study because of the relative ease of preparing functionally intact microsomes from these cells following their incubation with the hormone. In contrast to studies examining mitochondrial energy-linked reactions, the majority of studies with microsomes have involved administration of either glucagon or cAMP and its analogs (see Reinhart and Bygrave, 1981).

The principal action of glucagon administration on the intact tissue or isolated cells is stimulation of the initial rate of Ca^{2+} transport as measured *in vitro*. The degree of stimulation is greatest with the "heavy" populations of microsomes (Reinhart and Bygrave, 1981) and is dependent on the concentration of glucagon administered and the time for which the tissue or cells are incubated with the hormone. The earliest effects are seen 5 to

10 min after hormone administration to hepatocytes and can be prevented by coadministration of insulin (Taylor *et al.*, 1980b; Andia-Waltenbaugh *et al.*, 1980). Elevation of intracellular cAMP by inhibitors of phosphodiesterase also leads to a stimulation of Ca^{2+} transport (Taylor *et al.*, 1980b; Reinhart and Bygrave, 1981). When the microsomal membrane is incubated *in vitro* for short periods with small amounts of the supernatant fraction in the presence of ATP and cAMP, the initial rate of Ca^{2+} transport is stimulated approximately 40%. Exclusion of cAMP alone from the medium prevents any such stimulation (Reinhart and Bygrave, 1981). The evidence would suggest that the mechanism of stimulation of the initial rate of Ca^{2+} transport by glucagon involves a cAMP-dependent protein kinase that in turn phosphorylates a component on the microsomal membrane on or near the Ca^{2+} transport protein as appears to take place in sarcoplasmic reticulum (Tada *et al.*, 1979).

4. ACTION OF HORMONES ON PLASMA MEMBRANE Ca^{2+} TRANSPORT

Considerable experimental difficulties are associated with the study of Ca^{2+} fluxes across the plasma membrane (Racker, 1980). These partly relate to the perturbation of such fluxes by intracellular organelles. Thus, little definitive information is available regarding the action of hormones on plasma membrane Ca^{2+} transport. However, in a recent kinetic analysis of the effects of epinephrine on Ca^{2+} distribution in hepatocytes, Barritt *et al.* (1981) concluded that a major effect of the hormone is to increase the rate constant for influx of Ca^{2+} to a small intracellular compartment of exchangeable Ca^{2+}, a finding at variance with reports showing a rapid efflux of Ca^{2+} from perfused liver (Althaus-Saltzmann *et al.*, 1980; Blackmore *et al.*, 1979).

Reports that calmodulin activates plasma membrane Ca^{2+} pumps in some tissues (e.g., Lynch and Cheung, 1979; Pershadsingh *et al.*, 1980) raise the question whether this protein might be involved also in liver plasma membrane Ca^{2+} transport.

5. PHYSIOLOGICAL ROLE OF HORMONE-INDUCED CHANGES IN INTRACELLULAR Ca^{2+} FLUXES

Unlike glucagon, the α-adrenergic agonists are thought not to give rise to an increase in cAMP concentration. Instead the action of these hormones is considered to be mediated by a rise in intracellular Ca^{2+} (see Blackmore *et al.*, 1979), which in turn may stimulate the conversion of phosphorylase b to phosphorylase a and hence increase glycogenolysis. Independent studies from several laboratories have suggested that mitochondria are the source of this Ca^{2+} (Babcock *et al.*, 1979; Blackmore *et al.*, 1979; Taylor *et al.*, 1980a) although this view has been questioned (Poggioli *et al.*, 1980; Althaus-Saltzmann *et al.*, 1980). The recent kinetic analysis of Barritt *et al.* (1981) suggests all three membranes (mitochondrial, microsomal, and plasma membrane) are involved in the epinephrine-induced Ca^{2+} redistribution in liver.

A role for microsomal Ca^{2+} fluxes in the regulation of cytoplasmic Ca^{2+} is suggested from considering (1) the ability of the organelle to maintain a low steady-state concentration of Ca^{2+} (Becker *et al.*, 1980), (2) the lowering of the Ca^{2+} content in the microsomal fraction following glucagon or phenylephrine treatment (Blackmore *et al.*, 1979), and (3) the close correlation between glucagon-stimulated microsomal Ca^{2+}-flux activity and glycogenolysis during liver development (Reinhart and Bygrave, 1981). These data point strongly to a role of microsomal Ca^{2+} fluxes in the regulation of glycogen metabolism in liver.

Rapid cycling of Ca^{2+} across the inner mitochondrial membrane (Nicholls and Crompton, 1980) provides a means of controlling not only cytoplasmic Ca^{2+} but also intramitochondrial Ca^{2+} (Denton and McCormack, 1980). In this way changes in mitochondrial Ca^{2+} fluxes are suggested to control the activities of several Ca^{2+}-sensitive matrix dehydrogenases (Denton and McCormack, 1980). For example, Hems *et al.* (1978) consider that vasopressin may bring about an increase in pyruvate dehydrogenase through an increase in intramitochondrial Ca^{2+} (see also Sugden *et al.*, 1980).

The possible involvement of Ca^{2+}-regulated events in the control of hydrogen transfer between mitochondria and cytoplasmic compartments during the stimulation of gluconeogenesis by α-agonists and glucagon has been emphasized recently (Kneer *et al.*, 1979). Moreover, changes in Ca^{2+} fluxes may play a role in the hormonal control of gluconeogenesis through increases in the rate of mitochondrial ATP formation. The increases in mitochondrial Ca^{2+} fluxes (Yamazaki *et al.*, 1980; Taylor *et al.*, 1981), respiratory chain activity (Yamazaki *et al.*, 1980), and mitochondrial ATP content (Taylor *et al.*, 1981) brought about by the action of α-agonists all appear closely related events. Of additional interest are reports that shifts of the ATPase-inhibiting peptide to and from the ATPase (Tuena de Gomez-Puyou *et al.*, 1980; Carafoli *et al.*, 1980) may in turn regulate the rate at which mitochondria take up Ca^{2+}.

Some important questions that remain to be elucidated relate to the nature of any purported secondary messenger to the α-receptor and its interaction with any of the intracellular organelle membranes. Also, although the action of glucagon appears to involve the activation of cAMP-dependent protein kinase, the link between this activation and the observed alteration to cellular Ca^{2+} fluxes is not known.

One possible common focal point for both classes of hormones is their ability to phosphorylate the same 12 cytoplasmic proteins (Garrison, 1978). Presumably these must be capable of transferring information to the target organelles. Thus, there may exist a cAMP-independent protein kinase linked in some way to the generation of an α-agonist second messenger. It is of some interest that recently a cAMP-independent protein kinase linked to the turnover of phosphatidylinositol has been described (Kishimoto *et al.*, 1980). As the turnover of this phospholipid in tissues has been associated with α_1-receptors (Fain and Garcia-Sainz, 1980), the possibility is raised that this protein kinase is associated with such a second messenger and hence the physiological action in response to this hormone.

The role of Ca^{2+} redistribution in the molecular events linking the action of α-agonists and glucagon in liver tissue with their physiological responses poses an important challenge for the future.

REFERENCES

Althaus-Saltzmann, M., Carafoli, E., and Jakob, A. (1980). *Eur. J. Biochem.* **106**, 241–248.

Andia-Waltenbaugh, A. M., Lam, A., Hummel, L., and Friedmann, N. (1980). *Biochim. Biophys. Acta* **630**, 165–175.

Ash, G. R., and Bygrave, F. L. (1977). *FEBS Lett.* **78**, 166–168.

Babcock, D. F., Chen, J.-L. J., Yip, B. P., and Lardy, H. A. (1979). *J. Biol. Chem.* **254**, 8117–8120.

Barritt, G. J., Parker, J. C., and Wadsworth, J. C. (1981). *J. Physiol.*, **312**, 29–55.

Becker, G. L., Fiskum, G., and Lehninger, A. L. (1980). *J. Biol. Chem.* **255**, 9009–9012.

Blackmore, P. F., Dehaye, J.-P., and Exton, J. H. (1979). *J. Biol. Chem.* **254**, 6944–6950.

Bygrave, F. L. (1978a). *Biochem. J.* **170**, 87–91.

Bygrave, F. L. (1978b). *Biol. Rev. Cambridge Philos. Soc.* **53**, 43–79.

Carafoli, E., and Crompton, M. (1978). *Curr. Top. Membr. Transp.* **10**, 151–216.

Carafoli, E., Gavinales, M., Affolter, H., Tuena de Gomez-Puyou, M., and Gomez-Puyou, A. (1980). *Cell Calcium* **1**, 255–265.

Claret-Berthon, B., Claret, M., and Mazet, J. L. (1977). *J. Physiol.* **272**, 529–552.

Denton, R. M., and McCormack, J. G. (1980). *FEBS Lett.* **119**, 1–8.

Fain, J. N., and Garcia-Sainz, J. A. (1980). *Life Sci.* **26**, 1183–1194.

Garrison, J. C. (1978). *J. Biol. Chem.* **253**, 7091–7100.

Halestrap, A. P. (1978). *Biochem. J.* **172**, 399–405.

Hems, D. A., McCormack, J. G., and Denton, R. M. (1978). *Biochem. J.* **176**, 627–629.

Hughes, B. P., and Barritt, G. J. (1978). *Biochem. J.* **176**, 295–304.

Kishimoto, A., Takai, Y., Mori, T., Kikkawa, U., and Nishizuka, Y. (1980). *J. Biol. Chem.* **255**, 2273–2276.

Kneer, N. M., Wagner, M. J., and Lardy, H. A. (1979). *J. Biol. Chem.* **254**, 12160–12168.

Lynch, T. J., and Cheung, W. Y. (1979). *Arch. Biochem. Biophys.* **194**, 165–170.

Moore, L., Chen, T., Knapp, H. R., and Landon, E. J. (1975). *J. Biol. Chem.* **250**, 4562–4568.

Murphy, E., Coll, K., Rich, T. L., and Williamson, J. R. (1980). *J. Biol. Chem.* **255**, 6600–6608.

Nicholls, D. G., and Crompton, M. (1980). *FEBS Lett.* **III**, 261–268.

Pershadsingh, H. A., Landt, M., and McDonald, J. M. (1980). *J. Biol. Chem.* **255**, 8983–8986.

Poggioli, J., Berthon, B., and Claret, M. (1980). *FEBS Lett.* **115**, 243–246.

Prpic, V., and Bygrave, F. L. (1980). *J. Biol. Chem.* **255**, 6193–6199.

Prpic, V., Spencer, T. L., and Bygrave, F. L. (1978). *Biochem. J.* **176**, 705–714.

Racker, E. (1980). *Fed. Proc.* **39**, 2422–2426.

Reinhart, P. H., and Bygrave, F. L. (1981). *Biochem. J.,* **194**, 541–549.

Siess, E. A., and Wieland, O. H. (1980). *Eur. J. Biochem.* **110**, 203–210.

Sugden, M. C., Ball, A. J., Ilic, V., and Williamson, D. H. (1980). *FEBS Lett.* **116**, 37–40.

Tada, M., Okmori, F., Yamada, M., and Abe, H. (1979). *J. Biol. Chem.* **254**, 319–326.

Taylor, W. M., Prpic, V., Exton, J. H., and Bygrave, F. L. (1980a). *Biochem. J.* **188**, 443–450.

Taylor, W. M., Reinhart, P. H., Hunt, N. H., and Bygrave, F. L. (1980b). *FEBS Lett.* **112**, 92–96.

Taylor, W. M., Reinhart, P. H., and Bygrave, F. L. (1981). *Proc. Aust. Biochem. Soc.* **14**, 42.

Titheradge, M. A., Stringer, J. L., and Haynes, R. C. (1979). *Eur. J. Biochem.* **102**, 117–124.

Tuena de Gomez-Puyou, M., Gavilanes, M., Gomez-Puyou, A., and Ernster, L. (1980). *Biochim. Biophys. Acta* **592**, 396–405.

Yamazaki, R. K., Mickey, D. L., and Storey, M. (1980). *Biochim. Biophys. Acta* **592**, 1–12.

82

Na–Ca Countertransport in Cardiac Muscle

Harald Reuter

1. HISTORY AND THEORY OF Na–Ca EXCHANGE

The discovery of Na–Ca countertransport across cell membranes involved three major steps:

1. Wilbrandt and Koller (1948) and Lüttgau and Niedergerke (1958) suggested that Ca^{2+} and Na^+ ions compete for anionic groups at the membrane surface, either by distributing themselves according to a Donnan equilibrium (Wilbrandt and Koller, 1948) or by interacting more specifically with these groups (Lüttgau and Niedergerke, 1958). In each case, formation of a Ca–anion complex was assumed to be somehow responsible for activation of contraction of the frog heart. Application of the law of mass action predicted that the Ca–anion complex, and hence contraction, should depend on the ratio $[Ca^{2+}] : [Na^+]^2$ in the external medium. This prediction fitted the experimental data reasonably well. Both hypotheses also implied that Ca adsorption to surface membranes, and/or Ca uptake into intact tissues, should be inversely proportional to the external Na^+ concentration. This has been confirmed by various investigators (Niedergerke, 1963; Langer, 1964; Baker and Blaustein, 1968). This *Na–Ca antagonism,* however, could not account for *Na–Ca countertransport* across the membrane.

2. On the basis of ^{45}Ca flux measurement in guinea pig atria, Reuter and Seitz (1967, 1968) found that not only does Ca *influx* depend on external Na, but also Ca *efflux.* They suggested that a Na–Ca heteroexchange diffusion system exists in cardiac cell membranes that controls the intracellular Ca concentration. In such a "carrier-mediated" transport system, the downhill movement of Na^+ (or Ca^{2+}) ions can provide the free energy for uphill movement of Ca^{2+} (or Na^+) ions across the membrane. The original hypothesis of a Na–Ca countertransport system (Reuter and Seitz, 1968) suggested an electroneutral exchange of 2 Na^+ for 1 Ca^{2+}. The stoichiometry is under debate, but the basic concept of Na–Ca exchange has been confirmed by many studies in heart and other tissues (for re-

Harald Reuter • Department of Pharmacology, University of Bern, 3010 Bern, Switzerland. This chapter is dedicated to Professor Silvio Weidmann, Bern, on the occasion of his 60th birthday.

views see Baker, 1972; Reuter, 1974; Blaustein, 1974; Mullins, 1976; Sulakhe and St. Louis, 1980).

3. From experiments in squid axon, Baker *et al.* (1969) and Blaustein and Hodgkin (1969) came to a similar conclusion as Reuter and Seitz (1968). However, Blaustein and Hodgkin (1969) pointed out that an electroneutral exchange of 2 Na^+ from the external medium for 1 Ca^{2+} from the cytoplasm might not be able to keep the free cytoplasmic Ca^{2+} concentration at sufficiently low levels. They suggested a 3 Na^+ : 1 Ca^{2+} exchange, which would be electrogenic. Mullins (1976, 1979) even demanded a 4 Na^+ : 1 Ca^{2+} exchange. Although the exact stoichiometry of the Na–Ca exchange system has not been resolved so far for any kind of biological tissue, experiments in dialyzed squid axons and in vesicular membrane preparations from mammalian cardiac muscle (Mullins and Brinley, 1975; Reeves and Sutko, 1980) indicate membrane potential dependence of this transport system. This result favors a stoichiometry greater than 2 Na^+ : 1 Ca^{2+}.

A simple exchange of Na^+ for Ca^{2+} ions across the membrane requires the maintenance of a Na^+ gradient between extra- and intracellular spaces, in order to accomplish any net movement of Ca^{2+} in the opposite direction. In most cells an inwardly directed Na^+ gradient is established and maintained by the "Na pump," which is intimately related to the activity of the Na^+/K^+-activated ATPase in cell membranes. Under the assumption that the Na^+ gradient is the only direct energy source for driving a fully coupled Na–Ca countertransport, this system would be able to reduce the free intracellular Ca^{2+} ion activity (αCa_i^{2+}) to an equilibrium value given by

$$\alpha Ca_i^{2+} = \alpha Ca_0^{2+} \; \frac{(\alpha Na_i^+)^n}{(\alpha Na_o^+)^n} \; \exp \; \frac{(n-2)FV_m}{RT} \tag{1}$$

where n is in the number of Na^+ ions that exchange for one Ca^{2+} ion, V_m is the membrane potential, and F, R, and T have their usual meanings. Any Na^+ : Ca^{2+} stoichiometry different from 2 : 1 has to take into account the exponential term including V_m. Therefore, the direction of net fluxes of Ca^{2+} not only depends on the Na^+ gradient, but also on V_m. The equilibrium condition of the system is determined by (see Mullins, 1976)

$$n z_1 F(V_{Na} - V_m) - z_2 F(V_{Ca} - V_m) = 0 \tag{2}$$

where z_1 and z_2 are the electric valencies and V_{Na} and V_{Ca} are the equilibrium potentials for Na^+ and Ca^{2+}, respectively.

On the basis of such theoretical considerations, Mullins (1979) has recently proposed a model for ionic currents in the heart, where Na–Ca exchange serves as a current generator. The reversal potential (V_R), i.e., the membrane potential where the net current generated by the system is zero, can be calculated from

$$V_R = \frac{n V_{Na} - 2 V_{Ca}}{n - 2} \tag{3}$$

At membrane potentials positive to V_R, the current generated by Na–Ca exchange would be an outward current. In principle, the Na–Ca countertransport serves to extrude Ca^{2+} from the cell as long as an inwardly directed electrochemical Na^+ gradient exists, but changes in V_m may lead to transient changes in $[Ca^{2+}]_i$. For example, the cell will gain Ca^{2+} if $[n z_1 F(V_{Na} - V_m)] < [z_2 F(V_{Ca} - V_m)]$, as may occur during depolarization of the membrane to potentials positive to V_R. Although there are several experimental findings that would disagree with important details in Mullins' hypothesis, it is clear that the consequences of an electrogenic Na–Ca exchange for cell function cannot be ignored. Another system by which the cell gains Ca^{2+} during depolarization is the slow inward Ca current (see Reuter, 1974).

The role played by ATP in the Na–Ca exchange system is not completely understood. It seems that only the Michaelis constant (K_m), but not the maximal velocity (V_{max}) of Na^+ activation of Ca^{2+} efflux, depends on ATP (Baker and Glitsch, 1973; Di Polo, 1977; Blaustein, 1977; Jundt and Reuter, 1977). There is little information whether or not ATP is hydrolyzed for this purpose.

According to recent ion activity measurements with ion-sensitive electrodes in resting mammalian cardiac muscle (Lee and Fozzard, 1975; Deitmer and Ellis, 1978; Marban *et al.*, 1980; Lee *et al.* 1980), the following extra- and intracellular Na^+ and Ca^{2+} ion activities must be inserted into equation (1): $\alpha Na_o^+ = 114$ mM, $\alpha Na_i^+ = 5$–7 mM, $\alpha Ca_o^{2+} = 0.5$–0.6 mM, $\alpha Ca_i^{2+} \approx 0.1$ μM. With $V_m = -80$ mV, the coupling coefficient n in equation (1) has to be greater than 2 in order to reduce αCa_i^{2+} to the measured value. With $n = 2$, the calculated equilibrium value of αCa_i^{2+} would be about 1 μM, with $n = 3$ about 2.5 nM.

Following this general introduction, the remainder of the review will deal with experimental results in support of Na–Ca countertransport in the heart. Because space is limited, only a selected number of papers relevant to this subject can be discussed.

2. FLUX MEASUREMENTS

2.1. Intact Preparations

If Na^+ and Ca^{2+} compete for an anionic site at the outer surface of the cell membrane, ^{45}Ca uptake should increase when $[Na^+]_o$ is reduced. This was first demonstrated by Niedergerke (1963) in the frog heart and later confirmed by several investigators for various other preparations (e.g., Langer, 1964; Baker and Blaustein, 1968).

The concept of a Na–Ca countertransport system has been developed by Reuter and Seitz (1968) primarily on the basis of ^{45}Ca efflux measurements. They found that a minor fraction of ^{45}Ca efflux depended on $[Ca^{2+}]_o$, which was interpreted as Ca–Ca exchange diffusion (Reuter and Seitz, 1967, 1968). In the absence of $[Ca^{2+}]_o$, a much larger portion (more than 70%) of ^{45}Ca efflux could be inhibited by removal of $[Na^+]_o$, when external NaCl was replaced by LiCl, KCl, choline Cl, or sucrose. For this phenomenon, Reuter and Seitz (1968) suggested that the uphill transport of Ca_i^{2+} was coupled to a large extent to the downhill movement of Na_o^+. This heteroexchange diffusion had a Q_{10} of 1.35, and in the presence of normal $[Na^+]_o$ was not inhibited, but rather increased by metabolic poisons (Reuter and Seitz, 1968; Blaustein and Hodgkin, 1969). A prediction of this concept was that the net Ca concentration in the tissue should rise if $[Na^+]_o$ is lowered, and that it should decrease again upon readdition of $[Na^+]_o$. This prediction has been verified experimentally for cardiac muscle (Reuter and Seitz, 1968; Busselen and van Kerkhove, 1978) and squid axon (Requena *et al.*, 1979).

Figure 1A shows a normalized plot of the Na_o^+ dependence of ^{45}Ca efflux. The experimental results obtained from guinea pig auricles (Jundt and Reuter, 1977) have been fitted by the equation

$$v = \frac{V_{max}}{1 + \left\{ \dfrac{K_{Na}}{Na^+} \right\}^n} \tag{4}$$

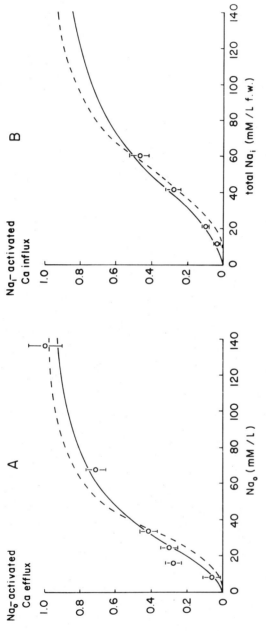

Figure 1. (A) Fraction of ^{45}Ca *efflux* that depends on the *external* Na concentration in guinea pig auricles. This fraction was estimated from changes in the rate coefficient of ^{45}Ca efflux when $[Na]_o$ was varied. Experimental points are means ($\pm$S.E.) of five auricles each (replotted from Jundt and Reuter, 1977). The data were fitted by equation (4) with $n = 2$ (solid line) or $n = 3$ (dashed line) and $K_{Na} = 39mM$. (B) Fraction of ^{45}Ca *influx* that depends on the *internal* Na concentration in guinea pig auricles. ^{45}Ca influx was measured during a 10-min incubation period in ^{45}Ca-containing Tyrode's solution after $[Na]_i$ had been changed by various means. The results have been replotted from Fig. 3 in Glitsch *et al.* (1970). The data were fitted by equation (4) with $n = 2$ (solid line) or $n = 3$ (dashed line) and $K_{Na} = 60$ mM.

where v is the rate of Na_o^+-activated Ca efflux in a Ca^{2+}-free medium with various Na^+ concentration ($V_{max} = 1$); K_{Nao} is the apparent half-saturating value of $[Na]_o$ ($= 39$ mM); n was assumed to be either 2 (solid line) or 3 (dashed line). It must be stressed that the experimental data do not discriminate between a stoichiometry of 2 or 3. Curve fittings like that in Fig. 1 have the inherent problem that they almost never provide an unambiguous answer as to the true stoichiometry. To solve this question one has to know, and keep constant, all forces that determine the flows through this system [see equation (2)]. So far, this has not been achieved in any biological preparation. Neither has the question been answered as to whether or not Na^+ ions act cooperatively on its transport site. In case of positive cooperativity, n would be a Hill coefficient providing a lower estimate of the degree of cooperativity. Although ATP seems to affect the affinity of Na and/or Ca binding (Baker and Glitsch, 1973; Di Polo, 1977; Blaustein, 1977; Jundt and Reuter, 1977), the way it does this is entirely unclear. All this becomes extremely important if one wants to define the molecular properties of this system. In intact cardiac preparations, no clear effect of membrane potential on Na_o^+-dependent ^{45}Ca efflux has been observed (Jundt *et al.*, 1975; Busselen and van Kerkhove, 1978; Reuter, unpublished). Possibly a release of Ca^{2+} from internal stores could have masked any depolarization-induced inhibition of ^{45}Ca efflux.

An important test of the Na–Ca exchange hypothesis was to show that an increase in $[Na^+]_i$, and hence a reduction or inversion of the Na^+ gradient, promotes Ca influx. This type of experiment was first done by Baker *et al.* (1969) in squid axon. Comparable data were obtained by Glitsch *et al.* (1970) in guinea pig auricles. Their results are replotted in Fig. 1B. The experimental data have been fitted again by equation (4) with $K_{Nai} = 60$ mM and $n = 2$ (solid line) or $n = 3$ (dashed line). Of course, all reservations concerning this curve-fitting procedure expressed above also apply to Fig. 1B. In addition, K_{Nai} is certainly subject to substantial error, as total Na_i instead of the relevant αNa_i^+ is plotted on the abscissa. However, αNa_i^+ is only a small fraction of total Na_i. Lee and Fozzard (1975) found an apparent activity coefficient for Na_i (γNa_i) of only 0.175 in rabbit papillary muscle. It is unknown whether γNa_i is constant when total Na_i is increased.

The results of Glitsch *et al.* (1970) shown in Fig. 1B have at least been qualitatively confirmed by Tillisch and Langer (1974) in rabbit myocardium, and very elegantly by Fosset *et al.* (1977) in embryonic cardiac tissue culture. Moreover, when the Na pump is inhibited by cardiac glycosides, a fraction of total ^{22}Na efflux becomes sensitive to $[Ca^{2+}]_o$ (Bridge *et al.*, 1981; Katzung and Reuter, unpublished; see also Deitmer and Ellis, 1978). This is to be expected if Na_i^+ does indeed exchange for Ca_o^{2+} as suggested by Fig. 1B.

It is of considerable functional significance for the heart that drugs and poisons that elevate $[Na^+]_i$ (cardiac glycosides, veratridine, sea anemone toxin ATX_{II}) also increase Ca influx (Courand *et al.*, 1976; Fosset *et al.*, 1977; Romey *et al.*, 1980). Drugs and poisons that release Ca^{2+} from internal stores and thereby increase Ca_i^{2+} (caffeine, theophylline, papaverine, cyanide, etc.) transiently stimulate Na-dependent Ca efflux (Jundt *et al.*, 1975; Busselen and van Kerkhove, 1978; Reuter, unpublished).

Although the quantitative accuracy of tracer flux experiments in intact cardiac muscle is certainly limited, the qualitative insight into Na–Ca countertransport gained by these methods has been considerable. The results discussed above can be summarized as follows:

1. Most of the coupled fluxes suggested by the Na–Ca exchange hypothesis have indeed been observed with tracer methods in various cardiac preparations.
2. The inwardly directed Na^+ gradient is the primary energy source for net extrusion of Ca^{2+} from cardiac cells through this system. If $[Na^+]_o$ is reduced or $[Na^+]_i$ is increased, the cells will accumulate Ca^{2+} to a new steady state. Thus, depending

on the size and the direction of the electrochemical Na^+ gradient, the cell can either lose or gain Ca^{2+} via this system.

3. ATP may modify the affinities of Na^+ (and Ca^{2+}) for the transport sites.
4. The exact coupling ratio (stoichiometry) of Na–Ca exchange is unclear, but probably greater than 2 Na^+ : 1 Ca^{2+}.
5. Drugs and poisons that release Ca^{2+} from internal stores stimulate Na_o-dependent Ca efflux, while those that increase αNa^+_i stimulate Na_i-dependent Ca influx.

2.2 Cardiac Sarcolemmal Vesicles

The recent development of suitable vesicular surface membrane preparations from mammalian cardiac muscle by several groups (for review see Sulakhe and St. Louis, 1980) provides an excellent opportunity for further advances in the experimental analysis of Na–Ca countertransport. So far, most of the basic findings obtained from flux measurements in intact cardiac preparations have been confirmed and extended in isolated sarcolemmal vesicles. Reeves and Sutko (1979) were able to show that sarcolemmal vesicles from rabbit heart rapidly accumulated Ca^{2+} in the absence of ATP if an outwardly directed Na^+ gradient was established. ^{45}Ca-loaded vesicles lost their Ca^{2+} only in the presence of an inwardly directed Na^+ gradient. Dissipation of the Na^+ gradients by nigericin or narasin stopped the Na-dependent Ca movements. Li^+ or K^+ could not replace Na^+ in this countertransport. The rate of Ca^{2+} uptake was half-saturated at a $[Ca^{2+}]_o$ of 18 μM, while half-maximal inhibition of Ca^{2+} uptake by external Na^+ occurred at a $[Na^+]_o$ of 16 mM. La^{3+} inhibited Na-dependent Ca fluxes.

Pitts (1979) demonstrated most clearly that ^{45}Ca efflux is coupled to ^{22}Na influx in sarcolemmal vesicles obtained from dog heart. Although he estimated a 3 Na^+ : 1 Ca^{2+} "stoichiometry" from the initial flux rates, close inspection of the corresponding curves leaves some doubts about his estimates. The true initial rates could not be determined very accurately, and the pertinent driving forces for the respective fluxes were hardly constant. Similar problems as discussed in relation to Fig. 1A also apply to Pitts' results. On the other hand, Pitts (1979) showed most convincingly that Na–Ca exchange occurred in the same population of vesicles that also exhibited ATP-dependent Na-pump activity. This would not have been expected if the observed Na–Ca exchange had to be attributed to mitochondria (Carafoli and Crompton, 1978) or sarcoplasmic reticulum.

An additional interesting observation has been made by Caroni and Carafoli (1980) in sarcolemmal vesicles from dog heart. Not only did they confirm the findings by Reeves and Sutko (1979) and Pitts (1979) on Na–Ca countertransport, but they also detected that even in the absence of an outwardly directed gradient vesicles accumulated Ca^{2+} when ATP was present. The K_m for this ATP-driven Ca^{2+} transport was between 0.2 and 0.6 μM. The accumulated Ca^{2+} could be rapidly released by the addition of external Na^+. They concluded from their results that in cardiac sarcolemma an ATP-driven Ca transport operates in parallel to Na–Ca exchange.

Similar results have been obtained by Trumble *et al.* (1980). Compared to Na–Ca exchange, the ATP-driven Ca pump had a very low transport capacity. Caroni *et al.* (1980) monitored net Ca^{2+} uptake via Na–Ca exchange by means of Ca^{2+}-sensitive electrodes, or with the metallochromic dye arsenazo III, and claimed a 10-fold lower K_m for Ca^{2+} (1.5 μM) than that reported by Reeves and Sutko (1979). If this is correct, the K_m for the two Ca transport systems would be different by a factor of 2–5.

The strongest evidence that Na–Ca exchange is electrogenic in cardiac muscle stems from the most recent work with sarcolemmal vesicles, by making use of tetraphenylphosphonium ions (TPP$^+$) as a membrane potential indicator. Both Reeves and Sutko (1980) and Caroni *et al.* (1980) reported an accumulation of TPP$^+$ within the vesicles that occurred together with a Ca^{2+} uptake driven by an outwardly directed Na$^+$ gradient. This indicates that Na–Ca exchange can, at least transiently (Caroni *et al.*, 1980), establish an inside negative potential within these vesicles.

3. ION-SENSITIVE ELECTRODES

Does Na–Ca countertransport indeed participate in regulating free Ca^{2+} activity (αCa_i^{2+}) in the myoplasm? The new development of Ca^{2+}-sensitive electrodes has provided some preliminary answers to this important question. In rabbit ventricular muscle, Lee *et al.* (1980) found a more than threefold increase in myoplasmic αCa_i^{2+} when they reduced [Na$^+$]$_o$ from 153 mM to 20 mM. Together with the rise in αCa_i^{2+}, there occurred an increase in resting tension and in force of contraction. Similarly, Marban *et al.* (1980) reported a rise in αCa_i^{2+} in ferret papillary muscles bathed in Na-free solution. However, in this preparation the change in αCa_i^{2+} was below the contracture threshold, and only in the presence of elevated [K$^+$]$_o$ (30 mM) did Na$^+$ removal cause a much larger increase in αCa_i^{2+} that led to a contracture. The results of both groups are compatible with the idea that αCa_i^{2+} is at least partially controlled by Na–Ca exchange.

Deitmer and Ellis (1978) have used Na$^+$-selective electrodes in sheep Purkinje fibers. They showed a reduction of αNa_i^+ when [Ca^{2+}]$_o$ was increased. The reduction was particularly prominent when the Na pump was inhibited by strophantidin. In fact, [Ca^{2+}]$_o$ seemed to be the main regulator of αNa_i^+ during Na-pump inhibition. Neither Mn^{2+} nor Mg^{2+} could replace Ca^{2+} in this function.

Further use of ion-sensitive electrodes will be of great help in defining the importance of Na–Ca exchange in the regulation of Ca$^{2+}_i$ and Na^+_i in the myoplasm of cardiac cells.

4. FUNCTIONAL TESTS

Initial evidence for an interaction between Na$^+$ and Ca^{2+} at the cell surface came from contractile tension measurements in the frog heart (Wilbrandt and Koller, 1948; Lüttgau and Niedergerke, 1958). This implies that Na–Ca exchange has a major impact on the mechanical function of the heart. However, although it is clear that αCa_i^{2+} is one of the most important factors in force development of the heart (for review see Chapman, 1979), the exact relationship between Na–Ca exchange and αCa_i^{2+} is not clear. Other important factors in the regulation of αCa_i^{2+} are voltage-sensitive Ca^{2+} channels that regulate the passive membrane permeability to Ca^{2+}ions, and ATP-driven Ca pumps in the sarcoplasmic reticulum and probably also in the surface membrane. Moreover, recent evidence (see Marban *et al.*, 1980) indicates that force is not a unique function of αCa_i^{2+}, but that the pCa–tension relationship could be modified by other factors, e.g., phosphorylation of troponin I.

Therefore, it seems that mechanical force measurements may not be very useful for further clarification of the behavior of the Na–Ca countertransport system. For example, if one wants to draw conclusions from force measurements on the stoichiometry and electrogenicity of Na–Ca exchange in the membrane (e.g., Horackova and Vassort, 1979), too many simplifying assumptions have to be made to be plausible. Force measurements have their place in functional tests where Na–Ca exchange could be important (see Chapman,

1979), but hardly in kinetic or even molecular investigations and interpretations of this transport system.

5. CONCLUSIONS

Na–Ca countertransport is a powerful exchange system that is involved in the regulation of the free intracellular Ca^{2+} (and Na^+) concentration in cardiac muscle and other biological tissues. While this transport system has so far been characterized in cardiac muscle in a qualitative sense, many quantitative questions are unanswered. These include (1) interaction (competition?, cooperativity?) of Na^+ and Ca^{2+} ions at the transport sites, (2) the coupling ratios and affinities of these ions, (3) the role played by ATP in regulating the affinities, (4) the contribution of the system to the electrical behavior of the membrane, and (5) its contribution to the mechanical behavior of cardiac cells. The development of suitable membrane preparations and new techniques has provided encouraging progress in this area. But there still is a long way to go until we understand the biophysical and biochemical behavior and ultimately the molecular structure of this transport system.

REFERENCES

Baker, P. F. (1972). *Prog. Biophys. Mol. Biol.* **24,** 177–223.
Baker, P. F., and Blaustein, M. P. (1968). *Biochim. Biophys. Acta* **150,** 167–170.
Baker, P. F., and Glitsch, H. G. (1973). *J. Physiol.* **233,** 44P–46P.
Baker, P. F., Blaustein, M. P., Hodgkin, A. L., and Steinhardt, R. A. (1969). *J. Physiol.* **200,** 431–458.
Blaustein, M. P. (1974). *Rev. Physiol. Biochem. Pharmacol.* **70,** 34–82.
Blaustein, M. P. (1977). *Biophys. J.* **20,** 79–111.
Blaustein, M. P., and Hodgkin, A. L. (1969). *J. Physiol.* **200,** 497–527.
Bridge, M. H. B., Cabeen, W. R., Langer, G. A., and Reeder, S. (1981). *J. Physiol.* **316,** 555–574.
Busselen, P., and van Kerkhove, E. (1978). *J. Physiol.* **282,** 263–283.
Carafoli, E., and Crompton, M. (1978). *Curr. Top. Membr. Transp.* **10,**151–216.
Caroni, P., and Carafoli, E. (1980). *Nature (London)* **283,** 765–767.
Caroni, P., Reinlib, L., and Carafoli, E. (1980). *Proc. Natl. Acad. Sci. USA* **77,** 6354–6358.
Chapman, R. A. (1979). *Prog. Biophys. Mol. Biol.* **35,** 1–52.
Courand, F., Rochat, H., and Lissitzky, S. (1976). *Biochim. Biophys. Acta* **433,** 90–100.
Deitmer, J. W., and Ellis, D. (1978). *J. Physiol.* **277,** 437–453.
Di Polo, R. (1977). *J. Gen. Physiol.* **69,** 795–813.
Fosset, M., De Barry, J., Lenoir, M.-C., and Lazdunski, M. (1977). *J. Biol. Chem.* **252,** 6112–6117.
Glitsch, H. G., Reuter, H., and Scholz, H. (1970). *J. Physiol.* **209,** 25–43.
Horackova, M., and Vassort, G. (1979). *J. Gen. Physiol.* **73,** 403–424.
Jundt, H., and Reuter, H. (1977). *J. Physiol.* **266,** 78P–79P.
Jundt, H., Porzig, H., Reuter, H., and Stucki, J. W. (1975). *J. Physiol.* **246,** 229–253.
Langer, G. A. (1964). *Circ. Res.* **15,** 393–405.
Lee, C. O., and Fozzard, H. A. (1975). *J. Gen. Physiol.* **65,** 695–708.
Lee, C. O., Uhm, D. Y., and Dresdner, K. (1980). *Science* **209,** 699–701.
Lüttgau, H. C., and Niedergerke, R. (1958). *J. Physiol.* **143,** 486–505.
Marban, E., Rink, T. J., Tsien, R. W., and Tsien, R. Y. (1980). *Nature (London)* **286,** 845–850.
Mullins, L. J. (1976). *Fed. Proc.* **35,** 2583–2588.
Mullins, L. J. (1979). *Am. J. Physiol.* **236,** C103–C110.
Mullins, L. J., and Brinley, F. J., Jr. (1975). *J. Gen. Physiol.* **65,** 135–152.
Niedergerke, R. (1963). *J. Physiol.* **167,** 515–550.
Pitts, B. J. R. (1979). *J. Biol. Chem.* **254,** 6232–6235.
Reeves, J. P., and Sutko, J. L. (1979). *Proc. Natl. Acad. Sci. USA* **76,** 590–594.
Reeves, J. P., and Sutko, J. L. (1980). *Science* **208,** 1461–1464.

Requena, J., Mullins, L. J., and Brinley, F. J. (1979). *J. Gen. Physiol.* **73,** 327–342.

Reuter, H. (1974). *Circ. Res.* **34,** 599–605.

Reuter, H., and Seitz, N. (1967). *Naunyn-Schmiedeberg's Arch. Exp. Pathol. Pharmakol.* **257,** 324.

Reuter, H., and Seitz, N. (1968). *J. Physiol.* **195,** 451–470.

Romey, G., Renaud, J. F., Fosset, M., and Lazdunski, M. (1980). *J. Pharmacol. Exp. Ther.* **213,** 607–615.

Sulakhe, P. V., and St. Louis, P. J. (1980). *Prog. Biophys. Mol. Biol.* **35,** 135–195.

Tillisch, J. H., and Langer, G. A. (1974). *Circ. Res.* **34,** 40–50.

Trumble, W. R., Sutko, J. L., and Reeves, J. P. (1980) *Life Sci.* **27,** 207–214.

Wilbrandt, W., and Koller, H. (1948). *Helv. Physiol. Pharmacol. Acta* **6,** 208–221.

83

Aspects of Gastric Proton-Transport ATPase

G. Sachs, H. R. Koelz, T. Berglindh, E. Rabon, and G. Saccomani

1. INTRODUCTION

The central role of H^+ transport in a variety of biological membranes has been firmly established by experiments of the last decade or so (Mitchell, 1966; Racker, 1976). Of direct interest to this review is the mechanism of proton transport by the mammalian gastric mucosa, a tissue where the H^+ gradient reaches a value of almost 10^7. Whether the mechanism in this cell has anything in common with other H^+ ion transport systems is not clear at present. It does seem that at least part of the process is mechanistically similar to the mechanism of Na^+ and K^+ transport in eukaryotic cells via the Na^+/K^+-ATPase (Taniguchi and Post, 1975) and hence is likely to be of general as well as special interest. At this time the only well-defined component of acidification is the H^+/K^+-ATPase initially described by Forte and his collaborators (Ganser and Forte, 1973a). It is this enzyme, and the evidence for its role in acid secretion that will be described in this brief review.

2. CELLULAR SITE OF ACID SECRETION

Based on morphological observations it has been suspected that the intracellular canaliculus of the parietal cell is the site of acid secretion. This was recently shown directly. An acidic intracellular compartment will trap a weak base such as aminopyrine. Because aminopyrine has a pK_a of 5, the base will accumulate only in compartments with a pH of less than 5. This excludes the possible contribution of other acidic compartments such as lysosomes. This technique, which uses trace concentrations of [^{14}C]aminopyrine, has become the standard method to measure acid secretion in multiple samples of isolated gastric

G. Sachs, H. R. Koelz, T. Berglindh, E. Rabon, and G. Saccomani • Laboratory of Membrane Biology, University of Alabama, Birmingham, Alabama 35294.

glands (Berglindh, 1978). The acid compartment can be visualized when the concentration of aminopyrine in the medium is raised to 1 mM, because accumulation now leads to osmotic swelling of this space. The identification of the acid-containing compartment was confirmed by the demonstration that a metachromatic fluorescent weak base such as acridine orange also accumulated in the intracellular canaliculus of the parietal cells (Berglindh et al., 1980a).

3. ROLE OF K^+

In the intact animal, in the in vitro mucosa, and in isolated gastric glands (Hirschowitz and Sachs, 1972; Harris and Edelman, 1960), the presence of K^+ in the medium is necessary for acid secretion. Whereas the simplest interpretation of this finding is that K^+ is required in the cytosol for acid secretion, another possibility is that as K^+ in the medium is reduced, the ratio of Na^+/K^+ in the cytoplasm increases and it is the increase of this cytosolic ratio that reduces acid secretion. Both possibilities have been shown to be the case. That intracellular Na^+ is in fact inhibitory to H^+ secretion is evidenced by the finding that acid secretion in isolated gastric glands is abolished by K^+ removal from the medium but restored by additional removal of Na^+ although even lower cellular K^+ levels are measured in the latter case (Berglindh et al., 1979). Thus, the question as to how much K^+ is actually required for H^+ transport can only be answered in the absence of Na^+ and using a method that allows adequate variation of intracellular K^+. Probably the most quantitative data on this question have been obtained in isolated gastric glands by treating the glands suspended in Na^+-free medium with ouabain and amphotericin B and varying medium K^+. Figure 1 shows the response of aminopyrine accumulation to variation of K^+ in the medium, in the absence and presence of histamine. The absolute requirement of K^+ for acid secretion is demonstrated by the finding that the aminopyrine accumulation ratio is close to unity with very low cellular K^+ and increases as the K^+ levels are raised. Without secretagogue, the apparent $K_{0.5}$ for K^+ is 18 mM; with 10^{-4} M histamine the $K_{0.5}$ is shifted to 11 mM. It should be noted that the effect of medium K^+ in isolated gastric glands may differ from the situations in the intact stomach. In isolated glands, K^+ may reach the acid secretory membrane from both the luminal and the cytosolic side whereas in the intact stomach it can be provided from the cytosol of the parietal cell only (Koelz et al., 1981). The complexity of the system will become clearer as we discuss properties of the K^+-activated ATPase in later sections.

4. ROLE OF ATP

The field of acid secretion has been furrowed by controversy between proponents of a redox-based proton pump vs. an ATP-based proton pump (Hersey, 1977). Indirect experiments whether using a battery of inhibitors (Sachs, 1968) or by measuring redox changes in the tissue cannot be interpreted exclusively (Hersey, 1977). The central role of ATP in the mechanism of acid secretion is shown from the following experiments using isolated gastric glands. Inhibition of mitochondrial ATP synthesis by amytal, azide, cyanide, or oligomycin essentially abolishes acid secretion. When glands are made permeable to trypan blue and ATP by either high-voltage shocking or by short exposure to digitonin (20 μg/ml),

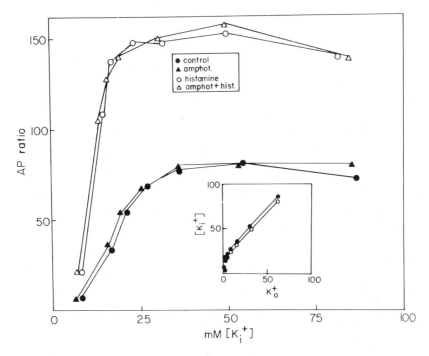

Figure 1. Aminopyrine (AP) accumulation ratio of rabbit gastric glands as a function of intracellular K^+ concentration. Glands were washed in Na^+K^+-free medium containing 150mM tetramethyl ammonium chloride, 10^{-4} M ouabain, and 10^{-5} M amphotericin B. The variation of cellular K^+ was obtained by altering K^+ concentrations in the incubation medium. Note that the apparent $K_{0.5}$ of cellular K^+ is shifted from 18 mM in control glands to 11 mM in histamine-stimulated (10^{-4} M) glands. The presence of amphotericin B in the incubation medium did not affect the results.

addition of ATP to the medium restores acid secretion to control values even in the presence of mitochondrial inhibitors (Berglindh *et al.*, 1980). ATP in digitonin-treated glands in the presence of oligomycin is also able to overcome anoxia (D. Malinowska *et al.*, 1981).

Thus, from the intact parietal cell, we learn that acid secretion occurs across the membrane of the secretory canaliculus, that it is K^+ and ATP dependent, and that it is unlikely that any redox system contributes to H^+ transport. Evidently therefore, the role of the K^+-activated ATPase mentioned above is of paramount interest.

5. LOCALIZATION OF THE ATPase

Centrifugal and free-flow electrophoretic techniques have allowed large-scale purification of the membrane containing the K^+-activated ATPase (Saccomani *et al.*, 1977). This membrane fraction has then been used to generate an antibody, which after purification provides a single band on rocket electrophoresis. This antibody binds selectively to the parietal cell where it can be visualized at the microvilli of the secretory canaliculus (Saccomani *et al.*, 1979a). Thus, the ATPase is appropriately located at the site of acid secretion.

6. CATALYTIC PROPERTIES OF THE ATPase

Even prior to the description of the K^+ activation of an ATPase in gastric mucosa, a K^+-activated pNPPase was described. This suggested that an enzyme with properties rather similar to the better known Na^+/K^+-ATPase might be responsible for acid secretion (Sachs and Hirschowitz, 1968). For example, both the Na^+/K^+- and the K^+-ATPase form a phosphorylated intermediate whose dephosphorylation is accelerated by the presence of K^+ (Ray and Forte, 1976; Skou, 1965). The gastric enzyme was then studied by transient kinetic techniques to determine the overall reaction pathway (Wallmark and Mardh, 1979; Wallmark *et al.*, 1980; Stewart *et al.*, 1981).

The reaction pathway determined thus far is shown in *Fig. 2*. Both transient and steady-state kinetics provide evidence for an occluded form of the enzyme, depicted as $E \cdot K_o^+$, which may be considered as the starting point of the reaction sequence. Binding of H_o^+ and ATP (in random order) is accompanied by release of K_o^+ from a low-affinity site. The active enzyme is thus the protonated form of the enzyme–substrate complex. Activation of the enzyme is inhibited by K^+ bound at the low-affinity site. The concentration of K^+ needed for inhibition rises with increasing ATP concentration and with decreasing pH at the external vesicle face. The occluded form of the enzyme can also be stabilized by Na^+, which suggests an explanation of the inhibitory role of cytoplasmic Na^+ on acid secretion in the intact parietal cell.

Phosphorylation of the enzyme is initiated by the addition of Mg^{2+}. If enzyme and ATP are premixed, before the addition of Mg^{2+}, the rate of phosphorylation is faster than if Mg^{2+} is prebound, or Mg^{2+} and ATP added simultaneously. This provides evidence that the rate-limiting step for phosphorylation of active enzyme is ATP binding.

The rate of phosphorylation by Mg ATP increases as the pH is reduced from 8 to 5.5. If the enzyme is preincubated with ATP in medium buffered at either pH 8 or pH 5.5 and phosphorylation initiated by adding Mg^{2+} and excess buffer at pH 5.5, the rate of phosphorylation is independent of the initial pH of the enzyme during the preincubation with ATP. Thus, protonation following the binding of ATP is at least as fast as the transphosphorylation reaction.

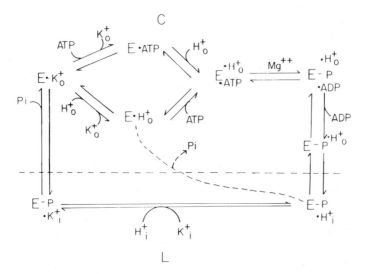

Figure 2. A schematic representation of the catalytic sequence of Mg^{2+}- and H^+/K^+-ATPase of gastric mucosa.

The phosphorylated group on the enzyme appears to be an acyl phosphate because of its sensitivity to high pH and hydroxylamine (Ray and Forte, 1976). We have so far not obtained direct evidence for an ADP-sensitive form of EP although it has been claimed that N-ethylmaleimide fixes EP in an ADP-sensitive form. On the other hand, the purified membrane fraction displays an ATP–ADP exchange insensitive to diadenosine pentaphosphate (Rabon and Sachs, 1981), suggesting the presence of this form during the reaction sequence. The generation of the ADP-insensitive form of EP is shown as occurring along with the appearance of the bound H^+ on the internal or *trans* face of the enzyme. The presence of K_i^+ on the *trans* face of the enzyme accelerates both loss of H_i^+ into the intravesicular space and dephosphorylation of the unstable E–P with a $K_{0.5}$ for K^+ at neutral pH of 200 μM. Decreasing pH progressively reduces the K_i^+ acceleration of dephosphorylation. Dephosphorylation of the E–$P \cdot K_o^+$ complex results in E $\cdot$ K$_o^+$, the occluded form of the enzyme, thus completing the catalytic cycle.

From the above reaction sequence it is evident that, depending on the concentration and localization with respect to the *cis* or *trans* face of the enzyme, both K^+ and H^+ may accelerate or inhibit the overall ATP hydrolysis. For example, with equal solutions on both faces of the enzyme, low K^+ levels stimulate ATP hydrolysis due to accelerated dephosphorylation of E–P whereas high levels are inhibitory by reducing the rate of enzyme phosphorylation. A significant decrease in the apparent affinity for ATP is stabilized by saturation of the low-affinity K^+ site.

The enzyme as mentioned above hydrolyzes phosphate esters such as acetyl phosphate or p-nitrophenyl phosphate. Hydrolysis of pNPP appears to be nonproductive in terms of transport of either H^+ or K^+, and also appears to occur on the same side of the enzyme as the ATP-binding site. The active sites for pNPP and ATP hydrolysis can be distinguished.

7. GROUPS INVOLVED IN CATALYSIS

A considerable effort has been expended in attempting to define the residues involved either in the hydrolysis of ATP or in the transport of cations in a variety of transport enzymes. Largely this has been done with relatively site-specific reagents.

SH Groups. p-Chloromercuribenzene sulfonic acid, for example, has been shown to inhibit ATPase activity (Sachs *et al.*, 1976). The SH reagent 5,5'-dithiobis-2-nitrobenzoic acid has been used to determine the role of SH in catalysis by the gastric ATPase. Based on differences in reactivity of the enzyme to the reagent in the presence and absence of Mg^{2+}, it has been suggested that the SH group involved is not at the catalytic site (Schrijen *et al.*, 1981).

Amino Groups. Reagents such as butanedione (Schrijen *et al.*, 1980) have shown that groups such as arginine may be involved in the active center of the enzyme. Inhibition by this reagent is prevented by ATP.

Histidine. The reagent diethylpyrocarbonate (DEPC) is relatively specific for this residue and again inhibition can be prevented by the presence of ATP. It is also of interest to note that DEPC has no effect on pNPPase activity. Hence, the DEPC-treated ATPase is converted into a simple phosphatase, and histidine catalysis is not involved in hydrolysis of the latter substrate (Saccomani *et al.*, 1980).

Carboxyl Groups. Both N,N'-dicyclohexylcarbodiimide and N-ethoxycarbonyl-2-ethoxy-1,2-dihydroquinoline (EEDQ) (Sachs *et al.*, 1976; Saccomani *et al.*, 1980) inhibit the ATPase and pNPPase activity. In contrast to the above reagents, ATP does not protect but exaggerates the inhibition found with EEDQ. Phosphorylation is inhibited, but what is particularly intriguing is that K^+ only on the vesicles' interior prevents EEDQ inhibition of the ATPase. The level of K^+ required is close to the K^+ concentration for activation of ATP hydrolysis.

All the above reagents, where investigated, appear to exert their inhibitory effects by reacting with a single type of group. It is obvious therefore that the catalytic center and the transport reaction involve the cooperation of several residues on the protein.

8. ENZYME COMPOSITION

When the enzyme-containing vesicles have been purified by free-flow electrophoresis, SDS gel electrophoresis shows that about 90% of the protein can be accounted for by a peptide at the 100,000-molecular-weight region. However, the peptide is heterogeneous as shown by two different lines of evidence.

Treatment of the enzyme with trypsin leads to peptide fragments along with inhibition of activity. After extensive tryptic hydrolysis, about one-third of the Coomassie blue staining material is left at the 100,000 region and is glycoprotein in nature (Saccomani *et al.*, 1979b). The presence of ATP protects the enzyme against inactivation, and now two-thirds of the protein is left at the 100,000 region based on SDS gel analysis. This can be interpreted as showing that the enzyme is composed of three nonidentical subunits: one glycoprotein subunit trypsin-insensitive, a second subunit that is trypsin-sensitive both in the presence and absence of ATP, and a third trypsin-sensitive subunit that is protected from digestion by ATP.

Isoelectric focusing shows that there is considerable heterogeneity of the isoelectric point. A major glycoprotein band occurs with a pH of 8.9, and a cluster of glycopeptides at a pH between 6.1 and 6.7. Second-dimension electrophoresis shows that the major bands have a molecular weight close to 100,000 (Sachs *et al.*, 1980).

An alternative means of investigating structure is to determine molecular weight by irradiation with electrons. From this the apparent molecular weight is 270,000 or thereabouts, suggesting an $\alpha\beta\gamma$, $\alpha_2\beta$, or $\alpha\beta_2$ type of arrangement in the membrane (Saccomani *et al.*, 1981). This can be contrasted to the $\alpha_2\beta_2$ arrangement of the Na^+/K^+-ATPase and the considerably more complex mitochondrial H^+ translocator (Esche and Allison, 1979; Kepner and Macey, 1968). It might appear to be a rule that the transport ATPases exist as dimers of at least the catalytic subunit in membrane-bound form. A so far insurmountable problem has been to prove what the role of the dimer is in transport or catalysis.

9. H⁺ TRANSPORT BY THE GASTRIC ATPase

A major advance in understanding acid secretion was the demonstration that H^+ uptake into dog gastric microsomes could be induced by the addition of ATP in the presence of K^+ (Lee *et al.*, 1974). The ionic requirements of H^+ transport by H^+/K^+-ATPase-containing vesicles are consistent with the catalytic cycle of the enzyme as discussed above. ATP-induced H^+ transport, like ATP hydrolysis, is absolutely dependent on the presence of K^+ in the vesicle interior (the *trans* side of the enzyme) and is inhibited by high external K^+ and Na^+ (Ganser and Forte, 1973b). Because the K^+ permeability of freshly prepared, "tight" vesicles is low (Rabon *et al.*, 1980), preincubation with K^+ is needed in order to detect H^+ transport. Also, due to the fact that protons enter the vesicle interior in exchange for K^+, sustained H^+ transport and maximal pH gradients can be achieved only if K^+ entry is facilitated by valinomycin. This indicates that the availability of internal K^+ is the rate-limiting step in this system. Under these conditions, determination of the $K_{0.5}$ of K^+ with initially equal K^+ concentrations in medium and intravesicular space gives a value of about 10 mM (Stewart *et al.*, 1981), which is close to the $K_{0.5}$ in the intact parietal cell (Koelz *et al.*, 1981). Thus, the correlation between the K^+ requirement, catalytic properties, the transport characteristics of the enzyme in vesicular form, and H^+ secretion by the isolated parietal cell provides important evidence for the central role of the H^+/K^+-ATPase in gastric acid secretion.

The magnitude of the maximal pH gradient reached in vesicles seems to be smaller as compared to the intact stomach or to isolated glands. Using the pH electrode technique, approximately 100 mmoles H^+/liter vesicle volume disappears from the medium upon addition of ATP, implying an internal pH close to 1. However, by using measures of internal vesicle pH, the maximal gradient developed at an external pH of 6.1 did not exceed 3–4 units, indicating that 90% or more of protons disappearing from the medium are buffered (Rabon *et al.*, 1978). It is not clear whether the expected gradient cannot be reached because of a deficient pump rate or because of a proton leak either conductively or as HCl. However, the deficiency in the pH gradient formed, whatever its basis, makes a measurement of the stoichiometry particularly important. As usual in the field of proton pump stoichiometry, there is disagreement. Whereas one group determined the stoichiometry to be unity (Reenstra *et al.*, 1980) neglecting Mg^{2+}-ATPase activity, we have published a stoichiometry of 3 to 4 (Sachs *et al.*, 1976) but ATPase activity and proton transport were not measured under absolutely identical conditions. Upon reexamining the question, two conditions were chosen: with no K^+ gradient and with an outward K^+ gradient as illustrated in Fig. 3, which is an Eadie–Hofstee plot of H^+ transport and ATPase activity at pH 6.1 at various ATP concentrations. From this, it seems that the stoichiometry is 2. Moreover, the high-affinity ATP state is evident with low K^+ at the vesicle external surface and the low-affinity state is found with high K^+ at this surface. Also, the high-affinity ATP state is competent in the transport reaction.

The finding that valinomycin increases H^+ transport when added subsequent to the addition of ATP to K^+-preequilibrated vesicles suggests either that the pump is electrogenic or that K^+ is depleted during the transport process or both.

10. CATION TRANSPORT BY ATPase

When gastric vesicles are preequilibrated with $^{86}Rb^+$, the addition of ATP induces a rapid efflux of isotope, hence creating an inward Rb^+ gradient. The ratio between Rb^+ efflux and H^+ uptake would appear to be unity, but it is difficult to compare H^+ flux measured with pH electrodes and Rb^+ flux measured by a filtration procedure (Schackmann *et al.*, 1977).

An alternative means of determining that a K^+ gradient is generated is to use a potential-sensitive probe such as ANS (Lewin *et al.*, 1977) or [^{14}C]-SCN$^-$ (Schackmann *et al.*, 1977). In the absence of an ionophore selective for K^+, no redistribution of the probes is observed. This implies that there is no generation of an internal positive potential during transport under these conditions, i.e., K^+ equilibration. If valinomycin is present, however,

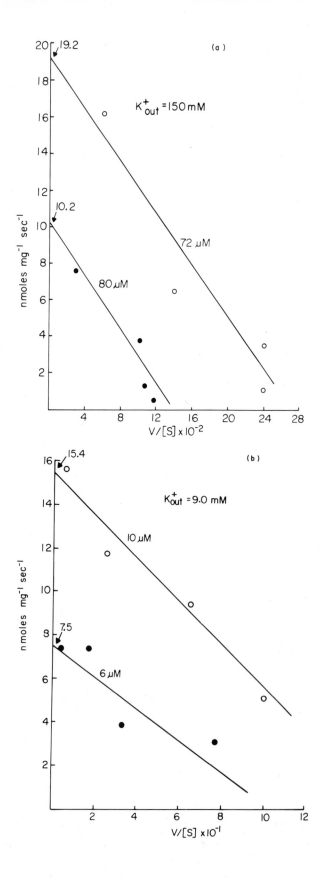

uptake of either probe is observed. This can be explained by the presence of the K^+ gradient induced by ATP causing an inwardly directed diffusion potential for K^+ when valinomycin is present.

If K^+ is exchanged for H^+ by the ATPase, as suggested by the above data, as well as by the catalytic properties of the enzyme, it is important to determine whether the pump displays any electrogenicity.

11. ELECTRICAL CHARACTERISTICS OF ATPase

Evidence exists that the transport catalyzed by the Na^+/K^+- and Ca^{2+}-ATPase is electrogenic as is the H^+ transport catalyzed by the ATPase of *Neurospora*. We have reviewed evidence that the gastric ATPase is not electrogenic in the direction of proton pumping in that we are unable to detect an internal positive potential.

The absence of a negative potential can be demonstrated by a lack of response of carbocyanine dyes during transport (Rabon *et al.*, 1978). However, as the converse of the valinomycin experiment discussed above, the addition of a protonophore such as TCS should result in the development of an outwardly directed diffusion potential for H^+. Indeed in the presence of TCS, accumulation of carbocyanine dye does result when H^+ transport is initiated by the addition of ATP.

The action of H^+- and K^+-selective ionophores is also instructive in determining whether the pump is electrogenic. If the pump is electrogenic in the proton transport direction, then TCS should short-circuit the pump and prevent accumulation of H^+ ion. Although partial inhibition occurs, this can readily be accounted for by an enhanced H^+ leak and enzyme inhibition. Equally the presence of valinomycin would enhance pump rate, which it does, but would also increase Rb^+ efflux, which it does not. The enhancement of rate would therefore be better explained by increased access of K^+ to the activating site of the pump.

The neutral exchange ionophore nigericin would be expected to enhance the potential of either an H^+ or a K^+ electrogenic transport. Instead no change in the potential characteristics is found. Admittedly, it is less easy to absolutely disprove the electrogenicity of a pump especially when it is not possible to design an experiment showing a lack of effect of applied voltage on pump rate. This type of experiment may have to wait for adequate reconstitution in planar bilayers, or patch-clamp studies.

12. PATHWAYS FOR K^+ AND Cl^-

From the above discussion K^+ has to reach the activating site of the ATPase. In the gastric vesicles, it appears to do this with an accompanying anion when H^+ transport is used as a measure. Thus, when sulfate is substituted for Cl^-, the time necessary for preincubation before H^+ transport is detected is greatly increased. When K^+ uptake is measured by reswelling of the vesicles following osmotic shrinkage due to application of a gradient of a K^+ salt (Cl^-, SO_4^{2-}, SCN^-), about 10% of the vesicles show high K^+ permeability

---

Figure 3. An Eadie–Hofstee plot of ATPase activity and H^+ transport as a function of ATP concentration either with 150 mM K^+ on both sides of the membrane (a), or with 150mM K^+ internal and 9 mM K^+ external to the vesicle (b). It appears that, at pH 6.1, the H^+/ATP stoichiometry is close to 2. It should also be noted that with high K^+ external, the apparent K_m for ATP is about 75 μM for both processes, and with low K^+ external, the K_m for ATP is about 8 μM.

that appears to be conductive because SCN^- increases the rate of reswelling. The rest have a lower permeability to KCl ($t_{1/2} = 56$ min). The significance of the permeable class of vesicles is not clear, but the slowly permeable majority are clearly the vesicles responsible for the H^+ transport signal detected (Rabon *et al.*, 1980). It would seem logical that regulation of activity of the pump could occur by the alteration of KCl permeability. Although an activating factor has been described in gastric homogenates, it does not appear to be acting on this parameter (Ray, 1978).

It is also surprising that no electrogenicity has been detected in gastric vesicles, for the intact gastric mucosa was one of the first epithelia to provide evidence for electrogenic transport. It is possible to construct models whereby the electrogenicity of the pump is retained, even given the properties described above (Sachs *et al.*, 1978). The Cl^- is transported with the H^+, and the K^+ is cycled back across the membrane. This model, if correct, requires the presence of a component lost in the vesicles studied to date. Perhaps a cytosolic factor is variably added or removed from the secretory membrane to provide this unusual pathway. Alternatively Cl^- permeability is added to accompany the ever present K^+ exchange permeability (Schackmann *et al.*, 1977; Malinowska and Sachs, 1982).

13. AVENUES FOR FUTURE RESEARCH

There are many unsolved problems in gastric secretion in mechanisms of the pump itself. Some of these have been touched upon above. The coupling of catalysis and transport is the central unsolved problem in pump biochemistry, and the gastric ATPase is no exception. The means of the physiological regulation is also quite unclear. Apparently both cAMP and Ca^{2+} may play a role (Chew *et al.*, 1980; Berglindh *et al.*, 1980c) but how is not known. Alteration of the kinetic characteristics of the occluded form of the enzyme may be one way. Another might be alteration in KCl permeability of the secretory membrane. We do not have sufficient information on the nature of the peptides comprising the ATPase, and the role of a possible dimer configuration of the catalytic subunit is completely unknown. The synthesis and assembly of the pump is also a timely problem, as well as its ontogeny. As a final problem, how unique is this pump in biology? Is it a derivative of plant root ATPases for example? Or similar to K^+-transport ATPases of bacteria? Clearly there is no lack of problems for future research in the area.

ACKNOWLEDGMENTS. The authors' research reported herein was supported by grants from the National Science Foundation (PCM 78-09208, PCM 80-08625) and the National Institutes of Health (AM15878, AM27606).

REFERENCES

Berglindh, T. (1978). *Acta Physiol. Scand. Special Suppl.* 55–68.
Berglindh, T., Helander, H. F., and Sachs, G. (1979). *Scand. J. Gastroenterol. Suppl.* **55**, 7–14.
Berglindh, T., Dibona, D. R., Ito, S., and Sachs, G. (1980a). *Am. J. Physiol.* **238**, G165–G176.
Berglindh, T., Dibona, D. R., Pace, C. S., and Sachs, G. (1980b). *J. Cell Biol.* **85**, 392–401.
Berglindh, T., Sachs, G., and Takeguchi, N. (1980c). *Am. J. Physiol.* **238**, G90–G94.
Chew, C. S., Hersey, S. J., Sachs, G., and Berglindh, T. (1980). *Am. J. Physiol.* **238**, G312–G320.
Esche, F. S., and Allison, W. S. (1979). *J. Biol. Chem.* **254**, 10740–10747.
Ganser, A. L., and Forte, J. G. (1973a). *Biochim. Biophys. Acta* **307**, 169–179.
Ganser, A. L., and Forte, J. G. (1973b). *Biochim. Biophys. Res. Commun.* **54**, 690–696.
Harris, J. B., and Edelman, I. S. (1960). *Am. J. Physiol.* **198**, 280–284.

Hersey, S. J. (1977). *Biochim. Biophys. Acta* **496,** 359–366.

Hirschowitz, B. I., and Sachs, G. (1972). *Am. J. Physiol.* **233,** 305–309.

Kepner, G. R., and Macey, R. I. (1968). *Biochim. Biophys. Acta* **163,** 183–203.

Koelz, H. R., Sachs, G., and Berglindh, T. (1981). *Am. J. Physiol.* **4,** G431–442.

Lee, J., Simpson, E., and Scholes, P. (1974). *Biochem. Biophys. Res. Commun.* **60,** 825–834.

Lewin, M., Saccomani, G., Schackmann, R., and Sachs, G. (1977). *J. Membr. Biol.* **32,** 301–318.

Malinowska, D. H., and Sachs, G. *Fed. Proc.,* in press.

Malinowska, D. H., Koelz, H. R., Hersey, S. J., and Sachs, G. (1981). *Proc. Natl. Acad. Sci. USA* **78,** 5908–5912.

Mitchell, P. (1966). *Biol. Rev.* **41,** 445–502.

Rabon, E., and Sachs, G. (1981). *Biochim. Biophys. Acta,* in press.

Rabon, E., Chang, H., and Sachs, G. (1978). *Biochemistry* **17,** 3345–3353.

Rabon, E., Takeguchi, N., and Sachs, G. (1980). *J. Membr. Biol.* **53,** 109–117.

Racker, E. (1976). In *A New Look at Mechanisms in Bioenergetics,* Academic Press, New York.

Ray, T. K. (1978). *FEBS Lett.* **92,** 49–52.

Ray, T. K., and Forte, J. G. (1976). *Biochim. Biophys. Acta* **443,** 451–467.

Reenstra, W., Lee, H. C., and Forte, J. G. (1980). In *Hydrogen Ion Transport in Epithelia* (I. Schulz, ed.), pp. 155–164, Elsevier/North-Holland, Amsterdam.

Saccomani, G., Stewart, H. B., Shaw, D., Lewin, M., and Sachs, G. (1977). *Biochim. Biophys. Acta* **465,** 311–330.

Saccomani, G., Helander, H. F., Cragon, S., Chang, H., Dailey, D. W., and Sachs, G. (1979a). *J. Cell Biol.* **83,** 271–283.

Saccomani, G., Dailey, D., and Sachs, G. (1979b). *J. Biol. Chem.* **254,** 2821–2827.

Saccomani, G., Barcellona, M. L., Rabon, E., and Sachs, G. (1980). In *Hydrogen Ion Transport in Epithelia* (I. Schulz, ed.), pp. 175–183, Elsevier/North-Holland, Amsterdam.

Saccomani, G., Sachs, G., Cuproletti, J., and Jung, C. Y. (1981). *J. Biol. Chem.* **256,** 7727–7729.

Sachs, G. (1968). *Biochim. Biophys. Acta* **162,** 210–219.

Sachs, G., and Hirschowitz, B. I. (1968). *Physiology of Gastric Secretion NATO* 186–202.

Sachs, G., Chang, H., Rabon, E., Schackmann, R., Lewin, M., and Saccomani, G. (1976). *J. Biol. Chem.* **251,** 7690–7698.

Sachs, G., Spenney, J. G., and Lewin, M. (1978). *Physiol. Rev.* 58, 106–173.

Sachs, G., Rabon, E., Stewart, H. B., Pierce, B., Smolka, A., and Saccomani, G. (1980). In *Hydrogen Ion Transport in Epithelia* (I. Schulz, ed.), pp. 135–143, Elsevier/North-Holland, Amsterdam.

Schackmann, R., Schwartz, A., Saccomani, G., and Sachs, G. (1977). *J. Membr. Biol.* **32,** 361–381.

Schrijen, J. J., Luyben, W. H. A. M., De Pont, J. J. H. M., and Bonting, S. L. (1980). *Biochim. Biophys. Acta* **597,** 331–344.

Schrijen, J. J., Loyben, W. H. A. M., De Pont, J. J. H. M., and Bonting, S. L. (1981). *Biochim. Biophys. Acta* **597,** 331–344.

Skou, J. C. (1965). *Physiol. Rev.* **45,** 596–617.

Stewart, H. B., Wallmark, B., and Sachs, G. (1981). *J. Biol. Chem.* **256,** 2682–2690.

Taniguchi, K., and Post, R. L. (1975). *J. Biol. Chem.* **250,** 3010–3018.

Wallmark, B., and Mardh, S. (1979). *Biol. Chem.* **254,** 11899–11902.

Wallmark, B., Stewart, H. B., Rabon, E., Saccomani, G., and Sachs, G. (1980). *J. Biol. Chem.* **255,** 5313–5319.

84

Phosphate Transport Processes of Animal Cells

Peter L. Pedersen and Janna P. Wehrle

1. FOREWORD

In this review an attempt is made to summarize as briefly as possible new developments concerned with the transport of P_i, not only across animal cell membranes, but across membranes of the bacterial and plant worlds as well. It may be helpful to the reader to first examine Fig. 1 in order to gain some appreciation of the importance of P_i transport processes in animal cells, and then Fig. 2 to obtain an overview of current thoughts on P_i transport processes across biological membranes. With this general overview at hand, more specific information concerning recent progress and unanswered questions about P_i transport processes can be obtained below. In particular, the literature cited emphasizes papers published between 1974 and 1980.

2. P_i TRANSPORT ACROSS THE MITOCHONDRIAL INNER MEMBRANE

Mitochondria contain at least two transport systems that catalyze the transport of P_i across the inner membrane (Figs. 1 and 2). One of these transport systems is referred to as the P^-_i/H^+ symport system (or P_i^-/OH^- antiport system) and was identified by the early work of Fonyó and Bessman (1966, 1968) and Chappell (1968). A second transport system, the $P_i^{2-}/dicarboxylate^{2-}$ antiport system, was identified by the work of Chappell and Haarhoff (1967). Of these two transport systems, the P_i^-/H^+ symport system has been studied most extensively in recent years and will be the major topic of discussion in this section. Specifically, we will focus on the following aspects of mitochondrial P_i transport: (1) identification and isolation of molecular components comprising the P_i^-/H^+ symport system, (2) identification of new inhibitors and activators of P_i transport, (3) establishment

Peter L. Pedersen and Janna P. Wehrle • Laboratory for Molecular and Cellular Bioenergetics, Department of Physiological Chemistry, The Johns Hopkins University School of Medicine, Baltimore, Maryland 21205.

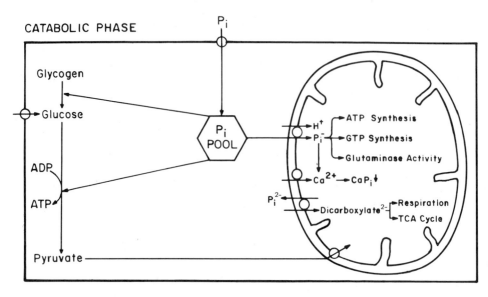

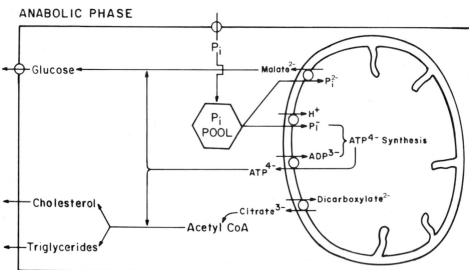

Figure 1. The importance of P_i and P_i transport processes to cell metabolism. The upper panel shows that during the catabolic phase of cell metabolism, P_i is necessary in the cytoplasm for glycogen breakdown (liver) and for glycolysis (most tissues). P_i can be obtained from the cell pool, which is most likely maintained by hydrolysis of ATP catalyzed by intracellular ATPases. However, during cell growth and division as well as during intense catabolic activity, a P_i transport system in the cell membrane is essential to maintain the P_i pool.

Specific transport systems make P_i available to the mitochondria of most tissues for ATP synthesis and GTP synthesis, and in the kidney for the activity of glutaminase, an enzyme involved in ammoniagenesis. In addition, P_i transport processes allow for entry of dicarboxylic acids into mitochondria of most tissues to help support respiration or the TCA cycle, and for entry of Ca^+ presumably to help maintain a low cytoplasmic level of this sometimes inhibitory cation.

The lower panel shows that during the anabolic phase of cell metabolism in liver, P_i stimulates exit of malate from mitochondria, which serves as the carbon source for glucose synthesis during gluconeogenesis. [Malate exit from mitochondria can occur via exchange with other dicarboxylic acids on the carrier indicated (dicarboxylate carrier) or via exchange with citrate on the tricarboxylate carrier. P_i-induced malate exit is therefore only one of the possible ways malate is made available for gluconeogenesis.] During the anabolic phase of cell metabolism, P_i is necessary also for ATP synthesis, which provides the energy source for synthesis of glucose, triglycerides, and cholesterol.

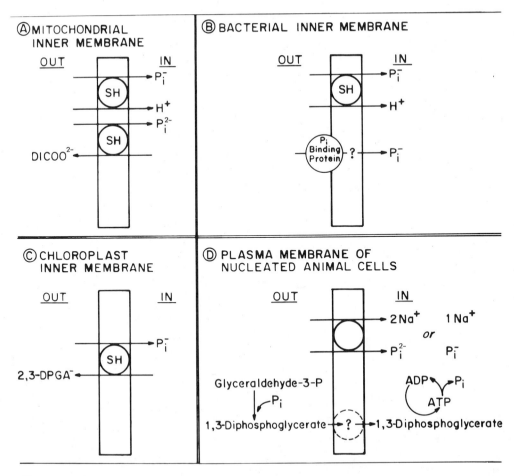

Figure 2. Transport systems identified or suggested to date for the movement of P_i into mitochondria, bacteria, chloroplasts, and into the cell cytoplasm. (A) In mitochondria there are at least two transport systems that catalyze the entry of P_i (a P_i^-/H^+ symporter or P_i^-/OH^- antiporter and a P_i^{2-}/dicarboxylate^{2-} antiporter). Work in this and other laboratories suggests that the P_i^-/H^+ symport and P_i^{2-}/dicarboxylate^{2-} antiport carriers may reside on the same protein complex and/or share a common polypeptide. The P_i^-/H^+ symport activity is specifically inhibited by NEM, and as many as five NEM-binding proteins have been detected under highly specific labeling conditions. One of these is a protein of 24,000–32,000 molecular weight. Crude preparations of this peptide have been incorporated into liposomes and shown to catalyze P_i movement across the membrane. (B) In bacteria, genetic evidence suggests the existence of as many as four P_i transport systems in the inner membrane, one of which has been identified as a P_i^-/H^+ symporter similar to that found in mitochondria, and another that involves a P_i-binding protein (~42,000 molecular weight) removable by osmotic shock. (C) In chloroplasts, a small protein system (~29,000 molecular weight) is present that exchanges P_i^- with 2,3-diphosphoglycerate (DPGA) rather than with OH$^-$. (D) In the plasma membrane of animal cells (other than in erythrocytes where a general anion transport system is operative), there have been reports of an Na$^+$/P_i^- symporter, the stoichiometry of which is thought to be different in intestine and kidney. It has been suggested also that P_i penetrates the plasma membrane of certain animal cells as 1,3-diphosphoglycerate by a process involving enzymatic steps on both sides of the membrane.

of the symmetry of the P_i^-/H^+ symporter, (4) elucidation of the energetics of P_i transport, (5) definition of the relationship of the P_i^-/H^+ symporter to other mitochondrial transport systems, and (6) definition of the relationship of P_i transport across the mitochondrial inner membrane to metabolic and/or physiological processes.

Table I. Summary of Molecular Information on Mitochondrial P_i Transport and/or Binding Systems

Investigator	System	Evidence for component(s) of P_i transport system(s)	Number	Molecular weight of major component	Amount in mitochondria (pmoles/mg)
Coty and Pedersen (1975a)	Liver	NEM binding	5	32,000	60
Hadvary and Kadenbach (1976)	Liver	NEM binding	1	27,000	30
Briand *et al.* (1976)	Liver	NEM binding	4	30,000	ND[a]
Kadenbach and Hadvary (1973)	Liver	P_i binding	Several	ND	32
Palmieri *et al.* (1974)	Liver	P_i binding	1	ND	ND
Banerjee *et al.* (1977); Banerjee and Racker (1979)	Heart	Isolation and reconstitution	Several	30,000	ND
Guérin and Napias (1978)	Yeast	P_i binding	1	10,000[b]	ND
Wohlrab (1980)	Heart Muscle	NEM binding	—	32,000	c
Blondin (1979)	Heart	P_i binding	2	7,000[b]	d

[a]ND, not determined.
[b]Suggested to be part of the H^+-ATPase complex.
[c]0.87 mole/mole cytochrome *a*.
[d]0.21–0.28 mole/mg ETP (electron transport particles).

2.1. Identification and Isolation of Molecular Components of the P_i^-/H^+ Symport System

N-Ethylmaleimide (NEM) is a rather specific inhibitor* of the P_i^-/H^+ symport system of liver mitochondria even though it is known to label SH groups of a number of inner-membrane proteins (Coty and Pedersen, 1975a; Briand *et al.*, 1976). NEM does not inhibit the ATP/ADP, P_i/dicarboxylate, or dicarboxylate/tricarboxylate antiport systems under conditions where it maximally inhibits the P_i^-/H^+ symporter (Meijer *et al.*, 1970; Klingenberg, 1970; Palmieri *et al.*, 1974; Coty and Pedersen, 1974). In rat liver mitochondria, a 24,000- to 32,000-molecular-weight peptide (and four larger peptides) is labeled with NEM under conditions where membrane SH groups unassociated with P_i transport are masked. The results of these studies and similar studies directed at identifying molecular components of the P_i^-/H^+ symport system of mitochondria are summarized in Table I.

Careful examination of the studies reported in Table I indicate that the 24,000- to 32,000-molecular-weight protein is the most likely candidate for the P_i^-/H^+ symporter. Perhaps the most compelling pieces of evidence for the role of this protein in P_i transport

---

*NEM under defined conditions inhibits, in addition to the P_i^-/H^+ carrier, the glutamate/aspartate carrier (Vignais and Vignais, 1973).

are the reconstitution studies of Wohlrab (1980) and of Racker's group (Banerjee *et al.*, 1977; Banerjee and Racker, 1979). Both laboratories have obtained detergent-solubilized preparations (using either Triton X-100 or octylglucoside) and incorporated the preparation into liposomal vesicles. In both cases the preparations are shown to catalyze P_i movement across liposomal membranes, which is inhibited by NEM. Particularly encouraging are the studies of Wohlrab (1980) reporting a very high P_i/P_i exchange rate (990 nmoles/min/mg) in the reconstituted systems. As yet, however, neither of these groups nor other research groups working on this problem have obtained a purified preparation. This is an essential point because, as will be noted below, there are reports of exit mechanisms for P_i in mitochondria. At this time it is not known whether these exit mechanisms are catalyzed by the P_i^-/H^+ symporter or a different P_i transporter.

It is also of interest to note in Table I that a protein of 7000–10,000 molecular weight that binds P_i has been isolated from both bovine heart and yeast mitochondria. This protein appears to be associated with the F_0 unit of the H^+-ATPase, which catalyzes ATP synthesis and ATP-dependent functions in mitochondria.

Whether this P_i-binding protein operates in conjunction with the P_i^-/H^+ symporter or with some other P_i transporter (i.e., the P_i/dicarboxylate carrier or a P_i exit carrier) is not known. It is possible that this P_i-binding protein functions exclusively in activities catalyzed by the H^+-ATPase and is unrelated to P_i transport processes altogether.

2.2. Inhibitors of P_i Transport

This subject has been treated extensively by Fonyó (1979) and only the most essential aspects will be covered here.

Table II summarizes a number of inhibitors of the P_i^-/H^+ symport system of mitochondria. Most of these are compounds that interact reversibly (organic mercurials or disulfides) or irreversibly (maleimides) with SH groups.

The test for reversibility is usually to add Cleland's reagent (dithiothreitol) after inhibition has been effected with a given compound and assess whether phosphate transport is restored.

Some of the inhibitors listed in Table II, like the maleimides, are rather specific for the P_i^-/H^+ symport system whereas others like the mercurials inhibit both the P_i^-/H^+ symport system and the P_i^{2-}/dicarboxylate^{2-} antiport system. Although all of these agents are thought to inhibit P_i transport by reacting with free, essential SH groups, it has not been shown unequivocally that these groups are the only functional groups involved. Cases in point concern the inhibitors *p*-diazobenzenesulfonate, 2-phenylindolone, and formaldehyde. *p*-Diazobenzenesulfonate is known to react with histidine, amino, and tyrosine groups as well as cysteine residues (Howard and Wild, 1957), whereas aldehydes and ketones are well known to form Schiff bases with amino residues.

Fonyó (1979) has emphasized that not every compound with an activated double bond and/or an ability to react with SH groups inhibits P_i transport into mitochondria. As an example, he points out that avenaciolide, a compound that reacts with certain SH groups in mitochondria, does not inhibit the P_i carrier (Meyer and Vignais, 1973). Consistent with this observation, we have found that sodium tetrathionate, also an SH-reactive compound, fails to inhibit P_i transport in mitochondria (Wehrle and Pedersen, unpublished observation).

It should be noted also that P_i transport activities of different mitochondrial types do not respond identically to inhibition by agents listed in Table II. Thus, either mersalyl or 6,6'-dithionicotinic acid completely inhibits P_i transport in porcine heart mitochondria,

Table II. Inhibitors of P$_i$ Transport in Mitochondria

Maleimide derivatives
  N-Ethylmaleimide (NEM)
  N-(N-acetyl-4-sulfamoylphenyl)maleimide (ASPM)
  N-Benzylmaleimide
  N-Cyclohexylmaleimide
Organic mercurials
  p-Hydroxymercuribenzoate (PMB)
  p-Chloromercuriphenylsulfonate (CMS)
  Mersalyl [sodium-σ-(3-hydroxymercuri-2-methoxypropyl) carbamoylpenoxyacetate]
  2-Chloromercuri-4,6-dinitrophenol
  Fluorescein mercuric acetate (FMA)
Disulfides
  5,5'-Dithiobis(2-nitrobenzoic acid) (DTNB or Ellman's reagent)
  6,6'-Dithionicotinic acid (carboxypyridine disulfide) (CPDS)
Other agents
  Fuscin (a mold metabolite)
  p-Diazobenzenesulfonate
  2-Phenylindolone
  Formaldehyde
  Diamide[a]
  Ethacrynic acid[a] [2,3-dichloro-4-(2'-methylene butyryl)phenoxyacetic acid]

---

[a] Not very effective inhibitors of P_i transport. Diamide inhibits only partially (Zoccarato *et al.*, 1977), and inhibition by ethacrynic acid requires prior incubation of the mitochondria with Mg^{2+} (Goldschmidt *et al.*, 1976).

whereas in rat liver mitochondria addition of n-butylmalonate (a specific inhibitor of the P_i^{2-}/dicarboxylate^{2-} antiport system) together with mersalyl is essential for complete inhibition (Abou-Khalil *et al.*, 1975).

Finally, it should be mentioned that the sulfhydryl oxidizing agent diamide has been reported to partially inhibit P_i transport in mitochondria (Zoccarato *et al.*, 1977). However, under most conditions diamide has no effect on phosphate transport nor does it alter inhibition of P_i transport by mersalyl (Siliprandi *et al.*, 1975). It does stimulate mitochondrial ATPase activity and may have a site of action in the H^+-ATPase complex.

2.3. Symmetry of the P_i^-/H^+ Symporter

Attempts have been made to distinguish the cytoplasmic from the matrix-facing surface of the P_i^-/H^+ symporter. Assuming that influx and efflux occur via the same system, a number of authors have attempted to use the differences in inhibitor sensitivity between influx and efflux to support the concept that the phosphate carrier has a single, reorienting active site, rather than symmetrical active sites, one at each surface. Guérin *et al.* (1970), Klingenberg *et al.* (1974), and Guérin and Guérin (1975) observed that DTNB and ASPM as well as NEM and mersalyl would inhibit P_i influx, whereas only the latter, supposedly more permeant, reagents could inhibit efflux. They interpreted this as supporting the existence of a single reorienting SH group, which could be protected from impermeant reagents if "pulled" to the matrix face by increasing internal phosphate levels. Siliprandi *et al.* (1975) reported that Mg^{2+} abolished the inhibition by mersalyl of P_i efflux, but left unaffected inhibition of P_i uptake. Rhodin and Racker (1974) reported results comparing differential inhibitor sensitivities for uptake by submitochondrial particles (supposedly inverted) compared to intact mitochondria. Unfortunately, although the investigators referred to above have agreed about asymmetry of the carrier, they have made different assumptions about the permeability characteristics of the SH reagents employed. For example, Guérin *et al*. (1970) assume that NEM is a permeant reagent and that mersalyl reacts only at the outside (Guérin and Guérin, 1975), whereas Rhodin and Racker (1974) assume that mersalyl penetrates the membrane and that NEM is a nonpenetrant.

We have recently readdressed the question of the membrane location of SH groups essential for P_i transport in rat liver mitochondria (Wehrle *et al*., 1978). Before beginning, we examined experimentally the permeability of the inner membrane to the reagents used, under the conditions used for transport studies. NEM, but not mersalyl, was found to have access to the matrix compartment. P_i uptake was studied both in intact mitochondria and in a highly purified and well-characterized population of inverted (>95%) inner-membrane vesicles. In both cases, either permeant or nonpermeant SH reagents fully inhibited P_i uptake. Thus, the critical SH group(s) is accessible from either surface. That there is only one (type of) SH group, which may be exposed at either surface, rather than two identical SH groups, one at either surface, may be inferred from the data of Coty and Pedersen (1975a), now that the reagent permeability has been ascertained. At that time we observed that prior treatment of intact mitochondria with mersalyl or PMB (nonpenetrant) could protect the phosphate transport activity from irreversible inactivation by NEM (penetrant), suggesting that when SH groups are reversibly blocked at the cytoplasmic face, no other critical SH groups remain at the matrix face. Fonyó *et al*. (1975) have proposed two critical SH groups per carrier, each of which must be blocked for inhibition to occur, but these are not imagined to be permanently at opposite membrane faces, as each can be inhibited by mersalyl, albeit at different concentrations.

In conclusion, all the evidence to date supports the existence of one class of critical SH group in the P_i^-/H^+ symporter, which can exist at either the matrix or the cytoplasmic surface.

2.4. Energetics—Relationship to Entry and Exit Pathways

It seems generally accepted today that the transport of many metabolites into mitochondria is driven by an electrochemical potential of hydrogen ions across the inner membrane established as a consequence of respiration. This is not to imply that chemical intermediates and/or conformational changes are not involved at some stage during the transport event (in fact, it is most likely that they are), but rather that a high-energy intermediate of the type X~I does not appear to be the primary driving force. The electrochemical potential of hydrogen ions $\Delta\tilde{\mu}_{H^+}$ is given by the expression

$$\Delta\tilde{\mu}_{H^+} = \Delta\psi + \frac{2.3RT}{F}\Delta pH \tag{1}$$

where F is the Faraday, $\Delta\psi$ is the electrical gradient across the membrane, R is the gas constant, T is the temperature (°K), and ΔpH is the chemical gradient of protons. Mitchell (1968) has rearranged this equation to give what he refers to as the protonmotive force (pmf) or Δp where

$$\Delta p = \Delta\psi - Z\Delta pH \tag{2}$$

With respect to P_i, it has been known for almost 10 years that movement of this anion into mitochondria proceeds predominantly via an effective P_i^-/H^+ symport (P_i^-/OH^- anti-

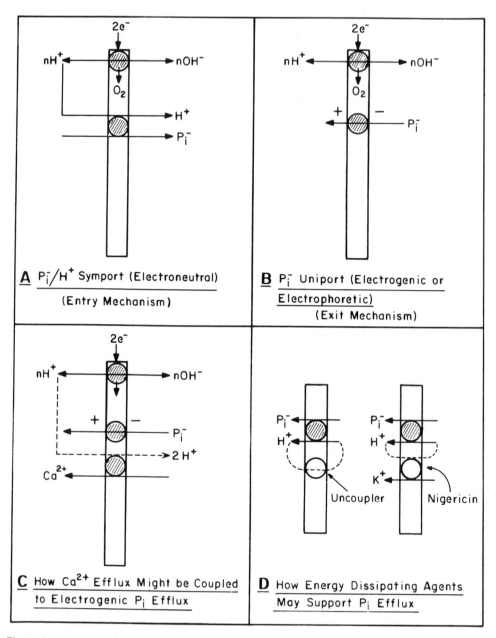

A P_i/H^+ Symport (Electroneutral)
(Entry Mechanism)

B P_i^- Uniport (Electrogenic or Electrophoretic)
(Exit Mechanism)

C How Ca^{2+} Efflux Might be Coupled to Electrogenic P_i Efflux

D How Energy Dissipating Agents May Support P_i Efflux

Figure 3. Comparison of energy-dependent entry (A) and exit (B, C, D) mechanisms for P_i in mitochondria. In C, it can be seen how Ca^{2+} efflux might be coupled to electrogenic efflux of P_i. [Fiskum and Lehninger (1980 for a review) believe that Ca^{2+} efflux in liver mitochondria proceeds via a $Ca^{2+}/2H^+$ antiport mechanism.] In D are mechanisms to account for P_i efflux under conditions where energy-dissipating agents like uncouplers or nigericin are present. Efflux under these conditions is best accounted for by assuming that the P_i^-/H^+ symporter is operating reversibly and that net efflux is "catalyzed" by proton cycling.

port) process as depicted in Fig. 3A. [Some P_i movement does take place in exchange with dicarboxylic acids, but this usually constitutes only a small fraction of the total P_i movement (Coty and Pedersen, 1974, 1975b).] Indirect evidence for a P_i^-/H^+ symport mechanism derived originally from the NH_4P_i swelling experiments of Chappell (1968), and direct evidence derived from experiments where P_i and proton movements were monitored simultaneously (Coty and Pedersen, 1975b). In terms of mechanism, however, the process appears to be HPO_4^{2-} translocation. Freitag and Kadenbach (1978) have shown that only P_i analogs with the -2 ionization state are translocated on the NEM-sensitive carrier and mimic P_i in their effects on dicarboxylate uptake. This implies that 2 H^+ must also be moved, as the net effect of the carrier appears electrically neutral in the NH_4P_i swelling assay. In more recent years, however, this laboratory (Wehrle *et al.*, 1978; Wehrle and Pedersen, 1979) and other laboratories (Azzone *et al.*, 1976; Lötscher *et al.*, 1979) have presented evidence that suggests that P_i movement across the mitochondrial inner membrane may occur not only in symport with H^+ in response to the pH gradient, but via a uniport process in response to the membrane potential as well (compare Fig. 3A and B).

That P_i may move electrogenically (or electrophoretically) in response to the membrane potential was inferred from the studies of Azzone *et al.* (1976) who showed that P_i can be extruded from mitochondria via an energy-dependent process. More direct evidence for electrogenic movement of P_i has been obtained from the reconstitution studies of Racker and his collaborators (Banerjee *et al.*, 1977), and from studies on P_i movements in inverted inner-membrane vesicles (Wehrle *et al.*, 1978; Wehrle and Pedersen, 1979; Lötscher *et al.*, 1979). The membrane potential of inverted inner-membrane vesicles of rat liver in the presence of respiratory substrate is from negative (outside) to positive (inside). P_i^- apparently moves in response to this potential rather than in symport with H^+, for the direction of the proton gradient is from alkaline (outside) to acid (inside). Whether electrogenic efflux of P_i takes place on the P_i^-/H^+ symporter or on a separate transport system remains to be established. However, it should be noted that both processes (P_i^-/H^+ symport and P_i^- uniport) are inhibited by the same SH reagents. Thus, a role for a single carrier in both processes does not seem unlikely.

Whether electrogenic efflux of P_i from mitochondria plays a role in the physiological movement of ions across the mitochondrial inner membrane remains to be established. We suggested almost a decade ago that phosphate efflux under certain physiological conditions may be coupled to efflux of calcium from mitochondria (Pedersen and Coty, 1972). As yet, this suggestion has not been put to experimental test in whole cells, but our original work (Pedersen and Coty, 1972) and our most recent studies (Wehrle and Pedersen, 1979) showing that calcium moves from the matrix to the cytoplasmic surface of the mitochondrial inner membrane in a phosphate-dependent manner (with electrogenic movement of P_i preceding movement of Ca^{2+}) have been supported by work in Carafoli's laboratory (Lötscher *et al.*, 1979). Figure 3C shows how electrogenic efflux of phosphate could be coupled to calcium efflux from mitochondria.

Finally, it should be pointed out that a number of laboratories have reported that phosphate efflux from mitochondria occurs under what might be best regarded as "deenergized" conditions (Papa *et al.*, 1970; Hoek *et al.*, 1971; Fonyó, 1968). For example, efflux of P_i from mitochondria can be induced by inhibition of respiration, by addition of uncouplers, or by addition of the K^+/H^+ exchange ionophore nigericin. As indicated in Fig. 3D, the latter two efflux processes could involve reversal of the P_i^-/H^+ symporter, thus distinguishing this type of P_i movement from electrogenic P_i^- uniport depicted in Fig. 3B.

2.5. *Relationship of the* P_i^-/H^+ *Symporter to Other Mitochondrial Transport Systems*

In the past it has been common to depict P_i^-/H^+ symport, P_i^{2-}/dicarboxylate^{2-} antiport, and the movement of calcium and adenine nucleotides in mitochondria as taking place on separate transport systems. In recent years, however, there have been reports that P_i^-/H^+ symport and P_i^{2-}/dicarboxylate^{2-} antiport may be associated with a single transport system, that the entry mechanism for Ca^{2+} may not be a uniport process in the presence of P_i, and that the ADP^{3-}/ATP^{4-} transport system may under conditions of ATP hydrolysis act as an ATP/ADP, P_i antiporter (or in conjunction with a P_i/ADP symporter).

We suggested that the P_i^-/H^+ symporter and the P_i^{2-}/dicarboxylate^{2-} antiporter may share a common subunit for P_i (Fig. 4B) because both activities were found to be inhibited by the same concentrations of PMB and both activities were found to have the same K_m

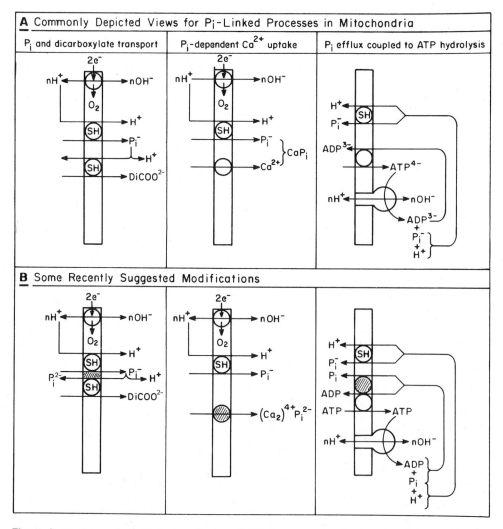

Figure 4. (A) Commonly depicted views for some P_i-linked processes in mitochondria and (B) some recently suggested modifications. Suggested modifications are indicated by cross-hatching. (See text for a discussion.)

for P_i (Coty and Pedersen, 1975a). Consistent with this view, experiments carried out by Lofrumento *et al.* (1974) and Lofrumento and Zanotti (1976) have been interpreted within the framework of a single protein complex specialized for P_i^-/H^+ symport and $P_i^{2-}/\text{dicarboxylate}^{2-}$ antiport. Specifically, these latter workers have shown that inhibition of the P_i^-/H^+ symport activity by NEM (a specific inhibitor of this activity) is prevented by low concentrations of *n*-butylmalonate, a specific inhibitor of $P_i/\text{dicarboxylate}$ activity. In addition, Kadenbach *et al.* (1978) have reported that a nontransportable phosphate analog (*n*-octylphosphate) can promote dicarboxylate uptake in precisely the same manner as can P_i. This contrasts with past conceptions of the mechanism of P_i-stimulated dicarboxylate uptake thought to be due to P_i uptake via the P_i^-/H^+ symporter, followed by $P_i^{2-}/\text{dicarboxylate}^{2-}$ exchange as illustrated in Fig. 4A. The effect of *n*-octylphosphate is inhibited by NEM, suggesting that it originates at the P_i^-/H^+ symporter protein rather than the dicarboxylate site, but that its effect can be transmitted directly to the dicarboxylate-transporting system. Thus, additional models not illustrated in Fig. 4 certainly deserve consideration.

Moyle and Mitchell (1977) have suggested that uptake of calcium phosphate by mitochondria may take place by a symport process (involving $Ca_3^{4+}-P_i^{2-}$) that need not involve the P_i^-/H^+ symporter directly (compare Figs. 4A and 4B). This view is supported by their finding that in the presence of P_i, Ca^{2+} can enter the mitochondria by a process that is inhibited by lanthanum but that is insensitive to NEM and mersalyl (inhibitors of the P_i^-/H^+ symporter as well as the $P_i^{2-}/\text{dicarboxylate}^{2-}$ antiporter). Import of Ca^{2+} via this transport system is stated to be stoichiometric with the import of P_i.

Fiskum *et al.* (1979) have challenged the view of Moyle and Mitchell (1977) that rat liver mitochondria transport calcium by a symport mechanism involving $Ca_2^{4+}-P_i^-$. Fiskum *et al.* (1979) report that NEM-insensitive symport of this type in which only one positive charge is carried per calcium ion transported could not be detected under any circumstances tested.

Finally, Reynafarje and Lehninger (1978) have suggested that during ATP-dependent processes in mitochondria, the ADP/ATP transport system may function in conjunction with an NEM- and mersalyl-insensitive P_i transport system separate and distinct from the P_i^-/H^+ symporter (Fig. 4B). Evidence for such a carrier is derived from the findings that during ATP hydrolysis in the presence of valinomycin, P_i is transported out of rat liver mitochondria in the presence of NEM or mersalyl. This P_i efflux process is inhibited by atractyloside, an inhibitor of the ADP/ATP transport system, suggesting that either P_i transport takes place on this carrier or, as indicated in Fig. 4B, on a carrier closely associated with the ADP/ATP transport system.

It is not clear from Reynafarje and Lehninger's studies (1978) whether NEM and mersalyl are actually completely inhibiting the P_i^-/H^+ symporter under the conditions of their experiments. Mg^{2+} has been shown under certain conditions to abolish mersalyl inhibition of P_i efflux from mitochondria (Siliprandi *et al.*, 1975). Along these lines, Fonyó *et al.* (1980) have recently challenged the claim of Reynafarje and Lehninger (1978) that mersalyl-insensitive P_i efflux from mitochondria occurs during ATP hydrolysis. Fonyó *et al.* (1980) find that if ATP is hydrolyzed in the presence of valinomycin and K^+, mersalyl inhibits the efflux immediately. However, these latter workers provide no explanation for the finding of Reynafarje and Lehninger (1978) that P_i efflux is partially inhibited by atractyloside.

In summary, it would appear that much additional work, preferably at the molecular level using purification and reconstitution approaches, will be necessary before workers in this field will be able to reach agreement on the choice of models depicted in Fig. 4.

2.6. Relationship to Metabolic and/or Physiological Processes

Although the necessity of mitochondrial P_i transport systems for ATP and GTP synthesis (P_i^-/H^+ symporter), for normal tricarboxylic acid cycle function, for entry of some respiratory substrates, and for gluconeogenesis (P_i^{2-}/dicarboxylate^{2-} antiporter) are now generally accepted, P_i transport processes in mitochondria may play important roles in at least two other processes. First, P_i transport mechanisms may be critical for ammoniagenesis in certain animals. Yu *et al.* (1976) have shown that P_i stimulates ammonia production in the rabbit. (Ammonia excretion provides a way for excreting H^+ ions while maintaining acid–base homeostasis.) These authors conclude that P_i-induced stimulation of ammoniagenesis in the rabbit kidney is mediated by removal of glutamate, the feedback inhibitor of P_i-dependent glutaminase. Significantly, glutamate removal is suggested to be linked to P_i-induced dicarboxylate exit across the mitochondrial membrane. A similar role of either a P_i/dicarboxylate carrier or a P_i/glutamine carrier in ammoniagenesis has been suggested by Kovaćević (1976).

Second, P_i transport may be essential for regulation of Ca^{2+}-dependent intracellular processes. Bygrave (1967) and more recently Lehninger (1971) have emphasized that mitochondria of animal cells may serve to maintain a low cytoplasmic Ca^{2+} level (~ 1 μM) in the face of a high plasma Ca^{2+} level (~ 2 mM). In view of the findings from several laboratories (Lehninger, 1974; Harris and Zaba, 1977; Moyle and Mitchell, 1977) that a permeant anion (usually P_i under physiological conditions) must accompany the entry of Ca^{2+} into mitochondria, it seems likely that P_i may be critical for regulating intracellular Ca^{2+} levels. The possibility that P_i transport may under certain physiological conditions promote Ca^{2+} efflux from mitochondria is also relevant to this discussion (see Section 2.4).

2.7. Other Aspects of Mitochondrial P_i Transport

Recent work that does not seem to appropriately fit into any of the above categories includes the studies of Briand *et al.* (1975) reporting that the reactivity of the SH groups of the P_i^-/H^+ symporter may be affected by the transmembrane pH; studies of Zimmer (1977) describing a procedure for distinguishing SH groups associated with P_i transport and those associated with oxidative phosphorylation; studies of Chateaubodeau *et al.* (1974) and Christiansen *et al.* (1974) demonstrating the presence of the P_i^-/H^+ symporter in yeast mitochondria (and promitochondria) and brown adipose tissue mitochondria, respectively. LaNoue *et al.* (1974) have shown that extramitochondrial P_i stimulates aspartate transport in rat liver and heart mitochondria in the absence of measurable P_i movement across the inner membrane. Finally, Homes *et al.* (1975) report that in certain cell types, vitamin E deficiency is accompanied by a decreased P_i uptake capacity (presumably mitochondrial).

3. P_i TRANSPORT ACROSS THE BACTERIAL INNER MEMBRANE

Although there have only been a few reports in the literature concerned with P_i transport across the bacterial inner membrane, it is of interest to note that two of these reports were by Peter Mitchell in the early 1950s (Mitchell, 1953, 1954). Mitchell showed that in *Staphylococcus aureus*, P_i can exchange across the osmotic barrier of the cell and that this exchange process is inhibited by mercurial agents and uncouplers of oxidative phosphorylation. More recent studies by Rosenberg and his colleagues (1975, 1977) using mutant strains of *E. coli* have shown that there may be at least two systems for P_i uptake in bacteria. Interestingly, one of the two systems appears to be similar to that described 21 years earlier by Mitchell (1954), i.e., it permits the complete exchange of intracellular P_i with extracellular P_i and is completely inhibited by uncouplers. This system does not require a "shock-sensitive" P_i-binding protein and operates in spheroplasts. A second transport system for P_i in *E. coli* is repressible by P_i concentrations above 1 mM, requires a P_i-binding protein for full activity, and does not operate in spheroplasts. It catalyzes very little exchange between internal and external P_i and is resistant to uncoupling agents. The P_i-binding protein required for this latter transport system has been purified and shown to have a molecular weight of 42,000 and to bind 1 mole of P_i per mole protein (Medveczky and Rosenberg, 1970; Gerdes *et al.*, 1977).

Rae and Strickland (1976) have also presented evidence suggestive of at least two transport systems for P_i in *E. coli*. These workers indicate that one of the two transport systems is dependent on respiration, which may drive P_i transport via an electrochemical gradient or "energized state." It is suggested that the other transport system may be dependent on ATP generated by glycolysis or, alternatively, an energy-rich intermediate of glycolysis. Recalling the earlier genetic studies of Willsky *et al.* (1973), Rae and Strickland (1976) also point out that the ability to utilize P_i in *E. coli* may be governed by at least four genetically distinct proteins. Therefore, the P_i transport problem in certain bacterial strains may be as complex as in mitochondria where as many as four transport systems for P_i have now been postulated.

In recent years, several investigators have provided evidence for the view that one or more of the bacterial P_i transport systems may be functionally analogous to mitochondrial P_i transport systems. Burnell *et al.* (1975) have shown that in membrane vesicles of *Paracoccus denitrificans*, P_i uptake can be driven by a pH gradient (alkaline inside) applied across the vesicle membrane. Moreover, they show that SH reagents like NEM inhibit P_i uptake. Consistent with these findings, Harold and Spitz (1975) have reported that P_i accumulation in *Streptococcus faecalis* occurs via a P_i^-/OH^- antiport system (P_i^-/H^+ symporter) that can be driven by ATP or by a metabolite driven by ATP. Also, Friedberg (1977) has shown in *Micrococcus lysodeikticus* that the electrochemical gradient of protons across the membrane is necessary for P_i uptake. Finally, Tsuchiya and Rosen (1975, 1976) show that P_i (or oxalate) is required for movement of Ca^{2+} into inverted inner-membrane vesicles of *E. coli*.

In summary, it would appear that at least two P_i transport activities observed in bacterial systems are similar to those observed in mitochondrial systems, namely, symport of P_i and H^+ and symport of P_i and Ca^{2+}.

4. P_i TRANSPORT ACROSS THE INNER MEMBRANE OF THE CHLOROPLAST ENVELOPE

Transport of P_i across the chloroplast inner membrane has received very little attention. Heldt and Rapley (1970) and Flügge and Heldt (1976) have shown that P_i enters chloroplasts in exchange for 3-phosphoglycerate (3-PGA), and suggest that this antiport activity may be important for the overall reaction of CO_2 fixation, enabling the export of carbon from chloroplasts in the form of triosephosphates. Similar to the P_i^-/H^+ symporter of mitochondria, this antiport activity is inhibited by *p*-diazobenzenesulfonate, which in radioactive form labels a 29,000-molecular-weight protein (Flügge and Heldt, 1977). Certainly additional work is essential in chloroplasts to define the number and types of P_i

transport activities present, and to establish whether any of these activities are bacterial-and/or mitochondrial-like.

5. P_i TRANSPORT ACROSS THE PLASMA MEMBRANE OF ANIMAL CELLS

The types of research activity in this area have varied in recent years from one cell type to the other. Studies of P_i transport (and the transport of other anions) across the plasma membrane of erythrocytes and spermatozoa have advanced to the molecular level. Studies of P_i transport across the plasma membranes of kidney and intestine have been concerned primarily with elucidating the cation-dependence and energetics of the P_i transport process. Finally, studies of P_i transport across the plasma membrane of rapidly growing cells have focused on the relationship between P_i uptake and cell growth. Below, transport of P_i across the plasma membranes of each of these cell types is considered under separate headings.

5.1. Erythrocytes and Spermatozoa

A number of investigators have studied the transport of P_i and other anions across the plasma membrane of the red cell (for reviews see Steck, 1974; Rothstein *et al.*, 1976; Cabantchik *et al.*, 1978). Unlike the P_i transport systems of mitochondria, the P_i transport system located in the red cell membrane has a very low affinity for anions [K_m (P_i) = 80 mM, K_m (Cl^-) = 26 mM], and is nonspecific, i.e., P_i, Cl^-, and SO_4^{2-} are translocated.

Ho and Guidotti (1975) have studied the P_i transport process of red cells in considerable detail. They have shown that P_i transport is a saturable process with an external K_m of 80 mM and a V_{max} of 3.8 moles per liter of red cells per minute. Using the sulfanilate anion, a specific inhibitor of anion transport, they were able to label the anion transport system and isolate a labeled glycoprotein of ~100,000 molecular weight. This glycoprotein belongs to the general class designated as component *a* by Bretscher (1971) and is thought to span the bilayer. Consistent with these studies is the more recent work of Wolosin *et al.* (1977). These workers have isolated vesicles of the red blood cell membrane and shown that the 100K protein is the main protein component of these vesicles. Because these vesicles display various characteristic properties of anion permeation closely resembling those of intact erythrocytes, these investigators conclude that the 100K protein is involved directly in permeation functions in the red cell.

A number of other types of studies of P_i transport in the red cell have provided additional information about this system. Ross and McConnell (1975) show that the spin-labeled P_i derivative "TEMPO phosphate" is readily taken up by red cells, that the rate of transport depends strongly on the transmembrane electrical potential, and that transport is inhibited by the specific inhibitor of anion transport, 4-acetoamido-4'-isothiocyanostilbene-2,2'-disulfonic acid. Studies on erythrocyte ghosts by Schrier (1970) show that saturation occurs with $H_2PO_4^-$, but that HPO_4^{2-} is distributed passively accordingly to the Donnan equilibrium. Finally, Murphy and Libby (1976) have shown that pressures of 200–1000 atm decrease significantly the amount of P_i transported into the rabbit erythrocyte. Because pressure did not result in morphological changes or cell disruption, it was concluded that a lipid phase transition is induced by high pressure and is propagated to the associated P_i transport system via a physical mechanism.

Perhaps the best direct evidence for a P_i transport system in the plasma membrane of an animal cell type, other than red blood cells, is that reported by Babcock *et al.* (1975) for bovine spermatozoa. These investigators have shown that P_i transport into the cell is inhibited by sulfhydryl reagents such as NEM and mersalyl, and that filipin, which interacts preferentially with the plasma membrane, enhances the permeability of spermatozoa to P_i. It is suggested by Babcock *et al.* (1975) that P_i entry into spermatozoa is controlled by a plasma membrane component similar to the P_i transport system in mitochondria.

5.2. Kidney and Intestine

In a very thorough study, Berner *et al.* (1976) present evidence that P_i may be transported across intestinal brush border vesicles via a P_i^-/Na^+ symport system. They suggest that this transport system is electroneutral at pH 7.4 and electrogenic at pH 6.0. K^+, Rb^+, or Cs^+ when replacing Na^+ resulted in much lower rates of P_i uptake.

In an analogous study, Hoffman *et al.* (1976) show that a sodium-dependent transport system for P_i is present in the brush border microvilli but absent from the basal-lateral plasma membranes. They provide a number of pieces of data indicating that entry of P_i across the brush border membrane occurs in symport with Na^+ either as an electroneutral $2Na^+/P_i^{2-}$ or as an electrogenic $2Na^+/P_i^-$.

Other work on phosphate transport in kidney and intestine has shown that transport may be vitamin D dependent and associated with the genetic disease hypophosphatemia. Peterlik and Wasserman (1978) report that P_i may enter intestinal mucosal tissue by a vitamin D-dependent uphill transport process. However, the transfer of P_i from tissue to serosal compartment appears to be by diffusion (possibly facilitated) and nonvitamin D dependent. In a previous study, Wasserman and Taylor (1973) showed that P_i uptake by the intestine of vitamin D_3-treated rachitic chicks is not directly dependent on calcium, although a direct relationship between serum calcium and the degree of P_i absorption is evident.

Hypophosphatemia is a genetic disease (X-linked dominant inheritance) that is sometimes referred to as familial vitamin D-resistant rickets. Affected individuals have essentially normal serum calcium levels, hypophosphatemia, impaired renal tubular reabsorption of phosphate anion, shortened stature, and vitamin D-nonresponsive rickets or osteomalacia. Recently a mouse model has been discovered, and using this model it has been shown that the mutant gene product of the *Hyp* gene is confined to the brush border membrane of the kidney (Eicher *et al.*, 1976; Tenenhouse and Schriver, 1978).

5.3. Rapidly Growing Cells

Early work in this area was carried out by Wu and Racker (1959) on Ehrlich ascites tumor cells. These workers showed that in the intact Ehrlich ascites tumor cell, glycolysis and mitochondrial oxidative phosphorylation contribute independently and almost equally to the transport of P_i in the cell. In these studies, P_i appeared to be a major rate-limiting factor in glycolysis, for high concentrations of external P_i markedly stimulated lactic acid production in intact cells.

The work of Wu and Racker (1959) was extended in some detail in the late 1960s and during the 1970s with the major objective of ascertaining whether P_i transport into the cell is one of the major rate-limiting factors in cell growth. To date, there does not seem to be any agreement on this point. Weber and Edlin (1971) showed that when cultures of mouse 3T3 cells become density inhibited, the rate of P_i transport across the cell membrane declines fivefold. In an analogous study, Cunningham and Pardee (1969) showed that P_i

transport is dependent on two serum factors of less than 40,000 molecular weight, which when added to confluent 3T3 cells result in a two to fourfold increase in P_i transport. Extending these studies, DeAsua *et al.* (1974) showed that stimulation of both P_i and uridine transport by addition of fresh serum to 3T3 cells is accompanied by a decrease in the intracellular concentration of 3' : 5'-cyclic AMP. From these and other studies, DeAsua *et al.* (1974) concluded that the increase in P_i transport is a primary event in the reinitiation of growth.

Follow-up studies from several laboratories suggest that although P_i transport may be related in some way to density-dependent inhibition of cell division, enhanced P_i transport is not essential for cell proliferation and does not accompany transformation of all cell types. Greenberg *et al.* (1977) have demonstrated by a number of different methods that the rapid increase in P_i uptake following addition of fresh serum to quiescent fibroblasts is not a necessary event for initiation of cell proliferation. These workers show that addition of dexamethasone, trypsin, or insulin stimulates proliferation but causes little or no change in P_i uptake. Weber *et al.* (1976), in examining a number of transport processes, conclude that P_i transport (as well as transport of nucleosides, amino acids, glucose, and K^+) is affected in the density-dependent inhibition of growth of chicken embryo fibroblasts, but only glucose transport is affected when cells are transformed by Rous sarcoma virus.

Because there does seem to be some relationship between density-dependent inhibition of cell division and P_i, Gray *et al.* (1976) examined this possibility in greater detail. They reported that prior to intracellular contact, the P_i concentration in 3T3 cells is 10 mM, but during critical contact this concentration is quickly reduced to approximately 2 mM and remains at this concentration to confluency. Similar alterations do not occur in Py 3T3 cells (polyoma virus-transformed 3T3 cells), which maintain a concentration of approximately 2 mM P_i regardless of cell density. It is suggested that the controlled modulation of P_i may regulate glycolysis and mitochondrial activity.

Two recent studies, in contrast to those noted above, have provided information about the effect of serum on the transport event (or carrier) for P_i in rapidly growing cells. Jullien and Harel (1976) summarize evidence suggesting that serum has two distinct effects, one on P_i transport and another on intracellular phosphorylation reactions presumably glycolysis and mitochondrial oxidative phosphorylation. Hilborn (1976) has shown that cAMP-stimulating drugs do not affect P_i transport in 3T3 cells and that the serum-stimulating effect on P_i transport is not due to a single serum fraction—more than one factor in the serum may be responsible.

Finally, several recent reports in the literature suggest that P_i accumulation by rapidly growing cells may involve monovalent cations, whereas another report implicates a role for glyceraldehyde-3-P. Nilsen-Hamilton and Hamilton (1976), Hamilton and Nilsen-Hamilton (1978), and Lever (1978) show that membrane vesicles derived from the plasma membrane and endoplasmic reticulum of mouse 3T3 cells transformed by simian virus 40 take up P_i by a mechanism that is dependent on sodium. Lever *et al.* (1976) show that in resting mouse fibroblast cultures, prostaglandin $F_{2\alpha}$ activates the Na^+/K^+-ATPase and stimulates P_i transport by a process partially coupled to the Na^+ pump. Mazumder and Wenner (1977) on the other hand find that ouabain, an inhibitor of Na^+/K^+-ATPase, stimulates the rate of incorporation of $^{32}P_i$ into cells and raises the net P_i level. Neihaus and Hammerstedt (1976) interpret $^{32}P_i$ labeling data on HeLa cells and human erythrocytes to suggest that P_i enters the cell as 1,3-diphosphoglycerate by reacting extracellularly with glyceraldehyde-3-P. Inside the cell, ATP is formed that is then assumed to be hydrolyzed to provide intracellular P_i.

ACKNOWLEDGMENT. This article was written while the authors were supported by NSF Grant PCM 76-11024.

REFERENCES

Abou-Khalil, S., SaBadie-Pialoux, N., and Gautheron, C. (1975). *Biochimie* **57**, 1087–1094.

Azzone, G. F., Massari, S., and Pozzan, T. (1976). *Biochim. Biophys. Acta* **423**, 15–26.

Babcock, D. F., First, N. L., and Lardy, H. A. (1975). *J. Biol. Chem.* **250**, 6488–6495.

Banerjee, R. K., and Racker, E. (1979). *Membr. Biochem.* **2**, 203–225.

Banerjee, R. K., Shertzer, H. G., Kanner, B. I., and Racker, E. (1977). *Biochem. Biophys. Res. Commun.* **75**, 772–778.

Berner, W., Kinner, R., and Murer, H. (1976). *Biochem. J.* **160**, 467–474.

Blondin, G. A. (1979). *Biochem. Biophys. Res. Commun.* **87**, 1087–1094.

Bretcher, M. S. (1971). *J. Mol. Biol.* **59**, 351–357.

Briand, Y., Debise, R., and Durand, R. (1975). *Biochimie* **47**, 787–796.

Briand, Y., Touraille, S., Debise, R., and Durand, R. (1976). *FEBS Lett.* **65**, 1–7.

Burnell, J. N., John, P., and Whatley, F. R. (1975). *Biochem. J.* **150**, 527–536.

Bygrave, F. L. (1967). *Nature (London)* **214**, 667–671.

Cabantchik, Z., Knauk, P., and Rothstein, A. (1978). *Biochim. Biophys. Acta* **515**, 239–302.

Chappell, J. B. (1968). *Brit. Med. Bull.* **24**, 150–157.

Chappell, J. B., and Haarhoff, K. (1967). In *Biochemistry of Mitochondria* (E. C. Slater, Z. Kaniuga, and L. Wojtczak, eds.), pp. 75–91, Academic Press, New York.

Chateaubodeau, G., Guérin, M., and Guérin, B. (1974). *FEBS Lett.* **46**, 184–187.

Christiansen, E. N., Grav, H. J., and Wojtczak, L. (1974). *FEBS Lett.* **46**, 188–191.

Coty, W. A., and Pedersen, P. L. (1974). *J. Biol. Chem.* **249**, 2593–2598.

Coty, W. A., and Pedersen, P. L. (1975a). *J. Biol. Chem.* **250**, 3515–3521.

Coty, W. A., and Pedersen, P. L. (1975b). *Mol. Cell. Biochem.* **9**, 109–124.

Cunningham, D. D., and Pardee, A. B. (1969). *Proc. Natl. Acad. Sci. USA* **64**, 1049–1056.

DeAsua, L. J., Rozenqurt, E., and Dulbecco, R. (1974). *Proc. Natl. Acad. Sci. USA* **71**, 96–98.

Eicher, E. M., Southard, J. L., Schriver, C. R., and Glorieux, F. H. (1976). *Proc. Natl. Acad. Sci. USA* **73**, 4667–4671.

Fiskum, G., and Lehninger, A. L. (1980). *Fed. Proc.* **39**, 2432–2436.

Fiskum, G., Reynafarje, B., and Lehninger, A. L. (1979). *J. Biol. Chem.* **254**, 6288–6295.

Flügge, U. I., and Heldt, H. W. (1976). *FEBS Lett.* **68**, 259–262.

Flügge, U. I., and Heldt, H. W. (1977). *FEBS Lett.* **82**, 29–33.

Fonyó, A. (1968). *Biochem. Biophys. Res. Commun.* **32**, 624–628.

Fonyó, A. (1979). *Pharmacol. Ther.* **7**, 627–645.

Fonyó, A., and Bessman, S. P. (1966). *Biochem. Biophys. Res. Commun.* **24**, 61–66.

Fonyó, A., and Bessman, S. P. (1968). *Biochem. Med.* **2**, 145–163.

Fonyó, A., Ligeti, E., Palmieri, F., and Quagliariello, E. (1975). In *Biomembranes, Structure and Function* (G. Gardos and I. Szasz, eds.), pp. 287–306, Akademial Kiado, Budapest.

Fonyó, A., Ligeti, E., and Vignais, P. V. (1980). In *First European Bioenergetics Conference*, pp. 291–292, Patron Editore, Bologna.

Freitag, H., and Kadenbach, B. (1978). *Eur. J. Biochem.* **83**, 53–57.

Friedberg, I. (1977). *FEBS Lett.* **81**, 264–266.

Gerdes, R. G., Strickland, K. P., and Rosenberg, H. (1977). *J. Bacteriol.* **131**, 512–518.

Goldschmidt, D., Gaudemer, Y., and Gautheron, D. (1976). *Biochimie* **58**, 713–722.

Gray, P. N., Cullum, M., and Griffin, M. J. (1976). *J. Cell. Physiol.* **89**, 225–234.

Greenberg, D. B., Barsh, G., Ho, T.-S., and Cunningham, D. D. (1977). *J. Cell. Physiol.* **90**, 193–210.

Guérin, B., Guérin, M., and Klingenberg, M. (1970). *FEBS Lett.* **10**, 265–268.

Guérin, M., and Guérin, B. (1975). *FEBS Lett.* **50**, 210–213.

Guérin, M., and Napias, C. (1978). *Biochemistry* **17**, 2510–2516.

Hadvary, P., and Kadenbach, B. (1976). *Eur. J. Biochem.* **67**, 573–581.

Hamilton, R., and Nilsen-Hamilton, M. (1978). *J. Biol. Chem.* **253**, 8247–8256.

Harold, F. M., and Spitz, E. (1975). *J. Bacteriol.* **122**, 266–277.

Harris, E. J., and Zaba, B. (1977). *FEBS Lett.* **79**, 284–289.

Heldt, H. W., and Rapley, L. (1970). *FEBS Lett.* **10**, 143–148.

Hilborn, D. (1976). *J. Cell. Physiol.* **87**, 111–122.

Ho, M. K., and Guidotti, G. (1975). *J. Biol. Chem.* **250**, 675–683.

Hoek, J. B., Lofrumento, N. E., Meijer, A. J., and Tager, J. M. (1971). *Biochim. Biophys. Acta* **226**, 297–308.

Hoffman, N., Thees, M., and Kinne, R. (1976). *Pfluegers Arch.* **362**, 147–156.

Homes, F. A., Masterbroek-Helder, D. J., and Molenaar, I. (1975). *Nutr. Metab.* **19**, 263–267.

Howard, A. N., and Wild, F. (1957). *Biochem. J.* **65**, 651–659.

Jullien, M., and Harel, L. (1976). *Exp. Cell Res.* **97**, 23–30.

Kadenbach, B., and Hadvary, P. (1973). *Eur. J. Biochem.* **39**, 21–26.

Kadenbach, B., Freitag, H., and Kolbe, H. (1978). *FEBS Lett.* **89**, 161–164.

Klingenberg, M. (1970). *FEBS Lett.* **6**, 145–154.

Klingenberg, M., Durand, R., and Guérin, B. (1974). *Eur. J. Biochem.* **42**, 135–150.

Kovaćević, Z. (1976). *Biochim. Biophys. Acta* **430**, 339–412.

LaNoue, K. F., Bryla, J., and Bassett, D. J. (1974). *J. Biol. Chem.* **249**, 7514–7521.

Lehninger, A. L. (1971). *Biochem. J.* **119**, 129–138.

Lehninger, A. L. (1974). *Proc. Natl. Acad. Sci. USA* **71**, 1520–1524.

Lever, J. E. (1978). *J. Biol. Chem.* **253**, 2081–2084.

Lever, J. E., Clingan, D., and DeAsua, L. J. (1976). *Biochem. Biophys. Res. Commun.* **71**, 136–143.

Lofrumento, N. E., and Zanotti, F. (1976). *FEBS Lett.* **63**, 129–133.

Lofrumento, N. E., Zanotti, F., and Papa, S. (1974). *FEBS Lett.* **48**, 188–191.

Lötscher, H. R., Schwerzmann, K., and Carafoli, E. (1979). *FEBS Lett.* **99**, 194–197.

Mazumder, A., and Wenner, C. E. (1977). *Arch. Biochem. Biophys.* **179**, 409–414.

Medveczky, N., and Rosenberg, H. (1970). *Biochim. Biophys. Acta* **211**, 158–168.

Meijer, A. J., Groot, G. S. P., and Tager, J. M. (1970). *FEBS Lett.* **8**, 41–44.

Meyer, J., and Vignais, P. M. (1973). *Biochim. Biophys. Acta* **325**, 375–384.

Mitchell, P. (1953). *J. Gen. Microbiol.* **9**, 273–287.

Mitchell, P. (1954). *J. Gen. Microbiol.* **11**, 73–82.

Mitchell, P. (1968). In *Chemiosmotic Coupling and Energy Transduction,* pp. 1–111, Glynn Research, Bodmin.

Moyle, J., and Mitchell, P. (1977). *FEBS Lett.* **77**, 136–140.

Murphy, R. B., and Libby, W. F. (1976). *Proc. Natl. Acad. Sci. USA* **73**, 2767–2769.

Neihaus, W. G., Jr., and Hammerstedt, R. H. (1976). *Biochim. Biophys. Acta* **443**, 515–524.

Nilsen-Hamilton, M., and Hamilton, R. (1976). *J. Cell. Physiol.* **89**, 795–800.

Palmieri, F., Genchi, G., Stipani, I., Francia, F., and Quagliariello, E. (1974). In *Membrane Proteins in Transport and Phosphorylation* (G. F. Azzone, M. E. Klingenberg, E. Quagliariello, and N. Siliprandi, eds.), pp. 245–256, North-Holland, Amsterdam.

Papa, S., Zanghi, M. A., Paradies, G., and Quagliariello, E. (1970). *FEBS Lett.* **6**, 1–4.

Pedersen, P. L., and Coty, W. A. (1972). *J. Biol. Chem.* **247**, 3107–3113.

Peterlik, M., and Wasserman, R. H. (1978). *J. Physiol.* **234**, E379–E388.

Rae, A. S., and Strickland, K. P. (1976). *Biochim. Biophys. Acta* **433**, 564–582.

Reynafarje, B., and Lehninger, A. L. (1978). *Proc. Natl. Acad. Sci. USA* **75**, 4788–4792.

Rhodin, T. R., and Racker, E. (1974). *Biochem. Biophys. Res. Commun.* **61**, 1207–1212.

Rosenberg, H., Cox, G. B., Butlin, J. D., and Gutowski, S. J. (1975). *Biochem. J.* **146**, 417–423.

Rosenberg, H., Gerdes, R. G., and Cheqwidden, K. (1977). *J. Bacteriol.* **131**, 505–511.

Ross, A. H., and McConnell, H. M. (1975). *Biochemistry* **14**, 2793–2798.

Rothstein, A., Cabantchik, Z. I., and Knauf, P. (1976). *Fed. Proc.* **35**, 3–10.

Schrier, S. L. (1970). *J. Lab. Clin. Med.* **75**, 422–434.

Siliprandi, D., Toninello, A., Zoccarato, F., and Bindoli, A. (1975). *FEBS Lett.* **51**, 15–17.

Steck, T. (1974). *J. Cell Biol.* **62**, 1–19.

Tenenhouse, H. S., and Schriver, C. R. (1978). *Can. J. Biochem.* **56**, 640–646.

Tsuchiya, T., and Rosen, B. P. (1975). *J. Biol. Chem.* **250**, 7687–7692.

Tsuchiya, T., and Rosen, B. P. (1976). *J. Biol. Chem.* **251**, 962–967.

Vignais, P. M., and Vignais, P. V. (1973). *Biochim. Biophys. Acta* **325**, 373–374.

Wasserman, R. H., and Taylor, A. N. (1973). *J. Nutr.* **103**, 586–599.

Weber, M. J., and Edlin, G. (1971). *J. Biol. Chem.* **246**, 1828–1833.

Weber, M. J., Hale, A. H., Yau, T. M., Buckman, T., Johnson, M., Brady, T. M., and LaRossa, D. D. (1976). *J. Cell. Physiol.* **89**, 711–722.

Wehrle, J., and Pedersen, P. L. (1979). *J. Biol. Chem.* **254**, 7269–7275.

Wehrle, J., Cintrón, N., and Pedersen, P. L. (1978). *J. Biol. Chem.* **253,** 8598–8603.

Willsky, G. R., Bennet, R. L., and Malamy, M. H. (1973). *J. Bacteriol.* **113,** 529–539.

Wohlrab, H. (1980). In *First European Bioenergetics Conference,* pp. 301–302, Patron Editore, Bologna.

Wolosin, J. M., Ginsburg, H., and Cabantchik, Z. I. (1977). *J. Biol. Chem.* **252,** 2419–2427.

Wu, R., and Racker, E. (1959). *J. Biol. Chem.* **234,** 1029–1035.

Yu, H. L., Giammarco, R., Goldstein, M. B., Stinebaugh, B. J., and Halperin, M. (1976). *J. Clin. Invest.* **58,** 557–564.

Zimmer, G. (1977). *Biochim. Biophys. Acta* **461,** 268–273.

Zoccarato, F., Rugolo, M., and Siliprandi, D. (1977). *J. Bioenerg. Biomembr.* **9,** 203–212.

85

Vitamin D and the Intestinal Absorption of Calcium and Phosphate

Robert H. Wasserman

1. INTRODUCTION

The transport of calcium and phosphate (P_i) across the intestinal epithelium has been studied extensively in recent years and a considerable number of these investigations have been concerned with the action of vitamin D. It has been known for several decades that vitamin D is required for the optimal absorption of calcium (Nicolaysen *et al.*, 1953) and, more recently, that there is also a direct effect of vitamin D on phosphate transport (Harrison and Harrison, 1961, 1963; Kowarski and Schachter, 1969). For many years, it was held that phosphate transport occurred secondarily to calcium transport, but this notion has been essentially discarded. Vitamin D has proven to be a useful variable in the study of these processes for a comparison can be made between the efficiency of calcium and phosphate transport in the presence and absence of vitamin D and, importantly, this allows an assessment of the molecular changes that coincide with vitamin D action (reviewed in Wasserman, 1980). Some of the recorded vitamin D-related macromolecular changes most probably bear on the stimulation of transport reactions and offer a means of gaining insight into the molecular basis of calcium and phosphate transport.

Vitamin D is known today to undergo biochemical transformations to yield biologically active metabolites, the most potent and most rapidly acting being 1,25-dihydroxycholecalciferol [$1,25(OH)_2D_3$]. The experimental use of $1,25(OH)_2D_3$ has added another dimension to investigations of calcium and phosphate transport.

For additional information on vitamin D metabolism and action, refer to the following monographs and reviews: Deluca (1979), Norman (1980), Lawson (1978).

Robert H. Wasserman • Department of Physiology, New York State College of Veterinary Medicine, Cornell University, Ithaca, New York 14853.

2. CALCIUM ABSORPTION AND VITAMIN D

Calcium is absorbed by two mechanisms: one is an active transport process and the other is diffusional in nature (Wasserman and Taylor, 1969). At low levels of intraluminal Ca^{2+}, the process that predominates is active transport, and at higher levels of Ca^{2+} (>5 mM), the diffusional mode becomes increasingly significant. Vitamin D, in our view, accelerates both types of processes and, further, it is possible that the diffusional absorption of Ca^{2+} occurs, at least in part, by way of the paracellular route. The active component of Ca^{2+} absorption assuredly is a transcellular event.

Recent studies also showed that microvilli isolated from the intestine of vitamin D-repleted chicks accumulate Ca^{2+} to a greater extent than those from rachitic chicks (Rasmussen *et al.*, 1979), which supports the hypothesis that a major site of Vitamin D action is at the brush border membrane (Wasserman and Taylor, 1969). Vitamin D also appears to increase the rate of transfer of Ca^{2+} across the basal-lateral membrane (Wasserman and Taylor, 1969). Thus, at least two sites of vitamin D action on the intestine appear likely.

Entrance of Ca^{2+} into the cell from the lumen is undoubtedly energetically independent because of a favorable downhill thermodynamic gradient. Extrusion of Ca^{2+} from the cell toward the lamina propria is against a thermodynamic gradient and therefore requires energy input. The latter is probably via a Ca^{2+}-ATPase identified on the basal-lateral membranes (Birge and Gilbert, 1974; Mircheff *et al.*, 1977) and/or by Na^+–Ca^{2+} exchange.

Vitamin D action has been considered to be analogous to the action of many steroid hormones, i.e., the induction of specific proteins that, in this case, are involved in Ca^{2+} and/or P_i translocation. One such product of the $1,25(OH)_2D_3$–gene interaction is the vitamin D-induced calcium-binding protein (CaBP). This has been shown by our group (Wasserman and Corradino, 1973; Wasserman *et al.*, 1978) and, most convincingly, by the elegant studies of Lawson and his colleagues (Emtage *et al.*, 1974; Spencer *et al.*, 1976).

Vitamin D-induced CaBP is a soluble protein that binds Ca^{2+} with high affinity (Wasserman *et al.*, 1978). Two varieties have been isolated and partially characterized. The avian type has a molecular weight of about 28,000 and binds 4 Ca^{2+} per molecule; the mammalian type has a molecular weight of about 10,000 and binds 2 Ca^{2+} per molecule. Both are acidic and have about 30–40% α-helicity. The pig (Hofmann *et al.*, 1979) and bovine (Fullmer and Wasserman, 1981) proteins have been sequenced, and show a high degree of homology with other calcium-binding proteins, such as calmodulin, troponin C, and parvalbumin (Kretsinger, 1980). CaBP is present in the other tissues in addition to the intestine, and these include the kidney, brain, bone, pancreas, parathyroid glands, and the shell gland (uterus) of the laying hen (Wasserman *et al.*, 1978; Christakos and Norman, 1980; Murray *et al.*, 1975). The avian type is present both in avian species and in certain tissues of the mammal. The mammalian type is found only in mammals, as far as we now know.

The possible involvement of CaBP in the vitamin D-dependent absorption of calcium has come from a wide variety of studies, demonstrating a high correlation between Ca transport and the concentration of CaBP in the intestinal mucosa (Wasserman and Corradino, 1973; Wasserman *et al.*, 1978). As one example, animals fed a normal Ca^{2+} diet and then fed a calcium-deficient diet have the capacity to increase their efficiency of calcium absorption and, concurrent with this change, there is an increase in CaBP levels (Morrissey and Wasserman, 1971; Bar *et al.*, 1978; Armbrecht *et al.*, 1980). These effects are mediated through an increase in the synthesis of $1,25(OH)_2D_3$ by the kidney-based hydroxylase system under the stimulus of calcium deprivation (DeLuca, 1979). Other experiments disclose an apparent dissociation between CaBP and calcium transport, particularly studies in which a single, acute dose of 1,25(OH)$_2$D$_3$ is given to vitamin D-deficient animals. According to some reports (Spencer *et al.*, 1978; Thomasset *et al.*, 1979), the appearance of CaBP in rachitic intestine lags behind changes in Ca^{2+} transport, although this has not been our experience (Wasserman *et al.*, 1977). However, it is agreed that, under these same conditions, CaBP is still present in the intestine after the 1,25(OH)$_2$D$_3$-stimulated absorption of Ca^{2+} has returned to the rachitic baseline. These and other studies suggest that other factors, in addition to the presence of CaBP, are required. In fact, the essentiality of new protein synthesis for the expression of vitamin D action has been questioned (Rasmussen *et al.*, 1979).

In this context, other vitamin D-responsive molecules and biochemical reactions have been identified in intestinal mucosa, and these include the stimulation of the synthesis of alkaline phosphatase (Holdsworth, 1970; Norman *et al.*, 1970; Haussler *et al.*, 1970), a Ca^{2+}-ATPase (Haussler *et al.*, 1970; Melancon and DeLuca, 1970), an 86,000 to 90,000-molecular-weight brush border protein (Rasmussen *et al.*, 1977; Lawson *et al.*, 1977), a 45,000-molecular-weight brush border protein (Lawson *et al.*, 1977), a high-molecular-weight particulate calcium-binding complex (Kowarski and Schachter, 1980), and an 18,500-molecular-weight calcium-binding protein (Miller *et al.*, 1979). Effects of vitamin D on adenylate cyclase activity (Neville and Holdsworth, 1969; Corradino, 1974) and sialotransferase activity (Moriuchi *et al.*, 1977) have been recorded, in addition to alterations in the degree of unsaturation of the acyl chain of membrane phospholipids (Max *et al.*, 1978). The relevance of these changes to vitamin D-mediated calcium and/or phosphate transport is yet to be resolved.

Deciphering the role of CaBP in the mechanism of calcium absorption has proven difficult, and one reason is the lack of unambiguous information on its cellular and subcellular localization. The recent study of Taylor (1980) suggests that it is cytoplasmic, and others place it more specifically in the terminal web/brush border region of the intestinal cell (Marche *et al.*, 1979; Roth *et al.*, 1980). Several ideas of the function of CaBP have been proposed, ranging from a role as an intracellular buffer, to a shuttle for the transcellular movement of Ca^{2+}, to an ability to confer Ca^{2+} sensitivity to another macromolecular complex, analogous to the action of calmodulin. The question of its mode of action has yet to be unequivocally defined.

3. PHOSPHATE ABSORPTION AND VITAMIN D

Earlier studies showed that the process of phosphate absorption is stimulated by vitamin D, that phosphate absorption has an active transport component, and that this active process is Na$^+$ dependent (Harrison and Harrison, 1961, 1963; Kowarski and Schachter, 1969). These early observations have been confirmed by others, and information on the process extended. Everted gut sac experiments revealed that the active transport site is associated with the brush border region of the cell, that phosphate after entering the cell does not apparently mix with the intracellular pool of phosphate, and that the release of phosphate toward the lamina propria is neither vitamin D dependent nor an active extrusion process (Peterlik and Wasserman, 1978). Studies with isolated brush border vesicles confirm the Na$^+$ dependency (Berner *et al.*, 1976; Fontaine *et al.*, 1979) and vitamin D dependency of the transfer of phosphate across this membrane (Fontaine *et al.*, 1979).

There is evidence that vitamin D-stimulated absorption of phosphate is dependent on protein synthesis (Peterlik and Wasserman, 1980). A macromolecule potentially involved in phosphate transport is a recently uncovered hydrophobic polypeptide derived from the intestinal brush border membranes (Wasserman and Brindak, 1979). This polypeptide is vitamin D responsive, has a molecular weight of about 81,000, transiently binds phosphate covalently, and the phosphate-binding reaction appears to be Na^+ dependent. A similar or identical protein was noted by Rasmussen *et al.* (1979) as a band on an SDS-acrylamide electrophoretic gel, and Lawson *et al.* (1977) reported the increased synthesis of a protein of similar molecular size in response to $1,25(OH)_2D_3$. There is a possible relationship between this polypeptide and alkaline phosphatase, although the nature of this association is not clear. The involvement of alkaline phosphatase in phosphate transport had been earlier proposed by Moog and Glazier (1971).

4. SUMMATION

There has been considerable information published on the nutritional and physiological aspects of calcium and phosphate transport, and the action of vitamin D thereon. Only part of this wealth of information could be included in this brief report, but despite these extensive efforts, major problems still remain in providing a complete description of these absorptive processes. Still to be described in unequivocal detail are the molecular basis of the transport of these ions across the brush border, the path taken in their movement through the cell, and the manner of their extrusion across the basal-lateral membrane. Several vitamin D-dependent molecular species and biochemical reactions have been reported, and the relationship of these (and yet-to-be-identified substances) to calcium and phosphate absorption requires further study.

ACKNOWLEDGMENTS. The studies from the author's laboratory were supported by NIH Grant AM-04652 and U.S. Department of Energy Contract EY-76-S-02-2792-004. The contributions of C. S. Fullmer, A. N. Taylor, R. A. Corradino, J. J. Feher, P. J. Bredderman, M. Peterlik, M. E. Brindak, and N. Jayne are recognized with thanks.

REFERENCES

Armbrecht, H. J., Zenser, T. V., Gross, C. J., and Davis, B. B. (1980). *Am. J. Physiol.* **239**, E322–E327.

Bar, A., Cohen, A., Edelstein, S., Shemesh, M., Montecuccoli, G., and Hurwitz, S. (1978). *Comp. Biochem. Physiol. B* **59**, 245–249.

Berner, W., Kinne, R., and Murer, H. (1976). *Biochem. J.* **160**, 467–474.

Birge, S. J., and Gilbert, H. R. (1974). *J. Clin. Invest.* **54**, 710–717.

Christakos, S., and Norman, A. W. (1980). In *Calcium-Binding Proteins: Structure and Function* (F. L. Siegel, E. Carafoli, R. H. Kretsinger, D. H. MacLennan, R. H. Wasserman, eds.), pp. 371–378, Elsevier/North-Holland, Amsterdam.

Corradino, R. A. (1974). *Endocrinology* **94**, 1607–1614.

DeLuca, H. (1979). *Vitamin D—Metabolism and Function*, Springer-Verlag, Berlin.

Emtage, J. S., Lawson, E. E. M., and Kodicek, E. (1974). *Biochem. j.* **140**, 239–247.

Fontaine, O., Matsumoto, T., Simoniescu, D. B. P., Goodman, D. B. P., and Rasmussen, H. (1979). In *Vitamin D: Basic Research and Its Clinical Application* (A. W. Norman, K. Schaefer, D. von Herrath, H.-G. Grigoleit, J. W. Coburn, H. F. DeLuca, E. B. Mawer, and T. Suda, eds.), pp. 693–701, de Gruyter, Berlin.

Fullmer, C. S., and Wasserman, R. H. (1981). *J. Biol. Chem.*, **256**, 5669–5674.

Harrison, H. E., and Harrison, H. C. (1961). *Am. J. Physiol.* **201**, 1007–1012.

Harrison, H. E., and Harrison, H. C. (1963). *Am. J. Physiol.* **205**, 107–111.

Haussler, M. R., Nagode, L. A., and Rasmussen, H. (1970). *Nature (London)* **228**, 1199–1201.

Hofmann, T., Kawakami, M., Hitchman, A. J. W., Harrison, J. E., and Dorrington, K. J. (1979). *Can. J. Biochem.* **57**, 737–748.

Holdsworth, E. S. (1970). *J. Membr. Biol.* **3**, 43–53.

Kowarski, S., and Schachter, D. (1969). *J. Biol. Chem.* **244**, 211–217.

Kowarski, S., and Schachter, D. (1980). *J. Biol. Chem.* **255**, 10834–10840.

Kretsinger, R. H. (1980). In *CRC Critical Reviews in Biochemistry* (G. D. Frasman, ed.), Vol. 8, pp. 119–174, CRC Press, Boca Raton, Fla.

Lawson, D. E. M. (ed.) (1978). *Vitamin D*, Academic Press, New York.

Lawson, D. E. M., Spencer, R., Charman, M., and Wilson, P. (1977). In *Vitamin D: Biochemical, Chemical and Clinical Aspects Related to Calcium Metabolism* (A. W. Norman, K. Schaefer, J. W. Coburn, H. F. DeLuca, D. Fraser, H.-G. Grigoleit, and D. von Herrath, eds.), pp. 265–275, de Gruyter, Berlin.

Marche, P., LeGuern, C., and Cassier, P. (1979). *Cell Tissue Res.* **197**, 69–77.

Max, E. E., Goodman, D. B. P., and Rasmussen, H. (1978). *Biochim. Biophys. Acta* **511**, 224–239.

Melancon, M. J., Jr., and DeLuca, H. F. (1970). *Biochemistry* **9**, 1658–1664.

Miller, A., Ueng, T.-H., and Bronner, F. (1979). *FEBS Lett.* **103**, 319–322.

Mircheff, A. K., Walling, M. W., Van Os, C. H., and Wright, E. M. (1977). In *Vitamin D: Biochemical, Chemical and Clinical Aspects Related to Calcium Metabolism* (A. W. Norman, K. Schaefer, J. W. Coburn, H. F. DeLuca, D. Fraser, H.-G. Grigoleit, and D. von Herrath, eds.), pp. 281–283, de Gruyter, Berlin.

Moog, F., and Glazier, H. S. (1971). *Comp. Biochem. Physiol. A* **42**, 321–326.

Moriuchi, S., Yoshizawa, S., and Hosoya, N. (1977). *J. Nutr. Sci. Vitaminol.* **23**, 497–504.

Morrissey, R. L., and Wasserman, R. H. (1971). *Am. J. Physiol.* **220**, 1509–1515.

Murray, T. M., Arnold, B. M., Kuttner, M., Kovacs, K., Hitchman, A. J. W., and Harrison, J. E. (1975). In *Calcium Regulating Hormones* (R. V. Talmage, M. Owen, and J. A. Parsons, eds.), pp. 371–375, Excerpta Medica, Amsterdam.

Neville, E., and Holdsworth, E. S. (1969). *FEBS Lett.* **2**, 313–316.

Nicolaysen, R., Eeg-Larsen, N., and Malm, O. J. (1953). *Physiol. Rev.* **33**, 424–444.

Norman, A. W. (1980). *Vitamin D: The Calcium Homeostatic Steroid Hormone*, Academic Press, New York.

Norman, A. W., Mircheff, A. K., Adams, T. H., and Spielvogel, A. (1970). *Biochim. Biophys. Acta* **215**, 348–359.

Peterlik, M., and Wasserman, R. H. (1978). *Am. J. Physiol.* **234**, E379–E388.

Peterlik, M., and Wasserman, R. H. (1980). *Horm. Metab. Res.* **12**, 216–219.

Rasmussen, H., Max, E. E., and Goodman, D. B. P. (1977). In *Vitamin D: Biochemical, Chemical and Clinical Aspects Related to Calcium Metabolism* (A. W. Norman K. Schaefer, J. W. Coburn, H. F. DeLuca, D. Fraser, H.-G. Grigoleit, and D. von Herrath, eds.), pp. 913–925, de Gruyter, Berlin.

Rasmussen, H., Fontaine, O., Max, E. E., and Goodman, D. B. P. (1979). *J. Biol. Chem.* **254**, 2993–2999.

Roth, S. I., Futrell, J. M., Wasserman, R. H., Brindak, M. E., Fullmer, C. S., Su, S. P. C., and Brakhop, N. H. (1980). *Fed. Proc.* **39**, 359.

Spencer, R., Charman, M., Emtage, J. S., and Lawson, D. E. M. (1976). *Eur. J. Biochem.* **71**, 399–409.

Spencer, R., Charman, M., Wilson, P. W., and Lawson, D. E. M. (1978). *Biochem. J.* **170**, 93–101.

Taylor, A. N. (1980). In *Calcium-Binding Proteins: Structure and Function* (F. L. Siegel, E. Carafoli, R. H. Kretsinger, D. H. MacLennan, and R. H. Wasserman, eds.), pp. 393–400, Elsevier/North-Holland, Amsterdam.

Thomasset, M., Cuisinier-Gleizes, P., and Mathieu, H. (1979). *FEBS Lett.* **107**, 91–94.

Wasserman, R. H. (1980). In *Pediatric Diseases Related to Calcium* (H. F. DeLuca, and C. S. Anast, eds.), pp. 107–132, Elsevier, Amsterdam.

Wasserman, R. H., and Brindak, M. E. (1979). In *Vitamin D: Basic Research and Its Clinical Application* (A. W. Norman, K. Schaefer, D. von Herrath, H.-G. Grigoleit, J. W. Coburn, H. F. DeLuca, E. B. Mawer, and T. Suda, eds.), pp. 703–710, de Gruyter, Berlin.

Wasserman, R. H., and Corradino, R. A. (1973). In *Vitamins and Hormones* (R. S. Harris, E. Diczfalusy, P. L. Munson, and J. Glover, eds.), Vol. 31, pp. 43–103, Academic Press, New York.

Wasserman, R. H., and Taylor, A. N. (1969). In *Mineral Metabolism, An Advanced Treatise* (C. L. Comar and F. Bronner, eds.), pp. 321–403, Academic Press, New York.

Wasserman, R. H., Corradino, R. A., Feher, J., and Armbrecht, H. J. (1977). In *Vitamin D: Biochemical, Chemical and Clinical Aspects Related to Calcium Metabolism* (A. W. Norman, K. Schaefer, J. W. Coburn, H. F. DeLuca, D. Fraser, H.-G. Grigoleit, and D. von Herrath, eds.), pp. 331–340, de Gruyter, Berlin.

Wasserman, R. H., Fullmer, C. S., and Taylor, A. N. (1978). In *Vitamin D* (D. E. M. Lawson, ed.), pp. 133–166, Academic Press, New York.

86

Alterations in Membrane Permeability to Ca^{2+} and Their Consequences during Maturation of Mammalian Spermatozoa

Donner F. Babcock and Henry A. Lardy

1. CALCIUM ACCUMULATION AND CALCIUM CONTENT

1.1. Comparison of the Ca^{2+} Content of Epididymal and Ejaculated Sperm

Work from this laboratory (Morton and Lardy, 1967a,b) established conditions for examination of mitochondrial function in bovine spermatozoa whose surface membranes are disrupted by shear forces, hypotonic shock, or by treatment with the polyene antibiotic filipin. Respiration of exposed sperm mitochondria, dependent upon the added oxidative substrates, supports extensive accumulation of Ca^{2+} (Babcock *et al.*, 1976). Uptake of Ca^{2+} also occurs in rabbit sperm mitochondria exposed by hypotonic shock (Storey and Keyhani, 1973) and in bovine sperm treated with digitonin (Singh *et al.*, 1981).

Mitochondria also represent the major site of Ca^{2+} sequestration in intact bovine epididymal sperm incubated with this cation *in vitro*. Uptake of Ca^{2+} is stimulated by the presence of phosphate (Babcock *et al.*, 1975), suppressed by the presence of bicarbonate (Singh *et al.*, 1978), and decreased by inhibitors of mitochondrial function (Babcock *et al.*, 1976; Singh *et al.*, 1978). Under carefully controlled conditions, the fluorescent chelate probe chlortetracycline monitors the mitochondrial Ca^{2+} content of sperm suspensions. Fluorescence increases during Ca^{2+} uptake, decreases during release of accumulated mitochondrial Ca^{2+}, and is insensitive to entry of Ca^{2+} into the extramitochondrial space of bovine sperm (Babcock *et al.*, 1976). More directly, electron microanalysis of individual cells (Babcock *et al.*, 1978) demonstrated that the calcium content of the midpiece region of bovine sperm (wherein the mitochondria are localized) increases and decreases in parallel with uptake and release of Ca^{2+}.

Donner F. Babcock and Henry A. Lardy • Institute for Enzyme Research and the Department of Biochemistry, University of Wisconsin, Madison, Wisconsin 53706.

Uptake of Ca^{2+} by epididymal sperm suspensions is not explained by a small subpopulation of damaged cells because the mitochondria of such cells presumably would require exogenous substrates, whereas the observed uptake is supported by endogenous energy sources (Babcock *et al.*, 1975, 1976). Electron microanalysis (Babcock *et al.*, 1978) indicates that, in fact, Ca^{2+} uptake capacity is distributed homogeneously within the sperm population.

Uptake of Ca^{2+} during incubation *in vitro* does not represent refilling of Ca^{2+} stores that are depleted by the washing procedure utilized in preparing epididymal sperm. Sperm loaded with Ca^{2+} in a preliminary incubation, then subjected again to the isolation procedure, retain 75% of accumulated Ca^{2+} (Babcock *et al.*, 1979). It was shown also that sperm subjected only to dilution of the epididymal contents have a low mitochondrial Ca^{2+} content. Therefore, bovine epididymal sperm *in situ* apparently contain approximately 5 nmoles mitochondrial Ca^{2+} per 10^8 cells and have the potential to increase this content by 5- to 10-fold during aerobic incubation *in vitro*.

The seminal plasma, to which bovine sperm are exposed at ejaculation, contains approximately 10 mM Ca^{2+}. However, ejaculated bovine sperm do not contain more Ca^{2+} than epididymal sperm (Babcock *et al.*, 1979). Only incomplete comparisons can be made for the epididymal and ejaculated sperm of other species. Epididymal sperm of the stallion also accumulate 25–30 nmoles Ca^{2+} per 10^8 cells during incubation *in vitro* (Tamblyn *et al.*, 1979). Preliminary experiments (Singh, Babcock, and Lardy, unpublished) indicate quantitatively similar accumulation by boar and ram epididymal sperm. In contrast, Peterson *et al.* (1979) found that ejaculated sperm of the boar and human accumulate less than 0.3 nmole Ca^{2+} per 10^8 cells under approximately comparable incubation conditions. The conclusion (Storey, 1975) that rabbit epididymal sperm do not accumulate Ca^{2+}, based upon respiratory measurements, may be invalid because bovine epididymal sperm display little if any respiratory enhancement during Ca^{2+} uptake (Babcock *et al.*, 1976). Uptake of Ca^{2+} by sperm collected at earlier stages of maturation has not been examined.

1.2. Ca^{2+} Transport Inhibitor in Seminal Plasma

The low Ca^{2+} content of ejaculated bovine sperm is apparently a consequence of their inability to accumulate extracellular Ca^{2+}. Uptake of Ca^{2+} by epididymal sperm also is prevented by addition of dialyzed seminal plasma. The responsible macromolecular component apparently operates to decrease plasma membrane permeability because simultaneous exposure of epididymal sperm to oxidative substrates, seminal plasma, and filipin restores mitochondrial uptake of Ca^{2+} (Babcock *et al.*, 1979).

We have purified a heat-stable, trypsin-sensitive component of bovine seminal plasma that is derived from secretions of the seminal vesicle and that is probably responsible for inhibition of Ca^{2+} uptake. The inhibitor is a single polypeptide of approximately 10,000 molecular weight. Its isoelectric point is greater than pH8, consistent with a determined amino acid composition rich in basic residues. The peptide in SDS-polyacrylamide gels does not stain for carbohydrates. The purified inhibitor itself does not have appreciable affinity for Ca^{2+} (Rufo *et al.*, in press). These properties are different from those of other known seminal proteins.

1.3. Ca^{2+} Uptake during Capacitation in Vitro

Between mating and fertilization, mammalian sperm undergo a series of obligatory maturational changes that are called, collectively, "capacitation." These apparently include "motility activation" and the exocytotic "acrosome reaction" that exposes hydrolytic enzymes that probably are required for egg penetration. Capacitation *in vitro* has been achieved for the sperm of a limited number of species.

Yanagimachi and Usui (1974) demonstrated a requirement for extracellular Ca^{2+} to achieve the acrosome reaction during *in vitro* capacitation of guinea pig sperm. A variety of membrane-directed agents promote the acrosome reaction in sperm of the guinea pig (Yanagimachi, 1975; Singh *et al.*, 1978, 1980; Green, 1978), hamster (Lui and Meizel, 1979), boar (Peterson *et al.*, 1978), ram (Shams-Borhan and Harrison, 1981), sea urchin (Decker *et al.*, 1976), and probably also for the rabbit and bull (for additional references see Triana *et al.*, 1980). Yanagimachi (1975) suggested that the common pharmacological mechanism is increased entry of extracellular Ca^{2+} and that similar entry normally occurs during capacitation. Singh *et al.* (1978) showed that indeed net uptake of Ca^{2+} parallels the time course of the acrosome reaction that takes place without pharmacological stimuli for guinea pig sperm incubated *in vitro*. Together, these studies indicate that influx of Ca^{2+} initiates the acrosome reaction for most, if not all, mammalian species.

Uptake of Ca^{2+} also accompanies the acrosome reaction that is induced by exposure of sperm of the sea urchins *L. pictus* and *S. purpuratus* to components that are released from eggs of the same or related species (Schackmann *et al.*, 1978; Kopf and Garbers, 1980). The responsible glycoprotein has been purified and partially characterized (Kopf and Garbers, 1980). Induced uptake is rapid and apparently results directly from alterations in membrane permeability.

The physiologically important mechanisms that presumably operate to increase mammalian sperm membrane permeability to Ca^{2+} during capacitation *in vivo* have not been elucidated. Unfortunately, study of capacitation *in vitro* runs a high risk of focusing upon mechanisms that operate only fortuitously to produce the end result. For example, capacitation of hamster (Lui and Meizel, 1977) and rat (Davis *et al.*, 1980) sperm by incubation in media containing serum albumin apparently results, in part, from removal of sperm membrane lipids. The oxidative metabolism required to produce capacitation of guinea pig sperm in chemically defined media (Rogers *et al.*, 1977) may also deplete membrane lipids that are the major endogenous energy stores of the sperm.

An incubation regimen that involves brief exposure to hypertonic conditions and subsequent prolonged incubation in fluid derived from ovarian follicles induces capacitation of rabbit sperm (Brackett and Oliphant, 1975; Akruk *et al.*, 1979). Similar treatment induces uptake of Ca^{2+} and release of the acrosomal enzyme, hyaluronidase (Triana *et al.*, 1980), and capacitation (Lorton and First, 1979; Brackett *et al.*, 1980) for bovine sperm. Hypertonic extraction alone releases protein from the sperm surface. Removal or modification of surface membrane proteins is also produced by incubation of sperm *in utero* (for references see Triana *et al.*, 1980; also see Esbenschade and Clegg, 1980). A testable hypothesis was formulated (Triana *et al.*, 1980) that envisions removal or modification of the Ca^{2+} transport inhibitor as an identifiable regulatory event resulting in enhanced membrane permeability to Ca^{2+} during capacitation.

2. REGULATION OF MOTILITY AND THE ACROSOME REACTION BY Ca²⁺

2.1. Role of Calmodulin and Its Localization

Calmodulin is the major acidic protein of the soluble fraction of bovine epididymal sperm (Brooks and Siegel, 1973) and also has been purified from sea urchin sperm (Garbers *et al.*, 1980). The content of calmodulin in sperm from several mammalian species was estimated (Jones *et al.*, 1978) at 10^5 molecules per cell and was associated primarily with the head fractions of sonicated cells. In more detailed studies (Jones *et al.*, 1980), calmodulin-specific immunofluorescence was found in the acrosomal and postacrosomal regions but also at the flagellar base and tip and to a lesser extent throughout the flagella of rabbit and guinea pig sperm. Similar localization of actin in the postacrosomal region by indirect immunofluorescence (Talbot and Kleve, 1978; Clarke and Yanagimachi, 1978; Tamblyn, 1980) suggests a possible role of calmodulin in contractile events associated with the acrosome reaction of mammalian sperm.

Pharmacological treatment produces Ca^{2+}-dependent activation of the motility of sperm of the guinea pig (Yanagimachi, 1975), bull (Babcock *et al.*, 1976), hamster (Lui and Meizel, 1979), and ram (Shams-Borhan and Harrison, 1981). Both motility and metabolism are stimulated by limited entry of Ca^{2+} into the extramitochondrial space of bovine sperm treated with ionophore A23187. With more extensive uptake, motility is suppressed (Babcock *et al.*, 1976). The local anesthetic nupercaine also activates bovine sperm motility by a mechanism that does not require external Ca^{2+} (Singh *et al.*, 1981) but which results in mobilization of internal stores of Ca^{2+} and an increase in the apparent free internal Ca^{2+} concentration (Babcock *et al.*, 1981). Mediation by calmodulin of the action of Ca^{2+} upon the contractile apparatus of the sperm flagella has not been established.

2.2. Ca^{2+} and the Cyclic Nucleotide Regulatory System

Increases in the cyclic AMP content of hamster (Morton *et al.*, 1974) and guinea pig (Hyne and Garbers, 1979) sperm apparently depend upon entry of extracellular Ca^{2+}. However, increases in intracellular Ca^{2+} concentrations, resulting from ionophore-induced uptake or from mobilization of internal stores by local anesthetics, do not activate cyclic AMP-dependent protein kinase of bovine sperm. Conversely, treatment with a phosphodiesterase inhibitor increases sperm cyclic AMP content and protein kinase activation ratios but does not affect the apparent free internal Ca^{2+} concentration (Babcock *et al.*, 1981).

The increase in cyclic AMP content that follows exposure of sea urchin sperm to egg components requires the presence of extracellular Ca^{2+}. However, neither calmodulin nor Ca^{2+}, added separately or together, activates solubilized sperm adenylate cyclase (Kopf and Garbers, 1980). Sperm cyclic nucleotide phosphodiesterase (Wells and Garbers, 1976) and guanylate cyclase (Garbers, 1976) also are unresponsive to additions of calmodulin and Ca^{2+}, respectively. Thus, the mechanisms that link Ca^{2+} and cyclic nucleotide metabolism apparently are indirect in both mammalian and invertebrate sperm.

2.3. Regulation by Other Ionic Species

Exposure of sea urchin sperm to egg components induces efflux of H^+ and also probably of K^+. A portion of the H^+ release apparently is coupled to Ca^{2+} uptake as both are prevented by treatment with the Ca^{2+} transport antagonist D600 (Schackmann *et al.*, 1978). Proton release also accompanies Na^+ uptake induced by the purified "speract" peptide that stimulates sperm respiration and motility (Hansbrough and Garbers, 1981). Tilney and co-workers (1976, 1978) provided convincing evidence that formation of the acrosomal process of invertebrate sperm is initiated by proton efflux that allows polymerization of non-filamentous actin contained in the acrosomal region. Mammalian sperm do not form a similar structure. However, some evidence indicates that the mammalian sperm acrosome is extremely acidic (Meizel and Deamer, 1978) and that elevation of the extracellular (and presumably the intracellular) pH induces the acrosome reaction for guinea pig sperm (Hyne and Garbers, 1981). It has been suggested that elevation of acrosomal pH may activate proteolytic zymogens that possibly are involved in the mammalian sperm acrosome reaction (for discussion see Meizel, 1978).

REFERENCES

Akruk, S. F., Faroqui, A. A., Williams, W. L., and Srivastava, P. N. (1979). *Gamete Res.* **2**, 1–13.

Babcock, D. F., First, N. L., and Lardy, H. A. (1975). *J. Biol. Chem.* **250**, 6488–6495.

Babcock, D. F., First, N. L., and Lardy, H. A. (1976). *J. Biol. Chem.* **251**, 3881–3886.

Babcock, D. F., Stammerjohn, D., and Hutchinson, T. (1978). *J. Exp. Zool.* **204**, 391–399.

Babcock, D. F., Singh, J. P., and Lardy, H. A. (1979). *Dev. Biol.* **69**, 85–93.

Babcock, D. F., Singh, J. P., and Lardy, H. A. (1981). In *Calcium Binding Proteins: Structure and Function* (F. Siegel, E. Carafoli, R. H. Kretsinger, O. H. MacLennan, and R. H. Wasserman, eds.), pp. 479–481, Elsevier, Amsterdam.

Brackett, B. G., and Oliphant, G. (1975). *Biol. Reprod.* **12**, 260–274.

Brackett, B. G., Oh, Y. K., Evans, J. F., and Donawick, W. J. (1980). *Biol. Reprod.* **23**, 189–205.

Brooks, J. C., and Siegel, F. L. (1973). *Biochem. Biophys. Res. Commun.* **55**, 710–716.

Clarke, G. N., and Yanagamachi, R. (1978). *J. Exp. Zool.* **205**, 125–132.

Davis, B. K., Byrne, R., and Beidgian, K. (1980). *Proc. Natl. Acad. Sci. USA* **77**, 1546–1550.

Decker, G. L., Joseph, D. B., and Lennarz, W. J. (1976). *Dev. Biol.* **53**, 115–125.

Esbenschade, K. L., and Clegg, E. D. (1980). *Biol. Reprod.* **23**, 530–537.

Garbers, D. L. (1976). *J. Biol. Chem.* **251**, 4071–4077.

Garbers, D. L., Hansbrough, J. R., Radany, E. W., Hyne, R. V., and Kopf, G. S. (1980). *J. Reprod. Fertil.* **59**, 377–381.

Green, D. P. L. (1978). *J. Cell Sci.* **32**, 153–164.

Hansbrough, J. R., and Garbers, D. L. (1981), *J. Biol. Chem.* **256**, 2235–2241.

Hyne, R. V., and Garbers, D. L. (1979). *Proc. Natl. Acad. Sci. USA* **76**, 5699–5703.

Hyne, R. V., and Garbers, D. L. (1981). *Biol. Reprod.*, **24**, 257–266.

Jones, H. P., Bradford, M., McRorie, R. A., and Cormier, M. J. (1978). *Biochem. Biophys. Res. Commun.* **82**, 1264–1272.

Jones, H. P., Lenz, R. W., Palevitz, B. A., and Cormier, M. J. (1980). *Proc. Natl. Acad. Sci. USA* **77**, 2772–2776.

Kopf, G. S., and Garbers, D. L. (1980). *Biol. Reprod.* **22**, 1118–1126.

Lorton, S. P., and First, N. L. (1979). *Biol. Reprod.* **21**, 301–308.

Lui, C. W., and Meizel, S. (1977). *Differentiation* **9**, 59–66.

Lui, C. W., and Meizel, S. (1979). *J. Exp. Zool.* **207**, 173–186.

Meizel, S. (1978). In *Development in Mammals* (M. H. Johnson, ed.), Vol. 3, pp. 1–64, Elsevier/North-Holland, Amsterdam.

Meizel, S., and Deamer, D. W. (1978). *J. Histochem. Cytochem.* **26**, 98–105.

Morton, B. E., and Lardy, H. A. (1967a). *Biochemistry* **6**, 50–56.

Morton, B. E., and Lardy, H. A. (1967b). *Biochemistry* **6**, 57–61.

Morton, B., Harrigan-Lum, J., Albagli, L., and Jooss, T. (1974). *Biochem. Biophys. Res. Commun.* **56**, 372–379.

Peterson, R., Russel, L., Bundman, D., and Freund, M. (1978). *Biol. Reprod.* **19**, 459–466.

Peterson, R. N., Seyler, D., Bundman, D., and Freund, M. (1979). *J. Reprod. Fertil.* **55**, 385–390.

Rogers, B. J., Ueno, M., and Yanagimachi, R. (1977). *J. Exp. Zool.* **199**, 129–136.

Rufo, G. A. R., Singh, J. P., Babcock, D. F., and Lardy, H. A. (1982). *J. Biol. Chem.*, in press.

Schackmann, R. W., Eddy, E. M., and Shapiro, B. M. (1978). *Dev. Biol.* **65**, 483–495.

Shams-Borhan, G., and Harrison, R. A. P. (1981). *Gamete Res.* **4**, 407–432.

Singh, J. P., Babcock, D. F., and Lardy, H. A. (1978). *Biochem. J.* **172**, 549–556.

Singh, J. P., Babcock, D. F., and Lardy, H. A. (1980). *Biol. Reprod.* **22**, 566–570.

Singh, J. P., Babcock, D. F., and Lardy, H. A. (1981). Submitted for publication.

Storey, B. T. (1975). *Biol. Reprod.* **13**, 1–9.

Storey, B. T., and Keyhani, E. (1973). *FEBS Lett.* **37**, 33–36.

Talbot, P., and Kleve, M. G. (1978). *J. Exp. Zool.* **204**, 131–134.

Tamblyn, T. (1980). *Biol. Reprod.* **22**, 727–734.

Tamblyn, T. M., Singh, J. P., Lorton, S. P., and First, N. L. (1979). *J. Reprod. Fertil. Suppl.* **27,** 31–37.

Tilney, L. G. (1976). *J. Cell Biol.* **69,** 73–89.

Tilney, L. G., Kiehart, D. P., Sardet, B., and Tilney, M. (1978). *J. Cell Biol.* **77,** 536–550.

Triana, L. R., Babcock, D. F., Lorton, S. P., First, N. L., and Lardy, H. A. (1980). *Biol. Reprod.* **23,** 47–59.

Wells, J. N., and Garbers, D. L. (1976). *Biol. Reprod.* **15,** 46–53.

Yanagimachi, R. (1975). *Biol. Reprod.* **13,** 519–526.

Yanagimachi, R., and Usui, N. (1974). *Exp. Cell Res.* **89,** 161–174.

Index